Springer Monographs in Mathematics

Springer
London
Berlin
Heidelberg
New York
Barcelona
Hong Kong
Milan
Paris
Santa Clara
Singapore
Tokyo

Seán Dineen

Complex Analysis on Infinite Dimensional Spaces

Springer

Seán Dineen, PhD, DSc
Department of Mathematics
University College Dublin
Belfield
Dublin 4
Ireland

ISBN-13: 978-1-4471-1223-5   e-ISBN-13: 978-1-4471-0869-6
DOI: 10.1007/978-1-4471-0869-6

British Library Cataloguing in Publication Data
Dineen, Seán, 1944-
   Complex analysis on infinite dimensional spaces. -
   (Springer monographs in mathematics)
   1. Holomorphic functions 2. Linear topological spaces
   I. Title
   515.7'3

Library of Congress Cataloging-in-Publication Data
Dineen, Seán, 1944-
   Complex analysis on infinite dimensional spaces / Seán Dineen.
      p. cm. -- (Springer monographs in mathematics)
   Includes bibliographical references and index.
                  (alk. paper)
   1. Holomorphic functions. 2. Linear topological spaces.
3. Functions of complex variables. I. Title. II. Series.
QA331.D636   1999   99-25273
515'.98—dc21   CIP

Mathematics Subject Classification (1991): 46G20, 46A32, 32E25, 32D05, 46B04, 58D12, 46A03

Typesetting: Camera-ready by Michael Mackey and Thomas Unger

12/3830-543210 Printed on acid-free paper SPIN 10677689

To
the memory
of

Paul Newbrough

Preface

Infinite dimensional holomorphy is the study of holomorphic or analytic functions over complex topological vector spaces. The terms in this description are easily stated and explained and allow the subject to project itself initially, and innocently, as a compact theory with well defined boundaries. However, a comprehensive study would include delving into, and interacting with, not only the obvious topics of topology, several complex variables theory and functional analysis but also, differential geometry, Jordan algebras, Lie groups, operator theory, logic, differential equations and fixed point theory. This diversity leads to a dynamic synthesis of ideas and to an appreciation of a remarkable feature of mathematics – its unity. Unity requires synthesis while synthesis leads to unity.

It is necessary to stand back every so often, to take an overall look at one's subject and ask "How has it developed over the last ten, twenty, fifty years? Where is it going? What am I doing?" I was asking these questions during the spring of 1993 as I prepared a short course to be given at Universidade Federal do Rio de Janeiro during the following July. The abundance of suitable material made the selection of topics difficult. For some time I hesitated between two very different aspects of infinite dimensional holomorphy, the geometric-algebraic theory associated with bounded symmetric domains and Jordan triple systems and the topological theory which forms the subject of the present book. I did not intend to write a book and, so, the choice of topic did not appear to have long term consequences. I had written a book[1] on locally convex structures on spaces of holomorphic functions some fifteen years previously and did not believe that the area has changed sufficiently to warrant another book. I took the apparently easy option of surveying recent developments, but while preparing my notes I was pleasantly surprised and, by the end of 1993, I knew that this book should be written and that it would take some time. It took almost six years and would never have been written without my biannual visits to Rio de Janeiro, kindly arranged at different times by Jorge Alberto Barroso, Luiza Moraes and Roberto Soraggi, where I had the opportunity to lecture on most of the material.

---

[1] *Complex Analysis on Locally Convex Spaces*, North Holland Math. Studies, 57, 1981.

The central theme in this book is the relationship on $H(U)$, the holomorphic functions on an open subset U of a locally convex space E, between the three topologies, τ_0 (the compact open topology), τ_ω (the ported or Nachbin topology) and τ_δ (the topology generated by the countable open covers). The portrayal of topologies as structures whose function is to define modes of convergence and continuity tends to obscure other roles they play. A topology is also a *selection* process which replaces the study of random subsets by *identifying* those likely to display interesting features. Different topologies give rise to different selections and reveal different features of the ambient space. The two extreme topologies considered here are τ_0 and τ_δ. The compact open topology has a function theoretic pedigree and is more likely to be useful, at least initially, with problems having their roots in several complex variables theory. It is, however, based on compact subsets of the underlying (domain) space and, in general, compact subsets of an infinite dimensional space are *small* and, consequently, less influential. On the other hand, open subsets are *large*, almost too large, and the τ_δ topology, based on such sets, can be a crude selection process when dealing with holomorphic functions. This mixture of ideas and concepts from different areas is, however, full of surprises and we find in Chapter 5 the following redeeming function theoretic property of the τ_δ topology:

> if Ω_1 and Ω_2 are domains spread over a Fréchet space and Ω_2 is an analytic extension of Ω_1 then the bijective mapping, defined by analytic continuation, is always a τ_δ isomorphism

(but may not be a τ_0 isomorphism).

Complications also arise on the linear side from the unavoidable presence of nuclearity in any reasonable topology on $H(U)$. This point is clearly and easily illustrated when E is a Banach space. In this case $(H(E), \tau_\delta)$ contains Fréchet nuclear spaces and Banach spaces as *complemented* subspaces – an unusual combination – intuitively avoided by the functional analyst. On the other hand the τ_0 topology on $H(E)$ has the useful property that its closed bounded sets are compact but has the drawback that it induces on E' (the continuous linear functions on E) a rather weak topology.

The ideal situation occurs when $\tau_0 = \tau_\delta$ and we have a topology which is acceptable in function theory and functional analysis. Examples are given in Chapter 4. These are important and include both $\mathcal{DFM}$ spaces and Fréchet nuclear spaces with (DN) but exclude *all* infinite dimensional Banach spaces. Between τ_0 and τ_δ we encounter the *intermediate* τ_ω topology, defined using neighbourhood systems of compact sets. It is sufficiently close to τ_0 to inherit some of its good behaviour and, yet, sufficiently removed, from τ_0 towards τ_δ, to potentially share properties with τ_δ. We may regard τ_ω as a compromise between the conflicting suggestions of several complex variables theory and linear functional analysis.

In spite of these conflicts and complications, or perhaps because of them, the results are surprisingly positive. Investigations, over many years, have shown that certain natural splittings, approximations and size restrictions need to be in place to obtain positive solutions to the topological equations $\tau_0 = \tau_\delta$ and $\tau_\omega = \tau_\delta$. Fortunately, these conditions are present in what are generally regarded as the most interesting locally convex spaces. The main results for these topologies are presented in the two most technically demanding chapters, four and five. The results for balanced domains – where the key to success is pointwise convergence of the Taylor series expansion at *all* points of the domain – are fairly complete and discussed in Chapter 4. The Taylor series expansion facilitates the employment of techniques from functional analysis but as we move from this setting there is a steady drift, especially in the final two chapters, towards non-linear methods and involvement with $H(U)$ as an algebra. The theory for arbitrary open sets, presented in Chapter 5, contains significant positive results and many open problems. Even when these topologies do not coincide they can be used in tandem to uncover what might otherwise have remained hidden, e.g. τ_δ, τ_ω and τ_0 combine to present $H(E)$ as a dual space.

Somehow, and surprisingly in view of its initial modest aims, the topological problems we consider capture the tension between the *finite* dimensional *holomorphic* theory and the *infinite* dimensional *linear* theory and, acting as a catalyst, fuse from them a topic with its own internal logic and intrinsic unity. Thus, in examining the basic definitions and considering fundamental topological questions we encounter in a natural and essential way such diverse topics as the bounded approximation property, finite dimensional decompositions, the Dunford–Pettis property, the Radon–Nikodým Property, the principle of local reflexivity, ultrapowers, (BB)-property, (DN)-property, the density condition, Arens regularity, hypocontinuity, spreading models, determining sets, the Levi problem, and meet new intrinsic concepts such as polarization constants, bounding sets, uniform factorization, compact non-polar sets, S-absolute decompositions, Taylor series completeness, entire functions of bounded type, etc. The answers that resulted from these topological questions (for instance in Section 5.2 we required almost the complete solution to the Levi problem in order to obtain $\tau_0 = \tau_\omega$ on open subsets of Fréchet–Schwartz spaces) and the presence of positive results suggest that infinite dimensional holomorphy will not be hindered by topological obstructions and, indeed, will be positively enriched when such considerations enter the picture.

This book is divided into six chapters, each devoted to a single theme. Chapters 1 and 3 introduce and cover the basic properties of polynomials and holomorphic functions over locally convex spaces respectively. With these two chapters as reference, the other four become almost independent self-contained units which complement one another and taken together add to the overall structure of the subject and book. The first two chapters are a self-contained study of polynomials (today an essentially independent field of investigation within linear functional analysis). This area has seen rapid development over the last ten years and the wide choice of material available obliged us to omit some interesting topics. In Chapter 1 we develop the basic theory, using tensor products, and discuss geometric properties of polynomials on Banach spaces. Chapter 2 is devoted to duality theory for different spaces of polynomials. Chapter 3 discusses Taylor and monomial expansions of Gâteaux and Fréchet holomorphic functions while Chapters 4 and 5 concentrate on relationships between the topologies τ_0, τ_ω and τ_δ. Chapter 6 examines the interplay between various concepts that were uncovered, in earlier chapters, as being intrinsic to infinite dimensional holomorphy.

Each chapter contains text, a set of exercises and a final section of notes. The exercises, notes, and appendix (which contains remarks on selected exercises) allowed us to insert material which, in the main text, would have interrupted the flow of essential material and led to the inclusion of excessive detail. In the notes and appendix we provide information on the history of the subject and references for the material presented. We have tried to be as careful as possible in this regard and take responsibility for the inevitable errors. Accurate and comprehensive records of this kind are not a luxury but essential background information in appreciating and understanding a subject and its evolution. Authors, who do not take this aspect of their work seriously, devalue their chosen subject and, ultimately, their own contribution.

We assume the reader has a basic knowledge of one complex variable theory and some experience with Banach space theory. The reader familiar with several complex variables and locally convex spaces will undoubtedly find the subject less difficult but we include definitions and results from these areas as required. We have tried to maintain a delicate balance between our desire to write a self-contained introduction for the non-expert and to provide a comprehensive summary for the expert.

It is a pleasure, and a relief, to arrive at the stage where I can thank those who helped me in this project. The many mathematicians who organized conferences and published proceedings over the years in this area performed a much appreciated service and facilitated my task enormously. The September 1994 Dublin conference on *"Polynomials and Holomorphic Functions over Infinite Dimensional Spaces"* occurred at a crucial time and the excellent survey lectures and the set of problems circulated at that conference played a key role in convincing me to continue writing this book. The participants at the weekly University College Dublin–Trinity College Dublin Analysis Seminar displayed remarkable patience, while I experimented with my presentation, and contributed with their honest and helpful advice. The analysis group at Universidade Federal do Rio de Janeiro, Roberto Soraggi, Luiza Moraes and Jorge Alberto Barroso, deserve to be mentioned in the introduction to each chapter for the wonderful hospitality and support they provided over the full period during which this book was written. The intensive courses in Complex Analysis, sponsored by the Erasmus Programme, organized initially by the Galois Network and Frank de Clerck (Ghent) and continued by Jaime

Carvalho de Silva (Coimbra) provided me with the opportunity to give short courses on some of the material in the delightful city of Coimbra. Financial support for some of these visits was provided by UFRJ (Universidade Federal do Rio de Janeiro), FAPERJ (Fundacaõ de Amparo a Pesquisa do Estado do Rio de Janeiro), CNPq (Conselho Nacional de Desenvolvimento Cientifica e Tecnologico), the Erasmus Programme of the European Union and the Faculty of Arts at University College Dublin. I am particularly grateful to the Modular Degree Programme at University College Dublin. The income from the night courses I gave in this programme was the only support I had to transform my handwritten notes into printed form.

Many individuals helped, with their technical advice and mathematical expertise, in the preparation of different chapters of this book but Jose Ansemil, Chris Boyd, Michael Mackey, Pilar Rueda and Thomas Unger gave unselfishly of their time with *all* chapters and I would like to single them out for special thanks. Their influence has been enormous and so pervasive that it is now impossible to detail. Raymundo Alencar, Richard Aron, Fernando Blasco, Yung Sung Choi, Veronica Dimant, Klaus Floret, José Isidro, Manolo Maestre, Pauline Mellon, Jorge Mujica, Yannis Sarantopoulos, Ray Ryan, Richard Timoney and Nacho Zalduendo provided specialized advice and encouragement when it mattered.

Finally, a special paragraph for Dana, soon to become Dr. Nicolau, who must have felt, at some points, that my revisions were not converging and that I was taking the "infinite" in the title too literally. Despite these reservations she did an excellent job in preparing this book for publication in the midst of a very busy period in her own studies. Thank you, Dana.

Susan Hezlet, recently of Springer-Verlag but now with the London Mathematical Society, is the type of editor that every author should have – helpful, realistic and encouraging. Her successor at Springer-Verlag, David Ireland, has been helpful and understanding during the final phase of this project.

Finally, a special word of thanks to the Dean of the Faculty of Arts, Professor Fergus D'Arcy, whose personal support and appreciation of scholarship and creativity does make a difference.

This book took a long time to write – much longer than planned. By reading it you will be thanking all those who helped me.

University College Dublin,
December 1998.

Contents

Chapter 4. Decompositions of Holomorphic Functions

Chapter 5. Riemann Domains

Chapter 6. Holomorphic Extensions

Chapter 1. Polynomials

Multilinear mappings, tensor products, restrictions to finite dimensional spaces and differential calculus may all be used to define polynomials over infinite dimensional spaces. All of these are useful, none should be neglected and, indeed, the different possible approaches and interpretations add to the richness of the subject. We adopt an integrated approach to the development of polynomials using multilinear mappings and tensor products. The philosophy of tensor products is easily stated: to exchange polynomial functions on a given space with *simpler* (linear) *functions* on a (possibly) more *complicated space*. As a typical example (see Section 1.2) we shall see that the space of continuous n-homogeneous polynomials on the locally convex space E can be realized as the space of continuous linear functions on the n-fold symmetric projective tensor product of E, $\widehat{\bigotimes}_{n,s,\pi} E$.

In Section 1.1 we begin by studying, at the algebraic level, the relationship between symmetric n-linear forms, symmetric n-tensors and n-homogeneous polynomials. Afterwards we refine this relationship to the continuous level by considering continuous polynomials between locally convex spaces and a further refinement arises, in Section 1.2, when we discuss topologies on spaces of polynomials. At the algebraic level we discuss linearization, duality and the polarization formula while, at the continuous level, we also meet factorization. This perspective leads to the recasting of certain results, e.g. (1.16) is just a translation of the polarization formula into the language of symmetric tensors, and to an examination, in Section 1.2, of the different topologies as uniform convergence over sets of tensors. Section 1.3, which begins as a more detailed analysis of the polarization inequality, ends up as a study of the interplay between polynomials and the *isometric* (or *geometric*) properties of Banach spaces – Chapter 2 is mainly devoted to the relationship between polynomials and *isomorphic* (or *linear topological*) properties of Banach spaces. As we proceed we require various concepts and results from locally convex space theory (Section 1.2) and Banach space theory (Section 1.3). These are introduced, without proof, as required, in a format deemed suitable from the perspective of infinite dimensional holomorphy.

1.1 Continuous Polynomials

$\mathbb{C}$, $\mathbb{R}$, $\mathbb{N}$ and $\mathbb{Z}$ denote respectively the complex numbers, the real numbers, the natural numbers and the integers. If A and B are sets, and $n, m \in \mathbb{N}$ then $A^n B^m$ will denote the Cartesian product of n copies of A and m copies of B and $x^n y^m$ will denote the element $(\underbrace{x, \ldots, x}_{n \text{ times}}, \underbrace{y, \ldots, y}_{m \text{ times}})$. E and F will denote vector spaces over $\mathbb{C}$.

For $n \in \mathbb{N}$ we let $\mathcal{L}_a(^n E; F)$ denote the *space of n-linear mappings* from E into F. The subscript a denotes *algebraic* since we do not assume any continuity properties. Hence, if $L \in \mathcal{L}_a(^n E; F)$, then L is an F-valued function defined on E^n which is linear in each variable when the remaining $(n-1)$ variables are fixed. Clearly, $\mathcal{L}_a(^n E; F)$ is a vector space over $\mathbb{C}$. 1-linear mappings are just linear mappings and in this case we use the notation $\mathcal{L}_a(E; F)$. 2-linear mappings are also called bilinear mappings and certain authors use the notation $\mathcal{B}_a(E; F)$ in place of $\mathcal{L}_a(^2 E; F)$. The notation $\mathcal{B}(E)$ is also used in the literature to denote the set of all bounded linear mappings from the Banach space E into itself. When $F = \mathbb{C}$ we write $\mathcal{L}_a(^n E)$ in place of $\mathcal{L}_a(^n E; \mathbb{C})$ and E^* in place of $\mathcal{L}_a(E; \mathbb{C})$. E^* is called the *algebraic dual of E*. When $n = 0$ we define $\mathcal{L}_a(^0 E; F)$ to be the set of constant mappings from E into F and this space can be identified with F in a natural fashion. If f is an F-valued function, F is a vector space over $\mathbb{C}$, α is a semi-norm on F and A is contained in the domain of f we let $\|f\|_{\alpha, A} = \sup_{x \in A} \alpha(f(x))$. If α is clearly understood from the context we just write $\|f\|_A$.

The following useful algebraic identities are easily established by induction.

Proposition 1.1 *If E, F and G are vector spaces over $\mathbb{C}$ and $m, n \in \mathbb{N}$ then the mappings I_m and J_m, defined as follows, are linear isomorphisms:*

$$I_m \colon \mathcal{L}_a(^{m+n} E; F) \longrightarrow \mathcal{L}_a(^m E; \mathcal{L}_a(^n E; F))$$

$$[I_m A(x)](y) := A(x, y),$$

where $A \in \mathcal{L}_a(^{m+n} E; F)$, $x \in E^m$ and $y \in E^n$,

$$J_m \colon \mathcal{L}_a(^m E; \mathcal{L}_a(^n F; G)) \longrightarrow \mathcal{L}_a(^n F; \mathcal{L}_a(^m E; G))$$

$$[J_m A(y)](x) := [A(x)](y),$$

where $A \in \mathcal{L}_a(^m E; \mathcal{L}_a(^n F; G))$, $x \in E^m$, $y \in F^n$.

A particular case, often used in the linear theory, is the isomorphism I_1 which gives, when $n = 1$ and $F = \mathbb{C}$,

$$\mathcal{L}_a(^2 E) \approx \mathcal{L}_a(E; E^*).$$

We now introduce *tensor products*. This approach is a *linearization* process for multilinear mappings and is important in presenting spaces of polynomials and holomorphic functions as dual spaces. The bilinear case illustrates the general method of constructing tensor products. If E is a vector space over $\mathbb{C}$ then the set, $E^{\{2\}}$, of all formal finite sums $\sum_i \lambda_i(x_i, y_i)$, where $\lambda_i \in \mathbb{C}$ and $(x_i, y_i) \in E^2$, can be endowed with the structure of a vector space with basis E^2 in an obvious way. Let I denote the subspace of $E^{\{2\}}$ generated by elements of the form

$$(x_1 + x_2, y) - (x_1, y) - (x_2, y), \ (x, y_1 + y_2) - (x, y_1) - (x, y_2),$$
$$(\lambda x, y) - \lambda(x, y), \ (x, \lambda y) - \lambda(x, y). \tag{1.1}$$

The 2-fold tensor product of E with itself, $\bigotimes_2 E$, is defined to be $E^{\{2\}}/I$. We let $x \otimes y := (x, y) + I$ and denote by i_2 the bilinear mapping from E^2 into $\bigotimes_2 E$ which maps (x, y) onto $x \otimes y$. If F is a vector space over $\mathbb{C}$ and $L \in \mathcal{L}_a(^2E; F)$ let $i_2^*(L)\left(\sum_i \lambda_i x_i \otimes y_i\right) = \sum_i \lambda_i L(x_i, y_i)$. Using (1.1) it is easily verified that $i_2^*(L)$ is well defined, and $i_2^*(L) \circ i_2 = L$. Moreover, $\left(\bigotimes_2 E, i_2\right)$ is uniquely determined by the above properties.

For arbitrary n we have the following result. There exists a vector space over $\mathbb{C}$, $\bigotimes_n E$, and $i_n \in \mathcal{L}_a(^nE; \bigotimes_n E)$ such that for any vector space F over $\mathbb{C}$ and any $L \in \mathcal{L}_a(^nE; F)$ there is a unique $i_n^*(L) \in \mathcal{L}_a(\bigotimes_n E; F)$ such that the diagram

$$\tag{1.2}$$

commutes. The pair $\left(\bigotimes_n E, i_n\right)$ is uniquely determined, by this property, in the category of vector spaces over $\mathbb{C}$ and $\mathbb{C}$-linear mappings. The *single* n-linear mapping, i_n, factors out the full non-linear content of *each* n-linear mapping L so that what remains, $i_n^*(L)$, is linear. We let $i_n(x_1, \cdots, x_n) = x_1 \otimes x_2 \cdots \otimes x_n$. Elements of $\bigotimes_n E$ are called tensors and the space $\bigotimes_n E$ is called the n-fold tensor product of E (with itself). It is easily seen that

$$i_n^*\colon \mathcal{L}_a(^nE; F) \longrightarrow \mathcal{L}_a\left(\bigotimes_n E; F\right)$$

is a linear isomorphism and, in particular, if $F = \mathbb{C}$, we have a representation of $\mathcal{L}_a(^nE)$ as a dual space, i.e.

$$\mathcal{L}_a({}^nE) \cong \left(\bigotimes_n E\right)^*. \tag{1.3}$$

Each element of $\bigotimes_n E$ has a representation of the form

$$\sum_{i=1}^{l} x_{i,1} \otimes x_{i,2} \cdots \otimes x_{i,n}.$$

However, this representation will never be unique. For instance, by (1.1), we have

$$(\lambda x) \otimes y \; = \; i_2(\lambda x, y) \; = \; i_2(x, \lambda y) \; = \; x \otimes (\lambda y)$$

and this lack of uniqueness features in many calculations on tensor products.

To define n-homogeneous polynomials we use the natural embeddings, called *diagonal mappings*, of E into E^n and $\bigotimes_n E$, denoted by Δ_n and δ_n respectively. These are defined as

$$\Delta_n : E \longrightarrow E^n$$
$$x \longrightarrow (x, x, \ldots, x)$$
$$\delta_n : E \longrightarrow \bigotimes_n E$$
$$x \longrightarrow x \otimes x \cdots \otimes x.$$

Clearly we have the commutative diagram

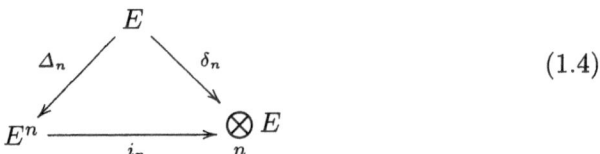

$$\text{(1.4)}$$

i.e. $i_n \circ \Delta_n = \delta_n$.

Definition 1.2 Let E and F be vector spaces over $\mathbb{C}$ and let $n \in \mathbb{N}$. A mapping from E into F which is the composition of Δ_n and an element of $\mathcal{L}_a({}^nE; F)$ is called an n-homogeneous polynomial from E into F.

We let $\mathcal{P}_a({}^nE; F)$ denote the vector space of all n-homogeneous polynomials from E into F. A polynomial from E into F is a finite sum of homogeneous polynomials from E into F. The vector space of all polynomials from E into F is denoted by $\mathcal{P}_a(E; F)$.

Hence $P \in \mathcal{P}_a({}^nE; F)$ if and only if $P = L \circ \Delta_n$ for some $L \in \mathcal{L}_a({}^nE; F)$. Since $L = i_n^*(L) \circ i_n$ and $\delta_n = i_n \circ \Delta_n$ we have

$$P = L \circ \Delta_n = i_n^*(L) \circ i_n \circ \Delta_n = i_n^*(L) \circ \delta_n$$

and, as i_n^* is a linear isomorphism, we have immediately the following result.

Proposition 1.3 *A mapping $P\colon E \to F$ is an n-homogeneous polynomial if and only if there exists $T \in \mathcal{L}_a(\bigotimes_n E; F)$, or equivalently $L \in \mathcal{L}_a(^nE; F)$, such that the diagram*

$$
\begin{array}{ccc}
E & \xrightarrow{\;\;\Delta_n\;\;} & E^n \\[2mm]
{\scriptstyle \delta_n}\big\downarrow & \;\;{\scriptstyle P} & \big\downarrow{\scriptstyle L} \\[2mm]
\bigotimes_n E & \xrightarrow[\;\;T\;\;]{} & F
\end{array}
\qquad (1.5)
$$

commutes.

In particular, $P(x) = L(x, \cdots, x) = T(x \otimes \cdots \otimes x)$ and $P(\lambda x) = \lambda^n P(x)$ for all $\lambda \in \mathbb{C}$ and $x \in E$. In many situations it is possible to present things in terms of both multilinear forms and tensor products. Initially, we will give both presentations but as we proceed we confine ourselves to the approach considered more suitable.

Example 1.4 If L is a 2-linear or bilinear form on $\mathbb{C}^n$, $n \in \mathbb{N}$, then it is well known that there exists an $n \times n$ matrix A such that $L(z, w) = zAw^t$ for all $z = (z_1, \ldots, z_n)$ and $w = (w_1, \ldots, w_n)$ in $\mathbb{C}^n$. If $A = (a_{ij})_{1 \le i, j \le n}$ then $L(z, w) = \sum_{1 \le i, j \le n} a_{ij} z_i w_j$. Hence any $\mathbb{C}$-valued 2-homogeneous polynomial, P, on $\mathbb{C}^n$ has the familiar form

$$
P(z) = L(z, z) = \sum_{1 \le i, j \le n} a_{ij} z_i z_j.
$$

If we replace the matrix A by its associated symmetric matrix $B := (A + A^*)/2$, where A^* is the *transpose*[2] of the matrix A, then $b_{ij} = (a_{ij} + a_{ji})/2$ and $zAz^t = zBz^t$ for every $z \in \mathbb{C}^n$. It follows that A and B define the same 2-homogeneous polynomial on $\mathbb{C}^n$.

---

[2] If $T : E \to F$ is a linear mapping between vector spaces the *transpose* mapping $T^* : F^* \to E^*$ is defined by letting $(T^*\phi)(x) = \phi(Tx)$ for all $x \in E$ and $\phi \in F^*$. The notation T^t is also used but we prefer the above except for vectors in $\mathbb{C}^n$ and use z^t to denote the $n \times 1$ column vector which is transpose to the $1 \times n$ row vector z. Note that the pairing given by $E^* \times E \to \mathbb{C}$, $(\phi, x) \to \phi(x)$ is (complex) *bilinear*. In operator theory one also uses the $*$ notation for the *adjoint* operator with respect to a *Hermitian form* (i.e. a real bilinear form which is complex linear in the first variable and conjugate linear in the second variable) and one should take care in distinguishing between these two notions. In this book we only consider the transpose as defined above. If $A := (a_{ij})_{1 \le i, j \le n}$ is an $n \times n$ matrix then $A^* = (a_{ji})_{1 \le i, j \le n}$.

By definition homogeneous polynomials are the restriction to the diagonal of multilinear mappings. The previous example, which is typical of the general situation, shows that different multilinear forms may define the same polynomial. To obtain a one to one correspondence we restrict ourselves to symmetric multilinear mappings and symmetric tensors.

An n-linear mapping from E into F is said to be symmetric if

$$L(x_1, \ldots, x_n) = L\left(x_{\sigma(1)}, \ldots, x_{\sigma(n)}\right)$$

for any $(x_1, \ldots, x_n) \in E^n$ and any permutation σ of the first n natural numbers. We let $\mathcal{L}_a^s({}^nE; F)$ denote the vector space of all symmetric n-linear mappings from E into F.

By averaging over all permutations we can associate, in a canonical fashion, a symmetric n-linear mapping with each n-linear mapping and in this way obtain a projection, s, from $\mathcal{L}_a({}^nE; F)$ onto $\mathcal{L}_a^s({}^nE; F)$. Formally, we achieve this in the following way. If $L \in \mathcal{L}_a({}^nE; F)$ let

$$s(L)(x_1, \ldots, x_n) = \frac{1}{n!} \sum_{\sigma \in S_n} L\left(x_{\sigma(1)}, \ldots, x_{\sigma(n)}\right)$$

where S_n is the set of all permutations of the first n natural numbers.

We call s the *symmetrization operator*. The following properties of s are easily verified:

(a) If $L \in \mathcal{L}_a({}^nE; F)$ then $s(L) \in \mathcal{L}_a^s({}^nE; F)$,
(b) $s(L) = L$ if and only if $L \in \mathcal{L}_a^s({}^nE; F)$,
(c) $s^2 = s$,
(d) s is a linear operator,
(e) If $L \in \mathcal{L}_a({}^nE; F)$ and $x \in E$ then $L(x^n) = s(L)(x^n)$.

Properties (a), (b), (c), (d) show that the mapping

$$s \colon \mathcal{L}_a({}^nE; F) \longrightarrow \mathcal{L}_a^s({}^nE; F)$$

is a surjective linear *projection*. Using (a), we obtain the commutative diagram

$$
\begin{array}{ccc}
\mathcal{L}_a({}^nE; F) & \xrightarrow{\quad s \quad} & \mathcal{L}_a^s({}^nE; F) \\
& \searrow{\scriptstyle L \to L \circ \Delta_n} \quad \swarrow{\scriptstyle L \to L \circ \Delta_n} & \\
& \mathcal{P}_a({}^nE; F) &
\end{array}
\tag{1.6}
$$

We denote the mapping, in the previous diagram, from $\mathcal{L}_a({}^nE; F)$ into $\mathcal{P}_a({}^nE; F)$ by $\widehat{}$ and write $\widehat{L} := L \circ \Delta_n$ for L in $\mathcal{L}_a({}^nE; F)$. This mapping is easily seen to be linear and, by (e) above, its restriction to $\mathcal{L}_a^s({}^nE; F)$, which we again denote by $\hat{}$, is surjective. As a consequence of the Polarization Formula (Corollary 1.7) we shall see that $\hat{}$ is injective on $\mathcal{L}_a^s(^nE;F)$ and thus obtain a canonical bijective linear mapping from $\mathcal{L}_a^s(^nE;F)$ onto $\mathcal{P}_a(^nE;F)$.

We consider the analogous situation for tensors. If E is a vector space over $\mathbb{C}$ and $x_1 \otimes x_2 \cdots \otimes x_n \in \bigotimes_n E$ let

$$s(x_1 \otimes \cdots \otimes x_n) = \frac{1}{n!} \sum_{\sigma \in S_n} x_{\sigma(1)} \otimes \cdots \otimes x_{\sigma(n)}.$$

The subspace of $\bigotimes_n E$ generated by $s(x_1 \otimes \cdots \otimes x_n)$, $x_i \in E$, is denoted by $\bigotimes_{n,s} E$ and called the n-fold symmetric tensor product of E. Elements of $\bigotimes_{n,s} E$ are called n-symmetric (or just symmetric) tensors. Clearly every tensor of the form $x \otimes \cdots \otimes x$ is a symmetric tensor. Moreover, each element θ in $\bigotimes_{n,s} E$ can be expressed as a finite (not necessarily unique) sum of the form

$$\sum_i x_i \otimes \cdots \otimes x_i.$$

We call such a sum a symmetric representation of θ. A direct verification is rather tedious – it is an extension to several variables of the following calculation –

$$s(x \otimes y) = \frac{1}{2}(x \otimes y + y \otimes x)$$
$$= \frac{1}{2}\left((x+y) \otimes (x+y) + (ix) \otimes (ix) + (iy) \otimes (iy)\right).$$

The result can also be obtained directly using universal properties of tensor products. The linear extension of s to $\bigotimes_n E$ gives a projection, which we again denote by s, of $\bigotimes_n E$ onto $\bigotimes_{n,s} E$. Symmetric tensors may also be introduced by the abstract approach used to obtain (1.2). It can be shown, using uniqueness of the pair constructed, that $\left(\bigotimes_{n,s} E,\ \delta_n\right)$ is characterized by the following property:

If F is any vector space over $\mathbb{C}$ and $P \in \mathcal{P}_a(^nE;F)$ then there exists a unique $j_n^*(P) \in \mathcal{L}_a\left(\bigotimes_{n,s} E; F\right)$ such that the diagram

(1.7)

commutes.

The mapping $j_n^*: P_a(^nE; F) \longrightarrow \mathcal{L}_a(\bigotimes_{n,s} E; F)$ is a linear isomorphism

and, in particular,

$$P_a(^nE) \cong \left(\bigotimes_{n,s} E\right)^*.$$

Formulae which recover the values of a symmetric n-linear form from its diagonal values or, equivalently, from the values of the associated n-homogeneous polynomial are known collectively as polarization formulae. Corollary 1.6 is the original *Polarization Formula* and can be proved directly. We prove, using probability theory, a more general formula which is useful later. We use complex-valued random variables and let $E[X]$ denote the expected value of the random variable X. A random variable X is said to be normalized if $E[X] = 0$ and $E[|X|^2] = 1$.

Proposition 1.5 Let $P \in P_a(^nE; F)$ where E and F are vector spaces over $\mathbb{C}$. Let $L \in \mathcal{L}_a^s(^nE; F)$ and suppose $\hat{L} = P$. If $r_1, \ldots, r_n$ are n independent normalized random variables on the probability space $(\Omega, \mathcal{F}, \mu)$ and $(x_1, \ldots, x_n) \in E^n$ then

$$L(x_1, \ldots, x_n) = \frac{1}{n!} E\left[\bar{r}_1 \cdots \bar{r}_n P\left(\sum_{i=1}^n r_i x_i\right)\right]. \qquad (1.8)$$

Proof. Consider the random variable

$$\Phi := \bar{r}_1 \cdots \bar{r}_n \cdot P\left(\sum_{i=1}^n r_i x_i\right).$$

We have

$$E[\Phi] = E\left[\sum_{\substack{0 \le j_i \le n \\ \sum_{i=1}^n j_i = n}} \binom{n}{j_1, \ldots, j_n} \bar{r}_1 r_1^{j_1} \cdots \bar{r}_n r_n^{j_n} \cdot L(x_1^{j_1}, \ldots, x_n^{j_n})\right]$$

$$= \sum_{\substack{0 \le j_i \le n \\ \sum_{i=1}^n j_i = n}} \binom{n}{j_1, \ldots, j_n} \cdot E[\bar{r}_1 r_1^{j_1}] \cdots E[\bar{r}_n r_n^{j_n}] L(x_1^{j_1}, \ldots, x_n^{j_n})$$

since the random variables are independent. If $j_k \ne 1$ for some k then $j_l = 0$ for some l and $E[\bar{r}_l r_l^{j_l}] = E[\bar{r}_l] = 0$. If $j_k = 1$ then

$$E[\bar{r}_k r_k^{j_k}] = E[|r_k|^2] = 1.$$

Hence $E[\Phi] = n! \, L(x_1, \ldots, x_n)$ and this completes the proof.

If $(b_i)_{i=1}^n$ are n independent Bernoulli random variables, which take the values ± 1 with probability $\frac{1}{2}$, then

$$E\left[b_1 \cdots b_n \cdot P\left(\sum_{i=1}^{n} b_i x_i\right)\right] = \sum_{\varepsilon_i = \pm 1} \frac{\varepsilon_1 \cdots \varepsilon_n}{2^n} P\left(\sum_{i=1}^{n} \varepsilon_i x_i\right)$$

and we have the following classical result.

Corollary 1.6 *Let* $P \in \mathcal{P}_a(^n E; F)$, *where* E *and* F *are vector spaces over* $\mathbb{C}$, *and let* $L \in \mathcal{L}_a^s(^n E; F)$. *If* $\hat{L} = P$ *then*

$$L(x_1, \ldots, x_n) = \frac{1}{2^n n!} \sum_{\varepsilon_i = \pm 1} \varepsilon_1 \cdots \varepsilon_n P\left(\sum_{i=1}^{n} \varepsilon_i x_i\right). \tag{1.9}$$

Corollary 1.7 *The mapping*

$$\hat{\ } : \mathcal{L}_a^s(^n E; F) \longrightarrow \mathcal{P}_a(^n E; F)$$

is a linear isomorphism.

We denote the inverse of this mapping by $\vee$. Thus we see that the spaces $\mathcal{L}_a^s(^n E; F)$, $\mathcal{L}_a(\bigotimes_{n,s} E; F)$ and $\mathcal{P}_a(^n E; F)$ are linearly isomorphic. For $P \in \mathcal{P}_a(^n E; F)$ and $x \in E$ we have

$$P(x) = \overset{\vee}{P}(x^n) = \langle j_n^*(P), x \otimes \cdots \otimes x \rangle \tag{1.10}$$

and for $L \in \mathcal{L}_a(^n E; F)$

$$L(x^n) = \hat{L}(x) = \langle j_n^*(\hat{L}), x \otimes \cdots \otimes x \rangle. \tag{1.11}$$

Note that $\hat{\ }$ is defined on $\mathcal{L}_a(^n E; F)$ and that its restriction to $\mathcal{L}_a^s(^n E; F)$, which we again denote by $\hat{\ }$, is bijective. When $F = \mathbb{C}$ we obtain isomorphisms between the spaces $\mathcal{L}_a^s(^n E)$, $(\bigotimes_{n,s} E)^*$ and $\mathcal{P}_a(^n E)$. To simplify our notation we write P in place of $j_n^*(\overset{\vee}{P})$ and, with this convention, we have

$$\langle P, x \otimes \cdots \otimes x \rangle = P(x) \tag{1.12}$$

for all $x \in E$ and $P \in \mathcal{P}_a(^n E)$.

The dual pairing between $\mathcal{P}_a(^n E)$ and $\bigotimes_{n,s} E$ is then given by the formula

$$\langle P, \theta \rangle = \sum_i P(x_i) \tag{1.13}$$

where $\sum_i x_i \otimes \cdots \otimes x_i$ is any symmetric representation of θ, $\theta \in \bigotimes_{n,s} E$, and $P \in \mathcal{P}_a(^n E)$.

We develop the above linearization and dual space representations at both the continuous (or topological) and holomorphic levels later. These representations may not appear remarkable in view of our way of introducing polynomials. But polynomials can also be defined, using the classical definition of polynomial on $\mathbb{C}^n$, by considering restrictions to finite dimensional subspaces (see Section 1.5). Using this definition one arrives at the same concept of polynomial although no hint of linearization or duality appears in this approach.

A classical inequality of Carleman states the following:

If $(a_n)_n$ is a sequence of positive real numbers then

$$\sum_n (a_1 \cdots a_n)^{1/n} \le e \sum_n a_n$$

and e is the best possible constant.

This inequality may be considered an extreme case (see Section 1.3) of a general *Polarization Inequality*, which we now derive from Corollary 1.6.

Proposition 1.8 *If E and F are vector spaces over $\mathbb{C}$, A is a convex balanced subset of E and α is a semi-norm on F then*

$$\|\widehat{L}\|_{\alpha,A} \le \|L\|_{\alpha,A^n} \le \frac{n^n}{n!} \|\widehat{L}\|_{\alpha,A} \tag{1.14}$$

for any $L \in \mathcal{L}_a^s(^nE;F)$.

Proof. The left hand side of this inequality is trivial since $\widehat{L}(A) \subset L(A^n)$. By Corollary 1.6 we have

$$\|L\|_{\alpha,A^n} \le \frac{1}{2^n n!} \sum_{\varepsilon_i = \pm 1} \sup_{x_i \in A} \alpha\left(\widehat{L}\left(\sum_{i=1}^n \varepsilon_i x_i\right)\right).$$

If $x_i \in A$ and $\varepsilon_i = \pm 1$ then $\dfrac{1}{n} \sum\limits_{i=1}^n \varepsilon_i x_i \in A$ since A is convex and balanced.

Hence

$$\alpha\left(\widehat{L}\left(\sum_{i=1}^n \varepsilon_i x_i\right)\right) = n^n \alpha\left(\widehat{L}\left(\frac{1}{n}\sum_{i=1}^n \varepsilon_i x_i\right)\right)$$

$$\le n^n \|\widehat{L}\|_{\alpha,A}$$

and

$$\|L\|_{\alpha,A^n} \le \frac{1}{2^n n!} \sum_{\varepsilon_i = \pm 1} n^n \|\widehat{L}\|_{\alpha,A} = \frac{n^n}{n!} \|\widehat{L}\|_{\alpha,A}.$$

This completes the proof.

In Section 1.3 we show that the constant $n^n/n!$ is best possible and more or less uniquely achieved.

To present a tensorial version of Proposition 1.8 we require some notation. If A is a subset of a vector space E let $\Gamma(A)$ denote the convex balanced hull (or absolutely convex hull) of A and if E is a locally convex space let $\overline{\Gamma}(A)$ denote the closed convex balanced hull of A. If n is a positive integer let

$$\bigotimes_{n} A := \{\underbrace{x_1 \otimes \cdots \otimes x_n}_{n \text{ times}} \ : \ x_i \in A \text{ for all } i\} = i_n(A^n) \quad \text{and}$$

$$\bigotimes_{n,s} A = \{\underbrace{x \otimes \cdots \otimes x}_{n \text{ times}} \ : \ x \in A\} = \delta_n(A).$$

If α is a semi-norm on F then, using (1.11), (1.14) implies

$$\|j_n^*(L)\|_{\alpha, \bigotimes_{n,s} A} \leq \|L\|_{\alpha, A^n} \leq \frac{n^n}{n!} \|j_n^*(L)\|_{\alpha, \bigotimes_{n,s} A} \tag{1.15}$$

for all $L \in \mathcal{L}_a^s({}^n E; F)$. Since $j_n^*(L) \circ j_n = L$ and $j_n(A^n) = s(\bigotimes_n A)$ the right hand side of (1.15) gives

$$\|j_n^*(L) \circ j_n\|_{\alpha, A^n} = \|j_n^*(L)\|_{\alpha, s(\bigotimes_n A)} \leq \|j_n^*(L)\|_{\alpha, \bigotimes_{n,s} \left(\frac{n}{(n!)^{1/n}} A\right)}$$

for all $L \in \mathcal{L}_a^s({}^n E; F)$. The Hahn–Banach Theorem, applied to finite subsets of A, implies

$$s\left(\bigotimes_n A\right) \subset \Gamma\left(\bigotimes_{n,s}\left(\frac{n}{(n!)^{1/n}} A\right)\right) = \frac{n^n}{n!}\Gamma\left(\bigotimes_{n,s} A\right). \tag{1.16}$$

So far we have confined ourselves to the algebraic theory of polynomials. Our main interest is continuous holomorphic mappings between locally convex spaces and, as a first step, we introduce continuous polynomials. We suppose that E and F are locally convex spaces over $\mathbb{C}$. Let $cs(E)$ denote the set of all continuous semi-norms on E. If α is a semi-norm on E let $B_\alpha(x, r) = B_E^\alpha(x, r) = \{y \in E : \alpha(x - y) < r\}$ and $\overline{B_\alpha(x, r)} = \{y \in E : \alpha(x - y) \leq r\}$. If $x = 0$ we sometimes write $B_\alpha(r)$ and $\overline{B_\alpha(r)}$ in place of $B_\alpha(0, r)$ and $\overline{B_\alpha(0, r)}$ respectively. If $\alpha \in cs(E)$ let $E_\alpha = (E, \alpha)/\alpha^{-1}(0)$ and π_α will denote the (canonical) mapping from E onto E_α. Note that E_α is a normed linear space, $\alpha(x) = 0$ if and only if $\pi_\alpha(x) = 0$ and $\pi_\alpha(B_\alpha(1))$ is the open unit ball of E_α. The topology on F is the initial topology arising from the mappings $(\pi_\alpha)_{\alpha \in cs(F)}$ and $f : E \to F$ is continuous if and only if $\pi_\alpha \circ f : E \to F_\alpha$ is continuous for each $\alpha \in cs(F)$. For this reason we frequently restrict ourselves to polynomials and holomorphic functions taking their values in a normed linear space.

We let $\mathcal{P}(^nE;F)$, $\mathcal{L}(^nE;F)$ and $\mathcal{L}^s(^nE,F)$ denote respectively the spaces of continuous n-homogeneous polynomials from E into F, the continuous n-linear mappings from E into F and the continuous symmetric n-linear mappings from E into F. In all cases E^n is given the product topology. We use the notation $\mathcal{P}(^nE)$, $\mathcal{L}(^nE)$ and $\mathcal{L}^s(^nE)$ in place of $\mathcal{P}(^nE;\mathbb{C})$, $\mathcal{L}(^nE;\mathbb{C})$ and $\mathcal{L}^s(^nE;\mathbb{C})$, respectively, and when $n=1$ we write E' in place of $\mathcal{P}(^1E) = \mathcal{L}(^1E)$. It is easily checked, using (1.5) and (1.6), that the restrictions

$$\wedge : \mathcal{L}(^nE;F) \longrightarrow \mathcal{P}(^nE;F)$$
$$\wedge : \mathcal{L}^s(^nE,F) \longrightarrow \mathcal{P}(^nE;F)$$
$$s : \mathcal{L}(^nE;F) \longrightarrow \mathcal{L}^s(^nE,F)$$

are well defined and, using the Polarization Formula, the mapping

$$\vee : \mathcal{P}(^nE;F) \longrightarrow \mathcal{L}^s(^nE,F)$$

is a well defined linear isomorphism. We next show that continuity of polynomials follows, as in the linear case, from apparently much weaker conditions. This leads to useful factorization results. For this we need two rather simple but useful preliminary results. The proofs use the Binomial Theorem and the one variable Maximum Modulus Theorem.

Lemma 1.9 *Let E and F be vector spaces over $\mathbb{C}$, $P \in \mathcal{P}_a(^nE;F)$ and $x, y \in E$.*

(a)

$$P(x+y) = \sum_{j=0}^{n} \binom{n}{j} \overset{\vee}{P}(x^j, y^{n-j})$$

$$(1.17)$$

$$P(x) - P(y) = \sum_{j=0}^{n-1} \binom{n}{j} \overset{\vee}{P}(y^j, (x-y)^{n-j})$$

(b) If $F = \mathbb{C}$ then

$$\sup_{|\lambda|\leq 1} |P(x+\lambda y)|^2 \geq |P(x)|^2 + |P(y)|^2.$$

Proof. The proof of (a) is a simple application of the Binomial Theorem. We prove (b). Let

$$g(\lambda) = P(x+\lambda y) = \sum_{j=0}^{n} a_j \lambda^j$$

for $\lambda \in \mathbb{C}$. Clearly $a_0 = P(x)$ and, since P is n-homogeneous, $a_n = P(y)$. If $i = \sqrt{-1}$ then

$$\sup_{|\lambda|\leq 1} |P(x+\lambda y)|^2 = \sup_{|\lambda|=1} |P(x+\lambda y)|^2$$

and hence,

$$
\sup_{|\lambda|\leq 1} |P(x+\lambda y)|^2 \geq \frac{1}{2\pi} \int_0^{2\pi} \left(\sum_{j=0}^n a_j e^{ij\theta}\right)\left(\sum_{k=0}^n \bar{a}_k e^{-ik\theta}\right) d\theta
$$

$$
= \sum_{0\leq j,k\leq n} a_j \bar{a}_k \frac{1}{2\pi} \int_0^{2\pi} e^{i(j-k)\theta} d\theta
$$

$$
= \sum_{j=0}^n |a_j|^2
$$

$$
\geq |a_0|^2 + |a_n|^2
$$

$$
= |P(x)|^2 + |P(y)|^2.
$$

This completes the proof.

If we consider k vectors $x_1,\ldots,x_k \in E$ then, by Lemma 1.9(b) and induction,

$$
\sup_{\substack{|\lambda_j|\leq 1 \\ j=1,\ldots,k}} \left|P\left(\sum_{j=1}^k \lambda_j x_j\right)\right|^2 = \sup_{\substack{|\lambda_j|\leq 1 \\ j=2,\ldots,k \\ |\lambda|\leq 1}} \left|P\left(x_1 + \lambda\sum_{j=2}^k \lambda_j x_j\right)\right|^2
$$

$$
\geq |P(x_1)|^2 + \sup_{\substack{|\lambda_j|\leq 1 \\ j=2,\ldots,k}} \left|P\left(\sum_{j=2}^k \lambda_j x_j\right)\right|^2 \tag{1.18}
$$

$$
\geq \sum_{j=1}^k |P(x_j)|^2.
$$

We obtain an l_1 version of this result later as a consequence of Lemma 1.57. Lemma 1.9(a) may also be interpreted as a Taylor series expansion of the polynomial P about the point x and, in anticipation of notation introduced in Chapter 3, we denote by $\dfrac{\widehat{d^j} P(x)}{j!}$, $x \in E$, and $\dfrac{\widehat{d^j} P}{j!}$, $0 \leq j \leq n$, respectively, the mappings

$$
y \in E \longrightarrow \binom{n}{j} \overset{\vee}{P}(x^{n-j}, y^j) \in F,
$$

$$
x \in E \longrightarrow \frac{\widehat{d^j} P(x)}{j!} \in \mathcal{P}(^j E; F).
$$

We have

$$
\frac{\widehat{d^j} P(x)}{j!} \in \mathcal{P}(^j E; F) \, , \quad \frac{\widehat{d^j} P}{j!} \in \mathcal{P}\left(^{n-j} E; \mathcal{P}(^j E; F)\right)
$$

and

$$P(x + y) = \sum_{j=0}^{n} \left[\frac{\widehat{d^j} P(x)}{j!} \right] (y)$$

for all $x, y \in E$. This is just the *Taylor series expansion* of P about the point x. The choice of notation will become clear in Chapter 3 when we discuss Taylor series expansions of holomorphic mappings between locally convex spaces.

If B is a convex balanced subset of the vector space E we define the *Minkowski functional* of B, $\| \ \|_B$, as follows:

$$\|x\|_B = \inf\{\lambda > 0 \ : \ x \in \lambda B\}$$

and $\|x\|_B = \infty$ if $x \notin \lambda B$ for all $\lambda > 0$. We have

$$\{x \ : \ \|x\|_B < 1\} \subset B \subset \{x \ : \ \|x\|_B \leq 1\}.$$

The function $\| \ . \ \|_B$ is a semi-norm on the subspace of E generated by B, E_B. Since B is convex and balanced,

$$E_B = \bigcup_{n>0} nB.$$

Lemma 1.10 *Let E and F be vector spaces over $\mathbb{C}$, $P \in \mathcal{P}_a(^nE; F)$, $B \subset E$ and let α denote a semi-norm on F.*

(a) If B is balanced and $x \in E$ then

$$\|P\|_{\alpha, B} \leq \|P\|_{\alpha, x+B}.$$

(b) If B is convex and balanced, $\lambda \in \mathbb{R}, \lambda > 0$ and $\lambda x \in B$ then

$$\|P\|_{\alpha, x+B} \leq \left(1 + \frac{1}{\lambda}\right)^n \|P\|_{\alpha, B}.$$

(c) If B is convex and balanced, x and $y \in B$, $\|y\|_B \leq \varepsilon$ and $\|x - y\|_B \leq 1$ then

$$\alpha\big(P(x) - P(y)\big) \leq \frac{n^n}{n!}(1 + \varepsilon)^n \|x - y\|_B \cdot \|P\|_{\alpha, B}.$$

Proof. (a) By the Maximum Modulus and Hahn–Banach Theorems we obtain

$$\|P\|_{\alpha, x+B} := \sup_{y \in B} \alpha\big(P(x + y)\big)$$

$$= \sup_{y \in B, \theta \in \mathbb{R}} \alpha\big(P(x + e^{i\theta}y)\big), \quad (B \text{ is balanced})$$

$$= \sup_{y \in B, \theta \in \mathbb{R}} \alpha\big(P(e^{-i\theta}x + y)\big), \quad (\text{homogeneity})$$

$$\geq \sup_{y \in B} \alpha\big(P(y)\big),$$

$$= \|P\|_{\alpha, B}.$$

(b) If $\lambda x \in B$ and $\lambda > 0$ then, since B is convex, $x + B \subset \frac{1}{\lambda}B + B = (\frac{1}{\lambda}+1)B$. By homogeneity, $\|P\|_{\alpha,x+B} \leq \|P\|_{\alpha,(\frac{1}{\lambda}+1)B} = (1 + \frac{1}{\lambda})^n \|P\|_{\alpha,B}$.

(c) By Lemma 1.9(a) and the Polarization Formula

$$\alpha(P(x) - P(y)) = \alpha\left(\sum_{r=0}^{n-1} \binom{n}{r} \overset{\vee}{P}(y^r, (x-y)^{n-r})\right)$$

$$\leq \sum_{r=0}^{n-1} \binom{n}{r} \|\overset{\vee}{P}\|_{\alpha,B^n} \cdot \|y\|_B^r \cdot \|x-y\|_B^{n-r}$$

$$\leq \frac{n^n}{n!} \|P\|_{\alpha,B} \|x-y\|_B \sum_{r=0}^{n-1} \binom{n}{r} \|y\|_B^r$$

$$\leq \frac{n^n}{n!} (1+\varepsilon)^n \|x-y\|_B \cdot \|P\|_{\alpha,B}.$$

This completes the proof.

The simple technique used in the above proof is often useful.

Proposition 1.11 *Let E be a locally convex space, F a normed linear space and suppose $P \in \mathcal{P}_a(^nE; F)$. The following are equivalent:*

(a) *P is locally uniformly continuous (i.e. for each $x \in E$ there exists a neighbourhood V of x such that $P|_V$ is uniformly continuous),*

(b) *P is everywhere continuous,*

(c) *P is continuous at some point,*

(d) *P is bounded on a neighbourhood of some point in E,*

(e) *P is a locally bounded function (i.e. every point in E contains a neighbourhood on which P is bounded).*

Proof. The implications $(a) \Rightarrow (b) \Rightarrow (c) \Rightarrow (d)$ are clear. By Lemma 1.10(a) and (b) we have $(d) \Leftrightarrow (e)$. If P is bounded on $x_0 + V$ where V is a convex balanced neighbourhood of zero then Lemma 1.10(a) implies $\|P\|_V < \infty$. Since $\| \cdot \|_V$ is a continuous semi-norm on E, Lemma 1.10(c) implies that P is uniformly continuous on αV for any $\alpha > 0$. Hence $(d) \Rightarrow (a)$. This completes the proof.

Corollary 1.12 *Let E and F be locally convex spaces over $\mathbb{C}$ and let $P \in \mathcal{P}_a(^nE; F)$. Then $P \in \mathcal{P}(^nE; F)$ if and only if P is continuous at one point.*

We now look at a useful factorization lemma. If E and F are locally convex spaces, $\alpha \in cs(E)$ and $P \in \mathcal{P}(^nE_\alpha; F)$ then

$$\pi_\alpha^*(P) := P \circ \pi_\alpha \in \mathcal{P}(^nE; F).$$

Hence $P(^nE_\alpha; F)$ may be identified with a subspace of $P(^nE; F)$. The factorization lemma tells us, in the case of a normed range, that the union of all such subspaces covers $P(^nE; F)$.

Lemma 1.13 (Factorization Lemma) *If E is a locally convex space and F is a normed linear space then*

$$P(^nE; F) = \bigcup_{\alpha \in cs(E)} \pi_\alpha^* \big(P(^nE_\alpha; F) \big)$$

for every positive integer n.

Proof. Let $P \in P(^nE; F)$. Since F is a normed linear space there exists a convex balanced neighbourhood of zero V such that $\|P\|_V < \infty$. Let $\alpha = \|\cdot\|_V$. If $x, y \in E$ and $\alpha(y) = 0$ then Lemma 1.10(c) implies

$$\|P(x+y) - P(x)\| \le \frac{n^n}{n!}(1 + \alpha(x))^n \alpha(y)\|P\|_V = 0$$

and hence $P(x+y) = P(x)$. This shows that the mapping $\widetilde{P}: E_\alpha \to F$ which sends x to $P(z)$ whenever $\pi_\alpha(z) = x$ is well defined. It is easily checked that $\widetilde{P} \in P_a(^nE_\alpha; F)$ and $\|\widetilde{P}\|_{\pi_\alpha(V)} = \|P\|_V$. Hence $\widetilde{P} \in P(^nE_\alpha; F)$. Since $P = \widetilde{P} \circ \pi_\alpha$ this completes the proof. $\quad\blacksquare$

Our next example shows that this result does not extend to arbitrary range spaces.

Example 1.14 The vector space of all sequences of complex numbers, $\mathbb{C}^{\mathbb{N}}$, endowed with the product topology or the topology of coordinate convergence, is a *Fréchet space* (i.e. a complete metrizable locally convex space). If $\alpha \in cs(\mathbb{C}^{\mathbb{N}})$ then $(\mathbb{C}^{\mathbb{N}})_\alpha$ is a finite dimensional normed linear space and hence linearly isomorphic to $\mathbb{C}^n$ for some $n \in \mathbb{N}$. We claim that

$$\mathcal{L}(\mathbb{C}^{\mathbb{N}}; \mathbb{C}^{\mathbb{N}}) \neq \bigcup_{\alpha \in cs(\mathbb{C}^{\mathbb{N}})} \pi_\alpha^* \big(\mathcal{L}((\mathbb{C}^{\mathbb{N}})_\alpha; \mathbb{C}^{\mathbb{N}}) \big).$$

Clearly the identity map I belongs to $\mathcal{L}(\mathbb{C}^{\mathbb{N}}; \mathbb{C}^{\mathbb{N}})$ and $I(\mathbb{C}^{\mathbb{N}}) = \mathbb{C}^{\mathbb{N}}$. However, if $L \in \pi_\alpha^* \big(\mathcal{L}((\mathbb{C}^{\mathbb{N}})_\alpha; \mathbb{C}^{\mathbb{N}}) \big)$ for some $\alpha \in cs(\mathbb{C}^{\mathbb{N}})$ then $L(\mathbb{C}^{\mathbb{N}})$ is a finite dimensional subspace of $\mathbb{C}^{\mathbb{N}}$. Hence

$$I \notin \bigcup_{\alpha \in cs(\mathbb{C}^{\mathbb{N}})} \pi_\alpha^* \big(\mathcal{L}((\mathbb{C}^{\mathbb{N}})_\alpha; \mathbb{C}^{\mathbb{N}}) \big).$$

To obtain a factorization result for polynomials with values in an arbitrary locally convex space we use projective limits. For each $\alpha \in A$, A a directed set, let E_α be a locally convex space and for $\alpha, \beta \in A, \alpha \leq \beta$, let $\pi_{\alpha,\beta} : E_\alpha \to E_\beta$ denote a continuous linear mapping. We suppose $\pi_{\alpha,\alpha}$ is the identity for all α and $\pi_{\alpha,\gamma} = \pi_{\alpha,\beta} \circ \pi_{\beta,\gamma}$ for $\alpha \leq \beta \leq \gamma$. We call $\{E_\alpha, \pi_{\alpha,\beta}\}_A$ a *projective system*. The subspace

$$E := \{(x_\alpha)_{\alpha \in A} : \pi_{\alpha,\beta}(x_\beta) = x_\alpha, \ \alpha \leq \beta\}$$

of the product $\prod_\alpha E_\alpha$ endowed with the induced topology is called the *projective limit* of $\{E_\alpha, \pi_{\alpha,\beta}\}_A$ and written $E = \varprojlim_\alpha E_\alpha$.

If E is a locally convex space, $\alpha, \beta \in \mathrm{cs}(E)$, $\alpha \leq \beta$, we define $\pi_{\alpha,\beta} :$ $E_\alpha \to E_\beta$ by letting $\pi_{\alpha,\beta}(\pi_\alpha(x)) = \pi_\beta(x)$ for all $x \in E$. The collection $\{E_\alpha, \pi_{\alpha,\beta}\}_{\mathrm{cs}(E)}$ is a projective system and E is a subspace of the projective limit of normed linear spaces $\varprojlim_{\alpha \in \mathrm{cs}(E)} E_\alpha$.

Proposition 1.15 *If E and $F := \varprojlim_\gamma F_\gamma$ are locally convex spaces and n is a positive integer then*

$$\mathcal{P}(^n E; F) = \varprojlim_\gamma \mathcal{P}(^n E; F_\gamma).$$

Combining Lemma 1.14 and Proposition 1.15 we see that for every $P \in$ $\mathcal{P}(^n E; F)$ and $\gamma \in \mathrm{cs}(F)$ there exists $\alpha \in \mathrm{cs}(E)$ and $\widetilde{P} \in \mathcal{P}(^n E_\alpha; F_\gamma)$ such that the diagram

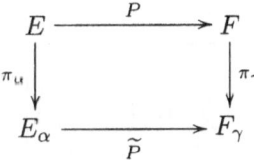

commutes. In view of Proposition 1.15 many results for polynomials taking their values in a normed linear space or Banach space can be readily extended to polynomials with values in an arbitrary locally convex space. For this reason we confine ourselves, in many cases, to Banach space valued polynomials and holomorphic functions. In fact, we often restrict ourselves even further by concentrating on the special Banach space $\mathbb{C}$. This avoids certain technical difficulties and allows a clearer presentation of the most important case. It also has two useful theoretical advantages since

(a) the space of $\mathbb{C}$-valued polynomials is an algebra;

(b) the space $\mathcal{P}(^n E)$ with its "natural" topology is a dual space (and this is not true for arbitrary Banach space valued polynomials).

Our next example is simple but shows some of the restrictions associated with the lack of local boundedness.

Example 1.16 Let $E = F \times F'_\beta$ where F is a locally convex space and F'_β is its *strong dual* (i.e. F' is the space of all continuous linear mappings on F and β is the topology on F' of uniform convergence on the bounded subsets of F). Let $A: E \times E \longrightarrow \mathbb{C}$ be defined by

$$A\big((x,x'),\ (y,y')\big) = \frac{1}{2}\big(x'(y) + y'(x)\big).$$

We have $A \in \mathcal{L}_a^s(^2E)$ and $\widehat{A}(x,x') = x'(x)$. Since $\widehat{A}$ is continuous if and only if it is locally bounded it follows that $\widehat{A}$ is continuous if and only if F is a normed linear space. From the definition of the β topology it follows that A and $\widehat{A}$ are continuous when restricted to the bounded, and hence compact, subsets of their domains.

We introduced continuous polynomials using multilinear forms and the product topology on E^n. To define continuous polynomials using tensors and to give a continuous version of (1.7) we need to place a locally convex topology on $\bigotimes_{n,s} E$. In fact, as we shall now see, this motivation leads us naturally to the projective topology on the symmetric tensors. Let $P \in \mathcal{P}_a(^nE)$, $\alpha \in cs(E)$ and $\theta \in \bigotimes_{n,s} E$. If $\sum_i x_i \otimes \cdots \otimes x_i$ is a symmetric tensor representation of θ then

$$
\begin{aligned}
|\langle P, \theta \rangle| &\leq \sum_i |\langle P, x_i \otimes \cdots \otimes x_i \rangle| \\
&= \sum_i |P(x_i)| \\
&= \sum_i \left| P\left(\frac{x_i}{\alpha(x_i)}\right) \right| \alpha(x_i)^n \\
&\leq \|P\|_{B_\alpha(1)} \sum_i \alpha(x_i)^n.
\end{aligned}
\tag{1.19}
$$

Since (1.19) holds for any representation of θ we have

$$|\langle P, \theta \rangle| \leq \|P\|_{B_\alpha(1)} \inf\left\{ \sum_i \alpha(x_i)^n \ :\ \theta = \sum_i x_i \otimes \cdots \otimes x_i \right\} \tag{1.20}$$

As polynomials are continuous if and only if they are locally bounded (1.20) suggests that we let

$$\pi_{\alpha,n}(\theta) = \inf\left\{ \sum_i \alpha(x_i)^n \ :\ \theta = \sum_i x_i \otimes \cdots \otimes x_i \right\}.$$

Clearly $\pi_{\alpha,n}$ is a semi-norm on $\bigotimes_{n,s} E$. We define the π or *projective topology* on $\bigotimes_{n,s} E$ as the locally convex topology generated by $\{\pi_{\alpha,n}\}_{\alpha \in cs(E)}$. Let $\bigotimes_{n,s,\pi} E$ denote the space $\bigotimes_{n,s} E$ endowed with the π topology and denote its completion by $\widehat{\bigotimes}_{n,s,\pi} E$. Inequality (1.20) now reads

$$|\langle P, \theta \rangle| \leq \|P\|_{B_\alpha(1)} \pi_{\alpha,n}(\theta)$$

and implies $\mathcal{P}(^nE) \subset \left(\bigotimes_{n,s,\pi} E \right)'$. On the other hand if $P \in \mathcal{P}_a(^nE)$ and $P \in \left(\bigotimes_{n,s,\pi} E \right)'$ then there exists $\alpha \in cs(E)$ such that $|\langle P, \theta \rangle| \leq 1$ for all $\theta \in \bigotimes_{n,s} E$ satisfying $\pi_{\alpha,n}(\theta) \leq 1$. If $\alpha(x) \leq 1$ then $\pi_{\alpha,n}(x \otimes \cdots \otimes x) \leq \alpha(x)^n \leq 1$. Hence $|P(x)| = |\langle P, x \otimes \cdots \otimes x \rangle| \leq 1$ and $\|P\|_{B_\alpha(1)} \leq 1$. By Proposition 1.11, $P \in \mathcal{P}(^nE)$ and, since the diagonal mapping $\delta_n : E \longrightarrow \bigotimes_{n,s,\pi} E$ is easily seen to lie in $\mathcal{P}(^nE; \widehat{\bigotimes}_{n,s,\pi} E)$, we have diagram (1.7) at the commutative level and have proved the following result.

Proposition 1.17 *If E is a locally convex space then*

$$\mathcal{P}(^nE) \cong \left(\widehat{\bigotimes}_{n,s,\pi} E \right)'$$

with duality given by $\langle P, x \otimes \cdots \otimes x \rangle = P(x)$.

Now suppose G_n is a complete locally convex space, $\varepsilon_n \in \mathcal{P}(^nE; G_n)$ and for each complete locally convex space F and each $P \in \mathcal{P}(^nE; F)$ there exists a unique $\tilde{P} \in \mathcal{L}(G_n; F)$ such that $P = \tilde{P} \circ \varepsilon_n$. By the Hahn–Banach Theorem $\varepsilon_n(E)$ generates a dense subspace of G_n. Using the definition of G_n and (1.7) we obtain $\theta \in \mathcal{L}\left(G_n; \widehat{\bigotimes}_{n,s,\pi} E\right)$ and $\psi \in \mathcal{L}\left(\widehat{\bigotimes}_{n,s,\pi} E; G_n\right)$ such that the diagrams

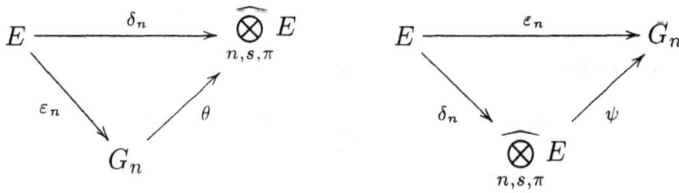

commute. Hence

$$\delta_n(x) = \theta(\varepsilon_n(x)) = x \otimes \cdots \otimes x,$$
$$\varepsilon_n(x) = \psi(\delta_n(x)) = \psi(x \otimes \cdots \otimes x),$$
$$\theta(\psi(x \otimes \cdots \otimes x)) = x \otimes \cdots \otimes x$$

and

$$\psi\Big(\theta\big(\varepsilon_n(x)\big)\Big) = \varepsilon_n(x).$$

This implies θ is a continuous linear isomorphism, with inverse ψ, between G_n and $\underset{n,s,\pi}{\bigotimes} E$ and the projective topology on $\underset{n,s}{\bigotimes} E$ is determined by the existence and commutativity of the diagram (1.7).

A similar construction is possible for $\underset{n}{\bigotimes} E$. If $\theta \in \underset{n}{\bigotimes} E$ and $\alpha \in \mathrm{cs}(E)$ let

$$\rho_{\alpha,n}(\theta) = \inf\Big\{\sum_i \alpha(x_{i,1}) \cdots \alpha(x_{i,n}) \; : \; \theta = \sum_i x_{i,1} \otimes \cdots \otimes x_{i,n}\Big\}$$

and denote the topology on $\underset{n}{\bigotimes} E$ generated by $\{\rho_{\alpha,n}\}_{\alpha \in \mathrm{cs}(E)}$ also by π. We write $\underset{n,\pi}{\bigotimes} E$ in place of $\big(\underset{n}{\bigotimes} E, \pi\big)$ and denote its completion by $\underset{n,\pi}{\widehat{\bigotimes}} E$.

If E is a locally convex space, F is a normed linear space and $L \in \mathcal{L}(^n E; F)$ then there exists an $\alpha \in \mathrm{cs}(E)$ such that

$$\|L(x_1, \ldots, x_n)\| \le \alpha(x_1) \cdots \alpha(x_n)$$

for all $x_1, \ldots x_n \in E$. Hence

$$\|i_n^*(L)(x_1 \otimes \cdots \otimes x_n)\| \le \alpha(x_1) \cdots \alpha(x_n)$$

and using tensor representations of $\theta \in \underset{n,\pi}{\bigotimes} E$ we see this implies

$$\|i_n^*(L)(\theta)\| \le \rho_{\alpha,n}(\theta)$$

for all $\theta \in \underset{n,\pi}{\widehat{\bigotimes}} E$. Since $\rho_{\alpha,n}(x_1 \otimes \cdots \otimes x_n) \le \alpha(x_1) \cdots \alpha(x_n)$ for all $x_1, \ldots, x_n \in E$ the mapping i_n in (1.2) is continuous and indeed the diagram (1.2) is commutative for continuous mappings. This shows that for any locally convex space we have (algebraically)

$$\mathcal{L}(^n E; F) \cong \mathcal{L}\Big(\underset{n,\pi}{\bigotimes} E; F\Big).$$

Similarly we have

$$\mathcal{L}^s(^n E, F) \cong \mathcal{L}\Big(\underset{n,s,\pi}{\bigotimes} E; F\Big) \cong \mathcal{P}(^n E; F).$$

In these identifications we can replace the spaces of tensors by their completions if F is complete. It is also easily seen for $\alpha \in \mathrm{cs}(E)$, E a locally convex space, that

$$\Big\{\theta \in \underset{n,s}{\bigotimes} E \; : \; \pi_{\alpha,n}(\theta) \le 1\Big\} = \overline{\Gamma}\Big(\underset{n,s}{\bigotimes} B_\alpha(1)\Big)$$

and

$$\left\{ \theta \in \bigotimes_{n} E \ : \ \rho_{\alpha,n}(\theta) \leq 1 \right\} = \overline{\Gamma}\left(\bigotimes_{n} B_\alpha(1) \right).$$

By the Polarization Formula, (1.16), we have

$$\bigotimes_{n,s} B_\alpha(1) \subset s\left(\bigotimes_{n} B_\alpha(1) \right) \subset \frac{n^n}{n!} \overline{\Gamma}\left(\bigotimes_{n,s} B_\alpha(1) \right).$$

This shows that

$$s \colon \widehat{\bigotimes_{n,\pi}} E \longrightarrow \widehat{\bigotimes_{n,s,\pi}} E$$

is a well defined continuous linear surjection. Hence $\widehat{\bigotimes}_{n,s,\pi} E$ is a complemented

subspace of $\widehat{\bigotimes}_{n,\pi} E$ for all n.

1.2 Topologies on Spaces of Polynomials

In this section we define, discuss and compare various locally convex topologies on $\mathcal{P}(^n E; F)$. The motivation for these topologies comes from point set topology, functional analysis and several complex variables theory.

Definition 1.18 Let E and F be locally convex spaces over $\mathbb{C}$ and let $n \in \mathbb{N}$. The compact open topology τ_0 on $\mathcal{P}(^n E; F)$ is the topology of uniform convergence on the compact subsets of E.

The τ_0 topology on $\mathcal{P}(^n E; F)$ is generated by the semi-norms $\| \cdot \|_{\alpha,K}$ where α ranges over $cs(F)$ and K over the compact subsets of E. We also use the notation E'_c in place of (E', τ_0).

We now define, motivated by the *strong topology* of linear functional analysis, the topology, on spaces of polynomials, of uniform convergence on bounded sets. First we recall that a subset B of a locally convex space E is *bounded* if it satisfies any of the following equivalent conditions:

–for each neighbourhood V of the origin there exists $\lambda > 0$ such that $B \subset \lambda V$,   (1.21)

–$\sup_{x \in B} \alpha(x) < \infty$ for each $\alpha \in cs(E)$,   (1.22)

–if $(x_n)_n$ is any sequence in B and V is any neighbourhood of the origin in E then there exists $\lambda > 0$ such that $x_n \in \lambda V$ for all n.   (1.23)

–if $(x_n)_n$ is any sequence in B and $(\lambda_n)_n$ is any null sequence in $\mathbb{C}$ then $(\lambda_n x_n)_n$ is a null sequence in E.   (1.24)

The bounded subsets of a locally convex space are closed under the operations of taking finite unions, vector sums, scalar multiples, absolutely convex hulls, subsets and, moreover, the image of a bounded set by a continuous linear mapping is also bounded.

Let E and F be locally convex spaces over $\mathbb{C}$ and let $P \in \mathcal{P}(^nE; F)$. If $\alpha \in \mathrm{cs}(F)$ then, by Proposition 1.11, there exists a neighbourhood V of 0 in E such that

$$\sup_{x \in V} \alpha(P(x)) := \|P\|_{\alpha,V} < \infty.$$

If B is a bounded subset of E then there exists $\lambda > 0$ such that $B \subset \lambda V$. Hence

$$\|P\|_{\alpha,B} \le \|P\|_{\alpha,\lambda V} = \lambda^n \|P\|_{\alpha,V} < \infty \tag{1.25}$$

and continuous polynomials map bounded sets onto bounded sets. This gives rise to the following definition.

Definition 1.19 Let E and F be locally convex spaces over $\mathbb{C}$ and let $n \in \mathbb{N}$. The bounded open topology on $\mathcal{P}(^nE; F)$, τ_b, is the topology of uniform convergence on the bounded subsets of E. If $n = 1$ and $F = \mathbb{C}$ we write E'_β in place of (E', τ_b) and call E'_β the strong dual of E.

The τ_b topology on $\mathcal{P}(^nE; F)$ is generated by the semi-norms $\| \cdot \|_{\alpha,B}$ where α ranges over $\mathrm{cs}(F)$ and B ranges over the bounded subsets of E. By homogeneity a fundamental system of semi-norms for τ_0 (respectively τ_b) is obtained by considering $\| \cdot \|_{\alpha,A}$ where α ranges over a fundamental system of semi-norms for F and A ranges over a collection $\mathcal{A}$ of compact (respectively bounded) subsets of E with the property that for each compact (respectively bounded) subset B of E there exists $\lambda \in \mathbb{C}$ and $A \in \mathcal{A}$ such that $B \subset \lambda A$.

Since each compact subset of a locally convex space is bounded we always have $\tau_0 \le \tau_b$ on $\mathcal{P}(^nE; F)$. If E is *quasi-complete* (i.e. if the closed bounded subsets of E are complete) and, in particular, if E is complete the Hahn–Banach Theorem implies that $\tau_b = \tau_0$ if and only if the closed convex hull of each bounded subset of E is compact. Motivated by Montel's Theorem in classical function theory a locally convex space in which each closed bounded set is compact is called *semi-Montel*.

Example 1.20 Let $(E, \| \cdot \|_E)$ and $(F, \| \cdot \|_F)$ denote normed linear spaces over $\mathbb{C}$. The τ_b topology on $\mathcal{P}(^nE; F)$ is generated by the norm

$$\|P\| := \sup_{\|x\|_E \le 1} \|P(x)\|_F.$$

Hence $\|P(x)\| \le \|P\| \|x\|^n$ for all $P \in \mathcal{P}(^nE; F)$ and all $x \in E$. In this case we write $\| \cdot \|$ in place of τ_b. The space $(\mathcal{P}(^nE; F), \| \cdot \|)$ is a normed linear space and, if F is complete, Proposition 1.11 shows that $(\mathcal{P}(^nE; F), \| \cdot \|)$ is a Banach space.

The two topologies we have introduced admit explicit fundamental systems of semi-norms but do not always enjoy good topological properties. We introduce a further topology originally motivated, in infinite dimensional holomorphy, by properties of analytic functionals in several complex variables theory (see Chapter 3). This topology also appears in functional analysis where it is called *the inductive topology*. It is perhaps more useful than either τ_0 or τ_b by virtue of its strong topological properties but is more difficult to characterize in a concrete fashion. This topology is called the τ_ω or *Nachbin topology*. We define it by using the τ_b topology when the range space is a normed linear space and use factorization results – Lemma 1.13 and Proposition 1.15 – for the general definition.

Definition 1.21 Let E be a locally convex space and let F denote a normed linear space. The τ_ω topology on $\mathcal{P}(^nE;F)$ is the inductive limit topology in the category of locally convex spaces and continuous linear mappings of the spaces $(\mathcal{P}(^nE_\alpha;F),\|\cdot\|)$ where α ranges over $\mathrm{cs}(E)$, i.e.

$$\left(\mathcal{P}(^nE;F),\tau_\omega\right) = \varinjlim_{\alpha\in\mathrm{cs}(E)} \left(\mathcal{P}(^nE_\alpha;F),\|\cdot\|\right).$$

Hence a semi-norm p on $\mathcal{P}(^nE;F)$ is τ_ω continuous if and only if for each neighbourhood V of zero in E there exists $c(V) > 0$ such that

$$p(P) \leq c(V)\|P\|_V \tag{1.26}$$

for all $P \in \mathcal{P}(^nE;F)$.

We will subsequently see that this amounts to saying that a semi-norm on $\mathcal{P}(^nE;F)$ is τ_ω continuous if and only if it is ported by the origin. For this reason τ_ω is also called the *ported topology*. For arbitrary range spaces we use the above definition and Proposition 1.15 to define τ_ω on $\mathcal{P}(^nE;F)$.

Definition 1.22 Let E and F be locally convex spaces and let $n \in \mathbb{N}$. The τ_ω topology on $\mathcal{P}(^nE;F)$ is defined as

$$\varprojlim_{\beta\in\mathrm{cs}(F)} \left(\varinjlim_{\alpha\in\mathrm{cs}(E)} \mathcal{P}(^nE_\alpha;F_\beta), \|\cdot\| \right) = \varprojlim_{\beta\in\mathrm{cs}(F)} \left(\mathcal{P}(^nE;F_\beta), \tau_\omega\right).$$

If $n = 1$ and $F = \mathbb{C}$ we write E'_i in place of (E',τ_ω) and call E'_i the *inductive dual* of E.

By (1.25) and (1.26) we have $\tau_\omega \geq \tau_b$ on $\mathcal{P}(^nE;F)$ for locally convex spaces E and F. Each of the topologies we have introduced defines a collection of bounded subsets of $\mathcal{P}(^nE;F)$. A further type of boundedness which arises in our studies is local boundedness. A set of functions $\mathcal{F}$ defined on an open subset Ω of a locally convex space and taking their values in a locally convex space F is said to be *locally bounded* if for each $x \in \Omega$ there exists a neighbourhood V_x of x, contained in Ω, such that $\bigcup_{f \in \mathcal{F}} f(V_x)$ is a bounded subset of F. If F is a normed linear space then $\mathcal{F}$ is locally bounded if and only if we can choose V_x such that

$$\sup_{f \in \mathcal{F}} \|f\|_{V_x} := \sup_{\substack{f \in \mathcal{F} \\ y \in V_x}} \|f(y)\| < \infty.$$

Proposition 1.11 shows that a locally bounded subset of $\mathcal{P}_a(^nE; F)$ already lies in $\mathcal{P}(^nE; F)$ and, indeed, is contained and locally bounded in $\pi_\alpha^*\big(\mathcal{P}(^nE_\alpha; F)\big)$ for some $\alpha \in \mathrm{cs}(E)$. This proves part (a) of the following lemma.

Lemma 1.23 *If E and F are locally convex spaces and $n \in \mathbb{N}$ then*

(a) the locally bounded subsets of $\mathcal{P}(^nE; F)$ are τ_ω bounded,
(b) τ_b and τ_0 define the same bounded subsets of $\mathcal{P}(^nE; F)$.

Proof. (b) We may suppose without loss of generality that F is a normed linear space. Since $\tau_b \geq \tau_0$ it follows that τ_b bounded sets are τ_0 bounded. Suppose the converse is not true. Then there exists a τ_0 bounded subset $\mathcal{A}$ of $\mathcal{P}(^nE; F)$ and a bounded subset B of E such that

$$\sup_{P \in \mathcal{A}} \|P\|_B = \infty.$$

Hence we can choose a sequence $(P_m)_m$ in $\mathcal{A}$ and a sequence $(x_m)_m$ in B such that $\|P_m(x_m)\| \geq m^{n+1}$ for all m. The sequence $(x_m/m)_{m=1}^\infty$ is a null sequence in E and $K := \{(x_m/m)_m \cup 0\}$ is a compact subset of E. Since $\|P_m(\frac{x_m}{m})\| = \frac{1}{m^n}\|P_m(x_m)\| \geq m$ it follows that $\sup_{P \in \mathcal{A}} \|P\|_K = \infty$. This contradicts our hypothesis and completes the proof.

In order to give examples and to introduce properties of the τ_ω topology we recall some definitions and results from linear functional analysis.

If E is a locally convex space with topology τ then the following conditions are equivalent:

– *if W is a convex balanced subset of E which absorbs all bounded subsets (i.e. if B is a bounded subset of E then there exists $\lambda > 0$ such that $B \subset \lambda W$) then W is a neighbourhood of zero;*   (1.27)

– *if α is a semi-norm on E and $\sup_{x \in B} \alpha(x) < \infty$ for each bounded subset B of E then $\alpha \in \mathrm{cs}(E)$;*   (1.28)

– *if F is a locally convex space and $T: E \to F$ is a linear mapping which maps bounded sets onto bounded sets then T is continuous;*   (1.29)

– *if τ' is a locally convex topology on E which has the same bounded sets as τ then $\tau \geq \tau'$;*   (1.30)

– $E = \varinjlim\limits_{\alpha} E_\alpha$ *for some collection of normed linear spaces in the*

category of locally convex spaces and continuous linear mappings. (1.31)

A locally convex space which satisfies any of these conditions is called *bornological*. Metrizable locally convex spaces are bornological and, by (1.31), $(\mathcal{P}(^nE;F), \tau_w)$ is a bornological space for all n and E, if F is normed. If (E, τ) is a locally convex space then there exists on E a finest locally convex topology, with the same bounded sets as (E, τ). This is called the *bornological topology associated with* τ and denoted by τ^{bor}.

Example 1.24 If E is a metrizable locally convex space and F is a normed linear space then the τ_0 bounded subsets of $\mathcal{P}(^nE;F)$ are locally bounded for any positive integer n. The argument used in the proof of Lemma 1.23(b) can be used to prove this. Hence τ_0, τ_b and τ_w define the same bounded sets and τ_w is the bornological topology associated with both τ_0 and τ_b. If E is a normed linear space then $\tau_b = \tau_w$ on $\mathcal{P}(^nE;F)$.

Example 1.25 Let $E = \varinjlim\limits_{n} E_n$ denote a countable inductive limit of

normed linear spaces in the category of locally convex spaces and continuous linear mappings. By (1.31), E is a bornological space and, moreover, it contains a fundamental system of convex, balanced, bounded sets $(B_n)_{n=1}^\infty$. We may suppose, without loss of generality, that $E_n \subset E_{n+1}$ for all n and that B_n is the closed unit ball of E_n. By (1.27), a set of the form

$$\sum_{m=1}^{\infty} \lambda_m B_m := \left\{ \sum_{j=1}^{k} \lambda_j b_j \; : \; b_j \in B_j, \; k \text{ arbitrary} \right\} \qquad (*)$$

is a neighbourhood of the origin if $(\lambda_m)_m$ is any sequence of positive real numbers. On the other hand if V is a convex balanced neighbourhood of the origin in E then for each integer m there exists $\lambda_m > 0$ such that $\lambda_m B_m \subset V$. Hence $\sum_{m=1}^{\infty} (\lambda_m/2^m) B_m \subset V$ and thus sets which have the form $(*)$ are a fundamental system of neighbourhoods of the origin.

Let n denote a positive integer and let F be a normed linear space. Suppose $\mathcal{F}$ is a subset of $\mathcal{P}_a(^nE;F)$ and for all m

$$\sup_{P \in \mathcal{F}} \|P\|_{B_m} := M_m < \infty.$$

Let $\lambda_1 = 1$. Suppose positive numbers $\lambda_2, \ldots, \lambda_m$ have been chosen so that

$$\sup_{P \in \mathcal{F}} \|P\|_{\sum_{i=1}^{m} \lambda_i B_i} \leq M_1 \left(\sum_{i=1}^{m} \frac{1}{2^{i-1}} \right).$$

If $x \in \sum_{i=1}^{m} \lambda_i B_i$, $y \in B_{m+1}$, $P \in \mathcal{F}$ and $\lambda > 0$ then

$$P(x + \lambda y) = P(x) + \sum_{j=1}^{n} \binom{n}{j} \overset{\vee}{P}(x^{n-j}, y^j)\lambda^j$$

and hence

$$\|P\|_{\sum_{i=1}^{m} \lambda_i B_i + \lambda B_{m+1}} \leq \|P\|_{\sum_{i=1}^{m} \lambda_i B_i} + \sum_{j=1}^{n} \binom{n}{j} \|P\|_m \lambda^j$$

where

$$\|P\|_m := \|\overset{\vee}{P}\|_{(\sum_{i=1}^{m} \lambda_i B_i + B_{m+1})^n}.$$

By Proposition 1.8 and our hypothesis there exists $c > 0$, independent of P in $\mathcal{F}$, such that $\|P\|_m \leq c$. Hence we can choose $\lambda := \lambda_{m+1} > 0$, sufficiently small, so that

$$\sup_{P \in \mathcal{F}} \|P\|_{\sum_{i=1}^{m+1} \lambda_i B_i} \leq M_1 \left(\sum_{i=1}^{m+1} \frac{1}{2^{i-1}} \right).$$

By induction there exists a sequence of positive real numbers $(\lambda_m)_m$ such that

$$\sup_{P \in \mathcal{F}} \|P\|_{\sum_{m=1}^{\infty} \lambda_m B_m} \leq 2M_1.$$

For single functions, this together with Proposition 1.11 and the proof of Lemma 1.23(b), shows that elements of $\mathcal{P}_a(^nE; F)$ which are bounded on either bounded or compact subsets of E are continuous. Hence $\mathcal{F} \subset \mathcal{P}(^nE; F)$ and the τ_0 bounded subsets of $\mathcal{P}(^nE; F)$ are locally bounded. This shows that τ_0, τ_b and τ_ω define the same bounded subsets of $\mathcal{P}(^nE; F)$. Since E has a fundamental system of bounded sets it follows that $(\mathcal{P}(^nE; F), \tau_b)$ is metrizable and hence bornological. Since $\tau_\omega \geq \tau_b$ it follows, by (1.29), that $\tau_\omega = \tau_b$ on $\mathcal{P}(^nE; F)$.

In discussing duality theory for Banach spaces one remains within the category of Banach spaces. However, the strong dual of a metrizable locally convex space is metrizable if and only if it is a normed linear space. Hence, in order to adequately discuss duality theory for metrizable spaces we need DF spaces.

Definition 1.26 A locally convex space E is a DF space if it contains a countable fundamental system of bounded sets and if the intersection of any sequence of absolutely convex neighbourhoods of zero which absorbs all bounded sets is itself a neighbourhood of zero.

While DF spaces do not fully characterize the strong duals of metrizable spaces we have the following relationships:

–If E is a DF space then E'_β is a Fréchet space.
–If E is a Fréchet space then E'_β is a DF space.

The collection of spaces used in Example 1.25, the countable inductive limits of normed linear spaces, coincides with the set of all bornological DF spaces. In particular, a countable direct sum of Banach spaces is a DF space.

A further collection, the *barrelled locally convex spaces*, are defined as those locally convex spaces which satisfy any of the following equivalent conditions:

$-$*if W is a closed convex balanced absorbing (if $x \in E$ there exists $\lambda > 0$ such that $\lambda x \in W$) subset of E then W is a neighbourhood of zero,* (1.32)

$-$*all lower semi-continuous semi-norms on E are continuous,* (1.33)

$-$*the pointwise bounded subsets of E' are locally bounded or equicontinuous.* (1.34)

Fréchet spaces (but not all metrizable locally convex spaces) are barrelled and arbitrary inductive limits of barrelled (respectively bornological) spaces are barrelled (respectively bornological). This means, in particular, that $\big(\mathcal{P}(^nE; F),\, \tau_\omega\big)$ is barrelled when E is any locally convex space and F is a Banach space.

A semi-Montel space E is *semi-reflexive* (i.e. $(E'_\beta)' = E$) while a barrelled semi-Montel space E (we call such spaces *Montel* spaces) is *reflexive* (i.e. $(E'_\beta)'_\beta = E$). A semi-Montel DF space is a Montel space and is called a $\mathcal{DFM}$ space. A locally convex space E is a $\mathcal{DFM}$ space if and only if E'_β is a Fréchet–Montel ($\mathcal{FM}$) space. $\mathcal{DFM}$ spaces are bornological DF spaces and thus included in Example 1.25. Finally a locally convex space E is *distinguished* if E'_β is a barrelled locally convex space (see Exercise 1.89).

To place these definitions in perspective it is worth comparing conditions (1.21), (1.27), and (1.32) with one another and similarly conditions (1.22), (1.28) and (1.33).

Example 1.27 In this example we consider the space $E := \mathbb{C}^{\mathbb{N}} \times \mathbb{C}^{(\mathbb{N})}$ where $\mathbb{C}^{\mathbb{N}}$ is the space of all complex sequences with the product topology and $\mathbb{C}^{(\mathbb{N})}$ is the space of all finite complex sequences with the direct sum topology.

Let $P_n: E \to \mathbb{C}$ be given by $P_n\big((x_m)_{m=1}^\infty,\, (y_k)_{k=1}^\infty\big) := x_n y_n$ and let $B = (P_n)_{n=1}^\infty$. Clearly $P_n \in \mathcal{P}(^2E)$ for all n. If K is a compact subset of E then there exist compact subsets of $\mathbb{C}^{\mathbb{N}}$ and $\mathbb{C}^{(\mathbb{N})}$ such that $K \subset K_1 \times K_2$. Since each compact subset of $\mathbb{C}^{(\mathbb{N})}$ is finite dimensional it follows that $\|P_n\|_K = 0$ for all n sufficiently large. Hence B is a τ_0 bounded subset of $\mathcal{P}(^2E)$.

Let $u_n = (0,\ldots,\underset{\underset{n^{\text{th}}\text{position}}{\llcorner}}{1},0,\ldots) \in \mathbb{C}^{\mathbb{N}}$ and let $v_n = (0,\ldots,\underset{\underset{n^{\text{th}}\text{position}}{\llcorner}}{1},0,\ldots) \in \mathbb{C}^{(\mathbb{N})}$. The sequences $(u_n)_{n=1}^\infty$ and $(v_n)_{n=1}^\infty$ are the standard bases for $\mathbb{C}^{\mathbb{N}}$ and $\mathbb{C}^{(\mathbb{N})}$ respectively. If $P \in \mathcal{P}(^2E)$ then there exists, by Lemma 1.13, $\alpha \in \mathrm{cs}(E)$

such that $P(x + y) = P(x)$ for all x, $y \in E$ satisfying $\alpha(y) = 0$. If $\alpha \in \mathrm{cs}(E)$ is given then $\alpha(u_n, 0) = 0$ for all n sufficiently large. Hence

$$P(nu_n, v_n) = P(0, v_n),$$

for all n sufficiently large and

$$p(P) = \sum_{n=1}^{\infty} |P(nu_n, v_n) - P(0, v_n)|$$

defines a semi-norm on $\mathcal{P}(^2E)$. For each n the mapping

$$P \in \mathcal{P}(^2E) \longrightarrow |P(nu_n, v_n) - P(0, v_n)|$$

defines a τ_0 and hence a τ_ω continuous semi-norm on $\mathcal{P}(^2E)$. Since we are considering $\mathbb{C}$-valued polynomials, $(\mathcal{P}(^2E), \tau_\omega)$ is an inductive limit of Banach spaces and thus barrelled. By (1.32) or (1.33) a semi-norm which is the pointwise limit of an increasing sequence of continuous semi-norms on a barrelled locally convex space is continuous. Hence p is τ_ω continuous. Since $p(P_n) = n$ for all n it follows that B is not a τ_ω bounded subset of $\mathcal{P}(^2E)$. We have thus shown that τ_0 and τ_ω do not define the same bounded subsets of $\mathcal{P}(^2E)$ and, in particular, that $\tau_0 \underset{\neq}{<} \tau_\omega$ on $\mathcal{P}(^2E)$. This also follows from general results given later for holomorphic functions on fully nuclear spaces with basis (Corollary 4.46). We see later (Exercise 3.118) that τ_ω is the barrelled topology associated with τ_0 on $\mathcal{P}(^2E)$.

The space $\mathbb{C}^{\mathbb{N}} \times \mathbb{C}^{(\mathbb{N})}$ is probably our most useful counterexample. Since $\mathbb{C}^{(\mathbb{N})} \cong (\mathbb{C}^{\mathbb{N}})'_\beta$ this space is a particular example of the space considered in Example 1.16. In later chapters we shall use the following properties of $\mathbb{C}^{\mathbb{N}} \times \mathbb{C}^{(\mathbb{N})}$: it is a reflexive nuclear space with absolute basis, it is a strict inductive limit of Fréchet nuclear spaces and an open compact surjective limit of $\mathcal{DFM}$ spaces. Fundamental neighbourhood systems at the origin and systems of compact sets are easily described for both $\mathbb{C}^{\mathbb{N}}$ and $\mathbb{C}^{(\mathbb{N})}$ and these can be combined to produce similar systems in $\mathbb{C}^{\mathbb{N}} \times \mathbb{C}^{(\mathbb{N})}$. Products of 1-dimensional closed bounded discs form a fundamental system of compact subsets of $\mathbb{C}^{\mathbb{N}}$ while every neighbourhood of the origin in $\mathbb{C}^{\mathbb{N}}$ contains a neighbourhood of the form $U \times \mathbb{C}^{\mathbb{N}-\{l\}}$ where $l \in \mathbb{N}$ and U is a neighbourhood of zero in $\mathbb{C}^l$. For fundamental systems in $\mathbb{C}^{(\mathbb{N})}$ we refer to Example 1.25. The space $\mathbb{C}^{\mathbb{N}} \times \mathbb{C}^{(\mathbb{N})}$ is barrelled, bornological, reflexive, Montel and distinguished but it is neither a Fréchet space nor a DF space. Holomorphic and polynomial properties of $\mathbb{C}^{\mathbb{N}} \times \mathbb{C}^{(\mathbb{N})}$ are discussed in Exercises 1.79, 3.107 and 3.118 and in Examples 3.8(g), 3.24(b) and (c), 3.50(d) and Corollary 4.46.

We shall shortly use the projective tensor product and linear analysis to discuss the various topologies defined on $\mathcal{P}(^nE; F)$. We first note, however, that it is possible to use the τ_0 topology and locally bounded sets – both of which can be defined without reference to tensor products – to define and obtain a concrete realization of the projective tensor product. This construction is modelled on the Banach space method of recovering preduals. Since the proof is a special case of the analogous result for holomorphic functions (Proposition 3.26) we do not provide details here.

Proposition 1.28 *Let E denote a locally convex space and endow*

$$\mathcal{Q}(^nE) := \{\phi \in \mathcal{P}(^nE)^* : \phi \text{ is } \tau_0 \text{ continuous on the locally bounded}$$
$$\text{subsets of } \mathcal{P}(^nE)\}$$

with the topology of uniform convergence on the locally bounded subsets of $\mathcal{P}(^nE)$. Then

$$\mathcal{Q}(^nE) \cong \widehat{\bigotimes_{n,s,\pi}} E. \tag{1.35}$$

The point evaluations[3] e_x defined by $e_x(P) = P(x)$, $P \in \mathcal{P}(^nE)$, belong to $\mathcal{Q}(^nE)$. Let $e \colon E \to \mathcal{Q}(^nE)$ be given by $e(x) = e_x$. The canonical isomorphism associated with (1.35) is the linear mapping which takes $e_x \in \mathcal{Q}(^nE)$ to $\underbrace{x \otimes \cdots \otimes x}_{n \text{ times}} \in \widehat{\bigotimes_{n,s,\pi}} E$.

The τ_0 and τ_b topologies on $\mathcal{L}(^nE; F)$ and $\mathcal{L}^s(^nE, F)$ are defined in an obvious manner and it is easily seen that the symmetrization operator s is a continuous projection from $\big(\mathcal{L}(^nE; F), \tau_0\big)$ onto $\big(\mathcal{L}^s(^nE, F), \tau_0\big)$ and from $\big(\mathcal{L}(^nE; F), \tau_b\big)$ onto $\big(\mathcal{L}^s(^nE, F), \tau_b\big)$. By the Polarization Formula we have

$$\big(\mathcal{L}^s(^nE, F), \tau_0\big) \cong \big(\mathcal{P}(^nE; F), \tau_0\big)$$

and

$$\big(\mathcal{L}^s(^nE, F), \tau_b\big) \cong \big(\mathcal{P}(^nE; F), \tau_b\big).$$

Hence we can identify $\big(\mathcal{P}(^nE; F), \tau\big)$, $\tau = \tau_0$ or τ_b, with a closed complemented subspace of $\big(\mathcal{L}(^nE; F), \tau\big)$.

Until now we have used the representation of $\mathcal{P}(^nE)$ as a space of functions on E to define and discuss the topologies τ_0, τ_b and τ_ω. Since all these are locally convex topologies and $\mathcal{P}(^nE)$ has $\widehat{\bigotimes_{n,s,\pi}} E$ as a predual it is of interest to display them as topologies of uniform convergence on subsets of $\widehat{\bigotimes_{n,s,\pi}} E$. In

---

[3] Both e_x (evaluation at the point x) and δ_x (the Dirac delta function at the point x) are used to denote the same mapping. The evaluation function comes from point set topology and function theory. The Dirac delta function arose in functional analysis, measure theory and distribution theory and reminds us that δ_x can be interpreted as a point mass and as a linear functional. Both are widely used and suggestive in different ways.

other words, we would like to extend the algebraic duality, in Proposition 1.17, by finding, for $\tau = \tau_0$, τ_b and τ_ω, collections of sets in $\widehat{\bigotimes_{n,s,\pi}} E$, $\mathcal{B}_\tau$, such that

$$\left(\mathcal{P}(^nE), \tau\right) \cong \left(\widehat{\bigotimes_{n,s,\pi}} E\right)'_{\mathcal{B}_\tau}$$

where $\left(\widehat{\bigotimes_{n,s,\pi}} E\right)'_{\mathcal{B}_\tau}$ is the space $\left(\widehat{\bigotimes_{n,s,\pi}} E\right)'$ endowed with the topology of uniform convergence over the sets in $\mathcal{B}_\tau$. Since the duality between $\mathcal{P}(^nE)$ and $\widehat{\bigotimes_{n,s,\pi}} E$ is the linear extension of

$$\langle P, \; x \otimes \cdots \otimes x \rangle = P(x)$$

it follows, for $A \subset E$, that

$$\|P\|_A = \sup_{x \in A} |\langle P, \; x \otimes \cdots \otimes x \rangle| = \|P\|_{\bigotimes_{n,s} A} = \|P\|_{\overline{\Gamma}(\bigotimes_{n,s} A)}. \qquad (1.36)$$

Hence

$$\mathcal{B}_{\tau_0} = \left\{ \overline{\Gamma}\left(\bigotimes_{n,s} K \right) \right\}_{K \text{ compact in } E}$$

and

$$\mathcal{B}_{\tau_b} = \left\{ \overline{\Gamma}\left(\bigotimes_{n,s} B \right) \right\}_{B \text{ bounded in } E}.$$

If V is a convex balanced neighbourhood of 0 in E we let

$$\mathcal{P}_V(^nE) := \{ P \in \mathcal{P}(^nE) \; : \; \|P\|_V < \infty \}$$
$$= \{ P \in \mathcal{P}(^nE) \; : \; \|P\|_{\overline{\Gamma}(\bigotimes_{n,s} V)} < \infty \}$$

and endow $\mathcal{P}_V(^nE)$ with the topology of uniform convergence on $\overline{\Gamma}\left(\bigotimes_{n,s} V \right)$. We have

$$\left(\mathcal{P}(^nE), \tau_\omega \right) = \varinjlim_V \mathcal{P}_V(^nE)$$

where the inductive limit is taken over any fundamental system of convex balanced neighbourhoods of 0 in E. Since sets of the form $\{ P \in \mathcal{P}(^nE) \; : \; \|P\|_V \leq c \}$ are the equicontinuous subsets of $\left(\widehat{\bigotimes_{n,s,\pi}} E \right)'$ it follows that $\left(\mathcal{P}(^nE), \tau_\omega \right)$ is the *inductive dual* of $\widehat{\bigotimes_{n,s,\pi}} E$, i.e.

$$\left(\mathcal{P}(^nE), \tau_\omega \right) \cong \left(\widehat{\bigotimes_{n,s,\pi}} E \right)'_i.$$

We have thus described the three topologies τ_0, τ_b and τ_ω on $\mathcal{P}(^nE)$ in terms of the duality between $\mathcal{P}(^nE)$ and $\widehat{\bigotimes_{n,s,\pi}} E$. The τ_0 and τ_b topologies may be considered *function space* topologies on $\mathcal{P}(^nE)$ while τ_ω exhibits both function theoretic and linear characteristics. From the descriptions above, it is not clear if we have included the most natural and useful *linear* dual space topology – the strong topology of uniform convergence on bounded sets. We define the *strong topology* β on $\mathcal{P}(^nE)$ as the topology of uniform convergence over the bounded subsets of $\widehat{\bigotimes_{n,s,\pi}} E$.

If $\mathcal{V}$ denotes a fundamental system of convex balanced neighbourhoods of 0 in E, then

$$\bigcap_{V \in \mathcal{V}} \overline{\Gamma}\left(\bigotimes_{n,s} \delta_V V\right)$$

forms a fundamental system of bounded subsets of $\widehat{\bigotimes_{n,s,\pi}} E$ as $(\delta_V)_{V \in \mathcal{V}}$ range over all collections of positive real numbers. If B is a bounded subset of E and V is a neighbourhood of the origin in E then there exists $\varepsilon_V > 0$ such that $B \subset \varepsilon_V V$. Hence

$$\overline{\Gamma}\left(\bigotimes_{n,s} B\right) \subset \overline{\Gamma}\left(\bigotimes_{n,s} \varepsilon_V V\right)$$

and $\overline{\Gamma}(\bigotimes_{n,s} B)$ is a bounded subset of $\widehat{\bigotimes_{n,s,\pi}} E$. This implies that $\tau_b \leq \beta$. Since the inductive dual topology is always finer than the strong topology we have $\beta \leq \tau_\omega$. This completes the list of locally convex topologies we consider on $\mathcal{P}(^nE)$. We have shown that

$$\tau_0 \leq \tau_b \leq \beta \leq \tau_\omega$$

on $\mathcal{P}(^nE)$ for all n.

Throughout this book we shall be concerned with conditions and examples under which some or all of these topologies, and the corresponding topologies on spaces of holomorphic functions, coincide. The problem of the coincidence of topologies has been a focal point for research over the last thirty years and is still a source of interesting open problems.

We now take a preliminary look at the problem of equality for these topologies.

$\tau_0 = \tau_b$:   If these two topologies coincide, even in the case $n = 1$, then by the Hahn–Banach Theorem, each bounded subset of E is contained in the closed convex hull of a compact set. Hence, if E is quasi-complete, then E is semi-Montel. If E is not quasi-complete the τ_0 topology may be quite pathological. Fortunately, the proof of Proposition 1.11 shows that each $P \in \mathcal{P}(^nE; F)$, E a locally convex space and F a normed linear space, is uniformly continuous on a neighbourhood of the origin. Hence, there exists a necessarily unique $\widetilde{P} \in \mathcal{P}(^n\widehat{E}; \widehat{F})$, where $\widehat{E}$ and $\widehat{F}$ denote the completions of E and F respectively, such that $\widetilde{P}\big|_E = P$. We shall see in Section 6.1 that the analogous property for holomorphic functions is not always valid. Thus, for many purposes, we can identify $\mathcal{P}(^nE; F)$ and $\mathcal{P}(^n\widehat{E}; \widehat{F})$. The τ_0 topology on $\mathcal{P}(^n\widehat{E}; \widehat{F})$ is finer, and often equal, to the topology of uniform convergence on the precompact subsets of E. To avoid uninteresting situations when discussing the compact open topology we prefer to assume that E and F are complete. While on the topic of completions we note that the equality $\widehat{\bigotimes_{n,s,\pi} E} = \widehat{\bigotimes_{n,s,\pi} \widehat{E}}$ is easily proved and implies

$$\left(\mathcal{P}(^nE), \beta\right) \cong \left(\mathcal{P}(^n\widehat{E}), \beta\right)$$

and

$$\left(\mathcal{P}(^nE), \tau_\omega\right) \cong \left(\mathcal{P}(^n\widehat{E}), \tau_\omega\right)$$

algebraically and topologically. Hence, for τ_b, β and τ_ω, there is no loss of generality in assuming that we are dealing with a complete locally convex domain space. Finally we note, by Lemma 1.23, that τ_0 and τ_b always define the same bounded subsets of $\mathcal{P}(^nE; F)$ and thus always have the same associated bornological topology.

$\underline{\tau_b = \beta} :$ By (1.36), $\tau_b = \beta$ on $\mathcal{P}(^nE)$ if and only if the sets $\overline{\Gamma}(\bigotimes_{n,s} B)$, B bounded in E, form a fundamental system of bounded subsets in $\widehat{\bigotimes_{n,s,\pi} E}$. This is the n-fold symmetric version of Grothendieck's "Problème des topologies".

Definition 1.29 A locally convex space E has the $(BB)_n$-property if each bounded subset of $\widehat{\bigotimes_{n,s,\pi} E}$ is contained in $\overline{\Gamma}(\bigotimes_{n,s} B)$ for some bounded subset B of E. If E has $(BB)_n$ for all n we say that E has $(BB)_\infty$.

Clearly we have $\tau_b = \beta$ on $\mathcal{P}(^nE)$ if and only if E has $(BB)_n$. By Example 1.25, we have $\tau_b = \beta = \tau_\omega$ on $\mathcal{P}(^nE)$ for all n when E is a bornological DF space and so these spaces have $(BB)_\infty$. In particular, every Banach space has $(BB)_\infty$. In fact, it is not difficult to extend the proof in Example 1.25 to show that every DF space has $(BB)_\infty$. We prefer, however, to deduce this in a different fashion (Example 1.32).

To discuss Grothendieck's problem we require the concept of the projective tensor product of a pair of locally convex spaces.

If E and F are locally convex spaces then there exists a unique pair $(E\bigotimes_\pi F, i_{E,F})$ consisting of a locally convex space, $E\bigotimes_\pi F$, and a continuous bilinear mapping $i_{E,F}: E \times F \to E\bigotimes_\pi F$ such that $i_{E,F}(E \times F)$ spans $E\bigotimes_\pi F$ and for each continuous bilinear mapping $b \in \mathcal{B}(E, F; G)$, G a locally convex space, there exists a unique $\tilde{b} \in \mathcal{L}(E\bigotimes_\pi F; G)$ such that the diagram

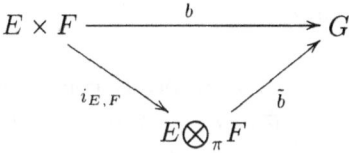

commutes. If $x \otimes y = i_{E,F}(x, y)$ then the π or *projective topology* on $E\bigotimes_\pi F$ is generated by

$$\rho_{\alpha,\beta}(\theta) := \inf\left\{\sum_i \alpha(x_i)\beta(y_i) \; : \; \theta = \sum_i x_i \otimes y_i \in E\bigotimes_\pi F\right\}$$

where $\alpha \in \mathrm{cs}(E)$ and $\beta \in \mathrm{cs}(F)$.

We denote the completion of $E\bigotimes_\pi F$ by $E \widehat{\bigotimes}_\pi F$. When $E = F$ we have $E \widehat{\bigotimes}_\pi E = \widehat{\bigotimes}_{2,\pi} E$. Grothendieck's "Problème des topologies" asks

> *"is every bounded subset of $E \widehat{\bigotimes}_\pi F$ contained in the closed convex hull of a set of the form $B_1 \otimes B_2$ where B_1 is a bounded subset of E and B_2 is a bounded subset of F?"*

When this is the case we say that the pair $\{E, F\}$ has the (BB)-property. Grothendieck's problem was posed in the mid-1950s and solved, in the negative, during the mid-1980s. Within the last ten years positive solutions were obtained for a number of interesting classes of Fréchet spaces and these motivated a number of results presented in Chapter 4. We are interested in $(BB)_n$, which involves symmetric tensors of degree n on the same space, while Grothendieck's problem involves arbitrary tensors of degree 2 on different spaces. Since $\widehat{\bigotimes}_{n,s,\pi} E$ is easily seen to be a complemented subspace of $\widehat{\bigotimes}_{n,\pi} E$ it is not surprising that solutions to the (BB)-problem should lead to solutions to the $(BB)_n$-problem.

Proposition 1.30 *If the pair $\{E, E\}$ has the (BB)-property then E has $(BB)_2$.*

Proof. Let B denote a bounded subset of $\widehat{\bigotimes}_{2,\pi,s} E$. Since $\widehat{\bigotimes}_{2,s,\pi} E$ is a subspace of $\widehat{\bigotimes}_{2,\pi} E$ bounded subsets of $\widehat{\bigotimes}_{2,s,\pi} E$ are also bounded subsets in $\widehat{\bigotimes}_{2,\pi} E$. By the (BB)-property for the pair $\{E, E\}$ there exist bounded subsets C_1 and C_2 of E such that $B \subset \overline{\Gamma}(C_1 \otimes C_2)$. If $C = \Gamma(C_1 \cup C_2)$ then C is a bounded subset of E and $B \subset \overline{\Gamma}(C \otimes C)$. Since $B \subset \widehat{\bigotimes}_{2,s,\pi} E$ we have, by (1.16),

$$B = s(B) \subset s\left(\overline{\Gamma}(\bigotimes_2 C)\right) = \overline{\Gamma}\left(s(\bigotimes_2 C)\right) \subset \overline{\Gamma}\left(\bigotimes_{2,s}(\sqrt{2}C)\right)$$

and this completes the proof.

By considering a collection $\mathcal{F}$ of locally convex spaces, stable under $\widehat{\bigotimes}_\pi$ (i.e. if E, $F \in \mathcal{F}$ then $E \widehat{\bigotimes}_\pi F \in \mathcal{F}$), such that $\{E, F\}$ has the (BB)-property for any pair E, F in $\mathcal{F}$ we obtain a stronger result.

Proposition 1.31 *Let $\mathcal{F}$ denote a collection of locally convex spaces which is stable under $\widehat{\bigotimes}_\pi$ and such that $\{E, F\}$ has the (BB)-property for any pair of spaces E and F in $\mathcal{F}$. Then each E in $\mathcal{F}$ has $(BB)_\infty$.*

Proof. We first prove, by induction, the following result: if $E \in \mathcal{F}$ and B is a bounded subset of $\widehat{\bigotimes}_{n,\pi} E$ then there exist bounded subsets $B_1, \ldots, B_n$ in E such that

$$B \subset \overline{\Gamma}(B_1 \otimes B_2 \ldots \otimes B_n). \tag{$*$}$$

When $n = 2$ this is our hypothesis. Suppose the result is true for $k \geq 2$. By induction $\widehat{\bigotimes}_{k,\pi} E \in \mathcal{F}$. Hence the pair $\{E, \widehat{\bigotimes}_{k,\pi} E\}$ has the (BB)-property. Thus, if B is a bounded subset of $\widehat{\bigotimes}_{k+1,\pi} E \cong E \widehat{\bigotimes}_\pi (\widehat{\bigotimes}_{k,\pi} E)$, there exist bounded subsets C_1 and C_2' in E and $\widehat{\bigotimes}_{k,\pi} E$, respectively, such that $B \subset \overline{\Gamma}(C_1 \otimes C_2')$. By our induction hypothesis there exist bounded subsets of E, $C_2, \ldots, C_{k+1}$ such that

$$C_2' \subset \overline{\Gamma}(C_2 \otimes C_3 \cdots \otimes C_{k+1}).$$

Hence

$$B \subset \overline{\Gamma}(C_1 \otimes \overline{\Gamma}(C_2 \otimes \cdots \otimes C_{k+1})).$$

Since it is easily verified that

$$\Gamma(C_1 \otimes \Gamma(C_2 \otimes \cdots \otimes C_{k+1})) = \Gamma(C_1 \otimes \cdots \otimes C_{k+1})$$

it follows that

$$B \subset \overline{\Gamma}(C_1 \otimes \cdots \otimes C_{k+1})$$

and this proves $(*)$.

If D is a convex balanced bounded subset of $\widehat{\bigotimes}_{n,s,\pi} E$ then the above shows that there exists $B_1, \ldots, B_n$ bounded subsets of E such that $D \subset \overline{\Gamma}(B_1 \otimes B_2 \otimes \cdots \otimes B_n)$. If $C = \Gamma\left(\bigcup_{i=1}^n B_i\right)$ then $D \subset \overline{\Gamma}(\bigotimes_n C)$. Since $D = s(D)$, (1.16) implies

$$B \subset \overline{\Gamma}\left(\bigotimes_{n,s}\left(\frac{n}{(n!)^{1/n}} C\right)\right).$$

Hence E has $(BB)_n$ and, since n was arbitrary, it follows that E has $(BB)_\infty$. This completes the proof.

Example 1.32 It is known that $E\bigotimes_\pi F$ is a DF space whenever E and F are DF spaces and that any pair of DF spaces has the (BB)-property. Hence, by Proposition 1.31, any DF space E has $(BB)_\infty$ and thus $\tau_b = \beta$ on $\mathcal{P}(^nE)$ for any positive integer n. We can show much more. Since $\widehat{\bigotimes_{n,s,\pi}} E$ is a comple-

mented subspace of $\widehat{\bigotimes_{n,\pi}} E$ it is a DF space and hence $(\mathcal{P}(^nE), \tau_b)$ is a Fréchet

space. Applying the known result, that every strongly bounded sequence in the dual of a DF space is equicontinuous, to the pair $(\widehat{\bigotimes_{n,s,\pi}} E, \mathcal{P}(^nE))$, we

see that τ_b bounded sequences in $\mathcal{P}(^nE)$ are locally bounded and hence τ_ω bounded. Since two locally convex topologies on the same space have the same bounded sets if and only if they have the same bounded sequences, β and τ_ω have the same bounded sets. As both $(\mathcal{P}(^nE), \beta)$ and $(\mathcal{P}(^nE), \tau_\omega)$ are bornological we have $\tau_\omega = \tau_b = \beta$ on $\mathcal{P}(^nE)$ for any n and any DF space E. This extends the result in Example 1.25 to arbitrary DF spaces.

We have considered two very different approaches to precisely the same problem in Examples 1.25 and 1.32. In Example 1.25 we used a function theory approach to $\mathcal{P}(^nE)$ while, in Example 1.32, we used a dual linear approach and applied results from functional analysis. This occurs often and it is advisable to consider, if possible, both methods in attacking problems.

We have already noted three differences between the symmetric tensor products, in which we are interested, and in the tensor products discussed in the linear functional analysis literature. This causes complications in the direct application of the extensive body of results on topological tensor products available. However, the fact that $\widehat{\bigotimes_{n,s,\pi}} E$ is a *complemented* subspace of

$\widehat{\bigotimes_{n,\pi}} E$ leads, as we have just seen, to a close relationship between these spaces and it was, for this reason, that we did not confine ourselves solely to the symmetric case.

The situation is even more favourable in view of the following result, which is based on the observation that a mapping $\phi\colon E^n \to \mathbb{C}$ which is linear in each variable may be regarded as an element of both $\mathcal{L}(^nE)$ and $\mathcal{P}(^n(E^n))$.

Proposition 1.33 *If E is a locally convex space over $\mathbb{C}$ then for each positive integer n, $(\mathcal{L}(^nE), \tau)$ is isomorphic to a complemented subspace of $(\mathcal{P}(^n(E^n)), \tau)$, $\tau = \tau_0$ or τ_b.*

Proof. For $1 \leq j \leq n$ let $\alpha_j\colon E \to E^n$ be given by

$$\alpha_j(x) = (0, \ldots, 0, \underset{\underset{j^{\text{th}} \text{ position}}{\vert}}{x}, 0, \ldots, 0) \ .$$

If $P \in \mathcal{P}(^n(E^n))$ then the *Monomial Theorem* (Lemma 1.9) implies

$$P(x_1, \ldots, x_n) = P\Big(\sum_{i=1}^{n} \alpha_i(x_i)\Big)$$

$$= \sum_{\substack{0 \le j_i \le n \\ \sum_{i=1}^{n} j_i = n}} \binom{n}{j_1, \ldots, j_n} \overset{\vee}{P}\big(\alpha_1(x_1)^{j_1}, \ldots, \alpha_n(x_n)^{j_n}\big)$$

for $(x_1, \ldots, x_n) \in E^n$. Each of the mappings

$$(x_1, \ldots, x_n) \in E^n \longrightarrow \binom{n}{j_1, \ldots, j_n} \overset{\vee}{P}\big(\alpha_1(x_1)^{j_1}, \ldots, \alpha_n(x_n)^{j_n}\big)$$

defines an element $P_{(j_1, \ldots, j_n)} \in \mathcal{P}(^n(E^n))$ and the mapping

$$P \longrightarrow P_{(j_1, \ldots, j_n)}$$

is a linear projection from $\mathcal{P}(^n(E^n))$ into $\mathcal{P}(^n(E^n))$.

If $j_1 = j_2 \ldots = j_n = 1$ we obtain a polynomial on E^n which is linear in each coordinate and hence the mapping

$$(x_1, \ldots, x_n) \longrightarrow n! \overset{\vee}{P}\big(\alpha_1(x_1), \ldots, \alpha_n(x_n)\big)$$

belongs to $\mathcal{L}(^n E)$. On the other hand, if $L \in \mathcal{L}(^n E)$, we define $\widetilde{L} \in \mathcal{P}(^n(E^n))$ by the formula

$$\widetilde{L}(x_1, \ldots, x_n) = L(x_1, \ldots, x_n).$$

We have

$$\widetilde{L}_{(1, \ldots, 1)}(x_1, \ldots, x_n) = n! \overset{\vee}{\widetilde{L}}\big(\alpha_1(x_1), \ldots, \alpha_n(x_n)\big)$$

$$= \frac{n!}{2^n n!} \sum_{\varepsilon_i = \pm 1} \varepsilon_1 \ldots \varepsilon_n \widetilde{L}\Big(\sum_{i=1}^{n} \varepsilon_i \alpha_i(x_i)\Big)$$

$$= \frac{1}{2^n} \sum_{\varepsilon_i = \pm 1} \varepsilon_1 \cdots \varepsilon_n L(\varepsilon_1 x_1, \ldots \varepsilon_n x_n)$$

$$= \frac{1}{2^n} \sum_{\varepsilon_i = \pm 1} \varepsilon_1^2 \cdots \varepsilon_n^2 L(x_1, \ldots, x_n)$$

$$= L(x_1, \ldots, x_n)$$

and thus shown that the range of the projection

$$P \longrightarrow P_{(1, \ldots, 1)}$$

can be linearly identified with $\mathcal{L}(^n E)$. Since it is easily checked that this is also a topological identification with respect to either τ_0 or τ_b this completes the proof.

When the locally convex space E has certain stability properties this result can be improved.

Proposition 1.34 *If E is a locally convex space and $E \cong E \times E$ then*[4] $(\mathcal{P}(^nE), \tau) \cong (\mathcal{L}(^nE), \tau), \tau = \tau_0, \tau_b$ *or* τ_ω.

The problem of topologies for spaces of polynomials on DF spaces turned out to be rather straightforward. For reasons, that we discuss in a wider context in Chapter 3, this is not surprising. What about the situation for Fréchet spaces? It is more interesting and complicated. It took considerable time, after the problem was first posed, before a counterexample was found showing that there exist pairs of Fréchet spaces, and indeed of Fréchet–Montel spaces, which did not have the (BB)-property. If K is a compact subset of a Fréchet space E then K is contained in the convex hull of a null sequence. From this it follows that $\overline{\Gamma}(\bigotimes_{n,s} K)$ and $\overline{\Gamma}(\bigotimes_{n} K)$ are each contained in the convex hull of a null sequence and hence both are compact. If, in addition, E is Montel and $\{E, E\}$ has the (BB)-property then every closed bounded subset of $\widehat{\bigotimes}_{2,\pi} E$ is compact. Since $\widehat{\bigotimes}_{2,\pi} E$ is always a Fréchet space it follows that it is Fréchet–Montel. By the same argument we see, if E is Fréchet–Montel and has $(BB)_n$, that $\widehat{\bigotimes}_{n,s,\pi} E$ is a Fréchet–Montel space. Hence, $\tau_0 = \tau_b = \beta$ on $\mathcal{P}(^nE)$. Moreover, $(\mathcal{P}(^nE), \beta)$ is a $D\mathcal{FM}$ space and hence is bornological. Since $\tau_\omega \geq \beta$ and, by Example 1.32, τ_ω and β have the same bounded sets it follows that $\tau_\omega = \beta$ on $\mathcal{P}(^nE)$.

Combining these results we obtain the following.

Proposition 1.35 *If E is a Fréchet–Montel space and n is a positive integer then the following are equivalent:*

(a) E has $(BB)_n$,

(b) $\widehat{\bigotimes}_{n,s,\pi} E$ is a Fréchet–Montel space,

(c) $\tau_0 = \tau_b = \beta$ on $\mathcal{P}(^nE)$,

(d) $\tau_b = \tau_\omega$ on $\mathcal{P}(^nE)$,

(e) $(\mathcal{P}(^nE), \tau_0)$ is a $D\mathcal{FM}$ space.

Proof. By the definition of $(BB)_n$ and the above remarks $(a) \Longleftrightarrow (c) \Longleftrightarrow (d)$. If (a) is satisfied then each bounded subset of $\widehat{\bigotimes}_{n,s,\pi} E$ is contained in a subset of the form $\overline{\Gamma}(\bigotimes_{n,s} K)$ where K is a compact subset of E. Hence

---

[4] A locally convex space E with this property is called stable.

each bounded subset of $\widehat{\underset{n,s,\pi}{\bigotimes}} E$ is relatively compact and $(a) \Rightarrow (b)$. Suppose

(b) holds. If B is a bounded subset of $\widehat{\underset{n,s,\pi}{\bigotimes}} E$ then the classical result of
Grothendieck, i.e. the $n = 1$ case, can be adapted to show that there exists
a null sequence $\{x_j\}_{j=1}^\infty$ in E such that $B \subset \overline{\Gamma}(\{x_j \otimes \cdots \otimes x_j\}_{j=1}^\infty)$. Hence E
has $(BB)_n$ and $(b) \Rightarrow (a)$.

If (a), (b), (c) and (d) hold then $\left(\widehat{\underset{n,s,\pi}{\bigotimes}} E\right)'_\beta$ is a $\mathcal{DFM}$ space and $\tau_0 =$
$\beta = \tau_\omega$. Hence $(\mathcal{P}(^nE), \tau_0)$ is a $\mathcal{DFM}$ space and $(a) \Rightarrow (e)$. If (e) holds then
τ_0 defines a bornological topology on $\mathcal{P}(^nE)$ and hence $\tau_0 = \tau_b = \tau_\omega$ and
$(e) \Rightarrow (d)$. This completes the proof.

We will give examples where the hypotheses of Proposition 1.35 are satisfied in Chapters 4 and 5 (see also Proposition 3.48). Now let F denote
a Fréchet–Montel space such that $\{F, F\}$ does not have the (BB)-property.
Then $(\mathcal{L}(^2F), \tau_b)$ is not a bornological space. By Proposition 1.33, $(\mathcal{L}(^2F), \tau_b)$
is a complemented subspace of $(\mathcal{P}(^2(F^2)), \tau_b)$ and hence $(\mathcal{P}(^2(F^2)), \tau_b)$ is also
not bornological and we have an example of a Fréchet–Montel space F^2 such
that $\tau_0 = \tau_b \underset{\neq}{<} \tau_\omega$ on $\mathcal{P}(^2(F^2))$. In fact, by Exercise 1.78, $\tau_b \neq \beta$ on $\mathcal{P}(^n(F^2))$
for all n.
So far we have obtained a good deal of information about the equality
$\tau_b = \beta$ on $\mathcal{P}(^nE)$ without discussing the intrinsic nature of the (BB)-property
or $(BB)_n$. From the definition we see that a pair of locally convex spaces has
the (BB)-property if every bounded subset of the completed π-projective
tensor product splits – modulo taking the absolutely convex hull. For this
reason it is to be expected that the existence of sufficiently many good splittings of the component spaces should lead to examples of pairs with the
(BB)-property. An ideal form of splitting is a projection. The mere existence
of projections is not sufficient, however, as the locally convex space structure
of a Fréchet space is defined by a sequence of semi-norms and it is necessary
to have some interaction between the projections and the semi-norms. In
Chapter 4 we discuss a collection of spaces, with sufficiently many splittings
(Definition 4.36), admitting a suitable interaction between the semi-norms
and the decompositions generated by the splittings. This leads to examples
of Fréchet–Montel spaces E for which $\tau_0 = \tau_\omega$ on $\mathcal{P}(^nE)$ for all n. The technique of using good estimates between sufficiently many different splittings
or projections appears to be fundamental as the first counterexample to the
(BB)-property was constructed so that the conditions required for the employment of this technique were absent.

$\underline{\beta = \tau_\omega}$:   Using preduals we see that $\beta = \tau_\omega$ on $\mathcal{P}(^nE)$ if and only if
the strong and the inductive duals of $\widehat{\underset{n,s,\pi}{\bigotimes}} E$ coincide. This approach to the
problem does not appear to lead to any concrete results and we prefer to

compare fundamental systems of zero neighbourhoods in $\mathcal{P}(^nE)$ for both topologies.

We have already noted that a fundamental system of neighbourhoods of zero in $(\mathcal{P}(^nE), \tau_\omega)$ is given by

$$\Gamma\left(\bigcup_{\substack{V \in \mathcal{V} \\ c_V > 0}} \{ P \in \mathcal{P}(^nE) : \|P\|_V \leq c_V \} \right) \tag{1.37}$$

where $\mathcal{V}$ is a fundamental neighbourhood system at the origin in E and $(c_V)_{V \in \mathcal{V}}$ ranges over all collections of positive real numbers indexed by $\mathcal{V}$.

From the definition of the strong topology it follows that a fundamental system of zero neighbourhoods in $(\mathcal{P}(^nE), \beta)$ is given by sets of the form

$$\left(\bigcap_{V \in \mathcal{V}} \Gamma\left(\bigotimes_{n,s} \delta_V V \right) \right)^\circ$$

where $\mathcal{V}$ is as above, $\delta_V > 0$ for all $V \in \mathcal{V}$ and $^\circ$ denotes the *polar* in $\mathcal{P}(^nE)$. By duality

$$\left(\bigcap_{V \in \mathcal{V}} \Gamma\left(\bigotimes_{n,s} \delta_V V \right) \right)^\circ = \overline{\Gamma}^{\sigma(\mathcal{P}, \otimes)}\left(\bigcup_{V \in \mathcal{V}} \left(\Gamma\left(\bigotimes_{n,s} \delta_V V \right) \right)^\circ \right)$$

where

$$\sigma(\mathcal{P}, \otimes) := \sigma\left(\mathcal{P}(^nE), \widehat{\bigotimes_{n,s,\pi}} E \right)$$

is the weak* topology on $\mathcal{P}(^nE)$ arising from the pairing $(\widehat{\bigotimes_{n,s,\pi}} E, \mathcal{P}(^nE))$.

We let $\sigma(\otimes, \mathcal{P})$ denote the weak topology on $\widehat{\bigotimes_{n,s,\pi}} E$ arising from the same pairing. Since

$$\left(\Gamma\left(\bigotimes_{n,s} \delta_V V \right) \right)^\circ = \{ P \in \mathcal{P}(^nE) : \|P\|_{\delta_V V} \leq 1 \}$$

$$= \left\{ P \in \mathcal{P}(^nE) : \|P\|_V \leq \frac{1}{(\delta_V)^n} \right\}$$

it follows that $(\mathcal{P}(^nE), \beta)$ has a fundamental system of zero neighbourhoods of the form

$$\overline{\Gamma}^{\sigma(\mathcal{P}, \otimes)}\left(\bigcup_{V \in \mathcal{V}} \{ P \in \mathcal{P}(^nE) : \|P\|_V \leq c_V \} \right) \tag{1.38}$$

where $\mathcal{V}$ and $(c_V)_{V \in \mathcal{V}}$ have the same meaning as in (1.37). Since fundamental neighbourhood systems in a locally convex space are preserved by taking closures, (1.37) and (1.38) imply that $\tau_\omega = \beta$ if and only if $(\mathcal{P}(^nE), \tau_\omega)$ has a fundamental neighbourhood system at the origin consisting of $\sigma(\mathcal{P}, \otimes)$-closed sets. Suppose the locally convex space E satisfies

$$\widehat{\bigotimes_{n,s,\pi}} E = \bigcup_K \overline{T}\left(\bigotimes_{n,s} K\right) \tag{1.39}$$

where the union is taken over all compact subsets of E.

Condition (1.39) may be considered a variant of $(BB)_n$. Fréchet spaces and some bornological DF spaces satisfy (1.39). If (1.39) is satisfied then $\sigma(\mathcal{P}, \bigotimes)$ and τ_0 are compatible topologies on $\mathcal{P}(^nE)$. By the Hahn–Banach Theorem the $\sigma(\mathcal{P}, \bigotimes)$-closure in (1.38) can be replaced by the τ_0-closure and we obtain the following result.

Proposition 1.36 *Let E denote a locally convex space satisfying (1.39). Then*

(a) τ_0, τ_b and β define the same bounded subsets of $\mathcal{P}(^nE)$,

(b) $\tau_\omega = \beta$ on $\mathcal{P}(^nE)$ if and only if $\mathcal{P}(^nE)$ has a τ_ω-neighbourhood basis at the origin consisting of τ_0-closed sets.

Proof. (a) Since $\widehat{\bigotimes_{n,s,\pi}} E$ is a complete locally convex space the weak and strong topologies on $\left(\widehat{\bigotimes_{n,s,\pi}} E\right)' = \mathcal{P}(^nE)$ generate the same bounded sets. By (1.39), $\sigma(\mathcal{P}, \bigotimes) \le \tau_0 \le \tau_b \le \beta$ and hence all define the same bounded subsets of $\mathcal{P}(^nE)$. This proves (a), and (b) follows from the remarks immediately preceeding the statement of the proposition.

The necessary and sufficient condition given in Proposition 1.36 is not very satisfactory as it is stated in terms of $\mathcal{P}(^nE)$ rather than E' or E. We look for a more concrete condition when E is Fréchet. By Example 1.24 we know that τ_0, τ_b, β and τ_ω all define the same bounded subsets of $\mathcal{P}(^nE)$. Since τ_ω is a bornological topology it follows that $\beta = \tau_\omega$ on $\mathcal{P}(^nE)$ if and only if $\left(\widehat{\bigotimes_{n,s,\pi}} E\right)'_\beta$ is bornological. Since E is Fréchet so also is $\widehat{\bigotimes_{n,s,\pi}} E$ and, by a result of Grothendieck, it follows that $\beta = \tau_\omega$ on $\mathcal{P}(^nE)$ if and only if $\widehat{\bigotimes_{n,s,\pi}} E$ is a *distinguished* Fréchet space. Since there exist Fréchet spaces which are not distinguished it follows that there are Fréchet spaces E such that $\beta \underset{\ne}{<} \tau_\omega$ on $E' \cong \mathcal{P}(^1E)$.

A useful related criterion which arose in the study of ultrapowers is the density condition. A Fréchet space E has the *density condition* if and only if bounded subsets of E'_β are metrizable. If E has $(BB)_n$ and the density condition then $\widehat{\bigotimes_{n,s,\pi}} E$ also has the density condition. Fréchet spaces with the density condition are distinguished. If $(x_n)_n$ is a dense sequence in a Fréchet space then

$$d(\phi, \psi) = \sum_{n=1}^{\infty} \frac{1}{2^n} \frac{|\phi(x_n) - \psi(x_n)|}{1 + |\phi(x_n) - \psi(x_n)|}$$

defines a continuous metric on E'_β. Since a continuous bijective mapping from a compact topological space onto a Hausdorff topological space is a homeomorphism, d induces the β topology on each compact subset of E'_β. In particular, if E is Fréchet–Montel then E is separable and, since closed bounded subsets of E'_β are compact, this implies that the bounded subsets of E'_β are metrizable and E has the density condition.

We complete this section by presenting a simple example involving the projective topology.

Example 1.37 $\left(\mathcal{P}(^n l_1),\ \tau_\omega\right) \cong l_\infty$. It is well known that $\widehat{\bigotimes}_{n,\pi} l_1 \cong l_1$ and that every infinite dimensional closed complemented subspace of l_1 is isomorphic to l_1 (a space with this property is said to be *prime*). Hence

$$\widehat{\bigotimes_{n,s,\pi}} l_1 \cong l_1$$

and

$$\left(\mathcal{P}(^n l_1),\ \tau_\omega\right) \cong \left(\widehat{\bigotimes_{n,s,\pi}} l_1\right)'_\beta \cong l'_1 \cong l_\infty.$$

1.3 Geometry of Spaces of Polynomials

The focus of our attention in this section is the fundamental inequality (1.14),

$$\|P\|_{\alpha,A} \le \|\stackrel{\vee}{P}\|_{\alpha,A^n} \le \frac{n^n}{n!}\|P\|_{\alpha,A}$$

where $P \in \mathcal{P}(^n E; F)$, A is a convex balanced subset of E and α is a seminorm on F.

It is fairly clear, by the Hahn–Banach Theorem that we may restrict our attention to the case where $F = \mathbb{C}$ and E is a normed linear space, since estimates can then be transferred to arbitrary locally convex spaces as required. The situation becomes more interesting when A is bounded and thus we suppose that A is the unit ball of E.

We begin by considering the first question that one asks about any inequality: is it best possible? Afterwards, we look at the standard consequences when and how the extremal constants are achieved and whether or not the estimates can be improved in special cases. We have taken the opportunity offered, by the unexpected directions in which our investigation proceeded, to display properties of homogeneous polynomials on l_p and c_0 and to show connections between properties of polynomials and the geometry of Banach spaces. In these cases we have sometimes given results which are special cases of later results and, at times, we have not presented results in their strongest known form (see the notes). We have tried to emphasize a direct approach and hope that the results and proofs will provide a bridge and some insight into the more technical results currently appearing in the literature.

So far we have given few examples of polynomials. The simplest and, in many cases, the most useful polynomials are obtained by multiplying linear functionals together. Thus if E is a locally convex space and $(\phi_i)_{i=1}^n$ is a finite subset of E' then the mapping

$$\phi_1 \cdots \phi_n \colon E \longrightarrow \mathbb{C}$$
$$x \longrightarrow \phi_1(x) \cdots \phi_n(x)$$

defines an element of $\mathcal{P}(^nE)$. To see this it suffices to define $L \colon E^n \longrightarrow \mathbb{C}$ by the formula

$$L(x_1, \ldots, x_n) = \phi_1(x_1) \cdots \phi_n(x_n)$$

and to note that $L \in \mathcal{L}(^nE)$ and $L \circ \Delta_n = \phi_1 \cdots \phi_n$. In particular, if $\phi \in E'$ then $\phi^n \in \mathcal{P}(^nE)$. The subspace of $\mathcal{P}(^nE)$ generated by $\{\phi^n\}_{\phi \in E'}$ is denoted by $\mathcal{P}_f(^nE)$ and elements of $\mathcal{P}_f(^nE)$ are called *polynomials of finite type*. Since E' is a vector space, $\mathcal{P}_f(^nE)$ contains all products of the form $\phi_1 \cdots \phi_n$, $\phi_i \in E'$. An important particular case occurs when E has a Schauder basis $(e_m)_{m=1}^\infty$. We let $(e_m')_{m=1}^\infty$ denote the coefficient functionals, i.e.

$$e_m'\Big(\sum_{n=1}^\infty z_n e_n\Big) = z_m$$

for all $m \in \mathbb{N}$ and $\sum_{n=1}^\infty z_n e_n \in E$. Finite products of coefficient functionals are called *monomials* relative to the basis $(e_m)_{m=1}^\infty$ or simply monomials. Hence a monomial is a polynomial of the form

$$(e_1')^{m_1}(e_2')^{m_2} \cdots (e_n')^{m_n}$$

for non-negative integers $m_1, \ldots, m_n$. A simplified notation has, however, become standard. Let $\mathbb{N}^{(\mathbb{N})}$ denote the set of all finite sequences of non-negative integers. We write $m \in \mathbb{N}^{(\mathbb{N})}$ in the form

$$m = (m_1, \ldots, m_n, 0, \ldots)$$

where each m_i is a non-negative integer and $m_i = 0$ for all i sufficiently large. If $z = \sum_{j=1}^\infty z_j e_j \in E$ then

$$(e_1')^{m_1}(e_2')^{m_2} \cdots (e_n')^{m_n}(z) = z_1^{m_1} z_2^{m_2} \cdots z_n^{m_n}.$$

If we let $m_i = 0$ for $i > n$ and $z^\circ = 1$ for all $z \in \mathbb{C}$ then

$$(e_1')^{m_1}(e_2')^{m_2} \cdots (e_n')^{m_n}\Big(\sum_{j=1}^\infty z_j e_j\Big) = \prod_{j=1}^\infty z_j^{m_j} =: z^m.$$

In accordance with the established notation of several complex variable theory we let z^m denote the monomial $(e_1')^{m_1} \cdots (e_n')^{m_n}$. If $m = (m_i)_{i=1}^\infty \in \mathbb{N}^{(\mathbb{N})}$ let

$$|m| = \sum_i m_i, \quad m! = m_1!m_2!\cdots m_n! \text{ and } m^m = m_1^{m_1}\cdots m_n^{m_n}.$$

We have $z^m \in \mathcal{P}(^{|m|}E)$. With this notation the Monomial Theorem has the following form:

If $P \in \mathcal{P}(^nE)$ and $(x_j)_{j=1}^k$ are k vectors in E then

$$P\left(\sum_{j=1}^k x_j\right) = \sum_{\substack{0 \leq m_i \leq n \\ \sum_{i=1}^k m_i = n}} \frac{n!}{m_1!\cdots m_k!} \overset{\vee}{P}(x_1^{m_1}, x_2^{m_2}, \ldots, x_k^{m_k})$$

$$= \sum_{\substack{m \in \mathbb{N}^k \\ |m| = n}} \frac{|m|!}{m!} \overset{\vee}{P}(x^m)$$

where $x = (x_1, \ldots, x_k)$ and $x^m = (x_1^{m_1}, \ldots, x_k^{m_k})$.

The space of polynomials on a finite dimensional space is spanned by the monomials relative to any basis.

For the remainder of this section E will denote a Banach space over $\mathbb{C}$ with norm $\|\cdot\|$ and open unit ball B. We shall use when convenient any one of the following:

$$\|P\| := \|P\|_E := \|P\|_B := \sup\{|P(x)| \ : \ \|x\| \leq 1\}.$$

On the finite product E^n we use the norm

$$\|(x_1, \ldots, x_n)\| = \sup_{1 \leq i \leq n} \|x_i\|.$$

In the previous section we used properties of E to derive properties of $\mathcal{P}(^nE)$. In this section we concentrate on deducing information about E from $\mathcal{P}(^nE)$. We first show that the constant $n^n/n!$ which appears in Proposition 1.8 is best possible.

Lemma 1.38 *If $E = l_p$, $1 \leq p \leq \infty$, and $m \in \mathbb{N}^{(\mathbb{N})}$ then*

$$\|z^m\|_{l_p} = \left(\frac{m^m}{|m|^{|m|}}\right)^{1/p}.$$

Proof. Let $m = (m_1, \ldots, m_n, 0, \ldots)$. We first suppose $p = 1$. By the classical inequality between the geometric and arithmetic means we have, for $z = (z_i)_i \in l_1$,

$$\frac{|z^m|}{m^m} = \left(\frac{|z_1|}{m_1}\right)^{m_1} \cdots \left(\frac{|z_n|}{m_n}\right)^{m_n}$$

$$\leq \left(\frac{|z_1| + |z_2| \cdots + |z_n|}{|m|}\right)^{|m|}$$

$$= \frac{\|z\|^{|m|}}{|m|^{|m|}}.$$

Hence

$$\|z^m\|_{l_1} \leq \frac{m^m}{|m|^{|m|}}.$$

On the other hand if $z_0 = \left(\frac{m_1}{|m|}, \frac{m_2}{|m|}, \ldots, \frac{m_n}{|m|}, 0, \ldots\right)$ then $\|z_0\| = \frac{\sum_i m_i}{|m|} = 1$
and

$$z_0^m = \left(\frac{m_1}{|m|}\right)^{m_1} \left(\frac{m_2}{|m|}\right)^{m_2} \cdots \left(\frac{m_n}{|m|}\right)^{m_n} = \frac{m^m}{|m|^{|m|}}.$$

Hence $\|z^m\|_{l_1} \geq m^m/|m|^{|m|}$ and this proves the result when $p = 1$.
If $1 < p < \infty$ then, letting $w_i = |z_i|^p$, we see that

$$\|z^m\|_{l_p} = \sup\{|z_1^{m_1} \ldots z_n^{m_n}| \; : \; \sum_i |z_i|^p \leq 1\}$$

$$= \sup\{|w_1^{m_1} \ldots w_n^{m_n}|^{\frac{1}{p}} \; : \; \sum_i |w_i| \leq 1\}$$

$$= \left(\frac{m^m}{|m|^{|m|}}\right)^{1/p}$$

by the $p = 1$ case. The case $p = \infty$ is easily seen to be true and this completes
the proof.

Example 1.39 The constant $n^n/n!$ in Proposition 1.8 is best possible. Let
$E = l_1$, $m = \underbrace{(1, 1, \ldots, 1}_{n \text{ times}}, 0, \ldots) \in \mathbb{N}^{(\mathbb{N})}$ and $P = z^m \in \mathcal{P}(^n l_1)$. By Lemma
1.38 we have $\|P\| = \|z^m\|_{l_1} = 1/n^n$. By the Polarization Formula

$$\overset{\vee}{P}(e_1, \ldots, e_n) = \frac{1}{n! 2^n} \sum_{\varepsilon_i = \pm 1} \varepsilon_1 \ldots \varepsilon_n P\left(\sum_{i=1}^n \varepsilon_i e_i\right)$$

$$= \frac{1}{n! 2^n} \sum_{\varepsilon_i = \pm 1} \varepsilon_1^2 \ldots \varepsilon_n^2$$

$$= \frac{1}{n!}.$$

Since $\|e_i\| = 1$ for all i this implies $\|\overset{\vee}{P}\| \geq 1/n!$. By Proposition 1.8

$$\|\overset{\vee}{P}\| \leq \frac{n^n}{n!} \|P\| = \frac{n^n}{n!} \cdot \frac{1}{n^n} = \frac{1}{n!} \leq \|\overset{\vee}{P}\|$$

and hence all these inequalities are equalities. In particular

$$\|\overset{\vee}{P}\| = \frac{n^n}{n!}\|P\|$$

and this proves our assertion.

We shall soon see that the space l_1^n and the polynomial z^m, $m = \underbrace{(1, 1, \ldots, 1, 0, \ldots)}_{n \text{ times}}$ provide the unique setting in which the supremum $n^n/n!$ is achieved. Since it is possible to improve the estimate for specific spaces we introduce the following definition.

Definition 1.40 If E is a Banach space over $\mathbb{C}$ and n is a positive integer let

$$c(n, E) = \inf\{\, M > 0 \, : \, \|\overset{\vee}{P}\| \leq M\|P\|, \text{ for all } P \in \mathcal{P}(^n E) \,\}.$$

We call $c(n, E)$ the nth *Polarization Constant* of the Banach space E and call

$$c(E) := \limsup_{n \to \infty} c(n, E)^{1/n}$$

the *Polarization Constant* of the space E.

By the Polarization Formula and Example 1.39 we have

$$1 \leq c(n, E) \leq \frac{n^n}{n!}$$

and the constant $n^n/n!$ is, in general, best possible. By Stirling's formula, $1 \leq c(E) \leq e$ for any Banach space E. If $\|\overset{\vee}{P}\| = c(n, E)\|P\|$, where $P \in \mathcal{P}(^n E)$ is non-zero, we call $\overset{\vee}{P}$ and P *extremal*. It is easily seen that $c(n, \cdot)$ is an isometric property of Banach spaces and that

$$c(n, E/F) \leq c(n, E)$$

for any closed subspace F of E. To exploit the isometric nature of $c(n, \cdot)$ we introduce the *Banach–Mazur distance* $d(\cdot, \cdot)$. If E and F are (linearly) isomorphic Banach spaces then there exists a continuous bijective linear mapping T from E into F. We have

$$E \xrightarrow{\ T\ } F \xrightarrow{\ T^{-1}\ } E$$

and

$$\|x\| = \|T^{-1}T(x)\| \leq \|T^{-1}\| \cdot \|T\| \cdot \|x\|.$$

Hence $\|T^{-1}\| \cdot \|T\| \geq 1$ and T is a scalar multiple of an isometry if and only if $\|T^{-1}\| \cdot \|T\| = 1$. By taking the infimum, over all isomorphisms, we obtain a quantitative measure of how close the spaces E and F are isometrically. We let

$$d(E, F) = \inf\{\|T\| \cdot \|T^{-1}\| \; : \; T \in \mathcal{L}(E; F) \text{ invertible}\}$$

and call $d(E, F)$ the Banach–Mazur distance between E and F. On any collection of *isomorphic* Banach spaces $\log d(\cdot, \cdot)$ is a pseudo-metric. If E and F are isometrically isomorphic Banach spaces then $d(E, F) = 1$ and the converse is true in certain cases, e.g. if E is finite dimensional.

If E and F are Banach spaces we say that F is finitely represented in E and write $F \stackrel{\circ}{\longhookrightarrow} E$ if for every finite dimensional subspace F_1 of F and every $\varepsilon > 0$ there exists a finite dimensional subspace E_1 of E such that

$$d(F_1, E_1) \leq 1 + \varepsilon. \tag{1.40}$$

Since the Banach–Mazur distance is only defined for isomorphic Banach spaces E_1 and F_1 have the same (finite) dimension.

Properties determined by the finite dimensional Banach subspaces of a Banach space are said to belong to the *local theory* (of Banach spaces). This area, which includes the topic of finite representability, has been extensively studied in recent years and many important results can be formulated as results in the local theory. In particular, we have the following:

$E \stackrel{\circ}{\longhookrightarrow} l_\infty$ for any Banach space E (we can even take $\varepsilon = 0$ in this case),

$E'' \stackrel{\circ}{\longhookrightarrow} E$ for any Banach space E *(Principle of Local Reflexivity)*,

$l_2 \stackrel{\circ}{\longhookrightarrow} E$ for any infinite dimensional Banach space E *(Dvoretzky's Spherical Sections Theorem)*.

We now return to the study of the isometric invariant $c(n, E)$. Let l_p^n denote $\mathbb{C}^n$ endowed with the l_p norm, $1 \leq p \leq \infty$.

Proposition 1.41 *If E is a Banach space and $c(n, E) = n^n/n!$ then $l_1^n \stackrel{\circ}{\longhookrightarrow} E$. If, moreover, E is an n dimensional Banach space then $c(n, E) = n^n/n!$ if and only if E is isometrically isomorphic to l_1^n.*

Proof. Let $0 < \varepsilon < 1$ be arbitrary and let

$$\widetilde{\varepsilon} = \frac{n^n}{2^n}\left[1 - \left(1 - \frac{\varepsilon}{n}\right)^n\right].$$

If $L \in \mathcal{L}^s(^n E)$ and $\|L\| > (n^n - \widetilde{\varepsilon})/n!\|\widehat{L}\|$ then there exists $x := (x_1, \ldots, x_n) \in E^n$, $\|x_i\| = 1$ for all i, such that

$$L(x) \geq \frac{n^n - \widetilde{\varepsilon}}{n!}\|\widehat{L}\|.$$

For $\beta = (\beta_j)_{j=1}^n \in \mathbb{R}^n$ let $e^{i\beta} = (e^{i\beta_j})_{j=1}^n$ and $e^{i\beta}x = (e^{i\beta_j}x_j)_{j=1}^n$. By the Polarization Formula, Corollary 1.6, we have

$$\frac{n^n - \tilde{\varepsilon}}{n!}\|\widehat{L}\| \leq L(x) = |L(e^{i\beta}x)|$$

$$\leq \frac{1}{2^n n!} \sum_{\varepsilon_j = \pm 1} \left| \widehat{L}\left(\sum_{j=1}^{n} \varepsilon_j e^{i\beta_j} x_j\right)\right|$$

$$\leq \frac{1}{2^n n!} \sum_{\varepsilon_j = \pm 1} \|\widehat{L}\| \cdot \left\|\sum_{j=1}^{n} \varepsilon_j e^{i\beta_j} x_j\right\|^n.$$

Hence

$$\sum_{\varepsilon_j = \pm 1} \left\|\sum_{j=1}^{n} \varepsilon_j e^{i\beta_j} x_j\right\|^n \geq 2^n (n^n - \tilde{\varepsilon})$$

for all $\beta \in \mathbb{R}^n$. Since

$$\left\|\sum_{j=1}^{n} \varepsilon_j e^{i\beta_j} x_j\right\|^n \leq \left(\sum_{j=1}^{n} \|\varepsilon_j e^{i\beta_j} x_j\|\right)^n \leq n^n$$

we have

$$\left\|\sum_{j=1}^{n} e^{i\beta_j} x_j\right\|^n \geq 2^n (n^n - \tilde{\varepsilon}) - (2^n - 1)n^n \tag{1.41}$$

$$= n^n \left(1 - \frac{\varepsilon}{n}\right)^n.$$

Let $a_j = r_j e^{i\beta_j}$ be n complex numbers. Without loss of generality we can assume $r_1 \geq r_j$ for all j. Then

$$\sum_{j=1}^{n} r_j \geq \left\|\sum_{j=1}^{n} r_j e^{i\beta_j} x_j\right\|$$

$$= \left\|r_1 \cdot \sum_{j=1}^{n} e^{i\beta_j} x_j + \sum_{j=2}^{n} e^{i\beta_j}(r_j - r_1)x_j\right\|$$

$$> r_1 \left\|\sum_{j=1}^{n} e^{i\beta_j} x_j\right\| - \sum_{j=2}^{n}(r_1 - r_j) \quad \text{(since } r_j \leq r_1 \text{ and } \|x_j\| = 1\text{)}$$

$$\geq r_1 n \left(1 - \frac{\varepsilon}{n}\right) - (n-1)r_1 + \sum_{j=2}^{n} r_j \quad \text{(by (1.41))}$$

$$= r_1(1 - \varepsilon) + \sum_{j=2}^{n} r_j$$

$$\geq (1 - \varepsilon) \sum_{j=1}^{n} r_j.$$

Translating this inequality into our original language we have

$$\sum_{j=1}^{n} |a_j| \geq \left\| \sum_{j=1}^{n} a_j x_j \right\| \geq (1-\varepsilon) \sum_{j=1}^{n} |a_j| \tag{1.42}$$

Let $T: l_1^n \longrightarrow F$ be defined by $T(e_i) = x_i$ and linearity, where F is the finite dimensional subspace of E spanned by $\{x_1, \ldots, x_n\}$. By (1.42),

$$\left\| \sum_{j=1}^{n} a_j x_j \right\| \leq \sum_{j=1}^{n} |a_j|$$

and hence $\|T\| \leq 1$. Moreover, since

$$(1-\varepsilon) \sum_{j=1}^{n} |a_j| \leq \left\| \sum_{j=1}^{n} a_j x_j \right\|$$

T is bijective and $\|T^{-1}\| \leq \dfrac{1}{1-\varepsilon}$. Since ε was arbitrary this shows $l_1^n \overset{o}{\hookrightarrow} E$. If E is an n dimensional space then $E = F$ and $d(E, l_1^n) \leq \dfrac{1}{1-\varepsilon}$ for all ε, $0 < \varepsilon < 1$. Hence $d(E, l_1^n) = 1$ and E and l_1^n are isometrically isomorphic. By Example 1.39 this completes the proof.

We now consider the form of extremal polynomials when $c(n, E) = n^n/n!$.

Proposition 1.42 *Suppose* $c(n, E) = n^n/n!$, $L \in \mathcal{L}^s(^n E)$, *and* $L(x) = (n^n/n!) \|\widehat{L}\|$ *where* $x = (x_1, \ldots, x_n)$ *and each* x_i *is a unit vector in* E. *Then there exists a complex number* c *such that*

$$\widehat{L}\left(\sum_{j=1}^{n} \alpha_j x_j \right) = c \cdot \alpha_1 \cdots \alpha_n$$

for all $(\alpha_j)_{j=1}^{n} \in \mathbb{C}^n$.

Remark. Our assumption is not only that L is extremal but, in addition, achieves its norm. If E is finite dimensional L always achieves its norm. Our conclusion shows that $\widehat{L}\big|_{(x_1,\ldots,x_n)}$ has the form of the extremal polynomial given in Example 1.39. If E is an n dimensional space this shows that, up to a scalar multiple, the extremal polynomial is unique.

Proof. Let $T = \{z \in \mathbb{C} : |z| = 1\}$ and let μ_n denote normalized Lebesgue measure on T^n. Let $r_j(e^{i\theta_1}, \ldots, e^{i\theta_n}) = e^{i\theta_j}$, $j = 1, \ldots, n$, denote n normalized independent random variables on $(T^n, \mathcal{M}, \mu_n)$. By Proposition 1.5 we have

$$\frac{n^n}{n!}\|\widehat{L}\| = L(x) = \frac{1}{n!}\left|\int_{T^n}\cdots\int e^{-i\theta_1}\cdot e^{-i\theta_2}\ldots e^{-i\theta_n}\widehat{L}\left(\sum_{j=1}^n e^{i\theta_j}x_j\right)d\mu_n(\theta)\right|$$

$$\leq \frac{1}{n!}\int_{T^n}\cdots\int\left|\widehat{L}\left(\sum_{j=1}^n e^{i\theta_j}x_j\right)\right|d\mu_n(\theta)$$

$$\leq \frac{\|\widehat{L}\|}{n!}\int_{T^n}\cdots\int\left\|\sum_{j=1}^n e^{i\theta_j}x_j\right\|^n d\mu_n(\theta)$$

$$\leq \frac{n^n}{n!}\|\widehat{L}\|.$$

Hence all the above inequalities are equalities. In particular,

$$\left\|\sum_{j=1}^n e^{i\theta_j}x_j\right\| = n \tag{1.43}$$

and

$$\left|\widehat{L}\left(\sum_{j=1}^n e^{i\theta_j}x_j\right)\right| = n^n\|\widehat{L}\| = n!L(x) \tag{1.44}$$

for all $\theta \in \mathbb{R}^n$. Let $\langle m,\theta\rangle = \sum_{i=1}^n m_i\theta_i$ for $m \in \mathbb{N}^{(\mathbb{N})}$ and $\theta \in \mathbb{R}^n$. By (1.44) we have

$$(n!L(x))^2 = (n^n\|\widehat{L}\|)^2 = \int_{T^n}\cdots\int\left|\widehat{L}\left(\sum_{j=1}^n e^{i\theta_j}x_j\right)\right|^2 d\mu_n(\theta)$$

$$= \int_{T^n}\cdots\int\left(\sum_{\substack{m\in\mathbb{N}^n\\|m|=n}}\frac{|m|!}{m!}e^{i\langle m,\theta\rangle}L(x^m)\right)\cdot\left(\sum_{\substack{m\in\mathbb{N}^n\\|m|=n}}\frac{|m|!}{m!}e^{-i\langle m,\theta\rangle}\overline{L(x^m)}\right)d\mu_n(\theta)$$

$$= \sum_{\substack{m\in N^n\\|m|=n}}\left(\frac{|m|!}{m!}\right)^2|L(x^m)|^2.$$

If $\tilde{m} = (1,1,\ldots,1)$ then

$$\frac{|\tilde{m}|!}{\tilde{m}!}\,|L(x^{\tilde{m}})| = n!L(x).$$

Hence $L(x^m) = 0$ if $m \in \mathbb{N}^n$ and $m \neq \tilde{m}$. If $\alpha = (\alpha_1,\ldots,\alpha_n) \in \mathbb{C}^n$ then

$$\widehat{L}\left(\sum_{j=1}^n \alpha_j x_j\right) = \sum_{\substack{m\in\mathbb{N}^n\\|m|=n}}\frac{|m|!}{m!}\alpha^m L(x^m)$$

$$= n!\,L(x)\alpha_1\ldots\alpha_n$$

as required. This completes the proof.

The extremal polynomial for l_1 can be used to construct non-isomorphic Banach spaces E and F such that $c(n, E) = c(n, F) = n^n/n!$ for all n. For example, if

$$E = l_2(\{l_1^n\}_{n=1}^\infty) = \left\{(x_n)_{n=1}^\infty \; : \; x_n \in l_1^n \text{ and } \|(x_n)_n\|^2 := \sum_{n=1}^\infty \|x_n\|^2 < \infty\right\},$$

then E is a reflexive Banach space and hence is not isomorphic to l_1. However, the extremal polynomial for l_1^n can be extended isometrically (i.e. without an increase of norm) to E since l_1^n is isometrically embedded in E and there exists a norm 1 (or contractive projection) from E onto l_1^n.

We turn now to the finite dimensional l_∞ case. We recall that the mapping

$$\phi(\lambda) := \frac{\lambda + \alpha}{1 + \overline{\alpha}\lambda} := \alpha + (1 - |\alpha|^2)\lambda + \sum_{j=2}^\infty a_j \lambda^j \tag{1.45}$$

is a biholomorphic mapping from the unit disc D in $\mathbb{C}$ onto itself when $|\alpha| < 1$.

Proposition 1.43 *If n and m are positive integers then*

$$c(n, l_\infty^m) \leq \frac{n^{n/2}(n + 1)^{(n+1)/2}}{2^n n!}. \tag{1.46}$$

Proof. Let $z = (z_j)_{j=1}^m$, $\|z\| = 1$, $w = (w_j)_{j=1}^m$, $\|w\| < 1$, denote elements in l_∞^m and let $P \in \mathcal{P}(^n(l_\infty^m))$. By (1.45) the mapping

$$g(\lambda) = P\left(\left(\frac{\lambda z_j + w_j}{1 + \overline{w}_j z_j \lambda}\right)_{j=1}^m\right)$$

is a holomorphic mapping of $\{\, \lambda \in \mathbb{C} \; : \; |\lambda| < 1 \,\}$ into $\{\, \lambda \in \mathbb{C} \; : \; |\lambda| \leq \|P\| \,\}$. By the Cauchy inequalities $|g'(0)| \leq \|P\|$. By Lemma 1.9(a) and (1.45)

$$g(\lambda) = P(w) + \sum_{j=1}^m n(1 - |w_j|^2)\overset{\vee}{P}(w^{n-1}, z_j e_j)\lambda + \sum_{k=2}^\infty b_k \lambda^k$$

where $e_j = (0, \ldots, 0, \underset{\underset{j^{\text{th}} \text{ position}}{\uparrow}}{1}, 0 \ldots 0) \in l_\infty^m$ for $1 \leq j \leq m$. Hence

$$\left|\sum_{j=1}^m n(1 - |w_j|^2)\overset{\vee}{P}(w^{n-1}, z_j e_j)\right| \leq \|P\|. \tag{1.47}$$

On replacing z_j by $e^{i\theta_j} z_j$ and using the n-linearity of $\overset{\vee}{P}$ we see that

$$n \cdot \sum_{j=1}^m (1 - |w_j|^2)|\overset{\vee}{P}(w^{n-1}, z_j e_j)| \leq \|P\|.$$

Hence

$$n\big(1 - \|w\|^2\big)\big|\overset{\vee}{P}(w^{n-1}, z)\big| = n(1 - \|w\|^2)\Big|\sum_{j=1}^{m} \overset{\vee}{P}(w^{n-1}, z_j e_j)\Big|$$

$$\leq n \sum_{j=1}^{m} (1 - |w_j|^2)\big|\overset{\vee}{P}(w^{n-1}, z_j e_j)\big|$$

$$\leq \|P\|$$

and this shows

$$n\big|\overset{\vee}{P}(w^{n-1}, z)\big| \leq \frac{\|P\|}{1 - \|w\|^2} \tag{1.48}$$

for any $P \in \mathcal{P}(^n(l_\infty^m))$, $w \in B_{l_\infty^m}$ and $z \in l_\infty^m$, $\|z\| = 1$. Replacing w by rx, $0 < r < 1$ and $\|x\| = 1$, in (1.48) yields

$$n\big|\overset{\vee}{P}(x^{n-1}, z)\big| \leq \inf_{0 < r < 1} \frac{\|P\|}{r^{n-1}(1 - r^2)}. \tag{1.49}$$

Since $\displaystyle\inf_{0 < r < 1} \frac{1}{r^{n-1}(1 - r^2)}$ occurs when $r = \left(\dfrac{n-1}{n+1}\right)^{1/2}$, (1.49) implies that

$$\big|n\overset{\vee}{P}(x^{n-1}, z)\big| \leq \frac{(n+1)^{(n+1)/2}}{2(n-1)^{(n-1)/2}} \|P\| \tag{1.50}$$

for any $P \in \mathcal{P}(^n(l_\infty^m))$, any n and $\|x\| = \|z\| = 1$. The mapping

$$x \in l_\infty^m \longrightarrow \overset{\vee}{P}(x^{n-1}, z)$$

is an $(n - 1)$-homogeneous polynomial for any fixed z and we may apply (1.50) to this polynomial. By iterating this process we eventually arrive at the inequality

$$\|\overset{\vee}{P}\| \leq \frac{n^{n/2}(n+1)^{(n+1)/2}}{2^n n!} \|P\|$$

and this completes the proof.

The same method of proof, i.e. the use of biholomorphic automorphisms of the unit ball can be used in any JB^* triple system E, in particular in any C^* algebra, and one obtains the estimate (1.47) which shows that

$$c(n, E) \leq \frac{n^{n/2}(n+1)^{(n+1)/2}}{2^n n!}.$$

By Stirling's Formula this implies $c(E) \leq e/2$. Since Hilbert spaces are JB^* triple systems, (1.46) gives an upper bound for $c(n, L_2(\mu))$. Our next result shows that this estimate can be greatly improved and that Hilbert spaces have the best possible polarization constants. To prove this result we use the following classical inequality of Bernstein:

If $T_n(\theta) = \sum_{k=-n}^{k=n} c_k e^{ik\theta}$ is a complex trigonometric polynomial of degree n, which satisfies $|T_n(\theta)| \leq 1$ for all real θ, then $|T_n'(\theta)| \leq n$ for all real θ.

Proposition 1.44 If H is a complex Hilbert space and n is a positive integer then $c(n, H) = 1$.

Proof. Let $P \in \mathcal{P}(^n H)$. For x, y in $\overline{B}_H$ put $T_n(\theta) = P(x \cos\theta + i\sigma y \sin\theta)$ where

$$\sigma = \begin{cases} 1, & \text{if } (x, y) = 0; \\ \dfrac{(x, y)}{|(x, y)|}, & \text{if } (x, y) \neq 0. \end{cases}$$

If θ is real then $|T_n(\theta)| \leq \|P\|$ and

$$T_n(\theta) = P\Big(x + i\sigma y\theta + \sum_{k \geq 2} a_k \theta^k\Big)$$

$$= P(x) + n \overset{\vee}{P}(x^{n-1}, i\sigma y)\theta + \sum_{k \geq 2} b_k \theta^k.$$

By Bernstein's Inequality this implies

$$|T_n'(0)| = n|\overset{\vee}{P}(x^{n-1}, i\sigma y)| \leq n\|P\|.$$

Since $|\sigma| = 1$ we get $|\overset{\vee}{P}(x^{n-1}, y)| \leq \|P\|$ and the inductive method used to complete the proof of Proposition 1.43 can be used to show $\|\overset{\vee}{P}\| \leq \|P\|$. This completes the proof.

It is also true that $\|P\| = \|\overset{\vee}{P}\|$ for real homogeneous polynomials on a real Hilbert space and, in fact, a real normed linear space E is isometrically isomorphic to a Hilbert space if and only if $\|P\| = \|\overset{\vee}{P}\|$ for all $P \in \mathcal{P}(^2 E)$. This result does not extend to polynomials over the complex field (see Exercise 1.92). An analysis similar to the above can be undertaken for the remaining $L_p(\mu)$ spaces. We present the following without proof.

Proposition 1.45

(a) When $n = 2^m$ and $1 \leq p \leq \infty$ then $c(n, L_p(\mu)) \leq (n^n/n!)^{|(p-2)/p|}$.

(b) If $1 \leq p \leq n/(n-1)$ then $c(n, L_p(\mu)) = n^{n/p}/n!$.

(c) If $n \leq p \leq \infty$, and $(1/p) + (1/p') = 1$, then $c(n, L_p(\mu)) \leq n^{n/p'}/n!$.

We have noted already that $c(n, E/F) \leq c(n, E)$ for any quotient E/F of E. What about subspaces? Since l_1^n is a subspace of l_∞, Example 1.39 and Proposition 1.43 show that $c(n, l_1^n) > c(n, l_\infty)$ so that in general we do not have $c(n, F) \leq c(n, E)$ for a subspace F of E. An examination of the result $c\big(n, l_2(\{l_1^n\}_{n=1}^\infty)\big) = n^n/n!$ shows that our proof used only the structure and position of certain finite dimensional subspaces. This leads easily to the following.

Lemma 1.46 *If F is a subspace of a Banach space and $\pi: E \to F$ is a continuous projection then $c(n, F) \le \|\pi\|^n c(n, E)$.*

Proof. If $P \in \mathcal{P}(^nF)$ then $P \circ \pi \in \mathcal{P}(^nE)$ and $P \circ \pi \big|_F = P$. We have

$$
\begin{aligned}
\|P \circ \pi\| &= \sup_{\substack{\|x\|\le 1 \\ x \in E}} |P(\pi(x))| \\
&\le \sup_{\substack{\|x\|\le 1 \\ x \in E}} \|P\| \cdot \|\pi(x)\|^n \\
&\le \|P\| \cdot \|\pi\|^n.
\end{aligned}
$$

By the Polarization Formula, $P \overset{\vee}{\circ} \pi \big|_{F^n} = \overset{\vee}{P}$. Hence

$$
\|\overset{\vee}{P}\| \le \|P\overset{\vee}{\circ}\pi\| \le c(n, E)\|P \circ \pi\| \le c(n, E)\|\pi\|^n\|P\|
$$

and $c(n, F) \le \|\pi\|^n c(n, E)$. This completes the proof.

Our main use of Lemma 1.46 will be in situations in which we have projections with norm estimates. We note in passing, however, that conversely Lemma 1.46 can also be used to estimate the norms of certain projections. For example, if $l_1^k \hookrightarrow l_\infty$ and if π is a projection from l_∞ onto l_1^k then (1.46) and Example 1.39 imply

$$
\|\pi\| \ge \frac{2\sqrt{n}}{(n+1)^{(n+1)/2n}}
$$

whenever $k \ge n$. Note that $(2\sqrt{n}) \big/ (n+1)^{(n+1)/2n} \longrightarrow 2$ as $n \to \infty$.

Clearly, if F is a 1-complemented subspace of E, Lemma 1.46 implies $c(n, F) \le c(n, E)$. Since 1-complemented subspaces are difficult to find and at the same time almost too well behaved we wish to weaken our hypothesis and draw the same conclusion. We require the following concept:

A Banach space E contains almost uniform copies of F if for every $\varepsilon > 0$ there exists a subspace F_ε of E and a projection $\pi_\varepsilon: E \longrightarrow F_\varepsilon$ such that

$$
d(F, F_\varepsilon) \le 1 + \varepsilon \quad \text{and} \quad \|\pi_\varepsilon\| \le 1 + \varepsilon.
$$

The existence of uniformly bounded projections is the difference between being represented and containing almost uniform copies. The following is now an easy extension of the method used to prove Lemma 1.46.

Proposition 1.47 *If a Banach space E contains almost uniform copies of F then*

$$c(n, F) \leq c(n, E).$$

We only required projections in Lemma 1.46 in order to extend polynomials and so with the aid of the following definition we get improved results.

Definition 1.48 Let F be a subspace of the Banach space E. The pair $\{F, E\}$ has the *Polynomial Extension Property* if for each $n \in \mathbb{N}$ and each $P \in \mathcal{P}(^n F)$ there exists $\widetilde{P} \in \mathcal{P}(^n E)$ such that

$$\widetilde{P}\big|_F = P \tag{1.51}$$

and

$$\|\widetilde{P}\|_{B_E} = \|P\|_{B_F}. \tag{1.52}$$

The Polynomial Extension Property is thus a polynomial version of the Hahn–Banach extension. Once more the proof of Lemma 1.46 can be adapted to prove the following result.

Proposition 1.49 *If the pair of Banach spaces $\{F, E\}$ has the Polynomial Extension Property then*

$$c(n, F) \leq c(n, E)$$

for all n.

Our results for $c(n, l_1)$ and $c(n, l_\infty)$ show that we do not always have the Polynomial Extension Property (see also Exercise 1.95). If F is a subspace of a finite dimensional Banach space E then the pair $\{F, E\}$ can be identified with $\{\mathbb{C}^k, \mathbb{C}^l\}$ for some positive integers k and l, $k \leq l$. Using this identification and the standard monomial representation of polynomials on finite dimensional spaces we see that each polynomial on F extends to a polynomial on E. Information regarding $c(n, E)$ and $c(n, F)$ can be used to find lower bounds on the norm of each possible extension.

The problem of extending polynomials arises in other contexts within infinite dimensional holomorphy such as the study of homomorphisms, the functional calculus, the equivalence of intrinsic metrics, the algebraic structure of the bidual of a Banach algebra and in various problems that we discuss later. A variety of methods are available for constructing extensions and, in general, extensions are not unique. We discuss two approaches here – one for Banach spaces and the other for arbitrary locally convex spaces. For Banach spaces we use *ultrapowers* and *the Principle of Local Reflexivity* and take the opportunity to introduce and discuss a useful general approach to a wide variety of problems. For reasons that we mention later the same approach leads to complications when applied to arbitrary locally convex spaces and in this case we use the original method of extending polynomials – the *Aron–Berner extension*. This proceeds by extending in turn each variable of the associated symmetric n-linear form using the classical Hahn–Banach Theorem.

Ultraproducts and powers in Banach spaces theory are a great relief to a general intrinsic problem in mathematics. Research mathematics is difficult. Techniques are learned slowly by trial and error usually after a great effort has been made to fully understand the underlying principle. By the time one is competent at applications one is regarded as an expert. Ultrapowers have, on the other hand, given us a few unexpected techniques which can be understood and applied almost immediately. I will spend some time presenting what I consider are the (A), (B) and (C) of applied ultrapowers.

First we define ultrapowers. Let E denote a Banach space and let I denote an index set. We let

$$l_\infty^I(E) = \left\{ (x_i)_{i \in I} \; : \; x_i \in E \quad \text{and} \quad \|(x_i)_{i \in I}\| := \sup_i \|x_i\| < \infty \right\}.$$

Clearly $l_\infty^I(E)$ is a Banach space. We have a natural isometric embedding of E in $l_\infty^I(E)$ by the *diagonal* mapping $x \in E \longrightarrow x^I \in l_\infty^I(E)$. If $\mathcal{U}$ denotes an ultrafilter on the set I we let

$$N_\mathcal{U} = \left\{ (x_i)_{i \in I} \in l_\infty^I(E) \; : \; \lim_\mathcal{U} \|x_i\| = 0 \right\}.$$

Clearly $N_\mathcal{U}$ is a closed subspace of $l_\infty^I(E)$ and we denote the quotient mapping by $\pi_\mathcal{U}$. The ultrapower of E over $\mathcal{U}$, $E_\mathcal{U}$ is defined to be $l_\infty^I(E)/N_\mathcal{U}$. The mapping $i_\mathcal{U} \colon x \in E \longrightarrow \pi_\mathcal{U}(x^I)$ is an isometric embedding of E in $E_\mathcal{U}$.

It is easy to check that

$$\left\| \pi_\mathcal{U}\left((x_i)_{i \in I} \right) \right\| = \lim_\mathcal{U} \|x_i\|.$$

The relationship between finite representability and ultrapowers is very direct:

We have $F \xrightarrow{\;\;\circ\;\;} E$ if and only if F is isometrically isomorphic to a subspace of some ultrapower of E.

We first give a reformulation of the *Principle of Local Reflexivity*. Let J_E denotes the canonical mapping of a Banach space into its bidual.

(A) There exists an ultrafilter $\mathcal{U}$, an isometric embedding $I_\mathcal{U}$ of E'' into $E_\mathcal{U}$ which extends J_E and a contractive projection $T_\mathcal{U}$ from $E_\mathcal{U}$ onto $I_\mathcal{U}(E'')$ such that the diagram

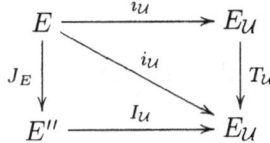

commutes. Moreover, $T_{\mathcal{U}}\big((x_i)_{i\in I} + N_{\mathcal{U}}\big) = I_{\mathcal{U}}(\omega^* - \lim_{\mathcal{U}} x_i)$.

For our second principle we need the concept of *superproperty*. If (P) is a property of infinite dimensional Banach spaces, e.g. reflexivity, cotype, etc. we say that the infinite dimensional Banach space E has *super* (P) if any infinite dimensional Banach space F, $F \xhookrightarrow{\ \circ\ } E$, has property (P). If (P) = super (P), i.e. if a Banach space has super (P) if and only if it has (P) we call (P) a superproperty.

Our second principle (B) states that the following conditions are equivalent for the property (P):

(i)   (P) is a superproperty,
(ii)   (P) is stable by taking ultrapowers and infinite dimensional subspaces,
(iii) (P) is determined by the finite dimensional Banach subspaces,
(iv)   A Banach space E has (P) if and only if any infinite dimensional Banach space which is finitely represented in E also has (P).

By (A) and (ii), or by (iv) and $E'' \xhookrightarrow{\ \circ\ } E \xhookrightarrow{\ \circ\ } E''$, it follows that superproperties are stable by taking biduals and so a Banach space E has super (P) if and only if E'' has super (P). Our third principle involves the extreme cases and here it is surprising that there are in fact *three* cases: the trivial superproperty, the first non-trivial superproperty and the strongest superproperty. Since the property of being infinite dimensional is clearly a superproperty we call it the trivial superproperty. This trivial superproperty would be less interesting were it not for the fact that there is a non-trivial superproperty which is implied by every other non-trivial superproperty. We call this the first non-trivial superproperty. Various concrete characterizations of this superproperty have been found, e.g. non-trivial cotype. By the Dvoretzky Spherical Sections Theorem we have $l_2^I \xhookrightarrow{\ \circ\ } F$ for any infinite dimensional Banach space. Hence, if *any* infinite dimensional Banach space satisfies a superproperty, then l_2^I also satisfies it for any infinite index set I. Since a Banach space is a Hilbert space if and only if all its finite dimensional subspaces are Hilbert spaces it follows by (iii) that the property of being a Hilbert space is a superproperty and so must be the strongest superproperty.

While the use of ultrapowers and superproperties is a sophisticated technique the above gives us a set of preliminary practical tests.

For a superproperty (P) we have the following:

(C_1) if c_0 (or l_∞) has (P) then every infinite dimensional Banach space has (P).

(C_2) if c_0 (or l_∞) does not have (P) then every space which satisfies (P) has the first non-trivial superproperty and so has, for instance, non-trivial cotype,

(C_3) if l_2 does not have (P) then no infinite dimensional Banach space has (P).

In practice the definition of ultrapower and (A) lead, as we shall see, to a simple constructive technique while (B) and (C) help to decide if a given property is a superproperty and often provides limits on the collection of spaces which satisfy a given superproperty. We include the following example solely to illustrate the ideas described above.

Example 1.50 Separability is not a superproperty.

Proof 1. If separability was a superproperty then it would be stable by taking biduals. But c_0 is separable while $c_0'' = l_\infty$ is not separable.

Proof 2. If F is any Banach space then $F \xrightarrow{\quad o \quad} c_0$. If separability was a superproperty then, since c_0 is separable, it would follow that every Banach space was separable. However, l_∞ is not separable.

Proof 3. Since there are infinite dimensional Banach spaces which are separable it follows that if separability was a superproperty then every Hilbert space would be separable. However, l_2^Γ, Γ uncountable, is a non-separable Hilbert space.

Proof 4. If separability was a superproperty then it would be determined by finite dimensional subspaces. In particular, if two Banach spaces had, isometrically, the same finite dimensional subspaces then both would either be separable or non-separable. However, l_2 and l_2^Γ, Γ uncountable, have the same finite dimensional subspaces but one is separable while the other is not.

We return now to polynomials.

Proposition 1.51 *For any Banach space E the pair $\{E, E''\}$ has the Polynomial Extension Property.*

Proof. If $P \in \mathcal{P}(^nE)$, x, $y \in E$, $\|y\| \le 1$ and $\|x - y\| \le 1$ then, by Lemma 1.10(c),

$$|P(x) - P(y)| \le \frac{(2n)^n}{n!} \|P\| \cdot \|x - y\| \qquad (1.53)$$

Let I denote an index set and let $\mathcal{U}$ denote an ultrafilter on I such that $E_\mathcal{U}$ contains E'' as a 1-complemented subspace. If $\|x_i\| \le M$ for all i and $\lim_\mathcal{U} \|x_i - y_i\| = 0$ then (1.53) implies that $\lim_\mathcal{U} P(x_i) = \lim_\mathcal{U} P(y_i)$. Hence the formula

$$\tilde{P}\left(\pi_\mathcal{U}\left((x_i)_{i \in I}\right)\right) = \lim_\mathcal{U} P(x_i)$$

gives a well defined mapping from $E_\mathcal{U}$ into $\mathbb{C}$ and, by the definition of the norm on $E_\mathcal{U}$, we have

$$\|\tilde{P}\|_{B_{E_\mathcal{U}}} = \|P\|.$$

Now let

$$A(x_1, \ldots, x_n) = \frac{1}{n!2^n} \sum_{\varepsilon_j = \pm 1} \varepsilon_1 \cdots \varepsilon_n \tilde{P}\left(\sum_{j=1}^{n} \varepsilon_j x_j\right)$$

for $x_1, \ldots x_n$ in $E_\mathcal{U}$. Then

$$A(x_1, \ldots, x_n) = \lim_{\mathcal{U}} \frac{1}{2^n n!} \sum_{\varepsilon_j = \pm 1} \varepsilon_1 \cdots \varepsilon_n P\left(\sum_{j=1}^{n} \varepsilon_j x_{j,i}\right)$$

where $(x_{j,i})_{i \in I} \in l_\infty^I(E)$, $j = 1, \ldots n$, and $\pi_\mathcal{U}\left((x_{j,i})_{i \in I}\right) = x_j$ for all j.

Since $P \in \mathcal{P}(^n E)$ it follows, by the Polarization Formula, that A is a symmetric n-linear form and $\hat{A} = \tilde{P}$. Thus $\tilde{P} \in \mathcal{P}(^n(E_\mathcal{U}))$, $\tilde{P}\big|_{T_\mathcal{U}(E_\mathcal{U})} \in \mathcal{P}(^n E'')$ and $\tilde{P} \circ J = P$. Hence $\tilde{P} \circ I_\mathcal{U}$ extends P to E'' and $\|\tilde{P} \circ I_\mathcal{U}\| = \|P\|$. This shows that $\{E, E''\}$ has the Polynomial Extension Property and completes the proof.

Corollary 1.52 *For any Banach space E and any positive integer n we have*

$$c(n, E) \le c(n, E'').$$

We now consider extending polynomials from an arbitrary locally convex space E to its bidual E''_e endowed with its natural topology, i.e. E''_e has $(V^{\circ\circ})_{V \in \mathcal{V}}$ as a fundamental neighbourhood basis where $\mathcal{V}$ is a fundamental neighbourhood basis for E and $V^{\circ\circ}$ is the bipolar of V in E''. We could use the Factorization Lemma (Lemma 1.13) and Proposition 1.51 to extend each individual polynomial but this might require the use of different ultrafilters on different neighbourhoods and the linearity of the extension operator, that we require in Chapter 4, becomes less obvious. The use of ultrapowers is also complicated due to the existence of two – the full and the bounded – possibly different ultrapowers on a locally convex space. The density of the bounded ultrapower in the full ultrapower gave rise to the density condition that we encountered earlier. Much more seriously, we do not have a satisfactory local reflexivity principle for arbitrary locally convex spaces.

Proposition 1.53 *If E is a locally convex space and n is a positive integer then there exists a linear extension operator (the Aron–Berner extension operator) $AB_n \colon \mathcal{P}(^n E) \to \mathcal{P}(^n E'')$ such that for every subset A of E and every convex balanced subset B of E*

$$\|AB_n(P)\|_{A+B^{\circ\circ}} \le \|P\|_{A+B}$$

for all $P \in \mathcal{P}(^n E)$.

Proof. Let $P \in \mathcal{P}(^n E)$. We extend $\overset{\vee}{P}$, beginning with the last variable and working backwards to the first, to an n-linear form $\tilde{P}$ on $(E'')^n$. Thus, if $1 \leq j \leq n$ and $\tilde{P}$ has been defined on $E^j \times (E'')^{n-j}$, we define $\tilde{P}$ on $E^{j-1} \times (E'')^{n-j-1}$ by using, for each fixed $(x_1, \ldots, x_{j-1}) \in E^{j-1}$ and $(x''_{j+1}, \ldots, x''_n) \in (E'')^{n-j}$, the second transpose of the linear mapping

$$x \in E \to \tilde{P}(x_1, \ldots, x_{j-1}, x, x''_{j+1}, \ldots, x''_n).$$

We define $AB_n(P)$ to be $\tilde{\tilde{P}}$. Clearly, AB_n is linear. It remains to show that the required estimate holds and that $(AB_n)(P)$ is continuous. We consider the case $A = \{0\}$. Minor adjustments yield the required result for arbitrary A. We may suppose $\|P\|_B < \infty$. Fix z in $B^{\circ\circ}$, the weak* closure of B in E'', and let $\varepsilon > 0$ be arbitrary. Given a positive integer l we claim there exists a subset $(x_j)_{j=1}^l$ of B such that

$$|\overset{\vee}{P}(x_{i_1}, \ldots, x_{i_n}) - \tilde{P}(z, \ldots, z)| < \frac{\varepsilon}{2} \qquad (*)$$

whenever $1 \leq i_j \leq l$ and $i_j \neq i_k$ if $j \neq k$. Since $\tilde{P}$ is a $\sigma(E'', E')$-continuous linear mapping on E'' in the jth variable whenever the first $j-1$ entries are fixed in E and the final $n-j$ entries are fixed in E'' we can choose, inductively, $(x_i)_i \subset B$ such that

$$|\tilde{P}(x_1, z, \ldots, z) - \tilde{P}(z, \ldots, z)| < \frac{\varepsilon}{2n}$$

$$|\tilde{P}(x_2, z, \ldots, z) - \tilde{P}(z, \ldots, z)| < \frac{\varepsilon}{2n}$$

$$|\tilde{P}(x_1, x_2, z, \ldots, z) - \tilde{P}(x_1, z, \ldots, z)| < \frac{\varepsilon}{2n}$$

$$\vdots$$

$$|\tilde{P}(x_{i_1}, \ldots, x_{i_r}, z, \ldots, z) - \tilde{P}(x_{i_1}, \ldots, x_{i_{r-1}}, z, \ldots, z)| < \frac{\varepsilon}{2n}$$

whenever $i_1 < i_2 < \ldots < i_r$. Let $x_{i_0} = z$. We then have

$$\left| \tilde{P}(x_{i_1}, \ldots, x_{i_n}) - \tilde{P}(z, \ldots, z) \right|$$

$$\leq \sum_{r=1}^n |\tilde{P}(x_{i_1}, \ldots, x_{i_r}, z, \ldots, z) - \tilde{P}(x_{i_1}, \ldots, x_{i_{r-1}}, z, \ldots, z)|$$

$$\leq n \cdot \frac{\varepsilon}{2n} = \frac{\varepsilon}{2}$$

whenever $i_1 < i_2 < \ldots < i_r$. Since $\overset{\vee}{P}$ is symmetric and $\tilde{P} = \overset{\vee}{P}$ on E^n this establishes our claim.

Now choose l such that

$$\frac{l^n - l(l-1) \cdots (l-n+1)}{l^n} \left[\frac{n^n}{n!} \|P\|_B + |\tilde{\tilde{P}}(z)| \right] < \frac{\varepsilon}{2}$$

and let $(x_i)_{i=1}^l$ satisfy $(*)$. Since B is convex $x := (1/l)\sum_{i=1}^l x_i \in B$. We have $[AB_n(P)](z) = \widehat{\widetilde{P}}(z) = \widetilde{P}(z,\dots,z)$ and

$$\left|P(x) - (AB_n(P))(z)\right| = \left|\overset{\vee}{P}\Big(\frac{1}{l}\sum_{i=1}^l x_i, \dots, \frac{1}{l}\sum_{i=1}^l x_i\Big) - \widetilde{P}(z,\dots,z)\right|$$

$$= \frac{1}{l^n} \sum_{i_1,\dots,i_r=1}^l \left|\overset{\vee}{P}(x_{i_1},\dots,x_{i_n}) - \widetilde{P}(z,\dots,z)\right|. \qquad (**)$$

The number of terms in the above series is l^n and $l(l-1)\cdots(l-n+1)$ of the n-tuples $\{(x_{i_1},\dots,x_{i_n})\}_{1\le i_j\le l, 1\le j\le n}$ have distinct subscripts. If $i_j \ne i_k$ for $j \ne k$ then $(*)$ implies $|\overset{\vee}{P}(x_{i_1},\dots,x_{i_n}) - \widetilde{P}(z,\dots,z)| < \varepsilon/2$. Otherwise, by the Polarization Inequality, we have

$$\left|\overset{\vee}{P}(x_{i_1},\dots,x_{i_n}) - \widetilde{P}(z,\dots,z)\right| \le \frac{n^n}{n!}\|P\|_B + \left|\widetilde{P}(z,\dots,z)\right|.$$

These two estimates and $(**)$ imply

$$\left|P(x) - (AB_n(P))(z)\right| \le \frac{l(l-1)\cdots(l-n+1)}{l^n} \cdot \frac{\varepsilon}{2}$$

$$+ \frac{l^n - l(l-1)\cdots(l-n+1)}{l^n}\Big[\frac{n^n}{n!}\|P\|_B + \big|\widetilde{P}(z)\big|\Big]$$

$$\le \frac{\varepsilon}{2} + \frac{\varepsilon}{2} = \varepsilon.$$

This proves the required estimate. Since $P \in \mathcal{P}(^nE)$ there exists a convex balanced neighbourhood of 0 in E, V, such that $\|P\|_V < \infty$. Hence $\|AB_n(P)\|_{V^{\circ\circ}} < \infty$ and $AB_n(P) \in \mathcal{P}(^nE'')$. The mapping $P \to (AB_n)(P)$ is easily seen to be linear and this completes the proof.

The Aron–Berner extension was defined using weak* continuity but may also be defined using ultrapowers. Indeed, if $E_{\mathcal{U}}$ is the ultrapower of E associated with the Principle of Local Reflexivity then it is not too difficult to show if $P \in \mathcal{P}(^nE)$ and $z \in E''$ that

$$\big[AB_n(P)\big](z) = \lim_{1,\mathcal{U}}\lim_{2,\mathcal{U}}\dots\lim_{n,\mathcal{U}} P(x_{i,1},\dots,x_{i,n})$$

where $(x_{i,j})_{i\in I} \in l_\infty^I$ and $\lim_{\mathcal{U}} x_{i,j} = x_j$ for $j = 1,\dots,n$. Thus we see that the extension obtained in Proposition 1.51 is obtained by taking a *joint limit* while the extension in Proposition 1.53 is derived by taking an *iterated limit*. Iterated limits along ultrafilters also arise in spreading models (see Section 2.4). Finally we note that $\big[(AB)_n(P)\big]\big(J_E(x)\big) = P(x)$ when $x \in E$ and J_E denotes the canonical embedding of E into E''.

We have completed our investigation of the polarization constants and devote the remainder of the chapter to results suggested by our excursion into ultrapowers. Our next result shows that c_0 is finitely represented in $\mathcal{P}(^nE)$ for any infinite dimensional Banach space E and any $n \geq 2$. It is clearly motivated by the *Dvoretzky Spherical Sections Theorem*. As a consequence we note that $\mathcal{P}(^nE)$, $n \geq 2$ and E infinite dimensional, satisfies no non-trivial superproperties such as cotype, superreflexivity etc. and places limits on the type of Banach space properties we should discuss for spaces of polynomials.

Proposition 1.54 *If E is an infinite dimensional Banach space and $n \geq 2$ then*

$$c_0 \lhook\joinrel\xrightarrow{\;\;o\;\;} \mathcal{P}(^nE).$$

Proof. By the linear Dvoretzky Theorem $l_2 \lhook\joinrel\xrightarrow{\;\;o\;\;} E'$. Hence, for any positive integer k and any $\varepsilon > 0$, there exists a sequence of vectors $(\phi_j)_{j=1}^k$ in E' such that for every $(\alpha_j)_{j=1}^k$ in $\mathbb{C}^k$

$$\sum_{j=1}^k |\alpha_j|^2 \leq \left\| \sum_{j=1}^k \alpha_j \phi_j \right\|^2 \leq (1+\varepsilon)^2 \cdot \sum_{j=1}^k |\alpha_j|^2.$$

By the Riesz Representation Theorem for finite dimensional Hilbert spaces we have

$$\sum_{j=1}^k |\phi_j(x)|^2 = \sup_{\sum_{j=1}^k |\alpha_j|^2 \leq 1} \left| \sum_{j=1}^k \alpha_j \phi_j(x) \right|^2 \leq (1+\varepsilon)^2$$

and in particular $\dfrac{|\phi_j(x)|}{1+\varepsilon} \leq 1$ for all j, $1 \leq j \leq k$, and all x, $\|x\| \leq 1$. Hence for $n \geq 2$ and $x \in E$, $\|x\| \leq 1$, we have

$$\left| \sum_{j=1}^k \alpha_j \phi_j^n(x) \right| \leq (1+\varepsilon)^n \sup_{j=1,\ldots,k} |\alpha_j| \cdot \sum_{j=1}^k \left| \frac{\phi_j(x)}{1+\varepsilon} \right|^n$$

$$< (1+\varepsilon)^n \sup_{j=1,\ldots,k} |\alpha_j| \cdot \sum_{j=1}^k \left| \frac{\phi_j(x)}{1+\varepsilon} \right|^2$$

$$\leq (1+\varepsilon)^n \sup_{j=1,\ldots k} |\alpha_j|.$$

and

$$\left\| \sum_{j=1}^k \alpha_j \phi_j^n \right\| = \sup_{\|x\| \leq 1} \left| \sum_{j=1}^k \alpha_j \phi_j^n(x) \right| \leq (1+\varepsilon)^n \cdot \sup_{j=1,\ldots,k} |\alpha_j|.$$

Since

$$\sum_{j=1}^k |\alpha_j|^2 = \sup_{\sum_{j=1}^k |\beta_j|^2 \leq 1} \left| \sum_{j=1}^k \alpha_j \beta_j \right|^2 \leq \sup_{\|x\| \leq 1} \left| \sum_{j=1}^k \alpha_j \phi_j(x) \right|^2$$

the Hahn–Banach Theorem implies that

$$\overline{B_{l_2^k}} \subset \overline{\{(\phi_j(x))_{j=1}^k : x \in E, \|x\| \le 1\}}. \tag{$*$}$$

Consider now a fixed $(\alpha_j)_{j=1}^k$. Let $\delta > 0$ and l, $1 \le l \le k$, be arbitrary. Since $(0, \ldots, 0, \underset{\underset{l^{\text{th}} \text{ position}}{\uparrow}}{1}, 0 \ldots 0) \in \overline{B_{l_2^k}}$, $(*)$ implies there exists $x \in E$, $\|x\| \le 1$, such that $|\phi_l(x) - 1| < \delta$ and $|\phi_{l'}(x)| < \delta$ if $l \ne l'$. Hence

$$\left| \sum_{j=1}^k \alpha_j \phi_j^n(x) \right| \ge |\alpha_l|(1 - \delta)^n - \sum_{l \ne l'} |\alpha_{l'}| \delta^n$$

and, since n is fixed and δ and l are arbitrary, this implies

$$\sup_{\substack{x \in E \\ \|x\| \le 1}} \left| \sum_{j=1}^k \alpha_j \phi_j^n(x) \right| \ge \sup_{j=1,\ldots,k} |\alpha_j|.$$

We have thus shown that for any $n \ge 2$

$$\sup_{j=1,\ldots,k} |\alpha_j| \le \left\| \sum_{j=1}^k \alpha_j \phi_j^n \right\| \le (1 + \varepsilon)^n \cdot \sup_{j=1,\ldots,k} |\alpha_j|.$$

Since $\varepsilon > 0$ was arbitrary this shows that $l_\infty^k \overset{\circ}{\hookrightarrow} \mathcal{P}(^n E)$ and since k was arbitrary we have $c_0 \overset{\circ}{\hookrightarrow} \mathcal{P}(^n E)$. This completes the proof.

With c_0 always finitely represented in $\mathcal{P}(^n E)$ it is natural to ask when c_0 is a subspace of $\mathcal{P}(^n E)$. The proof of Proposition 1.54 can easily be adapted to prove the following proposition and this leads to examples (Corollary 1.56). These results are placed in perspective in Chapters 2 and 4.

Proposition 1.55 *Suppose $\{\phi_j\}_{j=1}^k$ is a sequence of vectors in E', E a Banach space over $\mathbb{C}$, and there exist constants A and B and $p \ge 1$ such that*

$$A^p \cdot \sum_{j=1}^k |\alpha_j|^p \le \left\| \sum_{j=1}^k \alpha_j \phi_j \right\|^p \le B^p \cdot \sum_{j=1}^k |\alpha_j|^p \tag{1.54}$$

for any sequence of scalars $(\alpha_j)_{j=1}^k$. Then

$$A^n \cdot \sup_{j=1,\ldots,k} |\alpha_j| \le \left\| \sum_{j=1}^k \alpha_j \phi_j^n \right\| \le B^n \cdot \sup_{j=1,\ldots,k} |\alpha_j|$$

for all $n \ge q$ where $1/p + 1/q = 1$.

In particular, since $\mathcal{P}(^n E)$ is a dual space we have $l_\infty \hookrightarrow \mathcal{P}(^n E)$ if the above holds for all k.

Corollary 1.56 *If l_p is a quotient of E then $l_\infty \subset \mathcal{P}(^n E)$ for $n \geq p$.*

To complete this section we take one further look at the Polarization Formula. In proving Proposition 1.5 we used a sequence of independent normalized random variables $(r_j)_j$ but required only for real-valued random variables that the sequence satisfied

$$E\left[r_1^{j_1+1} \cdots r_n^{j_n+1}\right] = \begin{cases} 1 & \text{if } j_1 = j_2 = \ldots = j_n = 1, \\ 0 & \text{if } j_i = 0 \text{ for some } i. \end{cases}$$

Hence it follows that Proposition 1.5 holds for any sequence of random variables with this weaker property. This property of the random variables was used to identify a special term in an expansion involving n elements of an n-homogeneous polynomial. We now wish to identify a term in an expansion, involving k elements, of an n-homogeneous polynomial. For this we require a sequence of random variables $r_1, \ldots, r_k$ such that

$$E\left[r_{j_1} \cdots r_{j_n}\right] = \begin{cases} 1 & \text{if } j_1 = j_2 = \ldots = j_n, \\ 0 & \text{otherwise.} \end{cases}$$

A suitable sequence is provided by the *Generalized Rademacher Functions*. Let $\alpha_0 = 1, \alpha_1, \ldots, \alpha_{n-1}$ denote the n roots of unity. If $t \in [0,1]$ then t has an expansion (mod n) of the form

$$t = \sum_{j=1}^{\infty} \frac{\pi_j(t)}{n^j}$$

where $0 \leq \pi_j(t) \leq n - 1$. Let $s_k(t) = \alpha_{\pi_k(t)}$ for all k and t. The sequence $(s_k)_{k=1}^{\infty}$ is a sequence of random variables on $[0,1]$ endowed with normalized Lebesgue measure and, moreover, $E[s_{i_1} \cdots s_{i_n}] = 1$ if $s_{i_1} = s_{i_2} = \ldots = s_{i_n}$ and otherwise is equal to zero. The following proof is similar to the proof of Proposition 1.5.

Lemma 1.57 *For any k vectors $x_1, \ldots, x_k$ in the locally convex space E and any $P \in \mathcal{P}_a(^n E)$ we have*

$$\sum_{j=1}^{k} P(x_j) = E\left[P\left(\sum_{j=1}^{k} s_j x_j\right)\right].$$

Proof. We have

$$E\left[P\left(\sum_{j=1}^{k} s_j x_j\right)\right] = \int_0^1 P\left(\sum_{j=1}^{k} s_j(t) x_j\right) dt$$

$$= \int_0^1 \sum_{\substack{1 \le l \le n \\ 1 \le j_l \le k}} s_{j_1}(t) \cdots s_{j_n}(t) \overset{\vee}{P}(x_{j_1}, \ldots, x_{j_n}) dt$$

$$= \sum_{j=1}^k P(x_j).$$

This completes the proof.

Now suppose β_j, $j = 1, \ldots, k$ are scalars chosen so that $\beta_j^n P(x_j) = |P(x_j)|$ for all j. Since $|s_j(t)| = 1$ for all j and all $t \in [0, 1]$ we obtain from Lemma 1.57

$$\sum_{j=1}^k |P(x_j)| = \sum_{j=1}^k \beta_j^n P(x_j)$$

$$= E\left[P\left(\sum_{j=1}^k \beta_j s_j x_j \right) \right]$$

$$\le \sup_{|\lambda_j| \le 1} \left| P\left(\sum_{j=1}^k \lambda_j x_j \right) \right|.$$

The inequality

$$\sum_{j=1}^k |P(x_j)| \le \sup_{|\lambda_j| \le 1} \left| P\left(\sum_{j=1}^k \lambda_j x_j \right) \right| \tag{1.55}$$

is an l_1 version of (1.18). In particular, if $(x_j)_{j=1}^\infty$ is the standard unit vector basis $(e_j)_{j=1}^\infty$ for c_0 we get on letting $k \to \infty$ in (1.55)

$$\sum_{j=1}^\infty |P(e_j)| \le \|P\| \tag{1.56}$$

for all $P \in \mathcal{P}(^n c_0)$.

We may thus say that polynomials on c_0 have a *trace*. This result is not surprising since $c_0' = l_1$, the space of absolutely convergent sequences. The inequality (1.56) proves the $p = \infty$ case in the following proposition (we let $p/(p - n) = 1$ when $p = \infty$). A further proof is given in Example 2.25.

Proposition 1.58 *If* $1 \le p \le \infty$, $n \le p$ *and* $P \in \mathcal{P}(^n l_p)$ *then*

$$\left(P(e_j) \right)_{j=1}^\infty \in l_{p/(p-n)}.$$

and

$$\left\| (P(e_j))_{j=1}^\infty \right\|_{l_{p/(p-n)}} \le \|P\|.$$

Proof. For each j choose $\lambda_j \in \mathbb{C}$ such that $\lambda_j^n P(e_j) = |P(e_j)|^{p/(p-n)}$. By Lemma 1.57 we have, for any positive integer k,

$$\sum_{j=1}^{k} |P(e_j)|^{p/(p-n)} = \sum_{j=1}^{k} \lambda_j^n P(e_j) = \sum_{j=1}^{k} P(\lambda_j e_j)$$

$$= E\left[P\left(\sum_{j=1}^{k} \lambda_j s_j e_j\right)\right] = \int_0^1 P\left(\sum_{j=1}^{k} \lambda_j s_j(t) e_j\right) dt$$

$$\leq \|P\| \cdot \left(\sum_{j=1}^{k} |\lambda_j|^p\right)^{n/p}.$$

Since $|\lambda_j|^p = |P(e_j)|^{p/(p-n)}$ this implies

$$\left(\sum_{j=1}^{k} |P(e_j)|^{p/(p-n)}\right)^{1-(n/p)} = \left\|(P(e_j))_{j=1}^{k}\right\|_{l_{p/(p-n)}} \leq \|P\|$$

and since k was arbitrary this completes the proof.

We now apply Proposition 1.58 to show that continuous polynomials on c_0 enjoy a strong form of continuity. The proof of Proposition 2.34 can also be adapted to give an independent proof of this proposition.

Proposition 1.59 *Continuous polynomials on c_0 are weakly continuous on bounded sets.*

Proof. Since $c_0' = l_1$ is a separable Banach space the bounded subsets of c_0 are weakly metrizable. It is sufficient to consider homogeneous polynomials. We claim that it is enough to show that all continuous homogeneous polynomials on c_0 are weakly sequentially continuous at the origin. Indeed, suppose that we have proved this fact. Let $P \in \mathcal{P}(^n c_0)$ and suppose $x_j \to x$ weakly as $j \to \infty$. We have, for all j,

$$P(x_j) - P(x) = \sum_{k=0}^{n-1} \binom{n}{k} \overset{\vee}{P}(x^k, (x_j - x)^{n-k}). \tag{1.57}$$

For $0 \leq k \leq n - 1$ the mapping

$$y \in c_0 \longrightarrow \binom{n}{k} \overset{\vee}{P}(x^k, y^{n-k})$$

is an $(n - k)$-homogeneous polynomial and hence, by (1.57), it follows that $P(x_j) \to P(x)$ as $j \to \infty$.

Now suppose $P \in \mathcal{P}(^n c_0)$ and P is not weakly sequentially continuous at the origin. By using the norm continuity of P, the basis of c_0 and by taking subsequences, if necessary, we can find $\delta > 0$, a sequence $(u_j)_j \subset c_0$ and a strictly increasing sequence of positive integers $(k_j)_j$ such that $|P(u_j)| \geq \delta$,

$$u_j = \sum_{l=1}^{k_{j+1}} x_{l,j} e_l, \qquad \left\| \sum_{l=1}^{k_j} x_{l,j} e_l \right\| \leq \frac{1}{2^j}$$

and

$$\left\| \sum_{l=k_j+1}^{k_{j+1}} x_{l,j} e_l \right\| = 1$$

for all j where $(e_l)_{l=1}^{\infty}$ denotes the standard unit vector basis for c_0.

Let

$$v_j = \sum_{l=k_j+1}^{k_{j+1}} x_{l,j} e_l$$

for all j. The sequence $(v_j)_{j=1}^{\infty}$ consists of unit vectors with disjoint support and hence is a unit vector basis for a Banach subspace of c_0 which is isomorphic to c_0. By Proposition 1.58

$$\sum_{j=1}^{\infty} |P(v_j)| < \infty.$$

By Lemma 1.10(c)

$$\sum_{j=1}^{\infty} |P(u_j)| \leq \sum_{j=1}^{\infty} |P(v_j)| + \frac{n^n}{n!} 2^n \|P\| \cdot \sum_{j=1}^{\infty} \frac{1}{2^j} < \infty.$$

This contradicts the fact that $|P(u_j)| \geq \delta$ for all j and completes the proof.

The same method can be applied to show that continuous n-homogeneous polynomials on l_p, $n < p$, are weakly continuous on bounded sets (see Exercise 2.65).

1.4 Exercises

The following exercises develop topics encountered in later chapters and also certain material which we did not find convenient to include in the main body of the text. Consequently, some of these exercises are rather difficult. A serious attempt at solving them will, however, provide a good deal of insight into the theory – even if only as a means of identifying non-trivial problems. For the research worker they could easily lead to new techniques and worthwhile research projects. Starred exercises are commented on in the

Appendix. Unless otherwise stated all vector spaces are over the complex numbers although many of the results are also valid for real spaces.

Exercise 1.60 Show that $\mathcal{L}_a(^mE; F) = \mathcal{L}_a^s(^mE; F)$ if and only if either $m = 1$, $m = 0$, $\dim(E) \leq 1$ or $F = \{0\}$.

Exercise 1.61* If E is an infinite dimensional locally convex space, show that $\mathcal{P}_a(^nE) = \mathcal{P}(^nE)$ for all n if and only if $E \approx \mathbb{C}^{(\mathbb{N})}$.

Exercise 1.62* If E and F are vector spaces and $P: E \to F$ is a mapping such that $P\big|_G \in \mathcal{P}_a(^nG; F)$ for each finite dimensional subspace G of E show that $P \in \mathcal{P}_a(^nE; F)$. By considering the mapping $f: \mathbb{C}^{(\mathbb{N})} \to \mathbb{C}$ given by $f((z_n)_n) = \sum_{n=1}^{\infty} z_n^n$ show that this result does not extend from homogeneous polynomials to arbitrary polynomials.

Exercise 1.63* If E is a metrizable locally convex space, F an arbitrary locally convex space and $P \in \mathcal{P}_a(^nE; F)$, show that $P \in \mathcal{P}(^nE; F)$ if and only if $\phi \circ P \in \mathcal{P}(^nE)$ for all $\phi \in F'$.

Exercise 1.64* Let E and F be real Banach spaces and let f be a continuous mapping from E into F. Let $A_y f(x) = f(x + y) - f(x)$ for all x, y in E and define $A_{y_1} A_{y_2} \cdots A_{y_n} f(x)$ inductively. Show that f is a polynomial of degree $\leq n$ if and only if $A_{y_1} A_{y_2} \cdots A_{y_{n+1}} f = 0$ for all $y_1, \ldots, y_{n+1}$ in E. Show that this result is not valid for Banach spaces over $\mathbb{C}$.

Exercise 1.65* Let E denote a Fréchet space and suppose $P \in \mathcal{P}_a(^nE)$. Show that P is continuous if its restriction to a second category subset of E is continuous.

Exercise 1.66 Let $E = \sum_{m=1}^{\infty} E_m$ where each E_m is a Banach space. Let $P \in \mathcal{P}(^nE)$ and for each positive integer m let $P_m(x + y) = P(x)$ where $x \in \sum_{j=1}^{m} E_j$ and $y \in \sum_{j=m+1}^{\infty} E_j$. Show that $P_m \in \mathcal{P}(^nE)$ and that $P_m \to P$ as $m \to \infty$ uniformly on a neighbourhood of each point of E.

Exercise 1.67* If E is a metrizable locally convex space and n is a positive integer show that the compact open topology on $\mathcal{P}(^nE)$ is the finest locally convex topology on $\mathcal{P}(^nE)$ which agrees with the topology of pointwise convergence on equicontinuous (or equivalently locally bounded) subsets of $\mathcal{P}(^nE)$.

Exercise 1.68* If E is a complex normed linear space, $L \in \mathcal{L}^s(^mE)$, $1 \leq p \leq \infty$, and $(x_i)_{i=1}^k$ is a finite set of vectors in E such that

$$\left\| \sum_{i=1}^k z_i x_i \right\| \leq \left(\sum_{i=1}^k |z_i|^p \right)^{1/p} \quad \text{for all } (z_1, \ldots, z_k) \in \mathbb{C}^k \qquad (*)$$

show that

$$\left| L(x_1^{n_1}, \ldots, x_k^{n_k}) \right| \leq \frac{n_1! \cdots n_k! \, m^{m/p}}{\left(n_1^{n_1} \cdots n_k^{n_k}\right)^{1/p} m!} \|\widehat{L}\|$$

where $n_1, \ldots, n_k$ are positive integers with $n_1 + \cdots + n_k = m$. Show that the same estimate holds when $E = L^p(\mu)$, (*) is replaced by the condition $\|x_i\| = 1$ for all i and $1 \leq p \leq m'$ (where $\frac{1}{m} + \frac{1}{m'} = 1$).

Exercise 1.69* If E is a Fréchet space and F is a Banach space show that $B \subset \mathcal{P}(^n E; F)$ is τ_0-bounded if and only if $\sup_{P \in B} |P(x)| < \infty$ for every x in E. Construct an example which shows that this is not true on arbitrary locally convex spaces.

Exercise 1.70* If E and F are $\mathcal{DFM}$ spaces and $P \in \mathcal{P}_a(^n(E \times F))$ is separately continuous show that P is continuous.

Exercise 1.71* Let E denote a Banach space with basis $(e_m)_m$ and let F denote a Banach space with unconditional basis $(f_m)_m$. If n is a positive integer and $\sum_{m=1}^{\infty} P(e_m) f_m \in F$ for all $P \in \mathcal{P}(^n E)$ show that $l_\infty \hookrightarrow \mathcal{P}(^n E)'$.

Exercise 1.72* If $E = c_0(\Gamma)$, $F = l_p(\Gamma_1)$, $1 \leq p < \infty$, Γ and Γ_1 uncountable, and $P \in \mathcal{P}(^n E; F)$ show that there exists a countable subset Γ_2 in Γ and $\widetilde{P} \in \mathcal{P}(^n c_0(\Gamma_2); F)$ such that $P = \widetilde{P} \circ i_{\Gamma_2}$. ($i_{\Gamma_2}$ is the natural projection from $c_0(\Gamma)$ onto $c_0(\Gamma_2)$.) Show that $\{\phi^n : \phi \in c_0(\Gamma)'\}$ spans a dense subspace of $\left(\mathcal{P}(^n c_0(\Gamma)), \|\cdot\| \right)$.

Exercise 1.73* Let $K \in L^2([0,1]^{n+1})$ and suppose K is symmetric with respect to each of its coordinates. Let

$$[P(x)](t) = \int_0^1 \int_0^1 \cdots \int_0^1 K(t_1, \ldots, t_n, t) x(t_1) \cdots x(t_n) dt_1 \cdots dt_n$$

for every $x \in L^2([0,1])$. Show that $P \in \mathcal{P}(^n L^2[0,1]; L^2[0,1])$.

Exercise 1.74* If $4/3 \leq p \leq 2$ show that $C(4, l_p) \leq (8/3)^{(4-2p)/p}$.

Exercise 1.75* If H is a separable Hilbert space and $P \in \mathcal{P}(^n H; H)$ is compact (see Definition 2.5) show that there exists x in H, $\|x\| = 1$, such that $P(x) = \|P\|x$.

Exercise 1.76 Show that two locally convex topologies on the same vector space have the same bounded sets if and only if they have the same bounded sequences.

Exercise 1.77* If E is a locally convex space and $m \leq n$ show that $\bigotimes_{m,s,\pi} E$ is a complemented subspace of $\bigotimes_{n,s,\pi} E$.

Exercise 1.78* If m and n are positive integers, $m \leq n$, and E is a locally convex space show that $(\mathcal{P}(^m E), \tau)$ is isomorphic to a complemented subspace of $(\mathcal{P}(^n E), \tau)$ where $\tau = \tau_0$, τ_b or τ_ω.

Exercise 1.79* If the τ_0 bounded subsets of $\mathcal{P}(^n E)$ are locally bounded show that $(\mathcal{P}(^n E), \tau_0)$ is quasi-complete. Show that $\left(\mathcal{P}(^n(\mathbb{C}^{\mathbb{N}} \times \mathbb{C}^{(\mathbb{N})})), \tau_0\right)$ is not quasi-complete.

Exercise 1.80* If $P \in \mathcal{P}(^n E; F)$ where E and F are Banach spaces over $\mathbb{C}$ show that

$$\sup_{|\lambda_j| \leq 1} \left\| \sum_{j=1}^m \lambda_j P(x_j) \right\| \leq \sup_{|\lambda_j| \leq 1} \left\| P\left(\sum_{j=1}^m \lambda_j x_j \right) \right\|$$

for any finite sequence of vectors $(x_j)_{j=1}^m$ in E.

Exercise 1.81* Let E and F be real vector spaces and $E_{\mathbb{C}}$ and $F_{\mathbb{C}}$ their complexifications. If $P \in \mathcal{P}_a(^2 E; F)$ and

$$P_{\mathbb{C}}(x_1 + iy_1, x_2 + iy_2) := \overset{\vee}{P}(x_1, x_2) + i\overset{\vee}{P}(y_1, x_2) + i\overset{\vee}{P}(x_1, y_2) - \overset{\vee}{P}(y_1, y_2)$$

show that $P_{\mathbb{C}} \in \mathcal{L}_a(^2 E_{\mathbb{C}}; F_{\mathbb{C}})$ and that $P_{\mathbb{C}}\big|_E = \overset{\vee}{P}$. Generalize this formula to n-homogeneous polynomials. If $P \in \mathcal{P}(^n E_{\mathbb{C}})$ show that

$$P_{\mathbb{C}}(x + iy) = \frac{2^n}{2\pi} \int_0^{2\pi} P(x \cos\theta + y \sin\theta) e^{in\theta}\, d\theta.$$

If $\| \cdot \|$ is a norm on E show that

$$\gamma(x + iy) := \sup_{\phi \in \overline{B}_{E'}} \left(\phi(x)^2 + \phi(y)^2 \right)^{1/2} = \sup_{t \in \mathbb{R}} \| x \cos t - y \sin t \|$$

defines a complex norm on $E_{\mathbb{C}}$ with the following properties:

(a) $\gamma(\lambda(x + iy)) = |\lambda| \gamma(x + iy)$ for all x, $y \in E$, $\lambda \in \mathbb{C}$,
(b) $\gamma(x) = \|x\|$ for all $x \in E$,
(c) $\gamma(x) \leq \gamma(x + iy)$ for all x, $y \in E$,
(d) $\gamma(x + iy) = \gamma(x - iy)$ for all x, $y \in E$,
(e) if $P \in \mathcal{P}(^n E)$ then $\|P_{\mathbb{C}}\|_\gamma \leq 2^n \|P\|$.
    If β is any norm on $E_{\mathbb{C}}$ satisfying (a), (b) and (d) (i.e. with β in place of γ) show that $\gamma \leq \beta \leq 2\gamma$.

Exercise 1.82* If E and F are Banach spaces and $P \in \mathcal{P}(^n E; F)$ then we say that P is weakly compact if P maps bounded subsets of E onto relatively weakly compact subsets of F. If $P \in \mathcal{P}(^n E; F)$ show that the following are equivalent:

(i) P is weakly compact,

(ii) P^* (the adjoint of P) is weakly compact,

(iii) $P^{**}(E'') \subset F$,

(iv) $i_n^*(\overset{\vee}{P})$: $\underset{n,s,\pi}{\widehat{\bigotimes}} E \to F$ is a weakly compact operator.

Hence show, if E is reflexive and all continuous polynomials on E are weakly continuous on bounded sets, that $\mathcal{P}(^nE)$ is reflexive for all n.

Exercise 1.83* If E is a real reflexive Banach space and $n \geq 2$ show that $\theta \in \underset{n,s,\pi}{\bigotimes} E$ is an extreme point of the unit ball if and only if $\theta = \pm \underbrace{x \otimes \cdots \otimes x}_{n \text{ times}}$ for some unit vector x in E.

Exercise 1.84 Let E and F denote locally convex spaces over $\mathbb{C}$ with completions $\widehat{E}$ and $\widehat{F}$, respectively. If n is a positive integer and $P \in \mathcal{P}(^nE; F)$ show that there exists $\widetilde{P} \in \mathcal{P}(^n\widehat{E}; \widehat{F})$ such that $\widetilde{P}\big|_E = P$.

Exercise 1.85 Show that a subset B of a locally convex space E is bounded if and only if for some n, $\|P\|_B := \sup_{x \in B} |P(x)| < \infty$ for all $P \in \mathcal{P}(^nE)$.

Exercise 1.86* If Y is a subspace of a Banach space X show that $(Y \times Y', X \times Y')$ has the Polynomial Extension Property if and only if Y'' is a complemented subspace of X''.

Exercise 1.87* Show that $\underset{n,s,\pi}{\widehat{\bigotimes}} \mathbb{C}^I = \bigcup_{K \text{ compact}} \overline{\Gamma}\left(\underset{n,s}{\bigotimes} K\right)$ if and only if I is countable.

Exercise 1.88 Let E denote a locally convex space over $\mathbb{C}$. Show that the following conditions are equivalent:

(a) $E'_\beta = E'_i$,

(b) every $\sigma(E'', E')$-bounded subset of $E'' := (E'_\beta)'$ is contained in the $\sigma(E'', E')$ closure of $j_E(B)$ where B is a bounded subset of E and $j_E \colon E \to E''$ is the canonical embedding.

Show that both conditions are satisfied by E if E'_β is bornological and bounded sequences in E'_β are equicontinuous.

Exercise 1.89* If E is a Fréchet space show that the following are equivalent:

(1) E'_β is barrelled,

(2) E'_β is bornological,

(3) each $\sigma(E'', E')$ bounded subset of $E'' := (E'_\beta)'$ is contained in the $\sigma(E'', E')$ closure of $j_E(B)$ where B is a bounded subset of E and $j_E : E \to E''$ is the canonical embedding,

(4) $(E', \beta(E', E)) = (E', \beta(E', E''))$,

(5) if $E = \varprojlim_n E_n$ is a reduced projective limit of normed linear spaces

then $E'_\beta = \varinjlim_n E'_n$.

Exercise 1.90* Let τ_f denote the topology on $\mathcal{P}_a(^nE)$ of uniform convergence on the finite dimensional compact subsets of the vector space E and let $E^*_\sigma = (E^*, \sigma(E^*, E))$. Show that the mapping $(\mathcal{P}_a(^nE), \tau_f)' \to \mathcal{P}(^nE^*_\sigma)$ where $T \mapsto [\widetilde{T} : \phi \to T(\phi^n)]$ is a bijective linear isomorphism.

Exercise 1.91* Let $E = l_p$, $1 < p < \infty$. If $(x_j)_j \subset E$ and $P(x_j) \to 0$ as $j \to \infty$ for all $P \in \mathcal{P}(^nE)$, and all n, show that $\|x_j\| \to 0$ as $j \to \infty$.

Exercise 1.92* Let $E = H \times \mathbb{C}$ with the supremum norm where H is a Hilbert space. Show that $C(n, E) = 1$ for all n.

Exercise 1.93* If F is a closed subspace of the Banach space E and F'' is a 1-complemented subspace of E'' show that $\{F, E\}$ has the Polynomial Extension Property.

Exercise 1.94* By using the existence of discontinuous 2-homogeneous polynomials on $\mathbb{C}^{(I)}$, I uncountable, show that $(\mathcal{P}(^2\mathbb{C}^{(I)}), \tau_0)$ is not quasi-complete.

Exercise 1.95* Let $F = l^3_\infty$ and let $E = \{(x, y, z) \in F : x + y + z = 0\}$. If $P(\alpha + \beta, -\alpha, -\beta) = \alpha^2 + \alpha\beta + \beta^2$ for $\alpha, \beta \in \mathbb{C}$ show that $P \in \mathcal{P}(^2E)$ and $\|P\| = 1$. Show that any extension $\widetilde{P}$ of P to F must have norm strictly greater than 1.

Exercise 1.96* If F is a separable Fréchet space and $E = F'_c$ show that the compact subsets of E are metrizable (and hence separable). Show that each open subset of E contains a fundamental sequence of compact sets and hence deduce that each open subset of E is Lindelöf.

Exercise 1.97* If E is a real Banach space and P is a real-valued polynomial on E of degree n satisfying $\|P\| \le 1$ show that $\|dP(x)\| \le n^2$ for $\|x\| \le 1$.

Exercise 1.98* If E is a countable product of $\mathcal{DFN}$ spaces show that (E, E) has the (BB)-property if and only if $E \approx \mathbb{C}^{\mathbb{N}}$.

Exercise 1.99* If E is a countable direct sum of Fréchet nuclear spaces show that (E, E) has the (BB)-property if and only if E admits a continuous norm.

Exercise 1.100* Let E denote a Banach space and n a positive integer and let δ_x (the Dirac delta function) denote evaluation at the point x in E. Show that the mapping $\delta: E \to \mathcal{P}(^nE)'$, given by $\big(\delta(x)\big)(P) = \delta_x(P) = P(x)$ is a continuous n-homogeneous polynomial of norm 1. Show that $\big(\overset{\vee}{\delta}(x_1,\ldots,x_n)\big)(P) = \overset{\vee}{P}(x_1,\ldots,x_n)$.
If $\mathcal{P}(^nE)'$ is endowed with the topology of uniform convergence over the unit ball of E and the closed linear span of $\delta(E)$, $\widetilde{\delta(E)}$, is endowed with the induced topology show that $\widetilde{\delta(E)} \cong \underset{n,s,\pi}{\bigotimes} E$.

Exercise 1.101* If the vertices of the triangle Δ in $\mathbb{C}$ are the zeros of the polynomial P of degree 3 show that the zeros of P' are the foci of the ellipse of maximum area contained in Δ.

Exercise 1.102* If E is a Banach space, $P \in \mathcal{P}(^kE)$ and $Q \in \mathcal{P}(^lE)$ show that

$$\|P\|\|Q\| \le \frac{(k+l)^{k+l}}{k^k l^l} \|PQ\|.$$

Show that the constant may be replaced by $(k+l)!/k!\,l!$ when E is a Hilbert space.

Exercise 1.103 If $P: l_1 \to \mathbb{C}$ is defined by $P\big((x_j)_j\big) = \sum_{j=1}^{\infty} \lambda_j x_j^n$ show that $P \in \mathcal{P}(^nl_1)$ if and only if $(\lambda_j)_j \in l_{n/(n-1)}$ (where $l_{1/0}$ is identified with l_∞).

Exercise 1.104* If $L \in \mathcal{L}(^nc_0)$ and $(e_j)_j$ is the standard unit vector basis for c_0 show that

$$\sum_{\substack{k_i \in N \\ i=1,\ldots,n}} \big|L(e_{k_1},\ldots,e_{k_n})\big|^{2n/(n+1)} < \infty$$

and that the exponent $2n/(n+1)$ is best possible.

Exercise 1.105* If $E \cong E \times E$ show that $\mathcal{P}(^nE) \times \mathcal{P}(^nE) \cong \mathcal{P}(^nE)$ when each space of polynomials is endowed with either the τ_0, τ_b or τ_ω topologies.

Exercise 1.106* Let H denote a real Hilbert space and let $P: H \to \mathbb{R}$ be a polynomial of degree m satisfying $|P(x)| \le 1$ for all x, $\|x\| \le 1$. Show that

$$\big[(dP(x))(y)\big]^2 \le \frac{m^2\big(1 - \|x\|^2 + (x,y)^2\big)\big(1 - P(x)^2\big)}{1 - \|x\|^2}$$

for all x, $y \in H$, $\|x\| < 1$, $\|y\| = 1$.

Exercise 1.107* Let $\phi \colon X \times X \to \mathbb{C}$ denote a Hermitian form on the Banach space X (i.e. ϕ is real bilinear and $\phi(x,y) = \overline{\phi(y,x)}$ for all x, $y \in X$). If $A = \{x \in X : \phi(x,x) \geq 0,\ x \neq 0\}$, $P \in \mathcal{P}_a(^nE)$, $n > 1$, show that $\big(dP(x)\big)(y) \neq 0$ for all x, $y \in A$.

Exercise 1.108* Show that continuous polynomials between Banach spaces map weakly unconditionally convergent series onto weakly unconditionally convergent series. If E and F are Banach spaces and there exists $P \in \mathcal{P}(^nE; F)$ which does not map weakly unconditionally convergent series onto unconditionally convergent series show that E contains c_0.

Exercise 1.109* If E is a locally convex space show that $\bigotimes\limits_{n,s,\pi} E$ and $\bigotimes\limits_{n,\pi} E$ are Hausdorff spaces.

1.5 Notes

Mathematicians began exploring the concept of polynomial and holomorphic mapping in infinite dimensions at a time when ideas and theories such as the total derivative, point set topology and normed linear spaces, etc. were either still in their infancy or not yet discovered. Moreover, it appears that the search for fundamental concepts in infinite dimensional differential calculus stimulated much of the work which resulted in the satisfactory theory that we now know as *functional analysis*. These pioneers were motivated by many different considerations, and at times were not aware of one another's work. We provide here a brief outline of the development of polynomials over infinite dimensional spaces, a similar treatment of holomorphic functions is given in Section 3.6.

It is generally recognized that the definitive step in the creation of infinite dimensional analysis was taken by V. Volterra in 1887. In a series of notes [843, 844, 845, 846, 847], which later evolved into a book [848], he developed a theory of scalar-valued differentiable functions on $C[a,b]$ and obtained the following Taylor series expansion [843, p.105] for the real-valued analytic function y on $C[a,b]$

$$y\big|[\phi(x) + \psi(x)]\big| = y\big|[\phi(x)]\big| +$$

$$\sum_{1}^{\infty} \frac{1}{\Pi(n)} \int\int_{a}^{b} \cdots \int y^{(n)}\big|[\phi(x), t_1, t_2, \ldots, t_n]\big| \prod_{1}^{n} \psi(t_r)\, dt_1 \ldots dt_n$$

where ϕ, $\psi \in C[a,b]$. The nth term in the above expansion is an n-homogeneous polynomial on $C[a,b]$. Volterra did not, however, specifically discuss polynomial mappings.

The next step was taken by H. von Koch [520] in 1899 who, while discussing Cauchy's existence theorem for ordinary differential equations in n variables, asked

"Ce théorème subsiste-il si n croît au delà de toute limite?"

and replied

"Pour préciser cette question, qui sera l'objet principal de l'étude, il faut définir ce qu'on doit entendre par une fonction analytique d'une infinité de variables indépendentes."

Von Koch defined holomorphic functions of infinitely many variables and noted, but did not provide details, that many results from n complex variables theory extend to infinite dimensions. To von Koch each variable was a coordinate evaluation and he used a monomial expansion with absolute convergence on polydiscs. This is the approach we follow for Gâteaux-holomorphic functions in Section 3.1 and for Fréchet-holomorphic functions on fully nuclear spaces with basis in Section 3.3. Each holomorphic function Φ had, according to von Koch, the following Taylor series expansion:

$$\Phi = \varphi_0 + \varphi_1 + \varphi_2 + \cdots$$

where

$$\varphi_p = \sum_{(\mu),(\nu)} A^{(p)}_{\binom{\mu_1 \cdots \mu_k}{\nu_1 \cdots \nu_k}} x_{\mu_1}^{\nu_1} \cdots x_{\mu_k}^{\nu_k},$$

the μ_i are distinct numbers from the sequence $1,2,\ldots$, the ν_i are non-negative integers and the series converges absolutely on the set $|x_1| < R_1, |x_2| < R_2, \ldots$. It is clear that von Koch had a definite and precise concept of polynomial in mind. This was followed by D. Hilbert, who drew on the work of von Koch and investigations concerning quadratic forms in infinitely many variables, infinite matrices and linear equations with an infinite number of unknowns etc., by O. Toeplitz, E. Hellinger and E. Schmidt. Hilbert outlined a theory of holomorphic functions in infinitely many variables at the international congress in Rome in 1908 and published his results the following year [450]. The following expansion for ϕ is given in [450, p. 65]:

$$\phi(x_1, x_2, \ldots) = c + \sum_{(p)} c_p x_p + \sum_{(p,q)} c_{pq} x_p x_q + \sum_{(p,q,r)} c_{pqr} x_p x_q x_r + \cdots$$

$$= \sum c_{n_1 \cdots n_k} x_1^{n_1} \cdots x_k^{n_k}$$

the series converging absolutely on $|x_1| < |\varepsilon_1|, |x_2| < |\varepsilon_2|, |x_3| < |\varepsilon_3|, \ldots$ Monomial expansions over the unit ball of l_2 are also considered in [450].

During the same year, 1909, M. Fréchet published his first contribution [357] to the abstract theory of polynomials in infinitely many variables. Motivated by Cauchy's observation that any continuous real-valued function f of a real variable which satisfied the equation

$$f(x + y) - f(x) - f(y) = 0 \quad \text{for all } x, y \text{ in } \mathbb{R}$$

had to have the form $f(x) = Ax$ he gave an abstract "difference" characterization of real polynomials of one or several variables (see Exercise 1.64).
He then used this characterization to define real polynomials depending on
a countably infinite number of variables. His domain space was $\mathbb{R}^{\mathbb{N}}$ and on
it he defined, for continuity purposes, a metric which gives the usual coor-
dinatewise topology. The following year he used the same method in [358]
to define real polynomials on $\mathcal{C}[a, b]$ and showed that a real n-homogeneous
polynomial u on this space could be represented as

$$u_f = \lim_{m \to \infty} \int_a^b \cdots \int_a^b u_n^{(m)}(x_1, \ldots, x_n) f(x_1) \cdots f(x_n) \, dx_1 \cdots dx_n,$$

for $f \in \mathcal{C}[a, b]$, where $u_n^{(m)}$ is a sequence of n-homogeneous polynomials in
$x_1, \ldots, x_n$ independent of f and the limit is uniform over the compact subsets
of $\mathcal{C}[a, b]$. He also showed that any polynomial could be represented as a
finite sum of homogeneous polynomials. In a subsequent paper [359], Fréchet
obtained a Riesz Representation Theorem for bilinear forms on $\mathcal{C}[a, b]$.

The next step is due to R. Gâteaux. He made fundamental contributions
to infinite dimensional holomorphy (see Section 3.6) and his simple elegant
style makes pleasant reading. Gâteaux's work consists of three papers [383,
384, 385] which he wrote during the period 1912–1914. He died in 1914 and
most of his results, edited by P. Lévy, appeared in 1919 and 1922. Gâteaux
worked only on the spaces $\mathbb{K}^{\mathbb{N}}$ ($\mathbb{K} = \mathbb{R}$ or $\mathbb{C}$), l_2 and $\mathcal{C}[a, b]$. He noted
that Fréchet's definition of polynomial was inadequate for functions defined
on vector spaces over the field of complex numbers and proposed instead that
a continuous function P such that $P(\lambda z + \mu t)$ is a polynomial of degree n
in λ and μ for any vectors z and t in the domain be called a polynomial
of degree n. Gâteaux showed that his definition coincides with Fréchet's for
real-valued functions of real variables and went on to prove various results –
such as the relationship between the homogeneous parts and the "Gâteaux"
derivatives of a polynomial – with his definition. The development of the
concept of normed linear space and associated ideas between 1910 and 1929
allowed Fréchet to extend his definition of real polynomial to a rather general
setting in [360] and [362].

In 1931–1932, A.D. Michal, a student of Fréchet, gave a series of lectures at the California Institute of Technology in which he outlined the relationship between symmetric n-linear forms and homogeneous polynomials
(Corollary 1.7). This relationship had been noticed earlier for bilinear forms
and 2-homogeneous polynomials by M. Fréchet [359] and R. Gâteaux [384].
S. Mazur–W. Orlicz [599, 600] also established the connection between the
n-linear approach and the now classical approach of Fréchet and Gâteaux for
real Banach spaces. Further work on the definition of polynomial between
Banach spaces was carried out, during the following decade, by A.D. Michal
and his students A.H. Clifford, R.S. Martin, I.E. Highberg and A.E. Taylor
[569, 622, 621, 620, 818, 817, 819, 820], e.g. I.E. Highberg [449] clarified the

relationship between the different definitions and showed that Fréchet's difference method (Exercise 1.64) could be extended to the complex case if one added the hypothesis of complex Gâteaux differentiability.

The Polarization Formula (Corollary 1.6) and the Polarization Inequality (Proposition 1.8) are fundamental results which were known in 1930 but are constantly being rediscovered and published. The earliest references we have been able to locate are H.F. Bohnenblust–E. Hille [142, p.610], the unpublished thesis of R.S. Martin [569] (see [820]) and the two papers of S. Mazur–W. Orlicz [599, 600]. In L. Hörmander [465] (1954) and L. Gårding [381] (1959) we find the following:

$$M = \frac{1}{m!} \prod_k \left(\sum_i x_i^k \frac{\partial}{\partial x_i} \right) P(x)$$

where P is an m-homogeneous polynomial, M is its "polarized form", i.e. $M(x, \ldots, x) = P(x)$, $M = M(x^1, \ldots, x^m)$, $x^k = \{x_j^k\}_{j=1}^n \subset \mathbb{R}^n$. The proof in [465] uses differentiation and Euler's identity. A direct elementary proof from [137] is reproduced in [299, Theorem 1.5], while further proofs may be found in [64], [438], [714], [765] and [820] (see also the remarks on Problem 73 in *The Scottish Book* [595]). The proof given here is based on Proposition 1.5, and is inspired by the version of the Polarization Formula given in [765, Lemma 2].

During the period 1930–1960 many basic properties of polynomials and holomorphic functions appeared in conjunction with developments taking place in linear functional analysis and these appear at various places in the text and exercises. During the interval 1965–1985 there was an emphasis on the development of topics within infinite dimensional holomorphy which were motivated by several complex variables (e.g. holomorphy types, the $\overline{\partial}$ problem, the Levi problem, domains of holomorphy, etc.) – we refer to the notes on Chapters 2, 3 and 5 for details. Polynomial results were observed as special cases – mainly because of the central role of Taylor series expansions – or obtained as an intermediate step.

In this book we report on three important and overlapping lines of research with polynomials on infinite dimensional spaces as the *central* theme. The first of these concerns the interplay between the geometry of Banach spaces and polarization constants and we devote a good deal of Section 1.3 to this topic and discuss it again later in this section. The second involves connections between polynomials and the isomorphic theory of Banach spaces (and locally convex spaces). This is the main theme in Chapter 2 and we discuss its development in Section 2.6. The third, that we now discuss, follows from the introduction of tensor product methods to the study of polynomials. The theory of Banach space tensor products was initiated by R. Schatten [771] in the late 1940s and extended and considerably developed, in a locally convex space context, by A. Grothendieck [414, 415] during the early 1950s. During the mid-1960s the projective tensor product of Banach spaces was introduced by C.P. Gupta [421] into infinite dimensional holomorphy for duality purposes, but was only used sporadically during the 1970s, for example by A. Colojoară [240, 241] in constructing the symmetric algebra of certain locally convex spaces, although it could be said that it was implicit in many of the papers on holomorphy types and convolution equations from that era (see Section 2.6). During the same period, R.M. Aron [38] used the ε-tensor product (Definition 2.21) to study vector-valued holomorphic functions on Banach spaces, while M. Schottenloher [784] used the ϵ-product (which coincides with the ε-tensor product in the presence of the approximation property) to obtain formulae such as $\mathcal{H}(X \times Y) = \mathcal{H}(X) \; \varepsilon \; \mathcal{H}(Y)$. The 1980s saw the appearance of the thesis of R.A. Ryan [758] which introduced, in a systematic way, *symmetric tensor products* into infinite dimensional holomorphy and opened up the possibility of applying tensor product methods to a variety of polynomial problems, and the thesis of J. Taskinen [816], that focused attention on the (BB)-property, which in turn played an essential role in understanding the relationship between the τ_0 and τ_ω topologies on spaces of holomorphic functions. All these, together with the reduction of various holomorphic problems to spaces of homogeneous polynomials, led to the emergence of a research area in which the focus was *exclusively* on polynomials. The object was to study polynomial properties of Banach and Fréchet spaces. For Banach spaces this led to the results we discuss in Section 1.3 and Chapter 2 and the influence from the Fréchet (non-Banach side) appears mainly in Chapter 4. Together, these showed the importance of the fundamental relationship between polynomials and tensor products. The nature of polynomials led to the presentation of *symmetric* rather than general (or *unrestricted*) tensors as the natural setting in which to proceed. At the same time, for reasons we discuss in detail in Chapter 3, we felt it important to attempt a unified approach to the Banach and nuclear space cases. This perspective led to the recasting of certain known results in different ways, e.g. the introduction of the π topology prior to Proposition 1.17.

We now return to commenting on the text. The main duality result in Section 1.1 is Proposition 1.17, which is an extension due to R.A. Ryan [758] of a result in C.P. Gupta [421]. This is a topological version of the algebraic identity (1.7). The Factorization Lemma (and the corresponding result for holomorphic functions – see Chapter 3) has been implicit in the works of many authors, e.g. C.E. Rickart [743], A. Hirschowitz [454] and L. Nachbin [670]. A systematic study of this idea and its consequences is undertaken in S. Dineen [292] (see also [290]) and E. Ligocka [551] (see Section 3.6). Lemma 1.9(a) is just the binomial theorem, while Lemma 1.9(b), which is based on a well-known classical result of one complex variable, has been used by many authors. Lemma 1.10 is given in [299, Lemma 1.12].

Section 1.2 is devoted to the introduction and preliminary investigation of the various topologies we consider on spaces of homogeneous polynomials. The *compact open topology* and the *bounded open topology* are derived from point set topology and functional analysis respectively. The τ_ω topology (also called the *Nachbin topology* and the *ported topology*) which is given in Definitions 1.21 and 1.22 was introduced, initially for holomorphic functions and by inclusion for polynomials, by L. Nachbin [666] (we refer to Chapter 3 and the remarks on Exercise 3.85 for further details). The τ_ω topology in functional analysis (i.e. on E') is known as the inductive topology and had until relatively recently been somewhat neglected in the linear theory (see G. Köthe [522, p.400], P. Pérez Carreras–J. Bonet [720, p.260], K. Floret [352] and J.A. Berezanskiĭ [104]). Example 1.25 appears in [286] for a direct sum of Banach spaces, in [91] for open subsets of a $\mathcal{DFS}$ space, in [295] for $\mathcal{DFM}$ spaces and the tensor product approach to this problem (Example 1.32) is due to R.A. Ryan [758, Proposition 2.4] (see the remarks on Exercise 3.79). Example 1.27 is due to S. Dineen [286].

Initially we used symmetric tensors and duality to define homogeneous polynomials (Proposition 1.17). In Proposition 1.28 we reverse this and realize the space of symmetric tensors with the projective topology as a predual of $\mathcal{P}(^nE)$. Certain universal constructions must, and do, intervene in the proof of this proposition since, in general, dual spaces do not have unique preduals. This result is a special case of Proposition 3.26, and we refer the reader to our remarks on that proposition in Section 3.6 and to the notes on Exercise 1.100. Apart from Propositions 1.33 and 1.34 the remainder of this section follows a suggestion in [758] and examines, in tensor language, the different topologies on spaces of n-homogeneous polynomials – connections with the (BB)-property are established, the $(BB)_n$-property introduced (see [308]) and the differences between homogeneous polynomials on Fréchet and DF spaces observed. Proposition 1.35 amalgamates results due to J.M. Ansemil – S. Ponte [28] and S. Dineen [306, 308] (see also the notes on Chapter 4). Proposition 1.33 is due to A. Defant–M. Maestre [260] and arose in their study on connections between the (BB)-property and holomorphic functions on Fréchet–Montel spaces (the proof given here is different). Proposition 1.34 is due to A. Arias–J.D. Farmer [34] when $E = l_p$ and to J.C. Díaz–S. Dineen [266] for arbitrary E. In [34] one also finds the interesting result that $\mathcal{P}(^nl_p)$ is primary for $1 \leq p < \infty$ and $n \in \mathbb{N}$. Related results can be found in J.C. Díaz [263], who shows that Proposition 1.34 does not always extend to non-stable spaces, J. Bonet–A. Peris [166, Lemma 8], R. Alencar–K. Floret [10], J.M. Ansemil–K. Floret [24] and F. Cabello Sánchez–J. M. F. Castillo–R. García [201]. Using results of K.-D. Bierstedt–J. Bonet [116], H. Hüser [470] constructs an example of a distinguished Fréchet space E such that $\tau_\omega \neq \beta$ on $\mathcal{P}(^nE)$ for $n \geq 2$. A recent example of J.M. Ansemil–F. Blasco–S. Ponte [22] shows that $(BB)_2$ does not imply $(BB)_3$. Example 1.37 is due to R. M. Aron–M. Schottenloher [71] (see also [758, p.101] and [42]).

Monomials, which play an important role in later chapters, are introduced in Section 1.3. Lemma 1.38 is due to R.A. Ryan [759] (see [818, p.474] for a special case). The first part of this section examines the polarization constants on the classical spaces l_1, l_2 and l_∞. J. Kopeć–J. Musielak [519] showed in 1955 that $c(n, L^1[0,1]) = n^n/n!$ and thus that the constant $n^n/n!$ in Proposition 1.8 is best possible – a result which was found independently by B. Grünbaum [417] and which had previously been presented in a 1938 lecture at Princeton by A.E. Taylor [822]. Example 1.39 which recovers the result in [519] appears in L. Nachbin [666]. Propositions 1.41 and 1.42, which show that the extremal constant $n^n/n!$ is more or less uniquely achieved, are due to Y. Sarantopoulos [766] (see also [99, Proposition 6]). Proposition 1.43 and (1.50) are due to L.A. Harris [438] for C^*-algebras (a different proof is given in A. Tonge [828] who also provides evidence suggesting that the constants are the best possible) and for JB^*-triples to P. Mellon [613]. Proposition 1.44 appears in O.D. Kellogg [505] (1928) for real finite dimensional Hilbert spaces and is due to S. Banach [84, 85] (1933) for real separable Hilbert spaces (see also J.G. van der Corput–G. Schaake [254]). A proof for complex Hilbert spaces is given in [465] and also in [137] where the proof is attributed to S. Lojasiewicz. A different proof is given in L.A. Harris [438] and the proof here was motivated by [767, Proposition 1]. It is interesting to note (see [438]) that Proposition 1.44 for real Hilbert spaces is equivalent to the following classical inequality of Szegö:

if $T(\theta)$ is a real trigonometric polynomial of degree m and $|T(\theta)| \leq 1$ for all real θ then

$$m^2 T(\theta)^2 + T'(\theta)^2 \leq m^2.$$

Proposition 1.45(a) is again due to L.A. Harris [438] while (b), which improves (a), and (c) are due to Y. Sarantopoulos [765] (see also [820, p. 307]). Note that (a) is best possible when $p = 1$ (Example 1.39) and when $p = 2$ (Proposition 1.44) while it is not when $p = \infty$ (Proposition 1.43). The constants in Proposition 1.45 have been improved in many cases by Y. Sarantopoulos [765] and similar results for the non-commutative l_p-spaces – the c_p spaces – are given in [767]. The classical Bernstein and Markov polynomial inequalities have been generalized to Banach spaces (see [438], [768] and the remarks on Exercises 1.68, 1.74, 1.97, 1.106 and 1.107). Although many of the remaining results given in Chapter 1 were not directly motivated by the polarization constant the local and isometric nature of the polarization constant and the techniques employed in studying it for the spaces l_1, l_2 and l_∞ allowed us to present these topics as a logical development.

The Polynomial Extension Property was first investigated in 1978 by R.M. Aron–P. Berner [47]. They obtained extensions to certain superspaces of a given space but did not obtain the isometric extension (1.52) in the case of the bidual. The isometric nature of the Aron–Berner extension (Proposition 1.53) was obtained in 1989 by A.M. Davie–T.W. Gamelin [256]. The extension result in [47] was generalized to Fréchet spaces by N. V. Khue [506, 507] and the estimates in [256] extended in P. Galindo-D. García-M. Maestre [371] (see also [627, 628, 629, 632]). R.M. Aron-C. Boyd-Y.C. Choi investigate polynomials which admit "unique" extensions to the bidual in [48].

Independently, at the same time and for very different reasons, S. Dineen–
R.M. Timoney [328] and M. Lindström–R.A. Ryan [560] applied ultrapowers
to the extension problem and obtained alternative proofs of Proposition 1.51
(the proof given here is taken from [328]). For the sake of the non-specialist we
have included an informal introduction to ultrapowers and superproperties.

Following an idea of O. Nicodemi, [688, pp. 536–537], P. Galindo et al.
[372, 373] developed an inductive process for extending multilinear mappings
by letting

$$R_{n+1}A = I_m^{-1}\left[R_n \circ (R_1 \circ I_nA)^t\right]^t \qquad (1.58)$$

for $A \in \mathcal{L}(^{n+1}E; G)$ where I_m is defined in Proposition 1.1 and $R_1 \colon \mathcal{L}(E; G) \to$
$\mathcal{L}(F; G)$ is a given linear mapping which initiates the process. If R_1 is the
canonical embedding of E' into E''' this leads to the Aron–Berner extension.
This algebraic approach allows one to establish many different properties
by straightforward induction (see Exercise 6.59). In [856], I. Zalduendo obtains a topological characterization of polynomials in $\mathcal{P}(^nE'')$, E a Banach
space, which are Aron–Berner extensions of polynomials on E (Exercise 6.51)
and describes a vector-valued Aron–Berner extension (Example 6.15, [857],
[858]). J.A. Jaramillo–A. Prieto–I. Zalduendo [479] study a canonical map-
ping $\beta_n \colon \mathcal{P}(^nE)'' \to \mathcal{P}(^nE'')$ such that $\beta_n \circ J_{\mathcal{P}(^nE)} = AB_n$ where $J_{\mathcal{P}(^nE)}$
is the canonical mapping of $\mathcal{P}(^nE)$ into its bidual (see Section 2.4). For E
reflexive they show that $\mathcal{P}(^nE)$ is reflexive if and only if β_n is injective. Using the principle of local reflexivity M. Lindström–R.A. Ryan [560] prove
that $c(n, E) = c(n, E'')$ whenever E'' has the metric approximation property (see Section 2.1). As we remark in the text, extending polynomials (and
holomorphic functions) occurs in many contexts within infinite dimensional
holomorphy and features in Sections 2.4, 4.3, 5.2 and 6.3.

Propositions 1.54 and 1.55 are due to S. Dineen [309] while Corollary 1.56
is due to M. González–J. Gutiérrez [398] when $p = 2$ and to J. Gómez–
J.A. Jaramillo [392] and S. Dineen [309] for arbitrary p (see also the remarks
on Exercise 4.67, [274] and [404, Corollary 1.3]). The first result of this kind,
as noted in [758, p.107] is due to R.M. Aron who proved that $\mathcal{P}(^nl_p)$ contains
l_∞ for $n > p$ (see also [8]). We present further results on the containment of
l_∞ in $\mathcal{P}(^nE)$ in Sections 2.3 and 4.2.

The estimate (1.56) is due to K. John [484] and N. Th. Varopoulos [834]
when $n = 2$ and to R. M. Aron–J. Globevnik [59] for arbitrary n (see also
[72], [228], [859]). In [59] the authors introduced into infinite dimensional
holomorphy the generalized Rademacher functions ([216]) and these subsequently proved to be a useful technique in studying polynomials over infinite
dimensional spaces (see [10], [64], [172], [173], [174], [228], [274], [354], [404],
[405], [406], [589]). The inequality (1.55) is implicit in [64, Theorem 3] (where
it is attributed to A. Defant–J. Voigt) and [12, Theorem 3.10] and explicit
in [395] (see Exercise 1.108). Proposition 1.58 is due to I. Zalduendo [859]
(see also [22, Lemma 2.1 and Theorem 2.4], [64, Theorem 2], [228], [405,
Theorem 2.1] and the remarks on Exercise 1.104). This brings us to the final

result in this chapter, Proposition 1.59, which is due to J.E. Littlewood [562] when $n = 2$ and, independently, to W. Bogdanowicz [139] and A. Pełczyński [713] for arbitrary n. Many proofs of this result are now known (we provide two more in Chapter 2, see [64, Corollary 10], [167, Theorem 5], [377, Theorem 3.4.1] and [302, Proposition 2]) and the proof given here is also new. Proposition 1.59 is an appropriate closing statement for this chapter since it bridges the gap between the isometric and isomorphic theories and, moreover, weak continuity of polynomials on bounded sets plays a key role in the next chapter.

General references.

Banach space theory: S. Banach [83], A. Defant–K. Floret [258], J. Diestel [270], J. Diestel–H. Jarchow–A. Tonge [271], J. Diestel–J.J. Uhl, Jr. [272]. N. Dunford–J. Schwartz [337], S. Guerre-Delabrière [420], E. Hille [451], R.B. Holmes [461], J. Lindenstrauss–L. Tzafriri [555, 556].

Locally convex spaces: H. Goldman [391], A. Grothendieck [414, 415, 416], J. Horváth [467], H. Jarchow [480], G. Köthe [522, 523], P. Pérez Carreras–J. Bonet [720], A. Pietsch [724], H.H. Schaefer [770].

Several complex variables: L. Hörmander [466], J. Mujica [651].

Chapter 2. Duality Theory for Polynomials

The results of the previous chapter and, in particular, those in Section 1.3, give some idea of the interaction that is possible between polynomials and geometric concepts in Banach space theory. Recent and ongoing research is rapidly adding to our knowledge in this area and indeed the volume of material available and the pace of development obliged us to be selective in our choice of topics for presentation (see the end-of-chapter notes). We have tended to concentrate on polynomial results for which non-trivial holomorphic analogues are available.

The theme of this chapter is duality (and preduality) for spaces of polynomials. We describe situations in which the predual can itself be presented as a space of polynomials. This will be useful later when we present analytic functionals as holomorphic germs. We shall see in Chapters 3 and 4 that infinite dimensional holomorphy on Banach and nuclear spaces leads to the most important and interesting examples. It is thus appropriate to aim for a unified approach and, when this is not possible, to develop these two important cases in parallel.

In discussing duality we have found it convenient to refer to the following well known classical set of dual relationships:

X	X'	X''	
c_0	l_1	l_∞	commutative case
$\mathcal{K}(H)$	$\mathcal{N}(H)$	$\mathcal{L}(H; H)$	non-commutative case

where H is a Hilbert space, $\mathcal{K}(H)$ and $\mathcal{N}(H)$ are the spaces of compact and nuclear operators from H into H respectively. If we take the self-adjoint compact, nuclear and bounded linear mappings in the non-commutative case and translate into the language of polynomials we obtain the following:

X	X'	X''	
$\mathcal{K}(H)$	$\mathcal{N}(H)$	$\mathcal{L}(H;\,H)$	non-symmetric non-commutative case
$\mathcal{P}_A(^2H)$	$\mathcal{P}_N(^2H)$	$\mathcal{P}(^2H)$	symmetric non-commutative case

The notation $\mathcal{P}_A$ and $\mathcal{P}_N$ are explained in Section 2.1. Notice that the commutative case is obtained by restricting to the diagonal, i.e. the diagonal operators in $\mathcal{K}(H)$, $\mathcal{N}(H)$ and $\mathcal{L}(H;H)$ are isometrically isomorphic to c_0, l_1 and l_∞ respectively. We would also have arrived at this subspace following the procedure outlined in Propositions 1.50, 1.51 and Corollary 1.52. The first three sections of this chapter are devoted to obtaining analogous diagrams in more general situations. In the final section we give some applications.

2.1 Special Spaces of Polynomials and the Approximation Property

Important roles in our investigation are played by continuous polynomials which are

(i)   compact,
(ii) weakly continuous on bounded sets,
(iii) weakly uniformly continuous on bounded sets,
(iv) weakly sequentially continuous,
(v)   nuclear.

For linear operators between Banach spaces the first three classes coincide and for domain spaces, which do not contain l_1, these also coincide with (iv). We define these concepts now and show that the linear result does, in fact, extend to n-homogeneous polynomials. Having defined spaces of homogeneous polynomials, the corresponding spaces of n-linear forms and symmetric n-linear forms may be defined following the procedure outlined in Chapter 1 and will be assumed and used without further explanation.

A polynomial $P \in \mathcal{P}(^nE;F)$, E and F locally convex spaces, is said to be of *finite rank* if there exist finite subsets $\{\phi_i\}_{i=1}^l$ in E' and $(y_i)_{i=1}^l$ in F such that

$$P(x) = \sum_{i=1}^l \phi_i^n(x) y_i$$

for all x in E. We let $\mathcal{P}_f(^nE;F)$ denote the set of all n-homogeneous finite rank polynomials from E into F. When $F = \mathbb{C}$ the finite rank polynomials are just the polynomials of finite type introduced in Chapter 1.

The polynomials of finite type separate the points of E and collectively cover the full range space available. They form a relatively small and tractable subspace of the space of all continuous polynomials and are more likely to have properties closer to those of E' and F than $\mathcal{P}(^nE;F)$. It is often useful to initially consider problems on $\mathcal{P}_f(^nE;F)$ and to apply the experience accumulated to study the full space of polynomials. Since the polynomials of finite rank are rarely complete we consider their closure in $\big(\mathcal{P}(^nE;F),\tau_b\big)$.

Definition 2.1 We denote the closure of $\mathcal{P}_f(^nE;F)$ in $\big(\mathcal{P}(^nE;F),\tau_b\big)$ by $\mathcal{P}_A(^nE;F)$. Polynomials in $\mathcal{P}_A(^nE;F)$ are called approximable polynomials.

The term polynomial of compact type and the notation $\mathcal{P}_c(^nE;F)$ have previously been used in place of approximable polynomial and $\mathcal{P}_A(^nE;F)$ but we feel that this notation may be misleading. To obtain a more intrinsic description of approximable polynomials we introduce a further space of polynomials which coincides with $\mathcal{P}_A(^nE;F)$ when E' has the approximation property. This collection also features in biduality and reflexivity problems for $\mathcal{P}(^nE)$ (Section 2.4) and in studying the spectrum of $H_b(E)$ (Section 6.3).

Definition 2.2 We let $\mathcal{P}_\omega(^nE;F)$ denote the set of all continuous n-homogeneous polynomials which are weakly continuous on bounded sets.

Since $\mathbb{C}$-valued continuous linear functionals are weakly continuous and approximable polynomials are the uniform limit on bounded sets of finite rank polynomials we have

$$\mathcal{P}_f(^nE;F) \subset \mathcal{P}_A(^nE;F) \subset \mathcal{P}_\omega(^nE;F) \subset \mathcal{P}(^nE;F)$$

for any locally convex spaces E and F.

Before proceeding we compare at a very elementary level the definitions of continuity and uniform continuity in a locally convex space setting.

Let $f\colon A \subset E \to F$, E and F locally convex spaces. The mapping f is *continuous* if for each x in A and each neighbourhood W of 0 in F there exists a neighbourhood V of 0 in E such that $y \in A$ and $x - y \subset V$ imply $f(x) - f(y) \in W$.

The mapping f is *uniformly continuous* if for each neighbourhood W of 0 in F there exists a neighbourhood V of 0 in E such that $x, y \in A$ and $x - y \in V$ imply $f(x) - f(y) \in W$.

Clearly uniformly continuous mappings are continuous and, conversely, if f is linear then the identity

$$f(x) - f(y) = f(x - y) = f(x - x_0) - f(y - x_0)$$

shows that continuous linear mappings are uniformly continuous. We aim to prove a similar result for polynomials when E and F are Banach spaces, A is bounded and E has the weak topology. We proved this result in Proposition 1.11 for the norm topology. The result we now prove is rather more difficult.

Since a net $(x_\alpha)_\alpha$ in a locally convex space E is Cauchy if and only if for each neighbourhood V of zero in E there exists α_0 such that $x_\alpha - x_\beta \in V$ for all α, $\beta \geq \alpha_0$ it is clear that uniformly continuous mappings map Cauchy nets onto Cauchy nets. A subset A of a locally convex space E is *precompact* if every net in A contains a Cauchy subnet. Clearly uniformly continuous mappings map precompact sets onto precompact sets. Now suppose that A is precompact and $f: A \subset E \to F$ is not uniformly continuous. Then there exists a zero neighbourhood W in F such that in each zero neighbourhood V in E we can find x_V and y_V such that $x_V - y_V \in V$ and $f(x_V) - f(y_V) \notin W$. Let $\mathcal{V}$ denote the set of all neighbourhoods of zero in E. We order the index set $\mathcal{V} \times \mathbb{N}$ by $(V, i) \leq (U, j)$ if $V \supset U$ and $i \leq j$. For $(V, i) \in \mathcal{V} \times \mathbb{N}$ let

$$z_{V,i} = \begin{cases} x_V & \text{if } i \text{ is even,} \\ y_V & \text{if } i \text{ is odd.} \end{cases}$$

Since A is precompact we can suppose, if necessary by taking a subnet, that $(x_V)_V$ and $(y_V)_V$ are Cauchy nets. The net $(z_{V,i})_{(V,i) \in \mathcal{V} \times \mathbb{N}}$ is a Cauchy net in A but $\big(f(z_{V,i})\big)_{(V,i) \in \mathcal{V} \times \mathbb{N}}$ is not a Cauchy net in F. We have proved the following result.

Lemma 2.3 *If E and F are locally convex spaces and A is a precompact subset of E then $f: A \to F$ is uniformly continuous if and only if it maps Cauchy nets onto Cauchy nets.*

Let E denote a Banach space with closed unit ball B_E. The canonical linear mapping from E into its bidual shows that $\big(B_{E''}, \sigma(E'', E')\big)$ is the completion of $\big(B_E, \sigma(E, E')\big)$ and hence the unit ball of E is weakly precompact. Now suppose F is a Banach space and $T: E \to F$ is a continuous linear mapping. We claim that the following two conditions are equivalent:

T is weak to norm continuous on bounded sets, $\qquad\qquad\qquad$ (2.1)

$\overline{T(B_E)}$ is a (norm) compact subset of F. $\qquad\qquad\qquad\qquad$ (2.2)

If (2.1) holds then, since T is linear, it is uniformly continuous and hence $\overline{T(B_E)}$ is a compact subset of F, i.e. (2.2) is satisfied. Conversely suppose (2.2) holds. Let $(x_\alpha)_{\alpha \in \Gamma}$ denote a bounded net in E and suppose $x_\alpha \to x$ weakly as $\alpha \to \infty$. Since T is also weak to weak continuous, $T(x_\alpha) \to T(x)$ weakly as $\alpha \to \infty$. By (2.2), every subnet of $\big(T(x_\alpha)\big)_{\alpha \in \Gamma}$ contains a norm convergent subnet. By the Hahn–Banach Theorem each such subnet converges in norm to $T(x)$. Hence $\big(T(x_\alpha)\big)_{\alpha \in \Gamma}$ converges in norm to $T(x)$ and (2.1) holds. We have established the equivalence of (2.1) and (2.2). Our aim now is to prove the same result for polynomials.

Proposition 2.4 *Let $L \in \mathcal{L}(^n E; F)$ where E and F are Banach spaces. The following are equivalent:*

(a) *L is weakly continuous on bounded sets,*

(b) *if $(x_\alpha^i)_{\alpha \in \Gamma}$ are bounded weak Cauchy nets for $i = 1, \dots, n$ at least one of which is weakly null, i.e. converges weakly to zero, then $L(x_\alpha^1, \dots, x_\alpha^n)$*
   $$\longrightarrow 0 \quad as \quad \alpha \longrightarrow \infty,$$

(c) *if $(x_\alpha^i)_{\alpha \in \Gamma}$ are bounded weak Cauchy nets then $\left(L(x_\alpha^1, \dots, x_\alpha^n)\right)_\alpha$ is a Cauchy net in F,*

(d) *L is weakly uniformly continuous on bounded sets.*

Proof. We prove this result by induction. The case $n = 1$ is the linear case and this is immediate from our previous remarks.

We suppose (a), (b), (c) and (d) are equivalent for *all* continuous $(n-1)$-linear mappings. Suppose (a) is satisfied by the continuous n-linear mapping L but that (b) is false. Then there exists a neighbourhood W of 0 in F, n bounded weak Cauchy nets $(x_\alpha^i)_{\alpha \in \Gamma}$, $1 \le i \le n$, one of which, $(x_\alpha^1)_{\alpha \in \Gamma}$ say, converges to zero such that $L(x_\alpha^1, \dots, x_\alpha^n) \notin W$ for all α. For each fixed α in Γ the mapping

$$Lx_\alpha^n : E^{n-1} \longrightarrow F$$
$$(z_1, \dots, z_{n-1}) \longrightarrow L(z_1, \dots, z_{n-1}, x_\alpha^n)$$

is weakly continuous on bounded sets. Let V denote a neighbourhood of zero in F such that $V + V \subset W$. By our induction hypothesis there exists $\kappa(\alpha) \in \Gamma$, $\kappa(\alpha) \ge \alpha$, such that

$$Lx_\alpha^n(x_\beta^1, \dots, x_\beta^{n-1}) = L(x_\beta^1, \dots, x_\beta^{n-1}, x_\alpha^n) \in V$$

for all $\beta \ge \kappa(\alpha)$. Hence

$$L(x_{\kappa(\alpha)}^1, \dots, x_{\kappa(\alpha)}^{n-1}, x_{\kappa(\alpha)}^n - x_\alpha^n) \notin V$$

for all α and

$$\lim_{\alpha \to \infty} \left(x_{\kappa(\alpha)}^n - x_\alpha^n\right) = 0.$$

By repeating this argument $n - 2$ times we can produce n bounded weakly null nets which are not mapped, by L, onto a null net in F. This contradicts (a) and hence (a)$\Longrightarrow$(b).

Now suppose L satisfies (b). Let $(x_\alpha^i)_{\alpha \in \Gamma}$ denote n bounded Cauchy nets in E. For α, β in Γ

$$L(x_\alpha^1, \dots, x_\alpha^n) - L(x_\beta^1, \dots, x_\beta^n) = L(x_\alpha^1 - x_\beta^1, x_\alpha^2, \dots, x_\alpha^n)$$
$$+ L(x_\beta^1, x_\alpha^2 - x_\beta^2, \dots, x_\beta^n)$$
$$+ \dots + L(x_\beta^1, \dots, x_\beta^{n-1}, x_\alpha^n - x_\beta^n).$$

Since $z_{\alpha,\beta}^i := x_\alpha^i - x_\beta^i \to 0$ as α, $\beta \to \infty$ a re-indexing of nets and an application of (b) shows that $\left(L(x_\alpha^1, \ldots, x_\alpha^n)\right)_{\alpha \in \Gamma}$ is a Cauchy net in F. Hence (b)$\Longrightarrow$(c).

Suppose (c) holds. Let $(x_\alpha)_{\alpha \in \Gamma}$ denote a bounded weak Cauchy net in E^n. Let π_i denote the projection from E^n onto the ith coordinate E. Then $(x_\alpha^i)_{\alpha \in \Gamma} := \left(\pi_i(x_\alpha)\right)_{\alpha \in \Gamma}$ is a weak Cauchy net in E for each i. By (c), $\left(L(x_\alpha^1, \ldots, x_\alpha^n)\right)_{\alpha \in \Gamma}$ is a Cauchy net in F. By Lemma 2.3 this implies that L is weakly uniformly continuous on bounded sets and hence (c)$\Longrightarrow$(d). Clearly (d)$\Longrightarrow$(a) and this completes the proof.

By the Polarization Formula, $P \in \mathcal{P}(^nE; F)$ is weakly (respectively weakly uniformly) continuous on bounded sets if and only if $\overset{\vee}{P} \in \mathcal{L}_s(^nE; F)$ has the corresponding property. We can thus deduce from Proposition 2.4 the main result we were seeking and, moreover, by introducing compact polynomials obtain a full generalization of the linear result.

Definition 2.5 If $P \in \mathcal{P}(^nE; F)$, E and F locally convex spaces, then P is *compact* if there exists a neighbourhood V of 0 in E such that $\overline{P(V)}$ is a compact subset of F. We let $\mathcal{P}_K(^nE; F)$ denote the set of all compact n-homogeneous polynomials from E into F.

Proposition 2.6 *If $P \in \mathcal{P}(^nE; F)$, E and F Banach spaces, then the following are equivalent:*

(i) $P \in \mathcal{P}_\omega(^nE; F)$,

(ii) P *is weakly (uniformly) continuous on bounded sets,*

(iii) *for each k, $0 \le k \le n$, the mapping $\dfrac{\widehat{d}^k P}{k!}$ is weakly (uniformly) continuous on bounded sets,*

(iv) *for each k, $0 \le k \le n$, the mapping $\dfrac{\widehat{d}^k P}{k!}$ is compact,*

(v) $\dfrac{\widehat{d}^{n-1} P}{(n-1)!}$ *is a compact operator.*

Proof. By Proposition 2.4, (i) and (ii) are equivalent. Note also that (i) is equivalent to the $k = 0$ case in (iii). Now suppose (ii) is satisfied and that $1 \le k \le n$. Hence $\overset{\vee}{P}$ satisfies the conditions of Proposition 2.4. Let $(x_\alpha)_{\alpha \in \Gamma}$ denote a bounded net which converges weakly to x. Suppose there exists $\varepsilon > 0$ such that

$$\left\| \frac{\widehat{d}^k P}{k!}(x_\alpha) - \frac{\widehat{d}^k P}{k!}(x) \right\| \ge \varepsilon$$

for all α. Note that this implies $k < n$ since $\dfrac{\widehat{d}^n P}{n!}$ is a constant mapping. For each α there exists a unit vector y_α in E such that

$$\left\|\left(\frac{\widehat{d^k P}}{k!}(x_\alpha)\right)(y_\alpha) - \left(\frac{\widehat{d^k P}}{k!}(x)\right)(y_\alpha)\right\| \geq \frac{\varepsilon}{2}.$$

Since $(y_\alpha)_{\alpha \in \Gamma}$ is bounded it contains a weak Cauchy subnet which we suppose, without loss of generality, is the original net. Hence

$$\left(\frac{\widehat{d^k P}}{k!}(x_\alpha)\right)(y_\alpha) - \left(\frac{\widehat{d^k P}}{k!}(x)\right)(y_\alpha) = \binom{n}{k}\left[\overset{\vee}{P}(x_\alpha^{n-k}, y_\alpha^k) - \overset{\vee}{P}(x^{n-k}, y_\alpha^k)\right]$$

$$= \binom{n}{k}\sum_{j=0}^{n-k-1}\overset{\vee}{P}(x_\alpha^j, x_\alpha - x, x^{n-k-1-j}, y_\alpha^k)$$

$$\longrightarrow\!\!\!\!\!/\ \ 0 \text{ as } \alpha \to \infty$$

and this contradicts (b) of Proposition 2.4 and shows that (ii) $\Longrightarrow$ (iii). Hence the first three conditions are equivalent.

Since the unit ball of E is weakly precompact we see that (iii) $\Longrightarrow$ (iv). Clearly (iv) implies (v). If (v) is satisfied then $\dfrac{\widehat{d^{n-1} P}}{(n-1)!}$ is a compact linear operator and hence weak to norm continuous on bounded sets. If $(x_\alpha)_{\alpha \in \Gamma}$ is a bounded weakly convergent net in E, $x_\alpha \to x$ as $\alpha \to \infty$, then

$$\left\|\frac{\widehat{d^{n-1} P}}{(n-1)!}(x_\alpha) - \frac{\widehat{d^{n-1} P}}{(n-1)!}(x)\right\| = \sup_{\|y\| \leq 1}\left\|\left(\frac{\widehat{d^{n-1} P}}{(n-1)!}(x_\alpha - x)\right)(y)\right\|$$

$$= \sup_{\|y\| \leq 1}\left\|n\overset{\vee}{P}(x_\alpha - x, y^{n-1})\right\|$$

$$\longrightarrow 0 \text{ as } \alpha \to \infty.$$

By Proposition 2.4, $\overset{\vee}{P}$ and P are weakly continuous on bounded sets. Hence (v) $\Longrightarrow$ (i) and this completes the proof.

We now introduce various approximation properties.

Definition 2.7 A locally convex space E has the *approximation property* if for each compact subset K of E, each $\alpha \in cs(E)$ and each positive ε there exists $T = T_{\varepsilon,K} \in \mathcal{L}_f(E; E)$ such that

$$\alpha(x - Tx) \leq \varepsilon \tag{2.3}$$

for all x in K. If the collection $(T_{\varepsilon,K})_{\varepsilon,K}$, $\varepsilon > 0$ and K compact in E, form an equicontinuous subset of $\mathcal{L}_f(E; E)$ we say that E has the *bounded approximation property*.

A Banach space E has the bounded approximation property if there exists $\lambda \geq 1$ such that $\|T_{\varepsilon,K}\| \leq \lambda$ for all $\varepsilon > 0$ and all K compact in E. In this case we also say that E has the λ-*approximation property* and if E has the 1-approximation property we say that E has the *metric approximation property*. In *separable* Fréchet spaces we have the following useful characterization:

> *A separable Fréchet space E has the bounded approximation property if and only if there exists a sequence $(T_n)_n$ in $\mathcal{L}_f(E; E)$ such that $T_n(x) \to x$ as $n \to \infty$ for all x in E.*

If a locally convex space E admits a fundamental system of semi-norms $\mathcal{A}$ such that $E_\alpha := (E, \alpha) \, / \, \alpha^{-1}(0)$ has the approximation property for each $\alpha \in \mathcal{A}$ we say E has the *strict approximation property*. The strict approximation property is stronger than the approximation property. Every Fréchet space with a Schauder basis has the bounded approximation property and conversely every Fréchet space with the bounded approximation property is linearly isomorphic to a complemented subspace of a Fréchet space with a Schauder basis. If E has the approximation property then $\mathcal{P}_f(^nE; F)$ is dense in $\big(\mathcal{P}(^nE; F), \tau_0\big)$ for any locally convex space F. Many standard references on functional analysis contain a detailed study of the approximation property and so we confine ourselves to quoting the one linear result that we require. For spaces with the approximation property we get some idea, by comparing Definition 2.7 with (2.4), of the balance that must be maintained between the approximating topology (τ_0 or τ_ω) and the approximable functions (continuous or compact).

> *Let E denote a Banach space. Then E' has the approximation property if and only if for every Banach space F, every compact linear mapping T from E into F and every $\varepsilon > 0$ there exists $T_1 \in \mathcal{L}_f(E; F)$ such that*

$$\|T - T_1\| \le \varepsilon. \tag{2.4}$$

Proposition 2.8 *If E is a Banach space then E' has the approximation property if and only if*

$$\mathcal{P}_A(^nE; F) = \mathcal{P}_\omega(^nE; F)$$

for any n and any Banach space F.

Proof. Let B denote the open unit ball of E. If $T \in \mathcal{L}(E; F)$ is compact then Proposition 2.6 implies that $T \in \mathcal{P}_\omega(^1E; F)$. Hence if $\mathcal{P}_A(^nE; F) = \mathcal{P}_\omega(^nE; F)$ for all n and all F, (2.4) implies that E' has the approximation property.

Conversely suppose E' has the approximation property. Since we always have $\mathcal{P}_A(^nE; F) \subset \mathcal{P}_\omega(^nE; F)$ we must show that each P in $\mathcal{P}_\omega(^nE; F)$ can be uniformly approximated on the unit ball by finite rank polynomials. We prove this by induction. The case $n = 1$ follows from (2.4) and Proposition 2.6. Suppose the result is true for $\mathcal{P}_\omega(^mE; F)$, $m \le n - 1$. Let $P \in \mathcal{P}_\omega(^nE; F)$.

By Proposition 2.6, $Q := \dfrac{\widehat{d}^{n-1}P}{(n-1)!}$ is a compact linear mapping from E into $\mathcal{P}_\omega(^{n-1}E; F)$. By (2.4) there exists for every $\varepsilon > 0$, $(\phi_i)_{i=1}^l$ in $E' - \{0\}$, and $(Q_i)_{i=1}^l$ in $\mathcal{P}_\omega(^{n-1}E; F)$ such that

$$\sup_{x \in B} \left\| Q(x) - \sum_{i=1}^l \phi_i(x)Q_i \right\| \leq \varepsilon.$$

Since $\big(Q(x)\big)(x) = nP(x)$ it follows that

$$\sup_{x \in B} \left\| P(x) - \frac{1}{n} \sum_{i=1}^l \phi_i(x)Q_i(x) \right\| \leq \varepsilon.$$

By induction there exists for each i, $1 \leq i \leq l$, $\widetilde{Q}_i \in \mathcal{P}_f(^{n-1}E; F)$ such that

$$\|\widetilde{Q}_i - Q_i\| \leq \frac{\varepsilon}{l\|\phi_i\|}.$$

Hence

$$\left\| P - \frac{1}{n} \sum_{i=1}^l \phi_i \widetilde{Q}_i \right\|_B \leq 2\varepsilon.$$

This completes the proof.

To present duals and preduals as spaces of polynomials we introduce nuclear polynomials. These also appear in the predual of spaces of continuous polynomials and as the continuous polynomials on nuclear spaces.

Definition 2.9 If $P \in \mathcal{P}(^nE; F)$ where E and F are normed linear spaces then P is a *nuclear polynomial* if there exist bounded sequences $\{\phi_j\}_{j=1}^\infty \subset E'$, $(y_j)_{j=1}^\infty \in \widehat{F}$ (the completion of F) and $(\lambda_j)_{j=1}^\infty \in l_1$ such that

$$P(x) = \sum_{j-1}^\infty \lambda_j \phi_j^n(x)y_j \tag{2.5}$$

for all x in E. We let $\mathcal{P}_N(^nE; F)$ denote the space of all nuclear n-homogeneous polynomials from E into F and we norm this space by

$$\|P\|_N = \inf \left\{ \sum_{j=1}^\infty |\lambda_j| \|\phi_j\|^n \|y_j\| : P = \sum_{j=1}^\infty \lambda_j \phi_j^n y_j \right\}.$$

It is easily seen to be possible to specify two of the three quantities

$$\sum_{j=1}^\infty |\lambda_j|, \quad \sup_j \|\phi_j\| \quad \text{and} \quad \sup_j \|y_j\|$$

in (2.5) and, if required, one may also suppose that either $(\phi_j)_j$ or $(y_j)_j$ is a null sequence. If E and F are locally convex spaces we say that $P \in \mathcal{P}(^nE; F)$ is nuclear if for each $\beta \in \mathrm{cs}(F)$ there exists $\alpha \in \mathrm{cs}(E)$ and $\tilde{P} \in \mathcal{P}_N(^n\hat{E}_\alpha; \hat{F}_\beta)$ such that the following diagram commutes:

$$
\begin{array}{ccc}
E & \xrightarrow{\;\;P\;\;} & F \\
\pi_\alpha \downarrow & & \downarrow \pi_\beta \\
\hat{E}_\alpha & \xrightarrow[\;\;\tilde{P}\;\;]{} & \hat{F}_\beta
\end{array}
$$

where $\hat{E}_\alpha$ and $\hat{F}_\beta$ are the completions of the normed linear spaces E_α and F_β. This is clearly equivalent to saying that for each continuous semi-norm β on F there exist an equicontinuous sequence $(\phi_j)_j$ in E', $(\lambda_j)_j \in l_1$ and $(y_j)_j$, a bounded sequence in $\hat{F}$, such that

$$
P = \sum_{j=1}^{\infty} \lambda_j \phi_j^n y_j. \tag{2.6}
$$

Corresponding to various topologies on E' we can place topologies on $\mathcal{P}_N(^nE; F)$. If A is a subset of E and $\alpha \in \mathrm{cs}(F)$ let

$$
\pi_{N,A,\alpha}(P) = \|P\|_{N,A,\alpha} := \inf\Big\{ \sum_{j=1}^{\infty} |\lambda_j| \|\phi_j\|_A^n \alpha(y_j) \; : \; P = \sum_{j=1}^{\infty} \lambda_j \phi_j^n y_j \Big\}.
$$

for all $P \in \mathcal{P}_N(^nE; F)$. If $F = \mathbb{C}$ we just write $\|P\|_{N,A}$ or $\pi_{A,N}(P)$ or, if there is no danger of confusion, simply $\pi_A(P)$. As A ranges over the bounded sets of E we get the π_b topology and as A ranges over the compact subsets of E we get the π_0 topology on $\mathcal{P}_N(^nE; F)$. These correspond to the τ_b and τ_0 topologies on $\mathcal{P}(^nE; F)$. We also let

$$
\big(\mathcal{P}_N(^nE; F), \pi_\omega\big) = \varprojlim_{\gamma \in \mathrm{cs}(F)} \Big(\varinjlim_{\alpha \in \mathrm{cs}(E)} \big(\mathcal{P}_N(^nE_\alpha; F_\gamma), \pi_b\big)\Big)
$$

and obtain the analogue of the τ_ω topology on $\mathcal{P}_N(^nE; F)$. Our definitions of $\mathcal{P}_N(^nE)$ and $\widehat{\bigotimes}_{n,s,\pi} E'$ are rather similar and, using the approximation property, we shall connect these concepts. To apply the results of Section 1.2 note that $\bigotimes_{n,s} E'$ can be identified with $\mathcal{P}_f(^nE) \subset \mathcal{P}_N(^nE)$ as follows:

$$
\Big(\sum_{j=1}^{l} \phi_j \otimes \ldots \otimes \phi_j \Big)(x) := \sum_{j=1}^{l} \phi_j(x)^n. \tag{2.7}
$$

We may extend to the completion in certain cases. The following are true:

If E is a Banach space and E' has the approximation property then

$$\widehat{\bigotimes_{n,s,\pi}} E' \cong \left(\mathcal{P}_N(^nE), \|\cdot\|\right)$$

and if E is a locally convex space with the strict approximation property then

$$\varinjlim_{U \in \mathcal{U}} \left(\widehat{\bigotimes_{n,s,\pi}} E'_{U^\circ}\right) = \left(\mathcal{P}_N(^nE), \pi_\omega\right)$$

where $\mathcal{U}$ is a fundamental neighbourhood system at the origin in E consisting of convex balanced sets.

Next we introduce the Borel transform, B, by the formula

$$B \colon \mathcal{P}_f(^nE)' \longrightarrow \mathcal{P}_a(^nE') \tag{2.8}$$
$$BT(\phi) := T(\phi^n)$$

for all $\phi \in E'$. With this notation we easily obtain the following result.

Proposition 2.10 *Let E denote a locally convex space. The Borel transform is an isomorphism in each of the following cases.*

(a) If E'_β has the strict approximation property then

$$\left(\mathcal{P}_N(^nE),\ \pi_b\right)' \cong \mathcal{P}(^nE'_\beta)$$

and the equicontinuous subsets of the dual correspond to the locally bounded subsets of $\mathcal{P}(^nE'_\beta)$.

(b) If E'_c has the strict approximation property then

$$\left(\mathcal{P}_N(^nE),\ \pi_0\right)' \cong \mathcal{P}(^nE'_c)$$

and the equicontinuous subsets of the dual are the locally bounded subsets of $\mathcal{P}(^nE'_c)$.

(c) If E has a fundamental convex balanced neighbourhood system at the origin, $\mathcal{V}$, such that $(E_V)'$ has the approximation property for all $V \in \mathcal{V}$ then $\left(\mathcal{P}_N(^nE), \pi_\omega\right)' = \{P \in \mathcal{P}_a(^nE') : P \text{ is bounded on the equicontinuous subsets of } E'\}$. Moreover, the equicontinuous subsets of the dual correspond to sets of polynomials which are uniformly bounded on equicontinuous subsets of E'.

In the next two sections we discuss the implications of this proposition in two important cases – the Banach and nuclear space cases.

2.2 Nuclear Spaces

Spaces on which every continuous linear mapping is nuclear are particularly suited to the study of holomorphic functions. We introduce and study these spaces in this section. Since, however, they only include the finite dimensional Banach spaces they do not give the full story.

Definition 2.11 A locally convex space E is *nuclear* if $\mathcal{L}(E; F) = \mathcal{L}_N(E; F)$ for every locally convex space F. A locally convex space E is *dual nuclear* if E'_β is nuclear.

By definition E is nuclear if and only if for each $\gamma \in \operatorname{cs}(E)$ there exists $\alpha \in \operatorname{cs}(E)$, $\alpha \geq \gamma$, such that the canonical inclusion mapping

$$E_\alpha \longrightarrow E_\gamma$$

is nuclear (let $F = E_\gamma$ and apply the definition of nuclearity (Definition 2.9) to the linear mapping $E \to E_\gamma$). This means there exists $(\lambda_j)_{j=1}^\infty \in l_1$, $(\phi_j)_{j=1}^\infty$ a bounded sequence in $(E_\alpha)'$ and $(y_j)_{j=1}^\infty$ a bounded sequence in E_γ such that

$$x = \sum_{j=1}^\infty \lambda_j \phi_j(x) y_j \tag{2.9}$$

for all x in E where the series converges in E_γ. The series converges absolutely (and hence uniformly) over $B_\alpha(1)$.

By (2.9) every nuclear space has the approximation property and, indeed, the strict approximation property since it is known that a nuclear space has a fundamental system of semi-norms consisting of semi-inner products. If E is a quasi-complete dual nuclear space then it also has the strict approximation property. Since closed bounded subsets of quasi-complete nuclear and dual nuclear spaces are compact we only discuss the τ_0 and τ_ω topologies on such spaces.

Proposition 2.12 *If E is a nuclear space then*

$$\left(\mathcal{P}(^nE),\ \tau_\omega\right) = \left(\mathcal{P}_N(^nE),\ \pi_\omega\right) = \varinjlim_\alpha\ \mathcal{P}_N(^nE_\alpha).$$

Proof. For any Banach space F the inclusion mapping

$$\mathcal{P}_N(^nF) \longrightarrow \mathcal{P}(^nF)$$

is continuous. As we are comparing two inductive limits of **Banach** spaces, it suffices to show that for each $\gamma \in \operatorname{cs}(E)$ there exists $\alpha \in \operatorname{cs}(E)$, $\alpha \geq \gamma$, such that the mapping

$$\mathcal{P}(^nE_\gamma) \lhook\joinrel\longrightarrow \mathcal{P}_N(^nE_\alpha)$$

is well defined and continuous. Suppose (2.8) holds for α and γ. Let $P \in \mathcal{P}(^nE_\gamma)$ and $x = \sum_{k=1}^{\infty} \lambda_k \phi_k(x) y_k$. By the Polarization Formula

$$\sup_{k_j} |\overset{\vee}{P}(y_{k_1}, \ldots, y_{k_n})| \leq \frac{n^n}{n!} \|P\|_\gamma \left(\sup_k \gamma(y_k) \right)^n =: M_1 \|P\|_\gamma < \infty$$

and, by (1.16),

$$M_2 := \|s(\phi_{k_1} \cdots \phi_{k_n})\| \leq \frac{n^n}{n!} \|\phi_{k_1} \cdots \phi_{k_n}\|_{B_\alpha(1)}$$
$$\leq \frac{n^n}{n!} \left(\sup_k \|\phi_k\|_{B_\alpha(1)} \right)^n < \infty.$$

Let

$$M_3 := \sum_{\substack{k_j=1 \\ j=1,\ldots,n}}^{\infty} |\lambda_{k_1} \cdots \lambda_{k_n}| = \left(\sum_{k=1}^{\infty} |\lambda_k| \right)^n < \infty.$$

Hence the series

$$\sum_{\substack{k_j=1 \\ j=1,\ldots,n}}^{\infty} \lambda_{k_1} \cdots \lambda_{k_n} \phi_{k_1}(x) \cdots \phi_{k_n}(x) \overset{\vee}{P}(y_{k_1}, \ldots, y_{k_n}).$$

is absolutely convergent and equals $P(x)$. After applying the Polarization Formula and relabelling, we obtain

$$P = \sum_{k=1}^{\infty} \delta_k \psi_k^n \overset{\vee}{P}(z_k), \quad \psi_k \in E_\alpha', \quad z_k \in E_\gamma^n,$$

where

$$\sum_{k=1}^{\infty} |\delta_k| = M_3 < \infty, \quad \sup_k \|\psi_k\|_{B_\alpha(1)} \leq M_2, \quad \sup_k |\overset{\vee}{P}(z_k)| \leq M_1 \|P\|_\gamma$$

and the constants M_1, M_2 and M_3 are independent of P. Hence $P \in \mathcal{P}_N(^nE)$ and

$$\|P\|_{N,\alpha} \leq M_1 M_2 M_3 \|P\|_\gamma.$$

This completes the proof.

Proposition 2.13 *If E is a quasi-complete dual nuclear space then $\tau_0 = \pi_0$ on $\mathcal{P}_N(^nE)$ and $(\mathcal{P}(^nE), \tau_0)$ is a nuclear space.*

Proof. Clearly $\tau_0 \leq \pi_0$. Let K denote a compact convex balanced subset of E. Since E is dual nuclear and quasi-complete there exists a convex balanced compact subset K_1 in E, $K_1 \supset K$, such that the canonical mapping

$$(E', \|\cdot\|_{K_1}) \longrightarrow (E', \|\cdot\|_K)$$

is nuclear. The Banach space $(E', \|\cdot\|_{K_1})' = E_{K_1}$ has K_1 as its closed unit ball. Hence there exists $(\lambda_k)_{k=1}^\infty \in l_1$, $\{x_k\}_{k=1}^\infty \subset K_1$ and $(\phi_k)_{k=1}^\infty \subset E'$ with $\|\phi_k\|_K \le 1$ for all k such that

$$\phi = \sum_{k=1}^\infty \lambda_k \phi(x_k)\phi_k \qquad (2.10)$$

for all $\phi \in E'$ uniformly over K. Now suppose $P \in \mathcal{P}_N(^nE)$ and $\|P\|_{N,K} < \infty$. By the Hahn–Banach Theorem there exists $\theta \in (\mathcal{P}_N(^nE), \pi_0)'$ such that $\theta(P) = \|P\|_{N,K}$ and $\|\theta\|_{B_{N,K}(1)} = 1$. If $P = \sum_{i=1}^\infty \psi_i^n$ and $\sum_{i=1}^\infty \|\psi_i\|_K^n < \infty$ then, by (2.10),

$$\|P\|_{N,K} = \theta(P) = \sum_{i=1}^\infty \theta(\psi_i^n)$$

$$= \sum_{i=1}^\infty \sum_{k_1,k_2\ldots,k_n=1}^\infty \lambda_{k_1} \cdots \lambda_{k_n} \psi_i(x_{k_1}) \cdots \psi_i(x_{k_n}) \theta\big(s(\phi_{k_1} \cdots \phi_{k_n})\big)$$

$$= \sum_{k_1,k_2\ldots,k_n=1}^\infty \lambda_{k_1} \cdots \lambda_{k_n} \theta\big(s(\phi_{k_1} \cdots \phi_{k_n})\big)\Big\{\sum_{i=1}^\infty \psi_i(x_{k_1}) \cdots \psi_i(x_{k_n})\Big\}.$$

$$(2.11)$$

Since

$$\sum_{i=1}^\infty \psi_i(x_{k_1}) \cdots \psi_i(x_{k_n}) = \overset{\vee}{P}(x_{k_1}, \ldots, x_{k_n})$$

the Polarization Formula implies that

$$\Big|\sum_{i=1}^\infty \psi_i(x_{k_1}) \cdots \psi_i(x_{k_n})\Big| \le \frac{n^n}{n!}\|P\|_{K_1}$$

and a further application of the Polarization Formula implies

$$\big|\theta\big(s(\phi_{k_1} \cdots \phi_{k_n})\big)\big| \le \frac{n^n}{n!}\big(\sup_k \|\phi_k\|_K\big)^n \le \frac{n^n}{n!}.$$

Hence

$$\|P\|_{N,K} \le \Big(\sum_{k=1}^\infty |\lambda_k|\Big)^n \cdot \Big(\frac{n^n}{n!}\Big)^2 \cdot \|P\|_{K_1}.$$

This shows that π_0 and τ_0 agree on $\mathcal{P}_N(^nE)$.

For $k = (k_1, \ldots, k_n) \in \mathbb{N}^n$ let $\lambda_k = \lambda_{k_1} \cdots \lambda_{k_n}$, $L_k(P) = \overset{\vee}{P}(x_{k_1}, \ldots, x_{k_n})$ and $\phi_k = s(\phi_{k_1} \cdots \phi_{k_n})$. We have $\sum_{k\in\mathbb{N}^n} |\lambda_k| < \infty$, $(L_k)_k$ is a bounded sequence in $\left(\mathcal{P}_N(^nE), \|\cdot\|_{N,K_1}\right)'$, $(\phi_k)_k$ is a bounded sequence in $\left(\mathcal{P}_N(^nE), \|\cdot\|_{N,K}\right)$ and, by (2.11),

$$P = \sum_{k\in\mathbb{N}^n} \lambda_k L_k(P)\phi_k$$

in $\left(\mathcal{P}_N(^nE), \|\cdot\|_{N,K}\right)$. It follows that the canonical mapping

$$\left(\mathcal{P}_N(^nE), \|\cdot\|_{N,K_1}\right) \longrightarrow \left(\mathcal{P}_N(^nE), \|\cdot\|_{N,K}\right)$$

is nuclear and hence $\left(\mathcal{P}_N(^nE), \tau_0\right)$ is a nuclear space. Since E'_β has the approximation property $\mathcal{P}_N(^nE)$ is dense in $\mathcal{P}(^nE)$ for the compact open topology and $\left(\mathcal{P}(^nE), \tau_0\right)$ is nuclear.

A locally convex space E is infrabarrelled (or quasi-barrelled) if either of the following equivalent conditions are satisfied:

(a) *every closed convex balanced subset of E which absorbs bounded subsets is a neighbourhood of zero,*

(b) *strongly bounded, i.e. $\beta(E',E)$-bounded, subsets of E' are equicontinuous.*

If $J_E: E \to E'' := (E'_\beta)'_\beta$ denotes the canonical mapping into the bidual i.e. $(J_E x)(\phi) = \phi(x)$ for $x \in E$ and $\phi \in E'$, then E is infrabarrelled if and only if J_E is an isomorphism onto its image. If J_E is a surjective mapping we say that E is semi-reflexive. Thus

$$\text{infrabarrelled } + \text{ semi-reflexive } \Longleftrightarrow \text{ reflexive.}$$

Barrelled and bornological spaces are infrabarrelled.

Propositions 2.10, 2.12 and 2.13 motivate the introduction of the following interesting collection of locally convex spaces.

Definition 2.14 A locally convex space E is *fully nuclear* if E and E'_β are both complete infrabarrelled nuclear spaces.

If E is fully nuclear then it is reflexive, its closed bounded sets are compact (i.e. it is Montel) and its strong dual is also fully nuclear. Every Fréchet nuclear space, and hence every $\mathcal{DFN}$, is fully nuclear.

To identify certain dual spaces we must consider certain polynomials which are not necessarily continuous. The basic concepts – *k-spaces* and *hypocontinuity* – which come from point set topology will be useful later (see Chapters 4 and 5) and we introduce them here. A mapping between topological spaces is said to be *hypocontinuous* if its restriction to each compact set is continuous. A topological space X is a k-space if a mapping from X is continuous whenever its restrictions to compact subsets are continuous. Thus the topology of a k-space is *localized* on its compact sets and we have the following characterization:

a topological space (X, τ) is a k-space if and only if for any topology τ' on X such that $\tau'|_K = \tau|_K$ for every compact subset K of X we have $\tau \geq \tau'$.

Given a topological space (X, τ) there exists a unique k-space topology τ_k on X which agrees with τ on the compact subsets of X. We call (X, τ_k) the k-space associated with (X, τ). Clearly a topological space (X, τ) is a k-space if and only if $\tau = \tau_k$ on X. Metrizable spaces and $\mathcal{DFM}$ spaces are k-spaces.

We let $\mathcal{P}_{HY}(^n E; F)$ denote the set of all hypocontinuous n-homogeneous polynomials between the locally convex spaces E and F. If E is metrizable or a bornological DF space (Example 1.25) then $\mathcal{P}_{HY}(^n E; F) = \mathcal{P}(^n E; F)$. The compact open topology τ_0 on $\mathcal{P}_{HY}(^n E; F)$ is defined in an obvious fashion.

Lemma 2.15 *If E is a fully nuclear space then $P \in \mathcal{P}_a(^n E)$ is continuous on compact sets whenever it is bounded on compact sets and $\big(\mathcal{P}_{HY}(^n E), \tau_0\big)$ is the completion of $\big(\mathcal{P}(^n E), \tau_0\big)$.*

Proof. Since E is dual nuclear and the transpose of a nuclear mapping is nuclear it follows that for each convex balanced compact subset K of E there exists K_1, convex balanced compact in E, $K \subset K_1$, such that the inclusion mapping

$$E_K \longrightarrow E_{K_1}$$

is nuclear. Hence there exists $(\lambda_j)_j \in l_1$, $(\phi_j)_j \subset E_K'$, $\|\phi_j\|_K \leq 1$ for all j, $(y_j)_j \subset K_1$ such that for each x in K

$$x = \sum_{j=1}^{\infty} \lambda_j \phi_j(x) y_j \ . \tag{2.12}$$

and the series converges absolutely in E_{K_1}. In particular, K is a compact subset of E_{K_1}. Since the inclusion $E_{K_1} \longrightarrow E$ is continuous it follows that E_{K_1} and E induce the same topology on K. If $P \in \mathcal{P}_a(^n E)$ and P is bounded on the compact subsets of E then P is bounded on the bounded subsets of E_{K_1} and hence, by Proposition 1.11, $P\big|_{E_{K_1}} \in \mathcal{P}(^n E_{K_1})$. In particular $P\big|_{K_1}$ is continuous and this proves the first part of the lemma.

Since the completion of $\big(\mathcal{P}(^n E), \tau_0\big)$ is contained in $\big(\mathcal{P}_{HY}(^n E), \tau_0\big)$ for any locally convex space E we must show that for any $A \in \mathcal{L}_{HY}^s(^n E)$, the space of hypocontinuous symmetric n-linear forms on E^n, and any $\varepsilon > 0$ there exists $P \in \mathcal{P}(^n E)$ such that

$$\|\widehat{A} - P\|_K \leq \varepsilon.$$

Since $A\big|_{K_1^n}$ is continuous, (2.12) implies

$$\widehat{A}(x) = \sum_{k_1, \ldots, k_n = 1}^{\infty} \lambda_{k_1} \cdots \lambda_{k_n} \cdot A(y_{k_1}, \ldots, y_{k_n}) \phi_{k_1}(x) \cdots \phi_{k_n}(x)$$

for all x in K. Given $\varepsilon > 0$ there exists, since

$$\sum_{k_1,\ldots,k_n=1}^{\infty} |\lambda_{k_1} \cdots \lambda_{k_n}| |A(y_{k_1},\ldots,y_{k_n})| \cdot \|\phi_{k_1}\|_K \cdots \|\phi_{k_n}\|_K < \infty,$$

a finite subset J of $\mathbb{N}^n$ such that

$$\left\| \widehat{A} - \sum_{(k_1,\ldots,k_n)\in J} \lambda_{k_1} \cdots \lambda_{k_n} \cdot A(y_{k_1},\ldots,y_{k_n}) \cdot \phi_{k_1} \cdots \phi_{k_n} \right\|_K \le \varepsilon. \qquad (2.13)$$

Since E' is dense in E'_K we can approximate each ϕ_j uniformly on K by an element of E' and this combined with (2.13) completes the proof.

We now obtain a fairly complete duality theory for polynomials on fully nuclear spaces.

Proposition 2.16 *If E is a fully nuclear space and n is a positive integer then*

(a) $\left(\mathcal{P}_{HY}(^nE), \tau_0\right)'_\beta \cong \left(\mathcal{P}(^nE'_\beta), \tau_\omega\right),$

(b) $\left(\mathcal{P}(^nE), \tau_\omega\right)' = \mathcal{P}_{HY}(^nE'_\beta),$

(c) *$\tau_0 = \tau_\omega$ on $\mathcal{P}(^nE)$ if and only if $\mathcal{P}_{HY}(^nE'_\beta) = \mathcal{P}(^nE'_\beta)$ and the τ_0-bounded subsets of $\mathcal{P}(^nE'_\beta)$ are locally bounded.*

Proof. By Proposition 2.12, $\mathcal{P}(^nE) = \mathcal{P}_N(^nE)$ and, by Proposition 2.13, $\tau_0 = \pi_0$ on $\mathcal{P}(^nE)$. Hence, Proposition 2.10(a) or (b) and Lemma 2.15 imply

$$\left(\mathcal{P}_{HY}(^nE), \tau_0\right)' = \left(\mathcal{P}(^nE), \tau_0\right)' = \mathcal{P}(^nE'_\beta).$$

By Proposition 2.13 and Lemma 2.15, $\left(\mathcal{P}_{HY}(^nE), \tau_0\right)$ is a complete nuclear space and hence semi-reflexive. This implies that $\left(\mathcal{P}_{HY}(^nE), \tau_0\right)'_\beta$ is barrelled and has the Mackey topology. Since $\left(\mathcal{P}(^nE'_\beta), \tau_\omega\right)$ is also barrelled it follows that the two topologies on $\mathcal{P}(^nE'_\beta)$ will coincide if they have the same dual. Since $\left(\mathcal{P}_{HY}(^nE), \tau_0\right)$ is semi-reflexive we have

$$\left(\left(\mathcal{P}_{HY}(^nE), \tau_0\right)'_\beta\right)' = \mathcal{P}_{HY}(^nE).$$

By Propositions 2.10(c) and 2.12 and Lemma 2.15 we have

$$\left(\mathcal{P}(^nE'_\beta), \tau_\omega\right)' = \{P \in \mathcal{P}_a(^nE) \; : \; \|P\|_K < \infty \text{ for every compact subset}$$
$$K \text{ of } E\} = \mathcal{P}_{HY}(^nE).$$

This completes the proof of both (a) and (b).

If $\tau_0 = \tau_\omega$ on $\mathcal{P}(^nE)$ then (a) and (b) and Lemma 2.15 imply $\mathcal{P}_{HY}(^nE'_\beta) = \mathcal{P}(^nE'_\beta)$. Moreover, by Propositions 2.10(a) and 2.12, the equicontinuous subsets of $\big(\mathcal{P}(^nE),\tau_\omega\big)'$ are the subsets of $\mathcal{P}_a(^nE'_\beta)$ which are bounded on the equicontinuous subsets of E'. Since E' is fully nuclear the equicontinuous subsets of E' are precisely the relatively compact subsets of E' and so the equicontinuous subsets of $\big(\mathcal{P}(^nE),\tau_\omega\big)'$ are the τ_0-bounded subsets of $\mathcal{P}_{HY}(^nE'_\beta)$. By Proposition 2.10(a) the equicontinuous subsets of $\big(\mathcal{P}(^nE),\tau_0\big)'$ are the locally bounded subsets of $\mathcal{P}(^nE'_\beta)$. Hence the τ_0-bounded subsets of $\mathcal{P}(^nE'_\beta)$ $\big(=\mathcal{P}_{HY}(^nE'_\beta)\big)$ are locally bounded.

Conversely, if $\mathcal{P}_{HY}(^nE'_\beta) = \mathcal{P}(^nE'_\beta)$ and the τ_0-bounded subsets of $\mathcal{P}(^nE'_\beta)$ are locally bounded then $\big(\mathcal{P}(^nE),\tau_0\big)$ and $\big(\mathcal{P}(^nE),\tau_\omega\big)$ have the same dual by (a) and (b) and, Propositions 2.13, 2.10(b), 2.12 and 2.10(c), imply that the equicontinuous subsets of the dual coincide. This completes the proof of (c).

In Chapter 4 we give a holomorphic version of Proposition 2.16.

Corollary 2.17 *If E is a fully nuclear space and the τ_0-bounded subsets of $\mathcal{P}_{HY}(^nE)$ and $\mathcal{P}_{HY}(^nE'_\beta)$ are locally bounded then $\big(\mathcal{P}(^nE),\tau_0\big)$ and $\big(\mathcal{P}(^nE'_\beta),\tau_0\big)$ are both Montel nuclear spaces.*

Proof. Since the hypotheses apply equally to E and E'_β it suffices to prove the result for $\mathcal{P}(^nE)$. If $P \in \mathcal{P}_{HY}(^nE'_\beta)$ then $\{P\}$ is a τ_0-bounded subset of $\mathcal{P}_{HY}(^nE'_\beta)$. Hence P is locally bounded and continuous, i.e. $\mathcal{P}(^nE'_\beta) = \mathcal{P}_{HY}(^nE'_\beta)$. By Proposition 2.16(c), $\tau_0 = \tau_\omega$ on $\mathcal{P}(^nE)$ and hence $\big(\mathcal{P}(^nE),\tau_0\big)$ is barrelled. Similarly, $\mathcal{P}(^nE) = \mathcal{P}_{HY}(^nE)$ and hence, by Lemma 2.15, $\big(\mathcal{P}(^nE),\tau_0\big)$ is complete. By Proposition 2.13, $\big(\mathcal{P}(^nE),\tau_0\big)$ is nuclear. A complete barrelled nuclear space is Montel. This completes the proof.

Example 2.18 (a) If E is a Fréchet nuclear space then E'_β is a $\mathcal{DFN}$ space and, by Examples 1.24 and 1.25, the τ_0-bounded subsets of $\mathcal{P}(^nE)$ and $\mathcal{P}(^nE'_\beta)$ are locally bounded. Hence $\tau_0 = \tau_\omega$ on $\mathcal{P}(^nE)$ and $\mathcal{P}(^nE'_\beta)$ for all n. We have in fact proved this result for E'_β twice already (Example 1.25 and Proposition 1.33). Proposition 2.16(a) and (b) show, in particular, that $\big(\mathcal{P}(^nE),\tau_0\big)$ is a reflexive nuclear space when E is either a Fréchet nuclear or $\mathcal{DFN}$ space. (b) If $E = \mathbb{C}^{\mathbb{N}} \times \mathbb{C}^{(\mathbb{N})}$ then E is fully nuclear and $E'_\beta \cong E$. In Example 1.16 we gave an example of a 2-homogeneous polynomial on E which is easily seen to be bounded on compact sets but not continuous. Hence $\tau_0 \neq \tau_\omega$ on $\mathcal{P}(^2E)$. We have already given a proof of this result in Example 1.27. The result easily extends to arbitrary n.

Example 2.19 By Example 2.18(a), $\tau_0 = \tau_\omega$ on $\mathcal{P}(^2\mathbb{C}^{\mathbb{N}})$. We now show that $\tau_0 \neq \tau_\omega$ on $\mathcal{P}(^2\mathbb{C}^I)$ for any uncountable set I. Let $B := (b_i)_{i\in J}$ denote a subset of the real numbers indexed by an uncountable subset J of I and let $r(b_i,b_j) = \dfrac{1}{|b_i - b_j|}$ for $i \neq j$ and $r(b_i,b_i) = 0$ for all i. Suppose $t: B \to \mathbb{R}_+$

is a function such that

$$r(b_i, b_j) \leq t(b_i)t(b_j) \quad \text{for all} \ i, \ j \in J.$$

Since J is uncountable there exists a positive integer n such that $B_n := \{i \in J : t(b_i) \leq n\}$ is uncountable. Hence $(b_i)_{i \in B_n}$ has an accumulation point and there exist $i, \ j \in B_n$ such that $|b_i - b_j| < \dfrac{1}{n^2}$. On the other hand $t(b_i)t(b_j) \leq n^2$. This is impossible and we conclude that no such function exists.

If $P \in \mathcal{P}(^2\mathbb{C}^I)$ then the Factorization Lemma (Lemma 1.13) implies that there exists a finite subset F of I such that

$$\overset{\vee}{P}\big((z_i)_i, (w_i)_i\big) = \sum_{i,j \in F} \overset{\vee}{P}(e_i, e_j)z_i w_j$$

where e_i is the element of $\mathbb{C}^I$ which takes the value 1 at i and which is zero at all other points. This shows that $\phi(P) := \sum_{i,j \in J} \overset{\vee}{P}(e_i, e_j)r(b_i, b_j)$ is finite and $\phi \in \mathcal{P}(^2\mathbb{C}^I)^*$. Since the mapping $P \longmapsto \overset{\vee}{P}(e_i, e_j)$ is τ_0 continuous and τ_ω is a barrelled topology it follows that $\phi \in \big(\mathcal{P}(^2\mathbb{C}^I), \tau_\omega\big)'$. If $\phi \in \big(\mathcal{P}(^2\mathbb{C}^I), \tau_0\big)'$ there exists a set of positive real numbers $(t_i)_{i \in I}$ such that $|\phi(P)| \leq \sup\{|P((z_i)_{i \in I})| : |z_i| \leq t_i \ \text{for all} \ i\}$ for all $P \in \mathcal{P}(^2\mathbb{C}^I)$. If $i, \ j \in J$ and $P_{ij}\big((z_k)_{k \in I}\big) := z_i z_j$ then $|\phi(P_{ij})| = r(b_i, b_j) \leq t_i t_j$. This is not possible. Hence $\phi \notin \big(\mathcal{P}(^2\mathbb{C}^I), \tau_0\big)'$ and

$$\big(\mathcal{P}(^2\mathbb{C}^I), \tau_0\big)' \neq \big(\mathcal{P}(^2\mathbb{C}^I), \tau_\omega\big)'.$$

In proving results for nuclear spaces we frequently used the representations $\phi = \sum_{k=1}^\infty \lambda_k \phi(x_k)\phi_k$ and $x = \sum_{k=1}^\infty \lambda_k \phi_k(x)x_k$ where $\sum_k |\lambda_k| < \infty$ and the sequences $(\phi_k)_k$ and $(x_k)_k$ have certain boundedness properties. Since every compact subset of a Fréchet space is contained in the closed convex hull of a null sequence we obtain a similar representation for convex compact balanced subsets of Fréchet spaces. This is the hidden technique underlying the proof of the following proposition.

Proposition 2.20 *Let E be a Fréchet space. Then*

$$\big(\mathcal{P}(^nE), \tau_0\big)'_\beta \cong \widehat{\bigotimes_{n,s,\pi}} E.$$

If E has the density condition and $(BB)_n$ then

$$\big((\mathcal{P}(^nE), \tau_0)'_\beta\big)'_\beta \cong \big(\mathcal{P}(^nE), \tau_\omega\big).$$

If E has the strict approximation property then

$$\left(\mathcal{P}(^nE), \tau_0\right)'_\beta \cong \left(\mathcal{P}_N(E'_c), \pi_\omega\right).$$

Proof. For any Fréchet space E we have $(E'_c)'_\beta \cong E$. Since E Fréchet implies $\widehat{\bigotimes}_{n,\pi,s} E$ Fréchet and

$$\left(\mathcal{P}(^nE), \tau_0\right) = \left(\widehat{\bigotimes_{n,s,\pi}} E\right)'_c$$

it follows that

$$\left(\mathcal{P}(^nE), \tau_0\right)'_\beta \cong \left(\left(\widehat{\bigotimes_{n,s,\pi}} E\right)'_c\right)'_\beta \cong \widehat{\bigotimes_{n,s,\pi}} E.$$

If E has the density condition and $(BB)_n$ then

$$\left(\left(\mathcal{P}(^nE), \tau_0\right)'_\beta\right)'_\beta \cong \left(\widehat{\bigotimes_{n,s,\pi}} E\right)'_\beta \cong \left(\mathcal{P}(^nE), \tau_\omega\right).$$

Since every compact subset of $\widehat{\bigotimes}_{n,s,\pi} E$ is contained in the closed convex hull of $\{x_j \otimes \cdots \otimes x_j\}_{j=1}^\infty$ where $(x_j)_j$ is a null sequence in E and $\widehat{\bigotimes}_{n,s,\pi} E$ is Fréchet we have

$$\widehat{\bigotimes_{n,s,\pi}} E = \varinjlim_{K \in \mathcal{K}} \left(\widehat{\bigotimes_{n,s,\pi}} E_K\right)$$

where $\mathcal{K}$ denotes the set of all convex compact balanced subsets of E.

Now $E_K \cong (E'_{K^\circ})'_\beta$. Hence, if E has the strict approximation property, then

$$\widehat{\bigotimes_{n,s,\pi}} E = \varinjlim_{K \in \mathcal{K}} \left(\widehat{\bigotimes_{n,s,\pi}} (E'_{K^\circ})'\right)$$

$$= \varinjlim_{K \in \mathcal{K}} \left(\mathcal{P}_N(^nE'_{K^\circ}), \|\cdot\|_{K^\circ}\right)$$

$$= \left(\mathcal{P}_N(^nE'_c), \pi_\omega\right).$$

This completes the proof.

Nuclear spaces can also be introduced by means of the ε or *injective* tensor product.

Definition 2.21 If E and F are locally convex spaces then the ε or *injective* topology on $E \otimes F$ is the locally convex topology generated by the semi-norms

$$\varepsilon_{\alpha,\beta}\left(\sum_{i=1}^{l} x_i \otimes y_i\right) := \sup\left\{\left|\sum_{i=1}^{l} \phi(x_i)\psi(y_i)\right| \; : \; \phi \in E', \; \|\phi\|_{B_\alpha(1)} \le 1,\right.$$

$$\left.\psi \in F', \; \|\psi\|_{B_\beta(1)} \le 1\right\}$$

where α and β range over $cs(E)$ and $cs(F)$ respectively. We denote the completion of $E \otimes F$ with respect to the ε-topology by $E \widehat{\otimes}_\varepsilon F$.

Table 2.1

X	X'	X''	Conditions
$(\mathcal{P}(^nE), \tau_0)$	$(\mathcal{P}(^nE_{\beta'}), \tau_\omega)$	$\mathcal{P}_{HY}(^nE)$	E fully nuclear (Proposition 2.16)
$(\mathcal{P}(^nE), \tau_0)$	$(\mathcal{P}_N(^nE_c'), \tau_0)$	$(\mathcal{P}(^nE), \tau_\omega)$	E Fréchet with the strict approximation property, $(BB)_n$ and density condition (Proposition 2.20)

The ε-topology is always weaker than the π-topology and in a precise way, which we do not discuss, the π-topology – modelled on absolute sums and l_1 – may be regarded as the largest reasonable tensor product topology while the ε-topology – modelled on the supremum and c_0 – may be regarded as the weakest reasonable topology. On the other hand a locally convex space E is nuclear if and only if

$$E \otimes_\pi F = E \otimes_\varepsilon F$$

for every locally convex space F (see Sections 2.6 and 4.6).

For n-fold symmetric tensors we have the analogous definition.

Definition 2.22 If E is a locally convex space then the ε or injective topology on $\bigotimes_{n,s} E$ is the locally convex topology generated by the semi-norms

$$\varepsilon_{\alpha,n}\left(\sum_{i=1}^{l} x_i \otimes \ldots \otimes x_i\right) = \sup\left\{\left|\sum_{i=1}^{l} \phi(x_i)^n\right| \; : \; \phi \in E', \; \|\phi\|_{B_\alpha(1)} \le 1\right\}$$

where α ranges over $cs(E)$. The completion of $\bigotimes_{n,s} E$ with respect to this topology is denoted by $\widehat{\bigotimes}_{n,s,\varepsilon} E$.

Table 2.1 summarizes the duality results we have obtained.

2.3 Integral Polynomials and the Radon–Nikodým Property

In this section we consider preduals of spaces of polynomials on a Banach space. The classical case of $\mathcal{K}(H)$, $\mathcal{N}(H)$ and $\mathcal{L}(H, H)$ suggests that we begin by considering linear functionals on $\mathcal{P}_A(^nE)$. Using the ε-product we obtain another description of this space.

If $P = \sum_{j=1}^{l} \phi_j^n \in \mathcal{P}_f(^nE)$ then

$$\left\| \sum_{j=1}^{l} \phi_j \otimes \ldots \otimes \phi_j \right\|_\varepsilon = \sup_{x \in B_{E''}} \left| \sum_{j=1}^{l} \phi_j^n(x) \right|.$$

By Goldstine's Theorem, B_E is weak* dense in $B_{E''}$ and since all continuous linear functionals on E are weak* continuous on $B_{E''}$ we have

$$\left\| \sum_{j=1}^{l} \phi_j \otimes \ldots \otimes \phi_j \right\|_\varepsilon = \sup_{x \in B_E} \left| \sum_{j=1}^{l} \phi_j^n(x) \right| = \left\| \sum_{j=1}^{l} \phi_j^n \right\|.$$

In particular

$$\left(\mathcal{P}_f(^nE), \|\cdot\| \right) = \left(\mathcal{P}_f(^nE), \tau_\omega \right) = \bigotimes_{n,s,\varepsilon} E'$$

and

$$\mathcal{P}_A(^nE) \cong \widehat{\bigotimes_{n,s,\varepsilon}} E'.$$

Since each $P \in \mathcal{P}_A(^nE)$ is weakly uniformly continuous on bounded sets it has a unique extension to $\overline{B}_{E''}$ as a weak* continuous function; i.e. the mapping

$$P \in \mathcal{P}_A(^nE) \longrightarrow P\big|_{B_{E''}} \in \mathcal{C}\big(\overline{B}_{E''}, \sigma(E'', E')\big)$$

is an isometry. By the Hahn–Banach and the Riesz Representation Theorems there exists for each T in $\mathcal{P}_A(^nE)'$ a finite Borel measure μ on $\left(\overline{B}_{E''}, \sigma(E'', E') \right)$ such that

$$T(P) = \int_{\overline{B}_{E''}} P(x'')d\mu(x'') \tag{2.14}$$

and $\|T\| = \|\mu\|$. In particular, if we define the *Borel transform* B of T by

$$BT(\phi) := T(\phi^n) = \int_{\overline{B}_{E''}} x''(\phi)^n d\mu(x'')$$

for all $\phi \in E'$ then

$$\sup_{\|\phi\| \le 1} |T(\phi^n)| = \|BT\| \le \|\mu\|.$$

Since BT is easily seen to be an n-homogeneous polynomial, $BT \in \mathcal{P}(^nE')$. This motivates the following definition. We confine ourselves to the Banach space case and $\mathbb{C}$-valued situations although it is clear how to extend the definition to mappings between locally convex spaces.

Definition 2.23 If E is a Banach space then $P \in \mathcal{P}_a(^nE)$ is an *integral polynomial* or a *polynomial of integral type* if there exists a regular Borel measure μ of finite variation on $\left(\overline{B}_{E'}, \sigma(E', E)\right)$ such that

$$P(x) = \int_{\overline{B}_{E'}} \phi(x)^n d\mu(\phi) \tag{2.15}$$

for all x in E. We let $\|P\|_I = \inf\{\|\mu\| : \mu$ a regular Borel measure of finite variation satisfying (2.15)$\}$ and denote by $\mathcal{P}_I(^nE)$ the space of all integral n-homogeneous polynomials on E. If (2.15) holds we say that μ represents P.

From (2.15) it follows that $\sup_{\|x\| \le 1} |P(x)| \le \|P\|_I$. Hence $\mathcal{P}_I(^nE) \subset \mathcal{P}(^nE)$ and the inclusion mapping has norm 1. From (2.14) and (2.15) we obtain the following result.

Proposition 2.24 *For any Banach space E and any positive integer n*

$$\left(\mathcal{P}_A(^nE), \|\cdot\|\right)'_\beta \cong \left(\mathcal{P}_I(^nE'), \|\cdot\|\right)$$

with duality given by

$$BT(\phi) = T(\phi^n) \quad \text{for } T \in \mathcal{P}_A(^nE)' \text{ and } \phi \in E'.$$

This brief introduction and the Hilbert space case suggest that we investigate:

(A) the relationship between nuclear and integral polynomials on a Banach space,

(B) properties of polynomials which are weakly continuous on bounded sets.

We consider (A) and in Section 2.4 we build on the results in Section 2.1 to investigate (B). If $P \in \mathcal{P}_N(^nE)$ then there exists $(\lambda_j)_j \in l_1$, $\phi_j \in E'$, $\|\phi_j\| \le 1$, such that

$$P(x) = \sum_{j=1}^{\infty} \lambda_j \phi_j^n(x) \tag{2.16}$$

for all x in E and, moreover, $\|P\|_N \leq \sum_{j=1}^{\infty} |\lambda_j|$. If $\mu = \sum_{j=1}^{\infty} \lambda_j \delta_{\phi_j}$, where δ_{ϕ_j} is the Dirac delta function at ϕ_j, then μ represents P. Hence $\mathcal{P}_N(^nE) \subset \mathcal{P}_I(^nE)$ and $\|P\|_I \leq \|P\|_N$ for all $P \in \mathcal{P}_N(^nE)$. Comparing (2.15) and (2.16) we see that the main difference between nuclear and integral polynomials is that nuclear polynomials admit a representing measure with *countable support*. Our next example shows that these two classes of polynomials do not always coincide.

Example 2.25 $\mathcal{P}_N(^nl_1) \neq \mathcal{P}_I(^nl_1)$.
Let D denote the closed unit disc in $\mathbb{C}$. It is easily seen that

$$\left(\overline{B}_{l_\infty}, \; \sigma(l_\infty, l_1) \right) \cong D^{\mathbb{N}}$$

endowed with the product topology. For each m let μ_m denote a probability measure on D. If $\mu = \prod_m \mu_m$ on $D^{\mathbb{N}}$ then μ defines an integral polynomial $P_{\mu,n}$ on l_1 by the formula

$$P_{\mu,n}\left((x_m)_{m=1}^{\infty} \right) := \int_{D^{\mathbb{N}}} \left(\sum_{m=1}^{\infty} x_m y_m \right)^n d\mu((y_m)_m). \tag{2.17}$$

We make a particular choice of μ_m. Let M denote a subset of the positive integers. If $m \notin M$ let μ_m denote the point mass or Dirac measure at the origin. If $m \in M$ let μ_m denote a measure such that

$$\int_D z^k d\mu_m(z) = \begin{cases} 1 & \text{if } k = n, \\ 0 & \text{otherwise.} \end{cases}$$

By (2.17), we have

$$P_{\mu,M,n}\left((x_m)_m \right) := \sum_{k_1,\ldots,k_n=1}^{\infty} x_{k_1} \cdots x_{k_n} \int_{D^{\mathbb{N}}} y_{k_1} \cdots y_{k_n} d\mu((y_m)_{m=1}^{\infty})$$

$$= \sum_{m \in M} x_m^n.$$

We consider first the special case $M = \mathbb{N}$ and obtain the polynomial P

$$P\left((x_m)_{m=1}^{\infty} \right) = \sum_{m=1}^{\infty} x_m^n.$$

Let $(e_n')_{n=1}^{\infty}$ denote the unit vector basis in l_1 and let e_n'' denote the nth coordinate evaluation functional in l_∞. Suppose $P \in \mathcal{P}_N(^nl_1) \subset \mathcal{P}_\omega(^nl_1)$.

By Proposition 2.6 this implies that $\hat{d}P(B)$, B the unit ball of $(l_1)^{n-1}$, is a precompact subset of $\mathcal{P}(^1l_1) = l_\infty$. Since

$$\hat{d}P((x_m)_{m=1}^\infty) = n \sum_{m=1}^\infty x_m^{n-1} e_m''$$

we have $\|\hat{d}P(e_k') - \hat{d}P(e_l')\| = n\|e_k'' - e_l''\|$ for all positive integers k and l and this is impossible. Hence $P \notin \mathcal{P}_N(^nl_1)$ and

$$\mathcal{P}_N(^nl_1) \neq \mathcal{P}_I(^nl_1).$$

We use this example for one more observation and for this reason choose M arbitrary in $\mathbb{N}$. We proved in Proposition 1.59, using the trace, that all polynomials on c_0 are weakly continuous on bounded sets. This may also be proved independently by using the proof of Proposition 2.34. This implies

$$\mathcal{P}(^nc_0)' = \mathcal{P}_A(^nc_0)' = \mathcal{P}_I(^nl_1).$$

Here we use integral polynomials and the equality $\mathcal{P}_A(^nc_0) = \mathcal{P}(^nc_0)$ to show that $\mathcal{P}(^nc_0)$ has a trace. Let M denote an infinite subset of N and let $(M_j)_j$ denote an increasing sequence of finite subsets of M such that $\bigcup_j M_j = M$.

For all j we have $\|P_{\mu,M_j,n}\| \leq 1$ and, by duality, $\langle P_{\mu,M_j,n}, \phi^n \rangle = \sum_{k \in M_j} \phi^n(e_k)$ where $(e_k)_k$ is the unit vector basis in c_0. By continuity and duality we have

$$\langle P_{\mu,M_j,n}, P \rangle = \sum_{k \in M_j} P(e_k)$$

for all $P \in \mathcal{P}(^nc_0)$ and all M_j. Since $\langle P_{\mu,M_j,n}, P \rangle \longrightarrow \langle P_{\mu,M,n}, P \rangle$ for all $P \in \mathcal{P}(^nc_0)$ we have $\sum_{j \in M} P(e_j)$ convergent for all P in $\mathcal{P}(^nc_0)$. As M was arbitrary this implies $\sum_{j=1}^\infty |P(e_j)| < \infty$ for all $P \in \mathcal{P}(^nc_0)$. Moreover, since $\|P_{\mu,M_j,n}\| \leq 1$ for all P it follows that

$$\left| \sum_{j \in M} P(e_j) \right| \leq \|P\| \quad \text{for all} \quad P \in \mathcal{P}(^nc_0)$$

and hence

$$\sum_{j=1}^\infty |P(e_j)| \leq \|P\|.$$

Thus we see that in general the integral and nuclear polynomials do not coincide. To give conditions under which they do we introduce the Radon–Nikodým Property. As the name indicates this property originated in extending the Radon–Nikodým Theorem to Banach-valued measures. Geometric, analytic and probabilistic characterizations of this property have been found and the extensive literature on the topic is evidence of its important role in the modern theory on the geometry of Banach spaces. Here we just introduce the concepts we require in order to discuss nuclear and integral polynomials. We confine ourselves to the scalar-valued case and obtain a sufficient condition for the coincidence of nuclear and integral polynomials. By extending our investigation to the vector-valued case we could show that the same condition is also necessary but this would involve distinguishing between Pietsch and Grothendieck integrable functions and other technical complications that we prefer to avoid.

A Radon measure μ on a Hausdorff topological space X is an inner regular finite Borel measure on X. If E is a Banach space and $f: X \to E$ is a mapping then we say that f is μ-*Lusin measurable* if for each $\varepsilon > 0$ and each compact subset K in X there exists a compact subset K_ε in K such that $\mu(K - K_\varepsilon) < \varepsilon$ and $f \big|_{K_\varepsilon}$ is continuous. If f is μ-Lusin measurable for every Radon measure μ on X we say that f is *universally measurable*. With this definition we can introduce the Radon–Nikodým Property for dual Banach spaces.

Definition 2.26 Let E denote a Banach space. The space E' has the Radon–Nikodým Property if the identity mapping

$$I: \Big(E', \ \sigma(E', E) \Big) \longrightarrow \big(E', \ \| \cdot \| \big)$$

is universally measurable. A Banach space E is called an *Asplund space* if E' has the Radon–Nikodým Property.

An alternative proof and extension of the following result is given in Section 2.4.

Proposition 2.27 *If E is a Banach space and E' has the Radon–Nikodým Property then*

$$\mathcal{P}_N(^n E) = \mathcal{P}_I(^n E)$$

are isomorphic Banach spaces with norms satisfying

$$\| \cdot \|_I \leq \| \cdot \|_N \leq \frac{n^n}{n!} \| \cdot \|_I.$$

Proof. It suffices to show that each integral polynomial on E is nuclear. Let $P \in \mathcal{P}_I(^n E)$ and suppose μ is a representing measure for P. We suppose $\mu \geq 0$, the general case is obtained by taking the positive and negative real and imaginary parts of μ. By using inner regularity and Lusin measurability we can find a disjoint sequence of $\sigma(E', E)$-compact subsets of $B_{E'}$, $(K_j)_{j=1}^{\infty}$, such that the identity

$$I: \Big(E', \ \sigma(E', E) \Big) \longrightarrow \big(E', \ \| \cdot \| \big)$$

is continuous on each K_j and $\mu(B_{E'} \setminus S) = 0$ where $S = \bigcup_{j=1}^{\infty} K_j$. By uniform continuity on each K_j and inner regularity we see that I can be uniformly approximated on S by $(g_n)_n$, each of which has the form $\sum_{j=1}^{\infty} \phi_j \chi_{E_j}$ where $\|\phi_i\| \leq 1$ for all i and $(E_i)_{i=1}^{\infty}$ are pairwise disjoint Borel subsets of B_E (χ_A denotes the characteristic function of the set A).

Let $\varepsilon > 0$ be arbitrary. Choose $\varepsilon' > 0$ such that

$$\left(1 + \frac{\varepsilon'}{\mu(B_{E'})}\right)^n \mu(B_{E'}) \leq \mu(B_{E'}) + \varepsilon.$$

By choosing, if necessary, a subsequence we may suppose

$$\|\phi - g_1(\phi)\| \leq \frac{\varepsilon'}{2\mu(B_{E'})} \quad \text{and} \quad \|g_j(\phi) - g_{j-1}(\phi)\| \leq \frac{\varepsilon'}{2^j \mu(B_{E'})}$$

for all $j \geq 2$ and all $\phi \in S$. Let $g_j - g_{j-1} \equiv \sum_{k=1}^{\infty} \phi_{j,k} \chi_{E_{j,k}}$ where $\|\phi_{j,k}\| \leq 1$ for all j and k and $(E_{j,k})_{k=1}^{\infty}$ forms a sequence of pairwise disjoint measurable sets for each j. Let $g_0 \equiv 0$. For all ϕ in S we have

$$\phi = \sum_{j=1}^{\infty} \Big(g_j(\phi) - g_{j-1}(\phi)\Big) = \sum_{j=1}^{\infty} \sum_{k=1}^{\infty} \phi_{j,k} \chi_{E_{j,k}}(\phi)$$

and

$$\sum_{j=1}^{\infty} \sum_{k=1}^{\infty} \|\phi_{j,k}\| \chi_{E_{j,k}}(\phi) \leq \|\phi\| + \frac{\varepsilon'}{\mu(B_{E'})}. \tag{2.18}$$

For x in E and ϕ in S we have $\phi(x) = \sum_{j=1}^{\infty} \sum_{k=1}^{\infty} \phi_{j,k}(x) \chi_{E_{j,k}}(\phi)$ and, by (2.15),

$$P(x) = \int_{B_{E'}} \phi(x)^n \, d\mu(x)$$

$$= \int_S \phi(x)^n \, d\mu(x)$$

$$= \sum_{\substack{j_i, k_i = 1 \\ 1 \leq i \leq n}}^{\infty} \phi_{j_1, k_1}(x) \cdots \phi_{j_n, k_n}(x) \int_S \chi_{E_{j_1, k_1}}(\phi) \cdots \chi_{E_{j_n, k_n}}(\phi) \, d\mu(\phi).$$

Now

$$\sum_{\substack{j_i,k_i=1 \\ 1\leq i\leq n}} \|\phi_{j_1,k_1}\| \cdots \|\phi_{j_n,k_n}\| \cdot \left| \int_S \chi_{E_{j_1,k_1}}(\phi) \cdots \chi_{E_{j_n,k_n}}(\phi) d\mu(\phi) \right|$$

$$= \sum_{\substack{j_i,k_i=1 \\ 2\leq i\leq n}} \|\phi_{j_2,k_2}\| \cdots \|\phi_{j_n,k_n}\|$$

$$\cdot \left| \int_S \Big(\sum_{j_1,k_1=1}^{\infty} \|\phi_{j_1,k_1}\| \chi_{E_{j_1,k_1}}(\phi) \Big) \chi_{E_{j_2,k_2}}(\phi) \cdots \chi_{E_{j_n,k_n}}(\phi) d\mu(\phi) \right|$$

$$\leq \left(1 + \frac{\varepsilon'}{\mu(B_{E'})} \right) \sum_{\substack{j_i,k_i=1 \\ 2\leq i\leq n}}^{\infty} \|\phi_{j_2,k_2}\| \cdots \|\phi_{j_n,k_n}\|$$

$$\cdot \left| \int_S \chi_{E_{j_2,k_2}}(\phi) \cdots \chi_{E_{j_n,k_n}}(\phi) d\mu(\phi) \right| \qquad \text{(by 2.18)}$$

$$\leq \left(1 + \frac{\varepsilon'}{\mu(B_{E'})} \right)^n \int_S d\mu(\phi)$$

(by repeating the above argument $n-1$ more times)

$$\leq \mu(B_{E'}) \left(1 + \frac{\varepsilon'}{\mu(B_{E'})} \right)^n$$

$$\leq \mu(B_{E'}) + \varepsilon.$$

For $J := \{(j_1,k_1),\dots,(j_n,k_n)\} \in \mathbb{N}^{2n}$ there exists, by the Polarization Formula, a sequence $(\theta_{i,J})_i$ with $\sum_i \|\theta_{i,J}\|^n \leq \frac{n^n}{n!}$ such that

$$\frac{\phi_{j_1,k_1}(x)}{\|\phi_{j_1,k_1}\|} \cdots \frac{\phi_{j_n,k_n}(x)}{\|\phi_{j_n,k_n}\|} = \sum_i \theta_{i,J}(x)^n$$

for all x in E. Let

$$\lambda_J = \|\phi_{j_1,k_1}\| \cdots \|\phi_{j_n,k_n}\| \cdot \left| \int_S \chi_{E_{j_1,k_1}}(\phi) \cdots \chi_{E_{j_n,k_n}}(\phi) d\mu(\phi) \right|.$$

For all x in E we have

$$P(x) = \Big(\sum_J \lambda_J \Big) \cdot \Big(\sum_i \theta_{i,J}^n(x) \Big)$$

$$= \sum_{i,J} \left(\frac{\theta_{i,J}}{\|\theta_{i,J}\|} \right)^n (x) \cdot \|\theta_{i,J}\|^n \lambda_J$$

and, after reordering the above sum, we obtain a sequence $(\psi_k)_{k=1}^{\infty}$ in E', $\|\psi_k\| \leq 1$, and a sequence of real numbers $(\delta_k)_k$ such that

$$\sum_{k=1}^{\infty} |\delta_k| \leq \frac{n^n}{n!} \big(\mu(B_{E'}) + \varepsilon \big)$$

and

$$P(x) = \sum_{k=1}^{\infty} \delta_k \psi_k^n(x)$$

for all x in E. Hence P is nuclear and

$$\|P\|_N \le \frac{n^n}{n!} \Big(\mu(B_{E'}) + \varepsilon \Big).$$

Since ε was arbitrary and μ was any measure representing P it follows that $\|P\|_N \le \dfrac{n^n}{n!} \|P\|_I$. This completes the proof.

Since we have already defined the Radon–Nikodým Property for dual spaces we include, for the sake of completeness, the definition for arbitrary Banach spaces. In this proposition μ denotes Lebesgue measure on $[0,1]$.

Proposition 2.28 *A Banach space E has the Radon–Nikodým Property if it satisfies any of the following equivalent conditions:*

(1) *every function $f \colon [0,1] \to E$ of bounded variation is differentiable almost everywhere,*

(2) *every continuous linear operator $T \colon \big(L^1[0,1], \mu\big) \to E$ factors through l_1,*

(3) *each continuous linear operator $T \colon \big(L^1[0,1], \mu\big) \to E$ has the form $Tf = \int_0^1 f g d\mu$, where $g \colon [0,1] \to E$ is a bounded measurable function.*

In addition, if $E = F'$ (i.e. if E is a dual space) the above are equivalent to the following:

(4) *every non-empty closed bounded subset of E contains an extreme point of its closed convex hull,*

(5) *every separable subspace of F has a separable dual,*

(6) *every non-empty closed convex bounded subset of E is the closed convex hull of its extreme points.*

Reflexive spaces and separable dual spaces have the Radon–Nikodým Property. The spaces c_0 and $\mathcal{K}(H)$ do not have the Radon–Nikodým Property.

We have extended the classical linear duality theory to polynomials on Banach spaces. The polynomials which are weakly continuous on bounded sets replace the compact linear mappings in the linear theory and we require the approximation property and the Radon–Nikodým Property. We explore various consequences of this duality in the next section. We complete this section by presenting, in diagram form, the results we have obtained for a Banach space E.

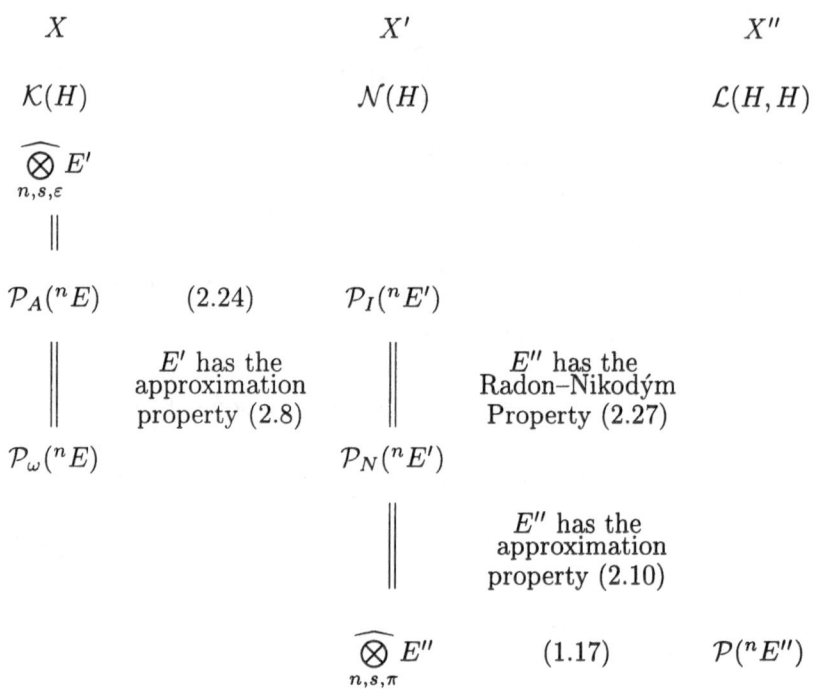

$$X \qquad\qquad X' \qquad\qquad X''$$

$$\mathcal{K}(H) \qquad\qquad \mathcal{N}(H) \qquad\qquad \mathcal{L}(H,H)$$

$$\underset{n,s,\varepsilon}{\widehat{\bigotimes}} E'$$

$$\|$$

$$\mathcal{P}_A(^nE) \qquad (2.24) \qquad \mathcal{P}_I(^nE')$$

$$\| \qquad \begin{array}{c} E' \text{ has the} \\ \text{approximation} \\ \text{property } (2.8) \end{array} \qquad \| \qquad \begin{array}{c} E'' \text{ has the} \\ \text{Radon--Nikodým} \\ \text{Property } (2.27) \end{array}$$

$$\mathcal{P}_\omega(^nE) \qquad\qquad \mathcal{P}_N(^nE')$$

$$\| \qquad \begin{array}{c} E'' \text{ has the} \\ \text{approximation} \\ \text{property } (2.10) \end{array}$$

$$\underset{n,s,\pi}{\widehat{\bigotimes}} E'' \qquad (1.17) \qquad \mathcal{P}(^nE'')$$

2.4 Reflexivity and Related Concepts

Now that we have developed a duality theory for polynomials – at least in the important nuclear and Banach space cases – it is natural to look at the problem of reflexivity. In the nuclear space case we have already obtained some results on this topic – Corollary 2.17 and Propositions 2.19 and 2.20 – and we use these as a basis for the development of a similar theory for holomorphic functions on fully nuclear spaces in Chapter 4. In this section we discuss reflexivity and related concepts for spaces of homogeneous polynomials on a Banach space. The following is immediate from the final table of the previous section.

Proposition 2.29 *If E is a Banach space and E'' has the approximation property and the Radon–Nikodým Property then for any positive integer n the canonical mapping of $\mathcal{P}_\omega(^nE)$ into its bidual identifies $\mathcal{P}_\omega(^nE)''$ with $\mathcal{P}(^nE'')$.*

Since reflexive spaces have the Radon–Nikodým Property this easily gives a criterion for reflexivity. In one direction, however, we do not need the approximation property and this is the content of the next proposition.

Proposition 2.30 *If continuous n-homogeneous polynomials on a reflexive Banach space E are weakly sequentially continuous then $\left(\mathcal{P}(^nE), \| \cdot \|\right)$ is reflexive.*

Proof. Since $\mathcal{P}(^nE)$ is a dual space it suffices by a theorem of James to show that each $P \in \mathcal{P}(^nE)$ attains its norm, i.e. there exists $x \in E$, $\|x\| \leq 1$ such that $P(x) = \|P\|$. Choose $(x_n)_n$, $\|x_n\| \leq 1$, such that $P(x_n) \to \|P\|$ as $n \to \infty$. By the Eberlein–Šmulian Theorem the closed unit ball of E is weakly sequentially compact. Hence $(x_n)_n$ contains a weakly convergent subsequence $(x_{n_j})_j$. If $x_{n_j} \to x$ weakly as $j \to \infty$ and P is weakly sequentially continuous then $P(x) = \|P\|$ and $\left(\mathcal{P}(^nE), \| \cdot \|\right)$ is reflexive. This completes the proof.

When E has the approximation property the converse is true.

Proposition 2.31 *Let E denote a reflexive Banach space with the approximation property. For any positive integer n the following are equivalent:*

(1) $\left(\mathcal{P}(^nE), \| \cdot \|\right)$ *is reflexive,*

(2) norm continuous n-homogeneous polynomials on E are weakly continuous on bounded sets,

(3) norm continuous n-homogeneous polynomials on E are weakly uniformly continuous on bounded sets,

(4) norm continuous n-homogeneous polynomial on E are weakly sequentially continuous,

(5) for each norm continuous n-homogeneous polynomial P on E there exists a unit vector x in E such that

$$P(x) = \|P\|.$$

Proof. (1) $\iff$ (2) follows from Propositions 2.29 and 2.30 and (2) $\iff$ (3) by Proposition 2.6. Clearly (3) $\implies$ (4) and (4) $\implies$ (5) follows from the proof of Proposition 2.30. By a theorem of James, (5) $\implies$ (1) and this completes the proof.

Proposition 2.31 reduces the problem of reflexivity to a property which must be satisfied individually by each polynomial. We would like to obtain examples, in terms of properties of E, when this is satisfied. To proceed we examine conditions (2)–(4) for *arbitrary* Banach spaces. We know that (2) and (3) are equivalent and clearly (3) $\implies$ (4). Condition (4), the weakest of the conditions, can be replaced by an apparently even weaker condition. This is the content of the following lemma (see also the proof of Proposition 1.59).

Lemma 2.32 *If E is a Banach space and n is a positive integer then all continuous polynomials of degree less than or equal to n are weakly sequentially continuous if and only if they are weakly sequentially continuous at the origin.*

Proof. One direction is clear. Suppose all continuous polynomials of degree less than or equal to n are weakly sequentially continuous at the origin. Let $P = \sum_{l=0}^{n} P_l$ where $P_l \in \mathcal{P}(^l E)$ for $0 \leq l \leq n$. If $x_j \to x$ weakly as $j \to \infty$ then, by Lemma 1.10,

$$\left| P(x_j) - P(x) \right| \leq \sum_{l=1}^{n} \sum_{k=1}^{l} \binom{l}{k} \left| \overset{\vee}{P}_l(x)^{l-k}(x_j - x)^k \right|$$

for all j. For each l and k, $1 \leq k \leq l$, the mapping

$$y \longrightarrow \binom{l}{k} \overset{\vee}{P}_l(x)^{l-k}(y)^k$$

is a k-homogeneous polynomial, $k \leq n$, and by hypothesis, weakly sequentially continuous at the origin. Hence, for $1 \leq k \leq l \leq n$,

$$\overset{\vee}{P}_l(x)^{l-k}(x_j - x)^k \to 0 \text{ as } j \to \infty.$$

This implies $P(x_j) \to P(x)$ as $j \to \infty$ and P is weakly sequentially continuous. This completes the proof.

Our next task is to find Banach spaces on which every continuous polynomial is weakly sequentially continuous at the origin (we already have one example – c_0 – by Proposition 1.59). Clearly this will be the case if weak and norm sequential convergence coincide and a well known classical result says that this occurs for l_1 (we say that l_1 has the *Schur Property*). We look for a weaker condition. Since $\|\phi\| = \sup_{\|x\| \leq 1} |\phi(x)|$ we can express this result for l_1 as follows:

A sequence $(x_j)_j \subset l_1$ is weakly null if and only if $\phi_j(x_j) \to 0$ as $j \to \infty$ for any bounded sequence $(\phi_j)_j \subset l_1' \cong l_\infty$.

Our next definition is obtained by replacing the bounded sequence by a weakly null sequence.

Definition 2.33 A Banach space E has the Dunford–Pettis Property if for any weakly null sequence $(x_j)_j$ in E and any weakly null sequence $(\phi_j)_j$ in E' we have $\phi_j(x_j) \to 0$ as $j \to \infty$.

The Banach spaces $\mathcal{C}(K)$ and $L^1(\mu)$ have the Dunford–Pettis Property.

Proposition 2.34 *If E is a Banach space with the Dunford–Pettis Property and $P \in \mathcal{P}(^n E)$ then P is weakly sequentially continuous.*

Proof. By Lemma 2.32, it suffices to show that all continuous homogeneous polynomials are weakly sequentially continuous at the origin. We prove this by induction. If $n = 1$ there is nothing to prove. We suppose the result is true for $n - 1$. Let $(x_j)_j$ denote a weakly null sequence in E. By our induction hypothesis $Q(x_j) \to 0$ as $j \to \infty$ for all $Q \in \mathcal{P}(^{n-1}E)$. Since $\left(\bigotimes\limits_{n-1,s,\pi} E \right)' \cong \mathcal{P}(^{n-1}E)$ this implies that $(\underbrace{x_j \otimes \ldots \otimes x_j}_{n-1 \text{ times}})_j$ is a weakly null sequence in $\bigotimes\limits_{n-1,s,\pi} E$.

Let $P \in \mathcal{P}(^n E)$ and let $\widetilde{P} \colon \bigotimes\limits_{n-1,s,\pi} E \to E'$ be defined by

$$\left(\widetilde{P}(\underbrace{x \otimes \cdots \otimes x}_{n-1 \text{ times}}) \right)(y) = \overset{\vee}{P}(x^{n-1}, y)$$

and linearity. Then $\widetilde{P} \in \mathcal{L}(\bigotimes\limits_{n-1,s,\pi} E; E')$ and $\left(\widetilde{P}(x \otimes \cdots \otimes x) \right)(x) = P(x)$ for all x in E. Since $\widetilde{P}$ is a continuous linear mapping it maps weakly null sequences onto weakly null sequences. Hence $\left(\widetilde{P}(\underbrace{x_j \otimes \cdots \otimes x_j}_{n-1 \text{ times}}) \right)_j$ is weakly null in E'. Since $(x_j)_j$ is weakly null in E and E has the Dunford–Pettis Property we have

$$\left(\widetilde{P}(\underbrace{x_j \otimes \cdots \otimes x_j}_{n-1 \text{ times}}) \right)(x_j) = P(x_j) \to 0 \text{ as } j \to \infty.$$

This completes the proof.

We now obtain a result similar to the previously established equivalence of (2.1) and (2.2). Since the converse is also true this proposition can be developed into a characterization of the Dunford–Pettis Property.

Proposition 2.35 *Let E denote a Banach space with the Dunford–Pettis Property and let F denote a reflexive Banach space. If $T \colon E \to F$ is a continuous linear operator then T maps weakly null sequences in E onto norm null sequences in F.*

Proof. Since $T \colon (E, \text{weak}) \to (F, \text{weak})$ is continuous it suffices to show, using the Hahn–Banach Theorem, that T maps weakly null sequences onto relatively compact subsets of E. We suppose this is not true and that $(x_j)_j$ is a weakly null sequence in E such that $(Tx_j)_j$ does not contain any norm convergent subsequence. Without loss of generality we may also suppose $\|T(x_j)\| \geq \delta > 0$ for all j. Choose $(\phi_j)_j \subset F'$, $\|\phi_j\| \leq 1$, such that

$$|\phi_j(Tx_j)| \geq \frac{\delta}{2}$$

for all j. Since F' is reflexive we may suppose $\phi_j \to \phi$ weakly as $j \to \infty$. If T^* is the adjoint of T then $T^*\phi_j \to T^*\phi$ weakly as $j \to \infty$ and, by the Dunford–Pettis Property, we have

$$\lim_{j \to \infty} \left(T^*(\phi_j - \phi)\right)(x_j) = \lim_{j \to \infty} (\phi_j - \phi)(Tx_j) = 0.$$

Since $(Tx_j)_j$ is weakly null we also have $\phi(Tx_j) \to 0$ as $j \to \infty$. Hence $\phi_j(Tx_j) \to 0$ as $j \to \infty$ and this contradicts the fact that $|\phi_j(Tx_j)| \geq \delta/2$ for all j. This completes the proof.

We complete our discussion on the conditions in Proposition 2.31 by connecting weak sequential continuity with weak continuity on bounded sets. To achieve this we use a result of Grothendieck on the canonical inclusion $i: l_1 \to l_2$. The classical Rademacher functions on $[0,1]$ are defined by

$$r_n(t) = \text{sgn}\big(\sin(2^n \pi t)\big), \quad t \in [0, 1].$$

The sequence $\big(r_n(t)\big)_{n=1}^{\infty}$ is an orthonormal sequence in $L^2[0,1]$ and forms a bounded sequence in $L^{\infty}[0,1]$. Let $(e_n)_n$ denote the standard basis for l_1 and let $s: l_1 \longrightarrow L^{\infty}[0,1]$ be defined by $s(e_n) = r_n$ and linearity. Since $\|r_n\| = 1$ the mapping s is continuous. Let $T: L^{\infty}[0,1] \to L^2[0,1]$ denote the canonical inclusion mapping. We define $R: L^2[0,1] \to l_2$ by mapping $(r_n)_n$ onto the standard orthonormal basis in l_2 and extending it by linearity. Both T and R are easily seen to be continuous and the diagram

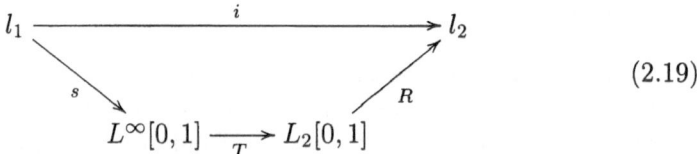

$$(2.19)$$

is commutative.

Proposition 2.36 *If E is a Banach space then $l_1 \longhookrightarrow\!\!\!\!\!/ \; E$ if and only if all weakly sequentially continuous polynomials on E are weakly (uniformly) continuous on bounded sets.*

Proof. We first suppose that $l_1 \longhookrightarrow\!\!\!\!\!/ \; E$. Let P denote a weakly sequentially continuous polynomial on E. By Rosenthal's l_1 Theorem every bounded sequence in E contains a weak Cauchy sequence. Moreover, any bounded subset of E is sequentially dense in its weak closure. If A is a bounded subset of E

and $x \in \overline{A}^{\sigma(E,E')}$ there exists $(x_j)_j$ in A such that $x_j \to x$ weakly as $j \to \infty$. Since P is weakly sequentially continuous we have $P(x_j) \to P(x)$ as $j \to \infty$. Hence

$$P\left(\overline{A}^{\sigma(E,E')}\right) \subset \overline{P(A)}$$

and P is weakly continuous on bounded sets.

Now suppose $l_1 \hookrightarrow E$. We complete the proof by constructing a weakly sequentially continuous polynomial on E which is not weakly continuous on bounded sets. By (2.19) we have the factorization

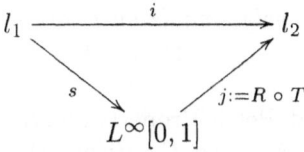

where s and j are continuous linear operators. Since $L^\infty[0,1]$ is injective, i.e. continuous linear mappings into $L^\infty[0,1]$ admit norm preserving extensions from subspaces, there exists a continuous linear mapping $U: E \to L^\infty[0,1]$ such that the following diagram commutes:

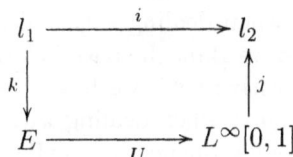

where k is the inclusion of l_1 in E.

Let $P\left((x_n)_{n=1}^\infty\right) = \sum_{n=1}^\infty x_n^m$ for $(x_n)_{n=1}^\infty \in l_2$ and $m \geq 2$. Then $P \in \mathcal{P}(^m l_2)$ and $Q := P \circ j \circ U \in \mathcal{P}(^m E)$. Since $L^\infty[0,1] \cong \mathcal{C}(K)$ has the Dunford–Pettis Property and l_2 is reflexive, Proposition 2.35 shows that j maps weakly convergent sequences in $L^\infty[0,1]$ onto norm convergent sequences in l_2. Hence Q is weakly sequentially continuous. We suppose that Q is weakly continuous on bounded sets. By Proposition 2.6, $\dfrac{\hat{d}^{m-1}Q}{(m-1)!}\Big|_{l_1}$ is a compact linear mapping from l_1 into $\mathcal{P}(^{m-1}l_1)$. Since

$$\left[\frac{\hat{d}^{m-1}Q}{(m-1)!}(e_n)\right](y) = m y_n^{m-1}$$

for $y = (y_j)_j \in l_1$ we have, for $a \neq b \in \mathbb{N}$,

$$\left\| \frac{\widehat{d}^{m-1} Q}{(m-1)!} (e_a - e_b) \right\| = m \sup_{\sum_{j=1}^{\infty} |y_i| \leq 1} \left| y_a^{m-1} - y_b^{m-1} \right| \geq m.$$

This is impossible. Hence Q is not weakly continuous on bounded sets and this completes the proof.

From Propositions 2.34 and 2.36 we obtain the following corollary.

Corollary 2.37 *If E is a Banach space with the Dunford–Pettis Property and $l_1 \not\hookrightarrow E$ then continuous polynomials on E are weakly (uniformly) continuous on bounded sets.*

Since c_0 has the Dunford–Pettis Property and $l_1 \not\hookrightarrow c_0$ we have yet another proof of Proposition 1.59. A space E satisfies the conditions of the above corollary if and only if weak and norm sequential convergence in E' agree, i.e. if and only if E' has the Schur Property.

From our analysis we can draw a number of conclusions. First, it is clear that l_1 and c_0 both play, in different ways, prominent roles in the relationships between continuity, weak continuity and weak sequential continuity. Second, although the Dunford–Pettis Property was helpful in clarifying the relationships between the conditions in Proposition 2.31 for general Banach spaces it will not be of use when dealing with reflexive Banach spaces since no infinite dimensional reflexive Banach space has the Dunford–Pettis Property. Finally, and on the positive side, we have Lemma 2.32 which tells us that in many cases, in particular when dealing with reflexive spaces, we may confine ourselves to weakly null sequences and call on the Bessaga–Pełczyński selection principle and spreading models.

We now look for positive and negative examples arising from Proposition 2.31. On the positive side we seek examples of Banach spaces E such that $\mathcal{P}(^n E)$ is reflexive for all n, while on the negative side we look for E with $l_\infty \hookrightarrow \mathcal{P}(^n E)$. If $\mathcal{P}(^n E)$ is reflexive then $l_\infty \not\hookrightarrow \mathcal{P}(^n E)$ and, in general, the converse is false (Example 2.43). In Corollary 1.56 we showed that $l_\infty \hookrightarrow \mathcal{P}(^n E)$ whenever l_p is a quotient of E and $n \geq p$. We start by improving this result with the aid of compact polynomials.

We require the following well known characterization of c_0.

A sequence $(x_j)_j$ in a Banach space E spans a closed subspace isomorphic to c_0 if and only if

(i)   *$(x_j)_j$ is a basic sequence,*

(ii) *$\sum_{j=1}^{\infty} |\phi(x_j)| < \infty$ for all $\phi \in E'$,*

(iii) *$\inf_j \|x_j\| > 0$.*

A series satisfying (ii) is called *weakly unconditionally Cauchy*. Note that (ii) and (iii) imply

$$0 < \inf_{j} \|x_j\| < \sup_{j} \|x_j\| < \infty.$$

Proposition 2.38 *Let E be Banach space. The following are equivalent.*

(a) $\mathcal{P}_K(^nE; l_p) = \mathcal{P}(^nE; l_p)$ for all p and n,

(b) If $(P_i)_i \subset \mathcal{P}(^nE)$ and $\sum_{i=1}^{\infty} |P_i(x)|^p < \infty$ for all x in E and some p, $1 \le p < \infty$, then $\|P_i\| \to 0$ as $i \to \infty$,

(c) $l_\infty \not\hookrightarrow \mathcal{P}(^nE)$ for all n.

Proof. Since the pointwise limit of a sequence of continuous Banach-valued n-homogeneous polynomials on a Banach space is continuous it follows that we can identify $\mathcal{P}(^nE; l_p)$ with the set of sequences

$$\left\{ (P_i)_{i=1}^{\infty} : P_i \in \mathcal{P}(^nE), \sum_{i=1}^{\infty} |P_i(x)|^p < \infty \right\}$$

by means of

$$P(x) = \sum_{i=1}^{\infty} P_i(x)e_i .$$

If $P \in \mathcal{P}_K(^nE; l_p)$ then clearly

$$\lim_{\substack{i \to \infty \\ \|x\| \le 1}} \sup \left(\sum_{j \ge i} |P_j(x)|^p \right)^{1/p} = 0$$

and hence $\|P_i\| \to 0$ as $i \to \infty$. On the other hand if $P = (P_i)_i \in \mathcal{P}(^nE; l_p)$ and

$$\limsup_{i \to \infty} \left(\sup_{\|x\| \le 1} \sum_{j \ge i} |P_j(x)|^p \right)^{1/p} = \delta > 0$$

then we can choose inductively a strictly increasing sequence of positive integers $(n_j)_j$, a sequence of unit vectors in E, $(x_j)_j$, and complex numbers $(\lambda_i)_i$, where $|\lambda_i| = 1$ for all i, if $p = 1$, and $\sum_{n_j+1}^{n_{j+1}} |\lambda_i|^q = 1$ for all j if $p > 1$ and $\frac{1}{p} + \frac{1}{q} = 1$, such that

$$\sum_{i=n_j+1}^{n_{j+1}} \lambda_i P_i(x_j) \ge \frac{\delta}{2}$$

for all j. Let $Q_j = \sum\limits_{i=n_j+1}^{n_{j+1}} \lambda_i P_i$ for all j. Then $Q_j \in \mathcal{P}(^n E)$ and $(Q_j)_j \in \mathcal{P}(^n E; l_p)$. Since $Q_j(x_j) \geq \dfrac{\delta}{2}$ we have $\|Q_j\| \geq \dfrac{\delta}{2}$. Hence $(Q_j)_j \notin \mathcal{P}_K(^n E; l_p)$. We have proved (a) $\Longleftrightarrow$ (b).

If (c) is not satisfied there exists a sequence $(P_j)_j$, which we may suppose consists of unit vectors, such that $\sum\limits_{i=1}^{\infty} |\phi(P_i)| < \infty$ for all $\phi \in \mathcal{P}(^n E)'$. In particular $\sum\limits_{i=1}^{\infty} |P_i(x)| < \infty$ for all $x \in E$ and (b) is not satisfied. Hence (b) $\Longrightarrow$ (c).

Now suppose there exist n, p and $(P_i)_{i=1}^{\infty} \subset \mathcal{P}(^n E)$ such that $\|P_i\| = 1$ for all i and $\sum\limits_{i=1}^{\infty} |P_i(x)|^p < \infty$ for all $x \in E$. Let $l = [p] + 1$ and $Q_i = P_i^l$ for all i. Then $(Q_i)_i \subset \mathcal{P}(^{nl} E)$, $\|Q_i\| = 1$ and $\sum\limits_{i=1}^{\infty} |Q_i(x)| < \infty$ for all $x \in E$. Since the pointwise limit of a sequence of continuous homogeneous polynomials on a Banach space is again a continuous polynomial it follows that $\sum\limits_{i=1}^{\infty} \lambda_i Q_i \in \mathcal{P}(^{nl} E)$ for all $(\lambda_i)_i \in l_{\infty}$. If $T \in \mathcal{P}(^{nl} E)'$ then $\sum\limits_{i=1}^{\infty} \lambda_i T(Q_i) < \infty$ and hence $\sum\limits_{i=1}^{\infty} |T(Q_i)| < \infty$ for all $T \in \mathcal{P}(^{nl} E)'$.

In particular $(Q_i)_i$ is a weakly null sequence of unit vectors and, by the Bessaga–Pełczyński selection principle, contains a basic sequence which we suppose, without loss of generality, is the original sequence. The sequence $(Q_i)_i$ spans a subspace of $\mathcal{P}(^{nl} E)$ isomorphic to c_0 and since $\mathcal{P}(^{nl} E)$ is a dual Banach space this implies $l_{\infty} \hookrightarrow \mathcal{P}(^{nl} E)$. Hence (c) is not satisfied and thus (c) $\Longrightarrow$ (b). This completes the proof.

A more precise relationship between p and n can be established by examining the proof of Proposition 2.38 (see Section 2.6).

If $\pi: E \to l_p$ is a quotient mapping and $\pi(x) = \big(\phi_i(x)\big)_{i=1}^{\infty}$ for all x in E then $\phi_i \in E'$ and $\sum\limits_{i=1}^{\infty} |\phi_i(x)|^p < \infty$ for all x in E. Hence (b) of Proposition 2.38 is not satisfied and we conclude that $l_{\infty} \hookrightarrow \mathcal{P}(^n E)$ for some n. This is an alternative proof of Corollary 1.56 and may be improved using the concept of upper p-estimates.

Definition 2.39 A sequence $(x_j)_j$ in a Banach space E has a lower q, $q < \infty$, (respectively an upper p, $p > 1$) estimate if there exists $c > 0$ such that for any sequence of scalars $(\lambda_j)_{j=1}^{\infty}$ we have

$$\Big\| \sum_{j=1}^{k} \lambda_j x_j \Big\|^q \geq c \cdot \sum_{j=1}^{k} |\lambda_j|^q$$

$$\text{(respectively } \Big\| \sum_{j=1}^{k} \lambda_j x_j \Big\|^p \leq c \cdot \sum_{j=1}^{k} |\lambda_j|^p)$$

for all k.

Corollary 2.40 *If E is a Banach space and E' contains a sequence of unit vectors satisfying an upper p-estimate for some $p > 1$ then there exists a positive integer n such that $l_\infty \hookrightarrow \mathcal{P}(^n E)$.*

Proof. If $(\phi_j)_j$ is the sequence of unit vectors in E' satisfying an upper p-estimate, there exists $c > 0$ such that

$$\Big\| \sum_{j=1}^{k} \lambda_j \phi_j \Big\|^p \leq c \cdot \sum_{j=1}^{k} |\lambda_j|^p$$

for all sequences of scalars $(\lambda_j)_j$ and all k. Using the duality $l'_p = l_q$, we see that $\sum_{j=1}^{\infty} |\phi_j^q(x)| < \infty$ for all $x \in E$. Hence (b) of Proposition 2.38 is not satisfied by $(\phi_j^n)_j$, $n > q$, and by (c) we conclude that $l_\infty \hookrightarrow \mathcal{P}(^n E)$ for some n.

We now turn to the problem of finding an infinite dimensional Banach space E such that $\big(\mathcal{P}(^n E), \| \cdot \| \big)$ is reflexive for all n. By Proposition 2.30 and Lemma 2.32 we must show that all continuous polynomials are weakly sequentially continuous at the origin. Proposition 2.38 and Corollary 2.40 give some necessary conditions -- and we shall see in Chapter 4 that, in certain circumstances, these are also sufficient. We know, from Corollary 2.40, that E must be a reflexive space which does not contain any l_p, $1 < p < \infty$, and this suggests that we examine Tsirelson's space. The original Tsirelson space, T^*, has an unconditional basis, $(t_j)_j$, with the property that

$$\sup_{n < j \leq 2n} |\alpha_j| \leq \Big\| \sum_{j=n+1}^{2n} \alpha_j t_j \Big\| \leq 2 \sup_{n \leq j \leq 2n} |\alpha_j|$$

for all n and any sequence of scalars $(\alpha_j)_j$. The proof of Proposition 1.59 can be adjusted to show directly that polynomials on T^* are weakly sequentially continuous and this implies that $\left(\mathcal{P}(^nT^*),\ \|\cdot\|\right)$ is reflexive for all n. It is also easily seen, from this estimate, that no weakly null sequence in T^* admits a lower q-estimate and, as we shall soon see, this also leads to the same result.

We now introduce *spreading models* by iterating ultrapowers. Let $\mathcal{U}$ denote a non-trivial ultrafilter on the natural numbers $\mathbb{N}$, i.e. $\bigcap_{F \in \mathcal{U}} F = \emptyset$. If E is a Banach space we define inductively E_n by letting $E_0 = E$ and $E_{n+1} = (E_n)_\mathcal{U}$ for all $n \geq 0$. Since each E_n is *isometrically* embedded in E_{n+1} in a canonical way we see that $\bigcup_{n \geq 0} E_n$ is a normed linear space and let E_∞ denote its completion. For each n let π_n denote the isometric embedding of E into E_n, i.e. π_n is the composition of the embeddings $E_0 \to E_1 \to \ldots \to E_n$ where $E_i \to (E_i)_\mathcal{U} = E_{i+1}$ is the canonical embedding of a space into an ultrapower of itself described in Chapter 1.

Let $(x_j)_j$ denote a weakly null sequence in E_0 which contains no Cauchy sequence. For any positive integer n the sequence $\left(\pi_n(x_j)\right)_j$ is a bounded sequence in E_n and hence determines an element of $(E_n)_\mathcal{U} = E_{n+1}$ which we denote by e_{n+1}. Since $(x_j)_j$ contains no Cauchy subsequence it is easily seen that $e_{n+1} \in E_{n+1} \setminus E_n$ for all $n \geq 0$. The closed subspace of E_∞ generated by $(e_j)_j$, $[(e_j)_j]$, is called the *spreading model* associated with $(x_j)_j$ and $(e_j)_j$ is called the *fundamental sequence* of the spreading model.

It is easily checked that

$$\Big\| \sum_{j=1}^k \alpha_j e_j \Big\| = \lim_{\mathcal{U},k} \lim_{\mathcal{U},k-1} \ldots \lim_{\mathcal{U},1} \Big\| \sum_{j=1}^k \alpha_j x_{n_j} \Big\| \qquad (2.20)$$

for any sequence of scalars $(\alpha_j)_j$ and any positive integer k. Now fix α_1 and α_2. For $\varepsilon > 0$ let

$$A^\varepsilon = \{ n \in \mathbb{N} \ : \ \big| \|\alpha_1 e_1 + \alpha_2 e_2\| - \|\alpha_1 e_1 + \alpha_2 x_n\| \big| < \varepsilon \}$$

and for $n \in A^\varepsilon$ let

$$B_n^\varepsilon = \{ m \in \mathbb{N} \ : \ \big| \|\alpha_1 e_1 + \alpha_2 x_n\| - \|\alpha_1 x_m + \alpha_2 x_n\| \big| < \varepsilon \}.$$

By (2.20), A^ε and B_n^ε belong to $\mathcal{U}$ for all $\varepsilon > 0$. Since $\mathcal{U}$ is a non-trivial ultrafilter we can choose by induction $(k_j)_j$, a strictly increasing sequence of positive integers, such that

$$k_1 \in A^{1/2}$$

$$k_2 \in A^{1/4} \cap B_{k_1}^{1/4}$$

$$k_3 \in A^{1/8} \cap B_{k_1}^{1/8} \cap B_{k_2}^{1/8}$$

$$\vdots$$

$$k_j \in A^{1/2^j} \cap B_{k_1}^{1/2^j} \cap B_{k_2}^{1/2^j} \ldots \cap B_{k_{j-1}}^{1/2^j}$$

for all j. Let $(x'_n)_{n=1}^{\infty} = (x_{k_j})_{j=1}^{\infty}$. Our construction shows that

$$\|\alpha_1 e_1 + \alpha_2 e_2\| = \lim_{n>m\to\infty} \|\alpha_1 x'_n + \alpha_2 x'_m\|.$$

This technique, separability of the complex numbers, a diagonal process and continuity of the norm allows us to choose a subsequence of $(x_n)_n$, $(x_{n_j})_j$, such that for any sequence of scalars $(\alpha_j)_j$ and any positive integer k we have

$$\left\|\sum_{j=1}^{k} \alpha_j e_j\right\| = \lim_{n_1>\ldots>n_k\to\infty} \left\|\sum_{j=1}^{k} \alpha_j x_{n_j}\right\|. \tag{2.21}$$

A subsequence of $(x_n)_n$ which satisfies (2.21) is called a *good subsequence*. Now suppose $P \in \mathcal{P}(^n E)$. In Chapter 1 we saw that there exists $P_1 \in \mathcal{P}(^n E_1)$ such that $P_1 \circ \pi_1 = P$ and $\|P_1\| = \|P\|$. By induction there exists, for every positive integer m, $P_m \in \mathcal{P}(^n E_m)$, $P_m \circ \pi_m = P$ and $\|P_m\| = \|P\|$ for all m. This defines a continuous polynomial on $\bigcup_{n\geq 0} E_n$ which, by uniform continuity, can be extended to an element $P_\infty \in \mathcal{P}(^n E_\infty)$ such that $\|P_\infty\| = \|P\|$.

The method used to construct a good subsequence can be applied in a straightforward fashion to find a subsequence of $(x_n)_n$, $(x_{n_j})_j$, such that

$$\lim_{n_1>\ldots>n_k\to\infty} P\left(\sum_{j=1}^{k} \alpha_j x_{n_j}\right) = P_\infty\left(\sum_{j=1}^{k} \alpha_j e_j\right) \tag{2.22}$$

for any sequence of scalars $(\alpha_j)_j$ and any positive integer k.

We suppose from now on that $(x_n)_n$ is a good, weakly null sequence of unit vectors in E. This implies also that $(e_j)_j$ is an unconditional basic sequence. If $(e_j)_j$ satisfies a lower q-estimate for some $q < \infty$, and $r > q$ then it is a non-trivial result that $(x_n)_n$ contains a subsequence which satisfies a lower r-estimate.

Proposition 2.41 *If no weakly null sequence of unit vectors in E admits a lower q-estimate for any $q > 1$ then all continuous $\mathbb{C}$-valued polynomials on E are weakly sequentially continuous.*

Proof. We suppose the result is not true. By Lemma 2.32, there exist a positive integer n, $P \in \mathcal{P}(^n E)$ and a weakly null sequence of unit vectors $(x_j)_j$ in E such that $|P(x_j)| \geq \delta > 0$ for all j. Let $[(e_n)_n]$ denote the spreading model associated with $(x_n)_n$. By taking a subsequence, if necessary, we may suppose that $(x_n)_n$ is a good sequence. Let P_∞ denote the extension of P to E_∞. By (2.22), $|P_\infty(e_n)| \geq \delta$ for all n. Let K denote the unconditionality constant of the basic sequence $(e_j)_j$. For any positive integer k and any sequence of scalars $(\alpha_j)_j$ we have

$$\delta \cdot \sum_{j=1}^{k} |\alpha_j|^n \le \sum_{j=1}^{k} |\alpha_j|^n |P_\infty(e_j)|$$

$$= \sum_{j=1}^{k} |P_\infty(\alpha_j e_j)|$$

$$\le \sup_{|\lambda_j| \le 1} \left| P_\infty \left(\sum_{j=1}^{k} \lambda_j \alpha_j e_j \right) \right|, \quad \text{by (1.55)},$$

$$\le \|P_\infty\| \cdot \sup_{|\lambda_j| \le 1} \left\| \sum_{j=1}^{k} \lambda_j \alpha_j e_j \right\|^n$$

$$\le K^n \|P\| \left\| \sum_{j=1}^{k} \alpha_j e_j \right\|^n.$$

Hence $(e_j)_j$ satisfies a lower n-estimate and $(x_j)_j$ admits a subsequence which has a lower q-estimate for $q > n$. This contradicts our hypothesis and, by Lemma 2.32, completes the proof.

It is easily seen that no weakly null sequence of unit vectors in c_0 satisfies a lower q-estimate and hence Proposition 2.41 is yet another proof that $\mathcal{P}_\omega(^n c_0) = \mathcal{P}(^n c_0)$ for all n (Proposition 1.59). The result also applies to T^* and shows that $\left(\mathcal{P}(^n T^*), \|\cdot\| \right)$ is reflexive for all n. More generally we have the following result.

Corollary 2.42 *If E is a reflexive Banach space and no weakly null sequence of unit vectors in E admits a lower q-estimate for any $q > 1$ then $\mathcal{P}(^n E)$ is reflexive for all n.*

Example 2.43 Let $(t_j)_j$ denote the standard unconditional basis for T^*. For any finite sequence of scalars $(\alpha_j)_j$ let

$$\|(\alpha_j)_j\|_{T_J^*} := \sup_{p_1 < \ldots < p_k} \left\| \sum_{j=1}^{k} (\alpha_{p_{2j-1}} - \alpha_{p_{2j}}) t_j \right\|_{T^*} \tag{2.23}$$

and let T_J^*, the Tsirelson–James space, denote the Banach space obtained by completing the space of finite sequences using the $\|\cdot\|_{T_J^*}$ norm. If we replace T^* by l_2 in (2.23) we obtain the classical James example of a quasi-reflexive non-reflexive Banach space. The space T_J^* has a shrinking monotone Schauder basis and $((T_J^*)''/T_J^*)$ is one-dimensional. Since no normalized block basic sequence in T_J^* satisfies a lower q-estimate for any $q < \infty$, Proposition 2.41 shows that $\mathcal{P}_\omega(^n T_J^*) = \mathcal{P}(^n T_J^*)$ for all n. The space T_J^* and all its higher duals have Schauder bases. Hence $(T_J^*)''$ has both the approximation property and the Radon–Nikodým Property. By Proposition 2.29,

$$\mathcal{P}_\omega(^nT_J^*)'' = \mathcal{P}(^nT_J^*)'' = \mathcal{P}(^nT_J^{*\prime\prime}).$$

Since $T_J^{*\prime}$ is separable and $\bigotimes\limits_{n,s,\varepsilon}(T_J^*)' = \mathcal{P}_\omega(^nT_J^*) = \mathcal{P}(^nT_J^*)$ it follows that $(\mathcal{P}(^nT_J^*), \|\cdot\|)$ is a separable dual space. Hence we have an example of a Banach space E such that $\mathcal{P}(^nE)$ is non-reflexive but $l_\infty \longleftrightarrow\!\!\!\!/\ \mathcal{P}(^nE)$ for any positive integer n.

We complete this section by taking a more general look at the biduals of spaces of polynomials. A canonical mapping from $\mathcal{P}(^nE)''$ into $\mathcal{P}(^nE'')$, E a Banach space, is obtained by taking the transpose, J_n^*, of the mapping

$$J_n \colon \bigotimes_{n,s,\pi} E'' \longrightarrow \mathcal{P}(^nE)'$$

where $\left[J_n\left(\underset{n}{\otimes}x\right)\right](P) = \left[AB_n(P)\right](x)$ for all $x \in E''$ and $P \in \mathcal{P}(^nE)$. It is easily seen that J_n is continuous and $\|J_n\| \le 1$.

Definition 2.44 A Banach space is Q-reflexive if J_n is an isomorphism for all n.

We confine ourselves for the remainder of this section to Banach spaces whose biduals have the approximation property. For such spaces we have the canonical isometries

$$\bigotimes_{n,s,\varepsilon} E' = \mathcal{P}_\omega(^nE) \quad \text{and} \quad \bigotimes_{n,s,\pi} E'' = \mathcal{P}_N(^nE') \tag{2.24}$$

and, moreover, the natural inclusion

$$\bigotimes_{n,s,\pi} E'' \longrightarrow \bigotimes_{n,s,\varepsilon} E'' \tag{$*$}$$

is a continuous *injection*. Since $\mathcal{P}_\omega(^nE) \subset \mathcal{P}(^nE)$ we may also define $J_{n,\omega}\colon \bigotimes\limits_{n,s,\pi} E'' \to \mathcal{P}_\omega(^nE)'$ by letting

$$J_{n,\omega}\left(\underset{n}{\otimes}x\right) = J_n\left(\underset{n}{\otimes}x\right)\Big|_{\mathcal{P}_\omega(^nE)}.$$

Proposition 2.45 *If E is a Banach space and E'' has the approximation property then E is Q-reflexive if and only if $\mathcal{P}_\omega(^nE) = \mathcal{P}(^nE)$ and $J_{n,\omega}$ is bijective for all n.*

Proof. We first suppose that E is Q-reflexive. By the Hahn–Banach Theorem $J_{n,\omega}$ is surjective and we are required to show that $J_{n,\omega}$ is injective.

Suppose $J_{n,\omega}\left(\sum_{i=1}^{\infty} \bigotimes_n x_i\right) = 0$ where $x_i \in E''$ and $\sum_{i=1}^{\infty} \|x_i\|^n < \infty$. We have

$$\left[J_{n,\omega}\left(\sum_{i=1}^{\infty} \bigotimes_n x_i\right)\right](\phi^n) = \sum_{i=1}^{\infty} x_i(\phi)^n = 0$$

for all $\phi \in E'$. By Goldstine's Theorem the unit ball of E' can be identified, via the canonical mapping into the bidual, with a $\sigma(E''', E'')$ dense subset of $B_{E'''}$. Since $\sum_{i=1}^{\infty} \|x_i\|^n < \infty$ this implies that $\sum_{i=1}^{\infty} \psi(x_i)^n = 0$ for all $\psi \in E'''$.

Hence $\left\|\sum_{i=1}^{\infty} \bigotimes_n x_i\right\|_{B_{E'''}} = 0$ and $\sum_{i=1}^{\infty} \bigotimes_n x_i = 0$ in $\widehat{\bigotimes_{n,s,\varepsilon}} E''$. The approximation property on E'' and $(*)$ imply that $\sum_{i=1}^{\infty} \bigotimes_n x_i = 0$ in $\widehat{\bigotimes_{n,s,\pi}} E''$. Hence $J_{n,\omega}$ is injective and consequently bijective.

Since $J_{n,\omega}$ is bijective we have $\mathcal{P}(^nE)' \cong \mathcal{P}_\omega(^nE)'$ and the Hahn–Banach Theorem implies $\mathcal{P}(^nE) = \mathcal{P}_\omega(^nE)$. This completes the proof in one direction. The converse is trivial since $\mathcal{P}_\omega(^nE) = \mathcal{P}(^nE)$ implies $J_{n,\omega} = J_n$. This completes the proof.

Corollary 2.46 *If E is a Q-reflexive Banach space and E'' has the approximation property then*

(a) $\mathcal{P}_I(^nE')$ and $\mathcal{P}_N(^nE')$ are isomorphic Banach spaces.
(b) $l_1 \nrightarrow E'$.

Proof. (a) By (2.24) and Proposition 2.45 the following identifications are implemented by isomorphisms of Banach spaces:

$$\mathcal{P}_N(^nE') = \widehat{\bigotimes_{n,s,\pi}} E'' = \left(\widehat{\bigotimes_{n,s,\varepsilon}} E'\right)' = \mathcal{P}_I(^nE').$$

(b) If $l_1 \hookrightarrow E'$ then $\widehat{\bigotimes_{n,s,\varepsilon}} l_1 = \mathcal{P}_\omega(^nc_0)$ is a closed subspace of $\mathcal{P}_\omega(^nE) = \widehat{\bigotimes_{n,s,\varepsilon}} E'$. By Proposition 2.24, $\mathcal{P}_\omega(^nc_0)' = \mathcal{P}_I(^nl_1)$ and hence, by the Hahn–Banach Theorem and Q-reflexivity, each $P \in \mathcal{P}_I(^nl_1)$ has an extension to $\tilde{P} \in \mathcal{P}_\omega(^nE)' = \widehat{\bigotimes_{n,s,\pi}} E''$. If $\tilde{P} := \sum_{j=1}^{\infty} \bigotimes_n x_j''$ where $x_j'' \in E''$ all j and $\sum_{j=1}^{\infty} \|x_j''\|^n < \infty$ then $P = \tilde{P}|_{\mathcal{P}_\omega(^nc_0)} = \sum_{j=1}^{\infty} \bigotimes_n (x_j''|_{l_1}) \in \mathcal{P}_N(^nl_1)$. By Example 2.25 this is not possible. Hence $l_1 \nrightarrow E'$ and this completes the proof.

We now prove an isometric extension of Proposition 2.27 and this leads to an isometric form of Q-reflexivity. For this we require the following results. Condition (2.26) may be compared with condition (6) in Proposition 2.28.

If $T: E \to F$ is a continuous linear mapping between Banach spaces and $B_F \subset \overline{T(B_E)}$ then $B_F \subset T((1 + \varepsilon)B_E)$ for every $\varepsilon > 0$ and T is surjective. (B_E and B_F are the closed unit balls of E and F.)

$$(2.25)$$

If E is a Banach space and $l_1 \not\hookrightarrow E$ then the closed unit ball of E', $B_{E'}$, is the closed convex hull of its extreme points,[5] $\mathrm{ext}(B_{E'})$ (see Proposition 2.26 and Exercise 2.60).

$$(2.26)$$

If E is a closed subspace of $C(K)$, K a compact Hausdorff space, then every extreme point ϕ of $B_{E'}$ has the form $\phi = \alpha\delta_q$ where $q \in K$, δ_q is evaluation at q, $\alpha \in \mathbb{C}$ and $|\alpha| = 1$.

$$(2.27)$$

Proposition 2.47 *If E is a Banach space, n is a positive integer and $l_1 \not\hookrightarrow \bigotimes\limits_{n,s,\varepsilon} E$ then $\mathcal{P}_I(^nE)$ and $\mathcal{P}_N(^nE)$ are isometrically isomorphic.*

Proof. Let B_I and B_N denote the closed unit balls of $\left(\mathcal{P}_I(^nE), \|\cdot\|_I\right)$ and $\left(\mathcal{P}_N(^nE), \|\cdot\|_N\right)$ respectively. It suffices to show $B_I = B_N$. The inclusion $\left(\mathcal{P}_N(^nE), \|\cdot\|_N\right) \longrightarrow \left(\mathcal{P}_I(^nE), \|\cdot\|_I\right)$ is a contraction and thus $B_N \subset B_I$. By hypothesis and (2.26) the closed unit ball of

$$\left(\bigotimes_{n,s,\varepsilon} E\right)' = \left(\mathcal{P}_I(^nE), \|\cdot\|_I\right)$$

is the closed convex hull of its extreme points. Since $\bigotimes\limits_{n,s,\varepsilon} E \subset C(B_{E'}, \sigma(E', E))$, isometrically, (2.27) implies that each extreme point of B_I has the form ϕ^n where $\phi \in E'$ and $\|\phi\| = 1$. Since $\|\cdot\|_I \le \|\cdot\|_N$ and $\|\phi^n\|_N = \|\phi^n\|_I$ for all $\phi \in E'$ this shows $\mathrm{ext}(B_I) \subset \mathrm{ext}(B_N)$. By our hypothesis and (2.25)

$$B_I = \overline{\Gamma}^{\left(\mathcal{P}_I(^nE), \|\cdot\|_I\right)}(\mathrm{ext}(B_I)) \subset \overline{\Gamma}^{\left(\mathcal{P}_I(^nE), \|\cdot\|_I\right)}(\mathrm{ext}(B_N)). \qquad (*)$$

---

[5] A point $x \in C$ is an extreme point of C if $y, z \in C$, $0 < \lambda < 1$ and $x = \lambda y + (1-\lambda)z$ imply $x = y = z$.

Hence $B_I \subset \overline{\Gamma}^{\left(\mathcal{P}_I(^nE),\, \|\cdot\|_I\right)}(B_N)$ and (2.25) implies $\mathcal{P}_I(^nE) = \mathcal{P}_N(^nE)$. By the open mapping theorem, $\mathcal{P}_I(^nE)$ and $\mathcal{P}_N(^nE)$ are isomorphic and again, by $(*)$,

$$B_I \subset \overline{\Gamma}^{\left(\mathcal{P}_N(^nE),\, \|\cdot\|_N\right)}(\text{ext}(B_N)) \subset B_N.$$

This completes the proof.

If E is Q-reflexive and E'' has the approximation property then

$$\widehat{\bigotimes_{n,s,\pi}} E'' = \Big(\widehat{\bigotimes_{n,s,\pi}} E\Big)''$$

and hence $\Big(\widehat{\bigotimes_{n,s,\pi}} E\Big)''$ has the approximation property. If $\widehat{\bigotimes_{n,s,\pi}} E$ is itself Q-reflexive, then Corollary 2.46 implies

$$l_1 \longleftrightarrow\!\!\!\!/ \ \Big(\widehat{\bigotimes_{n,s,\pi}} E\Big)' = \widehat{\bigotimes_{n,s,\varepsilon}} E'$$

and an application of Proposition 2.47 shows that $\mathcal{P}_I(^nE')$ and $\mathcal{P}_N(^nE')$ are isometrically isomorphic. A more transparent criterion appears in the following proposition.

Proposition 2.48 *If E is a Banach space such that E'' has the approximation property and the Radon–Nikodým Property, then $\mathcal{P}_I(^nE')$ and $\mathcal{P}_N(^nE')$ are isometrically isomorphic for all n.*

Proof. Since E'' has the Radon–Nikodým Property, E' and hence $\widehat{\bigotimes_{n,s,\varepsilon}} E'$ are Asplund spaces. This implies $l_1 \longleftrightarrow\!\!\!\!/\ \widehat{\bigotimes_{n,s,\varepsilon}} E'$ and an application of Proposition 2.47 completes the proof.

Combining the above we obtain the following result.

Proposition 2.49 *If E is a Banach space and E'' has the approximation property and the Radon–Nikodým Property then the following are equivalent:*

(a) *E is Q-reflexive, i.e. J_n is a linear isomorphism for all n,*

(b) *$\mathcal{P}_w(^nE) = \mathcal{P}(^nE)$,*

(c) *the canonical mapping $J_n^*: \mathcal{P}(^nE)'' \to \mathcal{P}(^nE'')$ is an isometric isomorphism for all n.*

Proof. By Proposition 2.45, (a) $\Rightarrow$ (b), and by Proposition 2.48, (b) $\Rightarrow$ (c). Since we always have (c) $\Rightarrow$ (a), this completes the proof.

The above result identifies the biduals of certain spaces of n-homogeneous polynomials. If E is Q-reflexive and E'' has the approximation property and the Radon–Nikodým Property then the canonical mapping into the bidual is given by

$$J_{\mathcal{P}(^n E)}\colon \left(\mathcal{P}(^n E),\ \|\cdot\|_B\right) \longrightarrow \left(\mathcal{P}(^n E''),\ \|\cdot\|_{B''}\right) \qquad (2.28)$$

where $J_{\mathcal{P}(^n E)}(x'') = \lim\limits_\alpha P(x_\alpha)$, $x'' \in E''$, $(x_\alpha)_\alpha$ is any bounded net in E such that $x_\alpha \to x''$ in the $\sigma(E'', E')$ topology, and B and B'' are the unit balls of E and E'' respectively. Moreover, $\left(\mathcal{P}(^n E),\ \|\cdot\|_B\right)''$ and $\left(\mathcal{P}(^n E''),\ \|\cdot\|_{B''}\right)$ are isometrically isomorphic and

$$J_{\mathcal{P}(^n E)}\left(\mathcal{P}(^n E)\right) = \mathcal{P}_{w^*}(^n E'')$$
$$:= \{P \in \mathcal{P}(^n E'') : P|_{B''} \text{ is } \sigma(E'', E') \text{ continuous}\}.$$

By Proposition 2.29 and Example 2.43, T_J^* is Q-reflexive. Since $c_0'' = l_\infty$ has the approximation property, Example 2.25 and Corollary 2.46(a) show that c_0 is not Q-reflexive.

In Section 6.3 we extend the above results to spaces of holomorphic functions. For entire functions the estimate in Proposition 2.27 suffices but for functions on the unit ball we require the more precise estimate given in Proposition 2.49.

2.5 Exercises

Exercise 2.50* Let $A : \mathcal{C}\left([0,1]\right) \longrightarrow \mathbb{C}$ be defined as

$$A(f) = \int_0^1 f(x)^2\, dm(x),$$

$m = $ Lebesgue measure. Show that $A \in \mathcal{P}_I(^2\mathcal{C}\,[0,1])$ but $A \notin \mathcal{P}_N(^2\mathcal{C}[0,1])$.

Exercise 2.51* If E is a Banach space show that the following conditions are equivalent.

(a) All continuous polynomials on E are weakly continuous on compact sets.

(b) All continuous polynomials on E are weakly sequentially continuous.

(c) For each n the diagonal mapping $\Delta : E \to \widehat{\bigotimes}_{n,s,\pi} E$, $x \to x \otimes \cdots \otimes x$, is weak to weak sequentially continuous.

(d) If $(x_k)_k$ is a weakly null sequence in E then $\underbrace{x_k \otimes \cdots \otimes x_k}_{n \text{ times}}$ is weakly null

in $\widehat{\bigotimes}_{n,s,\pi} E$.

(e) For each n the diagonal mapping in (c) is weak to weak continuous on weakly compact sets.

Exercise 2.52* Let $(x_m)_m$ denote an orthonormal subset in the Hilbert space H and suppose $n \geq 2$. Show that $P := \sum_{m=1}^{\infty} \lambda_m x_m^n \in \mathcal{P}_N(^n H)$ if and only if $\sum_{m=1}^{\infty} |\lambda_m| < \infty$. Show that $\|P\|_N = \sum_{m=1}^{\infty} |\lambda_m|$. What happens when $n = 1$?

Exercise 2.53* A sequence $(x_j)_j$ in a Banach space E is said to be *weakly p-summable*, $1 \leq p < \infty$, if $\sum_j |\phi(x_j)|^p < \infty$ for all $\phi \in E'$. Show that a sequence $(x_j)_j$ has an upper p-estimate if and only if it is weakly q-summable where $\frac{1}{p} + \frac{1}{q} = 1$.

Exercise 2.54* If $(x_j)_j$ is a sequence in a Banach space E and $n \leq p < \infty$ show that the following are equivalent:

(a) $(x_j)_j$ is weakly (p/n)-summable,
(b) if $(e_j)_j$ is the standard unit vector basis in $l_{p/n}$ then there exists $T \in \mathcal{L}(l_{p/n}; E)$ such that $T(e_j) = x_j$,
(c) if $(e_j)_j$ is the standard unit vector basis in l_p then there exists $P \in \mathcal{P}(^n l_p; E)$ such that $P(e_j) = x_j$ for all j.

Exercise 2.55* Let E and F denote reflexive Banach spaces with the approximation property. If n is a positive integer show that $\mathcal{P}(^n E; F)$ is reflexive if and only if
$$\mathcal{P}(^n E; F) = \mathcal{P}_A(^n E; F).$$
Show that $\mathcal{P}(^n l_p, l_q)$ is not reflexive if $nq > p$.

Exercise 2.56 Show that $P \in \mathcal{P}(^n E; F)$, E and F locally convex spaces, is weakly continuous if and only if it is of finite rank. Show that $(P_f(^n E; F), \tau_0)$ is separable for all n if and only if E'_c and F are both separable.

Exercise 2.57* Let $E = \varinjlim_m E_m$ be a strict inductive limit of Fréchet–Montel spaces. Show that the following are equivalent:

(a) Each E_m admits a continuous norm,
(b) $\mathcal{P}(^n E'_\beta) = \mathcal{P}_{HY}(^n E'_\beta)$ for some $n \geq 2$,
(c) $\mathcal{P}(^n E'_\beta) = \mathcal{P}_{HY}(^n E'_\beta)$ for all $n \geq 2$.

Exercise 2.58* Let X be a completely regular Hausdorff space and let $\mathcal{C}(X)$ denote the space of $\mathbb{C}$-valued continuous functions on X with the compact open topology. Show that $\mathcal{P}_{HY}(^n \mathcal{C}(X)) = \mathcal{P}(^n \mathcal{C}(X))$ if X is *Lindelöf*. If X is *paracompact* and $P \in \mathcal{P}_a(^n\mathcal{C}(X))$ is bounded on bounded sets show that P is continuous.

Exercise 2.59* If E is a Banach space such that no spreading model, built on a normalized weakly null sequence in E, has a lower q-estimate for any $q < \infty$ and E' has the approximation property show that E' has the Radon–Nikodým Property if and only if $\mathcal{P}(^nE)$ has the Radon–Nikodým Property for all n.

Exercise 2.60* Let E be a Banach space over $\mathbb{C}$ and let $E_{\mathbb{R}}$ denote E considered as a Banach space over $\mathbb{R}$. Show that $l_1 \hookrightarrow E$ if and only if $(l_1)_{\mathbb{R}} \hookrightarrow E_{\mathbb{R}}$.

Exercise 2.61* Show that the following conditions on a Banach space E are equivalent:

(a) weak* null sequences in E' are weakly null (spaces with this property are called *Grothendieck* spaces),
(b) for all (respectively one) positive integer n and $(P_m)_m \subset \mathcal{P}(^nE)$ we have $P_m(x) \to 0$ for all $x \in E$ if and only if $\tilde{P}_m(x'') \to 0$ for all $x'' \in E''$ ($\tilde{P}_m$ is the Aron–Berner extension of P_m to E''),
(c) if $P \in \mathcal{P}(^nE; c_0)$ then $\tilde{P} \in \mathcal{P}(^nE''; c_0)$.

Exercise 2.62* A quotient mapping between locally convex spaces $\pi : E \to F$ is said to lift bounded sets if for each bounded subset $B \subset F$ there exists a bounded set $A \subset E$ such that $B \subset \pi(A)$. If E is Banach show that quotient maps lift bounded sets. If π is a quotient mapping which lifts bounded sets and n is a positive integer show that the mapping

$$P \in (\mathcal{P}(^nF), \tau_b) \longrightarrow P \circ \pi \in (\mathcal{P}(^nE), \tau_b)$$

is an isomorphism onto its range. If $F = l_p$ and π lifts bounded sets show that $l_\infty \hookrightarrow (\mathcal{P}(^nE), \tau_b)$ for $n \geq p$. If E is quasi-normable and $l_p \cong E/F$ show that $(\mathcal{P}(^nE), \tau_b)$ is not reflexive for $n \geq p$.

Exercise 2.63* Let E be a Banach space. If $P \in \mathcal{P}_N(^nE)$ and $Q \subset \mathcal{P}_N(^mE)$ show that $P \cdot Q \in \mathcal{P}_N(^{n+m}E)$ and

$$\|PQ\|_N \leq \frac{(m+n)^{m+n}}{(m+n)!} \frac{n!m!}{n^n m^m} \|P\|_N \cdot \|Q\|_N.$$

Exercise 2.64* If E and F are Banach spaces, n is a positive integer, $P \in \mathcal{P}(^nE)$, and $0 < r < \infty$ show that the following are equivalent:

(i)   if $(x_i)_j$ is a sequence in E such that $\sum\limits_{j=1}^{\infty} |\phi(x_j)|^r < \infty$ for all $\phi \in E'$ then

$$\sum_{j=1}^{\infty} \|P(x_j)\|^{r/n} < \infty,$$

(ii) There exists $c > 0$ and μ a regular Borel measure on $(B_{E'}, \sigma(E', E))$ such that

$$\|P(x)\|^{r/n} \le c \int_{B_{E'}} |\phi(x)|^r \, d\mu(\phi)$$

for all $x \in E$.

Exercise 2.65* If $1 < p_i < \infty, i = 1, \ldots, m$ and $1 < q < \infty$ show that the following conditions are equivalent:

(a) $\sum\limits_{1=1}^{m} \dfrac{1}{p_i} < \dfrac{1}{q}$,

(b)   every continuous n-linear mapping $\Phi : l_{p_1} \times l_{p_2} \times \cdots \times l_{p_m} \longrightarrow l_q$ is compact,

(c)   every continuous n-linear mapping from $l_{p_1} \times l_{p_2} \times \cdots \times l_{p_m} \longrightarrow l_q$ is sequentially weak to norm continuous,

(d)   the space of continuous n-linear mappings from $l_{p_1} \times l_{p_2} \times \cdots \times l_{p_m} \longrightarrow l_q$ is reflexive.

Exercise 2.66* Let $E = \sum_{m=0}^{\infty} E_m$ where each E_m is a locally convex space. Let $(\phi_n)_{n=1}^{\infty} \subset E_0'$ and let $0 \ne \psi_n \in E_n'$ for all n. Show that $P = \sum_{n=1}^{\infty} \phi_n \psi_n \in \mathcal{P}_{HY}(^2 E)$ and that $P \in \mathcal{P}(^2 E)$ if and only if there exists a neighbourhood V of zero in E_0 such that $\|\phi_n\|_V < \infty$ for each n. Hence show that $(\mathcal{P}(^2 E), \tau_0)$ is not complete if E_0 is a non-normable metrizable locally convex space.

Exercise 2.67* If E is a Banach space and $P \in \mathcal{P}(^n E)$ show that P is weakly continuous on bounded sets if and only if there exists a compact subset K of E' such that

$$|P(x)| \le \sup\{|\phi(x)| : \phi \in K\}$$

for all x in E.

Exercise 2.68* If E is a Fréchet space show that E_c' is the k-space associated with $E_\sigma' := (E', \sigma(E', E))$.

Exercise 2.69* Show that open subsets of $\mathcal{DFM}$ spaces are k-spaces. If I is an uncountable set show that $\mathbb{C}^I$ is not a k-space. Show, however, that $\mathcal{P}(^n \mathbb{C}^I) = \mathcal{P}_{HY}(^n \mathbb{C}^I)$ for any positive integer n.

Exercise 2.70 If E, F and G are locally convex spaces show that

$$\left(E \widehat{\bigotimes}_\varepsilon F\right) \widehat{\bigotimes}_\varepsilon G = E \widehat{\bigotimes}_\varepsilon \left(F \widehat{\bigotimes}_\varepsilon G\right).$$

If $\widehat{\bigotimes}_{n,\varepsilon} E$ is defined inductively as $\left(\bigotimes_{n-1,\varepsilon} E\right) \widehat{\bigotimes}_\varepsilon E$ show that $\bigotimes_{n,s,\varepsilon} E$ is a closed complemented subspace of $\widehat{\bigotimes}_{n,\varepsilon} E$.

Exercise 2.71* If $(P_j)_j$ is a sequence in $\widehat{\bigotimes}_{n,s,\varepsilon} E'$ show that $(P_j)_j$ is weakly null if and only if $(\widetilde{P}_j)_j$ tends to zero pointwise in $\mathcal{P}(^n E'')$. ($\widetilde{P}_j$ is the unique extension of P_j to E'' as a weak* continuous polynomial on bounded sets – this coincides with the Aron–Berner extension.)

Exercise 2.72* Let $\mathcal{D}$ denote the space of $\mathcal{C}^\infty$-functions of compact support in $\mathbb{R}$. We endow $\mathcal{D}$ with its usual strict inductive limit topology. Let $\delta_a \in \mathcal{D}'$ be the Dirac delta function at the point a. Show that

$$f_m := \sum_{n=1}^m (\partial^n \delta_a) \cdot \delta_n \in \mathcal{P}(^2\mathcal{D})$$

for all m and that $(f_m)_{m=2}^\infty$ is a Cauchy sequence in $(\mathcal{P}(^2\mathcal{D}), \tau_0)$. Show that $P = \lim_{m\to\infty} f_m \in \mathcal{P}_{HY}(^2\mathcal{D})) \setminus \mathcal{P}(^2\mathcal{D})$.

Exercise 2.73* If E is a Banach space such that all spreading models over E are isomorphic to c_0 show that all polynomials on all subspaces of E are weakly sequentially continuous.

Exercise 2.74* A compact Hausdorff space X is said to be *scattered* or *dispersed* if every closed subset of X contains an isolated point. If X is dispersed show that $\left(\mathcal{P}(^n\mathcal{C}(X)), \|\cdot\|\right)$ has the approximation property.

Exercise 2.75* Let E denote a Banach space with the approximation property. If $(x_j)_j$ is a sequence in E which converges weakly to x and $x_j \otimes \cdots \otimes x_j \longrightarrow z \in \widehat{\bigotimes}_{n,s,\pi} E$ in the $\sigma\left(\bigotimes_{n,s,\pi} E, \mathcal{P}(^n E)\right)$ topology show that $z = \underbrace{x \otimes \cdots \otimes x}_{n \text{ times}}$.

Exercise 2.76* Let $x_1, \ldots, x_k$ be k points in the bidual E'' of the Banach space E. If $\sum_{j=1}^k AB_n(P)(x_j) = 0$ for all $P \in \mathcal{P}(^n E)$ show that $\sum_{j=1}^k P(x_j) = 0$ for all $P \in \mathcal{P}(^n E'')$.

Exercise 2.77* Show that $\mathcal{P}(^n l_1)'$ can be identified with the set of all finite symmetric Borel measures on $\beta(\mathbb{N}^n)$, the Stone–Čech compactification of $\mathbb{N}^n$.

Exercise 2.78* If E is a Banach space and E'' has the metric approximation property show that each $P \in \mathcal{P}(^n E'')$ is the pointwise limit of $\left(AB_n(P_\alpha) \right)_{\alpha \in \Gamma}$ where $(P_\alpha)_{\alpha \in \Gamma}$ is a bounded net in $\mathcal{P}(^n E)$.

Exercise 2.79* If the Banach space E has the approximation property show that $\mathcal{P}_f(^n E)$ is a dense subspace of $\left(\mathcal{P}(^n E), \tau_0 \right)$. If E has the compact approximation property show that $\mathcal{P}_\omega(^n E)$ is a dense subspace of $\left(\mathcal{P}(^n E), \tau_0 \right)$. (A Banach space E has the compact approximation property if each compact linear operator from E into itself can be uniformly approximated on compact sets by finite rank operators.)

Exercise 2.80* Show that $\tau_b = \tau_\omega$ on $\mathcal{P}(^n \lambda_{T^*}(A))$ where A is any Köthe matrix and T^* is the original Tsirelson space. Show that $\mathcal{P}(^n \lambda_{T^*}(A))$ is reflexive for all n.

2.6 Notes

It is informative, and not inaccurate, to consider the topics discussed in this chapter as having evolved essentially over two periods of time. The first period covered approximately the years 1965–1985 and concerns the first three sections and part of the final section. There were, however, important contributions prior to 1965, particularly from the Polish school of functional analysis, and we introduce these when relevant – see also the notes on Chapter 3. Today, we are in the middle of the second very active period which began around 1987. In writing a book over a five to six year period we felt it unwise to attempt to keep in touch with a rapidly changing subject and it will be some years before the definitive form of the current research emerges. Thus in Section 2.4 we only introduce topics that are related to the first three sections or to holomorphic results in later chapters. To obtain a better perspective of the internal growth of the subject we discuss, as far as possible, results grouped thematically and in chronological order. This is sometimes different to the arrangement in the text.

The affinity between bilinear forms and 2-homogeneous polynomials made it inevitable that any important concepts relating to the former would have implications for the latter. Thus it is no surprise that the linear and bilinear concepts introduced by A. Grothendieck [414, 415] during the 1950s, e.g. ε and π tensor products, nuclear spaces, nuclear and integral mappings between locally convex spaces and the approximation property, play a role in this chapter. The first period, characterized by the fact that the prime motivation of those involved was infinite dimensional analysis rather than polynomials, was initiated by L. Nachbin and C.P. Gupta in the mid-1960s. One particular consequence of this function theoretic approach was a desire to present duals and preduals as *function spaces*. This led first to the replacement of $\widehat{\bigotimes}_{n,s,\pi} E'$

([421]) by $\mathcal{P}_N(^nE)$ and, afterwards, to the spaces $\mathcal{P}_I(^nE)$ and $\mathcal{P}_A(^nE)$. This, of course, highlights the fact that one cannot, in general, sensibly add tensors of different degrees but this is possible if they admit realizations as functions on the *same* space. With function theory as the ultimate goal this is an important distinction. In the presence of the *approximation property* one can translate from one approach to the other. However, the different notation and ways of thinking have sometimes led to the significance of certain results being overlooked and go some way towards explaining the ubiquitous role of the approximation property in infinite dimensional holomorphy.

In his thesis, C.P. Gupta [421] (see also [422, 423, 665, 667, 684]) studied convolution operators on a Banach space and required nuclear polynomials, holomorphic functions of bounded type and a duality theory for spaces of polynomials. He defined nuclear polynomials (Definition 2.9), introduced the Borel transform and established the duality $\mathcal{P}_N(^nE)' \cong \mathcal{P}(^nE')$ for Banach spaces whose duals have the approximation property (Proposition 2.10). This had the advantage of presenting the predual as a space of polynomials. At the same time, L. Nachbin [665, 666], realizing the importance of both continuous and nuclear polynomials in a comprehensive theory, and influenced further by the various important classes of linear operators considered by A. Grothendieck [414], defined a general concept of *holomorphy type* with the aim of developing a unified theory which would include the nuclear and continuous cases as particular examples and which would incorporate holomorphic functions of bounded type by using weights (see the final remarks in [667]). We are still far – as we shall see in the next few chapters – from a single complete theory – but progress has been made and the motivation remains valid and helpful. In the intervening years holomorphy types (e.g. compact, integral, nuclear, Hilbert–Schmidt, etc. types) have been studied by many authors, e.g. L. Nachbin [665, 666, 667, 668], S. Dineen [276, 279], R.M. Aron [36, 37], T.A.W. Dwyer [338, 339, 340], P.J. Boland [145, 146, 147, 148], M.C. Matos [582, 583, 585], L.A. Moraes [625], M.C. Matos–L. Nachbin [591], O. Nicodemi [688] and, in recent years, this theory has begun to interact with the multilinear theory of operator ideals proposed by A. Pietsch [725] and currently being developed by G. Botelho [172, 173, 174], H. A. Braunss [194], H. Junek [495], M.C. Matos [587, 588, 589, 590] and Y. Meléndez–A. Tonge [612] (see also R. Alencar–M.C. Matos [12], H. A. Braunss–H. Junek [195, 196], S. Geiss [386], M. González–J.M. Gutiérrez [399], H. Junek–M.C. Matos [496], E. Plewnia [733, 734], B. Schneider [776], K. Floret–M.C. Matos [354] and Exercise 2.64). Since tensor products and operator ideals may be considered different approaches to the same subject matter this is not surprising and suggests that this direction will yield further interesting results.

We consider those developments of the first period which followed, directly or indirectly, from investigating holomorphy types and which are *related* to the *results presented here*. These mainly followed directions suggested by the nuclear, compact and continuous (Nachbin used the terminology current) types and dealt primarily with holomorphic functions. From these we extract the relevant polynomial results.

P.J. Boland initiated the study of holomorphic functions on nuclear and dual nuclear spaces and in a series of papers [146, 149, 150, 151] obtained many basic results. We present in Section 2.2 the polynomial versions of a number of these results and the corresponding holomorphic results are given in Section 4.4. The π_0 topology is introduced in [146] and Proposition 2.13 is given in [146, 149]. The algebraic part of Proposition 2.12 appears in [146] while the topological result is given in [299, Proposition 1.44]. Proposition 2.10 is, in the presence of the approximation property, essentially the preduality theory for polynomials discussed in Chapter 1. It appeared in this format for nuclear spaces in P.J. Boland [146], for Fréchet spaces in R.A. Ryan [758], and for locally convex spaces in [299, §§1.3–1.4]. Fully nuclear spaces were introduced by P.J. Boland–S. Dineen [152, 153]. Proposition 2.16, Corollary 2.17 and Example 2.18 appeared in [153] and are special cases of holomorphic results that we present later. In [247], J.F. Colombeau–R. Meise–B. Perrot developed a topological/bornological biduality theory for Silva holomorphic functions which, at the polynomial level, is similar to that given in Section 2.2. In the process they proved a general result which includes Lemma 2.15 – the special case for fully nuclear spaces with basis had previously appeared in [153] (see also [185]). Example 2.19, which compensates for the lack of nuclearity by using the compact open topology, is due to J. A. Barroso–L. Nachbin [93, Proposition 10] and J.A. Barroso [86, 87]. Further proofs are given in [93, 683]. Proposition 2.20 is essentially due to R.A. Ryan [758] and has been extended to locally convex spaces whose duals have the local Radon–Nikodým property – a concept due to A. Defant [257] – by C. Boyd [185]. The characterization of nuclearity using the coincidence of the ε and π topologies is due to A. Grothendieck [414]. Examples of G. Pisier [730], which settle a conjecture of Grothendieck, show that there exists an infinite dimensional Banach space E such that $\mathcal{P}_N(^2E) = \mathcal{P}(^2E)$. These examples were developed by H. Jarchow–K. John [481] (see K. John [485, 486]) to show that the same phenomena appear in non-nuclear Fréchet–Schwartz spaces (see also E. Mangino [567, remark 1.37] and Section 4.6 for further details).

The investigation of continuous polynomials on Banach spaces was under way at the same time and proceeded in two directions which eventually interacted. S. Dineen [276, 279] developed a duality theory for holomorphy types and in doing so introduced integral polynomials and proved $\mathcal{P}_A(^nE)' = \mathcal{P}_I(^nE')$ (Proposition 2.24). Connections with nuclear polynomials were established by R. Alencar in his thesis ([4, 5]). He proved $\mathcal{P}_I(^nE) = \mathcal{P}_N(^nE)$ when E' has the Radon–Nikodým Property (Proposition 2.27) and gave the first example where $\mathcal{P}_I \neq \mathcal{P}_N$ (Exercise 2.50) – the example given here, Example 2.25, is taken from [55]. Alencar also investigated the locally convex space situation in [4, Capítulo V] and showed that $\mathcal{P}_I(^nE'_c) = \mathcal{P}_N(^nE'_c)$ for all n whenever E is a quasi-complete locally convex space in which all compact sets are strictly compact (e.g. Fréchet spaces, $\mathcal{DFS}$ spaces, strict inductive limits of Banach spaces and spaces which satisfy the strict Mackey convergence criterion) and observed that $(\mathcal{P}(^nE), \tau_0)' \cong \mathcal{P}_I(^nE'_c)$ when E is a locally convex space with the approximation property. This may be regarded as a variation of Proposition 2.20. In [185], C. Boyd proves $(\mathcal{P}_N(^nE), \pi_\omega) = (\mathcal{P}_I(^nE), \tau_I)$ when E has a dual with the local Radon–Nikodým property and τ_I denotes the inductive limit topology on $\mathcal{P}_I(^nE)$ generated by the Banach spaces $\mathcal{P}_I(^n\hat{E}_\alpha)$, $\alpha \in cs(E)$, and extended the duality results of Alencar described above. We refer to J. Diestel–J.J. Uhl, Jr [272] for background information on the Radon–Nikodým Property. Condition (1) in Proposition 2.28 is usually taken as the definition of this property while, correctly speaking, Definition 2.26 is a theorem of L. Schwartz [795] (see also E. Saab [762, 763]). Our unorthodox introduction was chosen for convenience.

In a different direction R.M. Aron–M. Schottenloher [70, 71] initiated the study of compact polynomials and holomorphic functions, the approximation of holomorphic functions by polynomials of finite type, and the approximation property for spaces of holomorphic functions (see Exercise 3.124). Parallel to the holomorphic developments we have been discussing the theory of real differentiable functions over Banach spaces was under investigation in, perhaps, a less organized fashion. This subject led to different problems and phenomena, e.g. existence of partitions of unity, $\mathcal{C}^\infty$ functions of bounded support, separating polynomials (we refer to the books of J. Llavona [563] and A. Kriegl–P.W. Michor [530], and the survey article [425] for details) and included a number of results on the uniform approximation of differentiable functions. Motivated by one such result, essentially a version of Proposition 2.6 with $m = 1$ and E reflexive for functions which are weakly continuous and uniformly differentiable on bounded sets due to G. Restrepo [742], R.M. Aron–J.B. Prolla [68] proved the uniformly continuous version of Proposition 2.6 for arbitrary m and arbitrary Banach spaces and, in the same article, obtained Proposition 2.8 (see also [44, Proposition 4]). This led R.M. Aron [43], R.M. Aron–C. Hervés [61] and R.M. Aron–C. Hervés–M. Valdivia [62] to conduct a systematic and detailed investigation of polynomials and holomorphic functions which enjoyed various weak continuity properties (e.g. weak sequential continuity, weak continuity on bounded sets, weak uniform continuity on bounded sets, etc.). Proposition 2.4 is proved in [62] and has been generalized to different topologies and locally convex spaces in [399] by M. González–J. Gutiérrez. Another proof is given by T. Honda–M. Miyagi–M. Nishihara–M. Yosida in [463, 464]. The equivalence (2.1) $\iff$ (2.2), or (i) $\iff$ (v) in Proposition 2.6 for linear operators, is due to S. Banach [83, p.143]. The following non-trivial and interesting open problem would, if solved affirmatively, be a holomorphic version of Proposition 2.4.

If $f \in H(E)$ is weakly continuous on bounded sets is f bounded on bounded sets?

This has become known as "*the l_1-problem*" since a positive solution for $E = l_1$ would provide a positive solution for every Banach space. We refer to [62], [302] and [633] for further details, reformulations and positive and negative consequences of a solution. We found that Proposition 2.4, although the most recent of the results in the first three sections, offered itself as a natural starting point for our presentation and gave the chapter an improved sense of cohesion.

The next results we mention were motivated, not by holomorphic considerations, but by pre-1965 results of A. Pełczyński [715, 714] who introduced weakly compact polynomials. These are due to R.A. Ryan [757] who proved Proposition 2.34 and showed that the linear and polynomial Dunford–Pettis properties coincide for Banach spaces– a result re-proved in [156, 212, 377, 394]. Proposition 2.36 is taken from [62]. The first half is due to R.M. Aron–C. Hervés–M. Valdivia while the second part, involving the Rademacher functions, is due to J. Diestel (see [54, 351, 463, 464, 424, 633]). In our presentation of these results we have benefited from the lectures of J. Gutiérrez at University College Dublin during 1993–1994. The final contribution from this period was again initiated by R.A. Ryan [758] who, motivated by A. Pełczyński [714] and the theory of compact holomorphic mapping due to R.M. Aron–M. Schottenloher [70, 71], introduced weakly compact holomorphic mappings, established a connection between reflexivity and weak continuity on bounded sets and proved Proposition 2.30 and the equivalence of conditions (1), (2) and (5) in Proposition 2.31 (conditions (2) and (3) are equivalent by Proposition 2.4). This was followed by R. Alencar–R.M. Aron–S. Dineen [8] who initiated the study of polynomials on Tsirelson's space, T^*, and proved, using weak sequential continuity, that $\left(\mathcal{P}(^nT^*), \|\cdot\|\right)$ is reflexive for all n and, using lower p-estimates, showed that $\left(\mathcal{P}(^nT), \|\cdot\|\right)$ is not reflexive for $n \geq 2$. The results for T^*, introduced by B.S. Tsirelson [830] in 1974, were extended to vector-valued polynomials and to quotient spaces in [9] and [405] respectively. Lemma 2.32 is also given in [8]. Interestingly, A. Pełczyński [714] has shown – see also [405], [11] – that it suffices to consider bounded weakly *null* nets, when checking weak continuity on bounded sets for 2-homogeneous polynomials but, by R.M. Aron [43], this is not the case for 3-homogeneous polynomials. This phenomenon has recently been investigated by C. Boyd–R.A. Ryan [190] (see also [274, Proposition 1.5]) A detailed examination, which also answered several questions posed in [49], [275] and [401], of spaces on which all continuous polynomials are weakly sequentially continuous is undertaken in J.M.F. Castillo–R. García–R. Gonzalo [212]. Most of the above results have a holomorphic dimension and feature later in our study of holomorphic functions. Proposition 2.36 has been extended to Fréchet and DF spaces by C. Boyd [185].

This completes our summary of the first period of development which, although externally motivated, identified interesting concepts and results on polynomials over infinite dimensional spaces. This brings us to 1985 and the beginning of the second period. This did not represent a break in the scientific sense but was more of a psychological change of attitude notable for the appearance of polynomials as the *main object* of study – with holomorphic and differentiable functions an important but *secondary concern*. This moved the subject closer to linear functional analysis. Today this is an active research area containing many interesting results that we did not find possible to include in the main text (see our final remarks in this section).

J.D. Farmer [346] and J.D. Farmer–W.B. Johnson [348] introduced spreading models into the study of polynomials and proved Proposition 2.41, Corollary 2.42 and a result which contains Corollary 2.40 as a special case. This was taken a step further by R. Gonzalo [404] and R. Gonzalo–J.A. Jaramillo [405] who applied Ramsey theory and measures of weak summability to simplify, unify, quantify and extend many earlier results. We have followed the ultrafilter approach to spreading models given in S. Guerre–Delabrière [420] rather than the approach using nets in [405]. The extension (2.22) and the relationship between q-estimates on the weakly null sequence and its spreading model are taken from [405] and [406] respectively. We give one example of the type of result obtained in [405] (see also Section 4.6).

If E is a Banach space let

$$l(E) = \sup\{p \geq 1 : \text{every weakly null sequence of unit vectors in } E \text{ has a}$$
$$\text{subsequence with an upper } p\text{-estimate}\}$$

and

$$u(E) = \inf\{p \geq 1 : \text{every weakly null sequence of unit vectors in } E \text{ has a}$$
$$\text{subsequence with a lower } q \text{ estimate}\}$$

The indices $l(E)$ and $u(E)$ are related to type, cotype, hereditary p-Banach–Saks Property, Schur Property and loose rank. We have the following result:

If E and F are Banach spaces and $n \cdot u(F) < l(E)$ then every $P \in \mathcal{P}(^nE; F)$ is weakly sequentially continuous.

For the classical l_p spaces, $1 < p < \infty$, we have $l(l_p) = u(l_p) = p$ while l and u coincide with the upper and lower Boyd indices for reflexive Orlicz spaces. Extensions of this result to multilinear forms have been obtained by V. Dimant–I. Zalduendo [275] and R. Alencar–K. Floret [10]. Reflexivity of spaces of vector-valued homogeneous polynomials is investigated in [7], [9] and [405]. The relationship between Schauder bases and reflexivity for spaces of polynomials arises in Chapter 4. Proposition 2.38 and Corollary 2.40 are rearrangements of results in S. Dineen–M. Lindström [317] and are related to results in J. Gómez–J.A. Jaramillo [392]. R.M. Aron [42] had previously noted the relationship $\mathcal{P}(^nc_0; l_1) = \mathcal{P}_K(^nc_0; l_1)$ for all n.

Q-reflexive spaces (Definition 2.44) were introduced by R.M. Aron–S. Dineen [55] for Banach spaces E such that $\mathcal{P}_\omega(^nE) = \mathcal{P}(^nE)$. The definition was soon afterwards extended to arbitrary Banach spaces by M. González [393] who proved Proposition 2.45, which showed that the general definition extended that given in [55], and Corollary 2.46(b). While considering holomorphic functions of bounded type, P. Galindo–M. Maestre–P. Rueda [374] introduced isometric Q-reflexive Banach spaces. The space T_J^* is shown to be Q-reflexive in [55] and further examples of non-reflexive Q-reflexive spaces are given in [393] and [833]. An extensive study of the canonical mapping J_n for arbitrary Banach spaces is undertaken in J.A. Jaramillo–A. Prieto–I. Zalduendo [479]. Proposition 2.47 is due to C. Boyd–R.A. Ryan [191] and D. Carando–V. Dimant [203] and the three ingredients used (2.25), (2.26) and (2.27) may be found in J. Horváth [467, p.295], R. Haydon [442] and R. Arens–J.L. Kelley [33] respectively (see also the remarks on Exercise 2.60 for (2.26) and [337, p.441] for (2.27)). This result is an isometric version of Proposition 2.27 (see [203]). In [393], M. González showed (Corollary 2.46(b)) that $l_1 \hookrightarrow\!\!\!\!\!/\, E'$ if E is Q-reflexive and E'' has the approximation property (the proof given here is different). In [831, 832], the author shows that $\mathcal{P}_{\omega^*}(^nX')''$ is isometrically isomorphic to its τ_0-closure in $\mathcal{P}(^nX')$ if $l_1 \hookrightarrow\!\!\!\!\!/\, \mathcal{P}_{\omega^*}(^mX')$. This result and its clarifications in P. Rueda [753, Theorem 3.21] and P. Galindo–M. Maestre–P. Rueda [374] can be used in place of Proposition 2.47 in the proof of Proposition 2.49 (see also [377, Theorem 3.8.1]). Holomorphic properties of Q-reflexive Banach spaces are discussed in Section 6.3 and we refer to the remarks on Exercise 2.55 for vector-valued results. As noted by J. Diestel–J.J. Uhl, Jr. [272, p.215] there is an extremely fine dividing line between the conditions "$l_1 \hookrightarrow\!\!\!\!\!/\, E$" and "$E'$ has the Radon–Nikodým Property". An even finer line divides the conditions "$l_1 \hookrightarrow\!\!\!\!\!/\, \widehat{\bigotimes_{n,s,\varepsilon}} E'$" and "$E''$ has the Radon–Nikodým Property" but we conjecture that these properties do not coincide.

We conclude this section by listing briefly some of the topics in the theory of polynomials between Banach spaces that are currently under investigation. Current research on polynomials over Fréchet spaces is discussed in Chapters 1 and 4.

Topics concerning polynomials over Banach spaces currently under investigation (many of these topics are interrelated and a number also treat the holomorphic case):

(a) Geometric properties of spaces of polynomials, e.g. extreme points, exposed points, points of differentiability [63], [191], [192], [223], [227], [226], [350], [761].

(b) Topologies induced by polynomials, e.g. weak and bounded-weak polynomial topology, and related concepts such as polynomially continuous linear mappings, reciprocal Dunford–Pettis Properties, etc. [49], [124], [190], [207], [224], [394], [400], [403], [426], [436], [478].

(c) Polynomial analogues of various topological properties such as Property (V), Schur Property, Grothendieck Property, etc. and the classification of linear properties, e.g. Dunford–Pettis Property, using spaces of polynomials [124], [156], [348], [394], [397], [403], [426]. It is interesting to note that in the 1970s L. Nachbin initiated a programme similar to (c) with the aim of using holomorphy to classify locally convex spaces. Results from this endeavour appear at various places in this book (see Section 3.6).

(d) Polynomials which improve convergence, e.g. bounding, compact, weakly compact, weakly sequentially continuous, limited, Dunford–Pettis, unconditionally converging and (p, q, r)-summing polynomials, and the corresponding factorization results [10], [12], [57], [64], [124], [156], [172], [173], [174], [190], [212], [228], [348], [349], [354], [365], [370], [394], [395], [396], [397], [398], [401], [405], [478], [496], [587], [590], [612], [715], [714], [733], [734].

(e) Banach spaces which admit a "rich" supply of "good" polynomials, e.g. where all polynomials are weakly sequentially continuous, or where norm or numerical radius attaining polynomials are dense [2], [3], [56], [186], [212], [222], [225], [350], [483], [493], [494], [712], [769].

(f) Polynomials on spaces with additional structure, e.g. with basis or finite dimensional decomposition, etc. [263, 266, 274, 273].

(g) Extending and linearizing polynomials [48], [97], [201], [202], [204], [205], [212], [401], [479], [512], [532], [554], [598].

Chapter 3. Holomorphic Mappings between Locally Convex Spaces

In Section 3.1 we introduce Gâteaux holomorphic mappings between infinite dimensional spaces using only the classical concept of $\mathbb{C}$-valued holomorphic function of one complex variable. This very weak definition leads to monomial and Taylor series expansions and connections with derivatives. Holomorphic (or Fréchet holomorphic) mappings are defined as continuous Gâteaux holomorphic mappings. With these basic definitions in place we give various examples of holomorphic functions and, in doing so, encounter new concepts such as bounding sets and uniform holomorphy that play a role in later developments. In Section 3.2 we consider holomorphic analogues of the topologies defined on spaces of homogeneous polynomials in Chapter 1. The τ_0 and τ_ω topologies generalize quite readily, while the β-topology is obtained by using locally bounded sets and the compact open topology. Since τ_0 and τ_ω generally lack desirable topological properties such as barrelledness we introduce a further topology, the τ_δ topology, which may loosely be described as the topology generated by the countable open covers. This is the finest topology that we consider and coincides with the τ_ω topology on spaces of continuous homogeneous polynomials. In this section we motivate our investigation of these topologies and note the unavoidable and intrinsic role of nuclearity in the theory of holomorphic functions. In Section 3.1 we see that polydiscs and balanced domains are the natural domains of convergence for monomial and Taylor series expansions respectively. In Section 3.3 we study convergence of these expansions with respect to the topologies introduced in Section 3.2. Using Taylor series expansions we find that all these topologies lead to a strong form of decomposition – S-absolute decomposition. To obtain monomial expansions we consider holomorphic functions on open polydiscs in fully nuclear spaces with basis. We obtain an unconditional, and in many cases, an absolute monomial basis for the space of holomorphic functions with respect to the main topologies from Section 3.2.

Application of decompositions to obtain expansions of holomorphic mappings are the main tool used in Chapter 4 and in this context we may regard the Taylor series and monomial expansions as the two extreme cases. Taylor series expansions allow us to introduce a holomorphic version of the τ_b topology and this leads, in Chapter 4, to a projective description of the τ_ω topology. We use quasi-normability and completeness in Section 3.4 to show

how properties of homogeneous polynomials can be lifted to spaces of holomorphic functions.

3.1 Holomorphic Functions

We first introduce Gâteaux or $\mathcal{G}$-holomorphic functions from a finitely open subset of a vector space over $\mathbb{C}$ into a locally convex space by using holomorphic mappings from $\mathbb{C}$ into $\mathbb{C}$. The absence of a locally convex topology on the domain space obliges us to use finitely open sets but allows us to see clearly the finite or algebraic nature of $\mathcal{G}$-holomorphic mappings. This approach exposes the natural domain of convergence of monomial expansions – a polydisc – and the various modes of convergence of these expansions. The theory for holomorphic mappings on nuclear spaces is a natural development when a topology is introduced on the domain space. Using the monomial expansion we derive the Taylor series expansion and identify the natural domain of such expansions – balanced sets. Thus our study of $\mathcal{G}$-holomorphic mappings is a direct uncomplicated guide to a number of ideas which are refined and later developed within a topological setting.

A subset U of a vector space E over $\mathbb{C}$ is said to be *finitely open* if $U \cap F$ is open in the Euclidean topology of F for each finite dimensional subspace F of E.

Definition 3.1 A function $f: U \subset E \longrightarrow F$, where U is a finitely open subset of a vector space E over $\mathbb{C}$ and F is a locally convex space, is Gâteaux or $\mathcal{G}$-holomorphic if for each $\xi \in U$, $\eta \in E$ and $\phi \in F'$ the $\mathbb{C}$-valued function of one complex variable

$$\lambda \longrightarrow \phi \circ f(\xi + \lambda\eta)$$

is holomorphic on some neighbourhood of 0 in $\mathbb{C}$. We let $H_G(U; F)$ denote the set of all $\mathcal{G}$-holomorphic mappings from U into F and write $H_G(U)$ in place of $H_G(U; \mathbb{C})$.

Hartogs' Theorem says that separately holomorphic functions on $U \times V$, U and V open in $\mathbb{C}^m$ and $\mathbb{C}^n$ respectively, are holomorphic (as functions of $m + n$ complex variables). Hence $f: U \subset E \to F$ is $\mathcal{G}$-holomorphic if and only if $\phi \circ f\big|_{U \cap E_1}$ is holomorphic, in the classical sense, as a function of several complex variables, for each finite dimensional subspace E_1 of E. Consequently, one may use any of the finite dimensional criteria (e.g. Taylor series expansions, Cauchy–Riemann equations, existence of a $\mathbb{C}$-derivative, etc.) to define $\mathcal{G}$-holomorphic mappings. This implies, since the finite open topology is the inductive limit topology in the category of topological spaces given by the inclusion mappings $G \longrightarrow E$, where G with its Euclidean topology ranges over all finite dimensional subspaces of E, that $\mathcal{G}$-holomorphic mappings are continuous when the domain is endowed with the finite open topology and the range with the weak topology. In particular, $\mathcal{G}$-holomorphic functions map finite dimensional compact sets of their domain onto weakly compact subsets of their range. This suggests that we endow $H_G(U;F)$ with the (locally convex) topology of uniform convergence over the finite dimensional compact subsets of U.

We first suppose that E is a finite dimensional space with basis $(e_j)_{j=1}^k$ and that U is an open subset of E. If $f \in H_G(U;F)$ then, by the above remarks, $f(U)$ is a weakly separable subset of F. We may thus suppose, without loss of generality, that F is weakly separable and that $f(U)$ spans a weakly dense subspace of F. Let $(z^m)_{m \in \mathbb{N}^k}$ denote the set of monomials generated by the coordinate functionals associated with $(e_j)_{j=1}^k$ (see Section 1.3). For $\xi \in U$ we suppose $r := (r_j)_{j=1}^k$ is a finite sequence of positive real numbers such that

$$V_r := \xi + \left\{ \sum_{j=1}^k z_j e_j : |z_j| \le r_j,\ j = 1, \ldots, k \right\} \subset U.$$

If $\phi \in F'$ and $m = (m_j)_{j=1}^k \in \mathbb{N}^k$ let

$$a_{m,\xi}(\phi \circ f) = \frac{1}{(2\pi i)^k} \int \cdots \int_{\substack{|z_j|=r_j \\ j=1,\ldots,k}} \frac{\phi \circ f\left(\xi + \sum_{j=1}^k z_j e_j\right)}{z^m z_1 \cdots z_k} dz_1 \ldots dz_k. \qquad (3.1)$$

By the several variables Cauchy integral formula

$$\phi \circ f\left(\xi + \sum_{j=1}^k z_j e_j\right) = \sum_{m \in \mathbb{N}^k} a_{m,\xi}(\phi \circ f) z^m \qquad (3.2)$$

for $z = (z_j)_{j=1}^k$, $|z_j| \le r_j$, $j = 1, \ldots, k$. Now consider the mapping

$$a_{m,\xi}(f) : \phi \in F' \longrightarrow a_{m,\xi}(\phi \circ f).$$

Clearly $a_{m,\xi}(f)$ is linear and hence belongs to $(F')^*$. If W is a neighbourhood of zero in F then W° is a weak* compact subset of F'. Since F is weakly separable, W° is weak* separable and hence a weak* compact metrizable set. Since $f(V_r)$ is a bounded subset of F, the set $\left(\phi \circ f\big|_{V_r}\right)_{\phi \in W^\circ}$ is a uniformly bounded set of complex-valued functions on V_r. By (3.1) and the Lebesgue Dominated Convergence Theorem $a_{m,\xi}(f)$ is weak*-sequentially continuous on the equicontinuous subsets of F' and, as these sets are metrizable, $a_{m,\xi}(f)$ is weak* continuous on the equicontinuous subsets of F'. By Grothendieck's Completeness Theorem we may identify $a_{m,\xi}(f)$ with an element of $\hat{F}$ (the completion of F) and we have

$$\phi\big(a_{m,\xi}(f)\big) = a_{m,\xi}(\phi \circ f) \tag{$*$}$$

for all $\phi \in F'$. Since V_r is a compact subset of U there exists $s := (s_j)_{j=1}^r$, $s_j > r_j$ for all j, such that $V_s := \xi + \Big\{ \sum_{j=1}^{k} z_j e_j : |z_j| \le s_j \Big\} \subset U$. Moreover, (3.1) and (3.2) are still valid with r replaced by s. If $\alpha \in cs(F)$ and $m \in \mathbb{N}^k$, then, (3.1) and ($*$) imply

$$\alpha\big(a_{m,\xi}(f)\big) = \sup_{\substack{|\phi|\le\alpha \\ \phi\in F'}} \big|\phi\big(a_{m,\xi}(f)\big)\big| = \sup_{\substack{|\phi|\le\alpha \\ \phi\in F'}} \big|a_{m,\xi}(\phi \circ f)\big| \le \frac{\|f\|_{\alpha,V_s}}{s^m}. \tag{3.3}$$

Hence

$$\sum_{m\in\mathbb{N}^k} \|a_{m,\xi}(f)z^m\|_{\alpha,V_r} \le \sum_{m\in\mathbb{N}^k} \|f\|_{\alpha,V_s} \left(\frac{r}{s}\right)^m = \frac{\|f\|_{\alpha,V_s}}{\displaystyle\prod_{j=1}^{k}\left(1 - \frac{r_j}{s_j}\right)}$$

and the series $\sum_{m\in\mathbb{N}^k} a_{m,\xi}(f)z^m$ converges uniformly to a (continuous) function g on V_r. If $z = \sum_{j=1}^{k} z_j e_j$, $|z_j| \le r_j$ for all j, and $\phi \in F'$ then

$$\phi \circ g(\xi + z) = \sum_{m\in\mathbb{N}^k} \phi\big(a_{m,\xi}(f)\big)z^m$$

$$= \sum_{m\in\mathbb{N}^k} a_{m,\xi}(\phi \circ f)z^m$$

$$= \phi \circ f(\xi + z)$$

and the Hahn–Banach Theorem implies

$$f(\xi + z) = \sum_{m\in\mathbb{N}^k} a_{m,\xi}(f)z^m. \tag{3.4}$$

This shows that f is continuous on U and we may use Riemann's definition of the integral, (3.1) and the Hahn–Banach Theorem to conclude that

$$a_{m,\xi}(f) = \frac{1}{(2\pi i)^k} \underset{\substack{|z_j|=r_j \\ j=1,\dots,k}}{\int \cdots \int} \frac{f\big(\xi + \sum_{j=1}^{k} z_j e_j\big)}{z^m z_1 \cdots z_k}\, dz_1 \dots dz_k \tag{3.5}$$

for all $m = (m_j)_{j=1}^{k} \in \mathbb{N}^k$. By (3.5), $a_{m,\xi}(f) \in F$ if F is sequentially complete.

Now let E denote a vector space over $\mathbb{C}$ with algebraic (or Hamel) basis $(e_\alpha)_{\alpha\in\Gamma}$. Let $(e_\alpha^*)_{\alpha\in\Gamma}$ denote the set of coordinate functionals associated with this basis. A finite product of coordinate functionals is called a *monomial* (relative to the basis $(e_\alpha)_{\alpha \in \Gamma}$). If m_α is a non-negative integer for all $\alpha \in \Gamma$ and $m = (m_\alpha)_{\alpha \in \Gamma}$ let $s(m) = \{\alpha \in \Gamma : m_\alpha \neq 0\}$. Let $\mathbb{N}^{(\Gamma)} = \{(m_\alpha)_{\alpha \in \Gamma} : |s(m)| < \infty\}$. If $m = (m_\alpha)_{\alpha \in \Gamma} \in \mathbb{N}^{(\Gamma)}$ let $|m| = \sum_{\alpha \in \Gamma} m_\alpha$, $m! = \prod_{\alpha \in \Gamma} m_\alpha!$ and if Γ is an ordered indexing set let $l(m) = \sup\{\alpha : m_\alpha \neq 0\}$. We call $s(m)$ the *support*, $l(m)$ the *length* and $|m|$ the *modulus* of m. If $m = (m_\alpha)_{\alpha \in \Gamma} \in \mathbb{N}^{(\Gamma)}$ then the monomial $\prod_{\alpha \in s(m)} (e_\alpha^*)^{m_\alpha}$ is denoted by z^m (see Section 1.3). We have

$$z^m \left(\sum_{\alpha \in \Gamma} z_\alpha e_\alpha \right) = \prod_{\alpha \in \Gamma} z_\alpha^{m_\alpha}$$

for all $\sum_{\alpha \in \Gamma} z_\alpha e_\alpha \in E$ (we use the convention $0^0 = 1$). By (3.4) and (3.5), there exists, for each finite subset Γ_1 of Γ, a unique set of elements in $\widehat{F}$, $\left(a_{m,\xi}^{\Gamma_1}(f) \right)_{m \in \mathbb{N}^{\Gamma_1}}$, such that

$$f(\xi + z) = \sum_{m \in \mathbb{N}^{\Gamma_1}} a_{m,\xi}^{\Gamma_1}(f) z^m \tag{3.6}$$

whenever $z = \sum_{\alpha \in \Gamma_1} z_\alpha e_\alpha \in E$, $s(z) \subset \Gamma_1$ and $\xi + \sum_{\alpha \in \Gamma_1} w_\alpha e_\alpha \in U$ for all $|w_\alpha| \leq |z_\alpha|$, $\alpha \in \Gamma_1$. If Γ_1 and Γ_2 are finite subsets of Γ, $m \in \mathbb{N}^{(\Gamma)}$, and $s(m) \subset \Gamma_1 \cap \Gamma_2$, (3.5) implies that $a_{m,\xi}^{\Gamma_1}(f) = a_{m,\xi}^{\Gamma_2}(f)$ and we denote this common value by $a_{m,\xi}(f)$. Since (3.6) is valid for any finite subset of Γ we may combine these expansions to obtain the following result.

Let U denote a finitely open subset of a vector space E over $\mathbb{C}$ and let F denote a locally convex space over $\mathbb{C}$. Let $(e_\alpha)_{\alpha \in \Gamma}$ denote an algebraic or Hamel basis for E. If $\xi \in U$ let

$$P_\xi := \left\{ z = \sum_{\alpha \in \Gamma} z_\alpha e_\alpha \in E : \xi + \sum_{\alpha \in \Gamma} w_\alpha e_\alpha \in U \text{ whenever } |w_\alpha| \leq |z_\alpha| \text{ for all } \alpha \right\}.$$

If $f: U \to F$ then $f \in H_G(U; F)$ if and only if for each $\xi \in U$ there exists a unique set of elements in $\widehat{F}$ (the completion of F), $\left(a_{m,\xi}(f) \right)_{m \subset \mathbb{N}^{(\Gamma)}}$, such that

$$f(\xi + z) = \sum_{m \in \mathbb{N}^{(\Gamma)}} a_{m,\xi}(f) z^m \tag{3.7}$$

for all $z \in P_\xi$. If Γ_1 is a finite subset of Γ, $m = (m_\alpha)_{\alpha \in \Gamma}$, $s(m) \subset \Gamma_1$, $z = \sum_{\alpha \in \Gamma_1} z_\alpha e_\alpha \in P_\xi$ and $r_\alpha := |z_\alpha| > 0$ for all $\alpha \in \Gamma_1$ then

$$a_{m,\xi}(f) = \frac{1}{(2\pi i)^{|\Gamma_1|}} \int \cdots \int_{\substack{|z_\alpha| = r_\alpha \\ \alpha \in \Gamma_1}} \frac{f\left(\xi + \sum_{\alpha \in \Gamma_1} z_\alpha e_\alpha\right)}{z^m \prod_{\alpha \in \Gamma_1} z_\alpha} \prod_{\alpha \in \Gamma_1} dz_\alpha. \tag{3.8}$$

If $\beta \in cs(F)$ and $V_z := \xi + \left\{ \sum\limits_{\alpha \in \Gamma} w_\alpha e_\alpha \in E : |w_\alpha| \le |z_\alpha| \text{ for all } \alpha \right\} \subset P_\xi$ then

$\beta(a_{m,\xi}(f)z^m) \le \|f\|_{\beta,V_z}$ for all $m \in \mathbb{N}^{(\Gamma)}$, $s(m) \subset \{\alpha : z_\alpha \ne 0\}$, and

$$\sum_{m \in \mathbb{N}^{(\Gamma)}} \beta\left(a_{m,\xi}(f)z^m\right) < \infty.$$

We refer to (3.7) as the *monomial expansion* of f about ξ. Let

$$P_{n,\xi,f}(z) = \sum_{\substack{m \in \mathbb{N}^{(\Gamma)} \\ |m|=n}} a_{m,\xi}(f)z^m$$

for all $z \in E$ and all $n \in \mathbb{N}$. Since $P_{n,\xi,f}\big|_{E_1} \in \mathcal{P}(^nE_1; \widehat{F})$ for any finite dimensional subspace E_1 of E it is easily seen that $P_{n,\xi,f} \in \mathcal{P}_a(^nE; \widehat{F})$ for all n. By (3.7)

$$f(\xi + z) = \sum_{n=0}^{\infty} P_{n,\xi,f}(z) \tag{3.9}$$

for all $z \in P_\xi$. From the one dimensional identity principle for holomorphic mappings and the Hahn–Banach Theorem we see that $P_{n,\xi,f}$ does not depend on the choice of basis. If $z \ne 0$ belongs to E then there exists a basis for E, $(e_\alpha)_{\alpha \in \Gamma}$, such that $e_\alpha = z$ for some α. Hence (3.9) holds for all $z \in B_\xi := \{w \in E : \xi + \lambda w \in U, |\lambda| \le 1\}$ and from the finite dimensional theory we obtain the following result.

Proposition 3.2 *Let U denote a finitely open subset of a vector space E over $\mathbb{C}$ and let F denote a locally convex space over $\mathbb{C}$. If $\xi \in U$ let $B_\xi := \{z \in E : \xi + \lambda z \in U, |\lambda| \le 1\}$. If $f : U \to F$ then $f \in H_G(U; F)$ if and only if for every $\xi \in U$ there exists a unique sequence of homogeneous polynomials, $(P_{n,\xi,f})_{n=0}^{\infty}$, $P_{n,\xi,f} \in \mathcal{P}_a(^nE; \widehat{F})$, such that*

$$f(\xi + z) = \sum_{n=0}^{\infty} P_{n,\xi,f}(z) \tag{3.10}$$

for all $z \in B_\xi$. For each $n \in \mathbb{N}$ and $z \in B_\xi$

$$P_{n,\xi,f}(z) = \frac{1}{2\pi i} \int_{|\lambda|=1} \frac{f(\xi + \lambda z)}{\lambda^{n+1}} d\lambda. \tag{3.11}$$

If W is a balanced subset of E and $rW \subset B_\xi$, $r > 0$, then

$$\|P_{n,\xi,f}\|_{\beta,W} \le \frac{1}{r^n} \|f\|_{\beta,\xi+rW} \tag{3.12}$$

for all $\beta \in cs(F)$.

Formula (3.11) follows from (3.8) on taking a basis for E which contains $z := z_{\alpha_0}$ and noting that $P_{n,\xi,f}(z) = a_{m,\xi}(f)z^m$ where $m = (m_\alpha)_{\alpha \in \Gamma}$, $m_{\alpha_0} = n$ and $m_\alpha = 0$ for $\alpha \neq \alpha_0$. We refer to (3.12), which follows from (3.11), as the *Cauchy inequalities* or *estimates*, and call (3.10) the *Taylor series expansion* of f at ξ. From (3.10) and (3.11) we have

$$\left\| f(\xi + z) - \sum_{n=0}^{m} P_{n,\xi,f}(z) \right\|_{\beta,rW} \leq \frac{r^{m+1}\|f\|_{\beta,\xi+W}}{1 - r}$$

for any r, $0 < r < 1$, any positive integer m and any balanced set W contained in B_ξ. Since the right hand side of this inequality is finite for any finite dimensional compact subset of B_ξ the Taylor series expansion converges uniformly to f on the finite dimensional compact subsets of B_ξ. The set B_ξ is the maximal set in the collection of sets, $\mathcal{A} := \{A : A$ balanced, $\xi + A \subset U\}$. In particular, if we take $\xi = 0$ we see that U is balanced if and only if $U = B_0$. Thus, if U is balanced, we have

$$f = \sum_{n=0}^{\infty} P_{n,0,f}$$

in $H_G(U; \widehat{F})$ endowed with the topology of uniform convergence over the finite dimensional compact subsets of U.

We now establish a relationship between (Gâteaux) differentiable and $\mathcal{G}$-holomorphic mappings. If U is a finitely open subset of E, a vector space over $\mathbb{C}$, and F is a locally convex space over $\mathbb{C}$ then the mapping, $f: U \longrightarrow F$, is *Gâteaux differentiable* if for each $\xi \in U$ and each v in E

$$\lim_{\substack{\lambda \to 0 \\ \lambda \in \mathbb{C}}} \frac{f(\xi + \lambda v) - f(\xi)}{\lambda}$$

exists in $\widehat{F}$ (the completion of F). We denote this limit by $df(\xi)(v)$. It is easily checked that the mapping

$$df(\xi): E \longrightarrow \widehat{F}$$
$$v \longrightarrow df(\xi)(v)$$

is linear.

Lemma 3.3 *If U is a finitely open subset of a vector space E and F is a locally convex space then $f: U \longrightarrow F$ is Gâteaux differentiable if and only if it is $\mathcal{G}$-holomorphic.*

Proof. We assume the well known fact that this is true for finite dimensional spaces. First suppose that f is Gâteaux differentiable. If $\phi \in F'$ then $\phi \circ f$ is also Gâteaux differentiable and $d(\phi \circ f)(\xi) = \phi \circ df(\xi)$. By the finite dimensional result, $\phi \circ f \in H_G(U)$ and hence $f \in H_G(U; F)$.

Conversely, suppose $f \in H_G(U; F)$. If $\xi \in U$, $v \in E$, $\delta > 0$ and $\xi + \lambda v \in U$ for $|\lambda| \leq \delta$ then, by (3.10),

$$f(\xi + \lambda v) - f(\xi) = \sum_{n=1}^{\infty} P_{n,\xi,f}(\lambda v) = \sum_{n=1}^{\infty} \left(\frac{\lambda}{\delta}\right)^n P_{n,\xi,f}(\delta v)$$

for $|\lambda| \leq \delta$. If $\alpha \in cs(F)$ and $W := \{\lambda v : |\lambda| \leq \delta\}$ then

$$\|P_{n,\xi,f}\|_{\alpha,W} \leq \|f\|_{\alpha,\xi+W} =: M_\alpha$$

and

$$\lim_{\substack{\lambda \to 0 \\ \lambda \in \mathbb{C}}} \alpha \left(\frac{f(\xi + \lambda v) - f(\xi)}{\lambda} - P_{1,\xi,f}(v)\right) \leq \lim_{\lambda \to 0} \sum_{n=2}^{\infty} M_\alpha \delta \left(\frac{|\lambda|}{\delta}\right)^{n-1}$$

$$= \lim_{\lambda \to 0} \frac{\delta M_\alpha |\lambda|}{\delta - |\lambda|} = 0.$$

Hence f is Gâteaux differentiable and $df(\xi) = P_{1,\xi,f}$.

In proving this lemma we also showed

$$\alpha \left(\frac{f(\xi + \lambda v) - f(\xi)}{\lambda} - df(\xi)(v)\right) \leq 2|\lambda| \|f\|_{\alpha,\xi+W}$$

for $0 < |\lambda| < \delta/2$.

Example 3.4 If E is a vector space over $\mathbb{C}$, F is a locally convex space and $P \in \mathcal{P}_a(^nE; F)$ then

$$P(\xi + \lambda v) - P(\xi) = \sum_{j=1}^{n} \binom{n}{j} \overset{\vee}{P}((\lambda v)^j, \xi^{n-j})$$

$$= \sum_{j=1}^{n} \binom{n}{j} \overset{\vee}{P}(v^j, \xi^{n-j}) \lambda^j.$$

Hence P is $\mathcal{G}$-holomorphic and $[dP(\xi)](v) = n\overset{\vee}{P}(\xi^{n-1}, v)$.

We endow $\mathcal{L}_a(E; F)$ with the (locally convex) topology of uniform convergence on the finite dimensional compact subsets of E. If F is a complete locally convex space then $\mathcal{L}_a(E; F)$ is also complete. Suppose $f \in H_G(U; F)$ and $\xi \in U$. Consider the mappings

$$Q_{n,\xi,z} : v \in E \longrightarrow n\overset{\vee}{P}_{n,\xi,f}(z^{n-1}, v) = [dP_{n,\xi,f}(z)](v)$$

and

$$Q_{n,\xi} : z \in E \longrightarrow \left[v \in E \longrightarrow n\overset{\vee}{P}_{n,\xi,f}(z^{n-1}, v)\right].$$

It is easily seen that $Q_{n,\xi,z} \in \mathcal{L}_a(E; \widehat{F})$ for all $z \in E$ and $n \in \mathbb{N}$ and $Q_{n,\xi} \in \mathcal{P}_a(^{n-1}E; \mathcal{L}_a(E; \widehat{F}))$ for all $n \in \mathbb{N}$. Let $z \in B_\xi := \{w \in E : \xi + \lambda w \subset U, |\lambda| \leq 1\}$, $\alpha \in cs(F)$ and let V denote a finite dimensional compact subset of E. There exists $r > 0$ and $s > 1$ such that

$$W := \{s\gamma z + s\lambda v : |\gamma| \leq 1, |\lambda| \leq r, v \in V\} \subset B_\xi.$$

By Lemma 3.3, Example 3.4 and an application of the Cauchy estimates to $\overset{\vee}{P}_{n,\xi,f}$ and f

$$\sup_{v \in V} \alpha\big(n\overset{\vee}{P}_{n,\xi,f}(z^{n-1}, v)\big) \leq \frac{1}{r}\|P_{n,\xi,f}\|_{\alpha,z+rV} \leq \frac{1}{r} \cdot \frac{1}{s^n}\|f\|_{\alpha,\xi+W}. \qquad (3.13)$$

Hence $\displaystyle\sum_{n=1}^{\infty} \sup_{v \in V} \alpha(Q_{n,\xi,z}(v)) < \infty$ for all $\alpha \in cs(F)$ and every finite dimen-

sional compact subset V of E. This implies $\displaystyle\sum_{n=1}^{\infty} Q_{n,\xi,z} \in \mathcal{L}_a(E; \widehat{F})$ for all $z \in B_\xi$. By Lemma 3.3,

$$\alpha\Big(\frac{f(\xi + z + \lambda v) - f(\xi + z)}{\lambda} - \sum_{n=1}^{\infty} Q_{n,\xi,z}(v)\Big)$$

$$\leq \sum_{n=1}^{\infty} \alpha\Big(\frac{P_{n,\xi,f}(z + \lambda v) - P_{n,\xi,f}(z)}{\lambda} - Q_{n,\xi,z}(v)\Big)$$

$$\leq \sum_{n=1}^{\infty} 2|\lambda|\|P_{n,\xi,f}\|_{\alpha,z+\{\gamma v:|\gamma|\leq r\}},$$

$$\leq 2|\lambda| \sum_{n=1}^{\infty} \frac{\|f\|_{\alpha,\xi+W}}{s^n}, \quad \text{by (3.13)},$$

$$= 2|\lambda| \cdot \frac{s\|f\|_{\alpha,\xi+W}}{s - 1} \longrightarrow 0 \text{ as } \lambda \to 0$$

and

$$[df(\xi + z)](v) = \sum_{n=1}^{\infty} Q_{n,\xi,z}(v)$$

for all $z \in B_\xi$ and all $v \in E$. Hence

$$df(\xi + z) = \sum_{n=1}^{\infty} Q_{n,\xi,z} = \sum_{n=1}^{\infty} Q_{n,\xi}(z) \qquad (3.14)$$

for all $z \in B_\xi$. Since $Q_{n,\xi} \in \mathcal{P}_a(^{n-1}E; \mathcal{L}_a(E; \widehat{F}))$, $df \in H_G(U; \mathcal{L}_a(E; \widehat{F}))$. By the Hahn–Banach Theorem, Taylor series representations of $\mathcal{G}$-holomorphic functions are unique and hence (3.14) is the Taylor series expansion of df about ξ. By Lemma 3.3 and (3.14)

$$[d(df)](\xi) = Q_{2,\xi}.$$

Hence

$$\left(\left([d(df)](\xi)\right)(v)\right)(w) = (Q_{2,\xi,v})(w) = 2\overset{\vee}{P}_{2,\xi,f}(v,w) \in \widehat{F}.$$

Using the canonical isomorphism I_1 (Proposition 1.1) between $\mathcal{L}_a(^2E; F)$ and $\mathcal{L}_a(E; \mathcal{L}_a(E; \widehat{F}))$ we define

$$d^2 f(\xi)(v,w) := \left(\left((d(df))(\xi)\right)(v)\right)(w) = 2\overset{\vee}{P}_{2,\xi,f}(v,w).$$

Hence $d^2 f(\xi) = 2\overset{\vee}{P}_{2,\xi,f} \in \mathcal{L}_a^s(^2E; F)$ and

$$\widehat{d^2} f(\xi) := \widehat{d^2 f(\xi)} = 2P_{2,\xi,f} \in \mathcal{P}_a(^2E; \widehat{F}).$$

By induction we define $\widehat{d^n} f(\xi) := \widehat{d^n f(\xi)}$ where $d^n f(\xi) := d(d^{n-1} f)(\xi)$ and see that $\widehat{d^n} f \in H_G(U; \mathcal{P}_a(^nE; \widehat{F}))$ and, moreover, $\widehat{d^n} f(\xi)/n! = P_{n,\xi,f}$ for all n.

We have proved the following result.

Proposition 3.5 *If U is a finitely open subset of a vector space E, F is a locally convex space and $f \in H_G(U; F)$ then*

$$f(\xi + z) = \sum_{n=0}^{\infty} \frac{\widehat{d^n} f(\xi)}{n!}(z) \tag{3.15}$$

for all $\xi \in U$ and $z \in B_\xi$.

This completes our study of Gâteaux-holomorphic functions – which was achieved without the aid of a locally convex structure on the domain. Now we place a locally convex topology on the domain space and define holomorphic or Fréchet-holomorphic functions.

Definition 3.6 *If E and F are locally convex spaces over $\mathbb{C}$, U is an open subset of E then $f: U \to F$ is holomorphic if $f \in H_G(U; F)$ and f is continuous. We let $H(U; F)$ denote the set of all holomorphic mappings from U into F and write $H(U)$ in place of $H(U; \mathbb{C})$.*

Clearly $f \in H(U; F)$ if and only if $\pi_\alpha \circ f \in H(U; F_\alpha)$ for all $\alpha \in cs(F)$ and, in this case, we have

$$\pi_\alpha \circ \frac{\widehat{d^n} f(\xi)}{n!} = \frac{\widehat{d^n}(\pi_\alpha \circ f)}{n!}(\xi)$$

for $n \in \mathbb{N}$ and $\xi \in U$. Since $\pi_\alpha \circ f$ is continuous there exists, for each $\xi \in U$, a neighbourhood of 0 in E, V, such that $\xi + V \subset U$ and

$$\|\pi_\alpha \circ f\|_{\alpha,\xi+V} =: M < \infty.$$

By (3.12),

$$\left\|\pi_\alpha \circ \frac{\widehat{d^n} f(\xi)}{n!}\right\|_{\alpha,\delta V} = \left\|\frac{\widehat{d^n}(\pi_\alpha \circ f)}{n!}(\xi)\right\|_{\alpha,\delta V} \leq \delta^n M$$

for $n \in \mathbb{N}$ and $0 < \delta < 1$. Since α was arbitrary this implies

$$\widehat{d^n} f(\xi) \in \mathcal{P}(^n E; \widehat{F})$$

for all $n \in \mathbb{N}$ and all $\xi \in U$. By the above estimate the Taylor series of $\pi_\alpha \circ f$ at ξ converges uniformly on a neighbourhood of ξ. A slight modification of the above shows that locally bounded $\mathcal{G}$-holomorphic functions are holomorphic. When F is a normed linear space the converse is true. Since this elementary fact is so useful we list it as a proposition.

Proposition 3.7 *If U is an open subset of a locally convex space E and F is a normed linear space then $f \in H_G(U; F)$ is holomorphic if and only if it is locally bounded.*

In our next example we outline a number of situations in which the above properties characterize holomorphic functions and use the opportunity to introduce, in relatively simple situations, techniques, which are later refined, and to give examples of some essentially infinite dimensional holomorphic functions.

Example 3.8 (a) Let U denote a connected open subset of a Banach space E, let F denote a Banach space and suppose $f \in H_G(U; F)$. We claim that if there exists $\xi \in U$ such that $\widehat{d^n} f(\xi) \in \mathcal{P}(^n E; F)$ for all n then $f \in H(U; F)$. To establish this claim we may suppose, without loss of generality, that U is convex and balanced and that $\xi = 0$. Let

$$V_n = \left\{ x \in U : \left\|\frac{\widehat{d^m} f(0)}{m!}(x)\right\| \leq n \text{ for all } m \right\}$$

$$= \bigcap_{m=0}^{\infty} \left\{ x \in U : \left\|\frac{\widehat{d^m} f(0)}{m!}(x)\right\| \leq n \right\}.$$

Since $\widehat{d^m} f(0)$ is continuous, V_n is a closed subset of U for all n and as $f \in H_G(U; F)$, $\bigcup_{n=1}^{\infty} V_n = U$. By the Baire Category Theorem there exists n_0 such that V_{n_0} has non-empty interior. Choose $x_0 \in U$ and V a convex balanced neighbourhood of zero in E such that $x_0 + V \subset V_{n_0}$. We have

$$\sum_{n=0}^{\infty} \left\| \frac{\widehat{d}^n f(0)}{n!} \right\|_{\frac{1}{2}V} = \sum_{n=0}^{\infty} \frac{1}{2^n} \left\| \frac{\widehat{d}^n f(0)}{n!} \right\|_{V}$$

$$\leq \sum_{n=0}^{\infty} \frac{1}{2^n} \left\| \frac{\widehat{d}^n f(0)}{n!} \right\|_{x_0+V}, \quad \text{by Lemma 1.10(a),}$$

$$\leq 2n_0.$$

Hence f is continuous on a neighbourhood of the origin. If $y \in U$ then, using the Taylor series expansion of f at the origin, we obtain

$$[\widehat{d}^n f(y)](x) = \sum_{m=n}^{\infty} \left[\widehat{d}^n \left(\frac{\widehat{d}^m f(0)}{m!} \right)(y) \right](x).$$

Hence $\widehat{d}^n f(y)$, as the pointwise limit of a sequence of continuous functions on a Baire space, is of the first Baire class and thus admits points of continuity. By Proposition 1.11, $\widehat{d}^n f(y) \in \mathcal{P}(^n E; F)$ and the first part of the proof implies that f is continuous in a neighbourhood of y. Hence $f \in H(U; F)$.

(b) If E and F are Banach spaces, U is a connected open subset of E and $f \in H_G(U; F)$ then $f \in H(U; F)$ if f is continuous at a single point in U. To establish this just use the Cauchy estimates at the point of continuity ξ of f to show that $\widehat{d}^n f(\xi)$ is bounded on a neighbourhood of the origin. By Proposition 1.11 this implies that each $\widehat{d}^n f(\xi)$ is continuous and by (a) we obtain the desired conclusion.

(c) Let U and V denote open subsets of Banach spaces E and F, respectively, and suppose $f: U \times V \to \mathbb{C}$ is *separately holomorphic*, i.e. for each x in U the function $f_x: y \in V \to f(x, y)$ is holomorphic and for each y in V the function $f^y: x \in U \to f(x, y)$ is holomorphic. By the finite dimensional Hartogs' Theorem $f \in H_G(U \times V)$. Let x_0 denote a fixed point in U and for positive integers m and n let

$$A_{n,m} = \left\{ y \in V : |f(x, y)| \leq m \text{ whenever } x \in U \text{ and } \|x - x_0\| \leq \frac{1}{n} \right\}.$$

Since $f^y \in H(U)$ it follows that $\bigcup_{n,m} A_{n,m} = V$ and, since $f_x \in H(V)$, $A_{n,m}$ is closed for any pair (n, m). By the Baire Category Theorem there exists (n_0, m_0) such that A_{n_0, m_0} has an interior point (x_0, y_0). Hence there exists $\delta > 0$ such that $\|f(x, y)\| \leq m_0$ if $\|y - y_0\| \leq \delta$ and $\|x - x_0\| \leq 1/n_0$. This implies f is locally bounded at the point (x_0, y_0). By (a) or (b), $f \in H(U \times V)$.

(d) A locally convex space E is *superinductive* if it is the inductive limit of Banach spaces in the category of topological spaces and continuous mappings. Fréchet spaces and $\mathcal{DFS}$ spaces are superinductive spaces. We recall that a $\mathcal{DFS}$ (dual of Fréchet–Schwartz space) is a locally convex space which can be represented as $\varinjlim_{n} E_n$ where each E_n is a Banach space, $E_n \hookrightarrow E_{n+1}$ is compact and the inductive limit is taken in the category of locally convex spaces and continuous linear mappings. If E is superinductive, U is a connected open subset of E, F is a Banach space and $f \in H_G(U;F)$ then (a) and (b) imply $f \in H(U;F)$ if either all of the derivatives of f at a single point are continuous or if f is continuous at a single point.

Now suppose E and F are locally convex spaces and $E \times F'_\beta$ is superinductive. If U is an open subset of E and $f \in H(U;F)$ let $\widetilde{f}: U \times F'_\beta \to \mathbb{C}$ be given by the formula

$$\widetilde{f}(x, \phi) = \phi(f(x)).$$

Clearly $\widetilde{f}$ is separately holomorphic on $U \times F'_\beta$ and, by superinductivity, $\widetilde{f} \in H(U \times F'_\beta)$. Hence, if $\xi \in U$, there exists a neighbourhood V_ξ of ξ in U and a neighbourhood W of 0 in F'_β such that $\widetilde{f}$ is bounded on $V_\xi \times W$. Since neighbourhoods of 0 are absorbing it follows that $\|\phi(f(x))\|_{V_\xi} = \|\widetilde{f}(x, \phi)\|_{V_\xi} < \infty$ for all $\phi \in F'$. By Mackey's Theorem the set $\{f(x) : x \in V_\xi\}$ is a bounded subset of F and the function f is locally bounded. We have thus shown that $f \in H_G(U;F)$ is holomorphic if and only if it is locally bounded. This applies in particular to the pairs $\{E = \text{Fréchet}, F = DF\}$ and $\{E = \mathcal{DFS}, F = \text{Fréchet–Schwartz}\}$.

(e) Example (d) shows that $\mathcal{G}$-holomorphic functions on $\mathcal{DFS}$ spaces are continuous under apparently rather weak assumptions and it is natural to ask if a similar result holds for $\mathcal{DFM}$ spaces and indeed for any $\mathcal{DF}$ spaces. Using monomial expansions for holomorphic functions on l_1 it can be shown that the above results are still valid for entire $\mathcal{G}$-holomorphic functions on $\mathcal{DFM}$ spaces with an absolute basis but it is an open question if the result holds for all $\mathcal{DFM}$ spaces. Nevertheless, using the method of Example 1.25, we may prove a slightly weaker result for all $\mathcal{DFM}$ spaces. If E is a $\mathcal{DFM}$ space then E can be represented as an inductive limit, $\varinjlim_{m} E_m$, of Banach spaces in the category of locally convex spaces and continuous linear mappings so that $(B_m)_{m=1}^\infty$, $B_m := \{x \in E_m : \|x\|_{E_m} \leq 1\}$, forms a fundamental system of compact subsets of E. Let U denote an open subset of E and let F denote a Banach space. In view of later requirements we consider a set $\mathcal{F} \subset H_G(U;F)$ which is uniformly bounded on the compact subsets of U. Let $\xi \in U$ and suppose V is a convex balanced neighbourhood of 0 such that $\xi + V \subset U$. By the Cauchy inequalities $\left(\dfrac{\widehat{d^n} f(\xi)}{n!}\right)_{\substack{n=0,1\dots \\ f \in \mathcal{F}}}$ is uniformly bounded on the compact subsets of V. Let $\lambda_1 > 0$ be chosen so that $\lambda_1 B_1 \subset V$ and let

$$M = \sup_{\substack{n=0,1\dots \\ f \in \mathcal{F}}} \left\|\frac{\widehat{d^n} f(\xi)}{n!}\right\|_{\lambda_1 B_1}.$$ Suppose $\lambda_2, \dots, \lambda_m$ are positive scalars chosen

so that $\displaystyle\sum_{i=1}^{m} \lambda_i B_i$ is a compact subset of V and

$$\sup_{\substack{n=0,1\ldots \\ f\in\mathcal{F}}} \left\| \frac{\widehat{d^n} f(\xi)}{n!} \right\|_{\sum_{i=1}^{m} \lambda_i B_i} < 3M \cdot \sum_{i=1}^{m} \frac{1}{2^i}.$$

We can choose $\varepsilon > 0$ and $\lambda > 1$ such that $L(\varepsilon,\lambda) := \lambda\Big(\displaystyle\sum_{i=1}^{m} \lambda_i B_i + \varepsilon B_{m+1}\Big)$ is a compact subset of V. By the Cauchy inequalities

$$\sup_{\substack{n=0,1\ldots \\ f\in\mathcal{F}}} \left\| \frac{\widehat{d^n} f(\xi)}{n!} \right\|_{L(\varepsilon,\lambda)} \leq M' := \sup_{f\in\mathcal{F}} \|f\|_{\xi+L(\varepsilon,\lambda)}.$$

Since

$$\left\| \frac{\widehat{d^n} f(\xi)}{n!} \right\|_{L(\varepsilon,\lambda)} = \lambda^n \cdot \left\| \frac{\widehat{d^n} f(\xi)}{n!} \right\|_{\sum_{i=1}^{m} \lambda_i B_i + \varepsilon B_{m+1}}$$

there exists n_0 such that

$$\sup_{\substack{n\geq n_0 \\ f\in\mathcal{F}}} \left\| \frac{\widehat{d^n} f(\xi)}{n!} \right\|_{\sum_{i=1}^{m} \lambda_i B_i + \varepsilon B_{m+1}} \leq \frac{M'}{\lambda^{n_0}} < 3M \cdot \sum_{i=1}^{m+1} \frac{1}{2^i}.$$

By Example 1.25 we can choose for each n, $0 \leq n < n_0$, $\varepsilon_n > 0$ such that

$$\left\| \frac{\widehat{d^n} f(\xi)}{n!} \right\|_{\sum_{i=1}^{m} \lambda_i B_i + \varepsilon_n B_{m+1}} < 3M \cdot \sum_{i=1}^{m+1} \frac{1}{2^i}$$

for all $f \in \mathcal{F}$. Let $\lambda_{i+1} = \min(\varepsilon_1,\ldots,\varepsilon_m,\varepsilon)$. We have

$$\sup_{\substack{n=0,1\ldots \\ f\in\mathcal{F}}} \left\| \frac{\widehat{d^n} f(\xi)}{n!} \right\|_{\sum_{i=1}^{m+1} \lambda_i B_i} < 3M \cdot \sum_{i=1}^{m+1} \frac{1}{2^i}.$$

By induction we can find a sequence of positive numbers $(\lambda_n)_n$ such that $\left\| \dfrac{\widehat{d^n} f(\xi)}{n!} \right\|_{\sum_{i=1}^{\infty} \lambda_i B_i} \leq 3M$ for all n and all $f \in \mathcal{F}$. Hence

$$\sup_{f\in\mathcal{F}} \|f\|_{\xi+\frac{1}{2}\sum_{i=1}^{\infty} \lambda_i B_i} \leq \sup_{f\in\mathcal{F}} \sum_{n=0}^{\infty} \left\| \frac{\widehat{d^n} f(\xi)}{n!} \right\|_{\frac{1}{2}\sum_{i=1}^{\infty} \lambda_i B_i}$$

$$= \sup_{f\in\mathcal{F}} \sum_{n=0}^{\infty} \frac{1}{2^n} \left\| \frac{\widehat{d^n} f(\xi)}{n!} \right\|_{\sum_{i=1}^{\infty} \lambda_i B_i}$$

$$\leq 3M \cdot \sum_{n=0}^{\infty} \frac{1}{2^n}.$$

Since $\sum_{i=1}^{\infty} \lambda_i B_i$ is a neighbourhood of 0 in E, the collection $\mathcal{F}$ of $\mathcal{G}$-holomorphic functions is locally bounded. By Proposition 3.7, $\mathcal{F} \subset H(U; F)$. In particular, we note that every $\mathcal{G}$-holomorphic function on a $\mathcal{DFM}$ space which is bounded on compact sets is holomorphic and a collection of holomorphic functions which is uniformly bounded on compact sets is locally bounded or equicontinuous.

(f) Let E denote a superinductive space or a $\mathcal{DFM}$ space and let $(\phi_n)_n$ denote a weak* null sequence in E'. Then

$$f := \sum_{n=1}^{\infty} \phi_n^n \tag{3.16}$$

is easily seen to define a $\mathcal{G}$-holomorphic function on E with (3.16) as its Taylor series expansion at the origin. If E is superinductive then $f \in H(E)$, by (a) and (c), since $\dfrac{\hat{d}^n f(0)}{n!} = \phi_n^n$ is continuous. If E is a $\mathcal{DFM}$ space then $(\phi_n)_n$ tends to zero uniformly on the compact subsets of E. Hence, for any compact subset K of E there exists a positive integer n_0 such that $\|\phi_n\|_{2K} \leq 1$ for all $n \geq n_0$. By (e), the estimate

$$\|f\|_K \leq \sum_{n=1}^{\infty} \|\phi_n^n\|_K$$

$$= \sum_{n=1}^{\infty} \left(\|\phi_n\|_K\right)^n$$

$$\leq \sum_{n \leq n_0} \|\phi_n\|_K^n + \sum_{n_0+1}^{\infty} \frac{1}{2^n} \|\phi_n\|_{2K}^n < \infty$$

shows that $f \in H(E)$. The particular function given in (3.16) may be regarded as the simplest and most useful non-trivial holomorphic function. The criterion given to ensure that it is holomorphic is linear and thus provides a link between linear functional analysis and infinite dimensional holomorphy. It is employed as a "test" function for various problems and first arose in studying bounding sets. A subset A of a locally convex space E is said to be *bounding* if $\|f\|_A < \infty$ for all $f \in H(E)$. The Josefson–Nissenzweig Theorem says that the dual of any infinite dimensional Banach space E admits a weak*-null sequence of unit vectors, $(\phi_n)_n$. Let $f := \sum_{n=1}^{\infty} \phi_n^n$ and let B denote the unit ball of E. Fix $\varepsilon > 0$ and let $M = \|f\|_{(1+\varepsilon)B}$. By the Cauchy inequalities we have

$$\left\| \frac{\hat{d}^n f(0)}{n!} \right\|_{(1+\varepsilon)B} = (1+\varepsilon)^n \|\phi_n^n\|_B = (1+\varepsilon)^n \leq \|f\|_{(1+\varepsilon)B}$$

for all n. Hence $M = \infty$. Using translations and contractions we see that bounding sets in an infinite dimensional Banach space have empty interiors.

We use this result to show that the result in (e) does not extend to countable direct sums of infinite dimensional Banach spaces. Let $E = \sum_{m=1}^{\infty} E_m$ where each E_m is a Banach space and E_1 is infinite dimensional. Let $(\phi_j)_j$ denote a weak*-null sequence of unit vectors in $(E_1)'$ and for each positive integer k let $\psi_k \in E_k'$, $\|\psi_k\| = 1$. We identify E_k and E_k' with subspaces of E and E', respectively, in the usual fashion and hence ϕ_j and ψ_k may be identified with elements of E' for all j and k. For any positive integer k let $\theta_k(z) = kz$ where $z \in E$. Let $n_1 = 1$ and let $n_{j+1} = n_j + j + 1$ for all $j \geq 1$. It is easily seen that $(n_j)_{j=1}^{\infty}$ is strictly increasing and if $j \geq k$ and $j' \geq k'$ then $n_j + k = n_{j'} + k'$ if and only if $j = j'$ and $k = k'$. Consider the function

$$f\left(\sum_{j=1}^{\infty} x_j\right) := \sum_{k=2}^{\infty}\left\{\sum_{j=k}^{\infty} \phi_j^{n_j}(kx_1)\right\}\psi_k^k(x_k)$$

for $\sum_{j=1}^{\infty} x_j \in E$. Since $(\phi_j)_j$ is weak*-null we have

$$\sum_{j=k}^{\infty} \phi_j^{n_j}\Big|_{\sum_{i=1}^{l} E_i} \in H\left(\sum_{i=1}^{l} E_i\right)$$

for all l. Hence

$$f\Big|_{\sum_{j=1}^{l} E_j} = \sum_{k=2}^{l}\left\{\sum_{j=k}^{\infty} \phi_j^{n_j} \circ \theta_k\right\}\psi_k^k$$

is holomorphic. In particular, $f \in H_G(E)$ and f is continuous on the compact subsets of E. The condition $n_{j+1} = n_j + j + 1$ implies

$$\frac{\hat{d}^{(n_j+k)} f(0)}{(n_j + k)!} = \begin{cases} (\phi_j^{n_j} \circ \theta_k)\psi_k^k, & j \geq k \geq 2 \\ 0, & \text{otherwise.} \end{cases}$$

We claim that $f \notin H(E)$. Otherwise, the Cauchy inequalities would imply the existence of a sequence of positive real numbers $(\lambda_n)_{n\geq 1}$ such that

$$\|(\phi_j^{n_j} \circ \theta_k)\psi_k^k\|_{\sum_{n=1}^{\infty} \lambda_n B_n} \leq 1$$

for all $j \geq k$ (B_n denotes the unit ball of E_n for each n). Hence

$$\|(\phi_j^{n_j} \circ \theta_k)\psi_k^k\|_{\sum_{n=1}^{\infty} \lambda_n B_n} = \|\phi_j^{n_j} \circ \theta_k\|_{\lambda_1 B_1} \cdot \|\psi_k^k\|_{\lambda_k B_k}$$

$$= (\lambda_1 k)^{n_j}\|\phi_j\|^{n_j} \cdot \lambda_k^k \|\psi_k\|^k$$

$$= (\lambda_1 k)^{n_j} \lambda_k^k \leq 1$$

for all j and k. This implies

$$\limsup_{j \longrightarrow \infty} (\lambda_1 k)^{n_j/(n_j+k)} \cdot \lambda_k^{k/(n_j+k)} = k\lambda_1 \le 1$$

for all k. Since $\lambda_1 > 0$ this is impossible and proves our claim and as all derivatives of f at the origin are continuous this shows that the result in (a) does not extend to $E = \sum_{m=1}^{\infty} E_m$ unless each E_m is finite dimensional. If each E_m is finite dimensional then E is superinductive and by (d), (a) and (b) apply to E.

We let $H_{HY}(U; F)$ denote the set of all $\mathcal{G}$-holomorphic mappings from U into F which are continuous on the compact subsets of U. The elements of $H_{HY}(U; F)$ are called *hypoanalytic functions*. The function f considered above is hypoanalytic but not holomorphic. If each E_n in $\sum_{n=1}^{\infty} E_n$ is finite dimensional then $\sum_{n=1}^{\infty} E_n \cong \mathbb{C}^{(\mathbb{N})}$ and $H_G(E) = H_{HY}(E) = H(E)$ (Exercise 3.55). We have thus shown that hypoanalytic functions on a direct sum of Banach spaces are holomorphic if and only if the direct sum is isomorphic to $\mathbb{C}^{(\mathbb{N})}$. Since open subsets of Fréchet spaces and $\mathcal{DFM}$ spaces are k-spaces we have $H_{HY}(U; F) = H(U; F)$ for U open in either of these spaces and F any locally convex space.

(g) In the final part of this example we consider, as in the definition of $\mathcal{G}$-holomorphic functions, the composition of holomorphic functions and continuous linear mappings. If U is an open subset of a locally convex space E, F is a locally convex space, $f \in H(U; F)$ and $\phi \in F'$ then $\phi \circ f \in H(U)$. Conversely, if E is a metrizable space, $f \in H_G(U; F)$ and $\phi \circ f \in H(U)$ for all $\phi \in F'$ then f is weakly bounded on convergent sequences in U. By Mackey's Theorem f is bounded on each such sequence. Hence $\pi_\alpha \circ f$ is a locally bounded $\mathcal{G}$-holomorphic function and, by Proposition 3.7, $\pi_\alpha \circ f \in H(U; F_\alpha)$. Hence $f \in H(U; F)$. By (d), the same result is true for $\mathcal{DFS}$ spaces.

Now suppose E is a $\mathcal{DFM}$ space. If $\phi \circ f \in H(U)$ for all $\phi \in F'$ then f is weakly, and hence strongly, bounded on the compact subsets of U. Let $\alpha \in cs(F)$. Then $\mathcal{F} := (\phi \circ f)_{\substack{\phi \in F' \\ |\phi| \le \alpha}}$ is a subset of $H(U)$ which is uniformly bounded on the compact subsets of U. By (e), $\mathcal{F}$ is locally bounded. Hence, for each $\xi \in U$, there exists a neighbourhood V of ξ such that

$$\|f\|_{\alpha, V} = \sup_{\substack{\phi \in F' \\ |\phi| \le \alpha}} \|\phi \circ f\|_V < \infty$$

and f is locally bounded on U. Since $f \in H_G(U; F)$ it follows that $f \in H(U; F)$ and thus the same result applies to $\mathcal{DFM}$ spaces.

We extend this result to $E \times \mathbb{C}^{\mathbb{N}}$ where E is a $\mathcal{DFM}$ space. To obtain this extension we employ a factorization technique which has proved useful on more than one occasion.

Let U denote an open subset of $E \times \mathbb{C}^{\mathbb{N}}$ and let $f \in H_G(U; F)$ where F is an arbitrary locally convex space. Suppose $\phi \circ f \in H(U)$ for all $\phi \in F'$. Functions of this kind are said to be *weakly holomorphic*. We may suppose, without loss of generality, that F is a normed linear space and that $0 \in U$. Since $E \times \mathbb{C}^n$ is a $\mathcal{DFM}$ space for any positive integer n and sets of the form $V \times \mathbb{C}^{\mathbb{N}}$ where V is an open subset of $\mathbb{C}^n$, n finite, form a neighbourhood basis at the origin in $\mathbb{C}^{\mathbb{N}}$ we may suppose that $U = V \times \mathbb{C}^{\mathbb{N}}$ where V is a convex balanced neighbourhood of the origin in E. Let $\pi_n : \mathbb{C}^{\mathbb{N}} \to \mathbb{C}^{\mathbb{N}}$ denote the projection onto the first n coordinates. We claim that there exists a positive integer n_0 such that $f(x + y) = f(x)$ for all $x \in V$, all $y \in \mathbb{C}^{\mathbb{N}}$, $\pi_{n_0}(y) = 0$. If not there exists a sequence $(x_n)_n$ in V and a sequence $(y_n)_n$ in $\mathbb{C}^{\mathbb{N}}$ such that $\pi_n(y_n) = 0$ and $f(x_n + y_n) \neq f(x_n)$ for all n. For each n the function

$$g_n : \lambda \mapsto f(x_n + \lambda y_n) - f(x_n)$$

is a non-constant entire function, since $g_n(0) \neq g_n(1)$, and hence, by Liouville's Theorem and the Hahn–Banach Theorem, it is unbounded. For each n choose λ_n such that

$$\|f(x_n + \lambda_n y_n) - f(x_n)\| \geq n.$$

By Mackey's Theorem and by taking a subsequence, if necessary, we see there exists $\phi \in F'$ such that

$$|\phi \circ f(x_n + \lambda_n y_n) - \phi \circ f(x_n)| \geq n \qquad (3.17)$$

for all n. Since $\phi \circ f \in H(U)$ there exist neighbourhoods V_1 and W_1 of the origin in E and $\mathbb{C}^{\mathbb{N}}$, respectively, such that

$$\|\phi \circ f\|_{V_1 \times W_1} \leq M < \infty.$$

Since $\pi_{n_0}^{-1}(0) \subset W_1$ for some positive integer n_0 we have

$$|\phi \circ f(v + \lambda y_n)| \leq M$$

for all $v \in V_1$, all $\lambda \in \mathbb{C}$ and all $n \geq n_0$. Liouville's Theorem implies $\phi \circ f(v + \lambda y_n) = \phi \circ f(v)$ all $v \in W_1$, $\lambda \in \mathbb{C}$ and $n \geq n_0$ and, by the identity principle, the same equality is valid for all $v \in V$. This contradicts (3.17) and establishes our claim.

Let $\tilde{\pi}_{n_0} : E \times \mathbb{C}^{\mathbb{N}} \to E \times \mathbb{C}^n$ denote the canonical projection which is the identity on E and projects onto the first n_0 coordinates of $\mathbb{C}^{\mathbb{N}}$. By the above

$$f = f|_{\tilde{\pi}_{n_0}(E \times \mathbb{C}^{\mathbb{N}})} \circ \tilde{\pi}_{n_0}.$$

Since $f\big|_{\widetilde{\pi}_{n_0}(E\times\mathbb{C}^{\mathbb{N}})}$ is $\mathcal{G}$-holomorphic and $\psi\circ f\big|_{\widetilde{\pi}_{n_0}(E\times\mathbb{C}^{\mathbb{N}})}$ is holomorphic for all $\psi\in F'$ the result for $\mathcal{DFM}$ spaces shows that $f\big|_{\widetilde{\pi}_{n_0}(E\times\mathbb{C}^{\mathbb{N}})}$, and hence f, is holomorphic.

Finally we note that the space $\mathbb{C}^{\mathbb{N}}\times\mathbb{C}^{(\mathbb{N})}$, previously discussed in Example 1.27, is covered by this example. The space $\mathbb{C}^{(A)}$, A uncountable, admits weakly holomorphic functions which are not holomorphic (see the references for Exercise 3.114).

We now consider holomorphic versions of the Factorization Lemma proved for polynomials in Chapter 1 (Lemma 1.13). The earlier case was straightforward since polynomials are defined on the entire space and continuity at a single point implies continuity at all points (Corollary 1.12). This is not true of arbitrary holomorphic functions and thus topological and geometric properties of the domain have to be taken into consideration. Factorization results often reduce problems to simpler cases and have already appeared implicitly in our work, e.g. in Examples 1.27 and 3.8(g). The following lemma exposes the main principle in those two examples.

Lemma 3.9 *Let U denote a connected open subset of a locally convex space E, let F be a normed linear space and suppose $f\in H(U;F)$. There exists $\alpha\in \mathrm{cs}(E)$ such that for all $x\in U$, $y\in E$ for which $\alpha(y)=0$ and $\{x+\lambda y: 0\le\lambda\le 1\}\subset U$ we have $f(x+y)=f(x)$. If, in addition, U is balanced then $\pi_\alpha(U)$ is a finitely open subset of E_α and there exists $\widetilde{f}\in H_G\big(\pi_\alpha(U);F\big)$ such that $\pi_\alpha^*(\widetilde{f}):=\widetilde{f}\circ\pi_\alpha=f$.*

Proof. Let ξ denote a fixed point in U. Since F is a normed linear space and f is holomorphic there exists $\alpha\in\mathrm{cs}(E)$ such that $\xi+B_\alpha(1)\subset U$ and $\|f\|_{\xi+B_\alpha(1)}=:M<\infty$. If $x\in\xi+B_\alpha(1)$, $y\in E$ and $\alpha(y)=0$ then $x+\lambda y\in\xi+B_\alpha(1)$ for all $\lambda\in\mathbb{C}$. Hence the function $\lambda\in\mathbb{C}\to f(x+\lambda y)$ is a bounded entire function and, by Liouville's Theorem and the Hahn–Banach Theorem, is constant. This implies $f(x+y)=f(x)$ for all $x\in\xi+B_\alpha(1)$ and all $y\in E$ satisfying $\alpha(y)=0$. Now fix $y\in E$ satisfying $\alpha(y)=0$. Let $U_0=\{x\in U: \text{there exists }\varepsilon>0\text{ and a neighbourhood }V_x\text{ of }x\text{ such that }f(w+\beta y)=f(w)\text{ for all }w\in V_x\text{ and all }|\beta|<\varepsilon\}$. By definition U_0 is open and we have shown above that $\xi\in U_0$. Now suppose $x_\gamma\in U_0\to x\in U$ as $\gamma\to\infty$. Since U is open there exists an γ_0 such that the line segment, L, joining x_{γ_0} to x lies in U and as L is compact there exists a convex balanced neighbourhood W of 0 such that $L+2W\subset U$. Choose $\delta>0$ such that $\beta y\in W$ if $|\beta|<\delta$. The mapping $(z,\beta)\mapsto f(z+\beta y)-f(z)$ is defined on $(L+W)\times\{\beta\in\mathbb{C}: |\beta|<\delta\}$ and since $x_{\gamma_0}\in U_0$ it is zero on a neighbourhood of $(x_{\gamma_0},0)$. By the identity principle it is identically zero. Hence $f(z+\beta y)=f(z)$ for all z in a neighbourhood of x and all $\beta\in\mathbb{C}$, $|\beta|<\delta$ and $x\in U_0$. Since U is connected, $U=U_0$. Now suppose $x+\lambda y\in U$ for $0\le\lambda\le 1$. Since the mapping $\lambda\mapsto f(x+\lambda y)-f(x)$ is holomorphic and constant on a neighbourhood of zero the identity principle shows that $f(x + y) = f(x)$. This completes the proof of the first part of the proposition.

If U is balanced let $\xi = 0$. By the Cauchy estimates

$$\left\| \frac{\widehat{d^n} f(0)}{n!} \right\|_{B_\alpha(1)} \leq M \quad \text{for all } n.$$

By the Factorization Lemma for polynomials

$$\frac{\widehat{d^n} f(0)}{n!}(x + y) = \frac{\widehat{d^n} f(0)}{n!}(x) \tag{3.18}$$

for all $x, y \in E$, $\alpha(y) = 0$. Since π_α is surjective, $\pi_\alpha(U)$ is a finitely open subset of E_α. If $w \in \pi_\alpha(U)$ let

$$\widetilde{f}(w) = \sum_{n=0}^{\infty} \frac{\widehat{d^n} f(0)}{n!}(x) \tag{3.19}$$

where $x \in U$ satisfies $\pi_\alpha(x) = w$. By (3.18), $\widetilde{f}$ is well defined and clearly $f = \widetilde{f} \circ \pi_\alpha$. Using (3.18) and (3.19) one sees that $\widetilde{f}$ is $\mathcal{G}$-holomorphic and this completes the proof.

To obtain a factorization result for holomorphic functions we require $\pi_\alpha(U)$ to be open and $\widetilde{f}$ to be continuous. If π_α is an open mapping then clearly $\pi_\alpha(U)$ is open. Moreover, for any $x \in U$ there exists a neighbourhood V_x of x on which f is bounded. Since $\|\widetilde{f}\|_{\pi_\alpha(V_x)} = \|f\|_{V_x}$, $\widetilde{f}$ is continuous and hence holomorphic. This gives us our first holomorphic factorization result (Example 3.10(a)).

Example 3.10 (a) If E is a locally convex space and $cs(E)$ contains a fundamental directed set D of semi-norms such that π_α is an open mapping for all $\alpha \in D$ then, for any balanced open subset U of E and any normed linear space F,

$$H(U; F) = \bigcup_{\alpha \in D} \pi_\alpha^* \Big(H\big(\pi_\alpha(U); F\big) \Big).$$

(b) A Fréchet space E is called a *quojection* if it admits an increasing fundamental system of semi-norms $(\alpha_n)_n$ such that $E_n := E_{\alpha_n}$ is a Banach space for all n. Quojections have been characterized as the Fréchet spaces which are *quotients of countable products of Banach spaces*. However, there exist quojections which are not linearly isomorphic to a product of Banach spaces, these are called *twisted quojections*. By the Open Mapping Theorem for Banach spaces quojections are precisely the Fréchet spaces to which Example 3.10(a) applies. Hence, if E is a quojection, then

$$H(U; F) = \bigcup_{n} \pi_{\alpha_n}^* \Big(H\big(\pi_{\alpha_n}(U); F\big) \Big)$$

for any balanced open subset U of E and any normed linear space F. As a particular example we see that

$$H(\mathbb{C}^{\mathbb{N}}) = \bigcup_{n \in \mathbb{N}} \pi_n^*(H(\mathbb{C}^n))$$

where π_n is the canonical projection of $\mathbb{C}^{\mathbb{N}}$ onto its first n coordinates.

A similar approach yields the same result for any product of locally convex spaces and this was the method used in our discussion, in Example 3.8(g), of holomorphic functions on $E \times \mathbb{C}^{\mathbb{N}}$. We return to this later when discussing *surjective limits* (in Section 3.6).

The requirement that π_α be open is rather strong and does not apply if, for instance, E admits a continuous norm. By using a global in place of a local approach, we obtain a different type of factorization.

Example 3.11 Let U denote an open subset of the $\mathcal{DFM}$ space E. The set U is *Lindelöf*, i.e. each open cover of U admits a countable subcover, and *hemi-compact*, i.e. U contains a fundamental sequence of compact sets (see Exercise 1.96). For each $\xi \in U$ choose $\alpha_\xi \in cs(E)$ such that $\xi + B_{\alpha_\xi}(1) \subset U$. This implies $U = \bigcup_{\xi \in U} \{\xi + B_{\alpha_\xi}(1)\}$. By the Lindelöf Property there exists a sequence $(\xi_n)_{n=1}^\infty$ in U such that $\bigcup_{n=1}^\infty \{\xi_n + B_{\alpha_{\xi_n}}(1)\} = U$. Let $(B_n)_n$ denote an increasing fundamental system of bounded (or equivalently compact) subsets of E. For each n choose $c_n > 0$ such that $B_n \subset c_n B_{\alpha_{\xi_n}}(1) = B_{\alpha_{\xi_n}}(c_n)$ and let $V = \bigcap_n B_{\alpha_{\xi_n}}(c_n)$. If B is bounded in E then $B \subset B_{n_0}$ for some positive integer n_0. If $n \geq n_0$ then $B \subset B_n \subset B_{\alpha_{\xi_n}}(c_n)$. Hence V absorbs all bounded subsets of E and, by Definition 1.26, V is a neighbourhood of 0. For each n, $\alpha_{\xi_n} \leq c_n \alpha$ where $\alpha := \| \cdot \|_V \in cs(E)$. This implies $B_\alpha(1) \subset B_{\alpha_{\xi_n}}(c_n)$ for all n and hence U is an α-open subset of E. We have shown that each open subset of a $\mathcal{DFM}$ space is *uniformly open* with respect to a continuous semi-norm on E. Now let F denote a normed linear space and suppose $\mathcal{F}$ is a τ_0-bounded subset of $H(U; F)$. By Example 3.8(e), $\mathcal{F}$ is locally bounded and thus we can choose our α's so that $\sup_{f \in \mathcal{F}} \|f\|_{\xi + B_{\alpha_\xi}(1)} < \infty$ for all ξ. Hence $\mathcal{F}$ is a locally bounded subset of $H(U_\alpha; F)$, where U_α is the set U considered as an open subset of (E, α).

We can go further if $U = E$. In this case let $(K_n)_n$ denote a fundamental sequence of compact convex balanced subsets of E such that $nK_n \subset K_{n+1}$ for all n. For $\mathcal{F}$ a τ_0-bounded subset of $H(E; F)$ and $n \in \mathbb{N}$ choose $\alpha_n \in cs(E)$ such that $\sup_{f \in \mathcal{F}} \|f\|_{\overline{K_n + \alpha_n(1)}} < \infty$. Let

$$V = \bigcap_{n \in \mathbb{N}} \{K_n + \frac{1}{n} \overline{B_{\alpha_{n+1}}(1)}\}.$$

If $x \in E$ there exists n_0 such that $x \in K_{n_0} \subset K_n$ for all $n \geq n_0$. Since $K_n + \frac{1}{n}\overline{B_{\alpha_{n+1}}(1)}$ is a neighbourhood of zero there exists $\lambda \in \mathbb{C}$ such that $x \in \lambda\left(K_n + \frac{1}{n}\overline{B_{\alpha_{n+1}}(1)}\right)$ for all $n < n_0$. Hence V is absorbing and, as E is barrelled, V is a neighbourhood of zero. If $m < n$ then

$$mV \subset nV \subset nK_n + \overline{B_{\alpha_{n+1}}(1)} \subset K_{n+1} + \overline{B_{\alpha_{n+1}}(1)}$$

and hence

$$\sup_{f \in \mathcal{F}} \|f\|_{mV} \leq \sup_{f \in \mathcal{F}} \|f\|_{K_{n+1}+\overline{B_{\alpha_{n+1}}(1)}} < \infty.$$

By Lemma 3.9, there exists for each $f \in \mathcal{F}$, $\widetilde{f} \in H(E_\alpha)$, such that $f = \widetilde{f} \circ \pi_\alpha$ and, moreover, the set of holomorphic mappings $(\widetilde{f})_{f \in \mathcal{F}}$ is uniformly bounded on the bounded subsets of E_α. This result can be extended to separable $\mathcal{DFC}$ spaces (see Exercise 3.81).

The two previous examples show that on certain locally convex spaces each holomorphic function "depends" only on a single semi-norm. For balanced domains this leads to a factorization but this may not be the case for arbitrary open sets (see Section 3.6). We say that a locally convex space E *leads to uniform holomorphy* if for each U open in E, each Banach space F and each f in $H(U;F)$ there exists $\alpha \in cs(E)$ and an open covering $\mathcal{C}$ of U such that for each $W \in \mathcal{C}$ there exists an open subset V of E_α containing W and $f_V \in H(V;F)$ with $f_V\big|_W = f\big|_W$. Examples 3.10(b) and 3.11 show that quojections and $\mathcal{DFM}$ spaces lead to uniform holomorphy.

The study of Fréchet spaces which lead to uniform holomorphy establishes connections between factorizations, bounding sets and holomorphic extensions. If E is a Banach space let

$$H_b(E) = \{f \in H(E) : \|f\|_B < \infty \text{ for every bounded subset } B \text{ of } E\}$$
$$= \{f \in H(E) : \text{there exists a neighbourhood } V \text{ of } 0 \text{ such that}$$
$$\|f\|_{\alpha V} < \infty \text{ for all } \alpha \in \mathbb{R}\}.$$

The elements of $H_b(E)$ are called *holomorphic functions of bounded type* and endowed with the topology of uniform convergence on bounded sets $H_b(E)$ is a Fréchet space. In Example 3.8(f) we showed that $H_b(E)$ is a proper subspace of $H(E)$ for any infinite dimensional Banach space E. The two representations of $H_b(E)$ above are possible for a Banach space E since such spaces admit bounded open sets. For an arbitrary locally convex space E let

$$H_b(E) = \{f \in H(E) : \|f\|_B < \infty \text{ for every bounded subset } B \text{ of } E\}$$

and

$$H_{ub}(E) = \{f \in H(E) : \text{there exists a neighbourhood } V \text{ of } 0 \text{ such that}$$
$$\|f\|_{\alpha V} < \infty \text{ for all } \alpha \in \mathbb{R}\}.$$

The functions in $H_{ub}(E)$ are called *holomorphic functions of uniformly bounded type*. We always have the inclusions $H_{ub}(E) \subset H_b(E) \subset H(E)$ and, if E is semi-Montel, then $H_b(E) = H(E)$. By the Cauchy inequalities we have the following characterizations:

$$H_b(E) = \left\{ f \in H(E) : \lim_{n \to \infty} \left\| \frac{\widehat{d^n} f(0)}{n!} \right\|_B^{1/n} = 0 \text{ for every bounded} \right.$$

$$\left. \text{subset } B \text{ of } E \right\}$$

and

$$H_{ub}(E) = \left\{ f \in H(E) \colon \text{there exists a neighbourhood } U \text{ of zero} \right.$$

$$\left. \text{in } E \text{ such that } \lim_{n \to \infty} \left\| \frac{\widehat{d^n} f(0)}{n!} \right\|_U^{1/n} = 0 \right\}.$$

In Example 3.11 we show that $H_{ub}(E) = H(E)$ if E is a $\mathcal{DFM}$ space. The converse is true within the collection of DF spaces (see Exercise 3.78). If $H_{ub}(E) = H(E)$ then clearly E leads to uniform holomorphy (at least for entire functions).

Example 3.12 If E and F are locally convex spaces let $\mathcal{LB}(E; F)$ denote the set of all linear mappings which map a neighbourhood of zero onto a bounded subset of F. Clearly $\mathcal{LB}(E; F) \subset \mathcal{L}(E; F)$ and $\mathcal{L}(E; E) = \mathcal{LB}(E; E)$ if and only if E is a normed linear space.

If E is Fréchet and $T \in \mathcal{L}(E; H(\mathbb{C}))$ then $f(x, y) := [T(x)](y)$, $x \in E$, $y \in \mathbb{C}$, defines a separately holomorphic function on $E \times \mathbb{C}$. By Example 3.8(d), $f \in H(E \times \mathbb{C})$. If $H(E \times \mathbb{C}) = H_{ub}(E \times \mathbb{C})$, there exists a neighbourhood U of 0 in E such that for all $r > 0$

$$\|f\|_{x \in U, |y| < r} = \sup_{x \in U} \|T(x)\|_{\{y \in \mathbb{C}: |y| < r\}} < \infty.$$

Hence T maps a neighbourhood of zero in E onto a bounded subset of $H(\mathbb{C})$ and we have $\mathcal{LB}(E, H(\mathbb{C})) = \mathcal{L}(E; H(\mathbb{C}))$ whenever all holomorphic functions on $E \times \mathbb{C}$ are of uniformly bounded type. Since $H(\mathbb{C})$ is non-normable and $H(\mathbb{C}) \cong H(\mathbb{C}) \times \mathbb{C}$ this shows that $H(H(\mathbb{C})) \neq H_{ub}(H(\mathbb{C}))$.

This can also be seen directly by considering the holomorphic function

$$F : H(\mathbb{C}) \longrightarrow \mathbb{C}$$
$$F(f) = f(f(0)).$$

To prove that F does not factor continuously through $H(\mathbb{C})_\alpha$ for any $\alpha \in \mathrm{cs}(E)$ it suffices to show that $F^{-1}(0)$ is not closed with respect to any $\alpha \in \mathrm{cs}(E)$. We extend this result in Section 6.2 (see also Exercise 3.120).

3.2 Topologies on Spaces of Holomorphic Mappings

In this section we define the τ_0, τ_ω, τ_b, τ_δ and β topologies on $H(U; F)$, U an open subset of a locally convex space E and F a locally convex space, prove some general results and give elementary examples.

Definition 3.13 Let U denote an open subset of a locally convex space E and let F be a locally convex space. The compact open topology on $H(U; F)$ (or the topology of uniform convergence on the compact subsets of U) is the locally convex topology generated by the semi-norms

$$p_{\beta,K}(f) := \|f\|_{\beta,K} = \sup_{x \in K} \beta\big(f(x)\big)$$

where K ranges over the compact subsets of U and β over the continuous semi-norms on F. We denote this topology by τ_0.

We recall two results from several complex variables theory. The first of these provided Nachbin with one of his prime motivations for the introduction of the τ_ω topology on spaces of holomorphic functions over infinite dimensional domains, while the second motivated a sustained study of the same topology and the introduction of bounding sets (Example 3.8(f)). A linear functional on $H(U)$, U an open subset of $\mathbb{C}^n$, is called an *analytic functional*. If the analytic functional T on $H(U)$ is τ_0 continuous then there exists a compact subset K in U and $c > 0$ such that

$$|T(f)| \le c\|f\|_K \tag{3.20}$$

for all $f \in H(U)$. Since holomorphic functions are continuous the space $\{f|_K : f \in H(U)\}$ is a subspace of $\mathcal{C}(K)$, the Banach space of continuous functions on K endowed with the supremum norm over K. By (3.20), T is continuous when $H(U)$ has the topology induced by $\mathcal{C}(K)$. By the Hahn–Banach Theorem T has a norm preserving extension to $\mathcal{C}(K)$ and by the Riesz Representation Theorem there exists a finite Borel measure μ on K, $\|\mu\| \le c$, such that

$$T(f) = \int_K f(x)d\mu(x) \tag{3.21}$$

for all $f \in H(U)$. When (3.21) holds we call μ a representing measure for T. Since the compact set K which arises in (3.20) is not necessarily unique and, moreover, since Hahn–Banach extensions are not always unique, analytic functionals generally admit many representing measures. The Cauchy integral formula

$$f(0) = \frac{1}{2\pi i} \int_{|z|=1} \frac{f(z)}{z} dz$$

shows that δ_0 (the Dirac delta function at the origin) and $\dfrac{dz}{2\pi i z}\big|_{|z|=1}$ both represent the same analytic functional. While considering how to define the support of an analytic functional the following was observed; there exists an analytic functional T on $H(U)$ and a compact subset K of U such that T does not admit any finite Borel representing measure with support in K but every neighbourhood of K admits a finite Borel measure which represents T. One can say that T is almost *supported* by K and the terminology *ported* by K was introduced. In terms of estimates this says, again using the Riesz Representation Theorem, that for every neighbourhood V of K, $K \subset V \subset U$, there exists $c(V) > 0$ such that

$$|T(f)| \le c(V)\|f\|_V \tag{3.22}$$

for all $f \in H(U)$ but there is no estimate of the form

$$|T(f)| \le c\|f\|_K \tag{3.23}$$

for all $f \in H(U)$. These estimates suggested, since infinite dimensional locally convex spaces are not locally compact, studying analytic functionals which were not necessarily τ_0-continuous but satisfied (3.22). By considering semi-norms, rather than analytic functionals, one is led to the τ_ω topology (Definition 3.14). Interest in this topology was initially maintained by its possible application to another question that we now discuss.

A domain U in a locally convex space E is said to be *holomorphically convex* if for each compact subset K of U the set

$$\widehat{K}_{H(U)} := \{z \in U : |f(z)| \le \|f\|_K \text{ for all } f \in H(U)\}$$

is bounded away from the boundary of U. By the Hahn–Banach Theorem, convex domains are holomorphically convex. If E is quasi-complete the Hahn–Banach Theorem implies that $\widehat{K}_{H(U)}$, the *holomorphically convex hull* of K, is a relatively compact subset of E and, in this case, U is holomorphically convex if and only if $\widehat{K}_{H(U)}$ is compact for each compact subset K of U. In particular, we see that U is holomorphically convex if for each $(\xi_n)_n \subset U$, $\xi_n \to \xi \in \partial U$ there exists $f \in H(U)$ such that $\lim_{n \to \infty} |f(\xi_n)| = \infty$.

The converse is true for finite dimensional spaces and included in a result known as the *Cartan–Thullen Theorem*. The proof of this relies on a topological fact and an algebraic trick. In fact, suppose $\xi_n \in U \to \xi \in \partial U$ and $\sup_n |f(\xi_n)| < \infty$ for all $f \in H(U)$. Let

$$p(f) = \sup_n |f(\xi_n)| \tag{3.24}$$

for all $f \in H(U)$. By our hypothesis p is a semi-norm on $H(U)$. Since the mapping $f \to |f(\xi_n)|$ is τ_0 continuous for all n, p is a τ_0 lower semi-continuous semi-norm on $H(U)$. Since $\big(H(U), \tau_0\big)$ is a Fréchet space it is barrelled and p

is τ_0 continuous. Hence there exists $c > 0$ and a compact subset K of U such that $p(f) \leq c\|f\|_K$ for all $f \in H(U)$. Now $p(f^n) = p(f)^n$ and $\|f^n\|_K = \|f\|_K^n$ for all $f \in H(U)$. Hence

$$p(f)^n = p(f^n) \leq c\|f^n\|_K = c\|f\|_K^n$$

and taking n^{th} roots we obtain $p(f) \leq c^{1/n}\|f\|_K$ for all $f \in H(U)$. Letting $n \to \infty$ we see that $p(f) \leq \|f\|_K$ for all $f \in H(U)$. Hence $\xi_n \in \widehat{K}_{H(U)}$ for all n and U is not holomorphically convex.

The semi-norm defined by (3.24) motivated the introduction of bounding sets (Example 3.8(f)). It is easily seen (Exercise 3.85) that $(H(U), \tau_0)$ is not barrelled, if U is an open subset of an infinite dimensional Banach space, and so the above proof does not immediately generalize. If, however, $(H(U), \tau_\omega)$ is barrelled we can modify the proof as follows.

The semi-norm defined in (3.24) is τ_ω continuous (Definition 3.14 below) and so there exists a compact subset K of U such that for each open set V, $K \subset V \subset U$, there exists $c(V) > 0$ such that $p(f) \leq c(V)\|f\|_V$ for all $f \in H(U)$. We have

$$p(f)^n = p(f^n) \leq c(V)\|f^n\|_V = c(V)\|f\|_V^n$$

for all $f \in H(U)$. Hence

$$p(f) \leq \lim_{n \to \infty} c(V)^{1/n}\|f\|_V = \|f\|_V$$

and, as this holds for every neighbourhood V of K, we have

$$p(f) \leq \lim_{V \to K} \|f\|_V = \|f\|_K.$$

This also shows that the τ_0 and τ_ω multiplicative linear functions on $H(U)$ coincide for U an arbitrary open subset of a locally convex space.

The Cartan–Thullen Theorem is associated with concepts such as analytic continuation, envelopes of holomorphy and holomorphically convex domains. As mentioned above, it was initially hoped that the τ_ω topology would play a role in the development of these topics in an infinite dimensional setting. This hope did not materialize, and in fact the opposite happened – as we shall see in Chapter 5. Relationships between pseudo-convex domains, holomorphically convex domains and the envelope of holomorphy were developed independently of the τ_ω topology and these results led to a clarification of the relationship between the τ_0 and the τ_ω topologies on certain Fréchet spaces. Having motivated, and in fact used, the τ_ω topology we now formally introduce it.

Definition 3.14 Let U denote an open subset of a locally convex space E and let F be a normed linear space. A semi-norm p on $H(U; F)$ is ported by the compact subset K of U if for every open set V, $K \subset V \subset U$, there exists $c(V) > 0$ such that

$$p(f) \leq c(V)\|f\|_V$$

for all $f \in H(U; F)$. The τ_ω topology on $H(U; F)$ is the topology generated by the semi-norms ported by the compact subsets of U.

Definition 3.15 Let U denote an open subset of a locally convex space E and let F denote a locally convex space. The τ_ω topology on $H(U; F)$ is defined by

$$\big(H(U; F), \tau_\omega\big) = \varprojlim_{\alpha \in cs(F)} \big(H(U; F_\alpha), \tau_\omega\big).$$

We call τ_ω the *Nachbin* or *ported* topology on $H(U; F)$.

We also introduce here the τ_δ topology – the topology generated by the countable open covers. This helps us investigate the τ_ω topology and, at times, when the τ_ω topology does not have suitable properties we look to the τ_δ topology. We shall see, in Corollary 3.37, that, on balanced domains, τ_δ is the barrelled topology associated with τ_ω. The τ_δ topology enjoys good topological properties but like τ_ω, and unlike τ_0, its most common description includes a generating system of semi-norms described by means of *inequalities*. We always have $\tau_0 \leq \tau_\omega \leq \tau_\delta$ and apart from the obvious problem of investigating when some or all of these topologies coincide there are the specific problems of seeing which properties of τ_0 or τ_δ are common to τ_ω and of producing concrete sets of semi-norms which generate the τ_ω and the τ_δ topologies – this is sometimes referred to as the problem of finding a *projective description* of the topologies. Later using a predual of $\big(H(U; F), \tau_\delta\big)$ we define the β topology.

Definition 3.16 Let U denote an open subset of a locally convex space E and let F denote a normed linear space. A semi-norm p on $H(U; F)$ is τ_δ continuous if for each increasing countable open cover of U, $(V_n)_{n=1}^\infty$, there exist a positive integer n_0 and $c > 0$ such that

$$p(f) \leq c\|f\|_{V_{n_0}}$$

for every $f \in H(U; F)$. The τ_δ topology on $H(U; F)$ is the locally convex topology generated by the τ_δ continuous semi-norms. If F is an arbitrary locally convex space we let

$$\big(H(U; F), \tau_\delta\big) = \varprojlim_{\beta \in cs(F)} \big(H(U; F_\beta), \tau_\delta\big).$$

The relationship between the different topologies is given in the following lemma.

Lemma 3.17 *Let E and F be locally convex spaces and let U denote an open subset of E. On $H(U;F)$ we have*

$$\tau_0 \leq \tau_\omega \leq \tau_\delta.$$

Proof. We may suppose, without loss of generality, that F is a normed linear space. Since $\|f\|_K \leq \|f\|_V$ for every V containing K we have $\tau_\omega \geq \tau_0$. Now suppose p is a τ_ω continuous semi-norm on $H(U;F)$ ported by the compact subset K of U. Let $(V_n)_{n=1}^\infty$ denote an increasing countable open cover of U. Since K is compact there exists n_0 such that V_{n_0} is a neighbourhood of K. Hence there exists $c > 0$ such that $p(f) \leq c\|f\|_{V_{n_0}}$ for every $f \in H(U;F)$. This shows that p is τ_δ continuous and completes the proof.

A locally convex space is *ultrabornological* if it is an inductive limit of Banach spaces in the category of locally convex spaces and continuous linear mappings. Ultrabornological spaces are barrelled and bornological. Our next result shows that the τ_δ topology has good topological properties when the range space is a Banach space (a slightly less general result holds when the range space is a normed linear space). The proof contains an alternative description of the τ_δ topology.

Proposition 3.18 *Let U be an open subset of a locally convex space E and let F be a Banach space. Then $\big(H(U;F),\tau_\delta\big)$ is an inductive limit of Fréchet spaces and hence it is barrelled, bornological and ultrabornological.*

Proof. For each increasing countable open cover $\mathcal{V} := (V_n)_{n=1}^\infty$ of U, let $H_\mathcal{V}(U;F) = \{f \in H(U;F) : \|f\|_{V_n} < \infty \text{ for all } n\}$. We endow $H_\mathcal{V}(U;F)$ with the topology generated by the semi-norms $p_n(f) := \|f\|_{V_n}$ for all $f \in H_\mathcal{V}(U;F)$. $H_\mathcal{V}(U;F)$ is a metrizable locally convex space. It is, in fact, a Fréchet space since F is a Banach space and locally bounded $\mathcal{G}$-holomorphic functions are holomorphic. We claim that $H(U;F) = \bigcup_\mathcal{V} H_\mathcal{V}(U;F)$ where $\mathcal{V}$ ranges over all increasing countable open covers of U. If $f \in H(U;F)$ let $W_n = \{x \in U : \|f(x)\| < n\}$. As f is continuous W_n is open and hence $\mathcal{W} = (W_n)_n$ is an increasing countable open cover of U. Since $f \in H_\mathcal{W}(U;F)$ and $H_\mathcal{W}(U;F) \subset H(U;F)$ this establishes our claim. We let

$$\big(H(U;F),\tau_i\big) = \varinjlim_\mathcal{V} H_\mathcal{V}(U;F).$$

Since Fréchet spaces are ultrabornological and an inductive limit of ultrabornological spaces is ultrabornological, $\big(H(U;F),\tau_i\big)$ is ultrabornological. To complete the proof we show that $\tau_i = \tau_\delta$ on $H(U;F)$. Since the identity mapping from $H_\mathcal{V}(U;F)$ into $\big(H(U;F),\tau_\delta\big)$ is continuous for any countable increasing open cover $\mathcal{V}$ of U, $\tau_i \geq \tau_\delta$.

Let p denote a τ_i continuous semi-norm on $H(U;F)$. Suppose p is not τ_δ continuous. Then there exists an increasing countable open cover of U, $\mathcal{V} = (V_n)_{n=1}^\infty$ and a sequence $(f_n)_n$ in $H(U;F)$ such that $\|f_n\|_{V_n} \leq 1$ and $p(f_n) \geq n$ for all n. Let $W_n = \{x \in U : \|f_m(x)\| \leq n \text{ for all } m\}$ and $\mathcal{W} = (W_n^\circ)_{n=1}^\infty$ where W_n° denotes the interior of W_n. Since $\|f_m\|_{V_n} \leq 1$ for $m \geq n$ and each holomorphic function is locally bounded, $\mathcal{W}$ is an increasing countable open cover of U. Hence $p\big|_{H_\mathcal{W}(U;F)}$ is continuous and there exists a positive integer n_0 and $c > 0$ such that $p(f_n) \leq c\|f_n\|_{W_{n_0}}$ for all n. This implies

$$n \leq p(f_n) \leq c\|f_n\|_{W_{n_0}} \leq c\, n_0$$

for all n. Since this is impossible, p is τ_δ continuous and hence $\tau_i = \tau_\delta$. This completes the proof.

In proving Proposition 3.18 we showed that locally bounded subsets of $H(U;F)$ are τ_δ bounded sets. As previously noted in Chapter 1 the locally bounded sets are defined using the locally convex space stuctures of E and F and do not require a similar structure on $H(U;F)$. Nevertheless, it is often useful to be able to identify the locally bounded subsets of $H(U;F)$ with the bounded subsets of a locally convex space structure on $H(U;F)$. In the next proposition we see that this leads to a further description of the τ_δ topology. We recall that an inductive limit in the category of locally convex spaces and continuous linear mappings $E = \varinjlim_{\alpha} E_\alpha$ is *regular* if each bounded subset of E is contained and bounded in some E_α.

Proposition 3.19 *If U is an open subset of a locally convex space E and F is a Banach space then the following are equivalent:*

(a) the τ_δ bounded subsets of $H(U;F)$ are locally bounded,

(b) $\big(H(U;F),\tau_\delta\big) = \varinjlim_{\mathcal{V}} H_\mathcal{V}(U;F)$ is a regular inductive limit,

(c) there exists a locally convex topology τ on $H(U;F)$ such that the τ bounded subsets of $H(U;F)$ coincide with the locally bounded subsets of $H(U;F)$.

When (c) holds we have $\tau^{bor} = \tau_\delta$.

Proof. Since locally bounded sets are always τ_δ bounded and the bounded subsets of $H_\mathcal{V}(U;F)$ are locally bounded for any increasing countable open cover $\mathcal{V}$ of U the definition of regularity shows that (a) and (b) are equivalent. Clearly, (a)$\Longrightarrow$(c). Now suppose (c) is satisfied. Since τ and τ^{bor} have the same bounded sets, the τ^{bor} bounded subsets of $H(U;F)$ are locally bounded. Hence the identity mapping from $\big(H(U;F),\tau^{bor}\big)$ into $\big(H(U;F),\tau_\delta\big)$ maps bounded sets onto bounded sets and is continuous. This shows $\tau^{bor} \geq \tau_\delta$. If $\mathcal{V}$ is an increasing countable open cover of E then the inclusion mapping from $H_V(U;F)$ into $(H(U;F), \tau^{bor})$ maps bounded sets onto bounded sets and, since $H_V(U;F)$ is metrizable and hence bornological, it is continuous. By the definition of inductive limits the identity mapping

$$(H(U;F), \tau_\delta) = \varinjlim_V H_V(U;F) \longrightarrow (H(U;F), \tau^{bor})$$

is continuous. Hence $\tau_\delta \geq \tau^{bor}$. This shows that $\tau^{bor} = \tau_\delta$ and (c) implies (a). This completes the proof.

Example 3.20 (a) Let U denote an open subset of a metrizable locally convex space and let F denote a normed linear space. We claim that any τ_0 bounded subset $\mathcal{F}$ of $H(U;F)$ is locally bounded. If not, there exists $\xi \in U$ such that for every open set V, $\xi \in V \subset U$, we have $\sup_{f \in \mathcal{F}} \|f\|_V = \infty$. If $(V_n)_n$ is a fundamental system of neighbourhoods of ξ then for each $n \in \mathbb{N}$ there exists $\xi_n \in V_n$ and $f_n \in \mathcal{F}$ such that $\|f_n(\xi_n)\| \geq n$. The set $K := \{\xi_n\}_{n=1}^{\infty} \cup \{\xi\}$ is a compact subset of U. Since

$$\sup_{f \in \mathcal{F}} \|f\|_K \geq \sup_n \|f_n(\xi_n)\| = \infty$$

this is impossible and establishes our claim. Since $\tau_0 \leq \tau_\omega \leq \tau_\delta$ this shows that $\tau_0^{bor} = \tau_\omega^{bor} = \tau_\delta$ on $H(U;F)$.

(b) In Example 3.8(e) we proved that the τ_0 bounded subsets of $H(U;F)$, U an open subset of a $\mathcal{DFM}$ space, and F a normed linear space, are locally bounded. Hence τ_0, τ_ω and τ_δ define the same bounded subsets of $H(U;F)$. As every open subset of a $\mathcal{DFM}$ space admits a countable fundamental system of compact sets it follows that $(H(U;F), \tau_0)$ is a metrizable locally convex space. Since locally bounded $\mathcal{G}$-holomorphic functions are holomorphic, $(H(U;F), \tau_0)$ is quasi-complete and a Fréchet space and, as τ_0 and τ_δ are both bornological spaces with the same bounded sets, we have $\tau_0 = \tau_\omega = \tau_\delta$ on $H(U;F)$.

(c) (see Example 3.8(f)) Let A denote a closed bounding subset of a quasi-complete locally convex space E and let $p(f) = \|f\|_A$ for all $f \in H(E)$. Let $V = \{f \in H(E) : p(f) \leq 1\} = \bigcap_{x \in A} \{f \in H(E) : |f(x)| \leq 1\}$. The set V is τ_0, and hence τ_δ, closed and is clearly convex and balanced. Since A is bounding, V is absorbing. Since $(H(E), \tau_\delta)$ is barrelled, V is a τ_δ-neighbourhood of zero and hence p is τ_δ-continuous. If p is τ_ω-continuous there exists a convex balanced compact subset K of E such that for every V open, $K \subset V$, we can find $c(V) > 0$ such that $\|f\|_A \leq c(V)\|f\|_V$ for all $f \in H(E)$. An application of the algebraic structure of $H(E)$ implies, as in our discussion of holomorphic convexity, that $\|f\|_A \leq c(V)^{1/n}\|f\|_V$ for all n and hence $\|f\|_A \leq \|f\|_V$ for all $V \supset K$ and all $f \in H(E)$. Hence $\|f\|_A \leq \lim_{V \to K} \|f\|_V = \|f\|_K$ and, by the Hahn–Banach Theorem, $A \subset K$. This implies that p is τ_ω-continuous if and only if A is compact.

We now suppose that E is separable. If $(x_n)_{n=1}^\infty$ is a dense sequence in E and $\alpha \in \mathrm{cs}(E)$ then for every $\varepsilon > 0$ the sequence $(B_{\alpha,\varepsilon,n})_{n=1}^\infty$, where $B_{\alpha,\varepsilon,n} = \bigcup_{i=1}^n B_\alpha(x_i, \varepsilon)$, is an increasing countable open cover of E. Since p is τ_δ-continuous there exists $c > 0$ and a positive integer n_0 such that

$$p(f) = \|f\|_A \le c\|f\|_{B_{\alpha,\varepsilon,n_0}}$$

for all $f \in H(E)$. Hence

$$\|f\|_A = \|f^n\|_A^{1/n} \le c^{1/n}\|f^n\|_{B_{\alpha,\varepsilon,n_0}}^{1/n} = c^{1/n}\|f\|_{B_{\alpha,\varepsilon,n_0}}$$

for all $f \in H(E)$ and all n. This implies $\|f\|_A \le \|f\|_{B_{\alpha,\varepsilon,n_0}}$ for all $f \in H(E)$ and, by the Hahn–Banach Theorem, $A \subset \overline{\Gamma}(B_{\alpha,\varepsilon,n_0})$. Since $\overline{\Gamma}(B_{\alpha,\varepsilon,n_0})$ can be covered by a finite number of α-balls of radius 2ε it follows that A is compact and we have shown that closed bounding subsets of a quasi-complete separable locally convex space are compact.

In Example 3.8(g) we gave three examples where weakly holomorphic functions were holomorphic. An alternative proof for the first two follows from the above example and the following proposition.

Proposition 3.21 *Let U denote an open subset of a locally convex space such that τ_0 bounded subsets of $H(U)$ are locally bounded. If F is an arbitrary locally convex space then $H(U; F) = H(U; F_\sigma)$, $F_\sigma := \big(F, \sigma(F, F')\big)$.*

Proof. We may suppose, without loss of generality, that F is a normed linear space. Since $H(U; F) \subset H(U; F_\sigma)$ we consider $f \in H(U; F_\sigma)$. Let B denote the unit ball of F_β'. The set $\mathcal{F} := (\phi \circ f)_{\phi \in B}$ lies in $H(U)$ and, if K is a compact subset of U, then

$$\sup_{g \in \mathcal{F}} \|g\|_K = \sup_{\phi \in B} \|\phi \circ f\|_K = \|f\|_K < \infty.$$

Hence $\mathcal{F}$ is a τ_0 and, consequently, a locally bounded subset of $H(U)$. If $\xi \in U$ then there exists a neighbourhood V_ξ of ξ such that $\sup_{\phi \in B} \|\phi \circ f\|_{V_\xi} = \|f\|_{V_\xi} < \infty$. Hence f is locally bounded. This completes the proof.

Proposition 3.21 *cannot* be applied to the space $E := \mathbb{C}^{\mathbb{N}} \times \mathbb{C}^{(\mathbb{N})}$, which is a special case of the final example considered in Example 3.8(g), since, by Example 1.27, τ_0 and τ_ω do *not* define the same bounded subsets of $\mathcal{P}(^2E)$. Since the locally bounded subsets of $\mathcal{P}(^2E)$ are τ_ω bounded (Lemma 1.23) and $\tau_0 \le \tau_\omega$, there exists a τ_0 bounded subset $\mathcal{F}$ of $\mathcal{P}(^2E)$ which is not locally bounded. By Proposition 1.11 the set $(P|_U)_{P \in \mathcal{F}}$ is not locally bounded for any open subset U of E but clearly is τ_0 bounded.

The patient reader who has stayed with us may have noticed the apparently rather loose strands connecting some of the topics discussed and wondered if a more unified approach was possible or desirable. We pause here to make a number of heuristic remarks which may be regarded as part of the folklore of the subject. Hopefully, these will clarify our purpose and approach and, at the same time, justify our choice of material. We shall see that it is possible to provide a more unified approach but that this leads to severe restrictions on the scope of our work. Our aim is to study holomorphic functions on infinite dimensional spaces and one aspect of this study proceeds by considering locally convex structures on $H(U;F)$ even if, a priori, our main interest lies in a different direction. In many situations information on certain subspaces of a locally convex space leads to insight into the structure of the space itself – we have seen this already in Chapter 1 when we discussed the local theory of Banach spaces. This insight may be technical or intuitive, it may show us what is possible, likely or impossible and all of these may be helpful. We now look at certain locally convex spaces which always appear as complemented subspaces of the space of all holomorphic functions. For the sake of simplicity we confine ourselves to entire functions although similar remarks apply to the general case.

If E is a locally convex space then $(H(E),\tau)$, $\tau = \tau_0$, τ_ω or τ_δ, contains a number of special complemented subspaces and these have influenced our approach. If n is a positive integer then $(\mathcal{P}(^nE),\tau)$ and $(H(\mathbb{C}^n),\tau)$ are closed complemented subspaces of $(H(E),\tau)$ (Proposition 3.22). Moreover, $\tau_\omega = \tau_\delta$ on $\mathcal{P}(^nE)$ (Proposition 3.22(b)) and $\tau_0 = \tau_\omega = \tau_\delta$ on $H(\mathbb{C}^n)$ (Example 3.20(b)). Thus, in studying $(H(E),\tau)$, we are studying a superspace of

$$(H(\mathbb{C}^n),\tau_0), \ E'_c \text{ or } E'_\beta \text{ or } E'_i, \ \left(\bigotimes_{n,\pi,s} E\right)'_\tau \text{ or } (\mathcal{P}(^nE),\tau).$$

Since many properties of a locally convex space are inherited by complemented subspaces we expect $H(E)$ to possess only those properties shared by the above spaces. One is thus more likely to obtain positive results if these spaces have the *same* or *similar* structures. Since $(H(\mathbb{C}^n),\tau_0)$ is a Fréchet nuclear space we are first led to consider the case where E'_β, say, is also a Fréchet nuclear space. This is the case when E is a $\mathcal{DFN}$ space.

Results on the $\bar{\partial}$ problem, sheaf cohomology and the Levi problem on infinite dimensional spaces confirm what we have ourselves seen, for instance in Examples 1.25, 2.18 and 3.8, i.e. holomorphic functions on $\mathcal{DFN}$ spaces possess many good properties and, in looking for positive results in an infinite dimensional setting, one should first consider these spaces. A unified, simplified and self-contained approach to the theory is possible for such spaces since the only locally convex structures that arise in this context are $\mathcal{DFN}$ and Fréchet nuclear. Such a theory is significant and important but, by definition, omits the interesting theory of holomorphic functions on Banach spaces and we were not willing to accept such a restriction. Nevertheless, two useful general principles or, more precisely, guidelines, can be drawn from the above remarks. On a practical level, and this is borne out by the examples cited above, locally convex spaces with structures close to those of $\mathcal{DFN}$ spaces, for example $\mathcal{DFS}$ and $\mathcal{DFM}$ spaces, are more likely to yield positive results than spaces that might from the linear point of view alone be regarded as good spaces. In particular, this suggests a strong difference between the theories for Fréchet spaces and DF spaces. Secondly, we note that *any study of holomorphic functions will intrinsically involve nuclearity*. As a corollary one might expect the theory on nuclear and dual nuclear spaces to possess some special features and indeed this turns out to be the case.

The space of holomorphic functions on an infinite dimensional Banach space E, $\big(H(E), \tau_\omega\big)$, contains as complemented subspaces the Banach spaces $\mathcal{P}(^n E)$ and E' and the Fréchet nuclear spaces $H(\mathbb{C}^n)$. Despite the fact that the only locally convex spaces which are both Fréchet nuclear and Banach are finite dimensional an interesting theory has been developed for such spaces. Since the space of homogeneous polynomials on a Banach space, endowed with the τ_ω topology, is again a Banach space we could also have restricted ourselves to the study of polynomials and obtained another self-contained and interesting theory within the framework of Banach space theory. Sections 1.3, 2.1, 2.3 and 2.4, which are almost self-contained, indicate one important aspect of such a theory. Another aspect of infinite dimensional holomorphy which still retains the Banach space setting is the study of the space $H^\infty(U)$ – the bounded holomorphic functions on an open subset of a Banach space. Even within these purely Banach space settings we still do not manage to avoid nuclearity as we find nuclear mappings appearing in the duality theory for spaces of polynomials on a Banach space (Proposition 2.27).

Thus we see that the study of the full space of holomorphic functions on a Banach space leads inevitably to an interaction between nuclear and Banach space structures and we have chosen to present the theory in a framework which included both. This may give a somewhat disjointed appearance, at times, but we feel this is merely a temporary setback and that eventually a unified approach will appear – indeed this is already emerging – with the (BB)-property providing a link between Banach and Fréchet nuclear spaces.

We have listed separately the space E' as a complemented subspace since it has been more extensively studied within the linear framework and thus we often know much more about its structure than about the structure of spaces of homogeneous polynomials of degree greater than or equal to two. As a first approximation the space E' does provide a better indication than E of what to expect from $H(E)$ (see Exercise 6.59). For example, it is well known, and we implicitly used this in our local study of Banach spaces, that the presence of l_∞ often leads to negative linear results and counterexamples. Thus one might expect more extreme behaviour from $H(E)$ when E' contains l_∞, or when E contains l_1, and should not be too surprised to see that l_1 provided the largest possible polarization constant (Example 1.39 and Propositions 1.41

and 1.42) and that the non-containment of l_1 in E leads to more satisfactory results (Proposition 2.36). One, of course, has to be careful in accepting such general statements and we shall see in Proposition 4.56 that the monomials form an unconditional basis for $(H(l_1), \tau_0)$. Finally, within the circle of ideas we have just discussed, we mention that we shall soon see that $(H(E), \tau)$ itself often has the structure of a dual space and this fact can sometimes be used to advantage.

We return now to the study of topologies and begin by proving some of the assertions on which we based the above remarks.

Proposition 3.22 *Let U denote an open subset of a locally convex space E and let F denote a Banach space.*

(a) *For any positive integer n, $(P(^nE; F), \tau_0)$ is a closed complemented subspace of $(H(U; F), \tau_0)$.*

(b) *For any positive integer n, $(P(^nE; F), \tau_\omega)$ is a closed complemented subspace of $(H(U; F), \tau)$, $\tau = \tau_\omega$ or τ_δ, and in particular τ_ω and τ_δ coincide on $P(^nE; F)$.*

(c) *If $dimension(E) \geq n$, n a positive integer, then $(H(\mathbb{C}^n), \tau_0)$ is a closed complemented subspace of $(H(E; F), \tau)$, $\tau = \tau_0$, τ_ω or τ_δ.*

Proof. Since $(H(U; F), \tau) \cong (H(U - \xi, F), \tau)$ for any $\xi \in E$ and any of the topologies we are considering we may suppose $0 \in U$. Consider the mapping

$$\frac{\widehat{d^n}}{n!} : H(U; F) \longrightarrow H(U; F)$$

$$f \longrightarrow \frac{\widehat{d^n} f(0)}{n!}$$

This is a linear mapping and, since $\frac{\widehat{d^n} P(0)}{n!} = P$ for $P \in P(^nE; F)$, $\frac{\widehat{d^n}}{n!}$ is a projection from $H(U; F)$ into $H(U; F)$ with range $P(^nE; F)$ (we use the restriction mapping to identify $P(^nE; F)$ with a subspace of $H(U; F)$). Since uniform convergence on compact subsets of E is equivalent to uniform convergence on the compact subsets of a neighbourhood of the origin for elements of $P(^nE; F)$, $(H(U; F), \tau_0)$ induces the compact open topology on $P(^nE; F)$. Let V denote a convex balanced open subset of E which lies in U. If K is a compact balanced subset of V then the Cauchy inequalities, (3.12), imply

$$\left\| \frac{\widehat{d^n} f(0)}{n!} \right\|_K \leq \|f\|_K.$$

Hence $\frac{\widehat{d^n}}{n!}$ is continuous with respect to the compact open topology and this completes the proof of (a).

If p is a τ_ω-continuous semi-norm on $\mathcal{P}(^nE;F)$ then for every convex balanced neighbourhood of 0, W, contained in V there exists $c(W) > 0$ such that

$$p\left(\frac{\widehat{d^n}f(0)}{n!}\right) \le c(W)\left\|\frac{\widehat{d^n}f(0)}{n!}\right\|_W \le c(W)\|f\|_W$$

for every f in $H(U;F)$. Hence $\dfrac{\widehat{d^n}}{n!}$ is continuous with respect to the τ_ω topologies. To complete the proof of (b) it suffices to show that $\big(H(U;F),\tau_\delta\big)$ induces the τ_ω topology on $\mathcal{P}(^nE;F)$. Let p denote a τ_δ continuous semi-norm on $H(U;F)$ and let V_1 denote a convex balanced neighbourhood of zero in E. The sequence $(U \cap nV_1)_{n=1}^\infty$ is an increasing countable open cover of U. Hence there exist a positive integer m and $c > 0$ such that

$$\widetilde{p}(f) := p\left(\frac{\widehat{d^n}f(0)}{n!}\right) \le c\left\|\frac{\widehat{d^n}f(0)}{n!}\right\|_{U\cap mV_1} \le cm^n\left\|\frac{\widehat{d^n}f(0)}{n!}\right\|_{V_1} \le cm^n\|f\|_{V_1}$$

for all $f \in H(U;F)$. This shows that $\widetilde{p}$ is τ_ω continuous on $H(U;F)$ and, as we have already proved that $\dfrac{\widehat{d^n}}{n!}$ is a continuous projection with respect to the τ_ω topology, $p\big|_{\mathcal{P}(^nE;F)}$ is τ_ω continuous. This completes the proof of (b).

If n is a positive integer then we can find subspaces E_i and F_i of E and F, respectively, such that $E = E_1 \oplus E_2$, $F = F_1 \oplus F_2$, dimension$(E_1) = n$, dimension$(F_1) = 1$. Let q denote a continuous projection from F into F_1. If $f \in H(E;F)$ let $\pi f(x_1 + x_2) = q(f(x_1))$ for any $x_1 \in E_1$ and $x_2 \in E_2$. It is easily checked that π is a continuous projection from $\big(H(E;F),\tau\big)$ onto its range and that the range is linearly isomorphic to $\big(H(\mathbb{C}^n),\tau_0\big)$. This completes the proof of (c).

Proposition 3.22 allows us to lift certain results from spaces of homogeneous polynomials to spaces of holomorphic functions. For instance, Example 1.27 shows that $\tau_0 \underset{\neq}{<} \tau_\omega$ on $H(U)$ for any open subset U of $\mathbb{C}^{\mathbb{N}} \times \mathbb{C}^{(\mathbb{N})}$ and, since no infinite dimensional Banach space is Montel, $\tau_0 \underset{\neq}{<} \tau_\omega$ on $H(U;F)$ for any open subset U of an infinite dimensional Banach space and any locally convex space F (see Exercise 3.85). By Proposition 3.22(b) homogeneous polynomials do not distinguish between the τ_ω and τ_δ topologies and thus the results from Chapter 1 cannot be used to provide an example where $\tau_\omega \ne \tau_\delta$. We now give such an example. The key technique in the proof has already been used in Examples 1.27, 3.8(g) and 3.10 and in Lemma 3.9.

Definition 3.23 A sequence $(x_n)_n$ in a locally convex space E is *very strongly convergent* if $\lambda_n x_n \to 0$ in E as $n \to \infty$ for any sequence of scalars $(\lambda_n)_n$. The sequence is said to be *non-trivial* if $x_n \ne 0$ for each n.

The standard example of a non-trivial very strongly convergent sequence is the canonical unit vector basis for $\mathbb{C}^{\mathbb{N}}$ and, in fact, E admits a non-trivial very strongly convergent sequence if and only if E contains a basic sequence which is equivalent to the standard unit vector basis for $\mathbb{C}^{\mathbb{N}}$. If E is sequentially complete then E admits a non-trivial very strongly convergent sequence if and only if it contains $\mathbb{C}^{\mathbb{N}}$ as a complemented subspace (Exercise 3.87). In this case, following the proof of Proposition 3.22(c), we see that $(H(E), \tau)$ contains a complemented subspace isomorphic to $(H(\mathbb{C}^{\mathbb{N}}), \tau)$, $\tau = \tau_0$, τ_ω or τ_δ – a result which extends Proposition 3.22(c).

Example 3.24 (a) Let E denote a locally convex space which contains a non-trivial very strongly convergent sequence. By using a diagonal process and taking subsequences, if necessary, we can find a non-trivial very strongly convergent sequence $(x_n)_{n=0}^{\infty}$ in E and a sequence $(\phi_n)_{n=0}^{\infty}$ in E' such that $\phi_n(x_m) \neq 0$ if and only if $n = m$. Let $y = x_0$. For each f in $H(E)$ consider the sum

$$p(f) := \sum_{n=1}^{\infty} |f(ny + x_n) - f(ny)|.$$

If $f \in H(E)$ then, by Lemma 3.9, there exists an $\alpha \in \mathrm{cs}(E)$ such that $f(x + w) = f(x)$ for all x and w in E with $\alpha(w) = 0$. Since $(x_n)_n$ is very strongly convergent there exists a positive integer n_α such that $\alpha(x_n) = 0$ for all $n \geq n_\alpha$. Hence $f(ny + x_n) = f(ny)$ for all $n \geq n_\alpha$ and $p(f)$ is finite for every f in $H(E)$. The semi-norm

$$f \longrightarrow |f(ny + x_n) - f(ny)|$$

is τ_0 and hence τ_δ continuous on $H(E)$ and, since $(H(E), \tau_\delta)$ is barrelled, p is a τ_δ continuous semi-norm on $H(E)$.

We improve this result by showing that p is bounded on τ_0 bounded subsets of $H(E)$, i.e. p is a τ_0^{bor} continuous semi-norm on $H(E)$. Let B denote a τ_0 bounded subset of $H(E)$. We first show there exists a positive integer n_0 such that $f(\lambda y + x_n) = f(\lambda y)$ for all $\lambda \in \mathbb{C}$, all $f \in B$ and all $n \geq n_0$. If not, then by using subsequences if necessary, we can choose for each n, $\lambda_n \in \mathbb{C}$, and $f_n \in B$ such that $f_n(\lambda_n y + x_n) \neq f_n(\lambda_n y)$. For each n let

$$g_n(\lambda) = f_n(\lambda y + x_n) - f_n(\lambda y)$$

for $\lambda \in \mathbb{C}$. Then $g_n \in H(\mathbb{C})$ and since $g_n(\lambda_n) \neq 0$, $g_n \not\equiv 0$. By the identity principle for holomorphic functions of one complex variable we can select a sequence of complex numbers $(\beta_n)_n$ such that $|\beta_n| \leq 1/n^2$ and $g_n(\beta_n) \neq 0$ for all n. For each n and each w in $\mathbb{C}$ let

$$h_n(w) = f_n(\beta_n y + w x_n) - f_n(\beta_n y).$$

Then $h_n \in H(\mathbb{C})$ and $h_n(0) \neq h_n(1)$. By Liouville's Theorem we can choose a sequence of complex numbers, $(w_n)_n$, such that $|h_n(w_n)| > n + |f_n(\beta_n y)|$ for all n. Hence $|f_n(\beta_n y + w_n x_n)| > n$ for all n. Since $(x_n)_n$ is a very strongly convergent sequence, $(w_n x_n)_n$ is a null sequence in E and $|\beta_n| \leq 1/n^2$ for all

n implies $\beta_n y + w_n x_n \to 0$ as $n \to \infty$. Hence $K_1 := \{0\} \cup \{\beta_n y + w_n x_n\}_{n=1}^{\infty}$ is a compact subset of E and, for all n, we have

$$\|f_n\|_{K_1} \geq |f_n(\beta_n y + w_n x_n)| > n.$$

This contradicts the fact that B is a τ_0 bounded subset of E. Hence there exists a positive integer n_0 such that

$$f(\lambda y + x_n) = f(\lambda y)$$

for all $n \geq n_0$, all $\lambda \in \mathbb{C}$ and all f in B. This shows that

$$\sup_{f \in B} \sum_{n=0}^{\infty} |f(ny + x_n) - f(ny)| = \sup_{f \in B} \sum_{n=0}^{n_0} |f(ny + x_n) - f(ny)| < \infty$$

and p is bounded on τ_0-bounded subsets of $H(E)$.

We now show that p is not τ_ω continuous. Suppose otherwise and that p is ported by the compact subset K of E. Let $\phi_{n,m} = \phi_0^n \phi_m$ for any pair of positive integers n and m. Fix m and choose a neighbourhood V_m of K such that $\|\phi_0\|_{V_m} \leq \|\phi_0\|_K + \frac{1}{m} < \infty$ and $\|\phi_m\|_{V_m} < \infty$. By our assumption there exists $c_m > 0$ such that $p(f) \leq c_m \|f\|_{V_m}$ for all $f \in H(E)$. For all n we have

$$|\phi_{n,m}(my + x_m) - \phi_{n,m}(my)| = m^n |\phi_0(y)|^n |\phi_m(x_m)|$$
$$\leq p(\phi_{n,m})$$
$$\leq c_m \|\phi_0\|_{V_m}^n \|\phi_m\|_{V_m}.$$

Hence

$$m|\phi_0(y)| \cdot \lim_{n \to \infty} |\phi_m(x_m)|^{1/n} \leq \|\phi_0\|_{V_m} \cdot \lim_{n \to \infty} c_m^{1/n} \|\phi_m\|_{V_m}^{1/n}$$

i.e.

$$m|\phi_0(y)| \leq \|\phi_0\|_{V_m} \leq \|\phi_0\|_K + \frac{1}{m}.$$

Since $\phi_0(y) \neq 0$ and $\|\phi_0\|_K < \infty$ this is impossible. Hence p is not τ_ω-continuous. Letting $E = \mathbb{C}^{\mathbb{N}}$ we get an example of a Fréchet nuclear space with basis such that $\tau_\omega \neq \tau_\delta$ on $H(E)$.

The set $\{f : p(f) \leq 1\}$ is convex, balanced, absorbs all τ_0-bounded sets and is τ_0, and hence τ_ω, closed. We have thus shown that neither $(H(E), \tau_0)$ nor $(H(E), \tau_\omega)$ are infrabarrelled locally convex spaces. If we take $E = \mathbb{C}^{\mathbb{N}} \times \mathbb{C}^{(\mathbb{N})}$ then, by Example 1.27, on $H(E)$ we have $\tau_0 \underset{\neq}{<} \tau_\omega \underset{\neq}{<} \tau_\delta$. The above can easily be modified to give a similar result for $H(U; F)$, U an arbitrary subset of $\mathbb{C}^{\mathbb{N}} \times \mathbb{C}^{(\mathbb{N})}$ and F any locally convex space.

(b) Let $E = \prod_{n=1}^{\infty} E_n$ where each E_n is a $\mathcal{DFM}$ space and E_n' admits a continuous norm p_n. Then $p := \sum_{n=1}^{\infty} p_n$ is a continuous norm on $E_\beta' = \sum_{n=1}^{\infty} (E_n)_\beta'$ and $V := \{x \in E_\beta' : p(x) \leq 1\}^\circ$ is a convex balanced compact subset of E.

We prove by induction that V is a *determining set* for continuous polynomials on E, i.e. if $P \in \mathcal{P}(^n E)$ and $P|_V = 0$ then $P \equiv 0$. If $\phi \in \mathcal{P}(^1 E) = E'$ then $\phi|_V = 0 \iff p(\phi) = 0$ and the result follows since p is a norm on E'_β. Suppose the result is true for homogeneous polynomials of degree $\leq n$. If $P \in \mathcal{P}(^{n+1} E)$ and $P|_V = 0$ then, by the Polarization Formula, $\overset{\vee}{P}|_{V^n} = 0$. For each x in V the mapping $L_x : z \to \overset{\vee}{P}(x, z^n)$ is a n-homogeneous polynomial which vanishes on V. By our induction hypothesis $L_x \equiv 0$. Hence, for any $z \in E$, the continuous linear mapping

$$L^z : x \longrightarrow \overset{\vee}{P}(x, z^n)$$

vanishes on V and again, by induction, $L^z \equiv 0$. Since $L^z(z) = P(z)$ this implies that $P \equiv 0$ and, by induction, V is a determining set for polynomials. Using Taylor series expansions we see that V is a determining set for holomorphic functions on E.

For each n let π_n denote the canonical projection from E onto $F_n := \prod_{j=1}^n E_j$. Let B denote a τ_0-bounded subset of $H(E)$. We claim there exists a positive integer n_0 such that $f(x+y) = f(x)$ for all $f \in B$ and all $x, y \in E$ such that $\pi_{n_0}(y) = 0$. If not, we can find sequences $(f_n)_n \subset B$, $(x_n)_n \subset E$, $(y_n)_n \subset E$ such that $f_n(x_n + y_n) \neq f_n(x_n)$ for all n and $\pi_n(y_n) = 0$. Since V is determining we may suppose that $x_n \in V$ for all n and using the mappings

$$\lambda \longrightarrow f_n(x_n + \lambda y_n) - f_n(x_n)$$

and Liouville's Theorem we can choose λ_n such that

$$|f_n(x_n + \lambda_n y_n)| > n$$

for all n. The sequence $(y_n)_n$ is very strongly convergent and, as V is compact, the set $\{(x_n + \lambda_n y_n)_n\}$ is relatively compact in E. This is impossible since B is τ_0-bounded and we have established our claim. This implies that there exists n_0 such that for each $f \in B$ there exists $\widetilde{f} \in H(F_{n_0})$ such that $f = \widetilde{f} \circ \pi_{n_0}$. Let $\widetilde{B} = (\widetilde{f})_{f \in B}$. If $\widetilde{K}$ is a compact subset of F_{n_0} and 0 is the origin in $\prod_{j > n} E_j$ then $K := \widetilde{K} \times \{0\}$ is a compact subset of E and $\pi_{n_0}(K) = \widetilde{K}$. Hence

$$\sup_{\widetilde{f} \in \widetilde{B}} \|\widetilde{f}\|_{\widetilde{K}} = \sup_{f \in B} \|f\|_K < \infty$$

and $\widetilde{B}$ is a τ_0-bounded subset of the $\mathcal{DFM}$ space F_{n_0}. By Example 3.8(e), $\widetilde{B}$ is a locally bounded subset of F_{n_0}. If $\widetilde{B}$ is bounded on the open subset W of F_{n_0} then B is bounded on the open subset $W \times \prod_{n > n_0} E_n$ of E and we have proved that the τ_0 bounded subsets of $\prod_{n=1}^\infty E_n$ are locally bounded.

Since $\mathcal{D}'$ (the space of distributions on $\mathbb{R}^n$) is linearly isomorphic to $\prod_{n=1}^\infty E_n$ where $E_n \cong s$ (the rapidly decreasing sequences) the τ_0 bounded subsets of $H(\mathcal{D}')$ are locally bounded. We *cannot* drop completely the hypothesis of a continuous norm since, if $E_1 = \mathbb{C}^{(\mathbb{N})}$ and $E_n = \mathbb{C}$ for $n > 1$, then $\prod_{n=1}^{\infty} E_n \cong \mathbb{C}^{(\mathbb{N})} \times \mathbb{C}^{\mathbb{N}}$ and, by Example 1.27, τ_0 and τ_ω do not define the same bounded subsets of $H(\mathbb{C}^{(\mathbb{N})} \times \mathbb{C}^{\mathbb{N}})$. Since

$$\text{locally bounded} \implies \tau_\omega \text{ bounded} \implies \tau_0 \text{ bounded}$$

it follows that the τ_0-bounded subsets of $H(\mathbb{C}^{(\mathbb{N})} \times \mathbb{C}^{\mathbb{N}})$ are not locally bounded.

(c) In this example we keep the notation from (b) and show that we can remove the continuous norm condition in (b) by using the τ_ω topology. Let $E = \prod_{n=1}^{\infty} E_n$ where each E_n is a $\mathcal{DFM}$ space and let B denote a τ_ω-bounded subset of $H(E)$. We claim that there exists a positive integer n_0 such that B factors through F_{n_0}, i.e. $f(x + y) = f(x)$ for all $f \in B$, all $x, y \in E$ such that $\pi_{n_0}(y) = 0$. We first show that for each integer k there exists a positive integer n_k such that $\left(\widehat{d}^k f(0)\right)_{f \in B}$ factors through F_{n_k}. If not, then we can choose a sequence $(f_n)_n$ in B, $(x_n)_n \subset E$, $(y_n)_n \subset E$ such that

$$\widehat{d}^k f_n(0)(x_n + y_n) \neq \widehat{d}^k f_n(0)(x_n)$$

and $\pi_n(y_n) = 0$ for all n. Using the Polarization Formula, and taking a subsequence if necessary, we can find l, $l < k$, such that

$$d^k f_n(0)\left(x_n^l, y_n^{k-l}\right) \neq 0$$

for all n. On replacing y_n by $\beta_n y_n$, $\beta_n \in \mathbb{C}$, we may suppose

$$\left| d^k f_n(0)\left(x_n^l, (\beta_n y_n)^{k-l}\right) \right| > n$$

for all n. Since $(\beta_n y_n)_n$ is very strongly convergent and τ_ω is barrelled on $\mathcal{P}(^k E)$

$$p(f) := \sup_n \left| d^k f(0)\left(x_n^l, (\beta_n y_n)^{k-l}\right) \right|$$

is finite for all $f \in H(E)$ and defines a τ_ω-continuous semi-norm. This leads to a contradiction since $p(f_n) > n$ for all n and $(f_n)_n$ is τ_ω-bounded. This establishes the required result for $\left(\widehat{d}^k f(0)\right)_{f \in B}$.

If our claim is not true for B then we can choose strictly increasing sequences of positive integers $(s_n)_n$ and $(r_n)_n$, sequences $(x_n)_n$ and $(y_n)_n$ in E, and $(f_n)_n$ in B such that

$$\left| \frac{\widehat{d}^{s_n} f_n(0)}{s_n!}(x_n + y_n) - \frac{\widehat{d}^{s_n} f_n(0)}{s_n!}(x_n) \right| \neq 0$$

and $\pi_{r_n}(y_n) = 0$ for all n. By modifying, if necessary, the very strongly convergent sequence $(y_n)_n$ by scalar multiplication we can suppose

$$\left| \frac{\widehat{d}^{s_n} f_n(0)}{s_n!}(x_n + y_n) - \frac{\widehat{d}^{s_n} f_n(0)}{s_n!}(x_n) \right| > n \qquad (*)$$

for all n. Let

$$q(f) = \sum_{n=1}^{\infty} \left| \frac{\widehat{d}^{s_n} f(0)}{s_n!}(x_n + y_n) - \frac{\widehat{d}^{s_n} f(0)}{s_n!}(x_n) \right|$$

for all $f \in H(E)$. Since $(y_n)_n$ is very strongly convergent, $q(f) < \infty$ for all $f \in H(E)$. If V is a convex neighbourhood of 0 in E there exists a positive integer n_1 such that $\lambda y_n \in V$ for all $n \geq n_1$ and all $\lambda \in \mathbb{C}$. As the mapping

$$q_{n_1}(f) = \sum_{n=1}^{n_1} \left| \frac{\widehat{d}^{s_n} f(0)}{s_n!}(x_n + y_n) - \frac{\widehat{d}^{s_n} f(0)}{s_n!}(x_n) \right|$$

defines a semi-norm on $H(E)$ which is ported by the origin there exists $c(V) > 0$ such that

$$q_{n_1}(f) \leq c(V)\|f\|$$

for all $f \in H(E)$. If $f \in H(E)$ and $\|f\|_V < \infty$ then Lemma 3.9 implies that $\dfrac{\widehat{d}^n f(0)}{n!}(x + y_n) = \dfrac{\widehat{d}^n f(0)}{n!}(x)$ for all $x \in E$ whenever $n \geq n_1$. Hence $q(f) = q_{n_1}(f) \leq c(V)\|f\|_V$. If $\|f\|_V = \infty$ then $q(f) \leq c(V)\|f\|_V = \infty$ and we have proved that q is a τ_ω continuous semi-norm on $H(E)$. By $(*)$ $q(f_n) > n$ for all n and this contradiction establishes our claim. The final part of the proof of (b) can now be used to show that B is a locally bounded subset of $H(E)$.

We have shown that the τ_ω-bounded subsets of $H(\mathbb{C}^{(\mathbb{N})} \times \mathbb{C}^{\mathbb{N}})$ are locally bounded. In particular, the τ_δ-bounded subsets of $H(\mathbb{C}^{(\mathbb{N})} \times \mathbb{C}^{\mathbb{N}})$ are locally bounded and, using this result and Exercise 3.118, we have

$$\tau_0^{bor} \underset{\neq}{\leq} \tau_0^{bar} = \tau_\omega^{bor} = \tau_\delta$$

on $H(\mathbb{C}^{(\mathbb{N})} \times \mathbb{C}^{\mathbb{N}})$.

The space of symmetric tensors was defined in Chapter 1 in order to linearize single polynomials and to obtain a predual for spaces of homogeneous polynomials. To complete this section we carry out a similar programme for holomorphic functions. However, instead of following the direct approach in Chapter 1, we proceed along the lines suggested by the statement of Proposition 1.28. Indeed, Propositions 3.22(b) and 3.26 can be combined to form a proof of Proposition 1.28. We require the following lemma.

Lemma 3.25 *If U is an open subset of a locally convex space E and B is a locally bounded subset of $H(U)$ then τ_0 and the topology of pointwise convergence, τ_p, coincide on B and B is a relatively compact subset of $(H(U), \tau_0)$.*

Proof. By Ascoli's Theorem a subset of $\left(\mathcal{C}(U), \tau_0\right)$ – the continuous $\mathbb{C}$-valued functions on U endowed with the compact open topology – is relatively compact if and only if it is equicontinuous and pointwise bounded. Hence B is relatively τ_0-compact in $\mathcal{C}(U)$. By Montel's Theorem and Proposition 3.7, B is relatively τ_0-compact in $H(U)$. The identity mapping from $\left(\overline{B}^{\tau_0}, \tau_0\right)$ to $\left(\overline{B}^{\tau_0}, \tau_p\right)$ is a continuous bijective mapping from a compact space onto a Hausdorff space and hence is a homeomorphism. This completes the proof.

For U an open subset of a locally convex space let

$$G(U) = \{\phi \in H(U)^* : \phi \text{ is } \tau_0 \text{ continuous on the locally bounded}$$
$$\text{subsets of } H(U)\}.$$

It is easily seen that $G(U)$, endowed with the topology of uniform convergence on the locally bounded subsets of $H(U)$, is a complete locally convex space and

$$\left(H(U), \tau_0\right)' \subset G(U) \subset \left(H(U), \tau_\delta\right)'. \tag{3.25}$$

Consider the dual pairing $\langle G(U), H(U)\rangle$. Since point evaluations belong to $G(U)$ this is a separating duality. By Lemma 3.25, the $\sigma\left(H(U), G(U)\right)$-closure of a locally bounded subset of $H(U)$ is again locally bounded and this implies, by the Mackey–Arens Theorem, that $G(U)' = H(U)$. Since the topology of any locally convex space is uniform convergence over equicontinuous subsets of the dual the equicontinuous subsets of $G(U)'$ coincide with the locally bounded subsets of $H(U)$.

For B a convex balanced subset of $H(U)$ let $H(U)_B$ denote the subspace of $H(U)$ generated by B and normed by the Minkowski functional of B. Since a subset B of $H(U)$ is locally bounded if and only if it is contained and bounded in the Fréchet, and hence bornological, space $H_\mathcal{V}(U)$ for some increasing countable only cover $\mathcal{V}$ of U it follows (see the proof of Proposition 3.18) that

$$\left(H(U), \tau_\delta\right) = \varinjlim_{B \in \mathcal{B}} H(U)_B$$

where $\mathcal{B}$ denotes the set of all convex balanced locally bounded subsets of $H(U)$. As the inductive dual topology of $G(U)'$ is generated by $H(U)_B$ as B ranges over the equicontinuous subsets of $G(U)'$ the above identification of locally bounded and equicontinuous sets means we have proved the following proposition.

Proposition 3.26 *If U is an open subset of a locally convex space then* $G(U)'_i \cong \left(H(U), \tau_\delta\right).$

We denote by

$$J : f \in H(U) \longrightarrow Jf \in G(U)'$$

the isomorphism established in Proposition 3.26. Let $\delta_U : U \to G(U)$ denote the mapping which takes $x \in U$ to δ_x, the point evaluation or Dirac delta function at x. If $f \in H(U)$ then $\langle Jf, \delta_U(x) \rangle = \delta_x(f) = f(x)$ and hence the mapping

$$\lambda \in \mathbb{C} \longrightarrow \langle Jf, \delta_U(x + \lambda y) \rangle = f(x + \lambda y)$$

is holomorphic, for all $x \in U$ and $y \in E$, on a neighbourhood of the origin in $\mathbb{C}$. Hence $\delta_U \in H_G(U; G(U))$. If B is a locally bounded subset of $H(U)$ then for every x in U there exists a neighbourhood V_x of x such that $\sup\limits_{f \in B} \|f\|_{V_x} < \infty$.

Let $\|\cdot\|_B$ denote the semi-norm on $G(U)$ of uniform convergence over B. We have

$$\sup_{y \in V_x} \|\delta_U(y)\|_B = \sup_{\substack{y \in V_x \\ f \in B}} |\langle Jf, \delta_y \rangle| = \sup_{f \in B} \|f\|_{V_x} < \infty$$

and the mapping

$$\delta_U : U \longrightarrow (G(U), \|\cdot\|_B)$$

is locally bounded. Hence $\delta_U \in H(U; G(U))$.

We next claim that $\{\delta_x : x \in U\}$ generates a dense subspace of $G(U)$. If not, then, by the Hahn–Banach Theorem, there exists a non-zero $T := Jg$ in $G(U)'$ such that $T(\delta_x) = 0$ for all $x \in U$. We have $T(\delta_x) = \delta_x(g) = g(x) = 0$ for all $x \in U$. Hence $g \equiv 0$ and $T \equiv 0$. This contradiction establishes our claim.

We have thus shown that there exists a holomorphic mapping $\delta_U : U \to G(U)$ such that for all $f \in H(U)$, Jf is the unique element of $G(U)'$ for which the following diagram commutes:

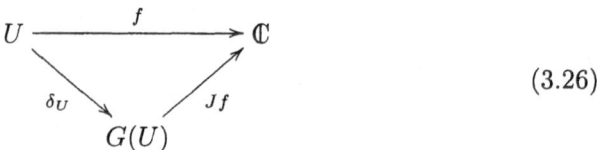

$$(3.26)$$

This is a scalar-valued holomorphic version of (1.5) and we may, in view of our next result, regard $G(U)$, algebraically and topologically, as the holomorphic analogue of $\widehat{\bigotimes}_{n,s,\pi} E$. We proceed to establish a vector-valued version of (3.26).

Proposition 3.27 *Let U denote an open subset of a locally convex space E and let F denote a complete locally convex space. For each $f \in H(U; F)$ there exists a unique $J_F f \in \mathcal{L}(G(U); F)$ such that the following diagram commutes:*

$$(3.27)$$

This property characterizes the pair $(G(U), \delta_U)$ up to linear topological isomorphism. The mapping $f \to J_F f$ establishes a linear topological isomorphism between the spaces $(H(U; F), \tau_\delta)$ and $(\mathcal{L}(G(U); F), \tau_\omega)$.

Proof. For fixed ψ in $G(U)$ consider the mapping

$$\phi \in F' \longrightarrow \langle J(\phi \circ f), \psi \rangle.$$

The collection $(\phi \circ f)_{\phi \in V^\circ}$ is a locally bounded subset of $H(U)$ whenever V is a neighbourhood of zero in F. If $(\phi_\alpha)_\alpha \in V^\circ$ is weak* null then $\phi_\alpha \circ f(z) \to 0$ for all $z \in U$ as $\alpha \to \infty$. By Lemma 3.25, $\psi(\phi_\alpha \circ f) \to 0$ as $\alpha \to \infty$ and, by Grothendieck's Completeness Theorem, there exists $J_F f(\psi) \in F$ such that $\phi(J_F f(\psi)) = \langle J(\phi \circ f), \psi \rangle$ for all $\phi \in F'$. The mapping $J_F f : \psi \in G(U) \to J_F f(\psi) \in F$ is easily seen to be linear. Since

$$\sup_{\substack{\phi \in V^\circ \\ \psi \in B^\circ}} |\phi(J_F f(\psi))| = \sup_{\substack{\phi \in V^\circ \\ \psi \in B^\circ}} |\langle J(\phi \circ f), \psi \rangle|$$

for any locally bounded subset B of $H(U)$, $J_F f$ is continuous. If $\phi \in F'$ and $x \in U$ then, by (3.26),

$$\phi(J_F f(\delta_U(x))) = \langle J(\phi \circ f), \delta_U(x) \rangle = \phi(f(x))$$

and by the Hahn–Banach Theorem, we obtain (3.27). Since $J(\phi \circ f)$ is unique for all $\phi \in F'$ the Hahn–Banach Theorem shows that $J_F f$ is also unique.

For each $\beta \in \mathrm{cs}(F)$ the spaces $(H(U; F_\beta), \tau_\delta)$ and $(\mathcal{L}(G(U); F_\beta), \tau_\omega)$ are inductive limits of normed linear spaces and by comparing these normed linear spaces, as in the proof of Proposition 3.26, we obtain the desired linear topological isomorphism.

Now suppose $\widetilde{G}(U)$ is a complete locally convex space, $\widetilde{\delta}_U \in H(U; \widetilde{G}(U))$ and for each $f \in H(U; F)$ there exists a unique $\widetilde{J}_F f \in \mathcal{L}(\widetilde{G}(U); F)$ such that $f = \widetilde{J}_F f \circ \widetilde{\delta}_U$. The uniqueness implies that $\widetilde{\delta}_U(U)$ spans a dense subspace of $\widetilde{G}(U)$. Applying (3.27) in turn to δ_U and $\widetilde{\delta}_U$ we obtain the commutative diagram

$$(3.28)$$

Since

$$\widetilde{J}_F \delta_U \circ J_F \widetilde{\delta}_U (\delta_U(x)) = \widetilde{J}_F \delta_U (\widetilde{\delta}_U(x)) = \delta_U(x)$$

and

$$J_F \widetilde{\delta}_U \circ \widetilde{J}_F \delta_U (\widetilde{\delta}_U(x)) = J_F \widetilde{\delta}_U (\delta_U(x)) = \widetilde{\delta}_U(x)$$

for all $x \in U$ density implies $\widetilde{J}_F \delta_U \circ J_F \widetilde{\delta}_U = I_{G(U)}$ and $J_F \widetilde{\delta}_U \circ \widetilde{J}_F \delta_U = I_{\widetilde{G}(U)}$. Hence $G(U) \cong \widetilde{G}(U)$, $\widetilde{\delta}_U = (\widetilde{J}_F \delta_U)^{-1} \circ \delta_U$ and the pair $(G(U), \delta_U)$ is determined up to linear isomorphism by (3.27). This completes the proof.

By considering n-homogeneous polynomials in place of holomorphic mappings in Proposition 3.27 we obtain a proof of Proposition 1.28 (see the remarks after Proposition 1.17).

If $U \xrightarrow{\ f\ } V$ is a holomorphic mapping between open subsets of locally convex spaces let $G(f) = J_{G(V)}(\delta_V \circ f)$. Since $G(f) \in \mathcal{L}(G(U); G(V))$ we have linearized the holomorphic mapping f. It is easily seen that $G(f \circ g) = G(f) \circ G(g)$ and $G(I_U) = I_{G(U)}$. If $i_{F,E}: F \longrightarrow E$ is the inclusion mapping from the subspace F of E into E then $G(i_{F,E}): G(F) \longrightarrow G(E)$ is also an inclusion mapping. In particular, if U is an open subset of a locally convex space, $G(U \cap F)$ may be identified with a subspace of $G(U)$ and using the natural inclusion order between subspaces of a given space we obtain the subspace $G_0(U)$ of $G(U)$ by letting

$$ G_0(U) = \bigcup_{\substack{F \subseteq E \\ dim(F) < \infty}} G(U \cap F). $$

Since point evaluations belong to $G_0(U)$, $G_0(U)$ is a dense subspace of $G(U)$. Hence $\delta_U \in H(U; G_0(U))$. If $f \in H_G(U; F)$ then $f|_{U \cap E_1} \in H(U \cap E_1; F)$ for any finite dimensional subspace E_1 of E. The corresponding linear mappings from $G(U \cap E_1)$ into F form a coherent family as E_1 ranges over the finite dimensional subspaces of E and it is easily seen that there exists $T_f \in \mathcal{L}_a(G_0(U); F)$ such that $T_f \circ \delta_U = f$. Moreover, T_f is continuous if and only if $f \in H(U; F)$. Thus we also have a linearization result for $\mathcal{G}$-holomorphic mappings. In fact, the approach outlined above can be adapted to obtain similar results for spaces such as $H_b(E; F)$, $H_{HY}(E; F)$ and $\mathcal{P}(^n E; F)$. The result for $\mathcal{P}(^n E)$ is Proposition 1.28. We use the predual $G(U)$ to obtain information about $H(U; F)$ and δ_U to study U. We state, without proof, some results of this type in the following example.

Example 3.28
(a) If U is an open subset of a locally convex space E then $G_0(U)$ is barrelled if and only if for each Banach space F, each subset B of $H(U; F)$ which is uniformly bounded on the finite dimensional compact subsets of U is locally bounded.

(b) An open subset U of a $\mathcal{DFM}$ space is holomorphically convex if and only if δ_U is a proper holomorphic mapping (i.e. $\delta_U^{-1}(K)$ is compact for each compact subset K of $G(U)$).

Montel's Theorem plays a crucial role in our study of $G(U)$ since it provides us with an abundance of *compact* sets. If we replace this theorem with a compactness hypothesis on the unit ball of a Banach space and consider linear, in place of holomorphic, mappings we obtain the standard method for obtaining a predual of a dual Banach space.

We are now in a position to define the β-topology on spaces of holomorphic functions.

Definition 3.29 Let U denote an open subset of a locally convex space E and let F denote a complete locally convex space. The β-topology on $H(U;F) = \mathcal{L}\big(G(U);F\big)$ is the topology of uniform convergence on the bounded subsets of $G(U)$.

Clearly, we have $\tau_0 \leq \beta \leq \tau_\delta$.

3.3 The Quasi-Local Theory of Holomorphic Functions

The topological theory of holomorphic mappings between locally convex spaces may be studied at three levels of increasing complexity, the most basic of these being the theory of *homogeneous polynomials*, considered in Chapters 1 and 2, and the most general, *the global theory*, i.e. the study of holomorphic functions on arbitrary open sets, is considered in Chapter 5. In between we find *the quasi-local theory*[6] – i.e. the theory for domains on which all holomorphic functions admit an expansion, into homogeneous polynomials, which converges pointwise at *all* points of the domain. This includes the study of Taylor series on balanced domains and monomial expansions on polydiscs. In contrast to the situation for G-holomorphic functions discussed in Section 3.1 the Taylor series expansion is the most general *topological* expansion available and applies to all locally convex spaces. The monomial expansion may be considered a refinement of the Taylor series expansion in which the homogeneous polynomials, of the Taylor series expansion, are themselves expanded using special assumptions on the space (full nuclearity and a basis) and uniform estimates obtained to give the required convergence on special domains (polydiscs). This point of view is further developed in Chapter 4. We adopt in this section however, a direct approach to monomial expansions in order to compare this topological situation with the algebraic monomial expansions of G-holomorphic functions given in Section 3.1 and also because an understanding of this special case is a suitable introduction and motivation for the more technical results in the next chapter.

In this section we discuss the quasi-local theory. This theory admits the direct application of well developed techniques from linear functional analysis such as Schauder decompositions and associated topologies. The global theory

---

[6] In classical several complex variables theory *"local theory"* refers to the study of holomorphic germs (see Section 4.3) and to avoid confusion we have adopted *quasi-local* as our terminology.

depends on both the quasi-local theory and, as we shall see in Chapter 5, techniques which originated in several complex variables theory.

We begin our discussion by considering abstract Schauder decompositions.

Definition 3.30 A sequence of subspaces $\{E_n\}_n$ of a locally convex space E is a Schauder decomposition of E if

(a) for each x in E there exists a unique sequence of vectors $(x_n)_n$, $x_n \in E_n$ for all n, such that

$$x = \sum_{n=1}^{\infty} x_n := \lim_{m \to \infty} \sum_{n=1}^{m} x_n$$

(b) the projections $(u_n)_{n=1}^{\infty}$ defined by

$$u_m\left(\sum_{n=1}^{\infty} x_n\right) := \sum_{n=1}^{m} x_n$$

are continuous.

If $\{E_n\}_n$ is a Schauder decomposition for a subspace of E we call $\{E_n\}_n$ a *basic decomposition*, i.e. $\{E_n\}_n$ is a Schauder decomposition for its closed linear span. If the sequence of projections $(u_n)_{n=1}^{\infty}$ are equicontinuous we say that the decomposition is *equi-Schauder*.

Lemma 3.31 *A Schauder decomposition of a barrelled locally convex space is an equi-Schauder decomposition.*

Proof. Let $\{E_n\}_n$ denote a Schauder decomposition of the barrelled locally convex space E and let $p \in \mathrm{cs}(E)$. Since $\lim_{n \to \infty} u_n(x) = x$ for all x in E we have $q(x) := \sup_n p(u_n(x)) < \infty$ and $p(x) \leq q(x)$ for all x in E.

Let $V = \{x \in E : q(x) \leq 1\} = \bigcap_n \{x \in E : p(u_n(x)) \leq 1\}$. Since each u_n is continuous, V is a closed convex balanced absorbing subset of E and, as E is barrelled, a neighbourhood of zero. Clearly $V = \bigcap_n u_n^{-1}(V)$.

Hence the sequence $(u_n)_n$ is equicontinuous and $\{E_n\}$ is an equi-Schauder decomposition. This completes the proof.

The proof of Lemma 3.31 shows that a locally convex space with an equi-Schauder decomposition, $\{E_n\}_n$, admits a fundamental system of semi-norms $\mathcal{F}$ such that

$$p\left(\sum_{n=1}^{\infty} x_n\right) = \sup_m p\left(\sum_{n=1}^{m} x_n\right) \tag{3.29}$$

for all $p \in \mathcal{F}$ and all $\sum_{n=1}^{\infty} x_n \in E$. By Lemma 3.31, every Schauder decomposition of a Banach space E is equi-Schauder and E can be renormed to satisfy (3.29). In this case the Schauder decomposition is called *monotone*. In Chapter 4 we shall need *unconditional decompositions*.

A Schauder decomposition $\{E_n\}_n$ is *unconditional* if, for each $x = \sum_{n=1}^{\infty} x_n$ in E, $x_n \in E_n$ for all n, we have

$$x = \sum_{n=1}^{\infty} x_{\pi(n)} \tag{3.30}$$

for any permutation π of the natural numbers. Various conditions equivalent to (3.30) are known and useful. For example (3.30) holds if and only if for each $p \in cs(E)$ and each $\varepsilon > 0$ there exists a finite subset J of $\mathbb{N}$ such that

$$p\left(x - \sum_{n \in J'} x_n\right) \leq \varepsilon$$

for all finite J', $J \subset J' \subset \mathbb{N}$. For this reason an unconditional decomposition is also called an *unordered decomposition*. If $\{E_n\}_n$ is an unconditional decomposition of the sequentially complete space E, $x = \sum_{n=1}^{\infty} x_n \in E$ and $|\theta_n| \leq 1$ for all n then $\sum_{n=1}^{\infty} \theta_n x_n \in E$. Moreover, if E is barrelled, then the proof of Lemma 3.31 can be modified, in an obvious way, to show that E has a fundamental system of semi-norms, each of which satisfies

$$p\left(\sum_{n-1}^{\infty} x_n\right) = \sup_{\substack{|\theta_n| \leq 1 \\ \theta_n \in \mathbb{C}}} p\left(\sum_{n=1}^{\infty} \theta_n x_n\right)$$

for all $\sum_{n=1}^{\infty} x_n \in E$, $x_n \in E_n$ all n.

A Schauder decomposition $\{E_n\}_n$ of E is an *absolute decomposition* if for each $p \in cs(E)$,

$$q\left(\sum_{n=1}^{\infty} x_n\right) := \sum_{n=1}^{\infty} p(x_n)$$

defines a continuous semi-norm on E. Absolute decompositions are unconditional and equi-Schauder.

If each E_n is one-dimensional and $\{E_n\}_n$ is a Schauder decomposition for E we say that E has a *Schauder basis*. In this case any sequence $(x_n)_n$, $x_n \neq 0$ and $x_n \in E_n$, is a Schauder basis for E. We say that the Schauder basis is

equi-Schauder, unconditional or absolute if the Schauder decomposition has
the corresponding property.

To study Taylor series expansions of holomorphic functions we require a
refinement of the concept of absolute decomposition – S-absolute decomposition. Let S denote the set of all scalar sequences $(\alpha_n)_{n=1}^{\infty}$ such that
$\limsup\limits_{n\to\infty} |\alpha_n|^{1/n} \le 1$.

Definition 3.32 A Schauder decomposition $\{E_n\}_n$ of a locally convex space
E is an S-absolute decomposition if for all $\alpha = (\alpha_n)_n \in S$ and $x = \sum_{n=1}^{\infty} x_n \in$
E, $x_n \in E_n$ all n,

$$\alpha \cdot x := \sum_{n=1}^{\infty} \alpha_n x_n \in E \tag{3.31}$$

and, for each $p \in \mathrm{cs}(E)$ and each $\alpha = (\alpha_n)_n$ in S,

$$p_\alpha\Big(\sum_{n=1}^{\infty} x_n\Big) := \sum_{n=1}^{\infty} |\alpha_n| p(x_n) \tag{3.32}$$

defines a continuous semi-norm on E.

In the following lemma we characterize bounded sets in spaces with S-
absolute decompositions. The method of proof is similar to that employed
earlier for nuclear spaces.

Lemma 3.33 *Let $\{E_n\}_n$ denote an S-absolute decomposition for the locally
convex space E and let $B := \Big\{\sum_{n=1}^{\infty} x_n^\lambda\Big\}_\lambda$ be a subset of E. The following are
all equivalent to the statement that B is bounded.*

$-\quad \sup\limits_{\lambda} p\Big(\sum_{n=1}^{\infty} x_n^\lambda\Big) < \infty$ *for all $p \in \mathrm{cs}(E)$,* $\tag{3.33}$

$-\quad \sum\limits_{n=1}^{\infty} \sup\limits_{\lambda} p(x_n^\lambda) < \infty$ *for all $p \in \mathrm{cs}(E)$,* $\tag{3.34}$

$-\quad \sum\limits_{n=1}^{\infty} |\alpha_n| \sup\limits_{\lambda} p(x_n^\lambda) < \infty$ *for all $\alpha = (\alpha_n)_n \in S$*
and all $p \in \mathrm{cs}(E)$, $\tag{3.35}$

$-\quad \sup\limits_{\lambda} p(x_n^\lambda) < \infty$ *for all n and $\limsup\limits_{n\to\infty}\big(\sup\limits_{\lambda} p(x_n^\lambda)\big)^{1/n} < 1$*
for all $p \in \mathrm{cs}(E)$, $\tag{3.36}$

$-\quad$ *for every $p \in \mathrm{cs}(E)$ there exists $\delta > 1$ such that*

$$\sum_{n=1}^{\infty} \delta^n \sup\limits_{\lambda} p(x_n^\lambda) < \infty \tag{3.37}$$

$$- \quad \sup_{n,\lambda} p(x_n^\lambda) < \infty \text{ for all } p \in \mathrm{cs}(E). \tag{3.38}$$

Proof. Since the sequence $(n^2)_n \in \mathcal{S}$, (3.32) and (3.33) imply that for each $p \in \mathrm{cs}(E)$, $\sup_\lambda \sum_{n=1}^\infty n^2 p(x_n^\lambda) =: M_p < \infty$. Hence

$$\sum_{n=1}^\infty \sup_\lambda p(x_n^\lambda) \leq \sum_{n=1}^\infty \frac{M_p}{n^2} < \infty$$

for every $p \in \mathrm{cs}(E)$ and (3.33) $\iff$ (3.34) (since trivially (3.34) $\implies$ (3.33)). By (3.32) we have (3.34) $\iff$ (3.35). Clearly (3.37) $\implies$ (3.36) $\implies$ (3.35). We can identify $\mathcal{S}$ with $H(D)$, D the open unit disc in $\mathbb{C}$, by means of the mapping

$$(\alpha_n)_n \in \mathcal{S} \longrightarrow \sum_{n=1}^\infty \alpha_n z^{n-1}.$$

Now $(H(D), \tau_0)$ is a Fréchet space and its strong dual under the pairing

$$\left\langle \sum_{n=0}^\infty a_n z^n, \sum_{n=0}^\infty b_n w^n \right\rangle := \sum_{n=1}^\infty a_n b_n$$

can be identified with $H(\overline{D})$, the holomorphic *germs*[7] defined on a neighbourhood of $\overline{D}$. Condition (3.35) says that $\left(\sup_\lambda p(x_n^\lambda) \right)_n$ forms the sequence of coefficients of an element of $H(D)' = H(\overline{D})$. Hence $\sum_n (\sup_\lambda p(x_n^\lambda)) z^n$ is holomorphic on an open disc about the origin in $\mathbb{C}$ of radius greater than 1 and (3.35) $\implies$ (3.37). Clearly (3.34) $\implies$ (3.38) and it remains to show that (3.38) $\implies$ (3.34). If (3.38) is satisfied and $p \in \mathrm{cs}(E)$ then

$$q\left(\sum_{n=1}^\infty x_n \right) := \sum_{n=1}^\infty n^2 p(x_n)$$

defines a continuous semi-norm on E. Hence $\sup_{n,\lambda} n^2 p(x_n^\lambda) =: M < \infty$ and

$$\sum_{n=1}^\infty \sup_\lambda p(x_n^\lambda) \leq \sum_{n=1}^\infty \frac{M}{n^2} < \infty$$

and (3.34) holds. This completes the proof.

We now prove a number of stability results for $\mathcal{S}$-absolute decompositions. We first note that the proof of Lemma 3.31 can be adjusted to show that a

---

[7] Holomorphic germs on arbitrary locally convex spaces are discussed later. We only require the one-dimensional case here.

Schauder decomposition of a barrelled locally convex space, satisfying (3.31), is an S-absolute decomposition.

Proposition 3.34 *Let $\{E_n\}_n$ denote an S-absolute decomposition for the locally convex space (E, τ). Then*

(a) $\{E_n, (\tau|_{E_n})^{bor}\}_n$ is an S-absolute decomposition for (E, τ^{bor}),

(b) If E_n is barrelled for each n then $\{E_n\}_n$ is an S-absolute decomposition for (E, τ^{bar}) and the τ^{bar} topology on E is generated by all semi-norms p on E which satisfy

(i) $p|_{E_n}$ is τ continuous,

(ii) $p\left(\sum\limits_{n=1}^{\infty} x_n \right) = \sum\limits_{n=1}^{\infty} p(x_n)$ for all $\sum\limits_{n=1}^{\infty} x_n \in E$.

(c) If $\{E_n\}_n$ is also an S-absolute decomposition for (E, τ^*) and $\tau^* \geq \tau$ then the following are equivalent for any τ^*-bounded net $(x^\beta)_{\beta \in \Gamma}$ in E,

(i) $x^\beta \longrightarrow 0$ as $\beta \to \infty$ in (E, τ^*),
(ii) $x^\beta \longrightarrow 0$ as $\beta \to \infty$ in (E, τ),

(iii) $x_n^\beta \longrightarrow 0$ as $\beta \to \infty$ for each n where $x^\beta = \sum\limits_{n=1}^{\infty} x_n^\beta$, $x_n^\beta \in E_n$ for

all β and n.

Proof. For locally convex spaces (F, τ_1) and (G, τ_2) it is easily seen that $(\tau_1 \times \tau_2)^b = \tau_1^b \times \tau_2^b$, where $b = bor$ or bar, on $F \times G$. Hence $(\tau|_{E_n})^{bor} = \tau^{bor}|_{E_n}$ in (a) and $\tau^{bar}|_{E_n} = \tau|_{E_n}$ in (b).

(a) If $x = \sum\limits_{n=1}^{\infty} x_n \in E$ then (3.35) implies $\left\{ n^2 \sum\limits_{m=n+1}^{\infty} x_m \right\}_n$ is a bounded sequence in E. If the semi-norm q on E is bounded on τ-bounded sets and

$$M := \sup_n q\left(n^2 \sum_{m=n+1}^{\infty} x_m \right) \text{ then}$$

$$q\left(x - \sum_{m=1}^{n} x_m \right) = \frac{1}{n^2} q\left(n^2 \sum_{m=n+1}^{\infty} x_m \right) \leq \frac{M}{n^2} \longrightarrow 0$$

as $n \to \infty$. Hence $x = \sum\limits_{n=1}^{\infty} x_n$ in (E, τ^{bor}). If $B := \left\{ \sum\limits_{n=1}^{\infty} x_n^\lambda \right\}_\lambda$ is a bounded subset of E and $(\alpha_n)_n \in S$ then (3.35) implies

$$\widetilde{B} := \left\{ \sum_{n=1}^{k} \theta_n \alpha_n x_n^\lambda \right\}_{|\theta_n| \leq 1, \, k=1,2,\dots}$$

is also a bounded subset of E. In particular, we have

$$\sup_{n,\lambda} n^2 |\alpha_n| q(x_n^\lambda) =: M_{q,\alpha} < \infty$$

for any semi-norm q on E which is bounded on τ-bounded sets. Since

$$\sup_\lambda \sum_{n=1}^\infty |\alpha_n| q(x_n^\lambda) \le M_{q,\alpha} \cdot \sum_{n=1}^\infty \frac{1}{n^2} < \infty$$

the semi-norm

$$\widetilde{q}\left(\sum_{n=1}^\infty x_n\right) := \sum_{n=1}^\infty |\alpha_n| q(x_n) \tag{3.39}$$

is bounded on τ-bounded sets and hence τ^{bor}-continuous. This completes the proof of (a).

(b) If the semi-norm p on E satisfies (i) and (ii) then it is easily seen (as in the proof of Lemma 3.31) that $V := \{x \in E : p(x) \le 1\}$ is τ-closed, convex, balanced and absorbing. Since $\tau^{bar} \ge \tau$ it is also τ^{bar}-closed and hence a neighbourhood of zero in (E, τ^{bar}).

To prove that (i) and (ii) are sufficient to describe τ^{bar} we use transfinite induction. Let $T_1 = \tau$. We define a topology T_β on E for every ordinal number β. If the ordinal β is not a limit ordinal, i.e. if β has an immediate predecessor β_p, let T_β be the locally convex topology on E with neighbourhood basis consisting of all T_{β_p}-closed convex balanced absorbing subsets of E. If β is a limit ordinal let

$$(E, T'_\beta) = \varprojlim_{\theta < \beta} (E, T_\theta)$$

and define T_β on E as the locally convex topology with neighbourhood base at the origin consisting of all T'_β-closed convex balanced absorbing subsets of E. By transfinite induction this defines T_β for every ordinal number β. Since the cardinality of all convex balanced absorbing subsets of E is less than or equal to $2^{|E|}$ there exists an ordinal θ, such that $T_\theta = T_{\theta_1}$ for all $\theta_1 \ge \theta$. By (1.32)–(1.34) a locally convex space is barrelled if and only if all closed convex balanced absorbing sets are zero neighbourhoods. Hence (E, T_θ) is a barrelled locally convex space and, as it is clearly the weakest barrelled topology finer than τ, $\tau^{bar} = T_\theta$. Since E_n is barrelled we have $\tau|_{E_n} = T_\beta|_{E_n}$ for every positive integer n and every ordinal number β. Hence $\tau|_{E_n} = \tau^{bar}|_{E_n}$ for all n.

We now prove, by transfinite induction, that $\{E_n\}_n$ is an $\mathcal{S}$-absolute decomposition for (E, T_β) for every ordinal number β. When $\beta = 1$ this is our hypothesis. If β is a limit ordinal and the result is true for all $\alpha < \beta$ then it is easily seen that $\{E_n\}_n$ is an $\mathcal{S}$-absolute decomposition for (E, T_β). Let $\tau' = T'_\beta$ if β is a limit ordinal and let $\tau' = T_{\beta_p}$ if β is not a limit ordinal. Thus the τ'-closed convex balanced absorbing subsets of E form a T_β-neighbourhood basis at the origin in E. For $\alpha = (\alpha_n)_n \in \mathcal{S}$ and $x = \sum\limits_{n=1}^{\infty} x_n \in E$ the set

$$V_{\alpha,x} := \left\{ \sum_{n=1}^{\infty} \delta_n x_n : \delta_n \in \mathbb{C}, |\delta_n| \leq |a_n| \text{ for all } n \right\} \tag{3.40}$$

is, by (3.31), a complete convex balanced bounded subset of (E, τ'), i.e. a Banach disc. If W is a T_β zero neighbourhood then the Banach–Mackey Theorem implies that W absorbs $V_{\alpha,x}$. If we let $\alpha_n = n^2$ for all n this implies there exists $c > 0$ such that $\sum\limits_{n=m}^{\infty} x_n \in \dfrac{c}{m^2} W$ for all m. Hence $x = \sum\limits_{n=1}^{\infty} x_n$ in (E, T_β). This implies that (E, T_β) satisfies (3.31).

Let p denote a T_β-continuous semi-norm on E and let $\alpha = (\alpha_n)_n \in \mathcal{S}$. Then $(n^2 \alpha_n)_n \in \mathcal{S}$ and by what we have just proved there exists $c > 0$ such that $n^2 \alpha_n x_n \in \{x : p(x) \leq c\}$ for all n. Hence

$$\sum_{n=1}^{\infty} |\alpha_n| p(x_n) \leq c \cdot \sum_{n=1}^{\infty} \frac{1}{n^2} < \infty.$$

Let

$$A = \left\{ \sum_{n=1}^{\infty} x_n \in E : \sum_{n=1}^{\infty} |\alpha_n| p(x_n) \leq 1 \right\}.$$

Since

$$A = \bigcap_{k=1}^{\infty} \left\{ \sum_{n=1}^{\infty} x_n \in E : \sum_{n=1}^{k} |\alpha_n| p(x_n) \leq 1 \right\}$$

and $p\big|_{E_n}$ is τ'-continuous, A is τ'-closed. Hence A is a T_β-neighbourhood of zero, $q := \sum_{n=1}^{\infty} |\alpha_n| p\big|_{E_n}$ is T_β continuous, and $\{E_n\}_n$ is an $\mathcal{S}$-absolute decomposition for (E, T_p) and, by induction, for (E, τ^{bar}).

If p is a τ^{bar}-continuous semi-norm on E then, since $\{E_n\}_n$ is a Schauder decomposition for (E, τ^{bar}), we have

$$p\left(\sum_{n=1}^{\infty} x_n \right) \leq p\left(\sum_{n=1}^{m} x_n \right) + p\left(\sum_{n=m+1}^{\infty} x_n \right)$$

$$\leq \sum_{n=1}^{m} p(x_n) + p\left(\sum_{n=m+1}^{\infty} x_n \right)$$

$$\leq \lim_{m \to \infty} \sum_{n=1}^{m} p(x_n) + \lim_{m \to \infty} p\left(\sum_{n=m+1}^{\infty} x_n \right) = \sum_{n=1}^{\infty} p(x_n) =: q\left(\sum_{n=1}^{\infty} x_n \right)$$

for all $\sum_{n=1}^{\infty} x_n \in E$. As q satisfies (i) and (ii) it is τ^{bar}-continuous and, since $p \leq q$, this completes the proof of (b).

(c) Clearly (i) $\Longrightarrow$ (ii) $\Longrightarrow$ (iii). Suppose (iii) is satisfied. Let p denote a τ^*-continuous semi-norm on E. Since $\{E_n\}_n$ is an S-absolute decomposition for (E, τ^*) the semi-norm

$$r\left(\sum_{n=1}^{\infty} x_n\right) := \sum_{n=1}^{\infty} n^2 p(x_n)$$

is τ^*-continuous. Hence

$$\sup_{\beta} \sum_{n=1}^{\infty} n^2 p(x_n^{\beta}) := M < \infty.$$

If $\varepsilon > 0$ is arbitrary we can choose n_0 such that

$$\sup_{\beta} \sum_{n=n_0}^{\infty} p(x_n^{\beta}) \leq M \cdot \sum_{n=n_0}^{\infty} \frac{1}{n^2} \leq \frac{\varepsilon}{2}.$$

By (iii) we can choose β_0 such that

$$\sum_{n=1}^{n_0-1} p(x_n^{\beta}) \leq \frac{\varepsilon}{2} \quad \text{for all } \beta \geq \beta_0.$$

If $\beta \geq \beta_0$ then

$$p(x^{\beta}) \leq \sum_{n=1}^{\infty} p(x_n^{\beta}) \leq \varepsilon$$

and $x^{\beta} \to 0$ as $\beta \to \infty$ in (E, τ^*). Hence (iii)$\Longrightarrow$(i) and this completes the proof.

A collection $\mathcal{B}$ of subsets of a locally convex space E with Schauder decomposition $\{E_n\}_n$ is S-*stable* if for each $B \in \mathcal{B}$ and each $\alpha = (\alpha_n)_n \in S$ the set

$$B(\alpha) := \left\{\sum_{n=1}^{\infty} \beta_n x_n : |\beta_n| \leq |\alpha_n| \text{ for all } n, \sum_{n=1}^{\infty} x_n \in B\right\}$$

is also in $\mathcal{B}$. Lemma 3.33 shows that the bounded subsets of a locally convex space E, with S-absolute decomposition $\{E_n\}_n$, form an S-stable collection. If $\{E_n\}_n$ is a Schauder decomposition for the locally convex space E and $T \in E'$ we let $T_n = T\big|_{E_n}$ for each n. If $x = \sum_{n=1}^{\infty} x_n \in E$ then

$$T\left(\sum_{n=1}^{\infty} x_n\right) = \sum_{n=1}^{\infty} T(x_n) = \sum_{n=1}^{\infty} T_n(x_n)$$

and $T = \sum\limits_{n=1}^{\infty} T_n$ in the weak* topology, $\sigma(E', E)$. A Schauder decomposition $\{E_n\}_n$ of a locally convex space E is *shrinking* if $\{(E_n)'_\beta\}_n$ is a Schauder decomposition for E'_β.

Proposition 3.35 *Let $\{E_n\}_n$ denote an S-absolute decomposition for the locally convex space E and let $\mathcal{B}$ denote an S-stable covering of E. If F is a subspace of E' let*

$$F_S = \Big\{ \sum_{n=1}^{\infty} \alpha_n T_n : (\alpha_n)_n \in S, \ \sum_{n=1}^{\infty} T_n \in F \Big\}.$$

Then F_S is a subspace of E' and $\{E'_n \cap F_S\}_n$ is an S-absolute decomposition for F_S, endowed with the topology of uniform convergence over the elements of $\mathcal{B}$. In particular, $\{(E_n)'_\beta\}_{n=1}^{\infty}$ is an S-absolute shrinking decomposition for E'_β.

Proof. Let $T = \sum\limits_{n=1}^{\infty} T_n \in F$. By continuity there exists $p \in cs(E)$ such that $|T(x)| \le p(x)$ for all $x \in E$. If $(\alpha_n)_n \in S$ and $x = \sum\limits_{n=1}^{\infty} x_n \in E$ then

$$\sum_{n=1}^{\infty} |\alpha_n T_n(x)| = \sum_{n=1}^{\infty} |\alpha_n T(x_n)| \le \sum_{n=1}^{\infty} |\alpha_n| p(x_n). \qquad (*)$$

Let $\Big(\sum\limits_{n=1}^{\infty} \alpha_n T_n \Big)(x) = \lim\limits_{k \to \infty} \sum\limits_{n=1}^{k} \alpha_n T_n(x)$. By $(*)$, $\sum\limits_{n=1}^{\infty} \alpha_n T_n \in E'$ for all $(\alpha_n)_n \in S$. Since F is a subspace of E' and the sum of two sequences in S is again in S, F_S is a subspace of E'. If $B \in \mathcal{B}$, $\alpha = (\alpha_n)_n \in S$ and $\beta_n = n^2 \alpha_n$ for all n then $B(\beta) \in \mathcal{B}$ and

$$\|T\|_{B(\beta)} = \sup_{\substack{\sum_{n=1}^{\infty} x_n \in B \\ |\gamma_n| \le |\alpha_n|}} \Big| \sum_{n=1}^{\infty} \gamma_n n^2 T_n(x_n) \Big|$$

$$= \sup_{\sum_{n=1}^{\infty} x_n \in B} \sum_{n=1}^{\infty} |\alpha_n| n^2 |T_n(x_n)|$$

$$\ge \sup_{\sum_{m=1}^{\infty} x_m \in B} |\alpha_n| n^2 |T_n(x_n)| \quad \text{(for any } n)$$

$$\ge |\alpha_n| n^2 \|T_n\|_B.$$

Hence

$$\Big\| \sum_{n=m}^{\infty} \alpha_n T_n \Big\|_B \leq \sum_{n=m}^{\infty} |\alpha_n| \|T_n\|_B \leq \Big(\sum_{n=m}^{\infty} \frac{1}{n^2} \Big) \|T\|_{B(\beta)} \tag{3.41}$$

and $\{E'_n \cap F_S\}_n$ is a Schauder decomposition for F_S. Moreover, if $(\alpha_n)_n \in S$ and $(\beta_n)_n \in S$ then $(\alpha_n \beta_n)_n \in S$ and hence $\{E'_n \cap F_S\}_{n=1}^{\infty}$ is a Schauder decomposition for F_S which satisfies (3.31).

If $\tilde{T} \in F_S$ then $\tilde{T} = \sum_{n=1}^{\infty} \alpha_n T_n$ where $\sum_{n=1}^{\infty} T_n \in F$ and $(\alpha_n)_n \in S$. If $B \in \mathcal{B}$ and $(\beta_n)_n \in S$ then $\sum_{n=1}^{\infty} \alpha_n \beta_n T_n \in F_S$. A calculation, similar to (3.41), shows

$$\sum_{n=1}^{\infty} |\beta_n| \|\alpha_n T_n\|_B = \sum_{n=1}^{\infty} \|\alpha_n \beta_n T_n\|_B \leq \Big(\sum_{n=1}^{\infty} \frac{1}{n^2} \Big) \Big\| \sum_{n=1}^{\infty} \alpha_n T_n \Big\|_{B(\tilde{\beta})}$$

where $\tilde{\beta} := (n^2 \beta_n)_n$. Hence $\{E'_n \cap F_S\}_n$ is an S-absolute decomposition. This completes the proof.

We apply the above results by showing that the Taylor series expansion at the origin gives rise to an S-absolute decomposition of the space of holomorphic functions on a balanced domain endowed with any of the topologies we have defined. Let U denote a balanced open subset of a locally convex space, let F denote a Banach space and suppose $f = \sum_{n=0}^{\infty} \frac{\widehat{d}^n f(0)}{n!} \in H_G(U; F)$. If V is a *balanced* subset of U, $1 < \lambda_1 < \lambda$, $(\alpha_n)_n \in S$, $|\alpha_n| \leq \lambda_1^n$ for $n \geq n_0$ and $\lambda V \subset U$ then

$$\sum_{n=0}^{\infty} \Big\| \alpha_n \frac{\widehat{d}^n f(0)}{n!} \Big\|_V = \sum_{n=0}^{n_0} \frac{|\alpha_n|}{\lambda^n} \Big\| \frac{\widehat{d}^n f(0)}{n!} \Big\|_{\lambda V} + \sum_{n=n_0+1}^{\infty} \Big(\frac{\lambda_1}{\lambda} \Big)^n \Big\| \frac{\widehat{d}^n f(0)}{n!} \Big\|_{\frac{\lambda \alpha_n^{1/n}}{\lambda_1} V}$$

$$\leq \Big(\sum_{n \leq n_0} \frac{|\alpha_n|}{\lambda^n} + \sum_{n=n_0+1}^{\infty} \Big(\frac{\lambda_1}{\lambda} \Big)^n \Big) \|f\|_{\lambda V}. \tag{3.42}$$

Proposition 3.36 *If U is a balanced open subset of a locally convex space E and F is a Banach space then*

(a) $\Big\{ \big(\mathcal{P}(^n E; F), \tau_0 \big) \Big\}_{n=0}^{\infty}$ *is an S-absolute decomposition for* $\big(H(U; F), \tau_0 \big)$,

(b) $\Big\{ \big(\mathcal{P}(^n E; F), \tau_\omega \big) \Big\}_{n=0}^{\infty}$ *is an S-absolute decomposition for* $\big(H(U; F), \tau_\omega \big)$
and $\big(H(U; F), \tau_\delta \big)$.

Proof. If $f = \sum_{n=0}^{\infty} \frac{\widehat{d}^n f(0)}{n!} \in H(U; F)$ and K is a compact balanced subset of U then there exists $\lambda > 1$ and a balanced open subset V of E such that

$\lambda K \subset \lambda V \subset U$ and $\|f\|_{\lambda V} < \infty$. If $(\alpha_n)_n \in S$ then, for any $\lambda_1, 1 < \lambda_1 < \lambda$, there exists n_0 such that $|\alpha_n| \leq \lambda_1^n$ for all $n \geq n_0$. By (3.42)

$$\sum_{n=0}^{\infty} \left\| \alpha_n \frac{\widehat{d^n} f(0)}{n!} \right\|_V < \infty$$

and hence

$$g := \sum_{n=0}^{\infty} \alpha_n \frac{\widehat{d^n} f(0)}{n!} \in H(U; F).$$

This shows that $\left\{ (\mathcal{P}(^nE; F), \tau) \right\}_{n=0}^{\infty}$, $\tau = \tau_0, \tau_\omega$ or τ_δ satisfies (3.31). Since $(H(U; F), \tau_\delta)$ is barrelled this implies $\left\{ (\mathcal{P}(^nE; F), \tau_\omega) \right\}_{n=0}^{\infty}$ is an S-absolute decomposition for $(H(U; F), \tau_\delta)$. Replacing V by K and letting

$$C = \left(\sum_{n \leq n_0} |\alpha_n| + \sum_{n=n_0+1}^{\infty} \left(\frac{\lambda_1}{\lambda} \right)^n \right) \text{ in (3.42) we obtain}$$

$$\sum_{n=0}^{\infty} |\alpha_n| \cdot \left\| \frac{\widehat{d^n} f(0)}{n!} \right\|_K \leq C \, \|f\|_{\lambda K}$$

for all $f = \sum_{n=0}^{\infty} \frac{\widehat{d^n} f(0)}{n!} \in H(U; F)$. Hence $\left\{ (\mathcal{P}(^nE; F), \tau_0) \right\}_{n=0}^{\infty}$ is an S-absolute decomposition for $(H(U; F), \tau_0)$.

Let p denote a τ_ω-continuous semi-norm on $H(U; F)$. Suppose p is ported by the compact balanced subset K. If V is a balanced open subset of U, $K \subset V$, there exists $C(V) > 0$ such that $p(f) \leq C(V)\|f\|_V$ for all $f \in H(U; F)$. Choose $\lambda > 1$ such that $\lambda K \subset U$. By (3.42)

$$\sum_{n=0}^{\infty} |\alpha_n| p \left(\frac{\widehat{d^n} f(0)}{n!} \right) \leq C(V) \cdot \sum_{n=0}^{\infty} |\alpha_n| \left\| \frac{\widehat{d^n} f(0)}{n!} \right\|_V$$

$$\leq C(V) \left(\sum_{n \leq n_0} \frac{|\alpha_n|}{\lambda^n} + \sum_{n=n_0+1}^{\infty} \left(\frac{\lambda_1}{\lambda} \right)^n \right) \|f\|_{\lambda V}$$

for all $f = \sum_{n=0}^{\infty} \frac{\widehat{d^n} f(0)}{n!} \in H(U; F)$. Since λV ranges over all neighbourhoods of the compact subset λK of U as V ranges over all neighbourhoods of K the semi-norm

$$q \left(\sum_{n=0}^{\infty} \frac{\widehat{d^n} f(0)}{n!} \right) := \sum_{n=1}^{\infty} |\alpha_n| p \left(\frac{\widehat{d^n} f(0)}{n!} \right)$$

is τ_ω-continuous. Hence $\left\{ (\mathcal{P}(^nE; F), \tau_\omega) \right\}_{n=0}^{\infty}$ is an S-absolute decomposition for $(H(U; F), \tau_\omega)$. This completes the proof.

The following corollary gives information on the relationship between the τ_ω and the τ_δ topologies on balanced domains.

Corollary 3.37 *Let U denote a balanced open subset of a locally convex space E and let F denote a Banach space.*

(a) The τ_δ topology on $H(U;F)$ is generated by all semi-norms on $H(U;F)$ satisfying

(i) $p\left(\sum_{n=0}^{\infty} \dfrac{\hat{d}^n f(0)}{n!}\right) = \sum_{n=0}^{\infty} p\left(\dfrac{\hat{d}^n f(0)}{n!}\right)$ *for all* $\sum_{n=0}^{\infty} \dfrac{\hat{d}^n f(0)}{n!} \in H(U;F)$,

(ii) $p\big|_{\mathcal{P}(^m E;F)}$ *is τ_ω-continuous for all m.*

(b) τ_δ is the barrelled topology associated with τ_ω on $H(U;F)$.

(c) If $(f_\beta)_{\beta \in \Gamma}$ is a τ_δ-bounded net in $H(U;F)$ then the following are equivalent:

(i) $f_\beta \longrightarrow 0$ *as $\beta \to \infty$ in* $\big(H(U;F), \tau_\delta\big)$

(ii) $f_\beta \longrightarrow 0$ *as $\beta \to \infty$ in* $\big(H(U;F), \tau_\omega\big)$

(iii) $\dfrac{\hat{d}^n f_\beta(0)}{n!} \longrightarrow 0$ *as $\beta \to \infty$ in* $\big(\mathcal{P}(^n E; F), \tau_\omega\big)$ *for all n.*

Proof. It suffices to apply Proposition 3.34(b) and (c) to Proposition 3.36(b). $\square$

It can be shown that the infrabarrelled and bornological topologies associated with τ_ω coincide but we do not know if these in turn coincide with τ_δ. We now obtain an $\mathcal{S}$-absolute decomposition for $G(U)$.

Proposition 3.38 *If U is a balanced open subset of a locally convex space E then $\left\{ \underset{n,s,\pi}{\widehat{\otimes}} E \right\}_{n=0}^{\infty}$ is an $\mathcal{S}$-absolute decomposition for $G(U)$. (We interpret $\underset{0,s,\pi}{\widehat{\otimes}} E$ as $\mathbb{C}$.)*

Proof. The space $G(U)$ is a subspace of $\big(H(U), \tau_\delta\big)'$ and is endowed with the topology of uniform convergence on the locally bounded subsets of $H(U)$. An application of (3.42) shows that the locally bounded subsets of $H(U)$ are $\mathcal{S}$-stable and, by Proposition 3.35, $\big\{\mathcal{P}(^n E)' \cap G(U)_\mathcal{S}\big\}_{n=0}^{\infty}$ is an $\mathcal{S}$-absolute decomposition for $G(U)_\mathcal{S}$. If $(f_\beta)_\beta$ is a locally bounded τ_0-null net in $H(U)$ and $(\alpha_n)_n \in \mathcal{S}$ then $\left(\sum_{n=0}^{\infty} \alpha_n \dfrac{\hat{d}^n f_\beta(0)}{n!}\right)_\beta$ is also a locally bounded τ_0-null net in $H(U)$. Hence, if $\sum_{n=0}^{\infty} T_n \in G(U)$ and $(\alpha_n)_n \in \mathcal{S}$, then

$$\left(\sum_{n=0}^{\infty} \alpha_n T_n\right)(f_\beta) = \left(\sum_{n=0}^{\infty} T_n\right)\left(\sum_{n=0}^{\infty} \alpha_n \dfrac{\hat{d}^n f_\beta(0)}{n!}\right) \longrightarrow 0$$

as $\beta \to \infty$. This implies $\sum\limits_{n=0}^{\infty} \alpha_n T_n \in G(U)$ and $G(U)_S = G(U)$. By Proposition 3.27 (see also Proposition 1.28)

$$\left(\mathcal{P}(^nE), \tau_\omega\right)' \cap G(U) \cong \bigotimes_{n,s,\pi} E.$$

This completes the proof.

The techniques used to prove Propositions 3.36 and 3.38 can easily be modified to show that Taylor series expansions at the origin lead to S-absolute decomposition for a number of other spaces of holomorphic functions, on balanced sets, such as $\left(H_{HY}(U; F), \tau_0\right)$, $\left(H_b(U), \tau_b\right)$ and $H(K)$. We omit the straightforward proofs.

The use of S-absolute decompositions is a first step in extending results from spaces of homogeneous polynomials to spaces of holomorphic functions. In cases where it does not immediately lead to a solution it often clarifies the problems that must be overcome and shows the (possibly) stronger results on spaces of homogeneous polynomials that are required.

Examples 3.8(f) and 3.20(c) show that holomorphic functions are not necessarily bounded on bounded sets. Consequently, the τ_b topology (of bounded convergence) does not generalize in a straightforward fashion from homogeneous polynomials to holomorphic functions. Using Taylor series expansions we define a suitable substitute on balanced domains.

A sequence of subsets, $(A_n)_n$, of a locally convex space E converges to a subset A of E if $A \subset A_n$ for all n and, for every open subset V of E which contains A, there exists a positive integer n_0 such that $A_n \subset V$ for all $n \geq n_0$.

Definition 3.39 If U is a balanced open subset of a locally convex space E and F is a Banach space we let τ_b denote the topology on $H(U; F)$ generated by the semi-norms

$$p\left(\sum_{n=0}^{\infty} \frac{\widehat{d}^n f(0)}{n!}\right) = \sum_{n=0}^{\infty} \left\|\frac{\widehat{d}^n f(0)}{n!}\right\|_{B_n} \tag{3.43}$$

for all $\sum\limits_{n=0}^{\infty} \dfrac{\widehat{d}^n f(0)}{n!} \in H(U; F)$ where $(B_n)_n$ ranges over all sequences of bounded subsets of U which converge to a compact subset of U.

The τ_b topology is defined on arbitrary open sets and for arbitrary range spaces by using projective limits. The following properties of τ_b on balanced domains are easily verified:

(a) $\tau_0 \leq \tau_b \leq \tau_\omega$, $\tau_b \leq \beta$ and $\tau_0 = \tau_b$ if and only if E is semi-Montel,

(b) $\left\{\left(\mathcal{P}(^nE; F), \tau_b\right)\right\}_{n=0}^{\infty}$ is an S-absolute decomposition for $\left(H(U; F), \tau_b\right)$

(for (b) it suffices to use $\alpha \|P\|_B = \|P\|_{\alpha^{1/n} B}$ for all $P \in \mathcal{P}(^n E; F)$ and all n and to note that $\alpha_n^{1/n} B_n \longrightarrow K$ whenever $(\alpha_n)_n \in \mathcal{S}$, $\alpha_n \geq 1$ all n, and $B_n \longrightarrow K$).

By (3.43), the τ_b topology, like the τ_0 topology, on balanced domains admits a nice projective description and our aim is to find general situations under which $\tau_b = \tau_\omega$. This will give us an explicit formula for a fundamental system of τ_ω semi-norms. We begin our investigations in this direction in the next section and continue in Chapter 4.

In Section 3.1 we showed that $\mathcal{G}$-holomorphic functions had monomials expansions with respect to any algebraic basis of the underlying domain space. This expansion has strong but very *local* convergence properties – it converges absolutely on *finite dimensional* polydiscs. Our aim now is to obtain a monomial expansion which converges absolutely with respect to the topologies introduced in Section 3.2. To obtain such an expansion we discuss holomorphic functions on fully nuclear spaces with a basis. We shall see later that nuclearity is, in many cases, necessary in order to obtain an absolute basis. The basis, or coordinate system, allows us to obtain fairly precise estimates and clarifies, without oversimplifying, the essential features of the situation. As a working hypothesis, the assumption of a basis, is extremely helpful although it is probable that it will eventually be weakened or removed entirely. It is noticeable, and not surprising, that many of the techniques used in this situation are similar to those used in the development of $\mathcal{S}$-absolute decompositions. In Chapter 4 we discuss holomorphic mappings on spaces endowed with a Schauder decomposition. The results in this section, although more special, are a good guide to the more technical aspects of Chapter 4.

Let E be a fully nuclear space with Schauder basis $(e_n)_n$. Since E is barrelled, Lemma 3.31 implies the basis is equicontinuous and as E is nuclear and quasi-complete it may be identified with a sequence space $\Lambda(P)$ where P is a collection of *weights* (non-negative sequences), i.e.

$$\Lambda(P) = \left\{ (x_n)_n \in \mathbb{C}^{\mathbb{N}} : \sum_{n=1}^{\infty} |x_n|\alpha_n < \infty \text{ for all } (\alpha_n)_n \in P \right\}.$$

The topology of $\Lambda(P)$ is generated by the semi-norms

$$\left\| (x_n)_n \right\|_{\alpha,1} := \sum_{n=1}^{\infty} |x_n|\alpha_n$$

where $(x_n)_n \in \Lambda(P)$ and $\alpha := (\alpha_n)_n \in P$. When we identify E with $\Lambda(P)$ we may take P to be $\left\{ (p(e_n))_n \right\}_{p \in cs(E)}$. The *Grothendieck–Pietsch criterion* for nuclearity says that $\Lambda(P)$ is nuclear if and only if for each $(\alpha_n)_n \in P$ there exists $(\beta_n)_n \in P$, $\beta_n \geq \alpha_n$ for all n, such that $\sum\limits_{n,\alpha_n \neq 0} \dfrac{\alpha_n}{\beta_n} < \infty$. If $\Lambda(P)$ is nuclear then its topology is generated by the semi-norms

$$\left\| (x_n)_n \right\|_{\alpha,\infty} := \sup_n |x_n| \alpha_n$$

where $(x_n)_n \in \Lambda(P)$ and $\alpha := (\alpha_n)_n \in P$. If $A \subset \Lambda(P)$ the *modular (or solid) hull of A, $\tilde{A}$*, is defined as follows:

$$\tilde{A} := \left\{ (y_n)_n \in \Lambda(P) : \text{ there exists } (x_n)_n \in A \text{ and } |y_n| \leq |x_n| \text{ for all } n \right\}.$$

If $A = \tilde{A}$ we say that A is *modularly decreasing* or *solid*. Subsets of $\Lambda(P)$ of the form

$$\left\{ (z_n)_n \in \Lambda(P) : \sup_n |z_n| \beta_n < 1 \right\}$$

or

$$\left\{ (z_n)_n \in \Lambda(P) : \sup_n |z_n| \beta_n \leq 1 \right\}$$

for some sequence $(\beta_n)_n$ in $[0, +\infty]$ are called *polydiscs*. We use the convention $a \cdot (+\infty) = +\infty$ if $a > 0$ and $0 \cdot (+\infty) = 0$. A polydisc of the first kind is a neighbourhood of zero if and only if $(\beta_n)_n \in P$, while polydiscs of the second kind are always closed. If $E \cong \Lambda(P)$ is a fully nuclear space with basis then E'_β is also a fully nuclear space with basis and $E'_\beta \cong \Lambda(P')$ where $P' = \left\{ \left(|x_n| \right)_n \right\}_{(x_n)_n \in \Lambda(P)}$.

If $U := \{ (z_n)_n : \sup_n |z_n| \alpha_n < 1 \}$ is an open polydisc in the fully nuclear space $\Lambda(P)$ then, by the Grothendieck–Pietsch criterion, we can find $(\alpha'_n)_n \in P$ such that $\alpha'_n > \alpha_n$ all n and $\sum\limits_{n=1}^{\infty} \dfrac{\alpha_n}{\alpha'_n} < \infty$. Let $\varepsilon_n = \alpha'_n / \alpha_n$ all n and let $V = \{ (z_n)_n : \sup_n |z_n| \alpha'_n < 1 \}$. Then V is an open polydisc in E and

$$\varepsilon V := \{ (\varepsilon_n z_n)_n : (z_n)_n \in V \} = U.$$

Definition 3.40 If $E \cong \Lambda(P)$ is a fully nuclear space and $A \subset E$ then

$$A^M := \left\{ (w_n)_n \in \Lambda(P') \cong E'_\beta : \sup_n |w_n z_n| \leq 1 \text{ for all } (z_n)_{n=1}^{\infty} \in A \right\}$$

is called the *multiplicative polar* of A.

It is immediate that A^M is a closed modularly decreasing subset of E'_β and that $A^{MM} := (A^M)^M$ is a closed subset of E which contains A. Multiplicative polars preserve the structure of polydiscs while retaining many properties of the usual polar $\circ$. This is evident in the following two technical lemmata.

Lemma 3.41 *Let U be an open polydisc in a fully nuclear space with basis $E \cong \Lambda(P)$. Then U^M is a compact polydisc in $E'_\beta \cong \Lambda(P')$. Furthermore, U*

contains a fundamental system of compact sets consisting of compact poly-discs and U^M has a fundamental neighbourhood system consisting of open polydiscs. The mapping $K \longrightarrow$ Interior (K^M) establishes a one-to-one correspondence between compact polydiscs in U and open polydisc neighbourhoods of U^M.

Proof. Let $U = \{(z_n)_n \in \Lambda(P) : \sup_n |z_n|\alpha_n < 1\}$ where $(\alpha_n)_n \in P$ and let

$$V := \{(z_n)_n \in \Lambda(P) : \sum_{n=1}^{\infty} |z_n|\alpha_n < 1\}. \text{ Then}$$

$$U^M = V^\circ = \Big\{(w_n)_n \in \Lambda(P') : \sup_n \Big|\frac{w_n}{\alpha_n}\Big| \leq 1\Big\}.$$

Since E is Montel, V° is a compact subset of E'_β and hence U^M is a compact polydisc in $\Lambda(P')$. Let K denote a compact subset of U and let $a_n = \sup_\lambda\{|z_n^\lambda| : (z_n^\lambda) \in K\}$. If $(\beta_n)_n \in P$ then

$$\sup_{z \in K}\big\|(z_n)_n\big\|_{\beta,\infty} = \sup_{(z_n^\lambda)_n \in K} |z_n^\lambda| \cdot \beta_n = \sup_n\Big[\big(\sup_\lambda |z_n^\lambda|\big) \cdot \beta_n\Big] < \infty.$$

Hence $(a_n)_n \in E$ and $K \subset [(a_n)_n] :=$ *modular hull of the point* $(a_n)_n$. Since $K \subset U$ there exists $\varepsilon < 1$ such that $\sup_n |z_n|\alpha_n \leq \varepsilon$ for all $(z_n)_n \in K$ and $[(a_n)_n]$ is a compact polydisc in U. This proves the first part of the lemma.

Now let W be a neighbourhood of U^M in E'_β. There exists $(\alpha'_n)_n \in P'$ such that

$$W \supset U^M + \{(w_n)_n \in \Lambda(P') : \sup_n |w_n|\alpha'_n < 1\}$$

$$= \{(w_n)_n \in \Lambda(P') : \sup_n |w_n|\gamma_n < 1\}$$

where

$$\gamma_n = \begin{cases} 0 & \text{if } \alpha'_n = 0, \\[2mm] \dfrac{1}{\alpha_n + \dfrac{1}{\alpha'_n}} & \text{if } \alpha'_n \neq 0. \end{cases}$$

Hence $(\gamma_n)_n \in P'$ and U^M has a fundamental neighbourhood system consisting of open polydiscs. If V is any open polydisc neighbourhood of U^M then we can choose β, $0 < \beta < 1$, such that βV is also an open polydisc neighbourhood of U^M. Hence $(\beta V)^M = \frac{1}{\beta} V^M \subset (U^M)^M = \overline{U}$. Since E'_β is also fully nuclear with basis, V^M is a compact polydisc in E which lies in $\beta\overline{U}$ and hence in U. If K is a compact polydisc in U then $K^\circ \subset K^M$ and hence K^M is a neighbourhood of zero in E'_β. Choose $\lambda > 1$ such that λK is also a compact subset of U. Then $(\lambda K)^M = \frac{1}{\lambda} K^M \supset U^M$ and

$K^M = \frac{1}{\lambda}K^M + \left(1 - \frac{1}{\lambda}\right)K^M \supset U^M + \left(1 - \frac{1}{\lambda}\right)K^M$ and hence K^M is a neighbourhood of U^M in E'_β. This shows that Interior (K^M) is an open polydisc neighbourhood of U^M. The one-to-one correspondence is now evident if one notices that for any closed polydisc B, $(\overline{B})^M = B^M$. This completes the proof.

Lemma 3.42 *Let U be an open polydisc in a fully nuclear space with basis, $E \approx \Lambda(P)$, and let K be a compact polydisc in U. There exists a sequence of real numbers $\delta = (\delta_n)_n$, $\delta_n > 1$ and $\sum_{n=1}^{\infty} \frac{1}{\delta_n} < \infty$, and V an open polydisc in $\Lambda(P)$ such that δK is a compact polydisc in U and $\delta(K + V) \subset U$.*

Proof. Let $K = \{(z_n)_n : \sup_n |z_n|\beta_n \le 1\}$. Choose $(\alpha_n)_n \in P$ and λ, $1 < \lambda < 2$, such that $U \supset \lambda K + 4W$ where $W := \{(z_n)_n : \sup_n |z_n|\alpha_n < 1\}$. By our previous remarks we can find $\varepsilon = (\varepsilon_n)_n$, $\varepsilon_n > 1$ all n and $\sum_n \frac{1}{\varepsilon_n} < \infty$ such that $V := \frac{1}{\varepsilon}W$ is an open polydisc in E. Since K is compact and V is open we have $(1/\beta_n)_n \in \Lambda(P)$ and $(\alpha_n/\varepsilon_n)_n \in P$. Hence $\sum_n \frac{\alpha_n}{\varepsilon_n\beta_n} < \infty$ and there exists a positive integer n_0 such that $\alpha_n < \varepsilon_n\beta_n$ for all $n \ge n_0$. If $(z_n)_n \in K$ this implies $(0, \ldots, 0, z_{n_0}, \ldots, z_n, \ldots) \in V$. Let $\delta_n = \lambda$ for $n < n_0$ and $\delta_n = \varepsilon_n$ for $n \ge n_0$. Then $\delta_n > 1$ all n and $\sum_n \frac{1}{\delta_n} < \infty$. If $(z_n)_n \in K$ and $(w_n)_n \in V$ then, since $\lambda < 2$,

$$\left(\delta_n(z_n + w_n)\right)_n \in \lambda K + \varepsilon V + 2V + \varepsilon V \subset \lambda K + 4W \subset U.$$

This completes the proof.

In Example 3.8(f) we introduced the space of hypoanalytic mappings $H_{HY}(U; F)$. It is easily seen that $\left\{\left(\mathcal{P}_{HY}(^nE; F), \tau_0\right)\right\}_{n=0}^{\infty}$ is an $\mathcal{S}$-absolute decomposition for $H_{HY}(U; F)$ endowed with the compact open topology where U is a balanced open subset of a locally convex space E and F is a Banach space. Since E is fully nuclear, the $\mathcal{S}$-absolute decomposition and Lemma 2.15 imply that $\mathcal{G}$-holomorphic functions which are bounded on bounded sets are hypoanalytic and that, moreover, $H(U)$ is τ_0-dense in $H_{HY}(U)$.

Let $f \in H_G(U)$ where $U := \{(z_n)_n \in E : \sup_n |z_n|\alpha_n < 1\}$ is an open polydisc in the fully nuclear space $E = \Lambda(P)$ with basis $(e_n)_n$. Since the (Schauder) basis in E can be extended to an algebraic basis we may apply (3.3), (3.7) and (3.8). If $m = (m_1, m_2, \ldots, m_n, \ldots) \in \mathbb{N}^{(\mathbb{N})}$ let

$$a_m(f) = \frac{1}{(2\pi i)^m} \int \cdots \int_{|z_i|=r_i} \frac{f\left(\sum_{i=1}^{n} z_i e_i\right)}{z^m z_1 \cdots z_n} dz_1 \ldots dz_n$$

where $0 < r_i < \frac{1}{\alpha_i}$ for $i = 1, \ldots, n$. By (3.7) and (3.8)

$$f(z) = \sum_{m \in \mathbb{N}^{(\mathbb{N})}} a_m(f) z^m \tag{*}$$

for all $z \in U$, $z \in \text{span}((e_n)_n)$. Let $K = \{(z_n)_n : \sup_n |z_n|\beta_n \le 1\}$ denote a compact polydisc in U. If $m = (m_n)_{n \in \mathbb{N}} \in \mathbb{N}^{(\mathbb{N})}$ and $\{n : \beta_n \ne \infty\} \supset \{n : m_n \ne 0\}$ then (3.3) implies

$$\left\| a_m(f) z^m \right\|_K \le \|f\|_K. \tag{3.44}$$

If $\beta_n = \infty$ for some n with $m_n \ne 0$ then $z^m = 0$ for all $z \in K$ and hence (3.44) holds for all $m \in \mathbb{N}^{(\mathbb{N})}$.

Now suppose $f \in H_{HY}(U)$. By Lemma 3.42 there exists a sequence of real numbers $(\delta_n)_n$, $\delta_n > 1$ and $\sum_n \dfrac{1}{\delta_n} < \infty$, such that δK is a compact polydisc in U. By (3.44)

$$\sum_{m \in \mathbb{N}^{(\mathbb{N})}} \left\| a_m(f) z^m \right\|_K = \sum_{m \in \mathbb{N}^{(\mathbb{N})}} \frac{1}{\delta^m} \left\| a_m(f) z^m \right\|_{\delta K} \le \|f\|_{\delta K} \cdot \sum_{m \in \mathbb{N}^{(\mathbb{N})}} \frac{1}{\delta^m}$$

$$= \frac{\|f\|_{\delta K}}{\displaystyle\prod_{n=1}^{\infty} \left(1 - \frac{1}{\delta_n}\right)} < \infty. \tag{3.45}$$

Since monomials are continuous and the uniform limit of continuous functions is continuous (3.45) implies that the series representation (*) is valid for all z in U. If J is any finite subset of $\mathbb{N}^{(\mathbb{N})}$ then, by uniqueness of the monomial coefficients in (*) we see, on applying (3.45) to $f - \sum_{m \in J} a_m(f) z^m$, that

$$\left\| f - \sum_{m \in J} a_m(f) z^m \right\|_K \le \sum_{m \in \mathbb{N}^{(\mathbb{N})} \setminus J} \| a_m(f) z^m \|_K \le \|f\|_{\delta K} \cdot \sum_{m \in \mathbb{N}^{(\mathbb{N})} \setminus J} \frac{1}{\delta^m}.$$

Hence the monomials form a Schauder basis for $\left(H_{HY}(U), \tau_0\right)$. The estimate (3.45) shows that the basis is absolute and, as $\left(H_{HY}(U), \tau_0\right)$ is complete, we have $\left(H_{HY}(U), \tau_0\right) \approx \Lambda(Q)$ where $Q := \left\{ \left(\|z^m\|_K \right)_{m \in \mathbb{N}^{(\mathbb{N})}} \right\}_{K \in \mathcal{K}(U)}$ and $\mathcal{K}(U)$ denotes the set of compact polydiscs in U. Since $\displaystyle\sum_{m \in \mathbb{N}^{(\mathbb{N})}} \frac{\|z^m\|_K}{\|z^m\|_{\delta K}} =$

$\displaystyle\sum_{m \in \mathbb{N}^{(\mathbb{N})}} \frac{1}{\delta^m} < \infty$ the Grothendieck–Pietsch criterion implies $\left(H_{HY}(U), \tau_0\right)$ is nuclear. We have thus proved most of the following proposition.

Proposition 3.43 *If U is an open polydisc in the fully nuclear space with basis $(e_n)_n$ then the monomials form an absolute basis for $\left(H_{HY}(U), \tau_0\right)$, $\left(H(U), \tau_0\right)$ and $\left(H(U), \tau_\omega\right)$. Moreover, $\left(H_{HY}(U), \tau_0\right)$ is a nuclear space.*

Proof. To complete the proof we must show that the monomials form an absolute basis for $(H(U), \tau_\omega)$. Let p denote a τ_ω-continuous semi-norm on $H(U)$ ported by the compact polydisc K in U. Let V denote an open subset of U which contains K. By Lemma 3.42 there exists a sequence of positive real numbers $\delta = (\delta_n)_n$, $\delta_n > 1$ and $\sum_n \dfrac{1}{\delta_n} < \infty$, and W an open polydisc in E such that δK is a compact polydisc in U and $\delta(K + W) \subset V$. Since p is ported by K there exists $c(W) > 0$ such that $p(f) \leq c(W)\|f\|_{K+W}$ for all $f \in H(U)$. By (3.44)

$$\sum_{m \in \mathbb{N}^{(\mathbb{N})}} p(a_m(f)z^m) \leq c(W) \sum_{m \in \mathbb{N}^{(\mathbb{N})}} \|a_m(f)z^m\|_{K+W}$$

$$= c(W) \sum_{m \in \mathbb{N}^{(\mathbb{N})}} \frac{1}{\delta^m} \|a_m(f)z^m\|_{\delta(K+W)}$$

$$\leq c(W)\|f\|_{\delta(K+W)} \cdot \sum_{m \in \mathbb{N}^{(\mathbb{N})}} \frac{1}{\delta^m}$$

$$\leq c(W)\|f\|_V \cdot \sum_{m \in \mathbb{N}^{(\mathbb{N})}} \frac{1}{\delta^m}.$$

Following the argument used for the τ_0 case we see that the monomials form a Schauder basis for $(H(U), \tau_\omega)$. The above also shows that

$$q(f) := \sum_{m \in \mathbb{N}^{(\mathbb{N})}} p(a_m(f)z^m)$$

is a τ_ω-continuous semi-norm on U (ported by K). Hence $\{z^m\}_{m \in \mathbb{N}^{(\mathbb{N})}}$ is an absolute basis for $(H(U), \tau_\omega)$ and this completes the proof.

To obtain nuclearity for the τ_ω topology we require a condition which may be regarded as a uniform Grothendieck–Pietsch criterion and which may also be compared with (3.32).

Definition 3.44 A locally convex space E with absolute basis $(e_n)_n$ is A-nuclear if there exists a sequence of positive real numbers $(\delta_n)_n$, $\sum_n \dfrac{1}{\delta_n} < \infty$, such that for each $p \in cs(E)$ the semi-norm

$$q\left(\sum_{n=1}^{\infty} x_n e_n\right) := \sum_{n=1}^{\infty} \delta_n p(x_n e_n)$$

is continuous.

Clearly A-nuclear spaces are nuclear and, if $E \approx \Lambda(P)$, then E is A-nuclear if and only if the diagonal mapping

$$\delta : \sum_n x_n e_n \in E \longrightarrow \sum_n \delta_n x_n e_n \in E$$

is a linear topological isomorphism.

Let E denote an A-nuclear space with absolute basis $(e_n)_n$ and fundamental system of bounded sets $\mathcal{B}(E)$. If $(e'_n)_n$ are the coefficient functionals associated with $(e_n)_n$ then, since the modular hull $\widetilde{B}$ of a bounded set B is also bounded,

$$\left\| \sum_n \alpha_n e'_n \right\|_B \le \left\| \sum_n \alpha_n e'_n \right\|_{\widetilde{B}} = \sum_n \| \alpha_n e'_n \|_{\widetilde{B}}$$

and E'_β has an absolute basis with system of weights $\left\{ \left(\| e'_n \|_B \right)_n \right\}_{B \in \mathcal{B}(E)}$. Since $\| e'_n \|_{\delta \widetilde{B}} = \delta_n \| e'_n \|_{\widetilde{B}}$ we have

$$\sum_n \frac{\| e'_n \|_{\widetilde{B}}}{\| e'_n \|_{\delta \widetilde{B}}} = \sum_n \frac{1}{\delta_n}$$

and the strong dual of an A-nuclear space is A-nuclear. In particular, we see that a reflexive A-nuclear space is a fully nuclear space with basis. We do not know of any fully nuclear space with basis that is not A-nuclear. Fréchet nuclear spaces with basis, their strong duals and direct sums of Fréchet nuclear spaces are all reflexive A-nuclear spaces. If E is a reflexive A-nuclear space then Proposition 3.43 and the diagonal mapping δ show that $\big(H_{HY}(E), \tau_0 \big)$ is A-nuclear.

Proposition 3.45 *If U is an open polydisc in the reflexive A-nuclear space E then $\big(H(U), \tau_\omega \big)$ is a nuclear space.*

Proof. Let p denote a τ_ω-continuous semi-norm on $H(U)$ ported by the compact polydisc K in U. Let $(\delta_n)_n$ denote the sequence which defines A-nuclearity. By Lemma 3.42 there exists a sequence of positive real numbers, $(\varepsilon_n)_n$, $1 < \varepsilon_n \le \delta_n$ and $\sum_n \frac{1}{\varepsilon_n} < \infty$, such that $\varepsilon^2 K := \varepsilon(\varepsilon K)$ is a compact polydisc in U. Let

$$q(f) = \sum_{m \in \mathbb{N}^{(\mathbb{N})}} \varepsilon^m |a_m(f)| p(z^m)$$

for all $f \in H(U)$ and suppose V is an open subset of U which contains $\varepsilon^2 K$. Since E is A-nuclear the diagonal mapping ε^2 is a linear isomorphism of E. Hence there exists an open polydisc W in E such that $\varepsilon^2 K + \varepsilon^2 W \subset V$. Since p is ported by K there exists $c(W) > 0$ such that

$$p(f) \le c(W) \| f \|_{K+W}$$

for all $f \in H(U)$. Hence

$$q(f) \leq \sum_{m \in \mathbb{N}^{(\mathbb{N})}} \varepsilon^m |a_m(f)| c(W) \|z^m\|_{K+W}$$

$$= c(W) \sum_{m \in \mathbb{N}^{(\mathbb{N})}} \|a_m(f) z^m\|_{\varepsilon(K+W)}$$

for all $f \in H(U)$. Since

$$\sum_{n \in \mathbb{N}^{(\mathbb{N})}} \frac{1}{\varepsilon^m} \|a_m(f) z^m\|_{\varepsilon^2(K+W)} \leq \|f\|_{\varepsilon^2(K+W)} \cdot \sum_{m \in \mathbb{N}^{(\mathbb{N})}} \frac{1}{\varepsilon^m}$$

$$\leq \left(\sum_{m \in \mathbb{N}^{(\mathbb{N})}} \frac{1}{\varepsilon^m} \right) \|f\|_V$$

this shows that q is τ_ω-continuous and since

$$\sum_{m \in \mathbb{N}^{(\mathbb{N})}} \frac{p(z^m)}{q(z^m)} = \sum_{m \in \mathbb{N}^{(\mathbb{N})}} \frac{1}{\varepsilon^m} < \infty$$

the Grothendieck–Pietsch criterion implies that $(H(U), \tau_\omega)$ is nuclear and completes the proof.

Finally we consider the τ_δ topology.

Proposition 3.46 *Let U denote an open polydisc in a fully nuclear space E with basis. The monomials $\{z^m\}_{m \in \mathbb{N}^{(\mathbb{N})}}$ form an unconditional equicontinuous Schauder basis for $(H(U), \tau_\delta)$. If E is A-nuclear then $(H(E), \tau)$, $\tau = \tau_0$, τ_ω, τ_0^{bor}, τ_ω^{bor} and τ_δ are A-nuclear spaces.*

Proof. Let $f \in H(U)$ be fixed. For each positive integer n let

$$V_n = \left\{ z \in U : \left| \sum_{m \in J} a_m(f) z^m \right| \leq n \text{ for every finite subset } J \text{ of } \mathbb{N}^{(\mathbb{N})} \right\}$$

and let W_n denote the interior of V_n. For each compact polydisc K in U there exists, by Proposition 3.43, a neighbourhood V of zero such that $\sum_{m \in \mathbb{N}^{(\mathbb{N})}} \|a_m(f) z^m\|_{K+V} < \infty$. Hence $(W_n)_n$ is an increasing countable open cover of U. If p is a τ_δ-continuous semi-norm on $H(U)$ there exists a positive integer n_0 and $c > 0$ such that

$$p(g) \leq c \, \|g\|_{W_{n_0}}$$

for all $g \in H(U)$. Hence

$$\sup_{\substack{J \subset \mathbb{N}^{(\mathbb{N})} \\ J \text{ finite}}} p\left(\sum_{m \in J} a_m(f) z^m \right) \leq c \, n_0$$

and $\left\{ \sum\limits_{m \in J} a_m(f) z^m \right\}_{\substack{J \subset \mathbb{N}^{(\mathbb{N})} \\ J \text{ finite}}}$ is τ_δ-bounded. By Proposition 3.37(c),

$$\sum_{\substack{m \in J \\ J \text{ finite}}} a_m(f) z^m \longrightarrow f \text{ as } J \to \infty \text{ in } \left(H(U), \tau_\delta\right)$$

if and only if for each non-negative integer n,

$$\sum_{\substack{m \in J, \, J \text{ finite} \\ |m| = n}} a_m(f) z^m \longrightarrow \frac{\widehat{d}^n f(0)}{n!} \text{ as } J \to \infty \text{ in } \left(\mathcal{P}(^n E), \tau_\omega\right).$$

By Proposition 3.43, the monomials form an absolute basis for $\left(H(U), \tau_\omega\right)$, and, since τ_ω and τ_δ coincide on $\mathcal{P}(^n E)$ by Proposition 3.22(b), this condition is satisfied and the monomials form an unconditional equicontinuous Schauder basis for $\left(H(U), \tau_\delta\right)$.

Now suppose E is A-nuclear with defining sequence $\delta = (\delta_n)_n$. Let $\widetilde{p}$ be a τ continuous semi-norm on $H(E)$, $\tau = \tau_0$ or τ_ω, and let

$$\widetilde{q}(f) = \sum_{m \in \mathbb{N}^{(\mathbb{N})}} \delta^m \widetilde{p}\left(a_m(f) z^m\right)$$

for all $f \in H(E)$. Since the diagonal isomorphism takes compact sets to compact sets and neighbourhood systems to neighbourhood systems, the proofs of Propositions 3.43 and 3.45 show that $\widetilde{q}$ is τ-continuous.

If p is τ_δ-continuous then, since $\sum\limits_{m \in \mathbb{N}^{(\mathbb{N})}} (\delta^2)^m a_m(f) z^m \in H(E)$ for all $f \in H(E)$, we have $\sup\limits_{m \in \mathbb{N}^{(\mathbb{N})}} (\delta^2)^m |a_m(f)| p(z^m) =: M < \infty$. Hence

$$q(f) := \sum_{m \in \mathbb{N}^{(\mathbb{N})}} \delta^m p\left(a_m(f) z^m\right) \leq M \sum_{m \in \mathbb{N}^{(\mathbb{N})}} \frac{1}{\delta^m} < \infty.$$

In this case q is the pointwise limit of a sequence of continuous semi-norms on a barrelled space and hence is continuous. This shows that the monomials form an absolute basis for $\left(H(E), \tau_\delta\right)$. Since

$$\sum_{m \in \mathbb{N}^{(\mathbb{N})}} \frac{p(z^m)}{q(z^m)} = \sum_{m \in \mathbb{N}^{(\mathbb{N})}} \frac{1}{\delta^m} < \infty$$

we see that $\left(H(E), \tau\right)$, $\tau = \tau_0$, τ_ω or τ_δ, are A-nuclear spaces. To complete the proof it suffices to prove and apply to $\left(H(E), \tau_0\right)$ and $\left(H(E), \tau_\omega\right)$ the following general result; if (E, τ) is complete A-nuclear with absolute basis $(e_n)_n$ then (E, τ^{bor}) is A-nuclear. This follows readily from a suitable modification of the proof of Proposition 3.34(a) or from the observation that the mapping

$$\delta : \sum_{n=1}^{\infty} x_n e_n \in E \longrightarrow \sum_{n=1}^{\infty} \delta_n x_n e_n \in E$$

is a linear isomorphism when $(\delta_n)_n$ is the sequence defining A-nuclearity. This completes the proof.

For fully nuclear spaces we confined our attention to the three topologies τ_0, τ_ω and τ_δ. Since complete nuclear spaces are semi-Montel we have $\tau_0 = \tau_b$. In the nuclear cases in which we are interested the spaces of holomorphic functions are semi-reflexive and the preduality theory coincides with the duality theory. We discuss duality in Section 4.4 using holomorphic germs.

3.4 Polynomials in the Quasi-Local Theory

In the previous section we saw that a holomorphic function on a balanced domain in a locally convex space may be regarded as a sequence of homogeneous polynomials which satisfies certain growth conditions. Locally convex topologies on the space of all holomorphic functions are a quantification of these conditions on aggregates of functions. Properties of this quantification are frequently obtained by combining, in an appropriate fashion, estimates on spaces of homogeneous polynomials. In this way we try and lift properties of homogeneous polynomials to spaces of holomorphic functions on balanced domains. In this section we give a number of examples of this general, if rather vague, principle. We have defined the following topologies on $H(U; F)$

$$\tau_0 \leq \tau_b \leq \overset{\tau_\omega}{\underset{\beta}{}} \leq \tau_\delta$$

and we may well ask what are we looking for. One answer is that we are looking for a topology with good linear properties (e.g. barrelled, bornological, etc), which admits a precise and useful projective system (i.e. it admits an explicit fundamental system of semi-norms) and which has applications. We have noted that the two topologies on the left, τ_0 and τ_b, have good projective systems, while the topology τ_δ has good linear topological properties. In general, however, we can say no more. In specific cases, as we have seen and shall see again, the most direct route to obtaining the kind of topology we hope for is to show $\tau_0 = \tau_\delta$ or $\tau_b = \tau_\delta$. This does happen, but it is often necessary to consider the intermediate $\tau_0 = \tau_\omega$, $\tau_b = \tau_\omega$ and $\tau_\omega = \tau_\delta$ problems. The τ_ω topology is in a crucial position and may be close enough to its neighbours on the left and right to inherit what we require, without actually coinciding with one or the other. It is worth noting that linear properties which are preserved under projective limits (e.g. semi-reflexivity, Schwartz Property, nuclearity, semi-Montel, etc.) are usually not too difficult to prove for the τ_0 topology, while properties preserved under inductive limits (e.g. infrabarrelledness, barrelledness etc) are true of the τ_δ topology.

We now consider and compare the τ_0, τ_b and τ_ω topologies on a balanced domain U in a Fréchet space E. Let F denote a Banach space and suppose p is a τ_ω-continuous semi-norm on $H(U; F)$. By Proposition 3.36(b) we may suppose

$$p\left(\sum_{n=0}^{\infty} \frac{\hat{d}^n f(0)}{n!}\right) = \sum_{n=0}^{\infty} p\left(\frac{\hat{d}^n f(0)}{n!}\right) \tag{3.46}$$

for all $\sum_{n=0}^{\infty} \frac{\hat{d}^n f(0)}{n!} \in H(U; F)$. Let p be ported by the compact balanced subset K of U and let $(W_m)_{m=1}^{\infty}$ denote a fundamental neighbourhood system of K. We suppose, without loss of generality, that each W_m is a balanced open subset of U and that the sequence $(W_m)_m$ is strictly decreasing. Since p is τ_ω-continuous there exists for each W_m, $c_m > 0$ such that

$$p(P_n) \le c_m \|P_n\|_{W_m}$$

for all n and all $P_n \in \mathcal{P}(^n E; F)$. Now choose inductively a strictly increasing sequence of positive integers $(k_m)_m$ such that $c_m^{1/n} \le 1 + \frac{1}{m}$ for all $n \ge k_m$. Let

$$V_n = \begin{cases} W_1, & 0 \le n < k_1 \\ \left(1 + \dfrac{1}{m}\right) W_m, & k_m \le n < k_{m+1}. \end{cases}$$

For $k_m \le n < k_{m+1}$ and $P_n \in \mathcal{P}(^n E; F)$ we have

$$p(P_n) \le c_m \|P_n\|_{W_m} = \|P_n\|_{c_m^{1/n} W_m} \le \|P_n\|_{(1+\frac{1}{m}) W_m} = \|P_n\|_{V_n}$$

and for $0 \le n < k_1$

$$p(P_n) \le c_1 \|P_n\|_{W_1} = c_1 \|P_n\|_{V_1}.$$

Combining these two estimates we see that the τ_ω topology on $H(U; F)$ is generated by a collection of semi-norms which satisfy (3.46) and for which there exists a fundamental neighbourhood system $(V_n)_n$ of a compact subset K of U such that

$$p\left(\frac{\hat{d}^n f(0)}{n!}\right) \le \left\|\frac{\hat{d}^n f(0)}{n!}\right\|_{V_n} \tag{3.47}$$

for all $f \in H(U; F)$ and all n.

If E is a Banach space then the sequence $(V_n)_{n=1}^{\infty}$ can be chosen to be a bounded sequence and hence we have proved the following result.

Proposition 3.47 *If U is a balanced open subset of a Banach space E, B is the open unit ball of E and F is a Banach space then the τ_ω topology on $H(U; F)$ is generated by the semi-norms*

$$p\left(\sum_{n=0}^{\infty} \frac{\widehat{d^n} f(0)}{n!}\right) = \sum_{n=0}^{\infty} \left\|\frac{\widehat{d^n} f(0)}{n!}\right\|_{K+\alpha_n B} \tag{3.48}$$

where K ranges over the compact balanced subsets of U and $(\alpha_n)_n$ over c_0. In particular, $\tau_\omega = \tau_b$ on $H(U;F)$.

The inequality (3.47) indicates an approach to the $\tau_0 = \tau_\omega$ and $\tau_b = \tau_\omega$ problems on balanced domains in arbitrary Fréchet spaces that we could use to prove results obtained by other means in Section 4.3. We confine ourselves here to convex balanced domains and give examples in the next chapter.

Proposition 3.48 *If U is a convex balanced open subset of a Fréchet–Montel space E then $\tau_0 = \tau_\omega$ on $H(U)$ if and only if E has $(BB)_\infty$.*

Proof. Since $\left(\mathcal{P}(^nE), \tau\right)$ is a complemented subspace of $\left(H(U), \tau\right)$, $\tau = \tau_0$ or τ_ω, for any convex balanced subset U of E, Proposition 1.36 implies that E has $(BB)_\infty$ if $\tau_0 = \tau_\omega$ on $H(U)$.

Conversely, suppose E has $(BB)_\infty$. Let p denote a τ_ω-continuous seminorm on $H(U)$, U balanced convex open in E. We may suppose, without loss of generality by (3.47), that

$$p(f) = \sum_{n=0}^{\infty} p\left(\frac{\widehat{d^n} f(0)}{n!}\right) \tag{3.49}$$

for all $f = \sum_{n=0}^{\infty} \frac{\widehat{d^n} f(0)}{n!} \in H(U)$, and that there exists a convex balanced compact subset K of U and $(W_n)_n$, a decreasing sequence of convex balanced open subsets of U converging to K, satisfying

$$p\left(\frac{\widehat{d^n} f(0)}{n!}\right) \leq \left\|\frac{\widehat{d^n} f(0)}{n!}\right\|_{W_n} \tag{3.50}$$

for all n and all $f \in H(U)$. Since E has $(BB)_\infty$, $\tau_0 = \tau_\omega$ on $\mathcal{P}(^nE)$ and since $p\big|_{\mathcal{P}(^nE)}$ is τ_ω-continuous, there exists, for all $n \geq 1$, a compact convex balanced subset K_n of E such that

$$p\left(\frac{\widehat{d^n} f(0)}{n!}\right) \leq \left\|\frac{\widehat{d^n} f(0)}{n!}\right\|_{K_n} \tag{3.51}$$

for all $f \in H(U)$. We now show that K_n can be chosen to lie in W_n and since $W_n \longrightarrow K$ this will complete the proof. Fix n and let $q := p\big|_{\mathcal{P}(^nE)}$. Let $B = \{\phi \in \mathcal{P}(^nE)' : |\phi| \leq q\}$. We have $q(P) = \|P\|_B$ for all $P \in \mathcal{P}(^nE)$. By the $\left(\bigotimes_{n,s,\pi} E, \mathcal{P}(^nE)\right)$ duality and (3.50)

$$q(P) = \|P\|_B \leq \|P\|_{\left(\bigotimes_{n,s} W_n\right)^{\circ\circ}}$$

for all $P \in \mathcal{P}(^n E)$. By the Hahn–Banach Theorem, $B \subset \left(\bigotimes_{n,s} W_n\right)^{\circ\circ}$ and by (3.51), $B \subset \left(\bigotimes_{n,s} K_n\right)^{\circ\circ}$. Hence B is a compact subset of the Fréchet–Montel space $\widehat{\bigotimes_{n,s,\pi}} E$ (Proposition 1.36) and there exists a null sequence, $\{x_j\}_{j=1}^\infty \subset \left(1 + \dfrac{1}{n}\right) W_n$, such that

$$B \subset \overline{\Gamma}\left(\{x_j \otimes \cdots \otimes x_j\}_{j=1}^\infty\right).$$

This implies

$$q(P) = \|P\|_B \leq \sup_j \|P(x_j)\|.$$

for all $P \in \mathcal{P}(^n E)$. Let $L_n = \{x_j\}_{j=1}^\infty \cup \{0\} \cup K$. Then L_n is a compact subset of $\left(1 + \dfrac{1}{n}\right) W_n$. Since $W_n \longrightarrow K$ as $n \to \infty$ it follows that $L_n \longrightarrow K$ as $n \to \infty$. We have

$$p\left(\sum_{n=0}^\infty \frac{\widehat{d^n} f(0)}{n!}\right) \leq \sum_{n=0}^\infty \left\|\frac{\widehat{d^n} f(0)}{n!}\right\|_{L_n}$$

for all $\sum_{n=0}^\infty \dfrac{\widehat{d^n} f(0)}{n!} \in H(U)$. By Proposition 3.36(a) we see that p is τ_0-continuous on $H(U)$. This completes the proof.

In Proposition 3.48 we succeeded in lifting the polynomial result (Proposition 1.36) to holomorphic functions on convex balanced domains. This required more precise estimates for polynomials than those needed in Proposition 1.36. We encounter, to a greater or lesser extent, similar problems in lifting other results from spaces of homogeneous polynomials to holomorphic functions on balanced domains. Proposition 3.48 is still true for balanced domains (Proposition 5.7) but the proof is quite different.

We now discuss completeness for spaces of holomorphic functions and in the process introduce a very weak form of completeness which is quite useful when combined with Taylor series expansions. Note that condition (3.31) in the definition of S-absolute decomposition is a completeness condition which was used in an essential way in the proof of Proposition 3.34(b).

Definition 3.49 Let $\{E_n\}_n$ denote an absolute Schauder decomposition for the locally convex space (E, τ). We say that $\{E_n\}_n$ is a T.S. (=Taylor series) τ-complete decomposition if for any sequence $(x_n)_n$, $x_n \in E_n$ for all n, $\sum_{n=1}^\infty p(x_n) < \infty$ for all $p \in cs(E)$ implies $\sum_{n=1}^\infty x_n \in E$.

A decomposition $\{E_n\}_n$ of a locally convex space is said to be boundedly complete if $x_n \in E_n$ all n and $\left\{\sum_{n=1}^{m} x_n\right\}_{m=1}^{\infty}$ bounded imply $\sum_{n=1}^{\infty} x_n \in E$. It is easily seen that an $\mathcal{S}$-absolute decomposition is T.S.-complete if and only if it is boundedly complete. When there is no danger of confusion we simply say that (E, τ) is T.S.-complete. In discussing spaces of holomorphic functions on balanced domains we always assume that we are dealing with the decomposition associated with the Taylor series expansion at the origin. If the locally convex space E is sequentially complete then it is T.S.-complete with respect to any $\mathcal{S}$-absolute decomposition.

If $\left\{(\mathcal{P}(^nE; F), \tau)\right\}_{n=0}^{\infty}$ is a Schauder decomposition for $(H(U; F), \tau)$, U a balanced open subset of E, $\tau \geq \tau_p$ (the topology of pointwise convergence), $P_n \in \mathcal{P}(^nE; F)$ for all n and $\sum_{n=0}^{\infty} p(P_n) < \infty$ for any τ-continuous seminorm p then $\sum_{n=0}^{\infty} P_n \in H_G(U; F)$. If, in addition, τ-bounded sequences in $H(U; F)$ are locally bounded then $\sum_{n=0}^{\infty} P_n \in H(U; F)$ and $(H(U; F), \tau)$ is T.S. τ-complete. This observation gives many examples and shows the close connection between local boundedness and T.S.-completeness.

Example 3.50

(a) If E is a Fréchet space or a $\mathcal{DFS}$ space and U is a balanced open subset of E and F is a Banach space then Examples 3.8(a) and (d) show that $H(U; F)$ is T.S. τ_p-complete.

(b) If E is a $\mathcal{DFM}$ space, U is a balanced open subset of E and F is a Banach space then Example 3.8(e) shows that $H(U; F)$ is T.S. τ_0-complete.

(c) If $\{E_m\}_{m=1}^{\infty}$ is a sequence of Banach spaces then Example 3.8(f) shows that $H\left(\sum_{m=1}^{\infty} E_m\right)$ is T.S. τ_0-complete if and only if each E_m is finite dimensional.

(d) Let $(P_n)_{n=1}^{\infty}$ denote the sequence of continuous 2-homogeneous polynomials defined on $E := \mathbb{C}^{\mathbb{N}} \times \mathbb{C}^{(\mathbb{N})}$ in Example 1.27. If K is a compact subset of E then $\|P_n\|_K = 0$ for all n sufficiently large. Hence $P_n^n \in \mathcal{P}(^{2n}E)$ and $\sum_{n=1}^{\infty} \|P_n^n\|_K < \infty$ for all K compact in E. This shows that $f := \sum_{n=1}^{\infty} P_n^n \in H_{HY}(E)$. It is easily seen, using very strongly convergent sequences in $\mathbb{C}^{\mathbb{N}}$, that f does not factor through $\mathbb{C}^n \times \mathbb{C}^{(\mathbb{N})}$ for any finite n. Hence $f \notin H(E)$ and $(H(\mathbb{C}^{\mathbb{N}} \times \mathbb{C}^{(\mathbb{N})}), \tau_0)$ is not T.S. τ_0-complete. This result also follows from Corollary 4.46. By Example 3.24(c), $H(\mathbb{C}^{\mathbb{N}} \times \mathbb{C}^{(\mathbb{N})})$ is T.S. τ_ω-complete.

The results in Example 3.50 can be extended to various other topologies on $H(U; F)$ in view of part (a) of the following proposition. We omit the proof which is immediate.

Proposition 3.51 *Let U denote an open subset of a locally convex space E and let F denote a Banach space.*

(a) If $\big\{\big(\mathcal{P}(^nE;F),\tau_i\big)\big\}_{n=0}^{\infty}$ is an S-absolute decomposition for $\big(H(U;F),\tau_i\big)$, $i=1,2$, and $\tau_1 \geq \tau_2$ then $H(U;F)$ $T.S.\,\tau_2$-complete implies $H(U;F)$ is $T.S.\,\tau_1$-complete.

(b) If $H(U;F)$ is $T.S.\,\tau_\omega$-complete then $\tau_\omega^{bor} = \tau_\delta$.

Proof. (b) We must show that τ_ω-bounded sets are τ_δ bounded. Let $(f_\alpha)_{\alpha\in\Gamma}$ denote a τ_ω-bounded subset of $H(U;F)$. By Lemma 3.33

$$\sum_{n=0}^{\infty} \sup_{\alpha} p\Big(\frac{\widehat{d^n} f_\alpha(0)}{n!}\Big) < \infty$$

for any τ_ω-continuous semi-norm p on $H(U;F)$. By T.S. τ_ω-completeness $\sum_{n=0}^{\infty}\dfrac{\widehat{d^n} f_{\alpha_n}(0)}{n!}$ is an element of $H(U;F)$ for any sequence $(\alpha_n)_n \subset \Gamma$. Since $\big\{\big(\mathcal{P}(^nE;F),\tau_\omega\big)\big\}_{n=0}^{\infty}$ is an S-absolute decomposition for $\big(H(U;F),\tau_\delta\big)$ we have $\sum_{n=0}^{\infty} q\Big(\dfrac{\widehat{d^n} f_{\alpha_n}(0)}{n!}\Big) < \infty$ for any τ_δ-continuous semi-norm q on $H(U)$ and any sequence $(\alpha_n)_n \subset \Gamma$. A simple approximation shows that

$$\sum_{n=0}^{\infty} \sup_{\alpha} q\Big(\frac{\widehat{d^n} f_\alpha(0)}{n!}\Big) < \infty$$

for any τ_δ-continuous semi-norm q on $H(U;F)$. By Lemma 3.33, the set $(f_\alpha)_{\alpha\in\Gamma}$ is τ_δ bounded. This completes the proof.

Proposition 3.52 *Let U denote a balanced open subset of a locally convex space E and let F denote a Banach space. If $\big\{\big(\mathcal{P}(^nE;F),\tau\big)\big\}_{n=0}^{\infty}$ is an S-absolute decomposition for $\big(H(U;F),\tau\big)$ then $\big(H(U;F),\tau\big)$ is complete (respectively quasi-complete, sequentially complete) if and only if $H(U;F)$ is $T.S.\,\tau$-complete and $\big(\mathcal{P}(^nE;F),\tau\big)$ is complete (respectively quasi-complete, sequentially complete) for all n.*

Proof. The conditions given are clearly necessary. We prove that they are also sufficient. We consider only the complete case as the others are handled in the same way. Let $(f_\alpha)_{\alpha\in\Gamma}$ denote a Cauchy net in $(H(U;F),\tau)$. Then $\Big(\dfrac{\widehat{d^n} f_\alpha(0)}{n!}\Big)_{\alpha\in\Gamma}$ is a Cauchy net in $\big(\mathcal{P}(^nE;F),\tau\big)$ for each n and, by our hypothesis, there exists $P_n \in \mathcal{P}(^nE;F)$ such that $\dfrac{\widehat{d^n} f_\alpha(0)}{n!} \longrightarrow P_n$ as $\alpha \to \infty$. Let p denote a τ-continuous semi-norm on $H(U;F)$. We may suppose, without loss of generality, that

$$p\left(\sum_{n=0}^{\infty} \frac{\hat{d}^n f(0)}{n!}\right) = \sum_{n=0}^{\infty} p\left(\frac{\hat{d}^n f(0)}{n!}\right)$$

for all $f \in H(U; F)$. Given $\varepsilon > 0$ we can find $\alpha_0 \in \Gamma$ such that

$$\sum_{n=0}^{\infty} p\left(\frac{\hat{d}^n f_\alpha(0)}{n!} - \frac{\hat{d}^n f_\beta(0)}{n!}\right) \leq \varepsilon$$

for all $\alpha, \beta \geq \alpha_0$. In particular for each positive integer k we have

$$\sum_{n=0}^{k} p(P_n) \leq \sum_{n=0}^{\infty} p\left(\frac{\hat{d}^n f_{\alpha_0}(0)}{n!}\right) + \varepsilon.$$

Hence $\sum_{n=0}^{\infty} p(P_n) < \infty$ and, since $H(U; F)$ is T.S. τ-complete, we have $f := \sum_{n=0}^{\infty} P_n \in H(U; F)$. By a standard argument we get

$$\sum_{n=0}^{\infty} p\left(P_n - \frac{\hat{d}^n f_\beta(0)}{n!}\right) \leq \varepsilon$$

for all $\beta \geq \alpha_0$. Hence $f_\alpha \longrightarrow f$ as $\alpha \to \infty$. This completes the proof.

Many questions regarding completeness can be resolved using the above techniques. We confine ourselves here to just one application but later we shall see that certain completeness conditions are equivalent to equalities of topologies.

Corollary 3.53 *If U is a balanced open subset of a Fréchet space then $(H(U), \tau_\omega)$ and $(H(U), \tau_\delta)$ are complete locally convex spaces.*

Proof. Since E is Fréchet, Example 3.50(a) shows that $H(U)$ is T.S. τ_p-complete and hence T.S. τ_ω and T.S. τ_δ-complete. Since τ_ω and τ_δ agree on $\mathcal{P}(^n E)$ it suffices, by Proposition 3.52, to show that $(\mathcal{P}(^n E), \tau_\omega)$ is complete. Now $(\mathcal{P}(^n E), \tau_\omega) = \left(\bigotimes_{n,s,\pi} E\right)'_i$ and $\bigotimes_{n,s,\pi} E$ is a Fréchet space. A result of Grothendieck says that the inductive dual of a Fréchet space is complete and hence $(\mathcal{P}(^n E), \tau_\omega)$ is complete. This completes the proof.

In our final example on the uses of $\mathcal{S}$-absolute decompositions we discuss quasi-normability of the space $(H(U), \tau_\omega)$. We have chosen this concept for a number of reasons. Our discussion in Section 3.2 suggests that we are more likely to be successful with properties that are common to Banach spaces and nuclear spaces and quasi-normability fits into this category. After proving our result we note that a number of other properties can be represented in a similar fashion and we shall take a certain feature of these representations as motivation for our approach in Chapter 4. A variation of quasi-normability also plays a role in Section 4.3.

A locally convex space E is *quasi-normable* if for every neighbourhood U of 0 in E there exists a neighbourhood V of 0 in E such that for every $\lambda > 0$ there exists a bounded subset M_λ in E satisfying

$$V \subset M_\lambda + \lambda U. \tag{3.52}$$

We can restate (3.52) in terms of semi-norms as follows: for each $p \in cs(E)$ there exists $q \in cs(E)$ such that for each $\lambda > 0$ there exists a bounded subset M_λ in E with

$$\{x : q(x) \le 1\} \subset M_\lambda + \{x : p(x) \le \lambda\}. \tag{3.53}$$

Proposition 3.54 *If U is a balanced open subset of a Fréchet space then* $\big(H(U), \tau_\omega\big)$ *is quasi-normable.*

Proof. Since countable inductive limits of Banach spaces are quasi-normable, $(\mathcal{P}(^nE), \tau_\omega)$ is quasi-normable for every positive integer n. Let p denote a τ_ω-continuous semi-norm on $H(U)$. Without loss of generality we may suppose

$$p\Big(\sum_{n=0}^{\infty} \frac{\widehat{d}^n f(0)}{n!}\Big) = \sum_{n=0}^{\infty} p\Big(\frac{\widehat{d}^n f(0)}{n!}\Big)$$

for all $f = \displaystyle\sum_{n=0}^{\infty} \frac{\widehat{d}^n f(0)}{n!} \in H(U)$.

For each n let $p_n := p\big|_{\mathcal{P}(^nE)}$. Since $\big(\mathcal{P}(^nE), \tau_\omega\big)$ is quasi-normable there exists $q_n \in cs\big(\mathcal{P}(^nE), \tau_\omega\big)$ such that (3.53) is satisfied by q_n and p_n. Let $(V_m)_{m=1}^{\infty}$ denote a decreasing fundamental neighbourhood system at the origin in E. For any pair of positive integers n and m there exists $c_n(V_m) > 0$ such that

$$q_n(P_n) \le c_n(V_m)\|P_n\|_{V_m}$$

for all $P_n \in \mathcal{P}(^nE)$. Without loss of generality we may suppose that the sequence $\big\{c_n(V_m)\big\}_{m=1}^{\infty}$ is increasing for each n. For all n let $\alpha_n = \dfrac{1}{c_n(V_n)}$ and, if $f \in H(U)$, let

$$q\Big(\sum_{n=0}^{\infty} \frac{\widehat{d}^n f(0)}{n!}\Big) := \sum_{n=0}^{\infty} \alpha_n q_n \Big(\frac{\widehat{d}^n f(0)}{n!}\Big) + \sum_{n=0}^{\infty} n^2 p\Big(\frac{\widehat{d}^n f(0)}{n!}\Big).$$

If $f \in H(U)$ and m is a positive integer then

$$\sum_{n=0}^{\infty} \alpha_n q_n \Big(\frac{\widehat{d}^n f(0)}{n!}\Big) \le \sum_{n=0}^{m-1} \frac{c_n(V_m)}{c_n(V_n)} \Big\|\frac{\widehat{d}^n f(0)}{n!}\Big\|_{V_m} + \sum_{n=m}^{\infty} \Big\|\frac{\widehat{d}^n f(0)}{n!}\Big\|_{V_m}$$

and thus

$$\sum_{n=0}^{\infty} \alpha_n q_n\left(\frac{\widehat{d}^n f(0)}{n!}\right) \le \sup_{1 \le n \le m}\left(\frac{c_n(V_m)}{c_n(V_n)}\right) \cdot \sum_{n=0}^{\infty}\left\|\frac{\widehat{d}^n f(0)}{n!}\right\|_{V_m}.$$

Hence q is τ_ω-continuous. To complete the proof we show that (3.53) is satisfied by p and q. Let $\lambda > 0$ be arbitrary. Choose a positive integer n_0 such that $\dfrac{1}{n_0^2} \le \dfrac{\lambda}{2}$. If $f \in H(U)$ and $q(f) \le 1$ then

$$n_0^2 \cdot \sum_{n=n_0}^{\infty} p_n\left(\frac{\widehat{d}^n f(0)}{n!}\right) \le \sum_{n=n_0}^{\infty} n^2 p_n\left(\frac{\widehat{d}^n f(0)}{n!}\right) \le q(f) \le 1.$$

Hence

$$p\left(\sum_{n=n_0}^{\infty} \frac{\widehat{d}^n f(0)}{n!}\right) \le \frac{1}{n_0^2} \le \frac{\lambda}{2}. \tag{3.54}$$

For each $n < n_0$ there exists a τ_ω-bounded subset of $\mathcal{P}(^n E)$, M_n, such that

$$\left\{P_n \in \mathcal{P}(^n E) : q_n(P_n) \le 1\right\} \subseteq M_n + \left\{P_n \in \mathcal{P}(^n E) : p_n(P_n) \le \frac{\alpha_n \lambda}{2^{n+2}}\right\}.$$

The set $M := \displaystyle\sum_{n=0}^{n_0-1} \frac{M_n}{\alpha_n}$ is a τ_ω-bounded subset of $H(U)$. Since $q(f) \le 1$ we have $q_n\left(\dfrac{\widehat{d}^n f(0)}{n!}\right) \le \dfrac{1}{\alpha_n}$ and hence

$$\frac{\widehat{d}^n f(0)}{n!} \in \frac{M_n}{\alpha_n} + \left\{P_n \in \mathcal{P}(^n E) : p_n(P_n) \le \frac{\lambda}{2^{n+2}}\right\}$$

for $n \le n_0$. Hence

$$\sum_{n=0}^{n_0-1} \frac{\widehat{d}^n f(0)}{n!} \in \sum_{n=0}^{n_0-1} \frac{M_n}{\alpha_n} + \left\{\sum_{n=0}^{n_0-1} P_n : P_n \in \mathcal{P}(^n E), \sum_{n=0}^{n_0-1} p_n(P_n) \le \frac{\lambda}{2}\right\}.$$

This together with (3.54) shows that

$$\{f \in H(U) : q(f) \le 1\} \subset M + \{f \in H(U) : p(f) \le \lambda\}$$

and completes the proof.

Several other important concepts in linear functional analysis can be expressed by set inclusions similar to (3.52). For example, a locally convex space E is Schwartz if and only if for each neighbourhood V of 0 in E there is a neighbourhood U of 0 such that for every $\lambda > 0$ there exists a *finite* subset F_λ of E satisfying

$$V \subset F_\lambda + \lambda U. \tag{3.55}$$

A quasi-complete locally convex space E is semi-Montel if and only if for each neighbourhood U of 0, each bounded subset B of E and each $\lambda > 0$ there exists a *finite* subset G_λ in E such that

$$B \subset G_\lambda + \lambda U. \tag{3.56}$$

These representations show that Schwartz spaces are quasi-normable, quasi-complete Schwartz spaces are semi-Montel and, moreover, a complete locally convex space is a Schwartz space if and only if it is semi-Montel and quasi-normable. In discussing these concepts for spaces of holomorphic functions on balanced domains the methods used in the proof of Proposition 3.54 are useful.

3.5 Exercises

Exercise 3.55* Show that $H_G(U; F) = H(U; F)$ for an open subset U of a locally convex space E if and only if $E \approx \mathbb{C}^{(\mathbb{N})}$.

Exercise 3.56* Show that the composition of holomorphic (respectively hypoanalytic) functions is holomorphic (respectively hypoanalytic).

Exercise 3.57* If E is an infinite dimensional Banach space show that $H_b(E)$ is a set of first category in $(H(E), \tau_\delta)$.

Exercise 3.58* If E is a locally convex space and F is a complete locally convex space show that the monomials form an unconditional generalized basis for $H_G(E; F)$ endowed with the topology of uniform convergence on the finite dimensional compact subsets of E.

Exercise 3.59* If E is a Fréchet space and $f \in H_G(E)$ show that $f \in H(E)$ if and only if f is a Borel measurable function.

Exercise 3.60* Let E and F denote Banach spaces and let U be a connected open subset of E. Suppose $f: U \to F^*$, $f(V) \subset F'$ for some non-empty open subset V of U and $\phi \circ f \in H(U)$ for every $\phi \in F$. Show that $f \in H(U; F'_\beta)$.

Exercise 3.61 Let U denote an open subset of a locally convex space E and let F denote a locally convex space. If $f \in H(U; F)$ show that the closed vector subspace of $\widehat{F}$ (the completion of F) generated by $f(U)$ is equal to the closed vector subspace of $\widehat{F}$ generated by

$$\bigcup_{\substack{n \in \mathbb{N} \\ \xi \in U, x \in E}} \widehat{d}^n f(\xi)(x).$$

Exercise 3.62* If $n \in \mathbb{N}$, F is a Fréchet space and $(f_m)_{m \in N} \subset H(\mathbb{C}^n, F)$ show that there exists a sequence of mutually orthogonal projections $(P_k)_k$ on F each of which has one-dimensional range such that for all m

$$f_m = \sum_{k=1}^{\infty} P_k \circ f_m$$

uniformly on compact subsets of $\mathbb{C}^n$.

Exercise 3.63 Let E, F and G denote Banach spaces and let U denote an open subset of E. Let $f \in H_G(U; \mathcal{L}(F, G))$. Show that f is holomorphic if and only if the mapping

$$z \in U \longrightarrow \phi\big(f(z)(\omega)\big).$$

is continuous for all $\omega \in F$ and $\phi \in G'$.

Exercise 3.64* If E is a Banach space with unit ball B, F is a Banach space and $f \in H(E; F)$ let

$$r_f(x) = \left(\limsup_{n \to \infty} \left\| \frac{\widehat{d^n} f(x)}{n!} \right\|^{1/n} \right)^{-1}.$$

Show that

$$r_f(x) = \sup\{\lambda > 0 : \|f\|_{x+\lambda B} < \infty\}$$

$$= \sup\left\{\lambda > 0 : \lim_{m \to \infty} \left\| f(x+z) - \sum_{n=0}^{m} \frac{\widehat{d^n} f(x)}{n!}(z) \right\|_{\{\|z\| \leq \lambda\}} = 0 \right\}$$

for each $x \in E$ (because of this result $r_f(x)$ is called the *radius of boundedness* or the *radius of uniform convergence* of f at x). Hence deduce that

$\lim_{n \to \infty} \left\| \dfrac{\widehat{d^n} f(0)}{n!} \right\|_K^{1/n} = 0$ for any compact subset K of E.

Exercise 3.65* Let U be an open subset of a Banach space E and let F be a Banach space with Schauder basis $(e_n)_n$. If f is a mapping from U into F then $f(x) = \sum_{n=1}^{\infty} f_n(x) e_n$ where $f_n : U \to \mathbb{C}$. The mapping f is said to be normal if, for each compact subset K of U and each positive δ, there exists a positive integer $n(K, \delta)$ such that

$$\left\| \sum_{n=m}^{\infty} f_n(x) e_n \right\|_K \leq \delta \text{ for all } m \geq n(K, \delta).$$

If $f_n \in H(U)$ for all n show that $f \in H(U; F)$ if and only if f is a normal mapping.

Exercise 3.66* Let $E = \sum_{n=1}^{\infty} E_n$ where each E_n is a locally convex space. Suppose $f_n \in H(E_n)$ and $g_n \in \mathcal{P}(\sum_{k=1}^{n} E_k)$ for each n. If there exists a neighbourhood U of zero in E such that $\|g_n\|_U < \infty$ for all n show that $f := \sum_{n=1}^{\infty} f_n g_n \in H(E)$.

Exercise 3.67 If U is an open subset of a locally convex space E, F is a locally convex space and f in $H_G(U; F)$ show that the following are equivalent:

(a) $f \in H(U; F)$ (respectively f is locally bounded)

(b) for each ξ in U, $\left(\dfrac{\widehat{d^n} f(\xi)}{n!} \right)_{n=0}^{\infty}$ is an equicontinuous (respectively locally bounded) sequence of mappings.

Exercise 3.68* If E is a locally convex Baire space, F is a locally convex metrizable Baire space, U is an open subset of $E \times F$ and f is a separately holomorphic function on U show that f is holomorphic.

Exercise 3.69* Let E denote a locally convex space and let $(\phi_n)_n$ denote an equicontinuous sequence in E'. If $f = \sum_{n=1}^{\infty} \phi_n^n$ show

(a) $f \in H(E)$ if and only if $(\phi_n)_n$ is weak*-null,
(b) $f \in H_b(E)$ if and only if $(\phi_n)_n$ is $\beta(E', E)$-null,
(c) $f \in H_{ub}(E)$ if and only if $\phi_n \to 0$ as $n \to \infty$ uniformly on a neighbourhood of the origin.

Exercise 3.70 Let E denote a Banach space and let $c_0(E')$ be the space of all weak*-null sequences in E' endowed with the sup norm topology. Let $L: c_0(E') \to (H(E), \tau_0)$ be defined by $L((\phi_n)_n) = \displaystyle\sum_{n=1}^{\infty} \phi_n^n$. Show that L is well defined and holomorphic.

Exercise 3.71* Let $g \in H(\mathbb{C})$ and let $E = F = \mathbb{C}^I$. Define $f: E \to F$ by $f((x_i)_{i \in I}) = (g(x_i))_{i \in I}$. Show that $f \in H(E; F)$. Show that for each x in E there exists a neighbourhood V_x of x such that the Taylor series expansion of f at x converges uniformly on V_x with respect to every $\beta \in \mathrm{cs}(F)$ if and only if g is a polynomial.

Exercise 3.72 Show that a G-holomorphic mapping $f: U \subset E \to F$, U an open subset of the locally convex E and F a locally convex space, is hypoanalytic in either of the following cases:

(a) $\widehat{d^n} f(\xi) \in \mathcal{P}_{HY}(^nE; F)$ for all $\xi \in U$ and all $n \in \mathbb{N}$ and f is bounded on the compact subsets of U,
(b) E is weakly separable, satisfies the Mackey convergence criterion and f is bounded on the compact subsets of U.

Exercise 3.73* Show that $\left(H(\mathbb{C}^{\mathbb{N}}), \tau_\delta\right) = \varinjlim_n \left(H(\mathbb{C}^n), \tau_0\right).$

Exercise 3.74* Let Γ denote an uncountable discrete set. If $f \in H\left(c_0(\Gamma)\right)$ show that there exists a countable subset Γ_1 of Γ and $\tilde{f} \in H(c_0(\Gamma_1))$ such that $f = \tilde{f} \circ \pi_{\Gamma_1}$ where π_{Γ_1} is the canonical surjection of $c_0(\Gamma)$ onto $c_0(\Gamma_1)$.

Exercise 3.75 If $f \in H\left(\prod_{\alpha \in A} E_\alpha\right)$ where each E_α is a locally convex space, show that there exists a finite subset A_1 of A and $\tilde{f} \in H\left(\prod_{\alpha \in A_1} E_\alpha\right)$ such that $f = \tilde{f} \circ \pi_{A_1}$ where π_{A_1} is the natural projection from $\prod_{\alpha \in A} E_\alpha$ onto $\prod_{\alpha \in A_1} E_\alpha$. If E_α is metrizable for each α in A show that $H_{HY}\left(\prod_{\alpha \in A} E_\alpha\right) = H\left(\prod_{\alpha \in A} E_\alpha\right)$.

Exercise 3.76* If X and Y are completely regular Hausdorff spaces and $\mathcal{C}(X)$ and $\mathcal{C}(Y)$, each with the compact open topology, are infrabarrelled show that $f \in H_G\left(\mathcal{C}(X), \mathcal{C}(Y)'_\beta\right)$ is holomorphic if and only if it is locally bounded.

Exercise 3.77* If X is a completely regular Hausdorff topological space show that the following are equivalent:

(a) $(\mathcal{C}(X), \tau_0)$ is barrelled,
(b) if $A \subset X$ and $\|f\|_A < \infty$ for all $f \in \mathcal{C}(X)$ then A is relatively compact in X,
(c) if $B \subset H(\mathcal{C}(X))$ is uniformly bounded on the finite dimensional compact subsets of U then B is locally bounded.

Exercise 3.78* If E is a DF space and F is a Banach space show that $f \in H_G(E, F)$ is continuous if and only if it is uniformly continuous on bounded sets. Show that $H_b(E; F) = H_{ub}(E; F)$ and hence deduce that each $f \in H_b(E)$ has an extension to E'' where it defines an element of $H_b(E'')$. Show that E is a DFM space if and only if $H(E) = H_{ub}(E)$.

Exercise 3.79* If $E = \varinjlim_n E_n$ is an inductive limit of normed linear spaces (or equivalently a bornological DF space) and $\mathcal{F} \subset H_G(E)$ is uniformly bounded on bounded sets show that $\mathcal{F}$ is a locally bounded subset of $H_b(E)$.

Exercise 3.80* Show that the equality $H_b(E; F) = H_{ub}(E; F)$ is inherited by subspaces and hence show that the Fréchet space E is a quojection if and only if $H_b(E; F) = H_{ub}(E; F)$ for every Fréchet space F with continuous norm.

Exercise 3.81* If F is a Fréchet space show that $E := F'_c$ leads to uniform holomorphy if and only if E is separable.

Exercise 3.82 If E is a complete locally convex space and the τ_0-bounded subsets of $H(E)$ are locally bounded show that E is barrelled.

Exercise 3.83* We say that a closed countable increasing cover $\mathcal{F} := (F_n)_{n=1}^{\infty}$ of a topological space U is *k-dominating* if each compact subset of U is contained in some F_n. Let U be an open subset of a locally convex space E and let

$$H_{\mathcal{F}}(U) = \{f \in H(U) : \|f\|_{F_n} < \infty \text{ for all } n\}.$$

$H_{\mathcal{F}}(U)$ is endowed with the locally convex topology generated by the seminorms $p_n(f) = \|f\|_{F_n}$. Let $\big(H(U), \tau_{\delta_K}\big) = \lim_{\overrightarrow{\mathcal{F}}} H_{\mathcal{F}}(U)$ where $\mathcal{F}$ ranges over all countable increasing k-dominating covers of U. Show that τ_{δ_k} is the bornological topology associated with τ_0 on $H(U)$. Show that $\big(H(U), \tau_{\delta_K}\big)$ is an ultrabornological space if U is a k-space.

Exercise 3.84* Let τ_p denote the topology of pointwise convergence. If E and F are locally convex spaces such that $H(E)$ and $H(F)$ are T.S. τ_p-complete and separately continuous polynomials on $E \times F$ are continuous show that separately holomorphic functions on $E \times F$ are holomorphic if and only if $H(E \times F)$ is T.S. τ_p-complete.

Exercise 3.85* Let U denote an open subset of a locally convex space E. For each $n \in \mathbb{N}$ let τ_n denote the topology on $H(U)$ generated by $p_{K,B}(f) := \sup_{\substack{0 \le i \le n \\ x \in K}} \|\widehat{d^i} f(x)\|_B$ where K ranges over the compact subsets of U and B over the bounded subsets of E. Show that $\tau_n = \tau_{n+1}$ for some (and hence for all n) if and only if each bounded subset of E is contained in the closed convex hull of a compact set. Let $\big(H(U), \tau_\infty\big) = \lim_{\overleftarrow{n}} \big(H(U), \tau_n\big)$. Show that $\big(H(U), \tau_n\big)$, $0 \le n \le \infty$, is a locally m-convex algebra. If E is an infinite dimensional Banach space show that $\tau_\infty \underset{\ne}{\le} \tau_\omega$ on $H(U)$.

Exercise 3.86* Let $E = \lim_{\overrightarrow{n}} E_n$ denote a strict inductive limit of Fréchet spaces and let $f \in H_G(E)$. If $(V_n)_n$ is a sequence of subsets of E where V_n is a neighbourhood of zero in E_n and $nV_n \subset V_{n+1}$ and, (*) $\sup_m \left\| \dfrac{\widehat{d^m} f(0)}{m!} \right\|_{V_n} < \infty$ for all n, show that $f \in H(E)$. If each E_n is a Banach space show that (*) is satisfied if and only if $f\big|_{E_n} \in H_b(E_n)$ for all n.

Exercise 3.87* Show that a sequentially complete locally convex space contains a non-trivial very strongly convergent sequence if and only if it contains $\mathbb{C}^{\mathbb{N}}$ as a complemented subspace.

Exercise 3.88* A net $(x_\alpha)_{\alpha \in \Gamma}$ is said to be very strongly convergent if $\lambda_\alpha x_\alpha \to 0$ as $\alpha \to \infty$ for any net of scalars $(\lambda_\alpha)_{\alpha \in \Gamma}$. If $x_\alpha \neq 0$ for an infinite number of α we said that the net is non-trivial. Show that a locally convex space contains a non-trivial very strongly convergent net if and only if it does not admit a continuous norm.

Let E denote a Banach space such that E' is not weak*-separable (for example a non-separable Hilbert space). Let τ denote the topology on E of uniform convergence on the weak*-separable bounded subsets of E'. Show that (E, τ) does not admit a continuous norm and does not contain any non-trivial very strongly convergent sequences.

Exercise 3.89* If E and F are Banach spaces, $f \in H(E; F)$ and $r_f(0) < \infty$ show that

$$\{\phi \in F' : r_{\phi \circ f}(0) > r_f(0)\}$$

is a set of first category in F'. If E is separable show that there exists $\phi \in F'$ such that $r_f = r_{\phi \circ f}$.

Exercise 3.90 If E is a barrelled locally convex space show that the bounding subsets of E are limited.

Exercise 3.91* If U is an open subset of a separable locally convex space and let $\mathcal{F} \subset H(U)$. Show that the following are equivalent:

(a) $\mathcal{F}$ is locally bounded,
(b) $\mathcal{F}$ is pointwise bounded and all sequences in $\mathcal{F}$ are locally bounded.

Exercise 3.92* Let (E, τ) denote a bornological locally convex space and suppose there exists a weaker topology τ_1 on E such that E contains a fundamental set of τ-bounded sets which are τ_1-compact. Let $F = \{\phi \in E' : \phi$ restricted to the τ-bounded subsets is τ_1-continuous$\}$ and endow F with the topology of uniform convergence on the τ-bounded subsets of E. Show that E is ultrabornological and $F_i' = (E, \tau)$. If E contains a τ-neighbourhood system at the origin consisting of τ_1-closed sets show that F is infrabarrelled and that $F_b' = F_i' = (E, \tau)$.

Exercise 3.93* If $\{E_n\}_n$ is a Schauder decomposition for the barrelled space E and $\sum_{n=1}^\infty \alpha_n x_n \in E$ whenever $(\alpha_n)_n \in S$ and $\sum_{n=1}^\infty x_n \in E$ show that $\{E_n\}_n$ is an S-absolute decomposition for E.

Exercise 3.94* Let U denote a balanced open subset of a locally convex space E and let F denote a Banach space. Show that sets of the form

$\left\{ \sum_{n=0}^{\infty} \left\| \frac{\widehat{d^n} f(0)}{n!} \right\|_{V_n} \leq \alpha \right\}$, where $\alpha > 0$ and $(V_n)_{n=1}^{\infty}$ is an increasing countable balanced open cover of U form a fundamental system of locally bounded subsets of $H(U; F)$.

Exercise 3.95* A locally convex space E is said to be countably barrelled if every convex balanced absorbing subset E which is the intersection of a sequence of closed convex zero neighbourhoods is a neighbourhood of zero. If U is a balanced open subset of a locally convex space E show that $\tau_\omega = \tau_\delta$ on $H(U)$ if and only if $(H(U), \tau_\omega)$ is countably barrelled.

Exercise 3.96 Let E denote a Banach space and suppose $(\phi_n)_n$ is a weak* null sequence of unit vectors in E'. Let $(k_n)_{n=1}^{\infty}$ be a strictly increasing sequence of positive integers and for each n let j_n be a non-negative integer with $0 \leq j_n \leq k_n$. Show that $\sum_{n=1}^{\infty} \phi_1^{j_n} \phi_n^{k_n - j_n} \in H(E)$ if and only if $\liminf_{n \to \infty} (k_n - j_n)/k_n > 0$.

Exercise 3.97 Show that $\left(\mathcal{P}_{HY}(^nE; F), \tau_0 \right)_{n=0}^{\infty}$ is an $\mathcal{S}$-absolute decomposition for $\left(H_{HY}(U; F), \tau_0 \right)$ for any balanced open subset of any locally convex space E and any complete locally convex space F.

Exercise 3.98* If U is a balanced open subset of a locally convex space E and F is a Banach space show that the following are equivalent:

(i) $\tau_\omega^{bor} = \tau_\delta$ on $H(U; F)$ or equivalently τ_δ and τ_ω define the same bounded sets,

(ii) τ_δ and τ_ω define the same compact subsets of $H(U; F)$,

(iii) τ_ω and τ_δ define the same topology on τ_ω-bounded sets,

(iv) τ_ω^{bor} is a barrelled topology on $H(U; F)$,

(v) If $T_n \in \left(\mathcal{P}(^nE; F), \tau_\omega \right)'$ for all n and $\sum_{n=0}^{\infty} T_n \left(\frac{\widehat{d^n} f(0)}{n!} \right)$ converges for every
$$f = \sum_{n=0}^{\infty} \frac{\widehat{d^n} f(0)}{n!} \text{ in } H(U; F) \text{ then } \sum_{n=0}^{\infty} T_n \in \left(H(U; F), \tau_{\omega, b} \right)'.$$

Exercise 3.99* If U is a balanced open subset of a locally convex space E and F is a Banach space show that τ_ω and τ_δ define the same convex balanced compact subsets of $H(U; F)$.

Exercise 3.100* A subset A of a locally convex space E is called a (holomorphic or analytic) *determining set* if $f \in H(E)$ and $f\big|_A = 0$ implies $f \equiv 0$. Show that E contains a compact determining set if and only if (E', τ_0) admits a continuous norm. Show that a metrizable locally convex space contains a compact determining set if and only if it is separable.

Exercise 3.101* If (E, τ) is a complete (respectively quasi-complete, sequentially complete) locally convex space show that (E, τ^{bar}) is complete (respectively quasi-complete, sequentially complete).

Exercise 3.102* Let U be an open polydisc in a fully nuclear space with basis E and let F be a complete locally convex space with equicontinuous basis $(e_n)_n \in \mathbb{N}$. Show that the monomials $(e_n z^m)_{n \in \mathbb{N}, m \in \mathbb{N}^{(\mathbb{N})}}$ form an equicontinuous basis for $H_{HY}(U; F)$.

Exercise 3.103* Let U be an open subset of a sequentially complete locally convex space E such that

(i)   every compact subset of E is contained in the closed balanced convex hull of a null sequence,

(ii) every null sequence is *Mackey-null*, i.e. if $x_n \to 0$ as $n \to \infty$ there exists $(\lambda_n)_n, \lambda_n \uparrow \infty$ such that $\lambda_n x_n \to 0$ as $n \to \infty$.
Show that $(H(U), \tau_0)$ is a Schwartz space.

Exercise 3.104 If U is a balanced open subset of a locally convex space E and F is a locally convex space show that $(H(U, F), \tau_0)$ is semi-Montel if and only if $H(U; F)$ is T.S. τ_0-complete and $(\mathcal{P}(^nE; F), \tau_0)$ is semi-Montel for all n.

Exercise 3.105* Let $\{E_n\}_n$ denote an $\mathcal{S}$-absolute decomposition of the locally convex space E. If each E_n is quasi-normable and for each sequence $(p_n)_n$, $p_n \in \operatorname{cs}(E_n)$, there exists a sequence of positive real numbers such that $p := \sum_{n=1}^{\infty} \alpha_n p_n \in \operatorname{cs}(E)$ show that E is quasi-normable. Hence show that $(H(U), \tau_\delta)$, U balanced in a Fréchet space, is quasi-normable.

Exercise 3.106* If $\{E_n\}_n$ is an $\mathcal{S}$-absolute decomposition for the locally convex space E show that E is semi-reflexive if and only if each E_n is semi-reflexive and E is T.S. τ-complete.

Exercise 3.107* If E is a fully nuclear space with basis show that the following are equivalent:

(1) the τ_δ bounded subsets of $H(E)$ are locally bounded,
(2) $G(E)$ (see Proposition 3.26) is bornological,
(3) $G(E)$ is barrelled,
(4) $G_0(E)$ (see Example 3.28(a)) is bornological,
(5) $G_0(E)$ is infrabarrelled.

Exercise 3.108* Let τ_0^* denote the finest locally convex topology on $H_b(E)$ that coincides with τ_0 on the τ_b-bounded subsets of $H_b(E)$. Show that $\{\mathcal{P}(^nE), \tau_0^*|_{\mathcal{P}(^nE)}\}_n$ is a $\mathcal{S}$-absolute decomposition $H_b(E)$. If E is a Fréchet space show that $\tau_0 \leq \tau_0^* \leq \tau_\delta$. If $H_b(E) = H(E)$ and $\tau_0 = \tau_\omega$ on $\mathcal{P}(^nE)$ all n

show that $\tau_0^* = \tau_\delta$. If E is Banach show that $\tau_0 = \tau_0^*$ if and only if E is finite dimensional.

Exercise 3.109* If p is a τ_w-continuous semi-norm on $\mathcal{P}(^n\mathbb{C}^I)$ and $p(z^m) = a_m$ for all $m \in I^{(I)}$ show that

$$q\Big(\sum_{m \in I^{(\mathbb{N})}} a_m z^m \Big) := \sum_{m \in I^{(\mathbb{N})}} |a_m| p(z^m)$$

defines a τ_w-continuous semi-norm on $\mathcal{P}(^n\mathbb{C}^I)$. Hence show that $\big(H(\mathbb{C}^I), \tau\big)$ is complete for $\tau = \tau_0$, τ_w or τ_δ.

Exercise 3.110* If E is a separable Fréchet space, $F := E'_c = (E', \tau_0)$ and U is an open subset of F show that $\tau_0 = \tau_\delta$ on $H(U)$.

Exercise 3.111 Let $(E_\alpha)_{\alpha \in \Gamma}$ denote a collection of locally convex spaces indexed by the directed set Γ and suppose $E_\alpha \subseteq E_\beta$ topologically if $\alpha \le \beta$. If $\{E_n\}_n$ is an $\mathcal{S}$-absolute decomposition for each E_α show that $\{E_n\}_n$ is also an $\mathcal{S}$-absolute decomposition for $\varinjlim_\alpha E_\alpha$.

Exercise 3.112* If E is an infinite dimensional reflexive Banach space and $F = \big(E, \sigma(E, E')\big)$ show that $(\mathcal{P}(^1F), \tau_w)' \approx (E')^*$ and that the ultra-bornological topology associated with τ_0 is strictly weaker than τ_w.

Exercise 3.113 If E is an infrabarrelled complete DF space show that $\big(H(E), \tau_0\big)$ is barrelled if and only if E is a Montel space.

Exercise 3.114* If A is an uncountable set show that $H\big(\mathbb{C}^{(A)}; l_2(A)\big) \ne H\big(\mathbb{C}^{(A)}, l_2(A)_\sigma\big)$ where σ denotes the weak topology.

Exercise 3.115* Let (E, τ) denote an A-nuclear space. Show that (E, τ^{bor}) is an A-nuclear space. Show that (E, τ) is a Mackey space and that (E, τ) is bornological if and only if it is infrabarrelled.

Exercise 3.116* Show that a Fréchet nuclear space with basis is an A-nuclear space.

Exercise 3.117* If E is a reflexive Banach space and $\mathcal{P}(^nE) = \mathcal{P}_w(^nE)$ for all n, show that $\big(H(U), \tau_w\big)$ is semi-reflexive for every open subset U of E and that $H_b(U)$ is a totally reflexive Fréchet space for any balanced open subset U of E. (Totally reflexive means that all quotients are reflexive.)

Exercise 3.118* If U is an open polydisc in a fully nuclear space with basis show that τ_δ is the barrelled topology associated with τ_0 on $H(U)$.

Exercise 3.119* Let U be an open polydisc in a hereditary Lindelöf fully nuclear space with basis. If $(H(U), \tau_\delta)$ is complete show that the monomials form an absolute basis for $(H(U), \tau_\delta)$.

Exercise 3.120* Let $\Lambda_\infty(\alpha)$ denote a power series space of infinite type where $\alpha = (\alpha_j)_j$ (see Example 4.13(a)). Show that the entire function

$$f((z_j)_j) = \sum_{j=2}^{\infty} z_j e^{z_1 \alpha_j}$$

is not of uniformly bounded type.

Exercise 3.121 If U is a holomorphically convex open subset of a $\mathcal{DFM}$ space and $(\xi_n)_n$ is a sequence in U which converges to $\xi \in \partial U$ show that there exists $f \in H(U)$ such that $\limsup_{n \to \infty} |f(\xi_n)| = \infty$.

Exercise 3.122* Let $\Lambda_\infty(\alpha)$ denote a power series space of infinite type (Example 4.13(a)). For $m = (m_i)_i \in \mathbb{N}^{(\mathbb{N})}$ let $(\alpha|m) = \sum_i m_i \alpha_i$. If $(a_m)_{m \in \mathbb{N}^{(\mathbb{N})}}$ is a set of complex numbers indexed by $\mathbb{N}^{(\mathbb{N})}$ show that

$$f(z) = \sum_{m \in \mathbb{N}^{(\mathbb{N})}} a_m z^m \in H(\Lambda_\infty(\alpha)'_\beta)$$

if and only if $\lim_{|m| \to \infty} |a_m|^{1/(\alpha|m)} = 0$.

Exercise 3.123* Show that $(H(s'), \tau_0) \cong s$.

Exercise 3.124* If E is a Banach space show

(a) $(H(E), \tau_0)$ has the approximation property if and only if E has the approximation property,
(b) $(H(E), \tau_\omega)$ has the approximation property if and only if $(\mathcal{P}(^n E), \| \cdot \|)$ has the approximation property for all n,
(c) if $(H(E), \tau_\omega)$ has the approximation property then $(H(E), \tau_0)$ also has the approximation property,
    Show that there exists a Banach space E such that $(H(E), \tau_0)$ has the approximation property but $(H(E), \tau_\omega)$ fails the approximation property.

Exercise 3.125* If E is a Banach space, $(\phi_n)_n \subset E'$ and $f = \sum_{n=1}^{\infty} \phi_n^n \in H(E)$ show that r_f is a constant mapping and find this constant. If $E = l_1$ and $g((z_n)_n) = \sum_{n=2}^{\infty} e^{-nz_1} z_n^n$ show that $g \in H(l_1)$ and

$$r_g((z_n)_n) = \begin{cases} 1 + \Re(z_1) & \text{when } \Re(z_1) \geq 0 \\ e^{\Re(z_1)} & \text{when } \Re(z_1) < 0. \end{cases}$$

If $E = c_0$ and $h((z_n)_{n=1}^{\infty}) = \sum_{n=1}^{\infty} (z_1 z_n)^n$ show that $h \in H(c_0)$ and

$$r_g\big((z_n)_{n=1}^\infty\big) = 2\big((|z_1|^2 + 4)^{1/2} + |z_1|\big)^{-1}$$

Exercise 3.126* If E is a Banach space and $f \in H(E)$ show that $|r_f(x) - r_f(y)| \le \|x - y\|$ for all $x, y \in E$. If E is uniformly convex show that

$$|r_f(x) - r_f(y)| < \|x - y\| \qquad (*)$$

for all $x \ne y$ in E. (A Banach space is *uniformly convex* if for every $\varepsilon > 0$ there exists $\delta > 0$ such that for any unit vectors x and y, $\|x + y\| \ge 2 - \delta$ implies $\|x - y\| \le \varepsilon$.) By using the function g in the previous exercise show that $(*)$ is not valid for l_1. Show, however, that $(*)$ holds for all $f \in H(c_0(\Gamma))$.

3.6 Notes

The historical notes of A.E. Taylor [823] are, probably, the most comprehensive guide to the development of infinite dimensional holomorphy up to the mid-1940s available and we have made extensive use of these notes in Section 1.5 and shall do so again in this section. In another paper, [824], A.E. Taylor documents the role of analyticity in operator and spectral theory and his history of the differential in the nineteenth and twentieth centuries [825] contains much of interest concerning the growth of infinite dimensional analysis. E. Hille–R.S. Phillips [452, Section 3.16 and Chapter 26] also provide a summary of the fundamentals known at that time – 1957 – together with a short bibliography on the subject (Chapter 26 was written in collaboration with M.A. Zorn). D. Pisanelli [728, 729] surveys most of the different concepts of holomorphic function currently in use and documents their origin. V.I. Averbukh–O.G. Smolyanov [79, §2] give a comprehensive account of the development of the concept of differential in topological vector spaces. They were mainly concerned with the real theory but naturally this has many consequences for the complex case. It is interesting to note the multiplicity of definitions that have been proposed in the real case – the authors list twenty five in [79] – and since a number of definitions were not seen to be equivalent and, indeed, various authors were not aware of one another's work, this led to a certain amount of chaos – or at least apparent chaos – which was rectified in [79]. Fortunately, differentiable mappings between complex locally convex spaces admit power series expansions and this led to fairly unanimous acceptance of Fréchet $\mathbb{C}$-differentiability as the standard definition with minor, but important, roles being played by Gâteaux holomorphic, Silva holomorphic and hypoanalytic functions.

Names associated with the development of real infinite dimensional differential calculus include V. Volterra, J. Hadamard, M. Fréchet, R. Gâteaux, P. Levy, A.D. Michal, L.M. Graves, E.W. Paxson, J.G. de Lamadrid, J. Sebastiaõ e Silva, H.H. Keller and G. Marinescu (see [79] for more detailed

information). In particular, we would like to emphasize the important contributions of J. Hadamard, M. Fréchet and A.D. Michal. Hadamard was one
of the first to recognize the importance of Volterra's work, he exercised a
deep influence on his students Fréchet and Gâteaux and insisted on linearity being incorporated into any definition of the derivative. M. Fréchet
and A.D. Michal both spent a considerable portion of their lives developing the theory – roughly forty years by Fréchet – and both opened fruitful
research directions. Readable accounts of the real theory are to be found
in P. Lévy ([550], both the 1922 and 1950 editions), A.D. Michal [619],
V.I. Averbukh–O.G. Smolyanov [78], A. Frölicher–W. Bucher [363] and S. Yamamuro [855]. A comprehensive recent account of real analysis over infinite
dimensional spaces is given in A. Krieg–P.W. Michor [530] (see also [531]).
Since the authors confine their presentation in [530] to the basic concepts
of infinite dimensional holomorphy, as defined by L. Fantappie [345], and to
results prior to 1970, the overlap with this book is rather slight. Surveys of
a more specialized nature, relevant to the topics discussed in this book, are
J.M. Ansemil [16], R.M. Aron [45], R.M. Aron–B. Cole–T.W. Gamelin [52],
K.-D. Bierstedt [115], K.-D. Bierstedt–R. Meise [122], J. Bonet–S. Dierolf
[159], P. Casazza [209], J.F. Colombeau–M.C. Matos [245], E. Dubinsky [334,
335], T.W. Gamelin [377], J.M. Gutiérrez–J.A. Jaramillo–J.G. Llavona [425],
J. Horváth [468, 469], B. Kramm [527], L. Lempert [544], P. Kreé [528, 529],
S. Dineen [296, 304, 305, 312], L.A. Moraes [628, 631], R. Meise–D. Vogt [611],
J. Mujica [656, 658], L. Nachbin [671, 672, 674, 678, 680, 681], M. Schottenloher [782, 783, 789], R.L Soraggi [810, 812], D. Vogt [837, 841] and I. Zalduendo
[860]. Books on infinite dimensional complex analysis include J.A. Barroso
[89], C.B. Chae [215], G. Coeuré [235], S. Dineen [299], M. Hervé [445, 446],
P. Mazet [597], J. Mujica [651], L. Nachbin [666] and Ph. Noverraz [696].

We now present our own brief history. We noted in Section 1.5 that the
definitive steps in the creation of a real and complex infinite dimensional calculus were taken by V. Volterra (1887), H. von Koch (1899), M. Fréchet and
D. Hilbert (1909). Prior to these, the work of J. Bernoulli on the curve of
quickest descent and L. Euler on the calculus of variations (see J. Hadamard
[430]) may also be mentioned as being of particular importance for later developments. H. von Koch [520] introduced a monomial approach to holomorphic
functions on infinite dimensional polydiscs which was developed and extended
by D. Hilbert [450]. M. Fréchet [357, 358] began at this time his development
of the concept of derivative and polynomial over both real and complex infinite dimensional spaces. Next came R. Gâteaux [383, 384, 385] who set
out to clarify and extend the works of Fréchet and Hilbert. In this he succeeded brilliantly. In [383, 384], he defined complex polynomials and noticed
their relationship with bilinear and quadratic forms, defined the "Gâteaux"
derivative, extended the Cauchy integral formula and the Cauchy inequalities, proved the identity theorem and convergence theorems for holomorphic
functions, established the now standard correspondence between derivatives

of a holomorphic function and homogeneous polynomials in its Taylor series expansion and obtained various results concerning analytic continuation and power series expansions – all in an infinite dimensional setting. His other paper [385] is also quite interesting and contains brief outlines of a number of topics which his untimely death prevented him from developing. Gâteaux's studies were confined to the spaces $\mathbb{C}^{\mathbb{N}}$, $\mathcal{C}[a,b]$ and l_2.

Von Koch and Hilbert's definition of holomorphic function of infinitely many variables was applied by H. Bohr [143] to investigate modes of convergence of Dirichlet series in the following rather ingenious fashion (see also Sections 1.5 and 4.6). Any positive integer n has a unique factorization $n = p_1^{v_1} p_2^{v_2} \cdots p_k^{v_k}$ where each v_i is a non-negative integer and $p_1 \leq p_2 \ldots$ denotes the set of all primes. Hence the Dirichlet series $f(s) = \sum_{n=1}^{\infty} a_n/n^s$ can be rewritten as

$$f(s) = \sum_{\substack{v_i \geq 0 \\ i=1,2,\ldots}} \frac{a(v_1, v_2, \ldots, v_k)}{(p_1^{v_1} \cdots p_k^{v_k})^s} = \sum_{\substack{v_i \geq 0 \\ i=1,2,\ldots}} a(v_1, \ldots, v_k) \left(\frac{1}{p_1^s}\right)^{v_1} \cdots \left(\frac{1}{p_k^s}\right)^{v_k}$$

where $a(v_1, \ldots, v_k) := a_n$ if $n = p_1^{v_1} \cdots p_k^{v_k}$. On replacing $1/p_i^s$ by y_i and letting y_i vary over some set of complex numbers we obtain, in monomial form, the holomorphic function of infinitely many variables,

$$F(y_1, y_2, \ldots) = \sum_{\substack{v_i \geq 0 \\ i=1,2,\ldots}} a(v_1, \ldots, v_k) y_1^{v_1} \cdots y_k^{v_k} = \sum_{v \in \mathbb{N}^{(\mathbb{N})}} a_v y^v.$$

We have $F(\frac{1}{p_1^s}, \frac{1}{p_2^s}, \ldots, \frac{1}{p_k^s}, \ldots) = f(s)$. Bohr showed that, although functions of a single variable s, the variables $y_i = p_i^s$ behave, in many ways, as if they were independent of one another. This is due, perhaps, as pointed out in [142, p.600] to the fact that the quantities $\log p_i$ are linearly independent over the rationals. The paper of H. Bohr [143] and subsequent papers by O. Toeplitz [826] and H.F. Bohnenblust–E. Hille [142] on the same topic contain interesting results on infinite dimensional holomorphy which would have played, if they had not been overlooked, a role in the development of holomorphic functions on c_0 and on fully nuclear spaces with basis. For instance we find, in our notation, the following result of H. Bohr (1913) in [143, p.462] (see also [142, p.613]).

If $\left| \sum_{m \in \mathbb{N}^{(\mathbb{N})}} a_m z^m \right| \leq c$ *for all finite sequences* $(z_i)_i$, $|z_i| \leq 1$ *for all* i,

then $\sum_{m \in \mathbb{N}^{(\mathbb{N})}} |a_m| \varepsilon^m < \infty$ *whenever* $\varepsilon = (\varepsilon_n)_n \in l_2$.

This result is best possible, i.e. it is not possible to replace l_2 by $l_{2+\varepsilon}$, and closely related to the essential ideas in Proposition 4.58. Refined versions for m-homogeneous polynomials are given in [142, Theorem III] (see Exercise 1.104).

In 1920, S. Banach gave, in his thesis, the axioms for a complete normed linear space (a Banach space) and in 1923, N. Wiener [852] who, incidentally, gave independently the same axioms three months after Banach, noted that the Cauchy integral formula extends to vector-valued holomorphic functions of one complex variable and in this setting, many classical results such as Morera's Theorem, Abel's Theorem and the residue theorem remain valid. L. Fantappié's long article [345] appeared in 1930. In it he proposed that a continuous function f defined on a domain D in a Banach space be called holomorphic if $f \circ \phi$ is holomorphic (as a function of one complex variable) for any holomorphic function ϕ from a domain in the complex plane into D. Fantappié's work served as motivation for A.E. Taylor [817] and was developed and put in a modern context using topological vector spaces by J. Sebastião e Silva [796, 797, 798, 799] (see D. Pisanelli [727, 728, 729], A. Kriegl–P.W. Michor [530]) and A. Kriegl–L.D. Nel [531].

In the 1930s an intensive study of the whole field of analysis and geometry over abstract spaces was carried out by A.D. Michal and his students R.S. Martin, A.H. Clifford, I.E. Highberg and A.E. Taylor. In 1932, R.S. Martin [569] developed the theory of holomorphic mappings for Banach spaces using a power series approach (see also A.D. Michal–A.H. Clifford [620]). The final step towards the current definition of holomorphic mapping was taken independently by L.M. Graves [408] and A.E. Taylor [817, 818]. Graves' treatment is rather brief, [408, pp. 649–653], but contains a lot of information, and in it he notes that Gâteaux differentiability plus continuity are equivalent to Fréchet differentiability (defined for normed linear spaces by Fréchet in 1925, [361]), when dealing with functions between complex normed linear spaces and observes that *balanced domains* are the natural domains of convergence for Taylor series expansions. Taylor's work is much more explicit. He defines a holomorphic function as a continuous function whose one-dimensional sections are holomorphic. This definition, a natural extension of the work of Gâteaux, allowed him to unify a number of previous results and to prove the following:

if f is a mapping between normed linear spaces (over $\mathbb{C}$) then the following are equivalent:

(a) *f is continuous and has a Gâteaux derivative at each point,*
(b) *f has a Fréchet derivative at each point,*
(c) *f has a power series expansion which converges uniformly on a neighbourhood of each point.*

Taylor goes on to generalize Riemann's Theorem on removable singularities, Mittag-Leffler's Theorem, Liouville's Theorem and the Cauchy–Riemann equations. In a further paper, [818], he gives an example of an entire function with finite radius of uniform convergence (see Exercise 3.64), discusses a number of interesting examples of holomorphic functions on l_p spaces and generalizes Hartogs' Theorem on separate analyticity (see Example 3.8(c) and the remarks on Exercise 3.68).

The Polish school of functional analysis, including S. Banach, S. Mazur and W. Orlicz, also contributed during this period (see the notes on Chapter 2) and although this traditional interest in non-linear functional analysis was maintained until the mid-1960s (see for instance A. Alexiewicz–W. Orlicz [14], W. Bogdanowicz [139, 140] and A. Pełczyński [713, 715, 714]), it was overshadowed by the rapid development of the linear theory. This dominance of linear functional analysis may be traced to the influence of S. Banach's book [83], published in 1932, although Banach himself hoped to develop a non-linear theory if we are to judge from his comment [83, p.231] on complex vector spaces:

"Ces espaces constituent le point de départ de la théorie des opérations linéaires complexes et d'une classe, encore plus vaste, des opérations analytiques, qui présentent une géneralization des fonctions analytiques ordinaires (cf.p. ex. L. Fantappié I funzionali analitici, Città di Castello, 1930). Nous nous proposons d'en exposer la théorie dans un autre volume."

See also the preface to [83].

M.A. Zorn [862, 863, 864] made a number of important contributions during the mid-1940s and we refer to these shortly, while J. Sebastião e Silva [796, 797, 798, 799] developed a theory of differentiation on arbitrary locally convex spaces during the early 1950s (and, in particular, on $\mathcal{DFS}$ spaces – now also referred to as Silva spaces). H.J. Bremermann [197, 198, 199, 200] initiated the study of pseudo-convex domains and tubular domains of holomorphy in the late 1950s and early 1960s (see Section 5.3). The mid-1960s saw the appearance of the thesis of A. Douady [332]. The fact that Douady required holomorphic functions over infinite dimensional Banach spaces (see [208]) attracted a lot of attention and motivated established mathematicians such as H. Cartan, H.J. Bremermann, M. Hervé, P. Lelong, A. Martineau, L. Nachbin, K. Stein to direct their students towards infinite dimensional holomorphy. This led to the rapid growth in the theory that we have seen over the last thirty years.

We return to commenting on the text. Most of the results in Section 3.1 up to and including Proposition 3.7 are well known and due to D. Hilbert, M. Fréchet, R. Gâteaux and others that we have previously mentioned. The arrangement of the material and, in particular, the initial use of monomial expansions with respect to a Hamel (or algebraic) basis *prior* to the introduction of Taylor series expansions appears to be new. Example 3.8(a) and (b) are due to M.A. Zorn [863], but we have followed the proof given in [13, §3]. The result in (a) shows that $\mathcal{G}$-holomorphic extensions of Fréchet holomorphic functions on Banach spaces are, in fact, continuous and hence Fréchet holomorphic (see [138, Theorem 6.1], [141] and [553]). Example 3.8(c) is due to A.E. Taylor [818] for $\mathbb{C} \times E$ and to M.A. Zorn [863] for an arbitrary pair of Banach spaces (see the remarks on Exercise 3.68). The extensions to pairs of Fréchet spaces and to pairs of $\mathcal{DFS}$ spaces given in Example 3.8(d) are due to Ph. Noverraz [690, 692] and A. Hirschowitz [459], respectively. Subsequently, it was found, by D. Pisanelli [728], J.F. Colombeau [242] and D. Lazet [533, 534], that these special cases could be obtained using superinductive limits of Banach spaces. Further generalizations of the first three parts of Example 3.8 can be found in S. Dineen [292], E. Ligocka [551] and Ph. Noverraz [690] (see the remarks on Exercise 3.68). Example 3.8(e) is due to J.A. Barroso–M.C. Matos–L. Nachbin [92] for $\mathcal{DFS}$ spaces and to S. Dineen [295] for $\mathcal{DFM}$ spaces and has been extended to bornological DF spaces by P. Galindo–D. García–M. Maestre [369, Proposition 15] (see also [162, Proposition 2.5] and the remarks on Exercises 3.78 and 3.79).

In his thesis on infinite dimensional analytic continuation H. Alexander [13] was led to the concept of *bounding set* (the analogue for continuous functions appears in Exercise 3.77). Clearly

$$relatively\ compact \implies bounding \implies bounded$$

in any Banach space. In [13], H. Alexander proved that

$$bounding \implies relatively\ compact$$

in a Hilbert space (see also [279]). S. Dineen [284] and A. Hirschowitz [453, 456], independently, extended this result by showing that bounding sets are relatively compact in any Banach space whose dual unit ball is weak* sequentially compact – we prove this result for separable spaces in Example 3.20(c) – and this was extended to weakly compactly generated Banach spaces by M. Schottenloher [777, 782, 789] (see Exercise 4.62). A linear characterization of these spaces cannot exist since B. Josefson [491] has shown that the closed convex hull of a bounding set need not be bounding. In [280], S. Dineen showed that l_∞ admits closed bounding non-compact subsets (Proposition 4.51) and B. Josefson [490] gave a complete linear characterization of the bounding subsets of l_∞ by proving that a subset A of l_∞ is bounding if and only if it is *weakly conditionally compact* (i.e. if and only if each sequence in A contains a weak Cauchy subsequence). In particular each bounded subset of $c_0 \hookrightarrow l_\infty$ is a bounding subset of l_∞ (see Sections 4.6 and 6.2). The Josefson–Nissenzweig Theorem was obtained independently in [488] and [689] (locally convex space generalizations are given in [163], [164], [557] and [561]). Josefson's motivation was to prove the result presented in the first part of Example 3.8(f), i.e. that $H_b(E)$ is a proper subspace of $H(E)$ for any infinite dimensional Banach space E.

A subset A of a Banach space E is *limited* if every weak*-null sequence in E' tends to zero uniformly on A. Limited sets are the linear analogue of bounding sets – in fact A is limited if and only if each holomorphic function, which has the form (3.16), is bounded on A (the Josefson–Nissenzweig Theorem says that the unit ball is not limited for any infinite dimensional Banach space). Bounding sets are limited and B. Josefson [491] and T. Schlumprecht [775] have given, independently, examples of limited sets which are not bounding (see also [492]). As a counterexample to the Levi problem, B. Josefson [487] constructed a polynomially convex domain in $c_0(\Gamma)$, Γ uncountable, which contains a closed precompact non-compact bounding set (we discuss this in Section 5.2). Further results on bounding and limited sets may be found in R.M. Aron [40], R.M. Aron–P. Berner [47], A. Bayoumi [95], M. Bianchini [112], J. Bonet–M. Lindström [163], J. Bourgain–J. Diestel [175], S. Dineen [284, 303], S. Dineen–L. Moraes [322], R. Haydon [443, 444], A. Hirschowitz [458], B. Josefson [492], M. Lindström [557, 558], M. Lindström–R.A. Ryan [559], J. Mujica [658], K. Rusek [754] and M. Schottenloher [789]. A polynomial version of limited sets is considered in P. Galindo [365] and M. González–J. Gutiérrez [394].

The function constructed on the direct sum in Example 3.8(f) was first considered in [303]. The result $H_{HY}(\sum_{i=1}^{\infty} E_i) = H(\sum_{i=1}^{\infty} E_i) \iff \dim(E_i) < \infty$ for all i is due to S. Dineen [286] and has been extended to strict inductive limits of Banach spaces in S. Dineen–L.A. Moraes [322]. Example 3.8(g) is due to J. Bonet–P. Galindo–D. García–M. Maestre [162] and generalizes the same result for $\mathbb{C}^{\mathbb{N}} \times \mathbb{C}^{(\mathbb{N})}$ which was announced in [677] and proved in [303].

With the intention of holomorphically classifying locally convex spaces, J.A. Barroso–M.C. Matos–L. Nachbin [91, 92] introduced holomorphic analogues of Mackey, infrabarrelled, barrelled and bornological locally convex spaces (holomorphically ultrabornological spaces are defined in [366]). In all cases it was found that the holomorphic property is *strictly* stronger than the linear property. A locally convex space E is holomorphically Mackey if $H(U; F) = H(U; F_\sigma)$ for any open subset U of E and any locally convex space F where $F_\sigma := (F, \sigma(F, F'))$. The result in Example 3.8(g) says that $E \times \mathbb{C}^{\mathbb{N}}$ is holomorphically Mackey for any $\mathcal{DFM}$ space E while Proposition 3.21, which is due to J.A. Barroso–M.C. Matos–L. Nachbin [92], says that a holomorphically infrabarrelled space is holomorphically Mackey. Using the universal linearization space $G_0(U)$ (see Example 3.28) J. Mujica–L. Nachbin [660] proved that a locally convex space E is holomorphically Mackey (respectively holomorphically barrelled) if and only if $G_0(U)$ is Mackey (respectively barrelled) for any open subset U of E (the barrelled result is Example 3.28(a)).

Factorization results, which follow from Liouville's Theorem on bounded entire functions, are implicit in A. Hirschowitz [454], C.E. Rickart [743] and L. Nachbin–J.A. Barroso [683, Example 4] and explicit in L. Nachbin [670]. Initially, they were introduced to extend holomorphic results from Banach spaces to Fréchet spaces, where one encounters the lack of a continuous norm (as for instance in dealing with $\mathbb{C}^{\mathbb{N}}$). S. Dineen [289, 290, 291, 292] and E. Ligocka [551] developed a general theory of factorization for holomorphic functions from these examples which has proved useful in generating spaces that preserve certain holomorphic properties. The basic concept is called *surjective limit*.

> *A locally convex space E is a surjective limit of the collection $(E_i, \pi_i)_{i \in A}$, E_i a locally convex space and π_i a continuous surjective linear mapping from E onto E_i for all i, if $\left(\pi_i^{-1}(V)\right)_{i \in A, V \in \mathcal{V}(E_i)}$ is a neighbourhood base (and not a subbase) for the topology of E as $\mathcal{V}(E_i)$ ranges over a neighbourhood basis for the space E_i, for all i.*

Of particular importance are surjective limits in which each π_i is open (open surjective limits) or each π_i lifts compact sets (compact surjective limits). The strong dual of a strict inductive limit of Fréchet–Montel spaces is an open and compact surjective limit of $\mathcal{DFM}$ spaces. A Fréchet space E which admits a fundamental directed set of semi-norms $(p_n)_n$ such that each $\left(E/p_n^{-1}(0), p_n\right)$ is a Banach space is, by the open mapping theorem, an open surjective limit of a sequence of Banach spaces. These spaces, first studied by A. Grothendieck [414], were called *quojections*, by S. Bellenot–E. Dubinsky [96] some years after surjective limits were introduced. This terminology has become universally accepted and we have adopted it. After the appearance of the influential article [637] by B.V. Moscatelli, quojections began to play an important role in the structural study of Fréchet spaces. We refer the reader to the excellent surveys, B.V. Moscatelli [638] and G. Metafune–B.V. Moscatelli [616] for further details.

In many cases a holomorphic property which is preserved by taking finite products can be extended to open surjective limits of one kind or another. Factorization methods and surjective limits have been applied to the Levi problem (Chapter 5) by J.F. Colombeau–J. Mujica [248], S. Dineen [287, 292], Ph. Noverraz [695, 703], M. Schottenloher [785], to the construction of the envelope of holomorphy by P. Berner [106, 107], L.A. Moraes–O.W. Paques–M.C.F. Zaine [635, 636], M. Schottenloher [785], in studying meromorphic functions and the Cousin I problem by V. Aurich [75] and S. Dineen [293], to the theory of convolution operators by P. Berner [109] and M.C. Matos [582], in studying locally convex topologies on spaces of holomorphic functions by P. Berner [105, 108], J. Bonet–P. Galindo–D. García–M. Maestre [162], S. Dineen [292], P. Galindo–D. García–M. Maestre [371] and Ph. Noverraz [706], and to germs of holomorphic functions by J. Mujica [644], Y.S. Choi [220] and R.L. Soraggi [811]. We have not undertaken a systematic study of surjective limits or quojections in this book and refer the reader to [292] and [299] for further information. However, a number of the results, exercises and examples presented here, in particular those involving products and direct sums, are special cases of general results obtained using surjective limits, e.g. Examples 3.24(b) and (c) are known to hold for open and compact surjective limits of $\mathcal{DFM}$ spaces (see [154, 626, 627] for (b), [108, 292] for (c), and, for related material, [299, Propositions 6.25 and 6.27] and the remarks on Exercises 3.80 and 4.74). Properties of $H(E)$, E a locally convex space, are often inherited by $H(F)$ when F is a *complemented* subspace of E (see Lemma 4.17

and Proposition 4.54) and this leads to new examples since complemented subspaces of products may be *twisted* (see [709] and Exercise 4.74).

Lemma 3.9 is given in [292] and Example 3.10 is due to L. Nachbin [670] (the space in Example 3.10(a) is an open surjective limit of normed linear spaces). A. Hirschowitz [454] and L. Nachbin [670] give examples of an open subset U in $\mathbb{C}^{\mathbb{N}}$ and $f \in H(U)$ which does not admit a global factorization through some finite dimensional subspace. This shows that we require some hypothesis, such as U balanced, in the final part of Lemma 3.9. The difficulties posed by these example were overcome by using Riemann domains (see Section 5.2) in P. Berner [106] and M. Schottenloher [780, 785, 787]. Example 3.11, which introduces a different kind of factorization result, is due to J.F. Colombeau–J. Mujica [248] who use it, together with the solution to the Levi problem on Hilbert spaces (see L. Gruman [411]), to solve positively the Levi problem on $\mathcal{DFN}$ spaces (see Section 5.2). This result may be combined with the extension theorem of R.M. Aron–P. Berner [47] for holomorphic functions of nuclear bounded type on a Banach space to give a further proof of P. Boland's holomorphic Hahn–Banach theorem [149] on $\mathcal{DFN}$ spaces (Exercise 4.77). The identity $\mathcal{L}(E,F) = \mathcal{LB}(E,F)$ has been extensively studied in the linear theory, see for instance [158, 161, 259, 839, 841].

Uniform holomorphy was introduced by L. Nachbin [670] who showed that $H(\mathbb{C})$ does not lead to uniform holomorphy (see [679], [681], [220], [221], [294] and [299, Examples 2.22 and 6.22(b)]) while holomorphic functions of uniformly bounded type were defined by J.F. Colombeau–M.C. Matos [244]. Example 3.12, which gives a necessary condition for uniform holomorphy, is due to R. Meise–D. Vogt [610] (see [371, Proposition 2.1]). Sufficient conditions for uniform holomorphy, also from [610], are discussed in Section 6.2 (see also L.M. Hai [432]). Topological vector space structures on $H_b(E)$ and $H_{ub}(E)$ are studied in [371], [507] and [610, Proposition 4.1]. If E is a Banach space then $H_b(E)$, endowed with the topology of uniform convergence on bounded sets, is a locally m-convex Fréchet algebra and thus has a less complicated locally convex space structure than any we have been able to associate with $H(E)$. For this reason it frequently yields results of interest and may be used as a stepping stone to $H(E)$ (see Section 6.3).

The topological vector space structure of $H(U)$, U an open subset of $\mathbb{C}^n$, has been studied by a number of authors including A. Grothendieck [413], G. Köthe [521] and A. Martineau [571]. The compact open topology on spaces of holomorphic functions in infinitely many variables was first investigated by H. Alexander [13] and L. Nachbin [666]. L. Nachbin [665, 666], motivated equally by properties of analytic functionals of several complex variables due to A. Martineau in [570, 572] and the classical definition (see Lemma 4.26) of the natural inductive limit topology on $H(K)$, K compact, introduced the τ_ω topology (Definition 3.14). The τ_δ (Definition 3.16) was first defined for holomorphic functions on separable Banach spaces by G. Coeuré [232] and, in general, by L. Nachbin [669]. Coeuré was motivated by problems of analytic continuation. Propositions 3.18 and 3.19 are given in [286] and [315] respectively. Example 3.20(b) is due to J.A. Barroso–M.C. Matos–L. Nachbin [91] for $\mathcal{DFS}$ spaces and to S. Dineen [295] for $\mathcal{DFM}$ spaces. The results in Proposition 3.22 for the τ_0 and τ_ω topologies are due to L. Nachbin [666] while the coincidence of τ_ω and τ_δ on $\mathcal{P}(^nE)$ is due to S. Dineen [286]. Very strongly convergent sequences were defined by C. Bessaga–A. Pełczyński [111] and introduced into infinite dimensional holomorphy by S. Dineen [285, 286]. They were applied in [286] to obtain Example 3.24(a) and are thoroughly investigated in M.A. Simões [803] (see Exercises 3.87, 3.88 and 5.77).

Presenting a $\mathbb{C}$-valued mapping on one space as a linear function between vector spaces is called *"linearizing"* the mapping. If linearizing can be realized in a systematic way for a collection of functions, $\mathcal{F}$, endowed with some topological structure, so that the resulting space of linear mappings is X' for some locally convex topological vector space X then it is reasonable to call X a *"predual"* of $\mathcal{F}$. We carried out this process in Chapter 1 using tensor products and universal constructions and, in Chapter 2, we followed the usual functional analysis approach to obtain preduals of $\mathcal{P}(^nE)$ etc. Linearizations and preduals were constructed for holomorphic functions over infinite dimensional spaces in 1972 by M. Schottenloher [784, Proposition 2.5]. Subsequently, R.A. Ryan [758, Propositions 1.3 and 2.6] obtained similar results for both homogeneous polynomials and holomorphic functions, and an abstract functorial approach was used in P. Mazet [597, Definition 3.4, Propositions 3.8 and 3.10]. All three authors, with different points of emphasis and using quite different approaches arrived at, more or less, the same results, Propositions 3.26 and 3.27. We have followed the approach of J. Mujica–L. Nachbin [660] who obtained a number of applications including Example 3.28 (Example 3.28(b) is due to P. Mazet [596] for finite dimensional spaces). It is interesting to note the evolution of certain ideas that led to the proof of Proposition 3.26 and to the concrete representation of $G(U)$. Exercise 2.100 of [299] asks one to prove that $\left(H^\infty(U), \|\cdot\|_U\right)$ is a dual Banach space. To prove this J. Mujica used the following result of J. Dixmier [330]–K.F. Ng [687]–L. Waelbroeck [849] (see R.B. Holmes [461, p.211]):

> *If E is a Banach space with closed unit ball B then E is a dual Banach space (i.e. there exists a Banach space F such that $E \cong F'_\beta$) if and only if there exists a Hausdorff locally convex topology τ on E such that (B, τ) is compact. Moreover $\{\phi \in E' : \phi|_B$ is τ continuous$\}$ is a predual of E.*

Mujica noted in 1980 that the completeness of $H(K)$, K a compact subset of a Fréchet space, was obtained in [300] by using the fact that the closed bounded subsets of $H(K)$ were compact with respect to the weaker τ_0 topology (which, of course, follows from Montel's Theorem). On combining these observations J. Mujica–L. Nachbin [660] (see also [647, 656]) were led to the following generalization of the Dixmier–Ng–Waelbroeck result:

Let E be a bornological locally convex space and suppose there exists a fundamental family $(B_\alpha)_\alpha$ of convex balanced bounded subsets of E and a Hausdorff locally convex topology τ on E such that each B_α is τ-compact then there exists a closed subspace F of E such that $E \cong F_i' \cong (F', \tau_\omega)$. $(F := \{u \in E^* : u|_{B_\alpha}$ is τ-continuous for each $\alpha\}.)$

A proof of this result (Exercise 3.92), together with further generalizations and applications, is given in K.-D. Bierstedt–J. Bonet [117]. This result led to the Mujica–Nachbin proof of Mazet's linearization theorem. To prove the completeness of $H(K)$, J. Mujica [647] (see also [656]) first proved an abstract completeness criterion for countable inductive limits of Banach spaces:

If $X = \varinjlim_j X_j$ is an LB space (i.e. a countable inductive limit of Banach spaces) and there exists a Hausdorff locally convex topology τ on X such that the closed unit ball of each X_j is τ-compact then X is the inductive dual of a Fréchet space and, in particular, it is complete.

A long standing open problem in linear functional analysis is whether or not *LB* spaces are complete. In the absence of a positive answer this result of Mujica has become one of the standard tools in the linear theory.

The next stage of development saw P. Galindo–D. García–M. Maestre [370] and J. Mujica [653, 654] prove linearization results for $H_b(U)$ and $H^\infty(U)$, respectively, and in a further paper, [655], J. Mujica clarified the relationship between the preduals constructed for those spaces. The articles [653, 654, 655] contain applications to such diverse topics as the metric approximation property, interpolating sequences, quasi-normability and the Mackey convergence criterion. Motivated by the above results C. Boyd [176, 177, 178, 179] showed that preduals of spaces of holomorphic mappings and germs, defined on subsets of E, have many properties in common with the space E. C. Boyd [177] also used preduals to discuss the relationship between the various topologies on $H(U)$ and $H(K)$ for U (open) and K (compact) in a Fréchet space (see Exercise 4.84).

In concluding our remarks on linearization we note that the predual $G(U)$ is now beginning to play the same role relative to $H(U)$ as $\widehat{\bigotimes}_{n,s,\pi} E$ has hitherto played relative to $\mathcal{P}(^nE)$ in Chapters 1 and 2 and, in retrospect, this is not at all surprising. An interesting feature of this development is the fact that all three of the "standard" topologies on $H(U)$, τ_0, τ_ω and τ_δ (and locally bounded sets) feature in different ways in the construction of $G(U)$, or to phrase it differently, $G(U)$ allows all three topologies to work together. This synthesis of properties led to results of interest which might not have been uncovered if each had been investigated in isolation.

Schauder decompositions of Banach spaces were first investigated by M.M. Grinblyum [409] and extended to topological vector spaces by C.W. Mc-Arthur–J.R. Rutherford [601]. A detailed study is undertaken in N.J. Kalton

[499, 500, 501]. They were first used systematically in infinite dimensional holomorphy by S. Dineen [286]. S-absolute decompositions (Definition 3.32) were introduced in [299] – decompositions which only satisfied (3.31) were called S-Schauder decompositions in [299] (see Exercise 3.93). Lemma 3.33 and Propositions 3.34(a), (c), and 3.36 collect and slightly improve some frequently used results from [299]. Corollary 3.37(a) is proved in [279] for Banach spaces and in [286] for locally convex spaces (see also [35, 39]). Corollary 3.37(b) is due to Ph. Noverraz [705, 707] while Corollary 3.37(c) is due to L. Nachbin [666] when E is a Banach space and to S. Dineen [286] for arbitrary E (a related result is due to M.C. Matos [579]). Proposition 3.38 is due to C. Boyd [178] while Propositions 3.34(b) is new and Proposition 3.35 is a generalization of [299, Proposition 3.13]. Y. Kōmura [524] was the first to discuss associated topologies in a (linear) locally convex space setting and while applications, especially of the associated bornological topology, appeared in various articles on infinite dimensional holomorphy the most systematic treatment is due to Ph. Noverraz [704, 705, 707] (see also [299, §3.1]). Ph. Noverraz [705] and P. Berner [105] proved that the infrabarrelled and bornological topologies associated with τ_ω coincide on $H(U)$ for U balanced (see Exercise 3.95) while Ph. Noverraz [705] showed that the barrelled and ultrabornological topologies associated with τ_ω coincide with τ_δ. It is an an open question if *all* these associated topologies coincide (see Proposition 3.51(b)). In proving Corollary 3.37(b), Ph. Noverraz [705, 707] used Corollary 3.37(a). We have followed a different route here and used Proposition 3.34(b). This result – with proof modelled on the standard proof using transfinite induction that the associated barrelled topology of a complete space is complete – was not previously available (see the remarks in [299, p.117] and the statement of Theorem 1 in [707]). The key is the very weak completeness built into (3.31). The τ_b topology on spaces of holomorphic functions was defined in [308] and is discussed in Chapter 4.

Holomorphic functions on nuclear and dual nuclear spaces were first investigated by P.J. Boland (see Section 2.6). In [149], P.J. Boland proved that $(H(E), \tau_0)$ is nuclear when E is a quasi-complete dual nuclear space (see also E. Nelimarkka [685]). This result was extended, independently, to arbitrary open sets by P.J. Boland [150] and L. Waelbroeck [851] (other proofs were later given by J.F. Colombeau–B. Perrot [251, 252] and N.V. Khue [507]) and contains the final part of Proposition 3.43 as a special case. Extensions to λ and s-nuclearity and to nuclear bornologies are due to K.-D. Bierstedt–B. Gramsch–R. Meise [119], K.-D. Bierstedt–R. Meise [120, 121, 122], J.F. Colombeau–R. Meise [246], J.F. Colombeau–B. Perrot [250, 251, 252], R. Meise–D. Vogt [606], and L. Waelbroeck [851]. For example the following result is proved in [246]:

Let E denote a quasi-complete locally convex space, then $(H(U; F), \tau_0)$ is an s-nuclear space for any open subset U of E if and only if E'_c and F are both s-nuclear spaces.

See Satz 1.12 of [119] for a preliminary result in this direction – s-nuclearity (also called strong nuclearity) is defined by replacing l_1 by s (Example 4.13(a)) in the Grothendieck–Pietsch criterion for nuclearity.

Applications of these results on nuclearity are given to the lifting of linear mappings (W. Kaballo [497]), to the classification of Stein algebras and uniform Fréchet algebras (B. Kramm [525, 526, 527] and H. Goldman [391]) and to the mathematical foundations of quantum field theory (P. Kreé [529], J.F. Colombeau [243] and J.F. Colombeau–B. Perrot [250]).

Fully nuclear spaces with basis were introduced by P.J. Boland–S. Dineen [152, 153] in order to study the absolute basis problem for $(H(E), \tau_0)$. Definition 3.40, Lemmata 3.41 and 3.42 and Propositions 3.43 and 3.45 are all given in [153]. A-nuclear spaces were defined and Proposition 3.46 proved in S. Dineen [301]. A duality theory between analytic functionals and holomorphic germs is developed in [153] and applied to the $\tau_0 = \tau_\omega$ problem in [153, 154, 155] – this is discussed in Section 4.3 – while the theory of A-nuclear spaces is developed in [154, 155, 298, 301, 307]. In [301], S. Dineen proved that $\tau_0 = \tau_\delta$ on $H(E)$ when E is a Fréchet nuclear space with basis and (DN). This result eventually evolved into Proposition 4.14 (see Sections 4.2 and 4.6).

A comprehensive study of holomorphic functions on Fréchet nuclear and $\mathcal{DFN}$ spaces with basis has been undertaken by M. Börgens–R. Meise–D. Vogt [169, 170, 171], R. Meise–D. Vogt [604, 605, 606, 607, 608, 610] and M. Scheve [772, 773, 773]. The research in these papers ranged over the interpretation of τ_0, τ_ω and τ_δ as normal topologies, the representation of spaces of holomorphic functions on $\mathcal{DFN}$ spaces with basis as Fréchet nuclear spaces with basis (see the remarks on Exercises 3.122 and 3.123), uniform holomorphy and holomorphic Hahn–Banach extensions (see Sections 6.1 and 6.2), counterexamples to the $\overline{\partial}$ problem, linear topological invariants of spaces of holomorphic functions and the holomorphic classification of infinite dimensional polydiscs. In these papers infinite dimensional holomorphy benefited from the structure theory for Fréchet nuclear spaces developed during the period 1976–1986. It is interesting to note how holomorphy led naturally to invariants, such as (DN) and $(\widetilde{\Omega})$ which also proved fundamental in the linear theory – the (DN)-property first arose in the work of D. Vogt [835] on the vector-valued $\overline{\partial}$ problem. Recent related results are given in L.M. Hai–T.T. Quang [435], N.M. Ha–L.M. Hai [427], N.M. Ha–N.V. Khue [428, 429]. Motivated by [153], A. Benndorf [100, 101] proved that $(H(E), \tau_0)$ admits a finite dimensional Schauder decomposition when E is a $\mathcal{DFS}$ space with a finite dimensional decomposition (see Exercise 4.68). We refer to [299, §6.4], [611] and Sections 6.1 and 6.2 for information on some of these topics.

Proposition 3.47 is due to S. Dineen [276, 279] for Banach spaces and to R.M. Aron [36, 37] for balanced open subsets of a Banach space (see also [113, 114]) – and motivated the introduction of the τ_b topology (see the introduction to [308]). The proof of Proposition 3.48 is taken from [308]. For balanced domains this result is due to J.M. Ansemil–S. Ponte [28] and their proof is based on results in J. Mujica [646]. Taylor Series completeness (Definition 3.49), Example 3.50, Propositions 3.51 and 3.52 and Corollary 3.53 are given in [286] and [299, §3.3]. A study of completeness in locally convex spaces with Schauder decompositions is undertaken in N.J. Kalton [500].

Quasi-normable (and the closely related Schwartz) spaces feature in many different places in this text (e.g. in Exercises 3.103 and 3.105, as motivation for Definition 4.36 and in Propositions 5.9 and 5.52). Proposition 3.54 is due to K.-D. Bierstedt–R. Meise [121, Proposition 16] for $\tau = \tau_\omega$ when E is Fréchet–Schwartz (see also E. Nelimarkka [685] for a proof using operator ideals), and to J.M. Isidro [475] for E Banach and $\tau = \tau_\omega$ or τ_δ. It can be shown that $\big(H_b(U), \beta\big)$, U balanced in a Fréchet space, is quasi-normable if and only if $\big(\mathcal{P}(^nE), \tau_b\big)$ is quasi-normable for each integer n ([307]) and this is the case if E is Banach and U convex balanced [25], E Banach and U balanced or if E has $(BB)_\infty$ ([17]) (see also [20], [308]). Further results on quasi-normability in infinite dimensional holomorphy may be found in [17, 25, 80, 120, 122, 166, 179, 188, 189, 307, 380, 474, 557, 558, 623, 647, 653, 655]. A useful linear characterization of quasi-normable spaces is given in R. Meise–D. Vogt [609].

Chapter 4. Decompositions of Holomorphic Functions

Until now we have developed the theories of holomorphic functions on Banach spaces and nuclear spaces in distinct but parallel fashions. In this chapter we partially integrate them by considering, as far as possible, a unified theory of holomorphic functions on Fréchet spaces. Our main technique is the use of various kinds of decompositions to obtain estimates.

Our motivation in Chapter 3 was to concentrate on properties common to Banach spaces and Fréchet nuclear spaces, e.g. the (BB)-property, quasi-normability etc. Now that we have obtained sufficiently many examples it is appropriate that we examine those which have shown themselves relevant and pliable to see if there are more intrinsic reasons for their appearance. Afterwards, we follow up some intuitive suggestions arising from this examination. We consider the following three properties of a locally convex space – we have already encountered the first two but also include (DN) as it played a significant role in the development of Fréchet nuclear spaces and was an important source of motivation for this chapter.

(1) *E has the $(BB)_\infty$-property.*

   If n is a positive integer and $\tilde{B}$ is a bounded subset of $\displaystyle\bigotimes_{n,s,\pi} E$ then there exists a bounded subset B of E such that

$$\tilde{B} \subset \overline{\Gamma}(\underbrace{B\otimes\cdots\otimes B}_{n \text{ times}})$$

(2) *E is quasi-normable.*

   For each zero neighbourhood U in E there exists a zero neighbourhood V in E such that for any $\lambda > 0$ there exists B_λ bounded such that

$$V \subset B_\lambda + \lambda U.$$

(3) *The Fréchet space E has (DN).*

   E admits a fundamental system of norms, $(\|\cdot\|_n)_n$, with $\|\cdot\|_1$ as the Dominant Norm, such that for each n there exist k and $c > 0$ such that

$$U_n \subset rU_1 + \frac{c}{r}U_{n+k}$$

for all $r > 0$, where $U_n := \{\phi \in E' : |\phi(x)| \leq 1 \text{ for all } x \in E, \ \|x\|_n \leq 1\} = \{x \in E : \|x\|_n \leq 1\}^\circ$.

We note, from the displayed inclusions, that all these properties split or decompose subsets of E. Why should this feature in the study of holomorphic functions on a locally convex space? The following is a naive explanation which may even have some merit.

We have been discussing *linear* topological structures on spaces of *holomorphic* and *polynomial mappings*. For a linear mapping ϕ we have the basic rule

$$\phi(x + y) = \phi(x) + \phi(y)$$

but for a polynomial from $\mathbb{C}$ to $\mathbb{C}$ we have, for instance,

$$(x + y)^n = x^n + y^n + \text{ other terms}$$

and for polynomials between locally convex spaces (e.g. see Lemma 1.9)

$$P(x + y) = P(x) + P(y) + \text{ other terms.}$$

If these other terms were not present the analysis might be much simpler (and less interesting). However, they are, and in many cases (e.g. Proposition 2.4) our main difficulty involves removing their influence. The approach we adopt to the "excess terms" in this chapter is to arrange things so that these terms split in a way that facilitates calculations and leads to estimates. The natural approach is to impose a splitting hypothesis on the original space. We do this by assuming the existence of a Schauder basis or decomposition or by requiring the existence of sufficiently many projections. These assumptions usually must be combined in some way with conditions of the type displayed in (1), (2) and (3) above. Such an approach has proved successful for Fréchet spaces as we shall see in Sections 4.2 and 4.3. In the long term it is probable that the basis and decomposition hypotheses on the underlying space will be replaced by more intrinsic conditions. In the meantime, results and the methods used to obtain them may uncover these intrinsic conditions (such a development took place in the linear theory for (DN) spaces). For these reasons we discuss in this chapter holomorphic functions on locally convex spaces which admit certain kinds of decompositions. The actual decomposition will depend on the problem under consideration. In Section 4.1 we discuss the general theory for normed linear spaces (we have already discussed this topic for fully nuclear spaces with a basis in Section 3.3) and show how decompositions of spaces of homogeneous polynomials may be found and lifted to the space of holomorphic functions on a balanced domain. Some of the results in this section may also be used to recover results such as Proposition 3.43 but the direct approach given in Section 3.3 is more efficient. In Section 4.2 we consider the topological problem $\tau_\omega = \tau_\delta$ and, using results from Section 2.4, discuss reflexivity of spaces of holomorphic functions on balanced domains in certain Banach spaces. In Section 4.3 we use holomorphic germs to discuss the $\tau_0 = \tau_\omega$ and the $\tau_b = \tau_\omega$ problems for balanced domains in Fréchet spaces with the strong localization property. In Section 4.4 we consider the necessity of our assumptions and limitation of the techniques used in earlier sections and give some counterexamples.

4.1 Decompositions of Spaces of Holomorphic Functions

In this section we combine Schauder decompositions of the domain space E with the S-absolute decomposition for spaces of holomorphic functions developed in Chapter 3 to obtain a decomposition of the space of holomorphic functions on a balanced domain.

We begin with an abstract result on absolute decompositions.

Proposition 4.1 *Let $\{E_n\}_{n=1}^\infty$ denote an absolute decomposition of the locally convex space E and for each n let $\{E_{n,m}\}_{m=1}^\infty$ denote a Schauder decomposition for the space E_n. With the order inherited from each E_n and unconditional order between the different E_n's, $\{E_{n,m}\}_{n,m=1}^\infty$ is a Schauder decomposition for E if and only if for each x in E, $x = \sum_{n=1}^\infty x_n$, $x_n = \sum_{m=1}^\infty x_{n,m}$, and each $p \in \mathrm{cs}(E)$*

$$\sum_{n=0}^\infty \sup_k p\left(\sum_{m=1}^k x_{n,m}\right) < \infty. \tag{4.1}$$

Proof. The mode of convergence we require is that for x in E, $p \in \mathrm{cs}(E)$ and $\varepsilon > 0$ there exist a finite subset J of $\mathbb{N}$ and $(k_n)_{n \in J}$ a finite set of positive integers such that for every J' finite, $J \subset J'$, and every set of positive integers $(k'_n)_{n \in J'}$, $k'_n \geq k_n$ for all $n \in J$, we have

$$p\left(x - \sum_{n \in J'} \sum_{m=1}^{k'_n} x_{n,m}\right) \leq \varepsilon.$$

If $\{E_{n,m}\}_{m=1}^\infty$ is a Schauder decomposition for E_n then the partial sums

$$\left\{ \sum_{n \in J} \sum_{m=1}^{k_n} x_{n,m} \right\}_{\substack{J \text{ finite in } \mathbb{N} \\ k_n \text{ arbitrary}}}$$

of any element of E form a bounded set. Hence condition (4.1) is necessary. We now show that it is sufficient. Let $p \in \mathrm{cs}(E)$. We may suppose, without loss of generality, that

$$p\left(\sum_{n=1}^{\infty} x_n\right) = \sum_{n=1}^{\infty} p(x_n)$$

for all $x = \sum_{n=1}^{\infty} x_n \in E$, $x_n \in E_n$ for all n. Now fix x in E and $\varepsilon > 0$. By (4.1) we can choose n_0 such that

$$\sum_{n=n_0}^{\infty} \sup_k p\left(\sum_{m=1}^{k} x_{n,m}\right) \le \varepsilon.$$

For each n, $1 \le n < n_0$, choose k_n such that

$$p\left(x_n - \sum_{m=1}^{k} x_{n,m}\right) \le \frac{\varepsilon}{2^n}$$

for all $k \ge k_n$. If $\{1,\dots,n_0 - 1\} \subset J$ and $(k'_n)_{n \in J}$ are positive integers such that $k'_n \ge k_n$ for $n < n_0$ then

$$p\left(x - \sum_{n \in J}\sum_{m=1}^{k'_n} x_{n,m}\right) = \sum_{n \in J} p\left(x_n - \sum_{m=1}^{k'_n} x_{n,m}\right) + \sum_{n \notin J} p(x_n)$$

$$\le \sum_{n=1}^{n_0-1} p\left(x_n - \sum_{m=1}^{k'_n} x_{n,m}\right) + \sum_{\substack{n \ge n_0 \\ n \in J}} p\left(\sum_{m > k'_n} x_{n,m}\right)$$

$$+ \sum_{n \notin J} p(x_n)$$

$$\le \sum_{n=1}^{n_0-1} \frac{\varepsilon}{2^n} + 4\varepsilon$$

$$\le 5\varepsilon.$$

Hence we have the required convergence for a Schauder decomposition. Since the projections $\pi_n\colon E \to E_n$ and $\pi_{n,m}\colon E_n \to E_{n,m}$ are continuous this shows that $\{E_{n,m}\}_{n,m=1}^{\infty}$ is a Schauder decomposition for E and completes the proof.

We apply Proposition 4.1 in the following way: we begin by considering a locally convex space E with Schauder decomposition and use monomials, relative to the decomposition, to obtain a Schauder decomposition for each space of homogeneous polynomials. This gives us the spaces $E_{n,m}$. We take a balanced domain U in E and consider on $H(U)$ a locally convex topology for which the homogeneous polynomials are an S-absolute decomposition. We then apply Proposition 4.1 to obtain a Schauder decomposition of $H(U)$.

Since there are delicate points involved in ordering the Schauder decomposition of the space of n-homogeneous polynomials we first discuss the situation informally. We consider a space E with a Schauder basis $(e_n)_n$ (we see later that it is preferable to actually begin with a Schauder decomposition but in this introduction we use a basis and relate it to the classical theory and known results). Since any Schauder basis in a locally convex space E can be extended to an algebraic (or Hamel) basis, (3.7) implies that each $f \in H(E)$ has a monomial expansion

$$\sum_{m \in \mathbb{N}^{(\mathbb{N})}} a_m z^m \tag{4.2}$$

where the series (4.2) converges to f at all points in the algebraic span of $(e_n)_n$, i.e. at all z of the form $\sum_{j=1}^{k} z_j e_j$ where k is an arbitrary positive integer. By Proposition 3.43 the series (4.2) converges absolutely for *all* $z = \sum_{j=1}^{\infty} z_j e_j \in E$ if E is a fully nuclear space. We prove in Section 4.4 that nuclearity is not only sufficient but also necessary to obtain absolute convergence when E is Fréchet–Montel. At the polynomial level we obtain the series

$$\sum_{\substack{m \in \mathbb{N}^{(\mathbb{N})} \\ |m|=n}} a_m z^m \tag{4.3}$$

and the same remarks apply. The problem with (4.2) and (4.3) is the unconditional or unordered convergence required. To obtain convergence at all points of E without nuclearity it is necessary to place a special order, that we now describe, on the method of summation.

If E is a Banach space with shrinking basis $(e_j)_j$ then $\{e_i \otimes e_j\}_{i,j=1}^{\infty}$ forms a basis with the square order for $\bigotimes_{2,\pi} E$. This order can be displayed in matrix fashion as follows:

$$
\begin{array}{ccccccc}
e_1 \otimes e_1 & & e_1 \otimes e_2 & \longrightarrow & e_1 \otimes e_3 & \cdots \\
\downarrow & & \uparrow & & \downarrow & \\
e_2 \otimes e_1 & \longrightarrow & e_2 \otimes e_2 & & e_2 \otimes e_3 & \cdots \\
& & & & \downarrow & \\
e_3 \otimes e_1 & \longleftarrow & e_3 \otimes e_2 & \longleftarrow & e_3 \otimes e_3 & \cdots \\
\downarrow & & & & & \\
\vdots & & \vdots & & \vdots &
\end{array}
$$

where we have indicated by arrows the order of the basis. This result has been extended to n-symmetric tensors and gives a Schauder basis for $\bigotimes_{n,s,\pi} E$. This basis is not, in general, shrinking and so does not lead to a basis for $\mathcal{P}(^nE)$. In special situations it does lead to a basis and connections with the reflexivity of $\mathcal{P}(^nE)$. This approach via tensors is linear – we have seen in Chapter 1 that tensor products are a means of linearizing polynomials – and we feel that the linear approach has led to a rather artificial presentation of the square order. If we consider polynomials from the function theory point of view it becomes more natural and shows why a Schauder decomposition – at least initially – is a more natural hypothesis on the underlying space than a Schauder basis. In the following discussion we identify $(z_j)_{j=1}^n$ with $\sum_{j=1}^n z_j e_j$ and $(z_j)_{j=1}^\infty$ with $\sum_{j=1}^\infty z_j e_j$.

Consider $f \in H(E)$ and $(z_j)_{j=1}^\infty \in E$. Then

$$f(z_1) + \sum_{j=1}^n \big(f(z_1,\ldots,z_{j+1}) - f(z_1,\ldots,z_j)\big) = f(z_1,\ldots,z_{n+1})$$

for all n. Since f is continuous

$$\lim_{n\to\infty} f(z_1,\ldots z_{n+1}) = f\big((z_j)_{j=1}^\infty\big)$$

and we obtain the series expansion for f

$$f\big((z_j)_{j=1}^\infty\big) = f(z_1) + \sum_{n=1}^\infty \big(f(z_1,\ldots,z_{n+1}) - f(z_1,\ldots,z_n)\big) \tag{4.4}$$

valid for all $(z_j)_{j=1}^\infty$ in E. We have obtained an expansion at all points – rather than at the points in the algebraic span of the basis – which does not require nuclearity. We establish a connection between (4.2) and (4.4), by rewriting (4.4) using the monomial expansion on each summand in (4.4). This is a valid operation since each summand involves only a finite number of variables and we may apply (3.7). We have

$$f\big((z_j)_{j=1}^\infty\big) = \sum_{n=0}^\infty a_n z_1^n + \sum_{\substack{n=0 \\ m=1}}^\infty a_{n,m} z_1^n z_2^m + \sum_{\substack{n_1,n_2=0 \\ n_3=1}}^\infty a_{n_1,n_2,n_3} z_1^{n_1} z_2^{n_2} z_3^{n_3} + \cdots$$

$$\parallel \qquad\qquad \parallel \qquad\qquad\qquad \parallel \tag{4.5}$$

$$f(z_1) \quad \big(f(z_1,z_2) - f(z_1)\big) \quad \big(f(z_1,z_2,z_3) - f(z_1,z_2)\big)$$

The expansions (4.2) and (4.5) contain precisely the same terms but, in (4.5), we have imposed a partial order on the sum so that, if we add the terms within each summand in (4.5) in any order and then add these together in the order given in (4.5), we obtain convergence at all points of E. Indeed, if π_n denotes the canonical projection from E onto span$\{e_1,\ldots,e_n\}$ then

$$f(z_1,\ldots,z_{n+1}) - f(z_1,\ldots,z_n) = (f \circ \pi_{n+1} - f \circ \pi_n)\big((z_j)_{j=1}^\infty\big)$$

and we obtain, in (4.4) and (4.5), uniform convergence on the compact subsets of E.

We now revert to $\mathcal{P}(^nE)$. The analysis we have just given clearly applies to (4.3) in place of (4.2) since we can take f to be an n-homogeneous polynomial in (4.5). For each positive integer k let

$$\mathcal{P}_k(^nE) = \left\{ \sum a_m z^m \ : \ m \in \mathbb{N}^{(\mathbb{N})}, \ |m| = n, \ l(m) = k \right\}.$$

The space $\mathcal{P}_k(^nE)$ is finite dimensional for each k and n and we have shown that $\left\{ (\mathcal{P}_k(^nE), \tau_0) \right\}_{k=1}^{\infty}$ is a finite dimensional Schauder decomposition for $(\mathcal{P}(^nE), \tau_0)$ whenever E has a Schauder basis. We would have arrived, as we shall see later, at the same conclusion if we had merely assumed that E had a finite dimensional decomposition rather than a basis.

To obtain a monomial basis for $\mathcal{P}(^nE)$ we must place an order on the monomials. We first say that $m < \overline{m}$ for $m, \overline{m} \in \mathbb{N}^{(\mathbb{N})}$, if $l(m) < l(\overline{m})$ and afterwards order the monomial basis in each $\mathcal{P}_k(^nE)$. It is probable that a number of different orderings lead, especially for the weaker topologies, to a Schauder basis. However, if the ordering was unimportant, especially for the τ_ω topology, we would end up with an unconditional basis for $\mathcal{P}(^nE)$ and, as we shall see later, this is not always possible. We order the monomial basis for $\mathcal{P}_k(^nE)$ by induction in such a way that the required convergence properties are easily verified. For $n = 1$ we have the natural order inherited from the basis in E. If $z^m \in \mathcal{P}_k(^{n+1}E)$ then there exists a unique m' such that $z^m = z^{m'} \cdot z_k$ and $z^{m'} \in \mathcal{P}_l(^nE)$ for some $l \leq k$. Suppose we have ordered the monomial basis in $\mathcal{P}_k(^nE)$ for all k. If z^m and $z^{\overline{m}} \in \mathcal{P}_k(^{n+1}E)$, $z^m = z^{m'} \cdot z_k$ and $z^{\overline{m}} = z^{\overline{m}'} \cdot z_k$ let $m \leq \overline{m}$ if $m' \leq \overline{m}'$. Thus the ordered monomial basis for $\mathcal{P}_k(^{n+1}E)$ is obtained from $\bigoplus_{l \leq k} \mathcal{P}_l(^nE)$ by multiplying each monomial by z_k and keeping the inherited order. We illustrate this by writing out a few cases.

		$k = 1$	$k = 2$	$k = 3$	
$n = 1$	E'	z_1	z_2	z_3	$\cdots$
$n = 2$	$\mathcal{P}(^2E)$	$\underbrace{z_1^2}_{\mathcal{P}_1(^2E)}$	$\underbrace{z_1z_2, \ z_2^2}_{\mathcal{P}_2(^2E)}$	$\underbrace{z_1z_3, \ z_2z_3, \ z_3^2}_{\mathcal{P}_3(^2E)}$	$\cdots$
$n = 3$	$\mathcal{P}(^3E)$	$\underbrace{z_1^3}_{\mathcal{P}_1(^3E)}$	$\underbrace{z_1^2z_2, \ z_1z_2^2, \ z_2^3}_{\mathcal{P}_2(^3E)}$	$\underbrace{z_1^2z_3, \ z_1z_2z_3, \ z_2^2z_3, \ z_1z_3^2, \ z_2z_3^2, \ z_3^3}_{\mathcal{P}_3(^3E)}$	$\cdots$

In fact we have arrived at the square ordering on the monomials. A formal definition of this order is the following:

If $m = (m_i)_{i=1}^{\infty}$ and $\overline{m} = (\overline{m}_i)_{i=1}^{\infty}$, $|m| = |\overline{m}| = n$ then $m < \overline{m}$ if

(1) $l(m) < l(\overline{m})$ or

(2) $l(m) = l(\overline{m})$ and for some $i \leq l(m)$, $m_i < \overline{m_i}$ and $m_j = \overline{m_j}$ for $j > i$.

This completes our informal introduction. We now proceed with a formal and more general discussion. Let $\{E_m\}_m$ denote a Schauder decomposition for the locally convex space E. We always assume in the expansion, $x = \sum_{j=1}^{\infty} x_j \in E$, that $x_j \in E_j$ for all j. We begin by defining certain concepts similar to those given in Section 3.3 for fully nuclear spaces with a basis. If $P \in \mathcal{P}(^n E)$ we say that P is k-homogeneous in the jth variable if

$$P\left(\lambda x_j + \sum_{i \neq j} x_i\right) = \lambda^k P\left(\sum_{i=1}^{\infty} x_i\right)$$

for all $\sum_{i=1}^{\infty} x_i \in E$ and all $\lambda \in \mathbb{C}$. If $m = (m_j)_j \in \mathbb{N}^{(\mathbb{N})}$ we denote by $\mathcal{P}(^m E)$ the set of all $P \in \mathcal{P}(^{|m|}E)$ which are m_j-homogeneous in the jth variable for all j. Hence $P \in \mathcal{P}(^m E)$ if and only if

$$P\left(\sum_{j=1}^{\infty} \lambda_j x_j\right) = \lambda^m P\left(\sum_{j=1}^{\infty} x_j\right)$$

for all $x = \sum_{j=1}^{\infty} x_j \in E$ and any sequence of scalars $(\lambda_j)_j$, such that $\sum_{j=1}^{\infty} \lambda_j x_j \in E$ (note that $\lambda^m = \prod_{j=1}^{\infty} \lambda_j^{m_j}$ converges since $\lambda_j^{m_j} = 1$ for all except a finite number of j). For n and k positive integers let $\mathcal{P}_k(^n E)$ denote, as before, the subspace of $\mathcal{P}(^n E)$ spanned by $\{P \in \mathcal{P}(^m E) : m \in \mathbb{N}^{\mathbb{N}}, |m| = n, l(m) = k\}$. If $P \in \mathcal{P}(^n E)$, $m \in \mathbb{N}^{(\mathbb{N})}$, $|m| = n$ and $k \geq l(m)$ let

$$P_m(x) = \frac{1}{(2\pi i)^n} \int \cdots \int_{\substack{|\lambda_i|=1 \\ i=1,\ldots,k}} \frac{P(\sum_{i=1}^{k} \lambda_i x_i)}{\lambda_1^{m_1+1} \cdots \lambda_k^{m_k+1}} d\lambda_1 \ldots d\lambda_k.$$

If $P \in \mathcal{P}(^n E)$, $x = \sum_{j=1}^{\infty} x_j \in E$ and k is a positive integer, then

$$\sum_{\substack{m \in \mathbb{N}^{(\mathbb{N})} \\ l(m)=k+1}} P_m\left(\sum_{j=1}^{\infty} x_j\right) = \sum_{\substack{m \in \mathbb{N}^{(\mathbb{N})} \\ l(m)=k+1}} P_m\left(\sum_{j=1}^{k+1} x_j\right)$$

$$= P\left(\sum_{j=1}^{k+1} x_j\right) - P\left(\sum_{j=1}^{k} x_j\right)$$

belongs to $\mathcal{P}_k(^n E)$. Moreover, we obtain the following formula, similar to (4.3) and (4.5),

$$P = \sum_{\substack{m\in \mathbb{N}^{(\mathbb{N})} \\ |m|=n}} P_m \tag{4.6}_a$$

$$= \sum_{k=1}^{\infty} \Big\{ \sum_{\substack{m\in \mathbb{N}^{(\mathbb{N})} \\ |m|=n \\ l(m)=k}} P_m \Big\} \tag{4.6}_b$$

where the series $(4.6)_a$ converges for all finite sums $\sum_{j=1}^{k} x_j \in E$ and the series $(4.6)_b$ converges for all x in E. For each m let $\pi_m\big(\sum_{j=1}^{\infty} x_j\big) = \sum_{j=1}^{m} x_j$ for $\sum_{j=1}^{\infty} x_j$ in E. Let $p \in cs(E)$ and suppose $p(x) = \sup_m p\big(\pi_m(x)\big)$ for all x in E. If $(P_k)_{k=1}^{\infty}$ is a sequence of n-homogeneous polynomials, $P_k \in \mathcal{P}_k(^nE)$ for all k, then for any positive integer k

$$\Big\| \sum_{j=1}^{k} P_j \Big\|_{B_p(1)} = \sup_{p\left(\sum_{m=1}^{\infty} x_m\right) \leq 1} \Big\| \sum_{j=1}^{k} P_j \Big(\sum_{m=1}^{\infty} x_m \Big) \Big\|$$

$$= \sup_{p\left(\sum_{m=1}^{k} x_m\right) \leq 1} \Big\| \sum_{j=1}^{k} P_j \Big(\sum_{m=1}^{k} x_m \Big) \Big\| \tag{4.7}$$

$$= \sup_{p\left(\sum_{m=1}^{k} x_m\right) \leq 1} \Big\| \sum_{j=1}^{k+1} P_j \Big(\sum_{m=1}^{k} x_m \Big) \Big\|$$

$$\leq \Big\| \sum_{j=1}^{k+1} P_j \Big\|_{B_p(1)}.$$

In the next two propositions we use the following well-known characterization of basic sequences in a Banach space:

A sequence of non-zero vectors $(w_j)_j$ in a Banach space is a basic sequence (i.e. a basis for its closed linear span) if and only if there exists $c > 0$ such that for any sequence of scalars $(\alpha_j)_j$ and any positive integers m and n

$$\Big\| \sum_{j=1}^{m} \alpha_j w_j \Big\| \leq c \cdot \Big\| \sum_{j=1}^{m+n} \alpha_j w_j \Big\|. \tag{4.8}$$

The infimum of all c satisfying (4.8) is called the *basis constant* of the sequence. By induction one easily sees that (4.8) is satisfied if one verifies (4.8) with $c = 1$ and $n = 1$ for all m. In this case the basic sequence is called *monotone*. If $\{E_n\}_n$ is a sequence of subspaces of E and any non-zero sequence $(x_n)_n$, $x_n \in E_n$, all n, is a basic sequence verifying (4.8) with c independent of the sequence then $\{E_n\}_n$ is a Schauder decomposition for its closed linear span.

Proposition 4.2 *Let $\{E_m\}_{m=1}^{\infty}$ denote an equi-Schauder decomposition for the locally convex space E. The sequence $\{\mathcal{P}_k(^nE)\}_{k=1}^{\infty}$ is a Schauder decomposition for the subspace*

$$G_n := \{P \in \mathcal{P}(^nE) : \text{there exists a neighbourhood } V \text{ of } 0 \text{ such that}$$
$$\|P - P \circ \pi_m\|_V \to 0 \ \text{as } m \to \infty\}$$

of $\left(\mathcal{P}(^nE), \tau_\omega\right)$.

Proof. Since the decomposition is equi-Schauder and local uniform convergence implies τ_ω convergence we may suppose that E is a Banach space. If $P \in G_n$ let $P_1 = P \circ \pi_1$ and $P_{k+1} = P \circ \pi_{k+1} - P \circ \pi_k$ for $k \geq 1$. For all k, $P_k \in \mathcal{P}_k(^nE)$, and by (4.7), the non-zero terms in $\{P_k\}_{k=1}^{\infty}$ form a basic sequence in the Banach space $\mathcal{P}(^nE)$. This shows that $\mathcal{P}_k(^nE)$ is a Schauder decomposition for G_n and completes the proof.

Corollary 4.3 *If $\{E_m\}_{m=1}^{\infty}$ is a finite dimensional shrinking Schauder decomposition for the Banach space E then $\{\mathcal{P}_k(^nE)\}_{k=1}^{\infty}$ is a finite dimensional Schauder decomposition for $\mathcal{P}_\omega(^nE)$.*

Proof. Since each E_m is finite dimensional and $\mathcal{P}_k(^nE)$ can be identified with a subspace of the finite dimensional space $\mathcal{P}(^n(\sum_{j=1}^{k} E_j))$ it follows that $\mathcal{P}_k(^nE)$ is finite dimensional for each k and n. By Proposition 4.2, $\{\mathcal{P}_k(^nE)\}_{k=1}^{\infty}$ is a Schauder decomposition for the subspace G_n of $\mathcal{P}(^nE)$. If $P \in \mathcal{P}(^nE)$ then $P \circ \pi_m$ is weakly continuous on bounded sets and hence $G_n \subset \mathcal{P}_\omega(^nE)$.

Now suppose $\phi \in E'$. Let $\psi_k = \phi - \phi \circ \pi_k$ for all k. Since $\{E_m\}_{m=1}^{\infty}$ is a shrinking basis $\|\psi_k\| \to 0$ as $k \to \infty$. From the identity $a^n - b^n = (a - b)\left(\sum_{j=1}^{n-1} a^{n-1-j}b^j\right)$ we see that

$$\|\phi^n - (\phi \circ \pi_k)^n\| = \sup_{\|x\|\leq 1} \left|\phi^n(x) - \phi^n(\pi_k(x))\right|$$

$$\leq \sup_{\|x\|\leq 1} \left|\phi(x) - \phi \circ \pi_k(x)\right| \cdot \left(\sum_{j=1}^{n-1} |\phi(x)|^{n-1-j}|\phi(\pi_k(x))|^j\right)$$

$$\leq n\|\psi_k\| \cdot \|\phi\|^{n-1} \longrightarrow 0 \ \text{as } k \to \infty.$$

Since $(\phi \circ \pi_k)^n = \phi^n \circ \pi_k$, $\phi^n \in G_n$ for all $\phi \in E'$ and $\overline{\mathcal{P}_f(^nE)} = \mathcal{P}_A(^nE) \subset G_n$. As $\{E'_m\}_m$ is a finite dimensional Schauder decomposition for E'_β, the space E'_β has the approximation property and, by Proposition 2.8, $\mathcal{P}_A(^nE) = \mathcal{P}_\omega(^nE)$. Hence $G_n = \mathcal{P}_\omega(^nE)$ and this completes the proof.

Proposition 4.4 *If the Banach space E has a shrinking basis then the monomials of degree n (relative to the basis) with the square order form a Schauder basis for $\left(\mathcal{P}_\omega(^nE), \|\cdot\|\right)$.*

Proof. We suppose, without loss of generality, that the basis for E is monotone. Since $\{\mathcal{P}_k(^nE)\}_{k=1}^{\infty}$ is a finite dimensional decomposition it suffices to show that the basis constants (for the monomial basis) of $\mathcal{P}_k(^nE)$ are, for each n, bounded with respect to k. We prove this by induction on n. Since the basis for E is shrinking the result is true for $n = 1$. Suppose the result is true for $\mathcal{P}_\omega(^nE)$ with c_n denoting the basis constant. Let $\{P_{k,j}^{n+1}\}_{j=1}^{t(k)}$ denote the monomial basis for $\mathcal{P}_k(^{n+1}E)$. We note that $P_{k,j}^{n+1} = Q_{k,j}z_k$ where $\{Q_{k,j}\}_{j=1}^{t(k)}$ is the monomial basis for $\bigoplus_{s=1}^{k} \mathcal{P}_s(^nE)$. Let $1 \leq n_1 \leq n_2 \leq l(k)$ and let $(a_{j,k})_{k,j}$ denote an arbitrary set of scalars. We have

$$\left\|\sum_{j=1}^{n_1} a_{k,j} P_{k,j}^{n+1}\right\| = \left\|\sum_{j=1}^{n_1} a_{k,j} Q_{k,j} z_k\right\|$$

$$\leq \left\|\sum_{j=1}^{n_1} a_{k,j} Q_{k,j}\right\| \cdot \|z_k\|$$

$$\leq 2\left\|\sum_{j=1}^{n_1} a_{k,j} Q_{k,j}\right\|$$

$$\leq 2c_n\left\|\sum_{j=1}^{n_2} a_{k,j} Q_{k,j}\right\|.$$

Thus we have to prove that there exists $c' > 0$ (independent of k and n_2) such that

$$\left\|\sum_{j=1}^{n_2} a_{k,j} Q_{k,j}\right\| \leq c'\left\|\sum_{j=1}^{n_2} a_{k,j} P_{k,j}^{n+1}\right\|.$$

We have

$$\left\|\sum_{j=1}^{n_2} a_{k,j} Q_{k,j}\right\| = \sup_{\substack{x \in E \\ \|x\| \leq 1}} \left|\sum_{j=1}^{n_2} a_{k,j} Q_{k,j}(x)\right|$$

$$= \sup_{\substack{x = \sum_{i=1}^{k} x_i e_i \\ \|x\| \leq 1}} \left|\sum_{j=1}^{n_2} a_{k,j} Q_{k,j}(x)\right|$$

$$= \left|\sum_{j=1}^{n_2} a_{k,j} Q_{k,j}(\omega)\right|$$

for some $\omega = \sum_{j=1}^{k} \omega_j e_j$, $\|\omega\| = 1$. If $|\omega_k| < 1/2$ let

$$g(\lambda) = \left(\sum_{j=1}^{n_2} a_{k,j} Q_{k,j}\right)(\omega + \lambda e_k).$$

The function g is a polynomial over $\mathbb{C}$ and, by the Maximum Modulus Theorem,

$$\sup_{|\lambda|=1} |g(\lambda)| \geq |g(0)| = \left\| \sum_{j=1}^{n_2} a_{k,j} Q_{k,j} \right\|.$$

Choose λ, $|\lambda| = 1$, where the supremum is achieved, and let $\tilde{\omega} = \omega + \lambda e_k$. Then

$$\left| \left(\sum_{j=1}^{n_2} a_{k,j} Q_{k,j} \right)(\tilde{\omega}) \right| \geq \left\| \sum_{j=1}^{n_2} a_{k,j} Q_{k,j} \right\|$$

and $\|\tilde{\omega}\| \leq \|\omega\| + \|\lambda e_k\| \leq 1 + s$ where $s = \sup_j \|e_j\|$. If $\tilde{\omega} = \sum_{j=1}^{k} \tilde{\omega}_k e_k$ then $|\tilde{\omega}_k| = |\omega_k + \lambda| \geq |\lambda| - |\omega_k| \geq 1/2$.

If $|\omega_k| \geq 1/2$ let $\tilde{\omega} = \omega$. In this case it is easily checked that $\tilde{\omega}$ also satisfies the properties listed for the first $\tilde{\omega}$. In either case we have

$$\left| \sum_{j=1}^{n_2} a_{k,j} P_{k,j}^{n+1}(\tilde{\omega}) \right| = \left| \sum_{j=1}^{n_2} a_{k,j} Q_{k,j}(\tilde{\omega}) \right| \cdot |\tilde{\omega}_k|$$

$$\geq \frac{1}{2} \left\| \sum_{j=1}^{n_2} a_{k,j} Q_{k,j} \right\|$$

and

$$\left| \sum_{j=1}^{n_2} a_{k,j} P_{k,j}^{n+1}(\tilde{\omega}) \right| \leq \left\| \sum_{j=1}^{n_2} a_{k,j} P_{k,j}^{n+1} \right\| \cdot \|\tilde{\omega}\|^{n+1}$$

$$\leq (1+s)^{n+1} \left\| \sum_{j=1}^{n_2} a_{k,j} P_{k,j}^{n+1} \right\|.$$

Hence

$$\left\| \sum_{j=1}^{n_2} a_{k,j} Q_{k,j} \right\| \leq 2(1+s)^{n+1} \left\| \sum_{j=1}^{n_2} a_{k,j} P_{k,j} \right\|$$

and this completes the proof.

Corollary 4.5 *If $\{E_m\}_{m=1}^{\infty}$ is a finite dimensional monotone Schauder decomposition for the reflexive Banach space E then $\mathcal{P}(^n E)$ is reflexive if and only if $\{\mathcal{P}_k(^n E)\}_{k=1}^{\infty}$ is a finite dimensional decomposition for $\mathcal{P}(^n E)$. If, in addition, $\dim(E_m) = 1$ for all m, i.e. if E has a basis, then $\mathcal{P}(^n E)$ is reflexive if and only if the monomials with the square order form a Schauder basis for $\mathcal{P}(^n E)$.*

Proof. A finite dimensional Schauder decomposition for a reflexive Banach space E is shrinking. Moreover, since E'_β also has a finite dimensional decomposition it has the approximation property. Applications of Propositions 2.31 and 4.4 and Corollary 4.3 complete the proof.

We apply Proposition 4.1 and the above results to obtain finite dimensional Schauder decompositions for spaces of holomorphic functions on certain convex balanced domains. We confine ourselves to a few examples. To verify (4.1) it is often necessary to place restrictions on the domain. A minimum requirement usually is that the domain be invariant under the projections $(\pi_n)_n$ associated with the given Schauder decomposition. When E is a Banach space with a monotone Schauder shrinking decomposition and we wish to apply Proposition 4.4 to holomorphic functions on U with the τ_ω topology it suffices to have $\overline{\bigcup_n \pi_n(K)}$ compact in U for any compact subset K of U. This condition is satisfied when $U = rB$, $r \in (0, \infty]$, and B is the open unit ball in E (we let $\infty \cdot B = E$). This leads to a finite dimensional Schauder decomposition. To obtain a Schauder basis is more complicated and it may be necessary, in order to verify (4.1), to obtain an upper bound for the basis constants, $(c_n)_n$, that appear in the proof of Proposition 4.4.

If U is a balanced open subset of a Banach space E let

$$H_\omega(U) = \left\{ f \in H(U) : \frac{\widehat{d^n} f(0)}{n!} \in \mathcal{P}_\omega(^nE) \text{ for all } n \right\}.$$

It is easily seen that $H_\omega(U)$ consists of all holomorphic functions which are locally weakly uniformly continuous, i.e.

$$H_\omega(U) = \{ f \in H(U) : \text{ for all } x \in U, \text{ there exists } V \text{ open}, x \in V \subset U$$
$$\text{such that } f\big|_V \text{ is weakly uniformly continuous}\}.$$

Moreover,

$$H_\omega(U) = H(U) \iff \mathcal{P}_\omega(^nE) = \mathcal{P}(^nE) \text{ for all } n.$$

Proposition 4.6 *Let E denote a Banach space with a monotone shrinking finite dimensional Schauder decomposition $\{E_m\}_{m=1}^\infty$ and let B denote the open unit ball of E. If $U = rB$, $r \in (0, \infty]$, then $\{\mathcal{P}_k(^nE)\}_{n=0,k=1}^{\infty,\infty}$ is a finite dimensional Schauder decomposition for $(H_\omega(U), \tau_\omega)$.*

Proof. Let $\widetilde{K}$ denote the closed convex hull of $\bigcup_n \pi_n(K)$, where K is a compact subset of U. Since the basis is monotone, $\widetilde{K}$ is a compact subset of U and, moreover, $\pi_n(\widetilde{K}) \subset \widetilde{K}$ for all n. Let $(\alpha_n)_n \in c_0$. If $f \in H_\omega(U)$ and $\dfrac{\widehat{d^n} f(0)}{n!}$ has the expansion $\sum_{k=1}^\infty P_{n,k}$, relative to the finite dimensional Schauder decomposition $\{\mathcal{P}_k(^nE)\}_{k=1}^\infty$ of $\mathcal{P}_\omega(^nE)$, then by (4.7),

$$\left\| \sum_{k=1}^m P_{n,k} \right\|_{K+\alpha_n B} \leq \left\| \frac{\widehat{d^n} f(0)}{n!} \right\|_{\widetilde{K}+\alpha_n B}$$

for all m and n. Since $\displaystyle\sum_{n=0}^{\infty}\left\|\frac{\widehat{d^n}f(0)}{n!}\right\|_{\widetilde{K}+\alpha_n B} < \infty$, Proposition 3.47 implies that (4.1) is satisfied. An application of Proposition 4.1 completes the proof.

Corollary 4.7 *If E is a Banach space with a monotone shrinking finite dimensional Schauder decomposition $\{E_m\}_{m=1}^{\infty}$ and $\mathcal{P}(^nE) = \mathcal{P}_\omega(^nE)$ for all n then $\{\mathcal{P}_k(^nE)\}_{n=0,k=1}^{\infty,\infty}$ is a finite dimensional Schauder decomposition for $\big(H(U),\ \tau_\omega\big)$ where $U = rB$, $0 < r \le \infty$, and B is the open unit ball in E.*

We remark that the hypothesis of a *monotone* Schauder decomposition in the last two results is not an essential restriction since a Banach space can always be renormed so that the Schauder decomposition becomes monotone and we obtain examples where U is a multiple of the unit ball in the renormed space. Corollary 2.37 and Proposition 2.41 give conditions under which $\mathcal{P}(^nE) = \mathcal{P}_\omega(^nE)$ for all n.

If E is

(a) c_0
(b) T^* (Tsirelson space)
(c) J_{T^*} (Tsirelson–James space)
(d) any Banach space which has the Dunford–Pettis Property, does not contain l_1 and admits a shrinking finite dimensional decomposition,

then $\big(H(E),\ \tau_\omega\big)$ has a finite dimensional Schauder decomposition.

The main result in this section, Proposition 4.2, essentially applies to Banach spaces since it treats each polynomial as a mapping which is continuous with respect to a single semi-norm on the space. To obtain non-trivial applications to polynomials and holomorphic functions on spaces other than Banach spaces it is necessary to use (and perhaps to assume) some interactions between the continuous semi-norms on the space. We shall see this occurring in the next two sections.

4.2 $\tau_\omega = \tau_\delta$ for Fréchet Spaces

In this section we discuss the $\tau_\omega = \tau_\delta$ problem for holomorphic functions on a collection of Fréchet spaces with finite dimensional Schauder decompositions. This is the key to extending polynomial results to spaces of holomorphic functions. We use a monomial expansion for holomorphic functions but, as we have seen in the previous section, it is necessary to find the correct setting for such expansions. Since we are considering a domain U in a Fréchet space the τ_δ topology is the bornological topology associated with τ_ω and we obtain $\tau_\omega = \tau_\delta$ by completing the following programme.

Let F denote a Banach space and let $T: H(U) \to F$ denote a linear mapping which is bounded on the τ_ω bounded subsets of $H(U)$ (or equivalently which is τ_δ-continuous).

Step 1:   Show that the monomials in $\{\mathcal{P}(^m E)\}_{m \in \mathbb{N}^{(\mathbb{N})}}$, relative to a given Schauder decomposition of E, form a Schauder decomposition for $(H(U), \tau_\delta)$.

Step 2:   Show that there exists a compact subset K of U such that for every V open, $K \subset V \subset U$, there exists $c(V) > 0$ such that

$$\|T(P)\| \le c(V) \|P\|_V$$

for all $P \in \mathcal{P}(^m E)$ and all $m \in \mathbb{N}^{(\mathbb{N})}$.

Step 3:   Show that for each neighbourhood V of K there exists a neighbourhood W of K and $c(V, W) > 0$ such that

$$\sum_{m \in \mathbb{N}^{(\mathbb{N})}} \|P_m\|_W \le c(V, W) \|f\|_V$$

for all $f = \sum_{m \in \mathbb{N}^{(\mathbb{N})}} P_m \in H(U)$.

Let E denote a Fréchet nuclear space with basis $(e_j)_j$ and increasing sequence of semi-norms $(\| \cdot \|_k)_{k=1}^\infty$. Let $a_{j,k} = \|e_j\|_k$ for all j and k and let A denote the Köthe matrix $(a_{j,k})_{1 \le j, k < \infty}$. By the Grothendieck–Pietsch criterion we may suppose $\sum_{j=1}^\infty \dfrac{a_{j,k}}{a_{j,k+1}} < \infty$ for all k. With the notation of Chapter 3 we have $E \approx \Lambda(P)$, $P = (a_k)_{k=1}^\infty$ and $a_k = (a_{j,k})_{j=1}^\infty$. We use here, however, the standard notation of Köthe sequence spaces, i.e. we let $E = \lambda(A)$ and $\|(x_j)_{j=1}^\infty\|_k = \sum_{j=1}^\infty a_{j,k} |x_j|$ for all k.

Now suppose $(E_j, \| \cdot \|_j')_{j=1}^\infty$ is a sequence of Banach spaces. Let

$$\lambda\left(A, \left\{(E_j, \| \cdot \|_j')\right\}_j\right) := \left\{(x_j)_j \; : \; x_j \in F_j, \; \|(x_j)_j\|_k := \sum_{j=1}^\infty a_{j,k} \|x_j\|_j' < \infty \right.$$

$$\left. \text{for all } k \right\}.$$

Since there is little danger of confusion we write $\| \cdot \|$ in place of $\| \cdot \|_j'$ for each j and $\lambda(\{E_j\}_j)$ in place of $\lambda(A, \{(E_j, \| \cdot \|_j')_j\})$. With the topology generated by $(\| \cdot \|_k)_{k=1}^\infty$ the space $\lambda(\{E_j\}_j)$ is a Fréchet space and $\{E_j\}_j$ is a Schauder decomposition of $\lambda(\{E_j\}_j)$.

Example 4.8
(a) If $E_1 = E$ and $E_j = \{0\}$ for $j > 1$ then $\lambda(\{E_j\}_j) = E$,
(b) If $\dim(E_j) = 1$ for all j then $\lambda(\{E_j\}_j) \cong \lambda(A)$,

(c) The space $\lambda(\{E_j\}_j)$ is Montel, or Schwartz, if and only if $\dim(E_j) < \infty$ for all j,

(d) Let $(b_j)_j$ denote an increasing sequence of positive integers and for each j let $E_j = l^\infty_{b_j}$. The space $\lambda(\{E_j\}_j)$ is a nuclear Fréchet space if and only if for each k there exists l such that

$$\sum_{j=1}^{\infty} \frac{b_j a_{j,k}}{a_{j,k+l}} < \infty.$$

In particular, if $a_{j,k} = j^{2k}$ for all j and k and $b_j = j^{2j}$ then $\lambda(\{E_j\}_j)$ is Schwartz but not nuclear.

If U_j is a balanced open subset of E_j for each j let

$$\bigoplus_j U_j = \text{Interior}\{(x_j)_j \in \lambda(\{E_j\}) \ : \ x_j \in U_j \text{ for all } j\}.$$

We call $\bigoplus_j U_j$ a *balanced open set of polydisc type*. The set $\bigoplus_j U_j$ will be *non-empty* if and only if there exists an integer k such that $\{x_j \in E_j \ : \ a_{j,k}\|x_j\| < 1\} \subset U_j$ for all j. When we consider $\bigoplus_j U_j$ we always assume it is non-empty. For each integer k the set $V_k := \{(x_j)_j \in \lambda(\{E_j\}_j) \ : \ \sup_j a_{j,k}\|x_j\| < 1\}$ is a balanced open set of polydisc type. In Example 4.8(a) every balanced open set is of polydisc type and in Example 4.8(b) the balanced open sets of polydisc type are the open polydiscs.

For each positive integer n let p_n denote the canonical projection from $E \cong \lambda(\{E_j\}_j)$ onto E_n. If $U := \bigoplus_j U_j$ is a balanced open set of polydisc type then $p_n(U) = U_n$. A subset K of E is compact if and only if $p_n(K)$ is a compact subset of E_n for each n and $\left(\sup_{x \in K} \|p_n(x)\|\right)_n \in \lambda(\{E_j\}_j)$.

Now suppose $f \in H(U)$ where $U := \bigoplus_j U_j$ is a balanced open set of polydisc type in $\lambda(\{E_j\}_j)$. Let $(P_m)_{m \in \mathbb{N}^{(\mathbb{N})}}$ denote the monomials associated with f by the decomposition $\{E_j\}_j$ of $\lambda(\{E_j\}_j)$ (see Section 4.1). By the local boundedness of f we see, as in the proof of Lemma 3.42, that there exists for each compact subset K of U a balanced open set of polydisc type in U, $V := \bigoplus_j V_j$, and a sequence of real numbers, $(\delta_j)_j$, $\delta_j > 1$ and $\sum_{j=1}^{\infty} 1/\delta_j < \infty$, such that $K \subset V \subset \delta V \subset U$ and $\|f\|_{\delta V} := M < \infty$ where $\delta V := \{(\delta_j x_j)_j \ : \ (x_j)_j \in V\}$. Hence $\|P_m\|_V \le M\delta^{-m}$ for all $m \in \mathbb{N}^{(\mathbb{N})}$ and for any finite subset J of $\mathbb{N}^{(\mathbb{N})}$ we have

$$\left\| f - \sum_{m \in J} P_m \right\|_V \le M \cdot \sum_{m \in \mathbb{N}^{(\mathbb{N})} \setminus J} \delta^{-m}.$$

This implies that $\left\{\sum_{m \in J} P_m\right\}_{\substack{J \subset \mathbb{N}^{(\mathbb{N})} \\ J \text{ finite}}}$ is a locally bounded, and hence a τ_ω and τ_δ bounded, subset of $H(U)$. We also have

$$\sum_{m \in \mathbb{N}^{(\mathbb{N})}} \|P_m\|_V \leq \left(\sum_{m \in \mathbb{N}^{(\mathbb{N})}} \delta^{-m} \right) \cdot \|f\|_{\delta V} \tag{4.9}$$

which was Step 3 in our programme. By Corollary 3.37(c) and the above we obtain the following proposition which completes Step 1.

Proposition 4.9 *If $\lambda(A)$ is a Fréchet nuclear space, $\{E_j\}_j$ is a sequence of Banach spaces and $E = \lambda(\{E_j\}_j)$ then $\left\{\mathcal{P}(^mE)\right\}_{m \in \mathbb{N}^{(\mathbb{N})}}$ is an absolute Schauder decomposition for $(H(U), \tau)$, $\tau = \tau_0$ or τ_ω, where U is a balanced open set in E of polydisc type. Moreover, $\left\{\mathcal{P}(^mE)\right\}_{m \in \mathbb{N}^{(\mathbb{N})}}$ is also an unconditional Schauder decomposition for $(H(U), \tau_\delta)$.*

There remains the technically more demanding second step. Let E denote a Banach space with finite dimensional Schauder decomposition $(E_j)_{j=1}^\infty$. We suppose, without loss of generality, that the decomposition is monotone, i.e. $\|x\| = \sup_n \|\pi_n(x)\|$ for all x in E where π_n denotes the canonical projection from E onto $F_n := E_1 + \cdots + E_n$. This implies, in particular, that $\|\pi_n\| \leq 1$ for all n. Let $(j_k)_{k=1}^\infty$ denote a strictly increasing sequence of positive integers and for each n let

$$\|x\|_{j_1,\ldots,j_n} = \left\| \pi_{j_1}(x) + \sum_{k=2}^n 2^{k-1} \left(\pi_{j_k}(x) - \pi_{j_{k-1}}(x) \right) + 2^n \left(x - \pi_{j_n}(x) \right) \right\|$$

for x in E. If $\|x\|_{j_1,\ldots,j_n} \leq 1$ then

$$\|\pi_{j_1}(x)\| \leq 1,$$

$$2^{k-1}\|\pi_{j_k}(x)\| \leq 2^{k-1}\|\pi_{j_{k-1}}(x)\| + 2, \quad 2 \leq k \leq n$$

$$2^n\|x\| \leq 2^n\|\pi_{j_n}(x)\| + 2.$$

By finite induction this implies $\|x\| \leq 3$. On the other hand

$$\|x\|_{j_1,\ldots,j_n} \leq \|x\| + \sum_{k=2}^{n+1} 2^k\|x\|.$$

Hence

$$\|x\| \leq 3\|x\|_{j_1,\ldots,j_n} \leq 3 \cdot 2^{n+2}\|x\| \tag{4.10}$$

for all x in E, all $j_1, \ldots, j_n$ and all n. From (4.10) we see that

$$B_{j_1,\ldots,j_n} := \{x \in E : \|x\|_{j_1,\ldots,j_n} \leq 1\}$$

is a closed subset of E. If $\|x\|_{j_1,\ldots,j_n} \leq 1$ then $\|\pi_{j_n}(x)\| \leq 3$ and $\|x - \pi_{j_n}(x)\| \leq 1/2^{n-1}$. Hence

$$B_{j_1,\ldots,j_n} \subset \{x \in F_{j_n} : \|x\| \leq 3\} + \left\{x \in E : \|x\| \leq \frac{1}{2^{n-2}}\right\}$$

and $B_{j_1,\ldots,j_n}$ may be covered by a finite number of balls of $\|\cdot\|$-radius less than or equal to $1/2^{n-2}$. Hence

$$B_{(j_k)_{k=1}^{\infty}} := \bigcap_{n=1}^{\infty} B_{j_1,\ldots,j_n}$$

is a compact subset of E. If V is a neighbourhood of $B_{(j_k)_{k=1}^{\infty}}$ then there exists $\varepsilon > 0$ such that

$$B_{(j_k)_{k=1}^{\infty}} + \{x \in E : \|x\| \le \varepsilon\} \subset V.$$

If $\delta > 1$ and $\|x\|_{j_1,\ldots,j_n} < \delta$ then $x = \frac{1}{\delta}\pi_{j_n}(x) + (1 - \frac{1}{\delta})\pi_{j_n}(x) + x - \pi_{j_n}(x)$, $\frac{1}{\delta}\pi_{j_n}(x) \in B_{(j_k)_{k=1}^{\infty}}$ and

$$\left\| \left(1 - \frac{1}{\delta}\right)\pi_{j_n}(x) + (x - \pi_{j_n}(x)) \right\| \le \left(1 - \frac{1}{\delta}\right) \cdot 3\delta + \frac{1}{2^{n-1}} < \varepsilon$$

for δ sufficiently close to 1 and n sufficiently large. This proves the following lemma.

Lemma 4.10 *If $\{E_n\}_{n=1}^{\infty}$ is a finite dimensional Schauder decomposition for the Banach space E then $c \cdot B_{(j_k)_{k=1}^{\infty}}$ forms a fundamental system of compact subsets of E as c ranges over all real numbers and $(j_k)_k$ over all strictly increasing sequences of positive integers. Moreover, if $(\delta_n)_{n=1}^{\infty}$ is any sequence of positive real numbers, $\delta_n > 1$ for all n and $\delta_n \to 1$ as $n \to \infty$ then the sequence $\{\delta_n B_{j_1,\ldots,j_n}\}_{n=1}^{\infty}$ forms a fundamental neighbourhood system for $B_{(j_k)_{k=1}^{\infty}}$.*

We require a similar description for $E := \lambda(\{E_n\}_n)$ where $\lambda(A)$ is a Fréchet nuclear space with basis and each E_n is a Banach space with a finite dimensional Schauder decomposition $\{E_{n,m}\}_{m=1}^{\infty}$. If $(K_{n,\alpha})_{\alpha \in \Gamma_n}$ is a fundamental system of compact subsets of E_n then a fundamental system of compact subsets of E is obtained by choosing for each n, $\alpha_n \in \Gamma_n$, such that

$$\left(\sup_{x \in K_{n,\alpha_n}} \|x\| \right)_n \in \lambda(A)$$

i.e. we require that there exists an increasing sequence of positive integers $(j_n)_n$, $j_n \to \infty$ as $n \to \infty$, such that

$$\sum_{n=1}^{\infty} a_{n,j_n} \left(\sup_{x \in K_{n,\alpha_n}} \|x\| \right) < \infty.$$

Sets K of the form $\{(x_n)_n \in E : x_n \in K_{n,\alpha_n} \text{ for all } n\}$ obtained in this way are a fundamental system of compact sets. Using the Banach space description and a diagonal process we can show that a fundamental system of compact sets and a fundamental neighbourhood system of each compact set in $\lambda(\{E_n\}_n)$ can be specified by means of a real number c, strictly increasing sequences of positive integers, $(j_n)_{n=1}^\infty$ and $(l_n)_{n=1}^\infty$, and a sequence of positive real numbers, $(\delta_n)_n$, $\delta_n > 1$ and $\delta_n \to 1$ as $n \to \infty$. Instead of describing the fundamental system of compact sets, as we did in Lemma 4.10 for Banach spaces, we give the inductive procedure which generates all compact sets. We do this in two steps for later convenience. Let

$$\left\| \sum_{n=1}^\infty x_n \right\|_{(1)} = \sum_{n=1}^\infty a_{n,1} \|x_n\|$$

for $\sum_{n=1}^\infty x_n \in \lambda(\{E_n\}_n)$. We suppose that $\|\cdot\|_{(k)}$ has been defined inductively starting with $\|\cdot\|_{(1)}$. Let

$$\left\| \sum_{n=1}^\infty x_n \right\|_{(k)}' = \left\| \sum_{n=1}^{j_k} x_n + \sum_{n=j_k+1}^\infty \frac{a_{n,k+1}}{a_{n,k}} x_n \right\|_{(k)} \tag{4.11}$$

for $\sum_{n=1}^\infty x_n \in \lambda(\{E_n\}_n)$, $x_n \in E_n$ for all n and let

$$\left\| \sum_{n=1}^\infty x_n \right\|_{(k+1)} = \left\| \sum_{n=1}^{j_k} \sum_{m=1}^{l_k} x_{n,m} + \sum_{n=1}^{j_k} \sum_{m=l_k+1}^\infty 2 x_{n,m} + \sum_{n=j_k+1}^\infty x_n \right\|_{(k)}' \tag{4.12}$$

where $x_{n,m} \in E_{n,m}$, $x_n = \sum_{m=1}^\infty x_{n,m}$ for all m. For example

$$\left\| \sum_{n=1}^\infty x_n \right\|_{(3)} = \sum_{n=1}^{j_1} a_{n,1} \|x_n\|_{l_1,l_2} + \sum_{n=j_1+1}^{j_2} a_{n,2} \|x_n\|_{l_2} + \sum_{n=j_2+1}^\infty a_{n,3} \|x_n\|.$$

Sets of the form

$$c \cdot K_{(j_n,l_n)_n} := c \cdot \bigcap_k \{x : \|x\|_{(k)} \le 1\}$$

form a fundamental system of compact subsets of E and a fundamental system of neighbourhoods of $c \cdot K_{(j_n,l_n)_n}$ is given by

$$c \cdot \bigcap_{j \le k} \{x : \|x\|_{(j)} \le \delta_k\} = c \cdot \delta_k \bigcap_{j \le k} \{x : \|x\|_{(j)} \le 1\}.$$

In proving our main result we require polynomials which are homogeneous in some but not in all variables. To simplify matters we distinguish between these variables by introducing the collection of homogeneous polynomials which are homogeneous in each *even* variable and, consequently, in the *combined odd* variables. For this we require a Schauder decomposition which is unconditional in the even variables. Since we will be considering decompositions of Fréchet spaces we may also suppose, without loss of generality, that we have a generating set of semi-norms which are monotone. Let

$\{E_n\}_n$ denote a Schauder decomposition of the locally convex space E with the following properties:

–if $\sum_{n=1}^{\infty} x_n \in E$ then $\sum_{n=1}^{\infty} x_{2n-1} \in E$ and $\sum_{n=1}^{\infty} \lambda_n x_{2n} \in E$ for any bounded sequence of scalars $(\lambda_n)_n$,

– the topology of E is generated by a collection of semi-norms each of which has the following property:

$$\left\| \sum_{n=1}^{\infty} x_n \right\| = \sup_{\substack{k \\ |\lambda_n| \leq 1}} \left\| \sum_{n=1}^{k} x_{2n-1} + \sum_{n=1}^{\infty} \lambda_n x_{2n} \right\|. \qquad (4.13)$$

If $m = (m_0, m_2, m_4, \ldots) \in \mathbb{N}^{(\mathbb{N})}$ we let $\mathcal{P}_e(^m E)$ denote the subspace of $\mathcal{P}(^{|m|}E)$ consisting of all polynomials which are homogeneous in each even variable, i.e. if $m_e := (m_2, m_4, \ldots)$ then $P \in \mathcal{P}_e(^m E)$ if and only if for any sequence of scalars $(\lambda_n)_{n=0}^{\infty}$ and any $\sum_{n=1}^{\infty} x_n \in E$ we have

$$P\left(\lambda_0 \cdot \sum_{n=1}^{\infty} x_{2n-1} + \sum_{n=1}^{\infty} \lambda_n x_{2n}\right) = \lambda^m P\left(\sum_{n=1}^{\infty} x_n\right)$$

where $\lambda^m = \prod_{n=0}^{\infty} \lambda_n^{m_{2n}}$.

The following is our fundamental technical result.

Proposition 4.11 *Let $\{E_n\}_n$ denote a Schauder decomposition of the Fréchet space E, U an open subset of E, F a Banach space and let*

$$T \colon (H(U), \tau_\delta) \longrightarrow F$$

denote a continuous linear mapping. Let $\beta_n \geq 1$ for all n, $\beta_{2n-1} = 2$ and suppose $\sum_{n=1}^{\infty} \beta_n^p x_n \in E$ for any $\sum_{n=1}^{\infty} x_n \in E$ and any $p > 0$. Let $\| \cdot \|$ denote a continuous norm on E satisfying (4.13). Suppose there exists $c > 0$ such that

$$\|T(P)\| \leq c\|P\| \qquad (4.14)$$

for all $P \in \mathcal{P}_e(^m E)$ and all $m \in \mathbb{N}^{(\mathbb{N})}$. Then, for any $\delta > 1$, there exists a positive integer j and $c_1 > 0$ such that

$$\|T(P)\| \leq c_1 \delta^{|m|} \|P\|_{\beta,j} \qquad (4.15)$$

for all $P \in \mathcal{P}_e(^m E)$ and all $m \in \mathbb{N}^{(\mathbb{N})}$, where

$$\|P\|_{\beta,j} := \sup\{|P(x)| \; : \; \|x\|_{\beta,j} \leq 1\}$$

and

$$\left\| \sum_{n=1}^{\infty} x_n \right\|_{\beta,j} := \left\| \sum_{n=1}^{j} x_n + \sum_{n=j+1}^{\infty} \beta_n x_n \right\|$$

all $\sum_{n=1}^{\infty} x_n \in E$, $x_n \in E_n$ for all n.

Proof. Suppose the result is not true. Then there exists $\delta > 1$ such that for each positive integer j there exists $m_j \in \mathbb{N}^{(\mathbb{N})}$ and $P_j \in \mathcal{P}_e(^{m_j}E)$ satisfying

$$\|T(P_j)\| > j\delta^{|m_j|}\|P_j\|_{\beta,2j}. \tag{4.16}$$

Since $\left\|\sum_{n=1}^{\infty} x_n\right\|_{\beta,j} = \lim_{m\to\infty}\left\|\sum_{n=1}^{j} x_n + \sum_{n=j+1}^{m} \beta_n x_n\right\|$, $\|\cdot\|_{\beta,j}$ is the limit of a sequence of continuous semi-norms on a Fréchet space and hence is continuous. Moreover, $\{x : \sup_j \|x\|_{\beta,j} \leq 1\}$ is closed convex balanced and absorbing and hence is a neighbourhood of zero in E. If $|m_j| = l$ for an infinite number of j, then $\left\{\dfrac{P_j}{\|P_j\|_{\beta,2j}}\right\}_{|m_j|=l}$ is a locally bounded, and hence a τ_δ-bounded subset of $\mathcal{P}(^l E)$, on which T is unbounded. This is impossible and we may suppose that $|m_j|$ is strictly increasing in j.

For $0 \leq l \leq |m_j|$, let

$$P_{j,l}\left(\sum_{n=1}^{\infty} x_n\right) = \binom{|m_j|}{l}\overset{\vee}{P_j}\left(\left(\sum_{n=1}^{j} x_{2n-1} + \sum_{n=1}^{\infty} x_{2n}\right)^{|m_j|-l}, \left(\sum_{n=j}^{\infty} x_{2n+1}\right)^l\right).$$

Then $P_{j,l} \in \mathcal{P}_e(^{m_j}E)$ and $\sum_{l=0}^{|m_j|} P_{j,l} = P_j$. By (4.16), for each j there exists l_j, $0 \leq l_j \leq |m_j|$, such that

$$\|T(P_{j,l_j})\| > \frac{j}{|m_j|+1}\delta^{|m_j|}\|P_j\|_{\beta,2j}. \tag{4.17}$$

Let $m_e^j = (m_2^j, m_4^j, \ldots)$ for all j and $\beta = (\beta_j)_{j=1}^{\infty}$. Since $\beta_j \geq 1$ for all j the following two cases cover all possibilities. (We may suppose, in Case 2, by taking a subsequence if necessary that the limit exists.)

<u>Case 1</u> $\lim_{j\to\infty}\left(\alpha^{l_j} \cdot \beta^{s_j}\right)^{1/|m_j|} = 1$ for all $\alpha \geq 1$.

<u>Case 2</u>   there exists $\alpha > 1$ such that $\limsup_{j\to\infty}\left(\alpha^{l_j} \cdot \beta^{s_j}\right)^{1/|m_j|} = \omega > 1$,

where $\beta^{s_j} = \prod_{i>j}\beta_{2i}^{m_{2i}^j}$ for all j.

We consider Case 1. By (4.14)

$$\limsup_{j\to\infty}\left(\frac{\|T(P_{j,l_j})\|}{\|P_{j,l_j}\|}\right)^{1/|m_j|} \leq 1. \tag{4.18}$$

Choose $0 < \varepsilon < 1$ such that $1 + 2\varepsilon < \delta$. By homogeneity

$$\|P_{j,l_j}\| = 2^{l_j}\beta^{s_j}\|P_{j,l_j}\|_{\beta,2j}. \tag{4.19}$$

If $x = \sum_{n=1}^{j} x_{2n-1}$, $y = \sum_{n=1}^{\infty} x_{2n}$ and $z = \sum_{n=j}^{\infty} x_{2n+1}$, $x_n \in E_n$ for all n, then

$$P_{j,l_j}(x + y + z) = \left(\frac{d^{l_j} P_j}{l_j!}(x + y) \right)(z)$$

and hence, by the Cauchy inequalities (3.12), we have

$$\|P_{j,l_j}\|_{\beta,2j} \leq \frac{1}{\varepsilon^{l_j}} \sup_{\substack{\|x+y+z\|_{\beta,2j} \leq 1 \\ |\lambda| \leq \varepsilon}} |P_j(x + y + \lambda z)|. \qquad (4.20)$$

Since $\| \cdot \|_{\beta,2j}$ also satisfies (4.13), $\|x + y + z\|_{\beta,2j} \leq 1$ implies $\|x + y\|_{\beta,2j} \leq 1$ and $\|z\|_{\beta,2j} \leq 2$. By (4.20) we have

$$\|P_{j,l_j}\|_{\beta,2j} \leq \frac{1}{\varepsilon^{l_j}} \sup_{\|x\|_{\beta,2j} \leq 1+2\varepsilon} |P_j(x)|$$

$$= \frac{(1 + 2\varepsilon)^{|m_j|}}{\varepsilon^{l_j}} \|P_j\|_{\beta,2j}. \qquad (4.21)$$

Hence

$$\liminf_{j \to \infty} \left(\frac{\|T(P_{j,l_j})\|}{\|P_{j,l_j}\|} \right)^{1/|m_j|}$$

$$\geq \liminf_{j \to \infty} \left(\frac{j}{|m_j| + 1} \cdot \frac{\delta^{|m_j|} \|P_j\|_{\beta,2j}}{\|P_{j,l_j}\|} \right)^{1/|m_j|} , \quad \text{(by (4.17))}$$

$$\geq \delta \liminf_{j \to \infty} \left(\frac{\varepsilon^{l_j}}{(1 + 2\varepsilon)^{|m_j|}} \cdot \frac{\|P_{j,l_j}\|_{\beta,2j}}{\|P_{j,l_j}\|} \right)^{1/|m_j|} , \quad \text{(by (4.21))}$$

$$\geq \frac{\delta}{1 + 2\varepsilon} \liminf_{j \to \infty} \left(\frac{\varepsilon^{l_j}}{2^{l_j} \beta^{s_j}} \right)^{1/|m_j|} , \quad \text{(by (4.19))}$$

$$= \frac{\delta}{1 + 2\varepsilon}, \quad \text{(taking } \alpha = 2/\varepsilon \text{ in Case 1).}$$

Since $\delta/(1 + 2\varepsilon) > 1$ this contradicts (4.18) and Case 1 is not possible.

We now consider Case 2. Choose $\alpha \geq 1$ such that $\lim_{j \to \infty} (\alpha^{l_j} \beta^{s_j})^{1/|m_j|} = \omega > 1$. Let $x = \sum_{n=1}^{\infty} x_n \in E$ and $p > 0$ be arbitrary. We have

$$y := \sum_{n=1}^{\infty} y_n := \sum_{n=1}^{\infty} \alpha^p x_{2n-1} + \sum_{n=1}^{\infty} \beta_{2n}^p x_{2n} \in E.$$

Since $(\| \cdot \|_{\beta,j})_j$ is an equicontinuous and monotone sequence of semi-norms there exists $M > 0$ such that $\sup_j \|x\|_{\beta,2j} \leq M$ and $\sup_j \left\| \sum_{n=1}^{\infty} y_n \right\|_{\beta,2j} \leq M$.

For any j we have

$$P_{j,l_j}\left(\sum_{n=1}^{\infty} x_n\right) = \frac{1}{\alpha^{pl_j}(\beta^{s_j})^p} P_{j,l_j}\left(\sum_{n\leq 2j} x_n + \sum_{n>2j} y_n\right).$$

Since

$$\left\|\sum_{n\leq 2j} x_n + \sum_{n>2j} y_n\right\|_{\beta,2j} \leq 3M$$

this implies

$$\limsup_{j\to\infty}\left(\frac{|P_{j,l_j}(\sum_{n=1}^{\infty} x_n)|}{\|P_{j,l_j}\|_{\beta,2j}}\right)^{1/|m_j|} \leq \frac{3M}{\omega^p}.$$

Since $\omega > 1$ and p is arbitrary we have

$$\lim_{j\to\infty}\left(\frac{|P_{j,l_j}(x)|}{\|P_{j,l_j}\|_{\beta,2j}}\right)^{1/|m_j|} = 0$$

for any x in E and, by Example 3.8(d), $\displaystyle\sum_{j=1}^{\infty}\frac{P_{j,l_j}}{\|P_{j,l_j}\|_{\beta,2j}} \in H(E)$. Proposition 3.36(b) implies

$$\sum_{j=1}^{\infty}\left\|\frac{T(P_{j,l_j})}{\|P_{j,l_j}\|_{\beta,2j}}\right\| < \infty.$$

This contradicts (4.17) and shows that Case 2 leads to a contradiction. This completes the proof.

To apply Proposition 4.11 to the space $\lambda(\{E_n\}_n)$ we require the (DN)-property already defined in the introduction to this chapter. For Fréchet nuclear spaces and, in particular, for Fréchet nuclear spaces with basis there are more concrete characterizations. We let s denote the set of all rapidly decreasing sequences. The space s is a Fréchet nuclear space isomorphic to $\lambda(A)$, $A = (a_{j,k})_{1\leq j,k<\infty}$, where $a_{j,k} := \|e_j\|_k = j^{2k}$ and $(e_j)_j$ is the usual basis for s. The space s has the following universal property:

A Fréchet space E is nuclear $\iff$ E is isomorphic to a subspace of $s^{\mathbb{N}}$.

For Fréchet nuclear spaces we have a similar characterization of the (DN)-property:

A Fréchet nuclear space E has (DN) $\iff$ E is isomorphic to a subspace of s.

When the Fréchet nuclear space has a basis we obtain a characterization of (DN) in terms of Köthe matrices.

Proposition 4.12 *The following are equivalent conditions on the Fréchet nuclear space with basis $E \approx \lambda(A)$.*

(a) E *has* (DN),

(b) the matrix $A = (a_{j,k})_{1 \le j,k < \infty}$ *can be chosen so that*

$$a_{j,k+1}^2 \le a_{j,k} a_{j,k+2}$$

for all j *and* k,

(c) the matrix A *can be chosen so that*

$$a_{j,k}^2 \le a_{j,1} a_{j,k+1}$$

for all j *and* k,

(d) the matrix $A = (a_{j,k})_{1 \le j,k < \infty}$ *can be chosen so that*

$$\beta_{j,k} := \frac{a_{j,k+1}}{a_{j,k}} \le \frac{a_{j,k+2}}{a_{j,k+1}} := \beta_{j,k+1}$$

for all j *and* k,

(e) for any fixed p, $1 < p < \infty$, *the matrix* $A = (a_{j,k})_{1 \le j,k < \infty}$ *can be chosen so that*

$$a_{j,k} \left(\frac{a_{j,k+1}}{a_{j,k}} \right)^p \le a_{j,k+p}$$

for all j *and* k,

(f) E is isomorphic to a closed subspace of s with basis.

Condition (e) in Proposition 4.12 has already appeared in a disguised form in Proposition 4.11 and is the reason for our interest in the (DN)-property.

Example 4.13 (a) Let $\alpha = (\alpha_j)_{j=1}^\infty$ be a strictly increasing sequence of positive real numbers such that $\sum_{j=1}^\infty q^{\alpha_j} < \infty$ for *some* q, $0 < q < 1$. Let $a_{j,k} = (1/q^{\alpha_j})^k$ for all j and k and let $A = (a_{j,k})_{1 \le j,k < \infty}$. Since

$$\sum_{j=1}^\infty \frac{a_{j,k}}{a_{j,k+1}} = \sum_{j=1}^\infty q^{\alpha_j} < \infty$$

the Grothendieck–Pietsch criterion implies $\lambda(A)$

is a nuclear Fréchet space with basis. Since $\beta_{j,k} = \frac{a_{j,k+1}}{a_{j,k}} = \frac{1}{q^{\alpha_j}} = \beta_{j,k+1}$ for all j and k, Proposition 4.12 implies that $\lambda(A)$ has (DN). If $\alpha_j = j$ for all j we obtain $H(\mathbb{C})$ with the compact open topology and if $\alpha_j = \log(j+1)$ for all j we obtain s. The space $\lambda(A)$ is usually denoted by $\Lambda_\infty(\alpha)$ and called a power series space of *infinite* type.

(b) Let $\alpha = (\alpha_j)_{j=1}^\infty$ be a strictly increasing sequence of positive real numbers and suppose $\sum_{j=1}^\infty q^{\alpha_j} < \infty$ for *all* q, $0 < q < 1$. Let $0 < p_0 < \infty$, let $(p_k)_k$ denote a strictly increasing sequence of positive numbers such that $\lim_{k \to \infty} p_k = p_0$ and let $a_{j,k} = p_k^{\alpha_j}$ for all j and k. For $k < k'$ we have

$$\sum_{j=1}^\infty \frac{a_{j,k}}{a_{j,k'}} = \sum_{j=1}^\infty \left(\frac{p_k}{p_{k'}} \right)^{\alpha_j} < \infty$$

and the Grothendieck–Pietsch criterion implies that $\Lambda_{p_0}(\alpha) := \lambda(\alpha)$ is a nuclear Fréchet space. $\Lambda_{p_0}(\alpha)$ is called a power series space of *finite* type. If $\alpha_j = j$ for all j and $p_0 = 1$ we obtain $H(D)$, the holomorphic functions on the open unit disc D in $\mathbb{C}$ with the compact open topology. From property (e) of Proposition 4.12 it follows easily that a power series space of finite type does not have (DN).

(c) It is possible to construct further examples of (DN)-spaces by choosing weights of even more rapid growth. For example let $a_{j,k} = 2^{j^k}$ for any positive integers j and k. Since

$$\sum_{j=1}^{\infty} \frac{a_{j,k}}{a_{j,k+1}} = \sum_{j=1}^{\infty} \frac{2^{j^k}}{2^{j^{k+1}}} = \sum_{j=1}^{\infty} \frac{1}{2^{(j-1)j^k}} < \infty$$

the Grothendieck–Pietsch criterion implies that $\lambda(A)$ is a nuclear Fréchet space. Since

$$\beta_{j,k} = \frac{2^{j^{k+1}}}{2^{j^k}} = 2^{(j-1)j^k} \le 2^{(j-1)j^{k+1}} = \beta_{j,k+1}$$

Proposition 4.12 implies that $\lambda(A)$ has (DN). It can be shown that $\lambda(A)$ is not isomorphic to any power series space.

The collection of Fréchet spaces with (DN) is the smallest collection of Fréchet spaces which contains all Banach spaces and s and which is closed with respect to $\otimes_\varepsilon$ and taking closed subspaces.

We now return to the $\tau_\omega = \tau_\delta$ problem.

Proposition 4.14 *Let* $E = \lambda(\{E_n\}_n)$, *where* $\lambda(A)$ *is a nuclear Fréchet space with basis and* (DN), *and each* E_n *is a Banach space with a finite dimensional Schauder decomposition* $\{E_{n,m}\}_{m=1}^{\infty}$. *If* U *is a balanced open subset of* E *of polydisc type then* $\tau_\omega = \tau_\delta$ *on* $H(U)$.

Proof. We will prove this result for the polydisc

$$U := \left\{ (x_n)_n \ : \ x_n \in E_n, \ \sum_{n=1}^{\infty} a_{n,1}\|x_n\| < 1 \right\}.$$

Our proof can be modified to obtain the result for balanced open sets of polydisc type. Let F denote a Banach space and let

$$T \colon \big(H(U), \tau_\delta\big) \longrightarrow F$$

denote a continuous linear mapping. We must show that

$$T \colon \big(H(U), \tau_\omega\big) \longrightarrow F$$

is continuous. The sequence $\left(\left(1 - \dfrac{1}{n}\right) U \right)_{n=2}^{\infty}$ is an increasing countable open cover of U. Hence there exists a positive integer n_0 and $c_1 > 0$ such that

$$\|T(f)\| \leq c_1 \|f\|_{(1-\frac{1}{n_0})U} \qquad (4.22)$$

for all $f \in H(U)$. Let $\delta_1 = 1 - \dfrac{1}{n_0}$ and let δ_k, $k \geq 2$, denote a sequence of real numbers, $\delta_k > 1$ for all k, such that $\prod_{k=1}^{\infty} \delta_k = \delta < 1$. Let $\left\| \sum_{n=1}^{\infty} x_n \right\|_{(1)} = \sum_{n=1}^{\infty} \alpha_{n,1} \|x_n\|$ for $\sum_{n=1}^{\infty} x_n \in E$. By (4.22), we have

$$\|T(P)\| \leq c_1 \delta_1^{|m|} \|P\|_{(1)} \qquad (4.23)$$

for all $P \in \mathcal{P}(^mE)$ and all $m \in \mathbb{N}^{(\mathbb{N})}$. Now suppose we have found $(c_i)_{i=2}^k$, a finite set of positive real numbers, and two strictly increasing finite sequences of positive integers $(j_i)_{i=1}^k$ and $(l_i)_{i=1}^k$ such that, for $i \leq k$,

$$\|T(P)\| \leq c_i (\delta_1 \ldots \delta_{2i})^{|m|} \|P\|_{(i)} \qquad (4.24)$$

for all $P \in \mathcal{P}(^mE)$ and all $m \in \mathbb{N}^{(\mathbb{N})}$ where

$$\|P\|_{(i)} := \sup\{|P(x)| \ : \ \|x\|_{(i)} \leq 1\}$$

and each $\| \cdot \|_{(i)}$ is constructed, using the sequences $(j_i)_{i=1}^k$ and $(l_i)_{i=1}^k$ and the process outlined in (4.11) and (4.12). Note that (4.24) reduces to (4.23) when $i = 1$.

Let $F_{2n} = E_n$ and $F_{2n-1} = \{0\}$ for all $n \geq 1$. The sequence $\{F_n\}_n$ is an unconditional Schauder decomposition for $\lambda(\{E_n\}_n)$ and, moreover, the *even* monomials for the $\{F_n\}_n$ decomposition coincide with the monomials for the decomposition $\{E_n\}_n$. Let $\beta_{2n} = \dfrac{\alpha_{n,k+1}}{\alpha_{n,k}}$ for all n. Since $\lambda(A)$ has (DN), $\sum_{n=1}^{\infty} \beta_n^p x_n \in E$ for $x = \sum_{i=1}^{\infty} x_n \in E$, $x_n \in F_n$ all n, and any positive integer p (recall that $\beta_{2n-1} = 2$ for all n but this is irrelevant here since $x_{2n-1} = 0$ for all n). The norm $\| \cdot \|_{(k)}$ satisfies (4.13) as we have assumed, without loss of generality, that the norm on each E_n is monotone with respect to the decomposition $\{E_{n,m}\}_{m=1}^{\infty}$. By (4.24), (4.14) of Proposition 4.11 is satisfied by the norm

$$\|| \cdot \|| := \delta_1 \cdots \delta_{2k} \| \cdot \|_{(k)}.$$

By Proposition 4.11, there exists $A > 0$ and $j_{k+1} > j_k$ such that

$$\|T(P)\| \leq A(\delta_1 \cdots \delta_{2k+1})^{|m|} \|P\|'_{(k)} \qquad (4.25)$$

for all $P \in \mathcal{P}_e(^m(\{F_n\}_n)) = \mathcal{P}(^m(\{E_n\}_n))$ where $\|P\|'_{(k)} := \sup\{|P(x)| \ : \ \|x\|'_{(k)} \leq 1\}$ and $\| \cdot \|'_{(k)}$ is constructed from $\| \cdot \|_{(k)}$ as in (4.11).

To go from $\|\cdot\|'_{(k)}$ to $\|\cdot\|_{(k+1)}$ it is necessary to apply finite induction and to use a different decomposition of $\lambda(\{E_j\})$. Let α_j, $0 \leq j \leq j_k$, denote a strictly increasing finite sequence where $\alpha_0 := \delta_1 \cdots \delta_{2k+1} < \alpha_j < \delta_1 \cdots \delta_{2k+2}$ for all

j. Let $0 \le k' \le j_{k+1}$. Suppose there exists $D_{k'} > 0$ and a positive integer $S_{k'}$ such that

$$\|T(P)\| \le D_{k'}\alpha_{k'}^{|m|}\|P\|_{(k,k')} \tag{4.26}$$

for all monomials with respect to the decomposition $\{E_n\}_n$ of E where

$$\|P\|_{(k,k')} := \sup\{|P(x)| : \|x\|_{(k,k')} \le 1\}$$

and

$$\left\|\sum_{n=1}^{k'}\sum_{m=1}^{\infty} x_{n,m} + \sum_{n=k'+1}^{\infty} x_n\right\|_{(k,k')}$$

$$:= \left\|\sum_{n=1}^{k'}\sum_{m=1}^{S_{k'}} x_{n,m} + \sum_{n=1}^{k'}\sum_{m=S_{k'}+1}^{\infty} 2x_{n,m} + \sum_{n=k'+1}^{\infty} x_n\right\|_{(k)}' \tag{4.27}$$

for $x_{n,m} \in E_{n,m}$, $\sum_{m=1}^{\infty} x_{n,m} \in E_n$, $n \le k'$, $x_n \in E_n$, $\sum_{n=k'+1}^{\infty} x_n \in E$. The norms $\|\cdot\|_{(k,k')}$ satisfy (4.13). Note that $\|\cdot\|_{(k,0)} := \|\cdot\|_{(k)}'$ and letting $D_0 = A$ we see, from (4.25), that (4.26) is satisfied when $k' = 0$.

Let $\phi: \mathbb{N} \to \mathbb{N} \setminus \{k'+1\}$ denote a bijective mapping. For each m let $G_{2m} = E_{\phi(m)}$ and $G_{2m-1} = E_{k'+1,m}$. The sequence $\{G_m\}_{m=1}^{\infty}$ is a Schauder decomposition for $\lambda(\{E_n\}_n)$ with generating system of norms satisfying (4.13). Moreover, the set of *even* monomials for the decomposition $\{G_m\}_{m=1}^{\infty}$ coincides with the set of all monomials for the decomposition $\{E_n\}_n$. Let $\beta_{2m} = 1$ for all m. By (4.26), (4.14) is satisfied by the norm $\alpha_{k'}\|\cdot\|_{(k,k')}$ for the decomposition $\{G_m\}_{m=1}^{\infty}$. By Proposition 4.11, there exists $D_{k'+1} > 0$ and a positive integer $S_{k'+1}$ such that

$$\|T(P)\| \le D_{k'+1}\alpha_{k'+1}^{|m|}\|P\|_{(k,k'+1)}$$

for all monomials with respect to the decomposition $\{E_n\}_n$ where $\|\cdot\|_{(k,k'+1)}$ is defined as in (4.27), using $k'+1$ and $S_{k'+1}$. We have shown, using the decomposition $\{G_m\}_{m=1}^{\infty}$ that (4.26) true for k' implies (4.26) true for $k'+1$. By finite induction we can find $D_{k'}$ and $S_{k'}$ for all $k' \le j_k$. Since we may increase S_k, without loss of generality, we may suppose $l_k := S'_{j_k} > l_{k-1}$. Then $\|\cdot\|_{(k,j_k)} = \|\cdot\|_{(k+1)}$. Since $\alpha_{j_k} < \delta_1 \cdots \delta_{2k}$ we have, with $c_{k+1} = D_{j_k}$, (4.24) for $i = k+1$. Hence, by induction, we can find sequences $(c_k)_k$, $(j_k)_k$ and $(l_k)_k$ for all k.

Let $K = K_{(j_k,l_k)_k}$. Since $\delta = \prod_{i=1}^{\infty} \delta_i < 1$, δK is a compact subset of U. By (4.24) we have shown that for every neighbourhood V of K there exists $c(V) > 0$ such that

$$\|T(P)\| \le c(V)\|P\|_V$$

for all $P \in \mathcal{P}(^m E)$ and all $m \in \mathbb{N}^{(\mathbb{N})}$. By (4.9) there exists for every V open, $K \subset V \subset U$, $c'(V) > 0$ such that

$$\|T(f)\| \le c'(V)\|f\|_V$$

for all $f \in H(U)$. Hence $T : (H(U), \tau_\omega) \to F$ is continuous and this completes the proof.

Combining Propositions 3.45, 3.47 and 4.14 we obtain the following corollaries.

Corollary 4.15 *If U is an open polydisc in a Fréchet nuclear space with basis and (DN) then $\tau_\omega = \tau_\delta$ on $H(U)$ and $(H(U), \tau_\omega)$ is a reflexive nuclear space.*

Corollary 4.16 *If U is a balanced open subset of a Banach space with Schauder basis then $\tau_b = \tau_\omega = \tau_\delta$ on $H(U)$.*

We improve Corollary 4.16 by using the following lemma, which bears comparison with Proposition 4.54, and the fact that a separable Banach space with the bounded approximation property (Definition 2.7) is a complemented subspace of a Banach space with a Schauder basis.

Lemma 4.17 *Let E denote a locally convex space such that $\tau_\omega = \tau_\delta$ on $H(U)$ for every balanced open subset U of E. If F is a complemented subspace of E then $\tau_\omega = \tau_\delta$ on $H(V)$ for every balanced open subset V of F.*

Proof. Let G denote a locally convex space such that $E = F \oplus G$. If V is a balanced open subset of F then $U := V + G$ is a balanced open subset of E. Let p denote a τ_δ continuous semi-norm on $H(V)$. If $f \in H(V)$ we define $\tilde{f}$ on U by the formula $\tilde{f}(x + y) = f(x)$ for $x \in V$ and $y \in G$. Clearly $\tilde{f} \in H(U)$ and $\tilde{f}\big|_V = f$. We define $\tilde{p}$ on $H(U)$ by $\tilde{p}(f) = p(f\big|_V)$ for all $f \in H(U)$. If $(V_n)_{n=1}^\infty$ is an increasing open cover of U then $(V_n \cap F)_{n=1}^\infty$ is an increasing open cover of V. Hence there exists $C > 0$ and n_0, a positive integer, such that

$$p(f) \le C\|f\|_{V_{n_0} \cap F}$$

for all $f \in H(V)$. If $f \in H(U)$ then

$$\tilde{p}(f) = p(f\big|_V) \le C\|f\big|_V\|_{V_{n_0} \cap F} \le C\|f\|_{V_{n_0}}.$$

Hence $\tilde{p}$ is τ_δ continuous and, by hypothesis, τ_ω-continuous on $H(U)$ and there exists K_1 compact in V and K_2 compact in G such that for every W_1 open in V, $K_1 \subset W_1 \subset V_1$, and every W_2 open in G, $K_2 \subset W_2$, there exists $C(W_1, W_2) > 0$ such that

$$\tilde{p}(f) \le C(W_1, W_2)\|f\|_{W_1 + W_2}$$

for all $f \in H(U)$. If $f \in H(V)$ then

$$p(f) = \tilde{p}(\tilde{f}) \le C(W_1, W_2)\|\tilde{f}\|_{W_1 + W_2} = C(W_1, W_2)\|f\|_{W_1}.$$

Hence p is τ_ω continuous and this completes the proof.

Corollary 4.18 *If U is a balanced open subset of a separable Banach space with the bounded approximation property then $\tau_\omega = \tau_\delta$ on $H(U)$.*

We are now in a position to extend results from Sections 2.4 and 4.1 on spaces of homogeneous polynomials to spaces of holomorphic functions. We recall that a locally convex space E is *semi-reflexive* if the canonical mapping from E into $E'' \cong \left(E', \ \beta(E', E) \right)'$ is surjective. A locally convex space is semi-reflexive if and only if it is quasi-complete for the $\sigma(E, E')$ topology. If $\{E_n\}_n$ is an S-absolute decomposition for the locally convex space (E, τ) then E is semi-reflexive if and only if each E_n is semi-reflexive and E is T.S. τ-complete (Exercise 3.106).

Since a separable reflexive Banach space has the approximation property if and only if it has the bounded approximation property and a locally convex space is reflexive if and only if it is semi-reflexive and barrelled we may apply Proposition 2.31 and Corollary 4.18 to obtain the following result.

Corollary 4.19 *If E is a separable reflexive Banach space with the approximation property then the following are equivalent:*

(a) $\left(H(U), \ \tau_\omega \right)$ *is reflexive for any balanced open subset U of E,*

(b) $\left(\mathcal{P}(^n E), \ \| \cdot \| \right)$ *is reflexive for all n,*

(c) *each norm continuous polynomial on E is weakly continuous on bounded sets,*

(d) *each norm continuous polynomial on E is weakly sequentially continuous.*

If E has a finite dimensional Schauder decomposition or Schauder basis we may use Corollaries 4.5 and 4.7 to obtain further equivalent conditions.

Corollary 4.20 *If $\{E_n\}_n$ is a finite dimensional Schauder decomposition for the reflexive Banach space E then the following conditions are equivalent:*

(a) $\left(H(U), \ \tau_\omega \right)$ *is reflexive for any balanced open subset U of E,*

(b) *for each n, $\left\{ \mathcal{P}_k(^n E) \right\}_{k=1}^{\infty}$ is a finite dimensional Schauder decomposition for $\mathcal{P}(^n E)$,*

(c) $\left\{ \left\{ \mathcal{P}_k(^n E) \right\}_{k=1}^{\infty} \right\}_{n=0}^{\infty}$ *is a finite dimensional Schauder decomposition for $\left(H(E), \ \tau_\omega \right)$,*

(d) $H_\omega(U) = H(U)$ *for any open subset U of E.*
     $\left(\mathcal{P}_1(^0 E) = \mathbb{C} \text{ and } \mathcal{P}_k(^0 E) = \{0\} \text{ for } k > 1 \right)$

If E has a Schauder basis then the above conditions are equivalent to

(e) *for each integer n the monomials of degree n with the square order form a Schauder basis for $\mathcal{P}(^n E)$.*

Example 4.21 If U is a balanced open subset of T^* (the original Tsirelson space) then $(H(U), \tau_\omega)$ is reflexive. For each n, $(\mathcal{P}(^nT^*), \|\cdot\|)$ has a Schauder basis and $(H(T^*), \tau_\omega)$ has a finite dimensional Schauder decomposition.

In Corollaries 1.56 and 2.40 and Proposition 2.38 we considered the containment of l_∞ (or equivalently c_0 since $\mathcal{P}(^nE)$ is a dual space) in $\mathcal{P}(^nE)$. We now consider the analogous problem for $(H(U), \tau_\omega)$. We begin by proving an abstract result for S-absolute decompositions.

Proposition 4.22 *Let* $\{E_n\}_n$ *denote an S-absolute decomposition, by Banach subspaces, of the locally convex space E. If the sequence $\{x_m :=$*
$$\sum_{n=1}^{\infty} x_{n,m}\}_{m=1}^{\infty}, \ x_{n,m} \in E_n \ \text{for all n and m, is equivalent to the unit vec-}$$
tor basis for c_0 then for each integer n either

(a) *$(x_{n,m})_{m=1}^{\infty}$ is a null sequence in E_n, or*
(b) *there exists a strictly increasing sequence of positive integers $(m_j)_j$ such that $(x_{n,m_{2j+1}} - x_{n,m_{2j}})_j$ is equivalent to the unit vector basis of c_0.*

Moreover, (b) occurs for some positive integer n, and hence $c_0 \hookrightarrow E$, if and only if $c_0 \hookrightarrow E_n$ for some positive integer n.

Proof. Fix n and let $\|\cdot\|_n$ denote the norm on E_n. Let $\|\sum_{m=1}^{\infty} \lambda_m x_m\|_0 = \sup_m |\lambda_m|$. By our hypothesis there exists $c := c(n)$ such that for any sequence of scalars $(\alpha_m)_m$ and any positive integer k

$$\left\|\sum_{m=1}^{k} \alpha_m x_{n,m}\right\|_n \leq c \left\|\sum_{m=1}^{k} \alpha_m x_m\right\|_0 = c \sup_{m \leq k} |\alpha_m|. \tag{4.28}$$

This implies in particular that $(x_{n,m})_{m=1}^{\infty}$ is bounded in $(E_n, \|\cdot\|_n)$.

We first claim that $(x_{n,m})_{m=1}^{\infty}$ does not contain a subsequence $(x_{n,m_j})_{j=1}^{\infty}$ equivalent to the unit vector basis for l_1. Otherwise, there would exist $c_1 > 0$ such that

$$c_1 \sum_{j=1}^{k} |\alpha_j| \leq \left\|\sum_{j=1}^{k} \alpha_j x_{n,m_j}\right\|_n \leq c \sup_{j \leq k} |\alpha_j|$$

for any sequence of scalars $(\alpha_j)_j$ and any positive integer k. This contradicts (4.28) for large k and establishes our claim. By Rosenthal's l_1 Theorem every subsequence of $(x_{n,m})_{m=1}^{\infty}$ contains a weak Cauchy subsequence. Let $(x_{n,m_i})_{i=1}^{\infty}$ denote a weak Cauchy subsequence of $(x_{n,m})_{m=1}^{\infty}$. We suppose that $(x_{n,m_i})_i$ is not a norm Cauchy sequence. By taking a subsequence, if necessary, we may suppose there exists $\delta > 0$ such that

$$\|x_{n,m_{2i+1}} - x_{n,m_{2i}}\|_n \geq \delta$$

for all i. Since $(x_{n,m_i})_{i=1}^{\infty}$ is a weak Cauchy sequence, $(x_{n,m_{2i+1}} - x_{n,m_{2i}})_i$ is a weakly null sequence and, by the Bessaga–Pełczyński selection principle, we may suppose, again by taking a further subsequence, that it is a basic sequence in E_n. For any sequence of scalars, $(\beta_i)_i$, and any positive integer k we have, by (4.28),

$$\left\| \sum_{i=1}^{k} \beta_i (x_{n,m_{2i+1}} - x_{n,m_{2i}}) \right\|_n \leq c \sup_{i \leq k} |\beta_i|.$$

Hence $(x_{n,m_{2i+1}} - x_{n,m_{2i}})_i$ is equivalent to the unit vector basis of c_0 and (b) is satisfied by the sequence $(x_{n,m_i})_{i=1}^{\infty}$. Otherwise, every subsequence of $(x_{n,m})_{m=1}^{\infty}$ contains a norm Cauchy subsequence. By choosing subsequences, if necessary, we may suppose there exists $x \in E_n$ such that

$$\|x_{n,m} - x\|_n \leq \frac{1}{2^m}$$

for all m. By (4.28)

$$c \geq \left\| \sum_{m=1}^{k} x_{n,m} \right\|_n \geq k\|x\|_n - \sum_{m=1}^{k} \frac{1}{2^m}$$

for all k. Hence $x = 0$. This implies that every subsequence of $(x_{n,m})_{m=1}^{\infty}$ contains a norm null subsequence and that $(x_{n,m})_{m=1}^{\infty}$ is a null sequence in E_n. Hence (a) is satisfied by the sequence $(x_{n,m})_{m=1}^{\infty}$ and we conclude that the sequence $(x_{n,m})_{m=1}^{\infty}$ always satisfies either (a) or (b). Clearly both (a) and (b) cannot be satisfied by the same sequence.

We now suppose that (a) is satisfied by $(x_{n,m})_{m=1}^{\infty}$ for all n. Since $(x_m)_{m=1}^{\infty}$ is a bounded sequence in E, Proposition 3.34(c), (ii) $\Longleftrightarrow$ (iii), implies that $(x_m)_{m=1}^{\infty}$ is a null sequence in E. Since however $(x_m)_{m=1}^{\infty}$ is equivalent to the unit vector basis of c_0 and the unit vector basis of c_0 is not a null sequence in c_0 condition (b) must be satisfied by some n. This completes the proof.

Corollary 4.23 *If U is a balanced open subset of a Banach space E then $c_0 \lhook\joinrel\longrightarrow (H(U), \tau_\omega)$ if and only if $c_0 \lhook\joinrel\longrightarrow (\mathcal{P}(^n E), \|\cdot\|)$ for some positive integer n.*

The following proposition can be combined with Propositions 4.19 and 4.20 to obtain further equivalent conditions.

Proposition 4.24 *Let E denote a Banach space with a finite dimensional shrinking unconditional Schauder decomposition $\{E_n\}_n$. The following conditions are equivalent:*

(a) $c_0 \lhook\joinrel\longrightarrow\kern-1.3em\diagup\; (H(U), \tau_\omega)$ for any balanced open subset U of E,

(b) $(H(U), \tau_\omega)$ is separable for any balanced open subset U of E,

(c) $\mathcal{P}_\omega(^n E) = \mathcal{P}(^n E)$ for all n,

(d) all continuous polynomials on E are weakly sequentially continuous.

Proof. By Corollary 4.23, $c_0 \not\hookrightarrow (H(U), \tau_\omega)$ if and only if $c_0 \not\hookrightarrow (\mathcal{P}(^nE), \|\cdot\|)$ for all n. Since $(\mathcal{P}(^nE), \|\cdot\|)$ is a dual Banach space (Proposition 1.28) this is equivalent to $l_\infty \longleftrightarrow\not\to \mathcal{P}(^nE)$ for all n. Hence $(b) \Longrightarrow (a)$. Since E' has the approximation property Proposition 2.28 implies $\mathcal{P}_\omega(^nE) = \overline{\mathcal{P}_f(^nE)}$ and, as the decomposition is shrinking, E' is separable. Hence $(\mathcal{P}_\omega(^nE), \|\cdot\|)$ is separable for all n. If (c) is true then it follows, using Taylor series expansions, that $(H(U), \tau_\omega)$ is separable and hence (c)$\Longrightarrow$(b).

Suppose (c) is not true. Let n denote the smallest positive integer for which $\mathcal{P}_\omega(^nE) \neq \mathcal{P}(^nE)$. Since $\mathcal{P}_\omega(^1E) = \mathcal{P}(^1E)$ for any Banach space we have $n > 1$. Let $P \in \mathcal{P}(^nE)\backslash\mathcal{P}_\omega(^nE)$. Since E' is separable the weak topology on bounded subsets of E is metrizable and hence P is not weakly sequentially continuous. By Lemma 2.32, there exists a weakly null sequence $(x_k)_k$ in E and $\delta > 0$ such that $|P(x_k)| \geq \delta$ for all k. Now fix a positive integer l and let $(e_i)_{i=1}^s$ denote a basis for the finite dimensional space spanned by $\{E_m\}_{m=1}^l$. Let F_i denote the one dimensional subspace of E spanned by e_i, $1 \leq i \leq s$, and let F_{s+1} denote the closed subspace of E spanned by $\{E_n\}_{n=l+1}^\infty$. If we take the monomial expansion of P with respect to the decomposition $\{F_i\}_{i=1}^{s+1}$ we obtain

$$P\Big(\sum_{i=1}^s \lambda_i e_i + x\Big) = P(x) + \sum_{\substack{m\in\mathbb{N}^l \\ 0<|m|<n}} a_m\lambda^m Q_m(x) + Q(\lambda_1, \ldots, \lambda_s)$$

where $\lambda_i \in \mathbb{C}$, $x \in F_{s+1}$, $Q_m \in \mathcal{P}(^{n-|m|}F_{s+1})$, $a_m \in \mathbb{C}$ and $Q := P\big|_{\sum_{m=1}^l E_m}$.

Let π_l denote the projection from E onto $\sum_{m=1}^l E_m$. If $y_k = \pi_l(x_k)$ and $z_k = x_k - \pi_l(x_k)$ then $\|y_k\| \to 0$ and $(z_k)_k$ is a weakly null sequence in E. Since $Q_m \in \mathcal{P}(^{n-|m|}F_{s+1})$ and $|m| > 0$ it follows that $Q_m(x_k) \to 0$ as $k \to \infty$. Hence $P(z_k) \longleftrightarrow\not\to 0$ as $k \to \infty$. Using this result for each l and a diagonal process we can generate a disjointly supported sequence of vectors $(w_j)_j$ such that $|P(w_j)| \geq \delta/2 > 0$ for all j. Clearly $(w_j)_j$ is a weakly null sequence in E. Let $w_j = \sum_{i=n_j+1}^{n_{j+1}} x_i$, $x_i \in E_i$ for all i, where $(n_j)_j$ is a strictly increasing sequence of positive integers. For each positive integer j let $P_j = P \circ (\pi_{n_{j+1}} - \pi_{n_j})$. Then $\|P_j\| \leq \|P\| \cdot \|\pi_{n_{j+1}} - \pi_{n_j}\|^n \leq c^n\|P\|$ for some $c > 0$ and $(P_j)_j \subset \mathcal{P}(^nE)$. Let $x = \sum_{j=1}^\infty x_j \in E$, $x_j \in E_j$ for all j. By (1.55)

$$\sum_{j=1}^k |P_j(x)| = \sum_{j=1}^k \Big|P\Big(\sum_{i=n_j+1}^{n_{j+1}} x_i\Big)\Big| \leq \sup_{|\lambda_j|\leq 1} \Big|P\Big(\sum_{j=1}^k \lambda_j\Big(\sum_{i=n_j+1}^{n_{j+1}} x_j\Big)\Big)\Big|$$

for any integer k. Since the decomposition is unconditional there exists $c' > 0$ such that

$$\Big\|\sum_{j=1}^k \lambda_j\Big(\sum_{i=n_j+1}^{n_{j+1}} x_i\Big)\Big\| \leq c'\Big\|\sum_{j=1}^\infty x_j\Big\|$$

for all k. Hence

$$\sum_{j=1}^{\infty} |P_j(x)| \le (c')^n \|P\| \cdot \|x\|^n.$$

Since $|P_j(w_j)| = |P(w_j)| \ge \delta/2$ and $(w_j)_j$ is bounded it follows that $\|P_j\| \longrightarrow\!\!\!\!\!/\ \ 0$ as $j \to \infty$. By Proposition 2.38, $l_\infty \hookrightarrow\!\!\!\!\!/\ \mathcal{P}(^k E)$ for some k. Hence (a) is not true and (a)$\Longrightarrow$ (c). In proving this last assertion we also showed $\sim$ (c) $\Rightarrow\sim$ (d) and, since clearly (c) $\Rightarrow$ (d) we have (c) $\Leftrightarrow$ (d). This completes the proof.

Corollary 4.25 *If E is a reflexive Banach space with an unconditional finite dimensional Schauder decomposition then the following are equivalent:*

(a) $c_0 \hookrightarrow\!\!\!\!\!/\ (H(U), \tau_\omega)$ *for any balanced open subset U of E,*

(b) $(H(U), \tau_\omega)$ *is reflexive for any balanced open subset of E,*

(c) $(\mathcal{P}(^n E), \|\cdot\|)$ *is reflexive for all n,*

(d) $c_0 \hookrightarrow\!\!\!\!\!/\ (\mathcal{P}(^n E), \|\cdot\|)$ *for any positive integer n.*

4.3 $\tau_b = \tau_\omega$ for Fréchet Spaces

In this section we discuss the coincidence of the τ_b and τ_ω topologies for balanced open subsets of a Fréchet space. This includes, for Fréchet–Montel spaces, results concerning the equality $\tau_0 = \tau_\omega$ but, for these spaces, a more delicate analysis in Section 5.1 leads to finer results. Two general, reasonably independent, approaches have led to fairly satisfactory positive results. Since the overlap between the results obtained by these methods is considerable, we did not feel it necessary to discuss both. The basic idea of the approach that we do not present is contained in the proof of Proposition 3.48 and amounts to showing that certain spaces have $(BB)_\infty$ *with bounds*. Results are obtained by a delicate, and often technical, analysis of estimates derived using splitting hypotheses on the underlying domain. The approach we follow is based on germs of holomorphic functions and the more manageable splitting condition – *the strong localization property* (Definition 4.36). Holomorphic germs also arise in Section 4.6 and the remaining two chapters.

We begin by introducing germs. Let K denote a compact subset of a locally convex space E and let F be a locally convex space. On $\bigcup_{K \subset V \text{ (open)}} H(V; F)$ we define an equivalence relationship $\sim$. Let $f \sim g$ if there exists a neighbourhood W of K on which f and g are both defined and agree, i.e. $f\big|_W = g\big|_W$. Let

$$H(K; F) = \bigcup_{\substack{K \subset V \\ V \text{ open}}} H(V; F)/\sim.$$

Elements of $H(K;F)$ are called *holomorphic germs* on K. If $f \in H(V;F)$, V open containing K, let f or $[f]_K$ denote the equivalence class in $H(K;F)$ containing f. We define natural topologies on $H(K;F)$ by defining

$$\big(H(K;F),\tau\big) = \varinjlim_{\substack{K \subset V \\ V \text{ open}}} \big(H(V;F),\tau\big)$$

where $\tau = \tau_0$, τ_b or τ_ω. Clearly $\tau_\omega \geq \tau_b \geq \tau_0$ on $H(K;F)$. The problems that arise in studying these topologies on germs are rather different to those arising in the function theory setting and may be traced back to differences between projective and inductive limits. Spaces of germs have many useful properties preserved by inductive limits and the main problems we encounter are those usually associated with inductive limits: completeness, characterization of bounded sets and a projective description of a fundamental system of semi-norms.

If F is a normed linear space and U is an open subset of the locally convex space E let

$$H^\infty(U;F) = \{f \in H(U;F) : \|f\| := \|f\|_U < \infty\}.$$

The space $H^\infty(U;F)$ is easly seen to be a normed linear space which is complete if F is a Banach space. Using the above equivalence, we find

$$H(K;F) = \bigcup_{\substack{K \subset V \\ V \text{ open}}} H^\infty(V;F)/\!\sim$$

and, moreover, we have the following topological relationship.

Lemma 4.26 *If K is a compact subset of a locally convex space E and F is a normed linear space then*

$$\big(H(K;F),\tau_\omega\big) = \varinjlim_{\substack{K \subset V \\ V \text{ open}}} \big(H^\infty(V;F),\|\cdot\|_V\big).$$

Proof. If V is an open subset of E which contains K then the natural injection from $H^\infty(V;F)$ into $\big(H(V;F),\tau_\omega\big)$ is continuous and hence the identity mapping from $\varinjlim_{\substack{V \supset K \\ V \text{ open}}} \big(H^\infty(V;F),\|\cdot\|_V\big)$ into $\varinjlim_{\substack{V \supset K \\ V \text{ open}}} \big(H(V;F),\tau_\omega\big)$ is con-

tinuous. Conversely, if p is a continuous semi-norm on

$$\varinjlim_{\substack{V \supset K \\ V \text{ open}}} \big(H^\infty(V;F),\|\cdot\|_V\big),$$

then for every V open, $V \supset K$, there exists $c(V) > 0$ such that $p(f) \leq c(V)\|f\|_V$ for every f in $H^\infty(V; F)$. If $f \in H(V; F) \setminus H^\infty(V; F)$ then $\|f\|_V = \infty$ and the same inequality holds. Hence the restriction of p to $H(V; F)$ is a τ_ω continuous semi-norm ported by the compact subset K of V. Thus p is also a continuous semi-norm on $\varinjlim_{\substack{V \supset K \\ V \text{ open}}} \left(H(V; F), \tau_\omega \right)$. This shows

$$\varinjlim_{\substack{K \subset V \\ V \text{ open}}} \left(H(V; F), \tau_\omega \right) = \varinjlim_{\substack{K \subset V \\ V \text{ open}}} \left(H^\infty(V; F), \|\cdot\|_V \right)$$

and completes the proof.

It follows that $H(K; F)$ is a bornological space if F is a normed linear space and an ultrabornological (and hence a barrelled) space if F is a Banach space. If E is a metrizable locally convex space and F is a Banach space then $H(K; F)$ is a countable inductive limit of Banach spaces and hence an ultrabornological DF space (Definition 1.26).

Holomorphic germs on a compact set K were defined using all open sets which contained K. This defined an inductive system. We now proceed in the opposite direction by considering the spaces of holomorphic germs on all compact sets inside a fixed open set. In this way we arrive at a projective system.

Lemma 4.27 *Let U be an open subset of the locally convex space E and let F denote a locally convex space. Then algebraically*

$$H(U; F) = \varprojlim_{K \in \mathcal{K}(U)} H(K; F)$$

where $\mathcal{K}(U)$ denotes the set of compact subsets of U directed by set inclusion.

Proof. The collection $(H(K), \pi_{K,L})_{K,L \in \mathcal{K}(U), K \subset L}$ where $\pi_{K,L}$ is the natural restriction mapping from $H(L)$ into $H(K)$ is a projective system. The canonical mapping

$$A: H(U) \longrightarrow \varprojlim_{K \in \mathcal{K}(U)} H(K)$$

where $A(f)(K) := [f]_K$ is the holomorphic germ on K induced by f, is linear and injective. We must show that A is surjective. Let $(f_K)_{K \in \mathcal{K}(U)} \in \varprojlim_{K \in \mathcal{K}(U)} H(K)$ be given. We define a function f on U by letting $f(x) = f_{\{x\}}(x)$ for all x in U. We claim that $f \in H(U)$ and $A(f) = (f_K)_{K \in \mathcal{K}(U)}$. If K is any compact subset of U then, since $(f_K)_{K \in \mathcal{K}(U)} \in \varprojlim_{K \in \mathcal{K}(U)} H(K)$,

$$\frac{\widehat{d}^n f_{\{x\}}(x)}{n!} = \frac{\widehat{d}^n f_K(x)}{n!}$$

for any n whenever $x \in K$. Hence if V is a convex balanced neighbourhood of 0, $x + V \subset U$, and $f_{\{x\}} \in H^\infty(x + V)$ then for any y in V we have

$$f(x + y) = f_{\{x+y\}}(x + y) = f_{[x,x+y]}(x + y).$$

This implies

$$f(x + y) = \sum_{n=0}^{\infty} \frac{\widehat{d}^n f_{[x,x+y]}(x)}{n!}(y)$$

$$= \sum_{n=0}^{\infty} \frac{\widehat{d}^n f_{\{x\}}(x)}{n!}(y)$$

where $[x, x + y] = \{x + \lambda y : 0 \le \lambda \le 1\}$ and shows that $f \in H(U)$. Moreover, if $K \subset \mathcal{K}(U)$, $x \in K$ and n is arbitrary then

$$\frac{\widehat{d}^n [f]_K(x)}{n!} = \frac{\widehat{d}^n f_{\{x\}}(x)}{n!} = \frac{\widehat{d}^n f_K(x)}{n!}$$

and $[f]_K = f_K$. Hence $A(f) = (f_K)_{K \in \mathcal{K}(U)}$. This completes the proof.

At the topological level we have the following result.

Lemma 4.28 *If U is an open subset of a locally convex space E and F is a locally convex space then*

$$\big(H(U; F), \tau_0\big) = \varprojlim_{K \in \mathcal{K}(U)} \big(H(K; F), \tau_0\big)$$

and the canonical mappings

$$\big(H(U; F), \tau_\omega\big) \longrightarrow \varprojlim_{K \in \mathcal{K}(U)} \big(H(K; F), \tau_\omega\big)$$

$$\big(H(U; F), \tau_b\big) \longrightarrow \varprojlim_{K \in \mathcal{K}(U)} \big(H(K; F), \tau_b\big)$$

are continuous.

To obtain $\tau_b = \tau_\omega$ using germs we require the canonical mapping above to be an isomorphism for the τ_ω topologies. This, as we shall see in Section 5.2, may be difficult for arbitrary open sets but for balanced domains it is rather straight-forward. Once this has been achieved we can reduce our main toplogical problem to showing that τ_b and τ_ω agree on spaces of germs. We concentrate on balanced compact sets and balanced open sets. Since a *compact balanced* subset of a locally convex space has a fundamental neighbourhood system consisting of *balanced open* sets one easily obtains the following result from Proposition 3.36.

Proposition 4.29 *If K is a compact balanced subset of a locally convex space E and F is a complete locally convex space then $\{(\mathcal{P}(^nE;F),\tau)\}_{n=0}^{\infty}$ is an S-absolute decomposition for $(H(K;F),\tau)$, $\tau = \tau_0$, τ_b and τ_ω.*

We now restrict ourselves to the case where E is a Fréchet space and $F = \mathbb{C}$ and obtain the following projective description (see Proposition 3.47).

Proposition 4.30 *If K is a compact balanced subset of a Fréchet space E then the following describe fundamental systems of semi-norms for $H(K)$.*

(a) $\tau = \tau_0$,

$$p(f) = \sum_{n=0}^{\infty} \left\| \frac{\widehat{d^n} f(0)}{n!} \right\|_{K+\alpha_n L}$$

where L is a compact subset of E and $(\alpha_n)_n \in c_0$.
(b) $\tau = \tau_b$,

$$p(f) = \sum_{n=0}^{\infty} \left\| \frac{\widehat{d^n} f(0)}{n!} \right\|_{K+\alpha_n L}$$

where L is a bounded subset of E and $(\alpha_n)_n \in c_0$.
(c) $\tau = \tau_\omega$,

$$p(f) = \sum_{n=0}^{\infty} p_n \left(\frac{\widehat{d^n} f(0)}{n!} \right)$$

where p_n is a τ_ω-continuous semi-norm on $\mathcal{P}(^nE)$ and for each balanced open subset V of E, $K \subset V$, there exists $c(V) > 0$ such that

$$p_n \left(\frac{\widehat{d^n} f(0)}{n!} \right) \le c(V) \left\| \frac{\widehat{d^n} f(0)}{n!} \right\|_V$$

for all $f = \displaystyle\sum_{n=0}^{\infty} \frac{\widehat{d^n} f(0)}{n!} \in H(V)$ and all n.

Proof. It is clear that each of the semi-norms defined above is continuous for the appropriate topology. We need to show that all semi-norms are dominated by semi-norms of the above type. Given a τ-continuous semi-norm p on $H(K)$ we may suppose by Proposition 4.29 that

$$p(f) = \sum_{n=0}^{\infty} p \left(\frac{\widehat{d^n} f(0)}{n!} \right)$$

for all $f = \sum_{n=0}^{\infty} \frac{\hat{d}^n f(0)}{n!} \in H(K)$. Let $(V_j)_{j=1}^{\infty}$ denote a decreasing fundamental convex balanced neighbourhood system at the origin such that $(j+1)^2 V_{j+1} \subset V_j$ for all j. If $\tau = \tau_0$ (respectively τ_b) then for each j there exist $c_j > 1$ and L_j compact (respectively bounded) and balanced in $K + V_j$ such that

$$p(P_n) \leq c_j \|P_n\|_{L_j}$$

for all $P_n \in \mathcal{P}(^n E)$ and for all n. If $R_j = \{y \in V_j : x+y \in L_j \text{ for some } x \in K\}$ then R_j is a balanced compact (respectively bounded) subset of V_j, $L_j \subset K + R_j$ and

$$p(P_n) \leq \|P_n\|_{c_j^{1/n} K + c_j^{1/n} R_j}$$

for all $P_n \in \mathcal{P}(^n E)$ and all n. For each j choose n_j such that $c_j^{1/n} \leq 1 + j^{-4}$ for all $n \geq n_j$. Without loss of generality we may suppose that $(n_j)_j$ is strictly increasing. Let

$$L = \left((c_1 - 1)K + c_1 R_1 \right) \cup \bigcup_{j=1}^{\infty} \left(\frac{1}{j^3} K + j \left(1 + \frac{1}{j^4} \right) R_j \right).$$

Since $\sum_{j=1}^{\infty} \frac{1}{j^3} < \infty$ and $jR_j \subset jV_j \subset \frac{1}{j}V_{j-1}$ for $j \geq j_0$, the set $\overline{L}$ is a compact (respectively bounded) subset of E. For all j, $\frac{1}{j^4}K + (1 + \frac{1}{j^4})R_j \subset \frac{1}{j}L$. Let $\alpha_n = 1$ for $n \leq n_2$ and $\alpha_n = \frac{1}{j}$ for $n_j < n \leq n_{j+1}$ and $j \geq 2$. Then $(\alpha_n)_n \in c_0$. For $n_j \leq n < n_{j+1}$ and $P_n \in \mathcal{P}(^n E)$ we have

$$p(P_n) \leq \|P_n\|_{K + (c_j^{1/n} - 1)K + c_j^{1/n} R_j} \leq \|P_n\|_{K + \alpha_n L}.$$

This completes the proof for $\tau = \tau_0$ and $\tau = \tau_b$. The proof for τ_ω follows immediately from (3.46).

Corollary 4.31 *If U is a balanced open subset of a Fréchet space then*

$$\left(H(U), \tau \right) = \varprojlim_{K \in \mathcal{K}(U)} \left(H(K), \tau \right)$$

for $\tau = \tau_0$, τ_b or τ_ω.

Corollary 4.32 *The following are equivalent conditions on the Fréchet space E:*

(a) $\left(H(U), \tau_\omega \right) = \left(H(U), \tau_b \right)$ for every balanced open subset U of E,
(b) $\left(H(K), \tau_\omega \right) = \left(H(K), \tau_b \right)$ for every balanced compact subset K of E.

Proof. It suffices to apply the definition of $\left(H(K), \tau \right)$ to obtain $(a) \implies (b)$ while $(b) \implies (a)$ follows from Corollary 4.31.

Corollaries 4.31 and 4.32 reduce the function space $\tau_0 = \tau_\omega$ problem to spaces of germs and we may now concentrate on $H(K)$. The following proposition will be proved later for arbitrary compact sets but since the proof is so simple for balanced sets we include it here.

Lemma 4.33 *If K is a compact balanced subset of a Fréchet space and $(V_j)_j$ is a fundamental neighbourhood system at the origin such that $(j+1)^2 V_{j+1} \subset V_j$ then*

$$\mathcal{K}_j := \{f \in H^\infty(K + V_j) : \|f\|_{K+V_j} \leq j\}$$

forms a fundamental system of bounded sets for $\big(H(K), \tau\big)$, $\tau = \tau_0$, τ_b and τ_ω.

Proof. By Lemma 4.26 each $\mathcal{K}_j$ is a τ_ω-bounded subset of $H(K)$. If the result is not true then, since $\tau_\omega \geq \tau_b \geq \tau_0$, we can find a bounded sequence $(f_j)_{j=1}^\infty$ in $\big(H(K), \tau_0\big)$ and $x_j \in V_j$ such that $\sum_{n=0}^\infty \left\| \dfrac{\widehat{d^n} f_j(0)}{n!} \right\|_{K+\{x_j\}} > j$ for all j. Since $jx_j \to 0$ as $j \to \infty$ we have $\sup_j p(f_j) = \infty$ where

$$p\left(\sum_{n=0}^\infty \frac{\widehat{d^n} f(0)}{n!} \right) := \sum_{n=0}^\infty \left\| \frac{\widehat{d^n} f(0)}{n!} \right\|_{K + \frac{1}{n} L}$$

and $L := \{jx_j\}_{j=1}^\infty \cup \{0\}$. This is impossible and the proof is complete.

We have proved much more than the statement of the lemma suggests, i.e. we have shown that $\big(H(K), \tau_0\big)$ is a semi-Montel space, that each of the three topologies admits a fundamental *sequence* of bounded sets and that each of the previously given inductive limit representations of $H(K)$ is regular.

Since we aim to show $\tau_\omega = \tau_b$ on $H(K)$ and $\big(H(K), \tau_\omega\big)$ is a DF-space (Definition 1.26) we will have to show eventually that $\big(H(K), \tau_b\big)$ is a DF-space. We proceed by deriving weaker properties and strengthening them.

A *locally convex space E has the* **countable neighbourhood property** *if for every sequence $(p_n)_n$ of continuous semi-norms on E there exists a continuous semi-norm p and a sequence of strictly positive scalars $(\varepsilon_n)_n$ such that $\varepsilon_n p_n \leq p$ for all n.*

This property, without a formal introduction, played an important role in Example 3.11.

Proposition 4.34 *If K is a compact balanced subset of a Fréchet space then $\big(H(K), \tau_b\big)$ has the countable neighbourhood property.*

Proof. Let

$$p_n(f) = \sum_{k=0}^\infty \left\| \frac{\widehat{d^k} f(0)}{k!} \right\|_{B_{k,n}}$$

where $(B_{k,n})_k$ is a sequence of balanced bounded subsets of E converging to K for each n. Since we can replace p_n by $\alpha_n p_n$, $\alpha_n > 0$, we can suppose, without loss of generality, that $B_{k,j} \subset K + V_n$ for all $j \leq n$ where $(V_j)_j$ is a neighbourhood basis at the origin consisting of convex balanced subsets such that $jV_j \subset V_{j-1}$ for all j. Let

$$B_k := \left(\bigcup_{j=1}^{k} B_{k,j} \right) \cup \left(\bigcup_{j>k} (B_{k,j} \cap V_j) \right)$$

for all k. The sequence $(B_k)_k$ consists of balanced bounded subsets of E which converges to K. Let

$$p(f) = \sum_{k=0}^{\infty} \left\| \frac{\hat{d}^k f(0)}{k!} \right\|_{B_k}$$

for all $f = \sum_{k=0}^{\infty} \dfrac{\hat{d}^k f(0)}{k!} \in H(K)$.

By construction $B_{k,n} \subset B_k$ for all $k \geq n$ and it is possible to find $\lambda_n > 1$ such that $B_{k,n} \subset \lambda_n B_k$ for $1 \leq k < n$. Choose ε_n such that $0 < \varepsilon_n < \lambda_n^{-n}$ for all n. Then $\| \cdot \|_{B_{k,n}} \leq \| \cdot \|_{B_k}$ for $k > n$ and on $\mathcal{P}(^k E)$, $1 \leq k \leq n$,

$$\varepsilon_n \| \cdot \|_{B_{k,n}} \leq \varepsilon_n \| \cdot \|_{\lambda_n B_k} = \varepsilon_n \lambda_n^k \| \cdot \|_{B_k} \leq \| \cdot \|_{B_k}.$$

Hence $\varepsilon_n p_n \leq p$ for all n and this completes the proof.

In the next stage we use the idea of a *localized* topology.

Definition 4.35 A locally convex space (E, τ) is a gDF (=generalized DF) space if it contains a fundamental sequence of bounded sets $(B_n)_n$ such that for any locally convex topology τ' on E, $\tau' \geq \tau$ and $\tau'\big|_{B_n} = \tau\big|_{B_n}$ all n imply $\tau = \tau'$.

The topology of a gDF space is localized, within the category of locally convex spaces, on its bounded sets. A further example of this type are the k-spaces, whose topology, within the category of topological spaces, is localized on its compact subsets (see Section 5.1). Continuity, of mappings within the category, need only be verified on the sets where the topology is localized. To show that $\tau_\omega = \tau_b$ on $H(K)$ it suffices to show $(H(K), \tau_b)$ is a gDF space, assume E has $(BB)_\infty$ and apply Proposition 3.34(c).

We find it convenient to use the following characterization of gDF spaces:

a locally convex space E with increasing fundamental sequence of convex balanced bounded sets $(B_n)_n$ is a gDF space if and only if for every sequence of neighbourhoods of zero in E, $(U_n)_n$, there exists a zero neighbourhood U such that $U \subset U_n + B_n$ for all n.

This leads us to a property more closely related to the type of splitting referred to in the introduction to this chapter.

Definition 4.36 A locally convex space E has the *strong localization property* if it contains an increasing fundamental sequence of bounded sets $(B_n)_n$ such that for any sequence of convex balanced neighbourhoods of zero, $(U_n)_n$ there exists a zero neighbourhood U and a sequence of continuous linear mappings $(L_n)_{n=1}^\infty$, $L_n \in \mathcal{L}(E; E)$, such that

$$L_n(U) \subset B_n \tag{4.29}$$

and

$$(I - L_n)(U) \subset U_n \tag{4.30}$$

for all n.

Since $U \subset L_n(U) + (I - L_n)(U) \subset U_n + B_n$ for all n we have

strong localization property $\Longrightarrow$ gDF $\Longrightarrow$ *countable neighbourhood property.*

Note that the sequence $(B_n)_n$, in Definition 4.36, can be replaced, without loss of generality, by any sequence $(B'_n)_n$, $B_n \subset B'_n$, and in verifying (4.29) and (4.30) we can replace $(U_n)_n$ by $(U'_n)_n$, if $U'_n \subset U_n$ for all n.

In proving our main result we use the Aron–Berner extension AB_n (Proposition 1.53) to extend polynomials from E to E''. We recall the following property of this extension.

If A is a subset of E and B is a convex balanced subset of E then

$$\|AB_n(P)\|_{A+B^{\circ\circ}} \le \|P\|_{A+B} \tag{4.31}$$

for all $P \in \mathcal{P}(^n E)$ and all n where $B^{\circ\circ}$ is the bipolar of B in E''.

Proposition 4.37 *If E is a Fréchet space and E'_β has the strong localization property then $(H(K), \tau_b)$ has the strong localization property for any compact balanced subset K of E.*

Proof. Let $(V_n)_n$ denote a fundamental decreasing neighbourhood system at the origin in E consisting of convex balanced sets such that the sequence $((nV_n)^\circ)_{n=0}^\infty$ satisfies the conditions for the strong localization property on E'_β. We take $(B_n)_{n=1}^\infty$ as our fundamental system of bounded sets in $H(K)$, where

$$B_n := \left\{ f \in H(K) : \sum_{k=0}^\infty \left\| \frac{\hat{d}^k f(0)}{k!} \right\|_{K+V_n} \le 1 \right\}.$$

Let $(W_n)_n$ denote a sequence of convex balanced neighbourhoods of zero in $H(K)$. By Proposition 4.30(b) and Lemma 4.34 we can find two decreasing null sequences of positive real numbers, $(\varepsilon_n)_{n=1}^{\infty}$ and $(\delta_k)_{k=0}^{\infty}$, $\varepsilon_n < 1/2$ for all n, and B bounded convex balanced in E containing K such that

$$\left\{ f \in H(K) : \sum_{k=0}^{\infty} \left\| \frac{d^k f(0)}{k!} \right\|_{K+\delta_k B} \leq \varepsilon_n \right\} \subset W_n$$

for all n. Let $\alpha = 2 + \max\{\delta_k : k \geq 0\}$. Since E'_β has the strong localization property we can find for the sequence of zero neighbourhoods, $((\varepsilon_n/\alpha^n)B^\circ)_{n=1}^{\infty}$, a convex balanced bounded subset $\tilde{B}$ of E, $\tilde{B} \supset B$, and a sequence of continuous linear mappings $(T_n)_n \subset \mathcal{L}(E'_\beta; E'_\beta)$ such that

$$T_n(\tilde{B}^\circ) \subset \frac{1}{n} V_n^\circ \tag{4.32}$$

and

$$(I - T_n)(\tilde{B}^\circ) \subset \frac{\varepsilon_n}{\alpha^n} B^\circ \tag{4.33}$$

for all n. Let Q_n denote the transpose of T_n. Since $\alpha > 2$ and $\varepsilon_n < 1/2$ we have $\varepsilon_n/\alpha^{n-1} \leq 1/n$. Hence, (4.32) and (4.33) imply

$$Q_n(K + V_n) \subset Q_n(K) + Q_n(V_n^{\circ\circ}) \subset K + (Q_n - I)(K) + \frac{1}{n}\tilde{B}^{\circ\circ} \subset K + \frac{2}{n}\tilde{B}^{\circ\circ}$$

and

$$(I - Q_n)(K + \delta_k B) \subset (I - Q_n)(\alpha B) \subset \frac{\varepsilon_n}{n}\tilde{B}^{\circ\circ}$$

for all k and n. Now choose inductively a strictly increasing sequence of positive integers $m(n)$ such that $\left(1 + \frac{1}{\sqrt{k}}\right)^{-k} < \varepsilon_n$ for all $k > m(n)$ and all $n \in \mathbb{N}$. Let $\tilde{\delta}_{m(n)} = \max\left(\frac{2}{n}, 2\delta_{m(n)} + \frac{1}{m(n)^{1/2}}\right)$ for all n and let $\tilde{\delta}_j = \tilde{\delta}_{m(n)}$ if $m(n) \leq j < m(n+1)$. Since $(\delta_j)_j$ is a null sequence it follows that $(\tilde{\delta}_j)_j$ is also a null sequence. If $k \leq m(n)$ then $\tilde{\delta}_k = \tilde{\delta}_{m(j)}$ for some $j \leq n$ and $\tilde{\delta}_k \geq 2/j \geq 2/n$. Hence, for $1 \leq k \leq m(n)$,

$$Q_n(K + V_n) \subset K + \frac{2}{n}\tilde{B}^{\circ\circ} \subset K + \tilde{\delta}_k\tilde{B}^{\circ\circ} \tag{a}$$

$$(I - Q_n)(K + \delta_k B) \subset \frac{\varepsilon_n}{n}\tilde{B}^{\circ\circ} \subset \varepsilon_n(K + \tilde{\delta}_k\tilde{B}^{\circ\circ}) \tag{b}$$

If $k \geq m(n)$ and $\lambda_k = 1 + \frac{1}{\sqrt{k}}$ then there exists $l \geq n$ such that $m(l) \leq k < m(l+1)$. Since $(\delta_k)_k$ is a decreasing sequence

$$\lambda_k(K + \delta_k B) \subset K + \left(\frac{1}{\sqrt{k}} + 2\delta_k\right)B$$

$$\subset K + \left(\frac{1}{\sqrt{m(l)}} + 2\delta_{m(l)}\right)B^{00}$$

$$\subset K + \tilde{\delta}_{m(l)}B^{00}$$

$$\subset K + \tilde{\delta}_k B^{00}. \tag{c}$$

We define the linear operators $R_n \colon H(K) \to H(K)$ by letting

$$R_n\Big(\sum_{k=0}^{\infty} \frac{\widehat{d^k} f(0)}{k!}\Big) = \sum_{k=0}^{m(n)} \Big[AB_k\Big(\frac{\widehat{d^k} f(0)}{k!}\Big)\Big] \circ Q_n\big|_E.$$

Let

$$W = \Big\{ f \in H(K) : \sum_{k=0}^{\infty} \Big\|\frac{\widehat{d^k} f(0)}{k!}\Big\|_{K+\tilde{\delta}_k \tilde{B}} \le 1 \Big\}.$$

If $f \in W$ then, by (a) and (4.31),

$$\sum_{k=0}^{m(n)} \Big\|\Big[AB_k\Big(\frac{\widehat{d^k} f(0)}{k!}\Big)\Big] \circ Q_n\big|_E\Big\|_{K+V_n} \le \sum_{k=0}^{m(n)} \Big\|\frac{\widehat{d^k} f(0)}{k!}\Big\|_{K+\tilde{\delta}_k \tilde{B}} \le 1.$$

Hence $R_n(W) \subset B_n$ for all n. Since $\Big[AB_0\Big(\frac{\widehat{d^0} f(0)}{0!}\Big)\Big] \circ (I - Q_n) = 0$ we have

$$(I - R_n)\Big(\sum_{k=0}^{\infty} \frac{\widehat{d^k} f(0)}{k!}\Big) = \sum_{k=1}^{m(n)} \Big[AB_k\Big(\frac{\widehat{d^k} f(0)}{k!}\Big)\Big] \circ (I - Q_n) + \sum_{k>m(n)} \frac{\widehat{d^k} f(0)}{k!}.$$

If $f \in W$ then (b), (c) and (4.31) imply

$$\sum_{k=1}^{m(n)} \Big\|\Big[AB_k\Big(\frac{\widehat{d^k} f(0)}{k!}\Big)\Big] \circ (I - Q_n)\big|_E\Big\|_{K+\delta_k B} + \sum_{k>m(n)} \Big\|\frac{\widehat{d^k} f(0)}{k!}\Big\|_{K+\delta_k B}$$

$$\le \sum_{k=1}^{m(n)} \Big\|\frac{\widehat{d^k} f(0)}{k!}\Big\|_{\varepsilon_n(K+\tilde{\delta}_k \tilde{B})} + \sum_{k>m(n)} \lambda_k^{-k}\Big\|\frac{\widehat{d^k} f(0)}{k!}\Big\|_{\lambda_k(K+\delta_k B)}$$

$$\le \varepsilon_n \sum_{k=1}^{m(n)} \Big\|\frac{\widehat{d^k} f(0)}{k!}\Big\|_{K+\tilde{\delta}_k \tilde{B}} + \sum_{k>m(n)} \lambda_k^{-k}\Big\|\frac{\widehat{d^k} f(0)}{k!}\Big\|_{K+\tilde{\delta}_k \tilde{B}}$$

$$\le \varepsilon_n \sum_{k=1}^{\infty} \Big\|\frac{\widehat{d^k} f(0)}{k!}\Big\|_{K+\tilde{\delta}_k \tilde{B}} \le \varepsilon_n$$

and $(I - R_n)(W) \subset W_n$ for all n.

Hence the neighbourhood of 0, W, and the sequence $(R_n)_n$ satisfy (4.29) and (4.30) with respect to the sequences $(W_k)_k$ and $(B_k)_k$. This shows that $(H(K), \tau_b)$ satisfies the strong localization property and completes the proof.

Corollary 4.38 *If E is a Fréchet space and E'_β has the strong localization property then $(H(K), \tau_b)$ is a gDF space.*

Corollary 4.39 *If E is a Fréchet space and E'_β has the strong localization property then the following are equivalent:*

(a) $\tau_\omega = \tau_b$ on $H(U)$ for every balanced open subset U of E,
(b) $\tau_\omega = \tau_b$ on $H(K)$ for every balanced compact subset K of E,
(c) E has the $(BB)_\infty$-property.

Proof. Clearly (a)$\Longrightarrow$(b)$\Longrightarrow$(c). Suppose (c) holds. Let K denote a compact balanced subset of E. By Corollary 4.38, $(H(K), \tau_b)$ is a gDF space. By Lemma 4.33, Proposition 3.34 and the $(BB)_\infty$-property, τ_ω and τ_b coincide on the bounded subsets of $(H(K), \tau_b)$. Hence $\tau_b = \tau_\omega$ on $H(K)$ and (c)$\Longrightarrow$(b). By Corollary 4.32, (b)$\Longrightarrow$(a) and this completes the proof.

Example 4.40 By Corollary 4.39, $\tau_\omega = \tau_b$ on $H(U)$, U a balanced open subset of any of the following spaces:

(a) Fréchet spaces with the density condition and topology generated by a sequence of semi inner-products,
(b) Fréchet–Schwartz spaces with the bounded approximation property,
(c) $H_b(F)$ where F is a Banach space and $H_b(F)$ has the topology of uniform convergence on the bounded subsets of F,
(d) $E = \lambda^p(A, (X_j)_j)$, $1 \le p < \infty$ or $p = 0$, i.e. $(X_j, \|\cdot\|_j)_j$ is a sequence of Banach spaces, $A = (a_{j,k})_{1 \le j, k < \infty}$ is a Köthe matrix and for $(x_j)_j$, $x_j \in X_j$,

$$\|(x_j)_j\|_{k,p} = \begin{cases} \left(\displaystyle\sum_{j=1}^{\infty} a_{j,k}\|x_j\|_j^p\right)^{1/p}, & 1 \le p < \infty \\ \sup_j a_{j,k}\|x_j\|_j, & p = 0 \end{cases}$$

and $\lambda^p(A, (X_j)_j) := \{(x_j)_j : x_j \in X_j$ and $\|(x_j)_j\|_{k,p} < \infty$ for all $k\}$ is endowed, for $1 \le p < \infty$, with the topology generated by $(\|\cdot\|_{k,p})_{k=1}^{\infty}$ and $\lambda^0(A, (X_j)_j) := \{(x_j)_j : (a_{j,k}\|x_j\|_j)_j \in c_0)$ for all $k\}$ is endowed with the topology generated by $(\|\cdot\|_{k,0})_{k=1}^{\infty}$.
(e) If U is an open polydisc in a Fréchet nuclear space E with basis and (DN) then, by combining Corollary 4.15 and (a), we obtain $\tau_0 = \tau_\delta$ on $H(U)$.

Example 4.40 leads to many spaces where $\tau_\omega = \tau_b$ and the method can be developed further, using vector-valued holomorphic functions, to linearly characterize the property $\tau_\omega = \tau_b$ within certain collections of Fréchet spaces (see Section 4.6). We have $\tau_\omega = \tau_0$ for Fréchet nuclear spaces by 4.40(a), for most Fréchet–Schwartz spaces by 4.40(b) and for many Fréchet–Montel spaces by 4.40(d).

Further results for these topologies on Fréchet nuclear spaces are given in Corollaries 4.44 and 4.46 and Propositions 4.48 and 4.55.

4.4 Examples and Counterexamples

In the previous two sections we obtained *positive* results concerning the relationship between the τ_0, τ_b, τ_ω and τ_δ topologies for locally convex spaces including many important Banach and Fréchet spaces. In proving these results we required various hypotheses and employed a variety of techniques. In this section we consider the necessity of these hypotheses and the scope of the techniques. We examine primarily the following:

- the restriction to Fréchet spaces in Sections 4.2 and 4.3;
- the requirement of a Schauder basis (Corollary 4.16);
- the (DN)-hypothesis (Corollary 4.15);
- the use of monomial expansions (Proposition 4.14).

To obtain positive results, as in Sections 4.2 and 4.3, without a Fréchet space assumption on the underlying domain it is natural, in view of our experience in previous chapters, to turn to the well behaved fully nuclear spaces with basis. For such spaces we find that a satisfactory duality theory can be developed using holomorphic germs. This leads to further results on the relationship between the topologies τ_0, τ_ω and τ_δ.

We define a Borel transform which is similar but not precisely the same as that given in Sections 2.2 and 2.3. For this reason we keep the same terminology but use a different notation. This transform is used, as in Chapter 2, to obtain a functional representation of analytic functionals, i.e. of linear functionals over spaces of holomorphic functions. If U is an open polydisc in the fully nuclear space with basis $E = \Lambda(P)$ and $T \in \big(H(U), \tau_\delta\big)'$ then the Borel transform of T, $\widetilde{B}T$, is defined on a subset of E'_β by the formula

$$\widetilde{B}T(w) := \sum_{m \in \mathbb{N}^{(\mathbb{N})}} T(z^m)w^m \qquad (4.34)$$

(we use $(z_n)_n$ and $(w_n)_n$ for elements of E and E'_β respectively with the usual duality between sequence spaces). The domain of definition of $\widetilde{B}T$ is the set of points where the right hand side of (4.34) converges unconditionally. On the finitely open interior of this subset of E', $\widetilde{B}T$ is $\mathcal{G}$-holomorphic. Since the monomials form an unconditional basis for $\big(H(U), \tau_\delta\big)$, $\widetilde{B}$ is an injective mapping. The relationship between $\widetilde{B}$ and the Borel transform B, defined in Chapter 2, can be seen by comparing (2.8) with (4.34);

$$BT(w) = \sum_{m \in \mathbb{N}^{(\mathbb{N})}} \binom{|m|}{m} T(z^m)w^m.$$

Lemma 4.41 *Let τ_1 and τ_2 be two locally convex topologies on E and suppose $(e_n)_n$ is an absolute basis for both. Then $\tau_1 = \tau_2$ in each of the following cases:*

(a) τ_1 and τ_2 are both barrelled topologies,

(b) $(E, \tau_1)' = (E, \tau_2)'$.

Proof. (a) Let $p\left(\sum_{n=1}^{\infty} x_n e_n\right) := \sum_{n=1}^{\infty} |x_n| p(e_n)$ for $\sum_{n=1}^{\infty} x_n e_n \in E$ denote a τ_2 continuous semi-norm on E. Since $(e_n)_n$ is an absolute basis for (E, τ_1) the semi-norms $p_m\left(\sum_{n=1}^{\infty} x_n e_n\right) := \sum_{n=1}^{m} |x_n| p(e_n)$ are τ_1 continuous for each m. Let

$$V := \left\{ x \in E : p(x) \le 1 \right\} = \bigcap_{m=1}^{\infty} \left\{ x \in E : p_m(x) \le 1 \right\}.$$

Then V is τ_1-closed, convex, balanced, and absorbing and, hence, since (E, τ_1) is barrelled, a τ_1-neighbourhood of zero. Hence p is τ_1-continuous and $\tau_2 \le \tau_1$. Interchanging τ_1 and τ_2 in the above argument gives $\tau_1 \le \tau_2$. Hence $\tau_1 = \tau_2$ and this proves (a).

(b) It is easily seen that each continuous semi-norm on a space with absolute basis has the form

$$p\left(\sum_{n=1}^{\infty} x_n e_n\right) = \sup_{\substack{\theta_n \in \mathbb{R} \\ m}} \left| \phi\left(\sum_{n=1}^{m} e^{i\theta_n} x_n e_n\right) \right|$$

for some continuous linear functional ϕ on E. If $(E, \tau_1)' = (E, \tau_2)'$ this implies $\tau_1 = \tau_2$.

We now consider the Borel transform of τ_0-analytic functionals.

Proposition 4.42 *Let U denote an open polydisc in a fully nuclear space E with basis. The Borel transform, $\tilde{B}$, is a linear topological isomorphism from $\left(H(U), \tau_0\right)'_\beta$ onto $\left(H(U^M), \tau_\omega\right)$. This isomorphism establishes a one to one correspondence between equicontinuous subsets of $(H(U), \tau_0)'$ and sets of germs which are defined and uniformly bounded on a neighbourhood of U^M.*

Proof. Let $T \in \left(H(U), \tau_0\right)'$. There exists $c > 0$ and a compact polydisc K in U such that $|T(f)| \le c \|f\|_K$ for every f in $H(U)$. By Lemma 3.42, there exists a sequence of positive real numbers $\delta = (\delta_n)_{n=1}^{\infty}$, $\delta_n > 1$ for all n and $\sum_{n=1}^{\infty} 1/\delta_n < \infty$, such that δK is a relatively compact subset of U. By Lemma 3.41, $(\delta K)^M$ is a neighbourhood of the compact polydisc U^M in E'_β and, for m in $\mathbb{N}^{(\mathbb{N})}$, we have

$$\|b_m w^m\|_{(\delta K)^M} \le c \|z^m w^m\|_{K \times (\delta K)^M} \le \frac{c}{\delta^m} \|z^m w^m\|_{\delta K \times (\delta K)^M} = \frac{c}{\delta^m}$$

where $b_m = T(z^m)$, $z = (z_n)_{n=1}^{\infty} \in E$ and $w = (w_n)_{n=1}^{\infty} \in E'_\beta$. Hence

$$\sum_{m \in \mathbb{N}^{(\mathbb{N})}} \|b_m w^m\|_{(\delta K)^M} \le c \sum_{m \in \mathbb{N}^{(\mathbb{N})}} \frac{1}{\delta^m} < \infty$$

and $\widetilde{B}T$ defines a holomorphic function on the interior of $(\delta K)^M$. We have thus shown that the Borel transform maps $\big(H(U), \tau_0\big)'$ into $H(U^M)$ and, moreover, since our bound on $\widetilde{B}T$ depends only on K and c, the image of an equicontinuous subset of $\big(H(U), \tau_0\big)'$ is a set of germs which is defined and uniformly bounded on a neighbourhood of U^M.

Conversely, given $g \in H(U^M)$, there exists, by Lemma 3.41, an open polydisc neighbourhood V of U^M and $\widetilde{g} \in H(V)$ such that g is the germ on U^M defined by $\widetilde{g}$. If $\widetilde{g}(w) = \displaystyle\sum_{m \in \mathbb{N}^{(\mathbb{N})}} b_m w^m$ for all $w \in V$, we may suppose without loss of generality, that

$$\sum_{m \in \mathbb{N}^{(\mathbb{N})}} \|b_m w^m\|_V := c' < \infty.$$

By Lemma 3.41, V and c' can be chosen uniformly for any family of germs which are defined and uniformly bounded on a fixed neighbourhood of U^M. Let

$$T_g\bigg(\sum_{m \in \mathbb{N}^{(\mathbb{N})}} a_m z^m \bigg) = \sum_{m \in \mathbb{N}^{(\mathbb{N})}} a_m b_m$$

for $\displaystyle\sum_{m \in \mathbb{N}^{(\mathbb{N})}} a_m z^m \in H(U)$. If $m \in \mathbb{N}^{(\mathbb{N})}$ and $\|w^m\|_V = +\infty$ then $b_m = 0$ and $|a_m b_m| \le c'\|a_m z^m\|_{V^M}$. If $\|w^m\|_V < \infty$ then $\|z^m w^m\|_{V^M \times V} = 1$ and

$$|a_m b_m| \le c' \frac{|a_m|}{\|w^m\|_V} = c' \|a_m z^m\|_{V^M}.$$

Hence

$$\sum_{m \in \mathbb{N}^{(\mathbb{N})}} |a_m b_m| \le c' \cdot \sum_{m \in \mathbb{N}^{(\mathbb{N})}} \|a_m z^m\|_{V^M}$$

and, since V^M is a compact polydisc in U, Proposition 3.43 implies that $T_g \in \big(H(U), \tau_0\big)'$. Since $\widetilde{B}T_g(w) = \widetilde{g}(w)$ for w in V the Borel transform is a linear isomorphism from $\big(H(U), \tau_0\big)'$ onto $H(U^M)$. As the monomials form an absolute basis for $\big(H(U), \tau_0\big)$ the semi-norm

$$p\bigg(\sum_{m \in \mathbb{N}^{(\mathbb{N})}} a_m z^m \bigg) = \sum_{m \in \mathbb{N}^{(\mathbb{N})}} \|a_m z^m\|_{V^M}$$

is τ_0 continuous and we have established the required result about equicontinuous subsets of $\big(H(U), \tau_0\big)'$.

Finally we show that $\widetilde{B}$ is a topological isomorphism. To prove this it suffices, by Lemma 4.41(a), to show that $\big(H(U^M), \tau_w\big)$ and $\big(H(U), \tau_0\big)'_\beta$ are both barrelled locally convex spaces with the monomials as absolute basis. By Lemma 4.26, $\big(H(U^M), \tau_w\big)$ is barrelled. By Proposition 3.43, the monomials form an absolute basis for $\big(H(V), \tau_w\big)$ as V ranges over a fundamental system of neighbourhoods of U^M and using the definition of inductive limits it follows easily that the monomials also form an absolute basis for $(H(U^M), \tau_\omega)$.

By Lemma 2.15, $H(U)$ is a dense subspace of $(H_{HY}(U), \tau_0)$. Moreover, if B is a bounded subset of $(H_{HY}(U), \tau_0)$ then (3.45) implies

$$\tilde{B} := \left\{ \sum_{m \in J} a_m e^{i\theta_m} z^m : \theta_m \in R, \sum_{m \in \mathbb{N}^{(\mathbb{N})}} a_m z^m \in B, J \text{ finite} \right\} \quad (4.35)$$

is a bounded subset of $(H(U), \tau_0)$ and $B \subset \tilde{B}$. Hence $(H(U), \tau_0)'_\beta = (H_{HY}(U), \tau_0)'_\beta$. By Proposition 3.43, $(H_{HY}(U), \tau_0)$ is nuclear and hence semi-reflexive. This implies that $(H(U), \tau_0)'_\beta$ is barrelled.

Let $w^m \left(\sum_{m \in \mathbb{N}^{(\mathbb{N})}} a_m z^m \right) = a_m$ for $m \in \mathbb{N}^{(\mathbb{N})}$ and $\sum_{m \in \mathbb{N}^{(\mathbb{N})}} a_m z^m \in H(U)$, i.e. $(w^m)_{m \in \mathbb{N}^{(\mathbb{N})}}$ is the dual basis to $(z^m)_{m \in \mathbb{N}^{(\mathbb{N})}}$. If $T \in (H(U), \tau_0)'$ let $b_m = T(z^m)$ for all $m \in \mathbb{N}^{(\mathbb{N})}$. If B and $\tilde{B}$ are as previously defined then

$$\|T\|_{\tilde{B}} = \sum_{m \in \mathbb{N}^{(\mathbb{N})}} \|b_m a_m\|_{\tilde{B}} \geq \|T\|_B. \quad (4.36)$$

By (4.36)

$$\lim_{\substack{J \to \infty \\ J \subset \mathbb{N}^{(\mathbb{N})} \\ J \text{ finite}}} \left\| T - \sum_{m \in J} b_m w^m \right\|_B \longrightarrow 0$$

and $\{w^m\}_{m \in \mathbb{N}^{(\mathbb{N})}}$ is a Schauder basis for $(H(U), \tau_0)'_\beta$. Moreover, (4.36) shows that the basis is absolute and completes the proof.

Corollary 4.43 *Let U be an open polydisc in a fully nuclear space E with basis. Consider the following conditions:*

(a) $(H(U), \tau_0)$ is bornological,

(b) $(H(U), \tau_0)$ is infrabarrelled,

(c) $(H(U^M), \tau_\omega) = \varinjlim_{\substack{V \supset U^M \\ V \text{ open in } E'_\beta}} (H^\infty(V), \|\cdot\|_V)$ is a regular inductive limit,

(d) $(H(U^M), \tau_\omega)$ is complete (quasi-complete, sequentially complete),

(e) τ_0 bounded linear functionals on $H(U)$ are τ_0 continuous.

Then (a) $\iff$ (b) $\iff$ (c) $\implies$ (d) $\iff$ (e) and all conditions are equivalent if E is A-nuclear and $U = E$.

Proof. In any locally convex space (a) $\implies$ (b) and (a) $\implies$ (e) $\implies$ (d). Suppose (b) holds. We show $\tau_0^{bor} = \tau_0$. By Propositions 3.46 and 3.43 the monomials form an unconditional basis for $(H(U), \tau_\delta)$ and an absolute basis for $\big(H(U), \tau_0\big)$. Since $\tau_\delta \geq \tau_0^{bor} \geq \tau_0$ they also form an unconditional basis for $\big(H(U), \tau_0^{bor}\big)$. If p is a τ_0^{bor} continuous semi-norm on $H(U)$ we may suppose

$$p\bigg(\sum_{m \in \mathbb{N}^{(\mathbb{N})}} a_m z^m \bigg) = \sup_{\substack{J \subset \mathbb{N}^{(\mathbb{N})} \\ J \text{ finite}}} p\bigg(\sum_{m \in J} a_m z^m \bigg)$$

for every $\displaystyle \sum_{m \in \mathbb{N}^{(\mathbb{N})}} a_m z^m \in H(U)$. The set $V := \{ f \in H(U) : p(f) \leq 1 \}$ is τ_0-closed, convex, balanced and absorbs all τ_0-bounded subsets of $H(U)$. By (b), V is a τ_0 neighbourhood of zero and p is τ_0-continuous. Hence $(b) \iff (a)$.

Since a locally convex space is infrabarrelled if and only if the equicontinuous and the strongly bounded subsets of the dual coincide, Proposition 4.42 shows that $(b) \iff (c)$.

Now suppose $\big(H(U^M), \tau_\omega\big)$ is complete. Since $\big(H(U^M), \tau_\omega\big)$ has an absolute basis it is complete if and only if it is sequentially complete. Hence all the conditions in (d) are equivalent. If $T \in \big(H(U), \tau_0^{bor}\big)'$ then, by (4.36), the partial sums in the (dual) monomial expansion of T form a Cauchy net in $\big(H(U), \tau_0\big)'_\beta$ and hence T is τ_0-continuous on $H(U)$. Hence $(d) \implies (e)$ and this completes the proof for arbitrary U.

If E is A-nuclear and $U = E$ the monomials form an absolute basis for $\big(H(U), \tau_0\big)$ and $\big(H(U), \tau_0^{bor}\big)$ and $(e) \implies (a)$ follows from Lemma 4.41(b). This completes the proof.

Corollary 4.44 *If U is an open polydisc in a Fréchet nuclear or $\mathcal{DFN}$ space with basis then $\tau_0 = \tau_\delta$ on $H(U)$ if and only if $\big(H(U^M), \tau_\omega\big)$ is a regular inductive limit.*

By Corollaries 4.15, 4.39 and 4.44 the (uncountable) inductive limit, which arises when E is a $\mathcal{DFN}$ space with basis,

$$\big(H(0_E), \tau_\omega\big) = \varinjlim_{\substack{0 \in V \\ V \text{ open}}} \big(H^\infty(V), \| \cdot \|_V \big).$$

is regular if E'_β has (DN). We see shortly that the converse is true, but in the meantime, note the partial converse; if E'_β does not admit a continuous norm then Example 3.24 and Corollary 4.44 imply that the inductive limit is not regular.

If E is a Fréchet space and K is a compact subset of E then, by Lemma 4.33,

$$\big(H(K), \tau_\omega\big) = \varinjlim_{\substack{V \supset K \\ V \text{ open}}} \big(H^\infty(V), \| \cdot \| \big)$$

is a regular inductive limit. By Corollary 4.44 this implies $\big(H(U), \tau_0\big)$ is bornological whenever U is an open polydisc in a $\mathcal{DFN}$ space with basis. However, this is only a special case of the result proved in Example 3.20(b).

We next characterize the Borel transform of τ_ω analytic functionals. For this we need hypoanalytic germs on a compact subset K of a locally convex space E. We define these, in the obvious way, by letting

$$H_{HY}(K) = \bigcup_{\substack{V \text{ open} \\ V \supset K}} H_{HY}(V)/\sim$$

where $f \sim g$ if f and g agree on a neighbourhood of K.

Proposition 4.45 *Let U be an open polydisc in a fully nuclear space with basis. The Borel transform, $\widetilde{B}$, is an algebraic isomorphism from $\left(H(U), \tau_\omega\right)'$ onto $H_{HY}(U^M)$. Moreover, under this isomorphism the equicontinuous subsets of $\left(H(U), \tau_\omega\right)'$ are in one to one correspondence with sets of hypocontinuous germs which are defined and uniformly bounded on the compact subsets of some neighbourhood of U^M.*

Proof. Let $T \in \left(H(U), \tau_\omega\right)'$. There exists a compact polydisc K in U such that for every open polydisc V, $K \subset V \subset U$, we have $c(V) > 0$ such that $|T(f)| \leq c(V)\|f\|_V$ for all $f \in H(U)$. Moreover, the set of all T which satisfy the above inequalities form an equicontinuous subset of $\left(H(U), \tau_\omega\right)'$. By Lemma 3.42, we can choose for each neighbourhood V of K a sequence of positive real numbers $\delta = (\delta_n)_{n=1}^\infty$, $\delta_n > 1$ for all n, $\sum_{n=1}^\infty 1/\delta_n < \infty$, and an open polydisc W in E such that δK is a relatively compact subset of U and $\delta(K + W) \subset V$.

Let $b_m = T(z^m)$ for all $m \in \mathbb{N}^{(\mathbb{N})}$ and let $\Gamma = \{m \in \mathbb{N}^{(\mathbb{N})} : \|z^m\|_V < \infty\}$. If $\|z^m\|_V = \infty$ then $\|w^m\|_{VM} = 0$. We have

$$\sum_{m \in \mathbb{N}^{(\mathbb{N})}} \|b_m w^m\|_{VM} \leq \sum_{m \in \Gamma} c(K + W)\|z^m\|_{K+W} \cdot \|w^m\|_{VM}$$

$$\leq c(K + W) \cdot \sum_{m \in \Gamma} \frac{1}{\delta^m}\|z^m\|_{\delta(K+W)} \cdot \|w^m\|_{VM}$$

$$\leq c(K + W) \cdot \sum_{m \in \Gamma} \frac{1}{\delta^m}\|z^m\|_V \cdot \|w^m\|_{VM}$$

$$\leq c(K + W) \cdot \sum_{m \in \mathbb{N}^{(\mathbb{N})}} \frac{1}{\delta^m}.$$

By Lemma 3.41, $\widetilde{B}T \in H_{HY}(U^M)$. By the uniformity of our bounds the Borel transform maps equicontinuous subsets of $\left(H(U), \tau_\omega\right)'$ onto subsets of $H_{HY}(U^M)$ which are defined and uniformly bounded on the compact subsets of some neighbourhood of U^M.

We now show that $\widetilde{B}$ is surjective. Let V be an open polydisc neighbourhood of U^M and let $\mathcal{C} = (c_K)_{K \in \mathcal{K}}$ denote a set of positive real numbers indexed by the family $\mathcal{K}$ of compact polydiscs in V. Let $H_{\mathcal{C}} = \{g \in$

$H_{HY}(V) : \|g\|_K \leq c_K$ for every K in $\mathcal{K}\}$. If $g \in H_C$ then $g = \displaystyle\sum_{m \in \mathbb{N}^{(\mathbb{N})}} b_m w^m$

in $\big(H_{HY}(V), \tau_0\big)$ and $\|b_m w^m\|_K \leq c_K$ for every K in $\mathcal{K}$ and every $m \in \mathbb{N}^{(\mathbb{N})}$.
If $f = \displaystyle\sum_{m \in \mathbb{N}^{(\mathbb{N})}} a_m z^m \in H(U)$ and $g \in H_C$ let

$$T_g(f) = T_g\Big(\sum_{m \in \mathbb{N}^{(\mathbb{N})}} a_m z^m \Big) = \sum_{m \in \mathbb{N}^{(\mathbb{N})}} a_m b_m.$$

For each open polydisc W in U containing V^M we can find a sequence of real numbers, $\delta = (\delta_n)_n$, $\delta_n > 1$, $\sum_{n=1}^{\infty} 1/\delta_n < \infty$ and W_1 a polydisc neighbourhood of the origin in E such that δV^M is a compact subset of U and $\delta(V^M + W_1) \subset W$. By Lemma 3.41, $(V^M + W_1)^M$ is a compact polydisc in V. Let $c = c_{(V^M+W_1)^M}$. If $m \in \mathbb{N}^{(\mathbb{N})}$

$$|a_m b_m| \leq \|a_m z^m b_m w^m\|_{(V^M+W_1) \times (V^M+W_1)^M}$$
$$\leq \frac{c}{\delta^m} \|a_m z^m\|_{\delta(V^M+W_1)}$$
$$\leq \frac{c}{\delta^m} \|f\|_W.$$

Hence

$$|T_g(f)| \leq \sum_{m \in \mathbb{N}^{(\mathbb{N})}} |a_m b_m| \leq c \cdot \|f\|_W \cdot \sum_{m \in \mathbb{N}^{(\mathbb{N})}} \frac{1}{\delta^m}$$

for every f in $H(U)$. Since V^M is a compact subset of U and W was arbitrary we have shown that $T_g \in \big(H(U), \tau_\omega\big)'$. As $\widetilde{B} T_g = g$, $\widetilde{B}$ is surjective, and since our bounds are uniform over g in H_C, $(T_g)_{g \in H_C}$ is an equicontinuous subset of $\big(H(U), \tau_\omega\big)'$. This completes the proof.

Combining Propositions 4.42 and 4.45 we obtain the following result.

Corollary 4.46 *If E is a fully nuclear space with basis then the following are equivalent:*

(a) *$\tau_0 = \tau_\omega$ on $H(U)$ for every open polydisc U in E,*

(b) *$\big(H(U), \tau_0\big)'_\beta = \big(H(U), \tau_\omega\big)'_\beta$ for every open polydisc U in E,*

(c) *$\big(H(U), \tau_0\big)' = \big(H(U), \tau_\omega\big)'$ for every open polydisc U in E,*

(d) *$H_{HY}(V) = H(V)$ for every open subset V of E'_β,*

(e) *$\big(H(V), \tau_0\big)$ is complete for every open subset V of E'_β,*

(f) *the bounded subsets of $\big(H(V), \tau_0\big)$ are locally bounded for every open subset V of E'_β.*

Proof. It is clear that (a)$\Longrightarrow$(b)$\Longrightarrow$(c) and Lemma 4.41(b) shows (c)$\Longrightarrow$(a). By Propositions 4.42 and 4.45, (c) and (d) are equivalent. Using partial sums

of the monomial expansion of hypoanalytic functions, Proposition 3.43, the
fact that hypoanalytic functions are locally defined and that E admits a
fundamental neighbourhood basis at the origin consisting of polydiscs we see
that (d), (e) and (f) are equivalent. This completes the proof.

If U is an open polydisc in a Fréchet nuclear or $\mathcal{DFN}$ space with basis then
$\tau_0 = \tau_\omega$ on $H(U)$. This follows from Corollary 4.46 since condition (d) is easily
seen to be satisfied in both cases. We have already proved this result for $\mathcal{DFN}$
spaces (Example 3.20(b)) and for Fréchet nuclear spaces (Example 4.40).
In fact, in both cases we previously obtained stronger results by different
methods. From Example 1.16 and either Proposition 2.16 or Example 3.50
and Corollary 4.46 ((a) $\Longleftrightarrow$ (d)) we have:

> if E is an infinite dimensional fully nuclear space with basis and U is
> an open subset of $E \times E'_\beta$ then $\tau_0 \neq \tau_\omega$ on $H(U)$.

The particular case, $E = \mathbb{C}^{(\mathbb{N})}$, is discussed in Examples 2.18(a) and 3.24.

Once again the introduction of nuclearity has had a positive effect and,
although the collection of fully nuclear spaces with bases has many special
properties the results and methods suggest an approach to duality and preduality for spaces of holomorphic functions on arbitrary locally convex spaces.
In Chapter 2 we identified preduals of spaces of homogeneous polynomials
on certain Banach spaces and in Proposition 3.26 constructed a predual for
$\big(H(U), \tau_\delta\big)$. A number of the results just presented may be interpreted in
this way. For example, by Proposition 4.42, $\big(H(U), \tau_0\big)$ is a strong predual
for $\big(H(U^M), \tau_\omega\big)$ whenever U is an open polydisc in a fully nuclear space
and we may use this result to construct a predual for $\big(H(U), \tau_\omega\big)$. In place
of stating a number of other immediate corollaries for fully nuclear spaces
we use the results in Sections 2.1 and 2.2 and S-absolute decompositions to
outline a preduality theory for holomorphic germs at the origin on *arbitrary*
locally convex space (see Section 6.3).

If A is a convex balanced subset of a locally convex space E and $f \in H_G(E)$ let

$$\pi_A(f) = \sum_{n=0}^{\infty} \pi_A \left(\frac{\widehat{d}^n f(0)}{n!} \right) := \sum_{n=0}^{\infty} \left\| \frac{\widehat{d}^n f(0)}{n!} \right\|_{N,A}$$

(see Section 2.1), and for V convex balanced and open let

$$H_N^\infty(V) = \Big\{ f \in H(V) : \widehat{d}^n f(0) \in \mathcal{P}_N(^n E) \text{ all } n \text{ and}$$

$$\pi_V(f) := \sum_{n=0}^{\infty} \pi_V \left(\frac{\widehat{d}^n f(0)}{n!} \right) < \infty \Big\}.$$

If $\mathcal{V} := (V_n)_{n=1}^{\infty}$ is an increasing countable convex balanced open cover of E
let

$$H_{N,\mathcal{V}}(E) := \{f \in H(E) : f|_{V_n} \in H_N^\infty(V_n) \text{ for all } n\} \, .$$

Definition 4.47 The entire functions of nuclear type on the locally convex space E, $H_N(E)$, is the space $\bigcup_{\mathcal{V}} H_{N,\mathcal{V}}(E)$ where $\mathcal{V}$ ranges over all increasing countable convex balanced open covers of E. The π_0 topology on $H_N(E)$ is generated by $(\pi_K)_{K \in \mathcal{K}}$ where $\mathcal{K}$ is the set of all compact convex balanced subsets of E.

A semi-norm p on $H_N(E)$ is π_ω continuous if there exists a compact convex balanced subset K of E such that for each V convex balanced open subset of E, $K \subset V$, there exists $c(V) > 0$ satisfying

$$p(f) \le c(V)\pi_V(f)$$

for all $f \in H_N(E)$. The π_ω topology is generated by the π_ω-continuous semi-norms.

The space $H_N(0_E)$ of nuclear holomorphic germs at the origin is defined (algebraically and topologically) as

$$\varinjlim_{V \in \mathcal{V}} \left(H_N^\infty(V), \pi_V\right)$$

where $\mathcal{V}$ is a fundamental neighbourhood system at the origin. If E is a Banach space with unit ball B let

$$H_{Nb}(E) = \left\{f \in H_N(E) : \lim_{n \to \infty} \pi_B\left(\frac{\widehat{d^n} f(0)}{n!}\right)^{1/n} = 0\right\}$$

and endow $H_{Nb}(E)$ with the Fréchet space topology generated by the semi-norms $(\pi_{nB})_{n=1,2,\dots}$. Functions in $H_{Nb}(E)$ are called holomorphic functions of bounded nuclear type.

The π_0 and π_ω topologies are the nuclear analogues on $H_N(E)$ of the τ_0 and τ_ω topologies on $H(E)$ (and we could also define the π_δ topology on $H_N(E)$). Clearly $H_N(E) \subset H(E)$, $\pi_0 \le \pi_\omega$, $\left(H_N^\infty(V), \pi_V\right)$ is a Banach space, $H_N(0_E)$ is a bornological DF space if E is a Fréchet space and $\{\mathcal{P}_N(^n E)\}_{n=0}^\infty$ with the induced topology is an S-absolute decomposition for $\left(H_N(E), \pi\right)$, $\pi = \pi_0$ or π_ω, $H_N(0_E)$ and $H_{Nb}(E)$. We omit the details since they are similar to those previously given for holomorphic functions in Chapter 3.

If $H_{(N)}$ is one of the above spaces and $T \in \left(H_{(N)}\right)'$ the Borel transform of T, BT, is defined as

$$BT = \sum_{n=0}^\infty BT_n$$

where $BT_n(\phi) := T_n(\phi^n) = T(\phi^n)$ for all $\phi \in E'$.

For our next proposition we require the following space of holomorphic functions:

if E is a locally convex space and V is a finitely open subset of E' containing the origin then

$$H_{\mathcal{E}}(V) := \left\{ f \in H_G(V) : \sum_{n=0}^{\infty} \left\| \frac{\widehat{d^n} f(0)}{n!} \right\|_A < \infty \text{ for} \right.$$

every equicontinuous subset A of V $\left. \vphantom{\sum} \right\}.$

Note that A is an equicontinuous subset of V if $A = W^{\circ}$ for some neighbourhood W of 0 in E.

Proposition 4.48 *Let E denote a locally convex space.*

(a) *If E'_c has the strict approximation property then the Borel transform is an isomorphism; $\left(H_N(E), \pi_0 \right)' \simeq H(0_{E'_c})$, and under this isomorphism equicontinuous subsets of $\left(H_N(E), \pi_0 \right)'$ correspond to subsets of $H(0_{E'_c})$ which are defined and bounded on some neighbourhood of 0 in E'_c.*

(b) *If E has a fundamental neighbourhood system at the origin $\mathcal{V}$ consisting of convex balanced open sets such that $(E_V)'$ has the approximation property for all $V \in \mathcal{V}$ then the Borel transform is an isomorphism*

$$\left(H_N(E), \pi_\omega \right)' \simeq H_{\mathcal{E}}(0_{E'_c}).$$

Under this isomorphism equicontinuous subsets of $\left(H_N(E), \pi_\omega \right)'$ correspond to subsets of $H_{\mathcal{E}}(0_{E'_c})$ which are defined and bounded on the equicontinuous subsets of some neighbourhood of 0 in E'_c.

(c) *If E is a fully nuclear space then $\tau_0 = \pi_0 = \pi_\omega = \tau_\omega$ on $H(E)$ when $H_{\mathcal{E}}(V) = H(V)$ and τ_0 bounded subsets of $H(V)$ are locally bounded for each open subset V of E'_β.*

Proof. (a) If $T \in \left(H_N(E), \pi_0 \right)'$ and $|T(f)| \leq \pi_K(f)$ for all $f \in H_N(E)$ then $BT = \sum_{n=0}^{\infty} T_n$ where $T_n(\phi) = T(\phi^n)$ for all $\phi \in E'$. By Proposition 2.10(b), $T_n \in \mathcal{P}(^n E'_c)$. Moreover,

$$\sum_{n=0}^{\infty} \|T_n\|_{(2K)^{\circ}} = \sum_{n=0}^{\infty} \frac{1}{2^n} \sup_{\phi \in K^{\circ}} |T_n(\phi)|$$

$$= \sum_{n=0}^{\infty} \frac{1}{2^n} \sup_{\phi \in K^{\circ}} |T(\phi^n)|$$

$$\leq \sum_{n=0}^{\infty} \frac{1}{2^n} \sup_{\phi \in K^{\circ}} \pi_K(\phi^n)$$

$$\leq \sum_{n=0}^{\infty} \frac{1}{2^n}$$

and $\sum_{n=0}^{\infty} T_n \in H(0_{E'_c})$. This shows that the Borel transform maps equicontinuous subsets of $\left(H_N(E), \pi_0\right)'$ onto subsets of $H(0_{E'_c})$ which are uniformly bounded on a neighbourhood of 0 in E'_c.

Conversely, if K is a convex balanced compact subset of E, $V =$ Interior(K°) and $g \in H^\infty(2V)$ let

$$T_n(\phi^n) = \frac{\widehat{d^n} g(0)}{n!}(\phi)$$

for all n and all $\phi \in E'$ ($\phi^0(x) = 1$ for all ϕ and x). By Proposition 2.10(b), T_n extends to define an element, that we again denote by T_n, in $\mathcal{P}_N(^n E)$, and by the definition of the nuclear norm, we have

$$|T_n(P_n)| \le \left\| \frac{\widehat{d^n} g(0)}{n!} \right\|_V \pi_K(P_n)$$

for all $P_n \in \mathcal{P}_N(^n E'_c)$ and all n. Hence $T := \sum_{n=0}^{\infty} T_n$ defines a linear functional on $H_N(E)$ and, for all $f \in H_N(E)$,

$$|T(f)| \le \sum_{n=0}^{\infty} |T_n(P_n)| \le \sum_{n=0}^{\infty} \left\| \frac{\widehat{d^n} g(0)}{n!} \right\|_V \pi_K(P_n)$$

$$\le \|g\|_{2V} \sum_{n=0}^{\infty} \pi_K(P_n)$$

$$\le \|g\|_{2V} \pi_K(f).$$

Since $BT = g$, the Borel transform is surjective and

$$B\{T \in H_N(E)' : |T(f)| \le \pi_K(f) \text{ all } f \in H_N(E)\}$$
$$\subset \{g \subset H(0_{E'_c}) : g \in H^\infty(2V), \|g\|_{2V} \le 1\}.$$

This completes the proof of (a). Since the proof of (b) is similar we leave it to the reader.

We now consider (c). An examination of the proofs of Propositions 2.12 and 2.13 and the use of Stirling's formula leads to uniform bounds on $\mathcal{P}_N(^n E)$ for all n which imply $\pi_\omega = \tau_\omega$ and $\pi_0 = \tau_0$ on $H_N(E) = H(E)$ and shows $H_N(0_{E'_c}) = H(0_{E'_c})$ when E is fully nuclear. We remark that the appearance of $n^n/n!$ in these propositions leads to estimates which oblige us to confine our attention to entire functions. By (a) and (b), $\pi_0 = \pi_\omega$ on $H_N(E)$ and this completes the proof.

In our next proposition we obtain a further representation of the predual of $H(0_{E'})$ by using functions of exponential type. This representation has proved useful in studying convolution operators on spaces of holomorphic functions over infinite dimensional spaces.

A function $f \in H(E)$, E a Banach space, is said to be of exponential type if there exist positive constants a and b such that

$$|f(z)| \le a \exp(b\|z\|)$$

for all $z \in E$. We let $\mathrm{Exp}(E)$ denote the set of all holomorphic mappings of exponential type on E. If $f \in H(E)$ then, by the Cauchy inequalities, $f \in \mathrm{Exp}(E)$ if and only if $\limsup_{n\to\infty} \|\widehat{d}^n f(0)\|^{1/n} < \infty$. From this we deduce that the mapping

$$f = \sum_{n=0}^{\infty} \frac{\widehat{d}^n f(0)}{n!} \in \mathrm{Exp}(E) \longrightarrow \sum_{n=0}^{\infty} \widehat{d}^n f(0)$$

is a bijective mapping from $\mathrm{Exp}(E)$ onto $H(0_E)$ and this leads to the following result.

Proposition 4.49 *If E is a Banach space and E' has the approximation property then*

$$H_{Nb}(E)' \simeq \mathrm{Exp}(E') \simeq H(0_{E'})$$

We now turn to the second of the hypotheses that we frequently used in Sections 4.2 and 4.3. The assumption that the Banach space E, in Corollary 4.16, had a Schauder basis was weakened in Corollary 4.18 to the hypothesis that E was a separable Banach space with the bounded approximation property. We show, using bounding sets, see Examples 3.8(f) and 3.20(c), that some hypothesis is necessary in this situation.

To give an example of a Banach space E such that $\tau_\omega \ne \tau_\delta$ on $H(E)$ it suffices, by Example 3.20(c), to find a closed bounding non-compact subset of E. We construct an example using the following characterization of bounding sets.

Lemma 4.50 *For A, a subset of a Banach space E, the following are equivalent:*

(a) A is a bounding subset of E,

(b) $\displaystyle\sum_{n=0}^{\infty} \left\| \frac{\widehat{d}^n f(0)}{n!} \right\|_A < \infty$ for all $f \in H(E)$,

(c) $\displaystyle\lim_{n\to\infty} \left\| \frac{\widehat{d}^n f(0)}{n!} \right\|_A^{1/n} = 0$ for all $f \in H(E)$.

Proof. Let A denote a bounding subset of E. By Example 3.20(c) and Corollary 3.37, $p(f) := \|f\|_A$ and

$$q(f) := \sum_{n=0}^{\infty} \left\| \frac{\widehat{d}^n f(0)}{n!} \right\|_A$$

are τ_δ-continuous. Hence (a)$\Longrightarrow$(b). If (b) holds then

$$\limsup_{n\to\infty}\left\|\frac{\widehat{d^n f}(0)}{n!}\right\|_A^{1/n} \le 1 \quad \text{for all } f \in H(E).$$

Since $\displaystyle\sum_{n=0}^\infty \frac{1}{\varepsilon^n}\frac{\widehat{d^n f}(0)}{n!} \in H(E)$ whenever $\varepsilon > 0$ and $f \in H(E)$ this implies

$$\limsup_{n\to\infty}\left\|\frac{\widehat{d^n f}(0)}{n!}\right\|_A^{1/n} \le \varepsilon$$

for all $\varepsilon > 0$ and $f \in H(E)$. Hence (b)$\Longrightarrow$(c). Clearly (c)$\Longrightarrow$(a) and this completes the proof.

Now let $u_n = (0,\ldots,\underset{\underset{n^{\text{th}}\text{position}}{\uparrow}}{1},0,\ldots)$ $\in l_\infty$ and let $A = \{u_n\}_{n=1}^\infty$. The set A is a closed non-compact subset of l_∞.

Proposition 4.51 A *is a bounding subset of* l_∞.

Proof. Suppose the result is not true. By Lemma 4.50 there exists $f \in H(l_\infty)$ and $(n_j)_j$, a strictly increasing sequence of positive integers, such that $\dfrac{\widehat{d^{n_j} f}(0)}{n_j!}(u_j) = 1$ for all j. Let $P_{n_j} = \dfrac{\widehat{d^{n_j} f}(0)}{n_j!}$ for all j.

We claim that for each $P \in \mathcal{P}(^n l_\infty)$, n arbitrary, and $\varepsilon > 0$ there exists an infinite subset S of $\mathbb{N}$ such that $\|P\big|_{l_\infty(S)}\| \le \varepsilon$. Otherwise, we could find a partition $(S_j)_j$ of $\mathbb{N}$ such that $\|P\big|_{l_\infty(S_j)}\| > \varepsilon$ for all j. By (1.18) or (1.55) this is impossible and establishes our claim.

Let $k_1 = 1$. Choose S_1 infinite, $k_1 \notin S_1$, such that

$$\sum_{0<r\le n_1}\binom{n_1}{r}\|\overset{\vee}{P}_1(u_1)^{n_1-r}\|_{B_{l_\infty(S_1)}} \le \frac{1}{n_1!}.$$

Suppose k_i and S_i have been chosen for $i \le m-1$ where $k_i \in S_{i-1} \setminus S_i$, S_i is infinite and contained in S_{i-1}. Let $k_m \in S_{m-1}$ and $C_m = \{k_1,\ldots,k_m\}$. Choose $S_m \subset S_{m-1}$, S_m infinite, $k_m \notin S_m$, such that

$$\sup_{y\in l_\infty(C_m)}\sum_{r>0}\binom{n_{k_m}}{r}\|\overset{\vee}{P}_{k_m}(y)^{n_{k_m}-r}\|_{B_{l_\infty(S_m)}} \le \frac{1}{n_k!}.$$

This is possible since C_m is finite and multilinearity implies that we are taking the supremum of a finite sum of continuous functions each of which can be made arbitrarily small by an appropriate choice of S_m. By induction we obtain an infinite set of positive integers $S := \{k_m\}_{m=1}^\infty$. By restricting to $l_\infty(S)$ we may suppose $k_m = m$ for all m. Let $Q_m = P_{k_m}\big|_{l_\infty(S)}$ and

$C'_m = S \backslash C_m$. For each positive integer m let $T_m(z) = Q_m(x)$ where $z = x+y$, $x \in l_\infty(C_m)$, $y \in l_\infty(C'_m)$. We have $T_m \in \mathcal{P}(^{n_{km}}l_\infty(S))$ for all m. Since $\|T_m - Q_m\| \le 1/m!$ and $\sum_{m=1}^{\infty} Q_m \in H(l_\infty(S))$ it follows that $\sum_{m=1}^{\infty} T_m \in H(l_\infty(S))$. By construction $|T_m(u_m)| = 1$ for all m. By Lemma 1.9(b) we can choose inductively a sequence of complex numbers $(\lambda_j)_j$, $|\lambda_j| \le 1$ such that $\|T_m(\sum_{j=1}^{m} \lambda_j u_j)\| \ge 1$ for all m. Then $u := \sum_{j=1}^{\infty} \lambda_j u_j \in l_\infty(S)$ and $T_m(u) = T_m(\sum_{j=1}^{m} \lambda_j u_j)$ for all m. Hence $\sum_{m=1}^{\infty} |T_m(u)| = \infty$. This contradicts the fact that $\sum_{m=1}^{\infty} T_m \in H(l_\infty(S))$. Hence A is a bounding subset of l_∞.

Corollary 4.52 $(H(l_\infty), \tau_\delta) \neq (H(l_\infty), \tau_\omega)$.

Since l_∞ has the bounded approximation property this corollary shows that the separability assumption in Corollary 4.18 cannot be removed without some substitute. Separability is also absent in a counterexample to the Levi problem (Example 5.55) which shows, in addition, that $c_0(\Gamma)$, Γ uncountable, contains an open set Ω such that $\tau_\omega \neq \tau_\delta$ on $H(\Omega)$.

We turn again to fully nuclear spaces for positive results and show, if we confine ourselves to entire functions, that the basis hypothesis in Proposition 4.41(b) is unnecessary. We require a result similar to Lemma 4.17 for closed rather than complemented subspaces. Nuclearity compensates for this weaker hypothesis. If F is a closed subspace of a locally convex space E let $H_E(F) = \{f \in H(F) : \text{there exists } g \in H(E) \text{ such that } g|_F = f\}$ and let $\mathcal{G} = \mathcal{G}(E; F) = \{f \in H(E) : f|_F = 0\}$. The set $\mathcal{G}$ is a τ_0 closed subspace of $H(E)$ and the mapping $f \in H(E) \to f|_F$ is a surjective linear mapping from $H(E)$ onto $H_E(F)$.

If E is fully nuclear and F is a closed subspace of E then F is fully nuclear and, by Proposition 2.12, $\mathcal{P}(^nF) = \mathcal{P}_N(^nF)$. Hence, for $P \in \mathcal{P}(^nF)$ there exists a convex balanced neighbourhood of 0 in E, V, and $(\phi_i)_i \subset F'$ such that $P = \sum_{i=1}^{\infty} \phi_i^n$ and $\sum_{i=1}^{\infty} \|\phi_i\|_{V \cap F}^n < \infty$. By the Hahn–Banach Theorem there exist $(\psi_i)_i \subset E'$ such that $\psi_i|_F = \phi_i$ and $\|\psi_i\|_V = \|\phi_i\|_{V \cap F}$ for all i. Hence $\tilde{P} = \sum_{i=1}^{\infty} \psi_i^n \in \mathcal{P}(^nE)$ and $\tilde{P}|_F = P$. We have shown $\mathcal{P}(^nF) \subset H_E(F)$ for all n.

Lemma 4.53 *If F is a closed subspace of a fully nuclear space then*

$$(H(E), \tau_0)/\mathcal{G} \cong (H_E(F), \tau_0).$$

Proof. Since the restriction mapping is continuous it suffices to prove the following: given K compact in E there exists L compact in F and $c > 0$ such that

$$q(f) := \inf_{g \in \mathcal{G}} \|f + g\|_K \le c\|f\|_L$$

for all f in $H(E)$. Since E is Montel, $E'_c/F^\perp \cong F'_c$, and for K compact in E, there exists L_1 compact in F, such that

$$p(\phi) := \inf\{\|\psi\|_K : \psi \in E', \psi\big|_F = \phi\big|_F\} \le \|\phi\|_{L_1}$$

for all $\phi \in E'$. If $\phi \in E'$ and n is a positive integer then

$$q(\phi^n) \le \inf\{\|\psi^n\|_K : \psi \in E', \psi\big|_F = \phi\big|_F\}$$
$$= \inf\{\|\psi\|_K : \psi \in E', \psi\big|_F = \phi\big|_F\}^n$$
$$= p(\phi)^n \le \|\phi\|_{L_1}^n.$$

Since $\mathcal{P}(^nE) = \mathcal{P}_N(^nE)$ this estimate implies

$$q(P) \le \|P\|_{N,L_1}$$

for all $P \in \mathcal{P}(^nE)$ and all n. Since F is dual nuclear, Proposition 2.13 implies there exists $c > 0$ and L_2 compact in F, $L_1 \subset L_2$, such that $\|P\|_{N,L_1} \le c\|P\|_{L_2}$ for all $P \in \mathcal{P}(^nF)$ and all n. If $f = \sum_{n=0}^\infty \dfrac{\widehat{d^n f}(0)}{n!} \in H(E)$ then

$$q(f) \le \sum_{n=0}^\infty q\left(\frac{\widehat{d^n f}(0)}{n!}\right) \le c \sum_{n=0}^\infty \left\|\frac{\widehat{d^n f}(0)}{n!}\right\|_{L_2}$$

and an application of Proposition 3.36(a) completes the proof.

Proposition 4.54 *If E is a fully nuclear space and $\tau_0 = \tau_\delta$ on $H(E)$ then $\tau_0 = \tau_\delta$ on $H(F)$ for any closed subspace F of E.*

Proof. Let p denote a τ_δ continuous semi-norm on $H(F)$. Let $\widetilde{p}(f) = p(f\big|_F)$ for all $f \in H(E)$. If $(V_n)_n$ is an increasing countable open cover of E then $(V_n \cap F)_{n=1}^\infty$ is an increasing countable open cover of F and there exists $c > 0$ and N a positive integer such that $p(g) \le c\|g\|_{V_N \cap F}$ for all $g \in H(F)$. Hence

$$\widetilde{p}(f) = p(f\big|_F) \le c\|f\big|_F\|_{V_N \cap F} \le c\|f\|_{V_N}$$

for all $f \in H(E)$ and $\widetilde{p}$ is τ_δ continuous on $H(E)$. Since $\tau_\delta = \tau_0$ on $H(E)$ and $\widetilde{p}\big|_{\mathcal{G}} = 0$ it follows, by Lemma 4.53, that $\widetilde{p}$ induces a continuous semi-norm q on $\big(H(E), \tau_0\big)/\mathcal{G} \cong \big(H_E(F), \tau_0\big)$. Hence there exists L compact in F such that $p(f) \le c\|f\|_L$ for all $f \in H_E(F)$. Since $\mathcal{P}(^nF) \subset H_E(F)$ for all n an application of Proposition 3.36 completes the proof.

Proposition 4.55 *If E is a Fréchet nuclear space with basis then $\tau_0 = \tau_\delta$ on $H(E)$ if and only if E has (DN).*

Proof. If E has (DN) then E is isomorphic to a closed subspace of s (Proposition 4.12) and s is a Fréchet nuclear space with basis and (DN). By Corollary 4.15 and either Example 4.40(a) or Corollary 4.43, $\tau_0 = \tau_\delta$ on $H(s)$ and hence, Proposition 4.54 implies $\tau_0 = \tau_\delta$ on $H(E)$.

We now suppose that $\tau_0 = \tau_\delta$ on $H(E)$. Let $(e_n)_n$ denote the basis for E and let F denote the closed subspace of E generated by $(e_n)_{n \geq 2}$. The space F is a Fréchet nuclear space with basis. If E_1 and E_2 are locally convex spaces with absolute bases $(f_n)_n$ and $(g_m)_m$ then it is well known (and not difficult to prove) that $(f_n \otimes g_m)_{n,m}$ is an absolute basis for $E_1 \otimes_\pi E_2$ with topology generated by

$$p \otimes_\pi q \left(\sum_{m,n=1}^\infty \alpha_{n,m} f_n \otimes g_m \right) = \sum_{n,m=1}^\infty |\alpha_{n,m}| p(f_n) q(g_m)$$

where p and q range over the continuous semi-norms of E_1 and E_2 respectively. If we apply this to $H(\mathbb{C})$ and F', identify $H(\mathbb{C}) \widehat{\otimes}_\pi F'$ with a subspace of $H(E)$ by extending the mapping

$$f \times \phi \longrightarrow [(z,w) \longrightarrow f(z)\phi(w)]$$

and use Proposition 3.43 we see that $H(\mathbb{C}) \widehat{\otimes}_\pi F'$ is isomorphic to a closed complemented subspace of $\big(H(E), \tau_0 \big)$. If $\tau_0 = \tau_\delta$ on $H(E)$ then $H(\mathbb{C}) \widehat{\otimes}_\pi F'$ is bornological. By Proposition 15, Chapter 2 of Grothendieck's thesis this implies that $F \cong \lambda(A)$ where A has the property that for all k and all $\varepsilon > 0$ there exists $k' > 0$ such that

$$\alpha_{j,k} \left(\frac{\alpha_{j,k}}{\alpha_{j,1}} \right)^{1/\varepsilon} \leq \alpha_{j,k'}$$

for all j. This is equivalent to condition (e) of Proposition 4.12 and completes the proof.

A version of Proposition 4.55, without a basis hypothesis, is given in Exercise 4.92. If $\lambda(A)$ is Fréchet nuclear and $\{E_n\}_n$ is a sequence of Banach spaces then any non-zero sequence of vectors $(x_n)_n$, $x_n \in E_n$ all n, generates a closed complemented subspace of $\lambda(\{E_n\}_n)$ isomorphic to $\lambda(A)$. Hence Lemma 4.17, modified for entire functions, and Propositions 4.55 show that $\lambda(A)$ has (DN) when $\tau_\omega = \tau_\delta$ on $H\big(\lambda(\{E_n\}_n)\big)$. We do not know if $\tau_\omega = \tau_\delta$ (or $\tau_b = \tau_\delta$) on $H(E)$, E a Fréchet space, implies that E has (DN). We do know (Example 3.24) that E must have a continuous norm but this condition is strictly weaker than (DN), even for Fréchet nuclear spaces, since $H(D)$, D the open unit disc in $\mathbb{C}$, admits a continuous norm but does not have (DN) (see Example 4.13(b)). On the other hand all Banach spaces have (DN) and hence, by Corollary 4.52, (DN) does not imply $\tau_\omega = \tau_\delta$ for Fréchet spaces.

A feature of earlier sections in this chapter was the employment of various expansions for polynomials and holomorphic functions (see Table 4.1). At one end of the spectrum of Fréchet spaces we have Banach spaces where we obtained a relatively weak expansion for the strong (norm) topology and a strong expansion for the relatively weak (compact open) topology. At the other end, we have Fréchet nuclear spaces where the strong and the compact open topologies coincide and there exists a strong expansion. It is natural to attempt to bridge the gap between these two extremes.

Table 4.1

	Domain space	Function space	Expansion
Proposition 4.4	E Banach with shrinking basis	$(\mathcal{P}_\omega(^nE), \|\cdot\|)$	Schauder basis with the square order
Exercise 4.71	E Banach with Schauder basis	$(\mathcal{P}(^nE), \tau_0)$	Schauder basis with the square order
Proposition 4.9	$E = \lambda(\{E_j\}_j)$ λ Fréchet-nuclear with basis, E_j Banach	$(H(E), \tau)$ $\tau = \tau_0, \tau_\omega$	absolute Schauder decomposition
Propositions 3.43 and 3.46	E Fréchet-nuclear with basis	$(H(E), \tau)$ $\tau_0, \tau_\omega, \tau_\delta$	absolute basis

The remaining results in this chapter, one positive and the other negative, may be regarded as efforts in this direction. In Proposition 4.56 we characterize the monomial coefficients of holomorphic functions on l_1 (see Exercise 3.122 for a similar result for infinite type power series spaces) and use it to show that the monomials form an unconditional basis for $H(l_1)$ endowed with the compact open topology. However, it is rarely, and perhaps never, the case that the monomials form an unconditional basis for $(\mathcal{P}(^nE), \tau_b)$ or $(\mathcal{P}(^nE), \tau_\omega)$ when E is an infinite dimensional Banach space and $n \geq 2$. If we move away from Banach spaces towards Fréchet–Montel spaces we see, in Propositions 4.58 and 4.59, that we do not obtain an absolute or unconditional monomial basis unless we go all the way to Fréchet nuclear spaces.

Let $rB_{c_0}^+ = \{(\xi_n)_n : 0 < \xi_n < r$ and $\lim_n \xi_n = 0\}$ for $0 < r \leq \infty$ and let B_{l_1} denote the open unit ball of l_1 with norm closure $\overline{B_{l_1}}$. Using the description outlined in Lemma 4.10 we see that $(rK_\xi)_{\xi \in B_{c_0}^+}$, $K_\xi := \{(\xi_n z_n)_n : (z_n)_n \in \overline{B_{l_1}}\}$, forms a fundamental system of compact sets in rB_{l_1}, $0 < r < \infty$. Moreover, if we let $\infty \cdot B_{l_1} := \bigcup_{r>0} rB_{l_1} = l_1$ then $(K_\xi)_{\xi \in c_0}$ is a fundamental system of compact sets in this case. It is easily verified that

$$\sup\{\|z\|_{l_1} : z \in K_\xi\} = \|\xi\|_{c_0}$$

for $\xi \in B_{c_0}^+$.

In earlier chapters we frequently used $\lim_{n \to \infty}(n^n/n!)^{1/n} = e$, but now need refined versions of this limit. These follow from *Stirling's formula*:

for any positive integer n there exists θ_n, $0 < \theta_n < 1$, such that

$$n! = \sqrt{2\pi n}\left(\frac{n}{e}\right)^n e^{\theta_n/12n}.$$

The two estimates we require are

$$n! \leq n^n \leq e^n n! \tag{4.37}$$

for all $n \in \mathbb{N}$, and for each η, $0 < \eta < 1$, there exists $a(\eta) > 0$ such that

$$(e\eta)^n n! \leq n^n \tag{4.38}$$

for all $n \geq a(\eta)$.

Proposition 4.56 *If $(a_m)_{m \in \mathbb{N}^{(\mathbb{N})}}$ is a set of complex numbers and $0 < r \leq \infty$ then $\sum_{m \in \mathbb{N}^{(\mathbb{N})}} a_m z^m$ is the monomial expansion of $f \in H(rB_{l_1})$ if and only if*

$$\lim_{|m| \to \infty} \frac{|a_m|\xi^m m^m}{|m|^{|m|}} = 0 \tag{4.39}_r$$

for all $\xi \in rB_{c_0}^+$. The monomial expansion of a holomorphic function f converges absolutely and pointwise to $f(z)$ and the convergence is uniform on compact subsets of l_1. The monomials form an equicontinuous unconditional basis for $\left(H(rB_{l_1}), \tau_0\right)$.

Proof. It suffices to prove the proposition for $r = 1$ since a rescaling reduces the finite r case to the case $r = 1$ and the $r = \infty$ can be handled by using $H(l_1) = \bigcap_{r>0} H(rl_1)$.

Let $f \in H(B_{l_1})$. The monomial coefficients of f, $(a_m)_{m \in \mathbb{N}^{(\mathbb{N})}}$, are defined in Sections 3.1 and 4.1. If $\xi \in B_{c_0}^+$ then we can choose $\delta > 1$ such that $K_{\delta\xi}$ is also a compact subset of B_{l_1}. If $m \in \mathbb{N}^{(\mathbb{N})}$ then (3.3) implies

$$\|a_m z^m\|_{K_{\delta\xi}} = |a_m|\xi^m \delta^{|m|} \|z^m\|_{\overline{B_{l_1}}} \leq \|f\|_{K_\xi}. \tag{4.40}$$

By Lemma 1.38, $\|z^m\|_{B_{l_1}} = m^m/|m|^{|m|}$. Hence

$$\lim_{|m|\to\infty}\left(\frac{|a_m|\xi^m m^m}{|m|^{|m|}}\right) \le \lim_{|m|\to\infty}\frac{\|f\|_{K_{\delta\xi}}}{\delta^{|m|}} = 0. \tag{4.41}$$

This proves the proposition in one direction.

Conversely, suppose we are given a collection $(a_m)_{m\in\mathbb{N}^{(\mathbb{N})}}$ satisfying $(4.39)_1$. We are required to show that for each $\theta = (\theta_n)_n \in B_{c_0}^+$ we can find $\xi = (\xi_n)_n \in B_{c_0}^+$, $\varepsilon = (\varepsilon_n)_n \in B_{c_0}^+$ and $c > 0$ such that

$$\frac{|m|^{|m|}}{m^m}\|z^m\|_{K_\theta} \le C\frac{|m|!}{m!}\xi^m\|z^m\|_{K_\varepsilon}$$

for all $m \in \mathbb{N}^{(\mathbb{N})}$. This requires two adjustments to the compact set K_θ, the first to produce ξ and the second to replace $|m|^{|m|}/m^m$ by $|m|!/m!$. Given $\theta = (\theta_n)_n \in B_{c_0}^+$ let $\xi_n = \theta_n^{1/2}$ for all n. Then $\xi = (\xi_n)_n \in B_{c_0}$ and K_ξ is a compact subset of B_{l_1}. Let $\eta > 0$ satisfy $\eta^2 = \sup\{\|z\| : z \in K_\xi\} = \|\xi\|$. Clearly $\eta < 1$. Since $\sup\{\sum_{n>k}|z_n| : (z_n)_n \in K_\xi\} = \sup_{n>k}|\xi_n| \to 0$ as $k \to \infty$ we can choose a positive integer l such that $|\xi_n| \le \eta/e$ for all $n \ge l$. Let $\varepsilon_n = \xi_n/\eta$ for $n < l$ and $\varepsilon_n = e\xi_n$ for $n \ge l$. Then $(\varepsilon_n)_n \in B_{c_0}^+$ and $\|(\varepsilon_n)_n\| \le \eta$.

Now fix an arbitrary element $m = (m_n)_{n=1}^\infty$ in $\mathbb{N}^{(\mathbb{N})}$. Let $F = \{n \in \mathbb{N} : n < l$ and $m_n \ge a(\eta)\}$ where $a(\eta)$ is chosen as in (4.38). We have

$$\frac{|m|^{|m|}}{m^m} = \frac{(\sum_n m_n)^{\sum_n m_n}}{\prod_n m_n^{m_n}} \le \frac{(\sum_n m_n)! \cdot \exp(\sum_n m_n)}{\prod_{n\notin F} m_n! \cdot \prod_{n\in F} m_n^{m_n}}, \text{ by (4.37)}$$

$$\le \frac{|m|! \cdot \exp(\sum_n m_n)}{\prod_{n\notin F} m_n! \cdot \prod_{n\in F}(e\eta)^{m_n} m_n!}, \text{ by (4.38)}$$

$$= \frac{|m|!}{m!} \cdot \exp\left(\sum_{n\notin F} m_n\right) \cdot \eta^{-\sum_{n\in F} m_n}$$

$$\le \frac{|m|!}{m!} \cdot \exp\left(\sum_{n\ge l} m_n + la(\eta)\right) \cdot \eta^{-\sum_{n<l} m_n}.$$

If $(z_n)_n \in K_\xi$ let $w_n = z_n/\eta$ for $n < l$ and $w_n = ez_n$ for $n \ge l$. By our construction $(w_n)_n \in K_\varepsilon$. If $z \in K_\xi$ then the above estimate implies

$$\frac{|m|^{|m|}}{m^m}|z^m| \le \frac{|m|!}{m!}e^{la(\eta)}\prod_{n\ge l}|ez_n|^{m_n} \cdot \prod_{n<l}\left|\frac{z_n}{\eta}\right|^{m_n} = \frac{|m|!}{m!}e^{la(\eta)}|w^m|$$

for all $m \in \mathbb{N}^{(\mathbb{N})}$. Hence

$$\frac{|m|^{|m|}}{m^m}\|z^m\|_{K_\xi} \leq \frac{|m|!}{m!}e^{la(\eta)}\|z^m\|_{K_\epsilon}. \tag{4.42}$$

If $z \in l_1$ then

$$\sum_{\substack{m \in \mathbb{N}^{(\mathbb{N})} \\ |m|=n}} \frac{|m|!}{m!}\|z^m\| = \sum_{\substack{m \in \mathbb{N}^{(\mathbb{N})} \\ |m|=n}} \frac{|m|!}{m!}|z_1|^{|m_1|} \cdots |z_n|^{|m_n|}$$

$$= \left(\sum_i |z_i|\right)^n = \|z\|^n.$$

This equality, (4.42) and $\sup\{\|z\| : z \in K_\epsilon\} = \|(\epsilon_n)_n\| < 1$ imply

$$\sum_{m \in \mathbb{N}^{(\mathbb{N})}} \|a_m z^m\|_{K_\theta} = \sum_{m \in \mathbb{N}^{(\mathbb{N})}} \frac{|a_m|\xi^m m^m}{|m|^{|m|}} \cdot \frac{|m|^{|m|}}{m^m}\|z^m\|_{K_\xi}$$

$$\leq \sup_{m \in \mathbb{N}^{(\mathbb{N})}} \frac{|a_m|\xi^m m^m}{|m|^{|m|}} \cdot \sum_{m \in \mathbb{N}^{(\mathbb{N})}} \frac{|m|!}{m!}e^{la(\eta)}\|z^m\|_{K_\epsilon}$$

$$\leq e^{la(\eta)} \sup_{m \in \mathbb{N}^{(\mathbb{N})}} \frac{|a_m|\xi^m m^m}{|m|^{|m|}} \cdot \sum_{n=0}^{\infty}\left(\sup_{z \in K_\epsilon}\|z\|\right)^n$$

$$\leq \frac{e^{la(\eta)}}{1 - \|(\epsilon_n)_n\|} \cdot \sup_{m \in \mathbb{N}^{(\mathbb{N})}} \frac{|a_m|\xi^m m^m}{|m|^{|m|}} < \infty.$$

Hence the series $\sum_{m \in \mathbb{N}^{(\mathbb{N})}} a_m z^m$ converges uniformly and absolutely on compact subsets of B_{l_1} and thus defines a continuous function f on B_{l_1}. Since it is easily seen that f is $\mathcal{G}$-holomorphic we have $f \in H(B_{l_1})$ and f has monomial expansion $\sum_{m \in \mathbb{N}^{(\mathbb{N})}} a_m z^m$. From the above

$$\left\|f(z) - \sum_{m \in J} a_m z^m\right\|_{K_\theta} \leq \frac{e^{la(\eta)}}{1 - \|(\epsilon_n)_n\|} \cdot \sup_{m \in \mathbb{N}^{(\mathbb{N})} \setminus J} \frac{|a_m|\xi^m m^m}{|m|^{|m|}} \tag{4.43}$$

for every finite subset J of $\mathbb{N}^{(\mathbb{N})}$.

Since $\dfrac{|a_m|\xi^m m^m}{|m|^{|m|}} \to 0$ as $|m| \to \infty$ the monomials form an unconditional basis for $(H(B_{l_1}), \tau_0)$. By (4.40) and (4.43) the τ_0 topology on $H(B_{l_1})$ is generated by the semi-norms

$$p_\xi\left(\sum_{m \in \mathbb{N}^{(\mathbb{N})}} a_m z^m\right) := \sup_{m \in \mathbb{N}^{(\mathbb{N})}} \frac{|a_m|\xi^m m^m}{|m|^{|m|}}$$

where ξ varies over $B_{c_0}^+$. Since

$$\sup_{\substack{J \subset \mathbb{N}^{(\mathbb{N})} \\ J \text{ finite}}} p_\xi \Big(\sum_{m \in J} a_m z^m \Big) = p_\xi \Big(\sum_{m \in \mathbb{N}^{(\mathbb{N})}} a_m z^m \Big)$$

the monomial basis is equicontinuous. This completes the proof.

If the monomial basis was $(H(B_{l_1}), \tau_0)$ was *absolute* then for every $\xi \in B_{c_0}^+$ we could find $\eta \in B_{c_0}^+$ and $c > 0$ such that

$$\sum_{m \in \mathbb{N}^{(\mathbb{N})}} \frac{|a_m| \xi^m m^m}{|m|^{|m|}} \le c \sup_{m \in \mathbb{N}^{(\mathbb{N})}} \frac{|a_m| \eta^m m^m}{|m|^{|m|}}$$

for all $\sum_{m \in \mathbb{N}^{(\mathbb{N})}} a_m z^m \in H(B_{l_1})$. This implies, in particular, that $\sum_{n=1}^k \xi_n \le c \cdot \sup_{n \le k} \eta_n \le c \|(\eta_n)_n\|$ for all k. If $\xi_n = 1/\sqrt{n}$ for all n this is impossible and implies that the monomial basis is not absolute.

For finite dimensional spaces the criterion established in Proposition 4.56 can, with some minor adjustments, be compared with the classical result

$$\sum_{n=1}^\infty a_n z^n \in H(D), \ D = \{z \in \mathbb{C} : |z| < 1\}, \ \textit{if and only if}$$

$$\limsup_{n \to \infty} |a_n|^{1/n} \le 1.$$

Corollary 4.57 *The series* $\displaystyle\sum_{n_1, n_2 = 0}^\infty a_{n_1, n_2} z_1^{n_1} z_2^{n_2}$ *defines a holomorphic function on* $\{(z_1, z_2) : |z_1| + |z_2| < 1\}$ *if and only if*

$$\limsup_{n_1 + n_2 \to \infty} \frac{\big(|a_{(n_1, n_2)}| n_1^{n_1} n_2^{n_2}\big)^{1/n_1 + n_2}}{n_1 + n_2} \le 1.$$

We now consider when the spaces of holomorphic functions on Montel or Fréchet–Montel spaces has an absolute or unconditional basis.

Let A_n denote a symmetric $n \times n$ matrix. We identify A_n with a symmetric bilinear form on $\mathbb{C}^n$ (see Example 1.4) in the following way:

$$A_n \big((z_j)_{j=1}^n, (w_k)_{k=1}^n \big) = \sum_{j,k=1}^n a_{jk} z_j w_k$$

where $A_n = (a_{j,k})_{1 \le j, k \le n}$. Let $\|A_n\|_p$ denote the norm of this bilinear form when $\mathbb{C}^n$ is given the l_p norm, i.e.

$$\|A_n\|_p = \sup \Big\{ \Big| \sum_{j=1}^n a_{jk} z_j w_k \Big| : \sum_{j=1}^n |z_j|^p \le 1, \sum_{k=1}^n |w_k|^p \le 1 \Big\}.$$

If $i = \sqrt{-1}$ and $a_{j,k} = \exp(2\pi i j k / n)$, $1 \le j$, $k \le n$, then by an inequality of Hardy–Littlewood there exists a universal constant α such that

$$\alpha \, n^{a(p)} \le \|A_n\|_p \le n^{a(p)} \tag{4.44}$$

where

$$a(p) = \begin{cases} \dfrac{3}{2} - \dfrac{2}{p} & \text{for } p \ge 2 \\[2mm] 1 - \dfrac{1}{p} & \text{for } 1 \le p \le 2 \end{cases}$$

Let E denote a Montel space and suppose the monomials form an absolute basis for $(H(E), \tau_0)$. In particular, E'_β is Montel with absolute basis and may be identified with a sequence space $\Lambda(P)$ (see Section 3.3). Since E'_β is Montel a fundamental system of compact subsets of E is obtained by taking polars of a fundamental system of neighbourhoods of zero in E'_β. The c_0–l_1 duality and the absolute basis in E'_β show that the sets

$$\left\{ \sum_{n=1}^{\infty} z_n e_n \in E : \sup_n \frac{|z_n|}{|a_n|} \le 1 \right\} \tag{4.45}$$

where $(a_n)_n \in P$ form a fundamental system of compact subsets of E (we adopt the usual convention when dividing by zero).

Proposition 4.58 *Let E denote a Montel space with basis $(e_n)_n$. If the monomials form an absolute basis for $(H(E), \tau_0)$ then E'_β is nuclear.*

Proof. By our hypothesis the monomials of degree 2 form an absolute basis for $(\mathcal{P}(^2 E), \tau_0)$. By (4.45) for each $(a_n)_n \in P$ there exists $(b_n)_n \in P$ and $c > 0$ such that for any set of scalars $(\alpha_{j,k})_{1 \le j,k < \infty}$, $\alpha_{j,k} = \alpha_{k,j}$, and any finite subset J of $\mathbb{N}$ we have

$$\sum_{j,k \in J} |\alpha_{j,k}| \cdot \sup_{j,k} \{ |z_j z_k| : |z_j| \le a_j \}$$

$$\le c \cdot \sup \left\{ \left| \sum_{j,k \in J} \alpha_{j,k} z_j z_k \right| : \sup_j \frac{|z_j|}{b_j} \le 1, \, \sup_k \frac{|z_k|}{b_k} \le 1 \right\}. \tag{4.46}$$

If $w_j = \dfrac{z_j}{b_j}$ then (4.46) becomes

$$\sum_{j,k \in J} |\alpha_{j,k}| \frac{a_j a_k}{b_j b_k} \le c \cdot \sup \left\{ \left| \sum_{j,k \in J} \alpha_{j,k} w_j w_k \right| : \sup_i |w_i| \le 1 \right\} \tag{4.47}$$

and, applying (4.44) with $p = \infty$, we get

$$\sum_{j,k \in J} \frac{a_j \, a_k}{b_j \, b_k} \le c \, n^{3/2} \tag{4.48}$$

for any finite subset J of $\mathbb{N}$. Let $(\beta_j)_j$ denote the non-increasing rearrangement of $(a_j/b_j)_j$. By (4.48),

$$(n\beta_n)^2 \leq \left(\sum_{j=1}^n \beta_j\right)^2 \leq c\, n^{3/2}.$$

for any positive integer n. Hence $\beta_n \leq \dfrac{c^{1/2}}{n^{1/4}}$ and $\displaystyle\sum_{n=1}^\infty \left(\frac{a_n}{b_n}\right)^8 < \infty$. A minor modification to the Grothendieck–Pietsch criterion shows that E'_β is nuclear and completes the proof.

Proposition 4.58 shows that nuclearity, in our approach to the topological problems in Sections 4.2 and 4.3, is not only convenient but unavoidable. The following improvement of Proposition 4.58 also holds (see Exercise 4.71).

Proposition 4.59 *If E is a DF-Montel space or a Fréchet–Montel space such that E'_β has an absolute basis and the monomials of degree 2 form an unconditional basis for $(\mathcal{P}(^2E), \tau_0)$ then E'_β (and hence E) is nuclear.*

4.5 Exercises

Exercise 4.60* Let E denote a Banach space and let $(n_j)_j$ denote a strictly increasing sequence of positive integers. If $P_j \in \mathcal{P}(^{n_j}E)$ is non-zero for all j show that $\{P_j\}_j$ is a basic sequence in $(H(E), \tau_\omega)$ and that the closed subspace of $(H(E), \tau_\omega)$ generated by $\{P_j\}_j$ is nuclear.

Exercise 4.61* If E is a Banach space with unit ball B, n is a positive integer, $(P_m)_{m>n}$ is a basic sequence of unit vectors in $\mathcal{P}(^nE), Q_m \in \mathcal{P}(^mE)$ for $m > n$ and $\|Q_m\|_B \to 0$ as $m \to \infty$ show that the closed subspace of $(H(E), \tau_\omega)$ generated by $\{P_m + Q_m\}_{m>n}$ is a Banach space isomorphic to the Banach subspace of $\mathcal{P}(^nE)$ generated by $(P_m)_{m>n}$.

Exercise 4.62* If E is a Banach subspace of $C(T)$, T a sequentially complete compact Hausdorff space, show that the bounding subsets of E are relatively compact. If E is a Banach space and E' has the Radon–Nikodým Property show that the bounding subsets of E are relatively compact.

Exercise 4.63* Let U be an open polydisc in a fully nuclear space with basis E. Show that the following are equivalent:

(a) $\left(H(U), \tau_\omega\right)'_\beta = H_{HY}(U^M) = \varinjlim_{\substack{V \supset U^M \\ V \text{ open}}} \left(H_{HY}(V), \tau_0\right).$

(b) $\left(H(U), \tau_\omega\right)'_\beta$ has the monomials as an absolute basis and $\left(H(U), \tau_\omega\right)$ is semi-reflexive.

(c) The τ_ω bounded subsets of $H(U)$ are locally bounded.

Moreover, if E is an A-nuclear space, show that the above are equivalent to:

(d) $\left(H(U), \tau_\omega\right)$ is semi-reflexive,

(e) $\left(H(U), \tau_\omega\right)$ is quasi-complete.

Exercise 4.64* If E is a Banach space with shrinking basis and $(n-1)u(E') < l(E)$ show that the monomials with the square order form a basis for $\mathcal{P}(^n E)$.

Exercise 4.65* Let E and F be Banach spaces and let U denote an open subset of E. Let $f \in H(U; F)$. If $\tilde{f} : U \longrightarrow H(0, F)$ (the space of F-valued holomorphic germs at 0 in E) is defined by

$$\tilde{f}(x) = \{\text{germ of } f \text{ at } x \text{ translated to the origin}\}$$

show that $\tilde{f} \in H_G(U; H(0; F))$. Is $\tilde{f}$ continuous?

Exercise 4.66* Let p denote a continuous semi-norm on $\left(H(0), \tau_\omega\right)$ where 0 is the origin in $\mathbb{C}^{(\mathbb{N})}$. Show that there exists a positive integer n such that if $f \in H(0)$ and $f\big|_{U \cap \mathbb{C}^n} = 0$ for some neighbourhood U of 0 in $\mathbb{C}^n$ then $p(f) = 0$.
Hence show that the sequence, $(f_n)_n$, defined by $f_n\left((z_m)_{m=1}^\infty\right) = \dfrac{z_n}{1 - nz_1}$ for all n is a bounded subset of $\left(H(0), \tau_\omega\right)$ and deduce that the inductive limit $\varinjlim_{\substack{V \text{ open} \\ 0 \in V}} \left(H^\infty(V), \|\cdot\|_V\right)$ is not regular.

Exercise 4.67* Let $\pi: E \to E/F$ denote the canonical quotient mapping where E is either a Banach space or a Fréchet–Schwartz space and let

$$\pi^* : \left(H\left(\pi(U)\right), \tau\right) \to \left(H(U), \tau\right)$$

where $\pi^*(f) = f \circ \pi$, $\tau = \tau_0$ or τ_ω, and U is a balanced open subset of E. Show that π^* is an embedding (i.e. a topological isomorphism onto its image).

Exercise 4.68* If $(E_n)_n$ is a finite dimensional decomposition of a Fréchet space E with sequence of projections $(\pi_n)_n$ show that E is a Schwartz space if and only if there exists a sequence of semi-norms $(\|\cdot\|_k)_k$ defining the topology of E such that

$$\limsup_{n \to \infty}\left\{\left\|x - \sum_{j=1}^n \pi_j(x)\right\|_k : \|x\|_{k+1} \le 1\right\} = 0$$

for all k.

If E is a Fréchet–Schwartz space with finite dimensional Schauder decomposition show that $\big(H(E_\beta'), \tau_0\big)$ has a finite dimensional Schauder decomposition.

Exercise 4.69* By using the estimate (4.42) in Section 4.6 show that a Fréchet–Schwartz Köthe sequence space $\lambda^p(A)$ is nuclear if and only if

$$\lambda^p(A) \,\widehat{\otimes}_\varepsilon\, \lambda^p(A) = \lambda^p(A) \,\widehat{\otimes}_\pi\, \lambda^p(A).$$

Exercise 4.70* Show that the bounding subsets of a Banach space E are relatively compact if and only if norm and weak holomorphic sequential convergence agree. (A sequence $(x_n)_n \subset E$ is weakly holomorphically convergent if $f(x_n) \to f(x)$ as $n \to \infty$ for all $f \in H(E)$.)

Exercise 4.71* Show that the monomials, with the square ordering, form a Schauder basis for $\big(\mathcal{P}(^nE), \tau_0\big)$ whenever E is a Banach space with Schauder basis.

Exercise 4.72* If E is a gDF space and F is a Banach space show that $H_b(E; F)$ endowed with the topology of uniform convergence on bounded sets is a Fréchet space. If E is complete show that $H(E) = H_b(E)$ if and only if $E = (F', \tau_0)$ for some Fréchet space E. Show that $H(E) = H_b(E)$ implies $H(E) = H_{ub}(E)$.

Exercise 4.73* If $\{F_j\}_j$ is a shrinking finite dimensional Schauder decomposition for the Banach space E and no disjointly supported sequence of vectors in E has a lower q estimate, $1 < q < \infty$, show that $\mathcal{P}_w(^nE) = \mathcal{P}(^nE)$ for all n.

Exercise 4.74* If $E = \varinjlim_n E_n$ is a strict inductive limit of Fréchet spaces and F has an unconditional basis show that $E \cong \sum_{n=1}^\infty F_n$ where each F_n is a Fréchet space with an unconditional basis.

Exercise 4.75* If $E = \sum_{n=1}^\infty E_n$ where each E_n is a Fréchet nuclear space with basis and (DN) show that $\tau_0 = \tau_\delta$ on $H(E)$.

Exercise 4.76* If E is a fully nuclear space such that $\big(H(E), \tau_0\big)$ is complete and F is a closed subspace of E show that each holomorphic function on F can be extended to a holomorphic function on E if and only if the quotient space $\big(H(E), \tau_0\big) \big/ \{f \in H(E) : f\big|_F = 0\}$ is a complete locally convex space.

Exercise 4.77* If F is a closed subspace of a $\mathcal{DFN}$ space E show that each holomorphic function on F admits a holomorphic extension to E.

Exercise 4.78* If E is a Fréchet nuclear space with (DN) show that $\widehat{\bigotimes}_{n,s,\pi} E$ is also a Fréchet nuclear space with (DN) for every positive integer n.

Exercise 4.79* By using bounding sets show that l_∞ does not contain any infinite dimensional separable complemented subspaces.

Exercise 4.80* If $f \in H(c_0)$ show that there exists $g \in H(l_\infty)$ such that $g\big|_{c_0} = f$ if and only if $r_f \equiv +\infty$.

Exercise 4.81* Let E and F be Banach spaces and let K be a compact determining set (Exercise 3.100) for holomorphic functions on E. If $f\colon K \to F$ is a mapping and for each ϕ in F' there exist an open neighbourhood V_ϕ of K and $f_\phi \in H(V_\phi)$ such that $f_\phi\big|_K = \phi \circ f$ show that there exists an open neighbourhood V of K and $\widetilde{f} \in H(V; F)$ such that $\widetilde{f}\big|_K = f$.

Exercise 4.82* If $f\colon l_2 \to \mathbb{C}$ is defined by $f\big((x_n)_n\big) = \sum_n x_n^2$ show that $f \in \mathcal{P}(^2 l_2)$. Since l_2 is separable it can be identified with a closed subspace of $\mathcal{C}([0,1])$, say $\pi(l_2)$. Show that there exists no holomorphic function on $\mathcal{C}([0,1])$ whose restriction to $\pi(l_2)$ coincides with f.

Exercise 4.83* Let $\pi\colon l_1 \to c_0$ be a continuous surjection. Show that the identity mapping from c_0 to c_0 cannot be lifted holomorphically to l_1.

Exercise 4.84* Let E denote a Fréchet space. Show that $\tau_0 = \tau_\omega$ on $H(U)$ for every balanced open subset U of E if and only if $G(U)$ is Montel for every balanced open subset U of E. Show that τ_0 and τ_ω are compatible topologies on $H(U)$, i.e. $\big(H(U), \tau_0\big)' = \big(H(U), \tau_\omega\big)'$, for every balanced open subset U of E if and only if $G(U)$ is reflexive for every balanced open subset U of E. Find a Banach space E such that τ_0 and τ_ω are compatible but do not coincide on $H(E)$.

Exercise 4.85* Let E denote a Fréchet space with the bounded approximation property. Use the Borel transform to show $(H(E), \tau_0)'_\beta \cong H_N(0_{E'_C})$.

Exercise 4.86 If $\{f_n\}_n$ is a sequence in $\big(H(0), \tau_\omega\big)$, 0 the origin in $\mathbb{C}^{(\mathbb{N})}$, and if for each n, there exists a neighbourhood of zero U_n in $\mathbb{C}^n$ such that $f_m\big|_{U_n} \equiv 0$ for all $m \geq n$ show that $(f_n)_n$ is a very strongly convergent sequence in $H(0)$.

Exercise 4.87* If U is a balanced open subset of a Banach space E show that
$$c_0 \hookrightarrow \big(H(U), \tau_\omega\big) \iff c_0 \hookrightarrow \big(H(U), \tau_\delta\big) \iff l_\infty \hookrightarrow \big(H(U), \tau_\omega\big).$$

Exercise 4.88* If F is a complex Banach space and $(a_m)_{m \in \mathbb{N}^{(\mathbb{N})}} \subset F$ show that $\sum_{m \in \mathbb{N}^{(\mathbb{N})}} a_m z^m \in H_b(l_1, F)$ if and only if

$$\lim_{|m| \to \infty} \left(\frac{\|a_m\| m^m}{|m|^{|m|}} \right)^{1/|m|} = 0.$$

Exercise 4.89* Show that the τ_ω topology on $H(l_1)$ is generated by the semi-norms

$$p_{\xi, \beta} \left(\sum_{m \in \mathbb{N}^{(\mathbb{N})}} a_m z^m \right) = \sup_{m \in \mathbb{N}^{(\mathbb{N})}} \frac{|a_m| m^m}{|m|^{|m|}} (\xi + \beta_{|m|})^m$$

where $\xi = (\xi_n)_n$ and $\beta = (\beta_n)_n$ range over c_0^+ $(= \bigcup_{r>0} r B_{c_0}^+)$.
Note: $(\xi + \beta_{|m|})^m = \prod_i (\xi_i + \beta_{|m|})^{m_i}$ for all $m = (m_i)_i$.

Exercise 4.90* Show that $H_N(E)$ is a translation invariant subalgebra of $H(E)$ for any locally convex space E.

Exercise 4.91* If U and V are open subsets of Fréchet spaces E and F, one of which has the approximation property, show that

$$H(U \times V) \simeq H(U; H(V)) \simeq H(U) \widehat{\otimes}_\varepsilon H(V)$$

when all spaces are endowed with the compact open topology.

Exercise 4.92* If E is a Fréchet nuclear space with basis and D is the open unit disc in $\mathbb{C}$ show that $\tau_0 = \tau_\delta$ on $H(D \times E)$ if and only if E has (DN).

4.6 Notes

The relationship between the τ_0, τ_ω, τ_b and τ_δ topologies on spaces of holomorphic functions has been an active research area for thirty years and, in the last five or six years, the theory for *balanced* domains has achieved a certain completeness. The results in this chapter represent the most important non-trivial positive results for these domains. We discuss, in Chapter 5, the same problem for arbitrary domains where the results are again highly non-trivial but far from being complete.

The square order on the unrestricted (i.e. the full) tensor product of two Banach spaces with Schauder bases was introduced in 1961 by B.R. Gelbaum-J.G. de Lamadrid [387] and extended, by R.A. Ryan [758] in 1980, to symmetric n-tensors and, by duality, to spaces of homogeneous polynomials. In his thesis, [758], R.A. Ryan essentially proved Proposition 4.4 and showed that the monomials form a basis for $\left(\mathcal{P}(^nE), \|\cdot\|\right)$ when E is a Banach space with shrinking basis and the Dunford–Pettis Property, e.g. c_0. In 1985, R. Alencar [6] proved Corollary 4.5 for reflexive Banach spaces with a basis (see also [275]). The remaining results presented in Section 4.1, i.e. Propositions 4.1, 4.2, 4.6 and Corollaries 4.3, 4.5 (for finite dimensional decompositions) and 4.7, were obtained by V. Dimant–S. Dineen [273]. For polynomials these are based on combining finite dimensional decompositions and the square ordering and, for holomorphic functions, $\mathcal{S}$-Schauder decompositions are used (see [299, Proposition 3.60 and Corollary 3.61]). Holomorphy on $\mathcal{DFS}$ and $\mathcal{DFN}$ spaces with finite dimensional decompositions was first considered, in 1980, by A. Benndorf [100, 101] (see Section 3.6).

In 1970, J.A. Barroso [86, 87] showed that $\tau_0 = \tau_\delta$ on $H(\mathbb{C}^I)$ if and only if I is finite and recently (1995) C. Boyd [180] proved that $\tau_0 = \tau_\omega$ or τ_δ on $H(\mathbb{C}^{(I)})$ if and only if I has cardinality less than the first measurable cardinal. The first positive $\tau_\omega = \tau_\delta$ results for balanced domains in a Banach space E were obtained by S. Dineen [283] in 1972 when E has an unconditional basis. This result was extended soon afterwards by G. Coeuré [233] to a collection of Banach spaces which included $L^1([0, 2\pi])$. The approach in [283] was extended in 1980 to Fréchet nuclear spaces and resulted in Corollary 4.15 ([301]). The (DN)-property was introduced by D. Vogt [835] in 1975 and has since featured prominently in the development and application of Fréchet spaces. We refer to [836] for Proposition 4.12 and to [837, 839, 841, 842] for further properties and linear characterizations of Fréchet spaces with (DN) (see also [429] and Exercises 4.68 and 4.78). In 1990, S. Dineen [310] combined the Banach and nuclear space cases and obtained Proposition 4.14 when the finite dimensional decompositions were unconditional. In 1995, S. Dineen and J. Mujica, independently and at the same time, removed the unconditionality hypothesis in the Banach space case and obtained Corollaries 4.16 and 4.18. The proofs were different with J. Mujica [657] obtaining Corollary 4.16 directly by following the Banach space approach in [283]. The author's Fréchet space proof, which includes Propositions 4.9, 4.11, 4.12 and 4.14, Lemmata 4.10 and 4.17 and Examples 4.8 and 4.13, was a further modification of [301]. This proof was outlined in [311] while the full proof, which is similar to the proof in [310], appears here for the first time. Lemma 4.17 is a typical example of how the bounded approximation property is employed within infinite dimensional holomorphy (see the remarks on Exercise 4.68 and Section 5.2). It is strictly stronger than the approximation property (Definition 2.7) even within the collection of Fréchet nuclear spaces ([333, 334, 840]). The equality $\tau_\omega = \tau_\delta$ is useful in extending results from spaces of homogeneous polynomials to holomorphic functions on balanced domains.

Corollary 4.20 is due to V. Dimant–S. Dineen [273]. The first part of Example 4.21 is due to R. Alencar–R. M. Aron–S. Dineen [8] while the second part appears in [273]. Propositions 4.22 and 4.24 and Corollaries 4.23 and 4.25

are due to V. Dimant–S. Dineen [273]. These were motivated by the following result of V. Dimant–I. Zalduendo [275]:

if E is a Banach space with a shrinking unconditional Schauder basis then $c_0 \not\hookrightarrow \mathcal{L}(^n E)$ if and only if the monomials with the square order form a basis for $\mathcal{L}(^n E)$.

In [275] the authors show, when E is a reflexive Banach space with unconditional basis, that $\mathcal{P}(^n E)$ is reflexive if $(n-1)u(E^*) < l(E)$ while $l_\infty \hookrightarrow \mathcal{P}(^n E)$ if $(n-1)l(E^*) > u(E)$ where l and u are the indexes of Gonzalo–Jaramillo (see [274, 404, 405] and Section 2.6). Reflexivity and non-containment of c_0 are, as we have seen in Chapter 2, closely related for spaces of polynomials. The remarks after Corollary 4.18 and the results in Proposition 4.22 and Corollary 4.23 reduce the analogous holomorphic problem to a problem on spaces of homogeneous polynomials. A general discussion on the lifting of polynomial results to spaces of holomorphic functions is given in the introduction to [273] and the following problem posed:

if E is a Banach space, F is a Banach subspace of $\big(H(E), \tau_\omega\big)$ and $(x_j)_j$ is a basic sequence in F does there exist a subsequence $(y_j)_j$ of $(x_j)_j$ and a positive integer n such that $(y_j)_j$ is equivalent to a basic sequence in $\big(\mathcal{P}(^n E), \|\cdot\|\big)$?

(See the remarks on Exercises 4.61.)

The theme of Section 4.3 has been an object of study for many years and provided the motivation for a number of other research directions within infinite dimensional holomorphy. Initially, the research focused on the $\tau_0 = \tau_\omega$ problem but in recent years this has been absorbed into the more general $\tau_b = \tau_\omega$ problem. Most of the positive solutions were obtained by localizing the problem to spaces of holomorphic germs or spaces of homogeneous polynomials. An exception was the Fréchet-nuclear case but even there germs entered the solution by duality.

Holomorphic germs on compact subsets of Banach spaces were first investigated by L. Nachbin [664, 666], and soon afterwards by S.B. Chae [213, 214] and A. Hirschowitz [458, 460]. Subsequently, J. Mujica [644] developed the theory on metrizable locally convex spaces (see also P. Avilés–J. Mujica [80], R.R. Baldino [82], K.-D. Bierstedt–R. Meise [120, 121, 122], S. Dineen [300], J. Mujica [639, 640], E. Nelimarkka [686], R.L. Soraggi [804, 805, 806, 807, 809, 810, 811], A.J.M. Wanderley [854] and Section 5.4). The τ_ω topology on spaces of holomorphic germs was defined by L. Nachbin [665, 666] while the τ_0 topology was introduced by O. Nicodemi [688]. A number of the results in Section 4.3 are special cases of results from Chapter 5 but we have included the much simpler proofs for balanced domains here, e.g. Lemma 4.33, which states that $H(K)$ is regular, is a special case of Proposition 5.4. This result is due to S.B. Chae [213, 214] and A. Hirschowitz [458] for the τ_ω topology on Banach spaces (although both proofs were originally incomplete) and to J. Mujica for the τ_0 [646] and the τ_ω topologies [644] on metrizable spaces. The application of $\mathcal{S}$-absolute decompositions to spaces of holomorphic germs is taken from [299, Theorem 3.25] while Corollary 4.31 is due to S.B. Chae [214] for Banach spaces and to J. Mujica [644] for Fréchet spaces. Corollary 4.32 is due to J.M. Ansemil–S. Ponte [28] for Fréchet–Montel spaces (see also S. Dineen [308]).

The result we present here for $\tau_b = \tau_\omega$ (Corollary 4.39) appeared at the end of a long sequence of results. The story began in 1970 with the first result in this direction, $\tau_0 = \tau_\omega$ on $H(\mathbb{C}^I)$, if and only if I is countable, by J.A. Barroso [86, 87]. This result was extended independently to arbitrary open subsets of $\mathbb{C}^{\mathbb{N}}$ by J.A. Barroso–L. Nachbin [93] and M. Schottenloher [787]. The proofs for $\mathbb{C}^{\mathbb{N}}$ use heavily the fact that holomorphic functions on open subsets of $\mathbb{C}^{\mathbb{N}}$ depend locally on a finite number of variables. The general approach adopted by Schottenloher of going to the envelope of holomorphy (see [299, Corollary 6.40]) is developed in Chapter 5 in the much more technically demanding setting of open subsets of Fréchet–Schwartz spaces. The first infinite dimensional representation theorem for holomorphic germs is due to J. Boland [151] who proved that $\big(H_N(U), \pi_0\big)'_\beta \equiv H(U^0)$, U a convex balanced open subset of a $\mathcal{DFN}$ space. Further representation theorems for analytic functionals on Banach and Fréchet spaces are due to J.M. Isidro [472] and R. Soraggi [810], respectively, while the classical theory for functions of one complex variable is due to A. Grothendieck [413], G. Köthe [521] and C.L. da Silva Diàs [801, 802]. A. Martineau [571] investigated the several variables case. In 1978, P.J. Boland–S. Dineen [153] used duality theory and holomorphic germs on $\mathcal{DFN}$ spaces to show that $\tau_0 = \tau_\omega$ on open polydiscs in Fréchet nuclear spaces with basis and on entire functions over arbitrary Fréchet nuclear spaces. This, as we have noted, is a consequence of Corollary 4.46 and the results needed to prove it – Propositions 4.42 and 4.45 – may also be found in [153]. Further results on the $\tau_0 = \tau_\omega$ problem on fully nuclear spaces (not necessarily metrizable) were obtained using duality, e.g. Corollaries 4.43 and 4.44 in [286] (see also [154, 155, 296, 298]). In 1981, R. Meise [603] showed that $\tau_0 = \tau_\omega$ on any balanced open subset of a Fréchet nuclear space and this was extended to balanced domains in Fréchet–Schwartz spaces in 1984 ([646]) and to polynomially convex domains in Fréchet nuclear spaces by J. Mujica [650] in 1986.

Mujica's results in [646] were based on a deep analysis of $\big(H(K), \tau_0\big)$ that we present in Chapter 5 and led J.M. Ansemil–S. Ponte [28] (see also [16]) to Proposition 5.7 and to show, using associated topologies, that $\tau_0 = \tau_\omega$ on $H(U)$ when U is a balanced open subset of a Fréchet–Montel space with absolute basis. J. Taskinen's 1986 paper, [816], contained positive and negative answers to Grothendieck's "Problème des topologies", and was the next important influence on the problem for balanced domains. From the negative answers J.M. Ansemil–J. Taskinen [30] showed that there exist Fréchet–Montel spaces E such that $\tau_0 \neq \tau_\omega$ on $H(E)$. The results in [816] drew attention (see Chapters 1 and 3) to the now obvious connection between the (BB)-property and the relationship between the different topologies on spaces of polynomials and holomorphic functions. Motivated by the positive results in [816] and [28], S. Dineen [306] showed that $\tau_0 = \tau_\omega$ on $H(U)$ for U balanced in a class of spaces which contained the Fréchet–Montel $\lambda^p(A)$ spaces (see Corollary 5.8) and afterwards P. Galindo–D. Garcia–M. Maestre [368] and A. Defant–M. Maestre [260] extended these results to a large class of Fréchet–Montel spaces including the Hilbertizable Fréchet–Montel spaces (see also J.C. Díaz–M.A. Miñarro [267]). Following the projective description for τ_ω on $H(U)$, U balanced in a Banach space (Proposition 3.47), and a similar characterization of the τ_0 topology on balanced domains (Proposition 3.30(a)), S. Dineen [308] introduced the τ_b topology (Definition 3.29) and showed, in 1992, that $\tau_b = \tau_\omega$ for a collection of Fréchet spaces which included the T-decomposable Fréchet spaces. The results in [308] include most previously known positive results on the equality $\tau_0 = \tau_\omega$ for scalar-valued holomorphic functions on balanced domains in Fréchet–Montel spaces. K.-D. Bierstedt–J. Bonet–A. Peris studied vector-valued holomorphic germs on $\mathcal{DFS}$ spaces in [118] and showed the existence of a Fréchet–Schwartz space E and a reflexive Banach space X such that $\tau_0 \neq \tau_\omega$ on $H(E, X)$ (see also C. Boyd [182]). This is in contrast to the result of J. Mujica for arbitrary Fréchet–Schwartz spaces (quoted above) and shows an unexpected difference between the scalar and vector-valued cases. The above summary brings us to the results we have presented in the main text which are due to C. Boyd–A. Peris [189] and appeared in 1996.

In his thesis, motivated by the concept of quasi-normability, A. Peris [721] defined (Qno) spaces (quasi-normable by operators) and developed a new approach to linearly classifying locally convex spaces. In [189] the authors studied vector-valued holomorphic functions and proved that $\tau_0 = \tau_\omega$ on $H(U, X)$ for every balanced open subset U of a fixed Fréchet–Schwartz space E and every Banach space X if and only if E has (Qno) (see the remarks on Exercise 4.91). Furthermore, they showed that $\tau_b = \tau_\omega$ on $H(U, X')$ for any balanced open subset U of a Fréchet space E and any Banach space X if and only if (E, X) has the (BB)-property for any Banach space X. These vector-valued results include many but not all of the scalar-valued results. To obtain these results Boyd–Peris introduced the class of generalized DF by operators, $(gDFop)$ spaces, that we have renamed, on artistic and humanitarian grounds, as spaces with the *strong localization property*. A detailed study of the countable neighbourhood property may be found in J. Bonet [157, 158]. To avoid certain technicalities we confined ourselves to the scalar-valued case. The results from [189], both explicit and implicit, included here are Propositions 4.34 and 4.37 and Corollaries 4.38 and 4.39 (which reduces to Proposition 5.7 when E is Montel). An important element in the proofs in [189] is the estimate (4.31) which is due to P. Galindo–D. García–M. Maestre [371].

Proposition 4.49 was proved in the 1966 thesis of C.P. Gupta [421, 423]. Gupta defined the Borel transform of an analytic functional T on $H_{Nb}(E)$,

E a Banach space, by letting

$$BT(\phi) = T(e^\phi) = \sum_{n=0}^{\infty} \frac{T(\phi^n)}{n!}$$

for all $\phi \in E'$. This is based on the classical definition used in studying partial differential equations (see [566]) and leads naturally to functions of exponential type. Holomorphic functions of exponential type over infinite dimensional spaces are discussed in J.M. Ansemil [15], P.J. Boland [146, 147, 151], T.A.W. Dwyer [338, 339, 340, 341], Y. Fujimoto [364], C.P. Gupta [422], P. Lelong [537, 540], M.C. Matos [577, 582, 583, 585], and L. Nachbin [667, 670]. At the n-homogeneous polynomial level the Borel transform gives, up to a factor $n!$, the definition given in Section 2.1 and in this setting there is no loss of generality in removing $n!$. Another variation, suitable for holomorphic functions on fully nuclear spaces, which again uses an adjustment of constants, is given in (4.34). Proposition 4.49 has been extended to $H_N(E)$ by C.P. Gupta [422], L. Nachbin [667] and L. Nachbin–C.P. Gupta [684] and served as motivation for many further studies of convolution equations over infinite dimensional spaces and for the development of holomorphic functions on nuclear and dual nuclear spaces (see Section 2.6). Proposition 4.48 is given in [299, §3.3].

In earlier chapters of this book we obtained directly a number of negative results (e.g. Example 3.24) on the coincidence problem for topologies on spaces of holomorphic functions. In the first three sections of this chapter we concentrated on obtaining positive results and our aim in Section 4.4 was to examine how far we could push the methods used to obtain these positive results before encountering counterexamples. At the moment it appears, intuitively, that apart from the standard topological considerations, discussed in introducing the different topologies in Chapter 3, the two main obstructions are "*size*" and "*approximation*". As noted by many authors (see the final section of [657]) the bounded approximation property appears frequently as a hypothesis for positive results. Whether or not this is due to intrinsic or convenient considerations is unclear at present. One major open problem is whether or not $\tau_\omega = \tau_\delta$ for (balanced) domains in *separable* Banach spaces. If the property "$\tau_\omega = \tau_\delta$" passed either to subspaces or quotients we would obtain a positive solution by using $\mathcal{C}([0,1])$ (for subspaces) and l^1 (for quotients) and Corollary 4.16 (see [67]). By Lemma 4.17 this property does pass to *complemented* subspaces and we used this to pass from Fréchet spaces with Schauder basis to separable Fréchet spaces with the bounded approximation property. In the case of Fréchet nuclear spaces we can go further and pass to arbitrary closed subspaces. This is the content of Proposition 4.54 ([298]) which allowed us to remove fully the basis hypothesis in Proposition 4.41(b) and to obtain Proposition 4.54 which is also due to S. Dineen [298] (see [630] for a study of quotient spaces of holomorphic functions over Banach spaces, [299, §5.4] where the problem of extending holomorphic results to products, subspaces and strict inductive limits is discussed, and Exercise 4.92 where the basis hypothesis in Proposition 4.54 is removed). Proposition 4.55 shows that the (DN)-property is necessary and sufficient for the equality $\tau_0 = \tau_\delta$ on Fréchet *nuclear* spaces with a basis but the role of (DN) in the $\tau_\omega = \tau_\delta$ problem for *arbitrary* Fréchet spaces is still unclear (see Examples 3.24 and 4.8(d) and Proposition 4.14). Proposition 4.55 appears in [299, pp.333 and 396] and related results may be found in [301]. The idea of using tensor products and the connection between this result and the result of Grothendieck, quoted in the proof of Proposition 4.55, was pointed out to the author by D. Vogt.

The other obstruction to positive results, "size", appears in the case of l_∞ (Corollary 4.52) and $c_0(\Gamma)$, Γ uncountable (Example 5.55). We refer to Section 3.6 for a discussion on bounding sets. Proposition 4.51 and Corollary 4.52 are due to S. Dineen [280]. Proposition 4.56 is due to R.A. Ryan [759] when $r = \infty$ and to L. Lempert [545] for arbitrary r. Ryan was motivated by topological considerations and applied his results to obtain lifting theorems for holomorphic mappings and to find a new description of the τ_ω topology on $H(l_1)$ (see Exercise 4.89). He had previously observed [758, p.102] the unconditionality of the monomial basis for $(\mathcal{P}(^n l_1), \tau_0)$. Lempert was motivated by the $\overline{\partial}$ problem. The Banach space $\overline{\partial}$ problem was an important research topic during the period 1970–1980 and partial solutions under additional hypotheses were obtained using Gaussian measures. These led to a complete solution on pseudo-convex domains in $\mathcal{DFN}$ spaces and to (DN) as a necessary condition for a solution in Fréchet nuclear spaces. During the last ten years the non-Banach space theory has been further developed by R. Soraggi [812, 813]. A counterexample of G. Coeuré [239] for $\mathcal{C}^1$ functions in 1978 pointed to intrinsic difficulties in the Banach space case and it is only recently that L. Lempert has made fundamental progress for such spaces. The programme outlined in [543] will be regarded as seminal for infinite dimensional holomorphy and a strong motivation for a renewed study of the $\overline{\partial}$ problem. The unrestricted $\overline{\partial}$ problem for $(0,1)$ forms on balls in l_1 is solved in [545] (see also [546, 547]). The solution uses Proposition 4.56 and does not involve Gaussian measures. We refer to P. Mazet [597] for further background on the infinite dimensional $\overline{\partial}$ problem.

The penultimate result in this chapter is due to S. Dineen–R.M. Timoney [326] and the final proposition may be found in A. Defant–L.A. Moraes [261] (see also [265] and [46, Remark 1.7]). These results together with the positive results for fully nuclear spaces, given in Section 4.3, show precisely the practical consequences for holomorphic functions in placing a nuclearity hypothesis on the domain (see our comments on nuclearity in Section 3.2). Although not apparent from the main text, the theme and proof of Proposition 4.58 have connections with a number of unexpected topics in complex and functional analysis. Among these we list:

(1) Banach space invariants,
(2) norms of random matrices,

(3) Pisier's counterexample to the ε–π conjectures of Grothendieck and Pietsch (see Section 2.6),
(4) Bohr's inequality,
(5) regions of uniform and absolute convergence of Dirichlet series.

The classical matrix inequality that we have used (4.44) can be found in G.H. Hardy–J.E. Littlewood [437] and J.E. Littlewood [562]. This was proved in the 1930s although an earlier calculation, along the same lines, with $p = \infty$, by O. Toeplitz [826] would have been sufficient for our purposes. In recent years (see [206] and [568] and their references) the estimate (4.44) has been placed in a modern context and shown to be relevant and sharp in calculating (either precisely or asymptotically) such *Banach space invariants* as type and cotype constants, approximation numbers, Banach–Mazur distance, maximum volume ellipsoids, projection norms, tensor norms etc.

To establish contact with these topics it is necessary to reinterpret (4.44) and (4.46). If we let $\tilde{A}_n$ denote the tensor naturally associated with A_n in (4.44), i.e. $\tilde{A}_n = \sum_{j,k=1}^{n} a_{jk} e_j \otimes e_k$, where $(e_j)_j$ is the usual unit vector basis for l_p, then (4.44) can be shown to imply without much difficulty

$$\|\tilde{A}_n\|_{l_p^n \otimes_\varepsilon l_p^n} \sim n^{a(p)}$$

and, using duality,

$$\|\tilde{A}_n\|_{l_p^n \otimes_\pi l_p^n} \geq n^{2-a(p')}$$

where $\frac{1}{p} + \frac{1}{p'} = 1$ and $a(p)$ is defined as in (4.44). These estimates imply

$$\|I_n : l_p^n \otimes_\varepsilon l_p^n \longrightarrow l_p^n \otimes_\pi l_p^n\| \geq n^{2-a(p)-a(p')} \qquad (4.49)$$

where I_n is the identity mapping.

Grothendieck's $\varepsilon - \pi$ *conjecture* asserts that $E \widehat{\otimes}_\varepsilon F = E \widehat{\otimes}_\pi F$ only when at least one of E or F is nuclear, while Pietsch conjectured that $E \widehat{\otimes}_\varepsilon E = E \widehat{\otimes}_\pi E$ if and only if E is nuclear. Clearly these have consequences for the relationship between the ε and π norms on well located subspaces of a Banach space and, indeed, using (4.49) one easily sees that a Banach space, which contains uniformly complemented $(l_p^n)_n$ is not a counterexample to *Pietsch's conjecture*. The Dvoretzky Theorem, that we discussed in Chapter 1, shows that any infinite dimensional Banach space contains almost isometric copies of $(l_2^n)_n$ but, in general, these will not be well located. Pisier's [730] counterexample $\tilde{E}$ to both conjectures has the property that there exists $c > 0$ such that if F is any finite dimensional subspace of $\tilde{E}$ and π is a projection from $\tilde{E}$ onto F then $\|\pi\| \geq c\sqrt{\text{Dimension }(F)}$. The monograph [731] is an extremely well-written but technically demanding study of these and other topics connected with the local structure and geometry of Banach spaces. The expository article [329] develops some of the above material for the non-expert.

Another approach to showing the existence of $n \times n$ matrices A_n with norms of a certain size is to consider the entries a_{ij} as random variables on some probability space – in this case A_n is called a *random matrix* – and to use probabilistic methods to estimate the *expected value* of the norm. For example, if we suppose $\alpha_{j,k} = \pm 1$ with equal probability and that the random variables $(\alpha_{j,k})_{1 \leq j,k \leq n}$ are *independent* then it can be shown that there exist positive constants A and B such that

$$An^{1/2} \leq E\left[\|A_n\|_2\right] \leq Bn^{1/2} \tag{4.50}$$

(see for instance [103] and [568]). Inequality (4.50) is clearly related to (4.44) when $p = 2$. In many cases one is interested in showing the existence of matrices (or bilinear forms, polynomials or tensors) for which there is a certain relationship between the coefficients (or individual terms) and the norm which depends on all coefficients. An estimate such as (4.50) shows that there exist matrices A_n with ± 1 as entries for which $\|A_n\| \sim n^{1/2}$.

In a different direction the authors in [326] noted a connection between absolute basis and the following result of H. Bohr [144] (see also P. Dienes [268, p.445]):

if the power series $\sum_{n=0}^{\infty} a_n z^n$ converges for all z in the unit disc in $\mathbb{C}$ and $\sup_{|z|<1} \left| \sum_n a_n z^n \right| < 1$ then $\sum_{n=0}^{\infty} |a_n||z^n| < 1$ for $|z| < 1/3$ and $1/3$ is best possible.

Using an n-tensor version of (4.50) given in [568, Theorem 1.1], S. Dineen–R.M. Timoney proved the following for $r = (r_1, \ldots, r_n)$, $r_j > 0$, $j = 1, \ldots, n$:

if $\displaystyle\sum_{m \in \mathbb{N}^n} |a_m| r^m \leq 1$ for all $(a_m)_{m \in \mathbb{N}^n}$ satisfying

$$\left\| \sum_{m \in \mathbb{N}^n} a_m z^m \right\|_{\{z = (z_j)_j : |z_j| < 1\}} < 1$$

then for all $\varepsilon > 0$ there exists $c(\varepsilon) > 0$ such that

$$\sum_{j=1}^{n} r_j \leq c(\varepsilon) n^{\frac{1}{2}+\varepsilon}. \tag{4.51}$$

The result in [568] is stated for *arbitrary* tensors but for random *symmetric* tensors the entries are not all *independent* and while the analogous result is true the constants involved change. The proof in [326], using the result in [568, Theorem 1.1], does give a correct proof of (4.51) but the claim in [326] that the proof and estimate extend to $\varepsilon = 0$ is incorrect. This was observed and kindly pointed out to the authors of [326] by H.P. Boas and D. Khavinson. In an interesting paper [134], H.P. Boas–D. Khavinson consider the n-dimensional Bohr inequality and using random trigonometric series in n variables proved directly the following improved version of (4.51):

$$\sum_{j=1}^{n} r_j \leq 2\sqrt{n \log n}$$

and showed that one cannot take $\varepsilon = 0$ in (4.51).

The story continues. In 1913, H. Bohr [143] studied the relationship between the half planes of uniform and absolute convergence of a Dirichlet series. It is surprising that the difference between such half-planes may contain a vertical strip of *positive* width. H. Bohr [143] proved that the width d of a maximal strip satisfies $d \leq 1/2$ and examples of O. Toeplitz [826] showed that $d \geq 1/4$. Bohr posed the problem of finding the actual value of d and this was settled in 1931 by H.F. Bohnenblust–E. Hille [142] who showed, using infinite dimensional holomorphy (see Section 3.6), that $d = 1/2$. In [327], S. Dineen–R.M. Timoney noted a connection between this problem and (4.51). Once more we used the incorrect $\varepsilon = 0$ case, but again the correct $\varepsilon > 0$ result and the method used are sufficient to show $d \geq 1/2 - \varepsilon$ for any $\varepsilon > 0$ and to give an independent solution to Bohr's problem. Recently, H.P. Boas has written an extremely informative and interesting article [133] aimed at making the results of Bohr on conditional, uniform and absolute convergence of Dirichlet series available to a wider audience.

Chapter 5. Riemann Domains

In comparing locally convex space structures on spaces of holomorphic functions in previous chapters we relied on expansions which converged everywhere, at least pointwise, to the functions under consideration. This confined our investigation to balanced domains but allowed us to make effective use of concepts and techniques developed in linear functional analysis. This approach is no longer adequate for arbitrary open sets and it is necessary to introduce ideas and methods of a non-linear kind which originated in several complex variables theory. These were associated in the finite dimensional theory with the Cartan–Thullen Theorem, the Levi problem and the $\bar{\partial}$ problem and were developed by exploiting the algebraic structure of the space of holomorphic functions and the principle of analytic continuation. Neither of these arise in studying linear functions or polynomials since polynomials can always be extended to the whole space and the space of n-homogeneous polynomials is not an algebra under pointwise multiplication. Our objective in this chapter, vis-à-vis the topological investigations in previous chapters, is to prove the following result:

$$\text{if } U \text{ is an arbitrary open subset of a Fréchet–Schwartz space with} \atop \text{the bounded approximation property then } \tau_0 = \tau_\omega \text{ on } H(U). \qquad (*)$$

We obtained this result for convex balanced domains in Example 4.40(b). The structure that we develop as we proceed towards this result is as important as the result itself and this chapter provides, for the functional analyst, a suitable introduction to non-linear aspects of infinite dimensional holomorphy.

In Section 5.1 we discuss $(H(K), \tau_0)$ for K an arbitrary compact subset of a Fréchet space. In this section we extend results given in Section 4.3. The methods are functional-analytic and the main result states that $(H(K), \tau_0)$ is a k-space when K is an arbitrary compact subset of a Fréchet space. Section 5.2 is devoted to an ab-initio study of Riemann domains.

We mention briefly how this fits in with our approach to the $\tau_0 = \tau_\omega$ problem on $H(U)$ by outlining, *without specifying the technical hypotheses needed* – these will appear in the course of the proofs – the main steps in establishing $(*)$.

We develop the theory of holomorphic functions on Riemann domains spread over a locally convex space. For (Ω, p) a Riemann domain we establish the following:

- the τ_0-spectrum of $H(\Omega)$, $S_c(\Omega)$, has the structure of a pseudo-convex Riemann domain,
- $S_c(\Omega)$ is a holomorphic extension of Ω,
- $S_c(\Omega)$ is holomorphically convex,
- $S_c(\Omega)$ has a fundamental system of compact sets each of which contains a fundamental neighbourhood system consisting of open sets satisfying a strong Oka–Weil approximation property,
- if K is a compact subset of $S_c(\Omega)$ then $\tau_0 = \tau_\omega$ on $H(K)$
- $\left(H\big(S_c(\Omega)\big), \tau_\omega \right) = \varprojlim_{K \in S_c(\Omega)} \big(H(K), \tau_\omega\big) = \big(H(\Omega), \tau_\omega\big)$ and
- $\left(H\big(S_c(\Omega)\big), \tau_0 \right) = \varprojlim_{K \in S_c(\Omega)} \big(H(K), \tau_0\big) = \big(H(\Omega), \tau_0\big)$.

It is clear, from the final result, that the general idea is to lift the $\tau_0 = \tau_\omega$ problem from a domain to its spectrum and to use the additional structure in the spectrum to solve the problem. Riemann domains are necessary since it is not always possible, even in the finite dimensional case, to identify the τ_0-spectrum of $H(U)$, U an open subset of E, with an open subset of E. We could, however, avoid the use of Riemann domains by confining our attention to pseudo-convex open subsets of a locally convex space but this approach, although quite general and not unnatural, would have avoided important aspects of infinite dimensional holomorphy. Before proceeding we remind the reader that *the above results are not true without certain hypotheses and these are presented at appropriate times throughout the chapter.* The problem of showing that pseudo-convex domains are domains of existence is known as the *Levi problem*.

5.1 Holomorphic Germs on a Fréchet Space

In Section 4.3 we discussed holomorphic germs on a compact *balanced* subset of a Fréchet space. In this section we examine in detail the space $\big(H(K), \tau_0\big)$ for arbitrary compact K. Our first goal is to obtain a projective description of $\big(H(K), \tau_0\big)$.

Lemma 5.1 *Let K denote a compact subset of the Fréchet space E. The semi-norms*

$$p(f) := \sum_{n=0}^{\infty} \varepsilon_n^n \sup_{x \in K} \left\| \frac{\widehat{d^n} f(x)}{n!} \right\|_L \tag{5.1}$$

and

$$q(f) := \sup_{k} \sup_{1 \le n \le n_k} 2^n \left| \sum_{m=0}^{n} \frac{\widehat{d^m} f(x_k)}{m!}(a_k) - \sum_{m=0}^{n} \frac{\widehat{d^m} f(y_k)}{m!}(b_k) \right| \tag{5.2}$$

where, L is a compact balanced subset of E, $(\varepsilon_n)_n$ is a non-negative sequence in c_0, $(x_k)_k$ and $(y_k)_k$ are sequences in K, $(a_k)_k$ and $(b_k)_k$ are null sequences in E, $x_k + a_k = y_k + b_k$ for all k and $(n_k)_k$ is a strictly increasing sequence of positive integers, are continuous on $\big(H(K), \tau_0\big)$.

Proof. We must show that both p and q restricted to $\big(H(K + V), \tau_0\big)$ are continuous for any convex balanced open subset V of E. We first consider p. There exists $r > 0$ such that $rL \subset V$. By the Cauchy inequalities, (3.12),

$$p(f) \le \sum_{n=0}^{\infty} \varepsilon_n^n \cdot \frac{1}{r^n} \|f\|_{K+rL} =: c\|f\|_{K+rL}$$

for all $f \in H(K + V)$. Since $K + rL$ is a compact subset of $K + V$ this completes the proof for p.

We now consider q. Since $(a_k)_k$ and $(b_k)_k$ are null sequences in E we may choose k_0 such that $2a_k$ and $2b_k$ lie in V for all $k \ge k_0$. Let

$$A = \{\lambda x : |\lambda| \le 1, \ x = a_k \text{ or } b_k, \ k \ge k_0\}.$$

Then A is a compact balanced subset of E and $2A \subset V$. If $f \in H(K+V)$, $k \ge k_0$ and $1 \le n \le n_k$ then

$$\left| \sum_{m=0}^{n} \frac{\widehat{d^m} f(x_k)}{m!}(a_k) - \sum_{m=0}^{n} \frac{\widehat{d^m} f(y_k)}{m!}(b_k) \right|$$

$$= \left| f(x_k + a_k) - \sum_{m=n+1}^{\infty} \frac{\widehat{d^m} f(x_k)}{m!}(a_k) - f(y_k + b_k) \right.$$

$$\left. + \sum_{m=n+1}^{\infty} \frac{\widehat{d^m} f(y_k)}{m!}(b_k) \right|$$

$$\le 2 \sum_{m=n+1}^{\infty} \sup_{x \in K} \left\| \frac{\widehat{d^m} f(x)}{m!} \right\|_A$$

$$= 2 \sum_{m=n+1}^{\infty} \frac{1}{2^m} \sup_{x \in K} \left\| \frac{\widehat{d^m} f(x)}{m!} \right\|_{2A}$$

$$\le 2 \left(\sum_{m=n+1}^{\infty} \frac{1}{2^m} \right) \|f\|_{K+2A}$$

On the other hand, if $B = \{\lambda x; |\lambda| \le 1, \ x = a_k \text{ or } b_k, \ k < k_0\}$, then for $f \in H(K + V)$, $k < k_0$ and $1 \le n \le n_k$ we have

$$\left| \sum_{m=0}^{n} \frac{\widehat{d^m f}(x_k)}{m!}(a_k) - \sum_{m=0}^{n} \frac{\widehat{d^m f}(y_k)}{m!}(b_k) \right| \leq 2 \sum_{m=0}^{n} \sup_{x \in K} \left\| \frac{\widehat{d^m f}(x)}{m!} \right\|_B$$

$$\leq 2 \left(\sum_{m=0}^{n} r^{-m} \right) \|f\|_{K+rB}$$

where $0 < r \leq 1$ is chosen so that $rB \subset V$. Let $C = 2A \cup rB$. We have shown

$$q(f) \leq 2\|f\|_{K+C} \cdot 2^{n_{k_0}} \cdot \sum_{m=0}^{n_{k_0}} r^{-m}$$

for all $f \in H(K + V)$. Since C is a compact subset of V this completes the proof.

For our next lemma we fix some notation. Let $(V_j)_j$ denote a fundamental system of neighbourhoods of the origin in the Fréchet space E consisting of convex balanced open sets with $jV_j \subset V_{j-1}$ for $j \geq 2$. Let K denote a compact subset of E and let

$$\mathcal{K}_j = \{ f \in H^\infty(K + V_j) : \|f\|_{K+V_j} \leq j \}.$$

By Lemma 3.25, $\mathcal{K}_j$ is a compact subset of $(H(K + V_j), \tau_0)$ and hence also a compact subset of $(H(K), \tau_0)$. If A is a compact balanced subset of E, n_1 is a positive integer and $\varepsilon > 0$ we let

$$U(A, n_1, \varepsilon) = \left\{ f \in H(K) : \sup_{x \in K} \left\| \frac{\widehat{d^n f}(x)}{n!} \right\|_A \leq \varepsilon \text{ for } 0 \leq n \leq n_1 \right\}.$$

If V is a neighbourhood of 0 in E a point $(x, a, y, b) \in (K \times V)^2$ is called a *point of ambiguity* if $x + a = y + b$. A subset S of $(K \times V)^2$ consisting of points of ambiguity is called a *set of ambiguity*. For each set of ambiguity S and each positive integer n_2 let

$$W(S, n_2) = \left\{ f \in H(K) : 2^n \left| \sum_{m=0}^{n} \frac{\widehat{d^m f}(x)}{m!}(a) - \sum_{m=0}^{n} \frac{\widehat{d^m f}(y)}{m!}(b) \right| \leq 1, \right.$$

$$\left. \text{for } (x, a, y, b) \in S \text{ and } n \leq n_2 \right\}.$$

By Lemma 5.1, $U(A, n_1, \varepsilon)$ and $W(S, n_2)$ are closed neighbourhoods of 0 in $(H(K), \tau_0)$.

Lemma 5.2 *Let G denote a subset of $H(K)$ containing the origin and suppose $G \cap \mathcal{K}_j$ is an open subset of $\mathcal{K}_j$ (with the induced topology from $(H(K), \tau_0)$) for all j. Then there exists:*

(a) $\varepsilon > 0$ and a strictly increasing sequence of positive integers $(n_j)_{j=0}^\infty$,

(b) a sequence of compact balanced subsets of E, $(A_j)_{j=0}^{\infty}$, with $A_j \subset 2V_j$ for $j \geq 1$,

(c) a sequence $(S_j)_{j=1}^{\infty}$ of finite sets of ambiguity, $S_j \subset (K \times V_j)^2$ for all $j \geq 1$, such that for $j \geq 0$

$$K_{j+1} \cap D_j \subset G \qquad (5.3)_j$$

where

$$D_j = \bigcap_{i=0}^{j} U(A_i, n_i, \varepsilon) \cap \bigcap_{i=1}^{j} W(S_i, n_i)$$

(for $j = 0$ the intersection $\bigcap_{i=1}^{j} W(S_i, n_i)$ is taken as $H(K)$).

Proof. The proof is by induction on j. We first consider $j = 0$. Since $G \cap K_1$ is τ_0 open in K_1 there exists a compact balanced subset $A \subset V_1$ and $0 < \varepsilon \leq 1/2$ such that

$$\{f \in K_1 : \|f\|_{K+A} \leq 3\varepsilon\} \subset G. \qquad (5.4)$$

Choose $r > 1$ such that $rA \subset V_1$. By the Cauchy estimates we have

$$\sup_{x \in K} \left\| \frac{\widehat{d^n f(x)}}{n!} \right\|_A \leq \frac{1}{r^n} \|f\|_{K+rA} \leq \frac{1}{r^n}$$

for all $f \in K_1$ and all $n \in \mathbb{N}$. Hence there exists $n_0 \geq 2$ such that

$$\sup_{f \in K_1} \sum_{n=n_0+1}^{\infty} \sup_{x \in K} \left\| \frac{\widehat{d^n f(x)}}{n!} \right\|_A \leq \varepsilon. \qquad (5.5)$$

Let $A_0 = n_0 A$. From (5.4) and (5.5) we see that

$$\left\{ f \in K_1 : \sup_{x \in K} \left\| \frac{\widehat{d^n f(x)}}{n!} \right\|_{A_0} \leq \varepsilon \text{ for } 0 \leq n \leq n_0 \right\}$$

$$\subset \left\{ f \in K_1 : \|f\|_K \leq \varepsilon \text{ and } \sup_{x \in K} \left\| \frac{\widehat{d^n f(x)}}{n!} \right\|_A \leq \frac{\varepsilon}{2^n}, 1 \leq n \leq n_0 \right\}$$

$$\subset \{f \in K_1 : \|f\|_{K+A} \leq 3\varepsilon\} \subset G.$$

Hence $K_1 \cap U(A_0, n_0, \varepsilon) \subset G$.

Next let $j \geq 1$ and suppose we have found $(A_i)_{i=0}^{j-1}$, $(n_i)_{i=0}^{j-1}$ and $(S_i)_{i=1}^{j-1}$ which satisfy $(5.3)_{j-1}$ but that we cannot find n_j, S_j and A_j to satisfy $(5.3)_j$. Then for each compact set $B \subset 2V_j$, each $n > n_{j-1}$ and each finite set of ambiguity S we have

$$K_{j+1} \cap D_{j-1} \cap U(B, n, \varepsilon) \cap W(S, n) \not\subset G.$$

The set $(K_{j+1} \setminus G) \cap D_{j-1}$ is a closed, and hence compact, subset of K_{j+1} and the sets $D_{j-1} \cap U(B, n, \varepsilon) \cap W(S, n)$ are closed non-empty subsets of D_{j-1} with the finite intersection property. Hence

$$(\mathcal{K}_{j+1} \setminus G) \cap D_{j-1} \cap \bigcap_{B,n,S} (U(B,n,\varepsilon) \cap W(S,n)) \neq \emptyset.$$

Let f belong to this intersection. Since $f \in \bigcap_{B,n} U(B,n,\varepsilon)$ for all $B \subset 2V_j$ we have

$$\sup_{x \in K} \left\| \frac{\widehat{d^n} f(x)}{n!} \right\|_{2V_j} \leq \varepsilon \qquad (5.6)$$

for all n and as $f \in \bigcap_{S,n} W(S,n)$ it follows that

$$\left| \sum_{m=0}^{n} \frac{\widehat{d^m} f(x)}{m!}(a) - \sum_{m=0}^{n} \frac{\widehat{d^m} f(y)}{m!}(b) \right| \leq \frac{1}{2^n} \qquad (5.7)$$

for all $n > n_{j-1}$ and all points of ambiguity $(x,a,y,b) \in (K \times V_j)^2$. By (5.6), we can define for each x in K, $f_x \in H(x + V_j)$, by the formula

$$f_x(x + a) = \sum_{n=0}^{\infty} \frac{\widehat{d^n} f(x)}{n!}(a)$$

and $\|f_x\|_{x+V_j} \leq 2\varepsilon < 1$. On the other hand if (x,a,y,b) is a point of ambiguity then (5.7) implies that $f_x(x+a) = f_y(y+b)$ and $f_x \equiv f_y$ on $(x+V_j) \cap (y+V_j)$ for all x,y in K. Hence $f \in H(K + V_j)$ and $\|f\|_{K+V_j} \leq 1$. This shows that $f \in \mathcal{K}_j \cap D_{j-1}$ but $f \notin G$. This contradicts $(5.3)_{j-1}$. Hence $(5.3)_{j-1}$ implies $(5.3)_j$ and, by induction, this completes the proof.

Proposition 5.3 *Let K denote a compact subset of a Fréchet space E. A subset G of $(H(K), \tau_0)$ is open if and only if $G \cap \mathcal{K}_j$ is open in $\mathcal{K}_j$ (with the induced topology from $(H(K), \tau_0)$) for all j. Moreover, the set of all seminorms satisfying (5.1) and (5.2) generate the topology of $(H(K), \tau_0)$.*

Proof. We keep the notation of Lemma 5.2. Suppose $G \cap \mathcal{K}_j$ is open in $(\mathcal{K}_j, \tau_0)$ for all j. We first suppose that $f_0 = 0 \in G$. Let $B_j = \bigcup_{i=j}^{\infty} A_i \cup \{0\}$ for $j \geq 0$. For each j, B_j is a compact subset of $2V_j$ and since $jV_j \subset V_{j-1}$ for all $j \geq 2$ it follows that

$$L := B_0 \cup \bigcup_{j=1}^{\infty} jB_j$$

is a compact subset of E. Let $n_{-1} = 0$ and for $j \geq 0$ let

$$R_j = \left\{ f \in H(K) : \sup_{x \in K} \left\| \frac{\widehat{d^n} f(x)}{n!} \right\|_{B_j} \leq \varepsilon \text{ for } n_{j-1} < n \leq n_j \right\}.$$

We have

$$\bigcap_{j=0}^{i} R_j \subset U(A_i, n_i, \varepsilon) \tag{5.8}$$

for $i \geq 0$. Let $\varepsilon_n = 1$ for $0 \leq n \leq n_0$, $\varepsilon_n = \dfrac{1}{j}$ for $n_{j-1} < n \leq n_j$, $j \geq 1$, and let

$$p(f) = \sum_{n=0}^{\infty} \varepsilon_n^n \sup_{x \in K} \left\| \widehat{\frac{d^n f(x)}{n!}} \right\|_L$$

for $f \in H(K)$. The semi-norm p satisfies (5.1) and hence is τ_0-continuous on $H(K)$. If $f \in H(K)$, $p(f) \leq \varepsilon$ and $n_{j-1} < n \leq n_j$ then

$$\sup_{x \in K} \left\| \widehat{\frac{d^n f(x)}{n!}} \right\|_{B_j} = \left(\frac{1}{j} \right)^n \sup_{x \in K} \left\| \widehat{\frac{d^n f(x)}{n!}} \right\|_{jB_j}$$

$$\leq \varepsilon_n^n \sup_{x \in K} \left\| \widehat{\frac{d^n f(x)}{n!}} \right\|_L, \quad \text{since } jB_j \subset L \text{ and } \varepsilon_n = \frac{1}{j},$$

$$\leq \varepsilon.$$

Hence $f \in R_j$ for all j and, by (5.8),

$$\{ f : p(f) \leq \varepsilon \} \subset \bigcap_{i=0}^{\infty} U(A_i, n_i, \varepsilon). \tag{5.9}$$

Let $l_0 = 0$. Now order the sequence of finite sets of ambiguity into a sequence $(x_k, a_k, y_k, b_k)_{k=0}^{\infty}$ where $S_i = (x_k, a_k, y_k, b_k)_{k=l_{i-1}}^{l_i}$, $i = 1, \ldots$. Let $m_k = n_i$ if $l_{i-1} < k \leq l_i$ for $i \geq 1$ and let

$$q(f) = \sup_{k} \sup_{1 \leq n \leq m_k} 2^n \left| \sum_{m=0}^{n} \widehat{\frac{d^m f(x_k)}{m!}} (a_k) - \sum_{m=0}^{n} \widehat{\frac{d^m f(y_k)}{m!}} (b_k) \right|$$

for $f \in H(K)$. Since q satisfies (5.2) it is τ_0 continuous on $H(K)$. Clearly

$$\{ f \in H(K) : q(f) \leq 1 \} \subset \bigcap_{i=1}^{\infty} W(S_i, n_i). \tag{5.10}$$

By Lemma 5.2, (5.9) and (5.10) we have

$$\mathcal{K}_{j+1} \cap \{ f \in H(K) : p(f) \leq \varepsilon \text{ and } q(f) \leq 1 \}$$

$$\subset \mathcal{K}_{j+1} \cap \bigcap_{i=0}^{\infty} U(A_i, n_i, \varepsilon) \cap \bigcap_{i=1}^{\infty} W(S_i, n_i) \subset G$$

for all j. Since $\bigcup_{j=1}^{\infty} \mathcal{K}_j = H(K)$ this shows that

$$\{ f \in H(K) : p(f) \leq \varepsilon \text{ and } q(f) \leq 1 \} \subset G$$

and G is a neighbourhood of $f_0 = 0$.

If f_0 is an arbitrary point in G then, given $j \in \mathbb{N}$, there exists $i \geq j$ such that $\mathcal{K}_j + f_0 \subset \mathcal{K}_i$. Since $G \cap \mathcal{K}_i$ is open in $\mathcal{K}_i$ there exists U open in $(H(K), \tau_0)$ such that $G \cap \mathcal{K}_i = U \cap \mathcal{K}_i$. Hence

$$
\begin{aligned}
(G - f_0) \cap \mathcal{K}_j &= \{G \cap (\mathcal{K}_j + f_0)\} - f_0 \\
&= \{U \cap (\mathcal{K}_j + f_0)\} - f_0 \\
&= (U - f_0) \cap \mathcal{K}_j
\end{aligned}
$$

and $(G - f_0) \cap \mathcal{K}_j$ is τ_0 open in $\mathcal{K}_j$. By the first part of the proof, $G - f_0$ is a neighbourhood of the origin in $(H(K), \tau_0)$ and G is open in $(H(K), \tau_0)$. Since the converse is trivial and we only required semi-norms of the form (5.1) and (5.2) in the proof, this completes the proof.

We recall that a topological space X is a *k-space* if its topology is localized on its compact sets, i.e. $U \subset X$ is open if and only if $U \cap K$ is open in K, with the induced topology, for each compact subset K of X. A mapping from a k-space into a topological space is continuous if and only if its restriction to each compact set is continuous. In the notation of Chapter 2 this says that hypocontinuous functions on a k-space are continuous.

Our next result shows the connection between the τ_0 and τ_ω topologies on $H(K)$ and extends Lemma 4.33.

Proposition 5.4 *If K is a compact subset of a Fréchet space E then the inductive limits*

$$
(H(K), \tau_\omega) = \varinjlim_{\substack{V \supset K \\ V \text{ open}}} (H(V), \tau_\omega) = \varinjlim_{\substack{V \supset K \\ V \text{ open}}} (H^\infty(V), \|\cdot\|_V)
$$

and

$$
(H(K), \tau_0) = \varinjlim_{\substack{V \supset K \\ V \text{ open}}} (H(V), \tau_0)
$$

are regular. The sets $\mathcal{K}_j = \{ f \in H(K + V_j) : \|f\|_{K+V_j} \leq j \}$ form a fundamental system of bounded sets for $(H(K), \tau_\omega)$ and $(H(K), \tau_0)$ and a fundamental system of compact sets for $(H(K), \tau_0)$ when $(V_j)_j$ ranges over a fundamental sequence of convex balanced open zero neighbourhoods in E.

Proof. Since $(H(K), \tau_\omega) = \varinjlim_{j} (H^\infty(K + V_j), \|\cdot\|_{K+V_j})$ is a bornological

DF space a fundamental system of bounded subsets of $(H(K), \tau_\omega)$ is given by $\{\overline{\mathcal{K}_j}^{(H(K),\tau_\omega)}\}_j$. By Montel's Theorem $\mathcal{K}_j$ is compact in $(H(K + V_j), \tau_0)$ and hence in $(H(K), \tau_0)$. Since $\tau_\omega \geq \tau_0$ it follows that $\mathcal{K}_j$ is closed in $(H(K), \tau_\omega)$ and this proves the result for $(H(K), \tau_\omega)$. To complete the proof it suffices to show that $(\mathcal{K}_j)_j$ is a fundamental system of bounded sets for $(H(K), \tau_0)$. If B is a bounded subset of $(H(K), \tau_0)$ then, using semi-norms satisfying (5.1) and arguing by contradiction, we can show that there exist $c_1 > 0$, $c_2 > 0$ and $j \in \mathbb{N}$ such that

$$\left\| \frac{\widehat{d^n} f(x)}{n!} \right\|_{V_j} \leq c_1 c_2^n$$

for all $f \in B$, $x \in K$ and $n = 0, 1 \dots$. Using these estimates and semi-norms which satisfy (5.2) it follows that there exists $\varepsilon > 0$ such that

$$\sum_{n=0}^{\infty} \frac{\widehat{d^n} f(x)}{n!}(y) = \sum_{n=0}^{\infty} \frac{\widehat{d^n} f(x_1)}{n!}(y_1)$$

for all $f \in B$, all x, $x_1 \in K$, y, $y_1 \in \varepsilon V_j$ such that $x + y = x_1 + y_1$. Hence $B \subset \mathcal{K}_j$ for some j and this completes the proof.

We rephrase the first part of Proposition 5.3 as follows.

Corollary 5.5 *If K is a compact subset of a Fréchet space then $(H(K), \tau_0)$ is a k-space.*

Corollary 5.6 *If K is a compact subset of a Fréchet space then $\tau_\omega = \tau_0$ on $H(K)$ if and only if $\tau_0 = \tau_\omega$ on the bounded subsets of $H(K)$.*

Proof. Since $\tau_\omega \geq \tau_0$ we have $\tau_0 = \tau_\omega$ if and only if the identity mapping

$$(H(K), \tau_0) \longrightarrow (H(K), \tau_\omega)$$

is continuous. An application of Proposition 5.3 completes the proof.

Our next result extends Proposition 3.48 to balanced domains and removes an assumption in Corollary 4.39 for Fréchet–Montel spaces.

Proposition 5.7 *If U is a balanced open subset of a Fréchet–Montel space E then $\tau_0 = \tau_\omega$ on $H(U)$ if and only if E has the $(DB)_\infty$-property.*

Proof. We have already noted, in Propositions 1.36 and 3.48, that $\tau_0 = \tau_\omega$ on $H(U)$ implies E has $(BB)_\infty$. Conversely, suppose E has $(BB)_\infty$. By Proposition 1.36, $\tau_0 = \tau_\omega$ on $\mathcal{P}(^n E)$ for all n. By Proposition 5.4, τ_0 and τ_ω define the same bounded subsets of $H(K)$ and, by Proposition 3.34(c), τ_0 and τ_ω coincide on these bounded subsets. An application of Corollary 5.6 completes the proof.

The Köthe sequence spaces $\lambda^p(A)$ (let $X_j = \mathbb{C}$ for all j in Example 4.40(d)) are Fréchet space generalizations of the classical l_p spaces (if $a_{j,k} = 1$ for all j and k then $\lambda^p(A) \cong l_p$) and have played an important role in the modern structure theory of Fréchet spaces. If A satisfies the Grothendieck–Pietsch criterion then $\lambda^p(A) \cong \lambda(A)$ is a Fréchet nuclear space. Similar criteria, which characterize the Montel and Schwartz $\lambda^p(A)$ spaces are known. If $\lambda^p(A)$ is not Montel then $\lambda^p(A)$ contains l_p as a complemented subspace and hence, by Corollary 1.56 or various results in Section 2.4, there exists a positive integer n such that $l_\infty \longleftrightarrow (\mathcal{P}(^n\lambda^p(A)), \tau_b)$. Conversely, if $\lambda^p(A)$ is Montel then, since $\lambda^p(A)$ is known to have $(BB)_\infty$, we have $\tau_0 = \tau_b = \tau_\omega$ by Proposition 5.7. In particular, $(\mathcal{P}(^n\lambda^p(A)), \tau_0)$ is reflexive and does not contain l_∞. We have proved the following result.

Corollary 5.8 *The following are equivalent conditions on the Köthe sequence space* $\lambda^p(A)$, $1 \le p < \infty$:

(a) $\tau_0 = \tau_\omega$ *on* $H(U)$ *for any balanced open subset* U *of* E,

(b) $(\mathcal{P}(^n(\lambda^p(A))), \tau_b)$ *is reflexive for all* n,

(c) $l_\infty \longleftrightarrow\!\!\!/\; (\mathcal{P}(^n(\lambda^p(A))), \tau_b)$ *for any positive integer* n,

(d) $\lambda^p(A)$ *is a Fréchet–Montel space.*

For Fréchet–Schwartz spaces we obtain the following result.

Proposition 5.9 *If* K *is a compact subset of a Fréchet–Schwartz space then* τ_0 *and* τ_ω *coincide on* $H(K)$.

Proof. Let V denote a convex balanced neighbourhood of zero in E. Since E is Fréchet–Schwartz there exists a convex balanced neighbourhood W of zero in E such that $2W \subset V$ and the canonical mapping $E_W \to E_V$ is precompact. Let $0 < \varepsilon < 1$ and $c > 0$ be arbitrary. Choose $0 < \delta < 1$ so that $\delta c \le \varepsilon(1 - \delta)$ and choose A finite in W such that $W \subset A + (\delta/2)V$. If $f \in H(K + V)$ and $\|f\|_{K+V} \le c$ then, by the Cauchy inequalities,

$$\|f\|_{K+W} \le \|f\|_{K+A+(\delta/2)V}$$

$$\le \|f\|_{K+A} + \sum_{n=1}^{\infty} \sup_{x \in K+A} \left\| \frac{\widehat{d^n} f(x)}{n!} \right\|_{(\delta/2)V}$$

$$\le \|f\|_{K+A} + \sum_{n=1}^{\infty} \delta^n \|f\|_{K+A+(1/2)V} \qquad (*)$$

$$\le \|f\|_{K+A} + \frac{\delta}{1-\delta} \|f\|_{K+V}$$

$$\le \|f\|_{K+A} + \frac{c\delta}{1-\delta} \le \|f\|_{K+A} + \varepsilon.$$

Let $\mathcal{K} = \{f \in H^\infty(K+V) : \|f\|_{K+V} \le c\}$. Consider the following sequence of canonical inclusions:

$$\left(H(K+W),\tau_0\right) \cap \mathcal{K} \xrightarrow[i_1]{} \left(H^\infty(K+W), \|\cdot\|_{K+W}\right) \cap \mathcal{K}$$
$$\xrightarrow[i_2]{} \left(H(K), \tau_\omega\right) \cap \mathcal{K} \xrightarrow[i_3]{} \left(H(K), \tau_0\right) \cap \mathcal{K}.$$

By $(*)$, i_1 is continuous and the definition of inductive limit implies i_2 is continuous. Since $\tau_\omega \geq \tau_0$, i_3 is continuous. By Montel's Theorem $\mathcal{K}$ is a compact subset of $\left(H(K+W),\tau_0\right)$ and since all the above topologies are Hausdorff they all agree on $\mathcal{K}$. By Corollary 5.5, $\tau_0 = \tau_\omega$ on $H(K)$. This completes the proof.

From the proof of this proposition we obtain the following.

Corollary 5.10 *If K is a compact subset of a Fréchet–Schwartz space and V is a convex balanced neighbourhood of the origin then there exists a neighbourhood of zero W, $W \subset V$, such that a bounded sequence $(f_n)_n \subset H^\infty(K+V)$ converges uniformly on compact subsets of $K+V$ to $f \in H^\infty(K+V)$ if and only if $\|f_n - f\|_{K+W} \to 0$ as $n \to \infty$.*

5.2 Riemann Domains over Locally Convex Spaces

The highlights of this section are

(a) the relationship between a Riemann domain, Ω, and its envelope of holomorphy $\Sigma(\Omega)$ (Corollary 5.50),
(b) the Oka–Weil approximation theorem (Proposition 5.47),
(c) the solution to the Levi problem (Proposition 5.44).

The routes to the proofs of all three are interwoven and the results themselves eventually combine to prove the topological result – which was our original motivation – Corollary 5.54.

This section uses many different results from several complex variables theory and we present these *without proof* as we proceed. Excellent accounts are available in the literature and we refer the reader to the notes for details. Assuming these finite dimensional results we present *with proof* the infinite dimensional results.

Definition 5.11 A Riemann domain (or surface) spread over a locally convex space E is a pair (Ω, p) consisting of a Hausdorff topological space Ω and a local homeomorphism $p \colon \Omega \to E$.

If U is an open subset of Ω and $p\big|_U \colon U \to E$ is a homeomorphism we call (U, p), or U, a *chart* in Ω and write p_U^{-1} in place of $\left(p\big|_U\right)^{-1}$. Thus a Riemann domain over E is a manifold modelled on E together with a *projection* p which globally defines *local coordinates* on Ω. The classical examples, and these should be kept in mind, are the Riemann surfaces associated with such functions as $\sqrt{z}$ and $\log z$, $z \in \mathbb{C}$. The complex structure of the locally convex space E endows Ω with the structure of a complex manifold and thus we can define *holomorphic* and *plurisubharmonic* functions on Ω.

We always identify an open subset U of E with the Riemann domain (U, i_U) where $i_U : U \to E$ is the canonical inclusion mapping. If (Ω, p) is a Riemann domain and p is an injective mapping we call Ω a *one-sheeted* or *univalent* Riemann domain. In this case we may identify (Ω, p) and $(p(\Omega), i_{p(\Omega)})$ as Riemann domains and regard (Ω, p) as an open subset of E. Using p we introduce, locally, notation suggested by the linear structure of the underlying space. This is extremely useful – in fact one could say "almost essential" – as a means of understanding what is happening and relating results on the Riemann domain to similar results on open subsets.

If $x \in \Omega$ and V is a convex balanced subset of E then there exists at most one subset of Ω containing x which is mapped by p homeomorphically onto $p(x) + V$. When such a set exists we denote it by $x + V$. If V is open then $(x + V, p)$ is a chart. If A is a subset of Ω, V is a convex balanced subset of E and $x + V$ exists for all $x \in A$ we let

$$A + V = \bigcup_{x \in A} \{x + V\}.$$

For $K \subset \Omega$ and $\alpha \in cs(E)$ we write $B_\Omega^\alpha(K, r)$ in place of $K + \{y \in E : \alpha(y) < r\}$ and if $a \in E$ we use the notation $D_\Omega(x, a, r)$ in place of $x + \{\lambda a, |\lambda| < r\}$. To avoid confusion with the image sets in E let $B_E^\alpha(x, r) = \{y \in E : \alpha(x - y) < r\}$ and $D_E(x, y, r) = \{x + \lambda y \in E : |\lambda| < r\}$ for x, $y \in E$ and $r \geq 0$. The mapping $p_{x+V}^{-1} : p(x) + V \to x + V$ is called a *section* of Ω. If $x \in \Omega$, $A \subset \Omega$, $a \in E$ and $\alpha \in cs(E)$ let

$$d_\Omega(x, a) = \sup\{r \geq 0 : D_\Omega(x, a, r) \text{ exists}\},$$
$$d_\Omega(A, a) = \inf_{x \in A} d_\Omega(x, a),$$
$$d_\Omega^\alpha(x) = \sup\{r \geq 0 : B_\Omega^\alpha(x, r) \text{ exists}\},$$

and

$$d_\Omega^\alpha(A) = \inf_{x \in A} d_\Omega^\alpha(x).$$

These functions may be considered distances to a virtual boundary. Clearly $d_\Omega(x, a) > 0$ for all $x \in \Omega$ and $a \in E$. It may happen that $d_\Omega^\alpha(x) = 0$ but, since Ω is locally isomorphic to E, there exists for each $x \in E$, $\alpha_x \in cs(E)$, such that $d_\Omega^{\alpha_x}(x) > 0$. On the other hand we may have $d_\Omega(x, a) = \infty$ and $d_\Omega^\alpha(x) = \infty$, e.g. this occurs when $(\Omega, p) = (E, i_E)$ for all x, a and α (see Exercises 5.57, 5.72 and 5.78). The *distance functions* d_Ω and d_Ω^α are related as follows: if $d_\Omega^\alpha(x) > 0$ then

$$d_\Omega^\alpha(x) = \inf\{d_\Omega(x, a) : \alpha(a) \leq 1\}.$$

The function $d_\Omega \colon \Omega \times E \to [0, \infty]$ is lower semi-continuous, while $d_\Omega^\alpha \colon \Omega \to [0, \infty]$ is continuous. Using a limit process one verifies easily that $B_\Omega^\alpha(x, d_\Omega^\alpha(x)) \subset \Omega$ and $D_\Omega(x, a, d_\Omega(x, a)) \subset \Omega$ for $x \in \Omega$, $a \in E$ and $\alpha \in \mathrm{cs}(E)$. The proof of the following simple result contains a standard approach to analysis on (Ω, p) and essentially captures the *germ* of a "topological vector space" structure on Ω.

Lemma 5.12 *If K is a compact subset of (Ω, p), $\alpha \in \mathrm{cs}(E)$, $d_\Omega^\alpha(K) > 0$ and L is a compact subset of $B_E^\alpha(0, d_\Omega^\alpha(K))$ then $K + L$ is a compact subset of Ω and*

$$d_\Omega^\alpha(K + L) \geq d_\Omega^\alpha(K) - \sup\{\alpha(x) \; : \; x \in L\}.$$

Proof. Clearly $K + L \subset \Omega$. Consider the nets $(x_\beta)_\beta \subset K$ and $(y_\gamma)_\gamma \subset L$. Since both K and L are compact we may suppose, without loss of generality, that $x_\beta \to x \in K$ as $\beta \to \infty$ and $y_\gamma \to y \in L$ as $\gamma \to \infty$. If $0 < \sup\{\alpha(z) : z \in L\} < r < s < d_\Omega^\alpha(K)$ then

$$p\big|_{B_\Omega^\alpha(x, s)} \colon B_\Omega^\alpha(x, s) \longrightarrow B_E^\alpha(p(x), s)$$

is a homeomorphism with inverse p_1. Choose β_0 such that $x_\beta \in B_\Omega^\alpha(x, s - r)$ for all $\beta \geq \beta_0$. Hence $\alpha\big(p(x_\beta) - p(x)\big) < s - r$ and $B_E^\alpha(p(x_\beta), r) \subset B_E^\alpha(p(x), s)$ for all $\beta \geq \beta_0$. Since $p(x_\beta) + y_\gamma \to p(x) + y$ in E as β and γ tend to ∞ it follows that

$$p_1(p(x_\beta) + y_\gamma) \longrightarrow p_1(p(x) + y)$$
$$\| \qquad\qquad\qquad \|$$
$$x_\beta + y_\gamma \qquad\qquad x + y$$

as β and $\gamma \to \infty$ and $K + L$ is a compact subset of Ω. If $z \in L$ then

$$p(x) + z + B_E^\alpha(0, s - r) \subset B_E^\alpha(p(x) + z, s - r) \subset B_E^\alpha(p(x), s)$$

and

$$p_1\big(B_E^\alpha(p(x) + z, s - r)\big) = x + z + B_\Omega^\alpha(0, s - r) - B_\Omega^\alpha(x + z, s - r).$$

Hence $B_\Omega^\alpha(x + z, s - r) \subset \Omega$ and $d_\Omega^\alpha(K + L) \geq s - r$. We complete the proof by letting $s \to d_\Omega^\alpha(K)$ and $r \to \sup\{\alpha(x) : x \in L\}$.

The above can easily be modified to show the following:

if $A \subset \Omega$, $x_\beta \in A \to x \in \Omega$ as $\beta \to \infty$, $y_\gamma \in E$, $y_\gamma \to y$ as $\gamma \to \infty$ and $\limsup_{\gamma \to \infty} \alpha(y_\gamma) < d_\Omega^\alpha(A)$ for some $\alpha \in \mathrm{cs}(E)$ then $x_\beta + y_\gamma \in \Omega$ for all β and γ sufficiently large and $x_\beta + y_\gamma \to x + y$ in Ω as $\beta, \gamma \to \infty$.

We define holomorphic functions on a Riemann domain over E by locally transferring the definition to E.

A function $f: (\Omega, p) \to F$ (a locally convex space) is holomorphic if, for each chart (U, p) in Ω, $f \circ p_U^{-1} \in H(p(U); F)$.

A mapping between Riemann domains $f: (\Omega, p) \to (\widetilde{\Omega}, \widetilde{p})$ is holomorphic if $\widetilde{p} \circ f$ is holomorphic on (Ω, p). Let $H(\Omega; \widetilde{\Omega})$ denote the set of all holomorphic mappings from (Ω, p) into $(\widetilde{\Omega}, \widetilde{p})$. If $\widetilde{\Omega} = \mathbb{C}$ we write $H(\Omega)$ in place of $H(\Omega; \mathbb{C})$. Uniqueness of Taylor series expansions implies that $f: \Omega \to \mathbb{C}$ is holomorphic if and only if for each $x \in \Omega$ there exists $\alpha \in cs(E)$ and a sequence of continuous homogeneous polynomials, $(P_n)_{n=0}^{\infty}$, $P_n \in \mathcal{P}(^n E)$, such that $d_\Omega^\alpha(x) > 0$ and for *some* r, $0 < r < d_\Omega^\alpha(x)$,

$$\lim_{m \to \infty} \left\{ \sup |f(y) - \sum_{n=0}^{m} P_n(p(y) - p(x))| \ : \ y \in B_\Omega^\alpha(x, r) \right\} = 0. \qquad (5.11)$$

Since the sequence $(P_n)_{n=0}^{\infty}$ in (5.11) is uniquely determined by f and x we write $P_n = \dfrac{\widehat{d^n} f(x)}{n!}$ for all n. In particular, this means that for any chart (U, p) containing x we have

$$\frac{\widehat{d^n} f(x)}{n!} = \frac{\widehat{d^n}(f \circ p_U^{-1})}{n!}(p(x)).$$

The Cauchy inequalities transfer easily and we have

$$\left\| \frac{\widehat{d^n} f(x)}{n!} \right\|_{B_E^\alpha(0,1)} \leq \frac{1}{\rho^n} \|f\|_{B_\Omega^\alpha(x, \rho)}$$

$$\left\| \frac{\widehat{d^n} f(x)}{n!} \right\|_{D_E(0, a, 1)} \leq \frac{1}{r^n} \|f\|_{D_\Omega(x, a, r)}$$

whenever $0 \leq \rho \leq d_\Omega^\alpha(x)$ and $0 \leq r \leq d_\Omega(x, a)$. The τ_0, τ_ω and τ_δ topologies are defined in the expected way and their standard properties on $H(\Omega)$ are obtained in a straightforward fashion.

We next define morphisms between Riemann domains, use analytic continuation to define *holomorphic extensions* and introduce the concepts of *domain* and *envelope of holomorphy*.

Definition 5.13 If (Ω, p) and $(\widetilde{\Omega}, \widetilde{p})$ are Riemann domains over the locally convex spaces E and F, respectively, and $T \in \mathcal{L}(E; F)$ then a continuous mapping $j: (\Omega, p) \to (\widetilde{\Omega}, \widetilde{p})$ is called a *T-morphism* if the diagram

$$\begin{array}{ccc} \Omega & \xrightarrow{\ \ j\ \ } & \widetilde{\Omega} \\ {\scriptstyle p}\big\downarrow & & \big\downarrow{\scriptstyle \widetilde{p}} \\ E & \xrightarrow[\ \ T\ \ ]{} & F \end{array}$$

commutes.

A T-morphism j is a T-*isomorphism*, if T is invertible, j is bijective and j^{-1} is a T^{-1}-morphism. If $E = F$ and T is the identity mapping then j is automatically continuous and we use the terms morphism and isomorphism in place of T-morphism and T-isomorphism, respectively. It is easily seen that morphisms are local isomorphisms.

Definition 5.14 Let (Ω, p) denote a Riemann domain over a locally convex space E and let $\mathcal{A} \subset H(\Omega)$.

(a) A morphism $j: (\Omega, p) \to (\widetilde{\Omega}, \widetilde{p})$ is called an $\mathcal{A}$-*extension* of (Ω, p) (or just an $\mathcal{A}$-extension) if for each $f \in \mathcal{A}$ there exists a *unique* $\widetilde{f} \in H(\widetilde{\Omega})$ such that $\widetilde{f} \circ j = f$.

(b) If each $\mathcal{A}$-extension of (Ω, p) is an isomorphism of Riemann domains we call (Ω, p) an $\mathcal{A}$-*domain of holomorphy*.

(c) An $\mathcal{A}$-extension $j: (\Omega, p) \to (\widetilde{\Omega}, \widetilde{p})$ is called an $\mathcal{A}$-*envelope of holomorphy* of (Ω, p) if for each $\mathcal{A}$-extension $j_1: (\Omega, p) \to (\Omega_1, p_1)$ there exists a morphism $j_2: (\Omega_1, p_1) \to (\widetilde{\Omega}, \widetilde{p})$ such that $j = j_2 \circ j_1$.

(d) If $\mathcal{A} = H(\Omega)$ $(\mathcal{A} = \{f\})$ we use the terms *holomorphic extension* (f-holomorphic extension), *domain of holomorphy* (domain of f-holomorphy), *envelope of holomorphy* (envelope or domain of existence) in place of $\mathcal{A}$-extension, $\mathcal{A}$-domain of holomorphy and $\mathcal{A}$-envelope of holomorphy, respectively.

Diagrams are useful in dealing with the concepts introduced above. Thus a continuous mapping $j: (\Omega, p) \to (\widetilde{\Omega}, \widetilde{p})$ of Riemann domains is a morphism if the following diagram commutes:

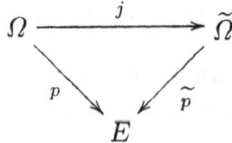

A morphism $j: (\Omega, p) \to (\widetilde{\Omega}, \widetilde{p})$ is a holomorphic extension if for each $f \in H(\Omega)$ there exists a *unique* $\widetilde{f} \in H(\widetilde{\Omega})$ such that the following diagram commutes:

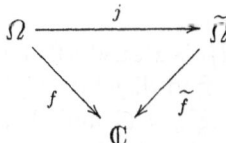

Uniqueness of the extension is easily seen to imply that $j(\Omega)$ has non-empty intersection with each connected component of $\widetilde{\Omega}$ and thus, if $(\Omega, p) = \bigcup_{i \in A}(\Omega_i, p|_{\Omega_i})$ is a partition of Ω into its connected components, then one readily sees that all holomorphic extensions of (Ω, p) are obtained by taking the disjoint union of holomorphic extensions of $(\Omega_i, p|_{\Omega_i})$, $i \in A$. Many of the problems we discuss can be solved by examining separately each component of (Ω, p) and we may often assume, without loss of generality, that (Ω, p) is connected. It is easily seen, using (c) of Definition 5.14, that any two envelopes of holomorphy of (Ω, p) are isomorphic as Riemann domains.

We use germs of holomorphic functions (see Section 4.3) to construct *the* envelope of holomorphy. Let I denote a fixed index set. For each point a in a locally convex space E consider the collection of pairs (U, ξ) where U is an open neighbourhood of a and $\xi = (\xi_i)_{i \in I} \subset H(U)$. Two such pairs (U, ξ) and (V, η) are *equivalent* if there exists a neighbourhood W of a, $W \subset U \cap V$, such that $\xi_i|_W = \eta_i|_W$ for all $i \in I$. Let H_a^I denote the set of all equivalence classes at the point a. The *germ* of (U, ξ) at a is defined to be the equivalence class containing (U, ξ) and is denoted either by ξ_a or $(U, \xi)_a$. If $|I| = 1$ we obtain the germ of a holomorphic function considered in Chapter 4.

The set H_a^I becomes an algebra when we let

$$\left(U, (\xi_i)_i\right)_a + \left(V, (\eta_i)_i\right)_a = \left(U \cap V, (\xi_i + \eta_i)_i\right)_a$$
$$\left(U, (\xi_i)_i\right)_a \cdot \left(V, (\eta_i)_i\right)_a = \left(U \cap V, (\xi_i \eta_i)_i\right)_a$$
$$\lambda\left(U, (\xi_i)_i\right)_a = \left(U, (\lambda \xi_i)_i\right)_a \text{ for } \lambda \in \mathbb{C}.$$

Consider the *disjoint* union of sets

$$H_E^I := \bigcup_{a \in E} H_a^I.$$

For each $\xi_a \in H_E^I$ let $\mathcal{N}(\xi_a)$ denote the collection of all sets $N(U, \xi)$ where

$$N(U, \xi) := \{\xi_b : b \in U\}$$

and (U, ξ) varies over all representatives of the germ ξ_a. We equip H_E^I with the unique topology such that $\mathcal{N}(\xi_a)$ is a neighbourhood base at ξ_a and let $\pi(\xi_a) = a$ for $\xi_a \in H_E^I$.

Proposition 5.15 (H_E^I, π) *is a Riemann domain over* E.

Proof. We first show that H_E^I is Hausdorff. Let ξ_a and η_b be distinct points in H_E^I. If $a \neq b$ then we can find disjoint neighbourhoods of a and b, U and V, respectively, such that $(U, \xi)_a = \xi_a$ and $(V, \eta)_b = \eta_b$. The sets $N(U, \xi)$ and $N(V, \eta)$ are disjoint neighbourhoods of ξ_a and η_b.

If $a = b$ choose a *connected* open neighbourhood W of a such that $\xi_a = (W, \xi)_a$ and $\eta_b = (W, \eta)_b$. If $N(W, \xi) \cap N(W, \eta) \neq \emptyset$ then there exists $c \in W$ such that $\xi_c = \eta_c$. By the identity principle for holomorphic functions $N(W, \xi) \equiv N(W, \eta)$ and, in particular, $\xi_a = \eta_a = \eta_b$. This contradicts our assumption that ξ_a and η_b are distinct. Hence H_E^I is Hausdorff. Clearly the mapping π is a local homeomorphism and this completes the proof.

Now let (Ω, p) denote a Riemann domain over a locally convex space E. Let $H(\Omega) = (f_i)_{i \in I}$ for some index set I and define $\tau \colon \Omega \to H_E^I$ by

$$\tau(x) := \left(p(U), (f_i \circ p_U^{-1})_{i \in I}\right)_{p(x)}$$

where (U, p) is a chart in Ω containing x. By the identity principle for holomorphic mappings τ is well defined, i.e. does not depend on the choice of chart. We denote by $\Sigma(\Omega)$ the union of those connected components of H_E^I which intersect $\tau(\Omega)$.

Proposition 5.16 $(\Sigma(\Omega), \pi)$ *is the envelope of holomorphy of* (Ω, p).

Proof. Since $\pi \circ \tau = p$ and p and π are morphisms the mapping τ is also a morphism. Let $f \in H(\Omega)$ and suppose $f = f_{i_0}$ where $i_0 \in I$. Let $g \colon \Sigma(\Omega) \to \mathbb{C}$ be defined by

$$g\left((V, (\xi_i)_{i \in I})_a\right) = \xi_{i_0}(a).$$

Since $g = \xi_{i_0} \circ \pi$ on a neighbourhood of $\left(V, (\xi_i)_{i \in I}\right)_a$, $g \in H(\Sigma(\Omega))$. If $x \in \Omega$ then

$$g \circ \tau(x) = g\left((p(U), (f_i \circ p_U^{-1})_{i \in I})_{p(x)}\right)$$
$$= f_{i_0} \circ p_U^{-1}(p(x))$$
$$= f_{i_0}(x)$$
$$= f(x).$$

Hence, each $f \in H(\Omega)$ can be extended to $\Sigma(\Omega)$ and $g \circ \tau = f$. Moreover, the extension is unique since each connected component of $\Sigma(\Omega)$ intersects $\tau(\Omega)$. This shows that $(\Sigma(\Omega), \tau)$ is a holomorphic extension of (Ω, p).

Let $j \colon (\Omega, p) \to (\widetilde{\Omega}, \widetilde{p})$ denote a holomorphic extension of (Ω, p). For each $i \in I$ there is a unique $h_i \in H(\widetilde{\Omega})$ such that $h_i \circ j = f_i$. If $y \in \widetilde{\Omega}$ let

$$\theta(y) = \left(\widetilde{p}(W), (h_i \circ \widetilde{p}_W^{-1})_{i \in I}\right)_{\widetilde{p}(y)}$$

where $(W, \widetilde{p})$ is a chart in $(\widetilde{\Omega}, \widetilde{p})$ which contains y. It is easily verified that θ is well defined and a morphism from $(\widetilde{\Omega}, \widetilde{p})$ into $(H_{E'}^I, \pi)$. If (U, p) is a chart in (Ω, p) containing x, $(W, \widetilde{p})$ is a chart in $(\widetilde{\Omega}, \widetilde{p})$ containing $j(x)$ and $j(U) = W$ then $p(U) = \widetilde{p}(W)$ and, for each $i \in I$,

$$f_i \circ p_U^{-1} = h_i \circ j \circ p_U^{-1} = h_i \circ \widetilde{p}_W^{-1}.$$

Hence

$$\theta \circ j(x) = \theta(j(x)) = \left(\widetilde{p}(W), (h_i \circ \widetilde{p}_W^{-1})_{i \in I}\right)_{\widetilde{p}(j(x))}$$
$$= \left(p(U), (f_i \circ p_U^{-1})_{i \in I}\right)_{p(x)}$$
$$= \tau(x).$$

Since each connected component of $\widetilde{\Omega}$ intersects $j(\Omega)$ we have $\theta(\widetilde{\Omega}) \subset \Sigma(\Omega)$. This completes the proof.

The proofs of Propositions 5.15 and 5.16 can easily be modified to show that each holomorphic function admits a unique, up to isomorphism of Riemann domains, envelope of holomorphy Ω_f. The classical Riemann surfaces of $\sqrt{z}$ and $\log z$ were constructed to obtain the natural domains on which to define these functions and are envelopes of holomorphy. It is clear that every domain of existence is a domain of holomorphy and, in an intuitive way, one may regard the domain of holomorphy of (Ω, p) as the intersection of Ω_f as f ranges over $H(\Omega)$ (see Exercise 5.80).

We have now defined and proved existence and uniqueness of the domains to which we will transfer the $\tau_0 = \tau_\omega$ problem. Our construction is rather abstract and to make further progress we require more concrete information. From our general results we know that each $f \in H(\Omega)$ can be extended uniquely to $\widetilde{f} \in H(\Sigma(\Omega))$ and, since we are interested in topological problems, we would like to know what happens collectively, that is to sets of functions. For instance, do τ-bounded subsets of $H(U)$ extend to τ-bounded subsets of $H(\Sigma(\Omega))$, $\tau = \tau_0$, τ_ω or τ_δ? The key is to first consider the points of $\Sigma(\Omega)$ as linear functionals on $H(\Omega)$. Points, viewed as linear functionals, are evaluations and have a special form which makes effective use of the algebraic structure of $H(\Omega)$. We begin shortly our investigation in this direction. Later we need to study special properties of $\Sigma(\Omega)$ such as pseudo-convexity and holomorphic convexity.

If $x \in \Sigma(\Omega)$ we define $\phi_x \colon H(\Omega) \to \mathbb{C}$ by letting

$$\phi_x(f) := \widetilde{f}(x)$$

where $\widetilde{f}$ is the unique extension of f to $\Sigma(\Omega)$ satisfying $\widetilde{f} \circ \tau = f$. By the principle of analytic continuation $\widetilde{f+g} = \widetilde{f} + \widetilde{g}$ and $\widetilde{fg} = \widetilde{f} \cdot \widetilde{g}$. Hence $\phi_x(f+g) = \phi_x(f) + \phi_x(g)$, $\phi_x(fg) = \phi_x(f) \cdot \phi_x(g)$ and ϕ_x defines a homomorphism from $H(\Omega)$ into $\mathbb{C}$. In general we do not know if these homomorphisms are continuous with respect to any of the topologies we consider on $H(\Omega)$ (see, however, the final two paragraphs in Section 5.4). We obtain certain results by placing restrictions on the underlying locally convex space E. For (Ω, p) a Riemann domain over a locally convex space E we let $S_b(\Omega)$ denote the set of *non-zero* τ_δ-continuous homomorphisms of $H(\Omega)$. The set $S_b(\Omega)$ is called the *bounded spectrum* of $H(\Omega)$ and elements of $S_b(\Omega)$ are called bounded homomorphisms. In some cases, e.g. when E is metrizable, the elements of $S_b(\Omega)$ are the homomorphisms which are bounded on the τ_0-bounded subsets of $H(\Omega)$.

A locally convex space E, which is also an algebra, is called a *locally m-convex* (m=multiplicative) *algebra* if it contains a neighbourhood basis at the origin $\mathcal{U} = (U_\alpha)_\alpha$ such that $U_\alpha \cdot U_\alpha = \{x \cdot y \ : \ x, y \in U_\alpha\} \subset U_\alpha$ for all α. The recent solution to the long standing *Michael Problem* states that multiplicative linear functionals on a locally m-convex Fréchet algebra over $\mathbb{C}$ are automatically continuous. This implies that all multiplicative linear functionals on $H(\Omega)$ are τ_δ-continuous and proves the following proposition. Instead of using this deep result, we prefer to give a direct self-contained proof which introduces a number of important techniques that are developed later in this chapter. Our proof requires a slightly refined inductive limit representation of $(H(\Omega), \tau_\delta)$. Let $(\alpha_n)_n$ denote a fundamental system of semi-norms for the metrizable locally convex space E and suppose $2\alpha_n \leq \alpha_{n+1}$ for all n. Let $\mathcal{C} := (C_n)_n$ denote an increasing countable open cover of Ω such that

$$C_n + B_E^{\alpha_{n+1}}(0, 2^{-n}) \subset C_{n+1} \tag{$*$}$$

for all n. The space

$$H_{\mathcal{C}}(\Omega) := \{f \in H(\Omega) \; : \; \|f\|_{C_n} < \infty \text{ for all } n\}$$

endowed with the topology of uniform convergence on each C_n, is a locally m-convex Fréchet algebra. We denote by $\mathcal{U}$ the collection of all increasing countable open covers of Ω satisfying $(*)$. Since the inclusion mapping $H_{\mathcal{C}}(\Omega) \longrightarrow (H(\Omega), \tau_\delta)$ maps bounded sets onto locally bounded and hence τ_δ bounded sets, it is continuous and the inclusion $\varinjlim_{\mathcal{C} \in \mathcal{U}} H_{\mathcal{C}}(\Omega) \longrightarrow (H(\Omega), \tau_\delta)$ is also continuous. If $\mathcal{F}$ is a locally bounded subset of $H(\Omega)$ let

$$A_n = \{x \in \Omega \; : \; d_\Omega^{\alpha_n}(x) > 2^{-n} \text{ and } \sup_{\mathcal{F}} \|f\|_{B_\Omega^{\alpha_n}(x, 2^{-n})} \leq n\}$$

and denote by B_n the interior of A_n for all n. Let $C_n = \bigcup_{i<n} B_i$. Since $\mathcal{F}$ is locally bounded $\mathcal{C} := (C_n)_n$ is an increasing open cover of Ω. Moreover, if $V_n = \{x \in E \; : \; \alpha_n(x) < 1\}$ then $C_n + \frac{1}{2^n} V_{n+1} \subset C_{n+1}$ for all n and $\mathcal{F}$ is a bounded subset of $H_{\mathcal{C}}(\Omega)$. Hence $\bigcup_{\mathcal{C} \in \mathcal{U}} H_{\mathcal{C}}(\Omega) = H(\Omega)$, $(H(\Omega), \tau_\delta) = \varinjlim_{\mathcal{C} \in \mathcal{U}} H_{\mathcal{C}}(\Omega)$, the inductive limit $\varinjlim_{\mathcal{C} \in \mathcal{U}} H_{\mathcal{C}}(\Omega)$ is regular and $(H(\Omega), \tau_\delta)$ is a locally m-convex algebra.

Proposition 5.17 *If (Ω, p) is a Riemann domain spread over a metrizable locally convex space E and $j \colon (\Omega, p) \to (\Sigma(\Omega), \pi)$ is a holomorphic extension to the envelope of holomorphy with transpose mapping $j^*(f) := f \circ j$ for all $f \in H(\Omega)$ then $\Sigma(\Omega) \subset S_b(\Omega)$ and*

$$j^* \colon \left(H(\Sigma(\Omega)), \tau_\delta \right) \longrightarrow (H(\Omega), \tau_\delta)$$

is an isomorphism of locally m-convex algebras.

Proof. If $\mathcal{C} := (C_n)_n \in \mathcal{U}$ let

$$S_{\mathcal{C}} := \{x \in \Sigma(\Omega) \; : \; \tilde{x}\big|_{H_{\mathcal{C}}(\Omega)} \text{ is continuous}\}$$

where $\tilde{x}(f) = \tilde{f}(x)$ and $\tilde{f}$ is the unique analytic extension of f to $\Sigma(\Omega)$. Clearly $\Omega \subset S_{\mathcal{C}}$. If $x \in S_{\mathcal{C}}$ there exists $C > 0$ and a positive integer n_0 such that

$$|\tilde{x}(f)| \leq C\|f\|_{C_{n_0}}$$

for all $f \in H_C(\Omega)$. For every positive integer k and $f \in H_C(\Omega)$ we have

$$|\tilde{x}(f)|^k = |\tilde{x}(f^k)| \leq C\|f^k\|_{C_{n_0}} = C\|f\|^k_{C_{n_0}}.$$

Hence $|\tilde{x}(f)| \leq C^{1/k}\|f\|_{C_{n_0}}$ and, letting k tend to infinity, we obtain $|\tilde{x}(f)| \leq \|f\|_{C_{n_0}}$ for all $f \in H_C(\Omega)$. Let

$$S_C(n) = \{x \in S_C \ : \ |\tilde{x}(f)| \leq \|f\|_{C_n} \text{ for all } f \in H_C(\Omega)\}.$$

If $x \in S_C(n_0)$ choose $m > n_0$ such that $d^{\alpha_m}_{\Sigma(\Omega)}(x) \geq 2 \cdot 2^{-m}$. Suppose $y \in E$ and $\alpha_m(y) < 2^{-m}$. By the Cauchy inequalities

$$\sum_{n=0}^{\infty} \left| \frac{\widehat{d^n \tilde{f}(x)}}{n!}(y) \right| \leq \sum_{n=0}^{\infty} \left\| \frac{\widehat{d^n f(\cdot)}}{n!}(y) \right\|_{C_{n_0}}$$

$$\leq \sum_{n=0}^{\infty} \frac{1}{2^n}\|f\|_{C_{n_0}+2\cdot2^{-(n_0+1)}V_{n_0+1}} \leq 2\|f\|_{C_{n_0+1}}$$

for all $f \in H_C(\Omega)$. Since

$$(x+y)(f) = \tilde{f}(x+y) = \sum_{n=0}^{\infty} \frac{\widehat{d^n \tilde{f}(x)}}{n!}(y)$$

for all $f \in H_C(\Omega)$, this implies $S_C(n_0)$ lies in the interior of $S_C(n_0 + 1)$ and hence $S_C = \bigcup_{n \geq 1} S_C(n)$ is an open subset of $\Sigma(\Omega)$. To show that S_C is closed it suffices to prove it is sequentially closed. Let $(x_n)_n \in S_C \to x \in \Sigma(\Omega)$ as $n \to \infty$. Since the sequence $(x_n)_n$ converges in $\Sigma(\Omega)$ and $H_C(\Omega)$ is barrelled the sequence $\{\delta_{x_n}\}_{n=1}^{\infty}$ is a weakly bounded and hence an equicontinuous subset of $H_C(\Omega)'$. Thus there exists a positive integer n_1 and $c > 0$ such that

$$|\tilde{f}(x_n)| \leq c\|f\|_{C_{n_1}}$$

for all n and all $f \in H_C(\Omega)$. This implies $|\tilde{f}(x)| \leq c\|f\|_{C_{n_1}}$ and $x \in S_C$. We have shown that S_C is an open and closed subset of $\Sigma(\Omega)$. Since $\Sigma(\Omega)$ is the envelope of holomorphy of Ω each connected component of $\Sigma(\Omega)$ intersects Ω. Hence $S_C = \Sigma(\Omega)$ and each $x \in \Sigma(\Omega)$ defines, by point evaluation, a multiplicative linear functional which is bounded on the bounded subsets of $H_C(\Omega)$. This shows that each $\tilde{x}$ is τ_δ-continuous and $\Sigma(\Omega) \subset S_b(\Omega)$.

Since $\Sigma(\Omega)$ is the envelope of holomorphy of Ω the mapping j^* is bijective and, by uniqueness of analytic continuation, j^* is a linear and algebraic isomorphism. The inverse mapping $(j^*)^{-1}$ is the restriction of holomorphic mappings on $\Sigma(\Omega)$ to Ω. Since the τ_δ-bounded subsets of $H(\Omega)$ and $H(\Sigma(\Omega))$ coincide with the locally bounded sets it follows that $(j^*)^{-1}$ is continuous with respect to the τ_δ topologies. To complete the proof we must show that τ_δ-bounded subsets of $H(\Omega)$ extend to τ_δ-bounded subsets of $H(\Sigma(\Omega))$.

If $\mathcal{F}$ is a τ_δ-bounded subset of $H(\Omega)$ then there exists $C := (C_n)_n \in \mathcal{U}$ such that $\mathcal{F}$ is contained and bounded in $H_C(\Omega)$. By the first part of the proof $\{\mathcal{S}_C(n)^\circ\}_{n=1}^\infty$ is an increasing countable open cover of $\Sigma(\Omega)$ (where $\circ$ denotes the interior). For each positive integer n,

$$\sup_{f \in \mathcal{F}} \|\tilde{f}\|_{\mathcal{S}_C(n)^\circ} = \sup\{|\tilde{f}(\tilde{x})| \, : \, \tilde{x} \in \mathcal{S}_C(n)^\circ, f \in \mathcal{F}\}$$

$$\leq \sup\{|\tilde{x}(f)| \, : \, \tilde{x} \in \mathcal{S}_C(n), f \in \mathcal{F}\}$$

$$\leq \sup\{\|f\|_{C_n} \, : \, f \in \mathcal{F}\}$$

$$< \infty.$$

Hence $\{\tilde{f} \, : \, f \in \mathcal{F}\}$ is a locally bounded and a τ_δ-bounded subset of $H(\Sigma(\Omega))$. This completes the proof.

To obtain the reverse inclusion, $\mathcal{S}_b(\Omega) \subset \Sigma(\Omega)$, we require additional hypotheses. It is more fruitful to start with a smaller set – the non-zero τ_0-continuous linear functionals – and to endow this set with the structure of a Riemann domain so that it becomes a holomorphic extension of Ω and afterwards to prove that this smaller set coincides with $\mathcal{S}_b(\Omega)$. It is, of course, also possible to consider τ_ω-continuous multiplicative linear functionals on $H(\Omega)$. If ϕ is ported by the compact subset K of Ω then it is easily seen, using powers of f, that $|\phi(f)| \leq \|f\|_K$ for all $f \in H(\Omega)$ so this case is already under observation when we consider τ_0 functionals.

Definition 5.18 If (Ω, p) is a Riemann domain over a locally convex space E, the τ_0-spectrum of $H(\Omega)$, $\mathcal{S}_c(\Omega)$, is the set of non-zero τ_0-continuous multiplicative linear functionals on $H(\Omega)$.

If $\phi \in \mathcal{S}_c(\Omega)$ there exists K compact in Ω and $c > 0$ such that

$$|\phi(f)| \leq c\|f\|_K$$

for all $f \in H(\Omega)$. Since $\phi(f^n) = \phi(f)^n$ and $\|f^n\|_K = \|f\|_K^n$ for all n it follows that

$$|\phi(f)| = |\phi(f^n)|^{1/n} \leq c^{1/n}\|f^n\|_K^{1/n} = c^{1/n}\|f\|_K$$

and letting $n \to \infty$ we obtain

$$|\phi(f)| \leq \|f\|_K \tag{5.12}$$

for all $f \in H(\Omega)$. When (5.12) holds we say that ϕ is *dominated* by K and write $\phi \prec K$. To endow $\mathcal{S}_c(\Omega)$ with the structure of a Riemann domain we assume that E is a quasi-complete locally convex space. We define a topology on $\mathcal{S}_c(\Omega)$ by specifying a neighbourhood basis about each point.

If $f \in H(\Omega)$ and $a \in E$ then the mapping

$$\left(\frac{d^n f}{n!}\right)_a := \frac{\widehat{d^n f_a}}{n!} : x \in \Omega \longrightarrow \frac{\widehat{d^n f(x)}}{n!}(a)$$

is a well defined holomorphic mapping on Ω. If $\phi \in \mathcal{S}_c(\Omega)$ is dominated by the compact subset K of Ω, $a \in E$ and $1 < \rho < d_\Omega(K, a)$ then the Cauchy inequalities imply

$$\left|\phi\left(\frac{\widehat{d^n f_a}}{n!}\right)\right| \le \sup_{x \in K}\left\|\frac{\widehat{d^n f(x)}}{n!}(a)\right\| \le \frac{1}{\rho^n}\|f\|_{D_\Omega(K,a,\rho)}$$

for all $n \in \mathbb{N}$ and all f in $H(\Omega)$. Hence

$$\sum_{n=0}^{\infty}\left|\phi\left(\frac{\widehat{d^n f_a}}{n!}\right)\right| \le \sum_{n=0}^{\infty}\frac{1}{\rho^n}\|f\|_{D_\Omega(K,a,\rho)} = \frac{\rho}{\rho-1}\|f\|_{\overline{D_\Omega(K,a,\rho)}} \qquad (*)$$

and ϕ_a defined by

$$\phi_a(f) := \sum_{n=0}^{\infty}\phi\left(\frac{\widehat{d^n f_a}}{n!}\right)$$

is a τ_0-continuous linear functional on $H(\Omega)$. By the product rule for differentiation $\phi_a(f \cdot g) = \phi_a(f) \cdot \phi_a(g)$ for all f, g in $H(\Omega)$. Since $\phi_a(1) = \phi(1) = 1$, ϕ_a is not identically zero and hence $\phi_a \in \mathcal{S}_c(\Omega)$. Using the multiplicative property of ϕ_a we can remove the constant $\rho/(\rho-1)$ in $(*)$ and letting $\rho \to 1$ we obtain

$$|\phi_a(f)| \le \|f\|_{\overline{D_\Omega(K,a,1)}}$$

for all f in $H(\Omega)$. By Lemma 5.12, $\overline{D_\Omega(K,a,1)}$ is a compact subset of Ω. If $a \in E$ is arbitrary and $0 < r < d_\Omega(K, a)$ then $d_\Omega(K, ra) > 1$ and $D_\Omega(K, ra, 1) = D_\Omega(K, a, r)$. We have proved the following:

> if $\phi \in \mathcal{S}_c(\Omega)$, $\phi \prec K$, $a \in E$ and $0 < r < d_\Omega(K, a)$ then $\phi_{ra} \in \mathcal{S}_c(\Omega)$ and $\phi_{ra} \prec \overline{D_\Omega(K, a, r)}$. $\qquad$ (5.13)

Standard manipulation of a power series within its domain of convergence shows that

$$(\phi_a)_b = \phi_{a+b}$$

for all $a, b \in E$ whenever there exists $\alpha \in \mathrm{cs}(E)$ such that $d_\Omega^\alpha(K) > 0$ and $\alpha(a) + \alpha(b) < d_\Omega^\alpha(K)$.

We take as a neighbourhood basis at ϕ all sets of the form $\{\phi_a : \alpha(a) < r\}$ where $\phi \prec K$, $\alpha \in \mathrm{cs}(E)$ and $0 < r < d_\Omega^\alpha(K)$. It is easily checked, using Lemma 5.12, that these sets form a neighbourhood system for a Hausdorff topology on $\mathcal{S}_c(\Omega)$ and, moreover, the mapping $a \in B_E^\alpha(0, r) \to \phi_a \in \{\phi_a : a < r\}$ is a homeomorphism. Next we define a continuous mapping from $\mathcal{S}_c(\Omega)$ into E and show that it is a local isomorphism. It is at this stage that we require E to be quasi-complete. For fixed ϕ in $\mathcal{S}_c(\Omega)$ consider the mapping

$$w_\phi : g \in E' \longrightarrow \phi(g \circ p).$$

Since $\phi \in \mathcal{S}_c(\Omega)$ and $g \circ p \in H(\Omega)$ there exists K compact in Ω such that

$$|w_\phi(g)| = |\phi(g \circ p)| \leq \|g \circ p\|_K = \|g\|_{p(K)}$$

for all g in E'. Since p is continuous, $p(K)$ is a compact subset of E and $w_\phi \in (E'_c)' = E$. Hence there exists a unique $q(\phi) \in E$ such that

$$w_\phi(g) = \phi(g \circ p) = g(q(\phi)) \tag{5.14}$$

for all $g \in E'$. The required mapping is

$$q: \mathcal{S}_c(\Omega) \longrightarrow E$$
$$\phi \longmapsto q(\phi).$$

If $x, y \in \Omega$ and $g \in E'$ then $g \circ p(y) = g(p(x)) + g(p(y) - p(x))$. By (5.11) and uniqueness of Taylor series expansions

$$\left(\frac{\widehat{d^n}(g \circ p)_a}{n!}\right)(x) = \begin{cases} g \circ p(x) & \text{if } n = 0 \\ g(a) & \text{if } n = 1 \\ 0 & \text{if } n > 1. \end{cases}$$

If $\phi \prec K$, K compact in Ω, and $d_\Omega(K, a) > 1$ then, by (5.14),

$$g(q(\phi_a)) = \phi_a(g \circ p) = \phi(g \circ p) + \phi(g(a))$$
$$= g(q(\phi)) + g(a)\phi(1_\Omega)$$

for all $g \in E'$. Since $\phi(1_\Omega^2) = \phi(1_\Omega)^2$ and, ϕ is non-zero, $\phi(1_\Omega) = 1$. Hence $g(q(\phi_a)) = g(q(\phi) + a)$. By the Hahn–Banach Theorem

$$q(\phi_a) = q(\phi) + a. \tag{5.15}$$

On replacing a by ra, $0 < r < d_\Omega(K, a)$, we obtain

$$q(\{\phi_a \; : \; \alpha(a) < r\}) = B_E^\alpha(q(\phi), r)$$

and q is a local homeomorphism. We have established the following result.

Proposition 5.19 *If (Ω, p) is a Riemann domain over a quasi-complete locally convex space E then $(\mathcal{S}_c(\Omega), q)$ is a Riemann domain over E. Moreover, if $\phi \in \mathcal{S}_c(\Omega)$, $\phi \prec K$ for K compact in Ω, $\alpha \in \mathrm{cs}(E)$ and $d_\Omega^\alpha(K) > 0$ then*

$$d_{\mathcal{S}_c(\Omega)}^\alpha(\phi) \geq d_\Omega^\alpha(K). \tag{5.16}$$

If (Ω, p) is a Riemann domain over a quasi-complete locally convex space E then each point x of Ω defines a point evaluation $\delta_x: H(\Omega) \to \mathbb{C}$, $\delta_x(f) = f(x)$, and consequently an element of $\mathcal{S}_c(\Omega)$. This induces the mapping

$$\delta: \Omega \longrightarrow \mathcal{S}_c(\Omega)$$
$$x \longmapsto \delta_x.$$

By (5.14), for all $g \in E'$ we have

$$g(q(\delta_x)) = \delta_x(g \circ p) = g \circ p(x) = g(p(x))$$

and, by the Hahn–Banach Theorem, $q \circ \delta = p$. If $x \in \Omega$, $\alpha \in cs(E)$ and $0 < r < d^\alpha_\Omega(x)$, (5.11) implies

$$\delta_y(f) = f(y) = \sum_{n=0}^{\infty} \frac{\widehat{d^n} f(x)}{n!} (p(y) - p(x))$$

$$= \sum_{n=0}^{\infty} \frac{\widehat{d^n} f_{p(y)-p(x)}}{n!} (x)$$

$$= \sum_{n=0}^{\infty} \delta_x \left(\frac{\widehat{d^n} f_{p(y)-p(x)}}{n!} \right)$$

$$= (\delta_x)_{p(y)-p(x)}(f)$$

for all $f \in H(\Omega)$, $y \in B^\alpha_\Omega(x, r)$. Hence

$$\delta(B^\alpha_\Omega(x, r)) = \{\delta_y \; : \; y \in B^\alpha_\Omega(x, r)\}$$
$$= \{(\delta_x)_y \; : \; \alpha(y) < r\}$$
$$= B^\alpha_{\mathcal{S}_c(\Omega)}(\delta_x, r) \tag{5.17}$$

and δ establishes a correspondence between neighbourhoods of $x \in (\Omega, p)$ and neighbourhoods of $\delta_x \in (\mathcal{S}_c(\Omega), q)$. This shows that δ is a local isomorphism and, in particular, a morphism. The mapping δ is injective if and only if for each pair of distinct points x, $y \in \Omega$ there exists $f \in H(\Omega)$ such that

$$f(x) = \delta_x(f) \neq \delta_y(f) = f(y).$$

A Riemann domain (Ω, p) is called *holomorphically separated* if each pair of distinct points in Ω can be separated by a holomorphic function on Ω. We have proved the following result.

Proposition 5.20 *If (Ω, p) is a Riemann domain over a quasi-complete locally convex space then the mapping*

$$\delta: \Omega \longrightarrow \mathcal{S}_c(\Omega)$$
$$x \longmapsto \delta_x$$

is a morphism from (Ω, p) into $(\mathcal{S}_c(\Omega), q)$. If (Ω, p) is holomorphically separated then δ is injective.

Proposition 5.21 *Let (Ω, p) denote a Riemann domain over a quasi-complete locally convex space E. For $f \in H(\Omega)$ define $\widetilde{f} : S_c(\Omega) \to \mathbb{C}$ by $\widetilde{f}(\phi) = \phi(f)$. Then*

(a) $\widetilde{f} \in H(S_c(\Omega))$,
(b) $S_c(\Omega)$ is holomorphically separated,
(c) $\widetilde{f} \circ \delta = f$ for all $f \in H(\Omega)$,
(d) for each compact subset $\widetilde{K}$ of $S_c(\Omega)$ there exists a compact subset K in Ω such that
$$\|\widetilde{f}\|_{\widetilde{K}} \leq \|f\|_K$$

for all f in $H(\Omega)$.

Proof. (a) Let $f \in H(\Omega)$ and $\phi \in S_c(\Omega)$ be arbitrary. Choose K compact in Ω such that $\phi \prec K$ and $\alpha \in cs(E)$ such that $d_\Omega^\alpha(K) > 1$ and $\|f\|_{B_\Omega^\alpha(K,s)} =: M(s) < \infty$ for some s, $1 < s < d_\Omega^\alpha(K)$. The mapping

$$Q_n : a \in E \longrightarrow \phi\left(\frac{\widehat{d}^n f_a}{n!}\right)$$

defines a continuous n-homogeneous polynomial on E. By (5.15),

$$\left|\widetilde{f}(\phi_a) - \sum_{n=0}^m Q_n\big(q(\phi_a) - q(\phi)\big)\right| = \left|\phi_a(f) - \sum_{n=0}^m Q_n(a)\right|$$

$$= \left|\sum_{n=m+1}^\infty \phi\left(\frac{\widehat{d}^n f_a}{n!}\right)\right| \leq \sum_{n=m+1}^\infty \frac{1}{s^n} M(s)$$

whenever $\alpha(a) \leq 1$. Hence

$$\sup_{\substack{a \in E \\ \alpha(a) \leq 1}} \left|\widetilde{f}(\phi_a) - \sum_{n=0}^m Q_n\big(q(\phi_a) - q(\phi)\big)\right| \longrightarrow 0 \quad \text{as } m \to \infty$$

and, by (5.11), $\widetilde{f} \in H(S_c(\Omega))$. This establishes (a).

If $\phi, \psi \in S_c(\Omega)$, $\phi \neq \psi$ there exists $f \in H(\Omega)$ such that $\phi(f) \neq \psi(f)$. Hence $\widetilde{f}(\phi) \neq \widetilde{f}(\psi)$, $S_c(\Omega)$ is holomorphically separated and (b) holds.

Since $\widetilde{f}(\delta_x) = \delta_x(f) = f(x)$ we have the commutative diagram

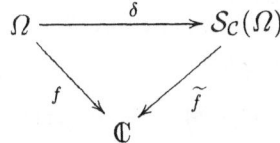

for all $f \in H(\Omega)$ and $\widetilde{f} \circ \delta = f$. This proves (c).

If $\widetilde{K}$ is a compact subset of $(\mathcal{S}_c(\Omega), q)$ then we can find $\phi_1, \ldots, \phi_k \in \widetilde{K}$, $K_1, \ldots, K_k$ compact in Ω, $\alpha \in cs(E)$ and positive numbers $r_1, \ldots, r_k$ such that $\phi_i \prec K_i$, $d^\alpha_{\mathcal{S}_c(\Omega)}(K_i) > r_i$ for all i and

$$\widetilde{K} \subset \bigcup_{i=1}^{k} B^\alpha_{\mathcal{S}_c(\Omega)}(\phi_i, r_i).$$

Since q is continuous the sets

$$A_i := \left(q(\widetilde{K}) - q(\phi_i)\right) \cap \overline{B^\alpha_E(0, r_i)}$$

are compact subsets of E for $i = 1, \ldots, k$. If $\phi \in \widetilde{K}$ choose i, $1 \leq i \leq k$, and $a_i \in \overline{B^\alpha_E(0, r_i)}$ such that $\phi = (\phi_i)_{a_i}$. By (5.15), $q(\phi) = q(\phi_i) + a_i$. Hence $a_i \in A_i$ and

$$\widetilde{K} \subset \bigcup_{i=1}^{k} \{(\phi_i)_a : a \in A_i\}.$$

By Lemma 5.12

$$K := \bigcup_{i=1}^{k} (K_i + A_i)$$

is a compact subset of Ω. If $1 \leq i \leq k$, $a \in A_i$ and $f \in H(\Omega)$, (5.13) implies

$$\left|(\phi_i)_a(f)\right| \leq \|f\|_{K_i + A_i} \leq \|f\|_K.$$

Hence $\|\widetilde{f}\|_{\widetilde{K}} \leq \|f\|_K$ for all $f \in H(\Omega)$. This establishes (d) and completes the proof.

We confine ourselves to the connected Riemann domain (Ω, p). This, neither immediately nor obviously, implies that $\mathcal{S}_c(\Omega)$ is connected – it would be very surprising if this was not the case and we do not know of any counterexamples (see Corollary 5.50 for a positive result) – and let $\widehat{\Omega}_c$ denote the connected component of $\mathcal{S}_c(\Omega)$ containing $\delta(\Omega)$. To obtain a holomorphic extension which yields a bijective mapping between spaces of holomorphic functions we consider $\widetilde{f}\big|_{\widehat{\Omega}_c}$ for each $f \in H(\Omega)$. Let $\delta^*: H(\widehat{\Omega}_c) \to H(\Omega)$ denote the transpose mapping $\delta^*(f) = f \circ \delta$.

Proposition 5.22 *Let (Ω, p) denote a connected holomorphically separated Riemann domain over a quasi-complete locally convex space E. The mapping $\delta: \Omega \to \widehat{\Omega}_c$ is an injective morphism and a holomorphic extension and*

$$\delta^*: \left(H(\widehat{\Omega}_c), \tau\right) \longrightarrow \left(H(\Omega), \tau\right)$$

is a topological isomorphism and homomorphism for both $\tau = \tau_0$ and τ_ω. Furthermore, if $(\widetilde{\Omega}, \widetilde{p})$ is a connected and holomorphically separated Riemann domain and

$$j_1 \colon (\Omega, p) \to (\widetilde{\Omega}, \widetilde{p})$$

is a holomorphic extension such that

$$j_1^*(f) := f \circ j_1$$

defines a τ_0-isomorphism from $\big(H(\widetilde{\Omega}), \tau_0\big)$ onto $\big(H(\Omega), \tau_0\big)$ then there exists an injective holomorphic extension

$$j_2 \colon (\widetilde{\Omega}, \widetilde{p}) \to (\widehat{\Omega}_c, q)$$

such that the following diagram commutes:

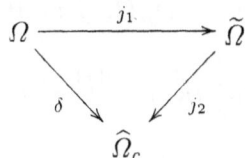

Hence $\widehat{\Omega}_c$ is the connected holomorphic extension of Ω which is characterized by being maximal with respect to inducing a τ_0-isomorphism. Furthermore, $\widehat{\Omega}_c$ can be identified with an open subset of $\Sigma(\Omega)$.

Proof. By Propositions 5.20 and 5.21, $\delta \colon (\Omega, p) \to (\widehat{\Omega}_c, q|_{\widehat{\Omega}_c})$ is a holomorphic extension. We have already noted that δ is an injective open mapping when Ω is holomorphically separated. Hence $\delta \colon \Omega \to \delta(\Omega)$ is a topological isomorphism. If (U, p) is a chart in Ω, then $(\delta(U), q)$ is a chart in $(\widehat{\Omega}_c, q)$ and $q_{\delta(U)}^{-1} = \delta \circ p_U^{-1}$. If $f \in H(\widehat{\Omega}_c)$, then $f \circ q_{\delta(U)}^{-1} \in H\big(q(\delta(U))\big)$ and $f \circ \delta \circ p_U^{-1} \in H(p(U))$. This implies $\delta^*(f) \in H(\Omega)$ for all $f \in H(\widehat{\Omega}_c)$. Clearly δ^* is a linear mapping. If $\delta^*(f) = 0$ then $f\big|_{\delta(\Omega)} = 0$ and, since $\widehat{\Omega}_c$ is connected, the principle of analytic continuation implies $f = 0$. Hence δ^* is injective. If $f \in H(\Omega)$, Proposition 5.21(a) implies, $\widetilde{f}\big|_{\widehat{\Omega}_c} \in H(\widehat{\Omega}_c)$. For all $x \in \Omega$,

$$\big[\delta^*\big(\widetilde{f}\big|_{\widehat{\Omega}_c}\big)\big](x) = \big(\widetilde{f}\big|_{\widehat{\Omega}_c} \circ \delta\big)(x) = \widetilde{f}\big(\delta(x)\big) = \widetilde{f}(\delta_x) = \delta_x(f) = f(x).$$

Hence $\delta^*\big(\widetilde{f}\big|_{\widehat{\Omega}_c}\big) = f$ and δ^* is surjective. We have shown that δ^* is a linear isomorphism and

$$H(\widehat{\Omega}_c) = \{\widetilde{f}\big|_{\widehat{\Omega}_c} \colon f \in H(\Omega)\}.$$

If $f, g \in H(\widehat{\Omega}_c)$ and $x \in \Omega$ then

$$\big(\delta^*(f \cdot g)\big)(x) = (f \cdot g)\big(\delta(x)\big) = f\big(\delta(x)\big) \cdot g\big(\delta(x)\big) = \big(\delta^*(f)\big)(x) \cdot \big(\delta^*(g)\big)(x).$$

Hence $\delta^*(f \cdot g) = \delta^*(f) \cdot \delta^*(g)$ and δ^* is a homomorphism. Since δ is a homeomorphism from Ω onto an open subset of $\widehat{\Omega}_c$ the identity

$$\|\delta^*(f)\|_A = \|f \circ \delta\|_A = \|f\|_{\delta(A)}$$

for $f \in H(\widehat{\Omega}_c)$ and $A \subset \Omega$ shows that δ^* is continuous when both spaces are endowed with either the τ_0 or τ_ω topologies. By Proposition 5.21(d), δ^* is an open mapping for the τ_0 topologies and hence a topological isomorphism.

If α is a τ_ω-continuous semi-norm on $H(\widehat{\Omega}_c)$ ported by the compact subset $\widetilde{K}$ of $\widehat{\Omega}_c$, let $\widetilde{\alpha}(f) = \alpha(\widetilde{f}\big|_{\widehat{\Omega}_c})$ for all $f \in H(\Omega)$. By Proposition 5.21(d), there exists a compact subset K of Ω such that $\|\widetilde{f}\|_{\widetilde{K}} \le \|f\|_K$ for all $f \in H(\Omega)$. If $\phi \in \widetilde{K}$ then $|\phi(f)| = |\widetilde{f}(\phi)| \le \|f\|_K$ for all $f \in H(\Omega)$ and hence $\phi \prec K$. If $\beta \in cs(E)$, $d_\Omega^\beta(K) > 0$, $a \in E$ and $0 < r < d_\Omega^\beta(K)$, Proposition 5.19 implies $B_{\widehat{\Omega}_c}^\beta(\widetilde{K}, r) \subset \widehat{\Omega}_c$. By (5.13), $\phi_{ra} \prec \overline{D_\Omega(K, a, r)}$. Hence

$$\|\widetilde{f}\|_{B_{\widehat{\Omega}_c}^\beta(\widetilde{K}, r)} \le \|f\|_{B_\Omega^\beta(K, r)}$$

for all $f \in H(\Omega)$. Since α is ported by $\widetilde{K}$ there exists $c := c(r) > 0$ such that

$$|\widetilde{\alpha}(f)| = |\alpha(\widetilde{f}\big|_{\widehat{\Omega}_c})| \le c\|\widetilde{f}\|_{B_{\widehat{\Omega}_c}^\beta(\widetilde{K}, r)} \le c\|f\|_{B_\Omega^\beta(K, r)}$$

for all $f \in H(\Omega)$. This implies $\widetilde{\alpha}$ is τ_ω-continuous and δ^* is an open mapping and an isomorphism for the τ_ω topologies.

Let $j_1 : (\Omega, p) \to (\widetilde{\Omega}, q)$ denote a holomorphic extension such that

$$j_1^* : (H(\widetilde{\Omega}), \tau_0) \to (H(\Omega), \tau_0), \quad j_1^*(f) = f \circ j_1$$

is a topological isomorphism. The mapping $(j_1^*)^{-1}$, which implements the holomorphic extension from Ω to $\widetilde{\Omega}$ is a homomorphism and, by uniqueness of analytic continuation,

$$(j_1^*)^{-1}\left(\left(\frac{\widehat{d^n f}}{n!}\right)_y\right) = \left(\frac{\widehat{d^n}}{n!}((j_1^*)^{-1} f)\right)_y \tag{$*$}$$

for all $f \in H(\Omega)$, all n and all $y \in E$. If $x \in \widetilde{\Omega}$ we define $j_2(x)$ on $H(\Omega)$ by

$$[j_2(x)](f) := [(j_1^*)^{-1}(f)](x).$$

Since j_1^* is inverse to a bijective homomorphism it is also a homomorphism and $j_2(x)$ defines a non-zero multiplicative linear functional on $H(\Omega)$. As j_1^* is a linear isomorphism with respect to the compact open topologies, $j_2(x)$ is τ_0-continuous and belongs to $\mathcal{S}_c(\Omega)$. Hence $j_2 : \widetilde{\Omega} \to \mathcal{S}_c(\Omega)$. If $x \in \Omega$ and $f \in H(\Omega)$ then

$$\begin{aligned}
[j_2(j_1(x))](f) &= [(j_1^*)^{-1}(f)](j_1(x)) \\
&= [(j_1^*)^{-1}(f) \circ j_1](x) \\
&= [j_1^*((j_1^*)^{-1}(f))](x) \\
&= f(x) = [\delta(x)](f)
\end{aligned}$$

and hence $j_2 \circ j_1 = \delta$. We now show that j_2 is a morphism. If $x \in \tilde{\Omega}$, K is compact in Ω, $j_2(x) \prec K$, $\alpha \in cs(E)$, $d_{\tilde{\Omega}}^{\alpha}(x) > 1$ and $d_{\Omega}^{\alpha}(K) > 1$ then,

$$
(j_2(x) + y)(f) = \sum_{n=0}^{\infty} j_2(x)\left(\frac{\widehat{d^n f_y}}{n!}\right) = \sum_{n=0}^{\infty} \left[(j_1^*)^{-1}\left(\frac{\widehat{d^n f_y}}{n!}\right)\right](x)
$$
$$
= \sum_{n=0}^{\infty} \frac{\widehat{d^n\left((j_1^*)^{-1}(f)\right)_y}}{n!}(x), \quad (\text{by } (*)),
$$
$$
= \left[(j_1^*)^{-1}(f)\right](x + y)
$$
$$
= j_2(x + y)(f)
$$

for $\alpha(y) < 1$ and $f \in H(\Omega)$. Hence $j_2(x) + y = j_2(x + y)$ and j_2 is a local isomorphism and a morphism. In particular, j_2 is continuous and, since $\tilde{\Omega}$ is connected, j_2 is a morphism from $\tilde{\Omega}$ into $\hat{\Omega}_c$. If $f \in H(\tilde{\Omega})$ then $\widetilde{f \circ j_1}\big|_{\hat{\Omega}_c} \in H(\hat{\Omega}_c)$ and

$$
\widetilde{f \circ j_1}\big|_{\hat{\Omega}_c}(j_2(x)) = j_2(x)(f \circ j_1) = \left[(j_1^*)^{-1}(f \circ j_1)\right](x)
$$
$$
= \left[(j_1^*)^{-1}\left(j_1^*(f)\right)\right](x)
$$
$$
= f(x)
$$

for all $x \in \Omega$. Hence $\widetilde{f \circ j_1}\big|_{\hat{\Omega}_c} \circ j_2 = f$. Since $\hat{\Omega}_c$ and Ω are connected the principle of analytic continuation implies that $\widetilde{f \circ j_1}\big|_{\hat{\Omega}_c}$ is the unique holomorphic extension of f to $\hat{\Omega}_c$. Hence $j_2 \colon (\tilde{\Omega}, \tilde{p}) \to (\hat{\Omega}_c, q)$ is a holomorphic extension and, since $(\tilde{\Omega}, \tilde{p})$ is holomorphically separated, j_2 is an injective mapping. This completes the proof.

By Proposition 5.22, $S_c\big(S_c(\Omega)\big) \cong S_c(\Omega)$ and $\widehat{(\hat{\Omega}_c)}_c = \hat{\Omega}_c$ for any holomorphically separated Riemann domain (Ω, p) over a quasi-complete locally convex space.

Before proceeding we note that Proposition 5.22 lifts the problem of topologies to the connected τ_0 spectrum.

Corollary 5.23 *If (Ω, p) is a connected holomorphically separated Riemann domain spread over a quasi-complete locally convex space then $\tau_0 = \tau_\omega$ on $H(\Omega)$ if and only if $\tau_0 = \tau_\omega$ on $H(\hat{\Omega}_c)$.*

Proof. It suffices to consider the following commutative diagram in which the horizontal mappings are the identity mappings:

$$\begin{array}{ccc}
\big(H(\Omega), \tau_\omega\big) & \xrightarrow{\quad I \quad} & \big(H(\Omega), \tau_0\big) \\
{\scriptstyle \delta^*}\big\uparrow & & \big\uparrow{\scriptstyle \delta^*} \\
\big(H(\widehat{\Omega}_c), \tau_\omega\big) & \xrightarrow{\quad I \quad} & \big(H(\widehat{\Omega}_c), \tau_0\big)
\end{array}$$

Hence the identity mapping $\big(H(\Omega), \tau_\omega\big) \to \big(H(\Omega), \tau_0\big)$ is an isomorphism, i.e. $\tau_0 = \tau_\omega$ on $H(\Omega)$, if and only if the identity mapping $\big(H(\widehat{\Omega}_c), \tau_\omega\big) \to \big(H(\widehat{\Omega}_c), \tau_0\big)$ is also an isomorphism, i.e. if and only if $\tau_0 = \tau_\omega$ on $H(\widehat{\Omega}_c)$.

This completes the first stage of our investigation. In it we used the methods of *functional analysis* to construct a new Riemann domain $\widehat{\Omega}_c$ canonically from the given Riemann domain Ω and set up a system of transferring "holomorphy" from Ω to $\widehat{\Omega}_c$. We have shown

$$\widehat{\Omega}_c \subset \Sigma(\Omega) \subset \mathcal{S}_b(\Omega)$$

and it is interesting to note that in the process of proving $\tau_0 = \tau_\omega$ on $H(\Omega)$ we will eventually show $\mathcal{S}_c(\Omega) = \mathcal{S}_b(\Omega)$, i.e. τ_0 and τ_δ define the same *multiplicative* linear functionals on $H(\Omega)$. In the next stage we use concepts associated with the theory of several complex variables, e.g. plurisubharmonic functions and pseudo-convexity, to examine what might be called the holomorphic-metric structure of $\widehat{\Omega}_c$. Our treatment is brief and aimed solely at providing the necessary background for the main results in this section (Proposition 5.44 and Corollaries 5.45 and 5.53).

Definition 5.24 A function $u\colon (\Omega, p) \to [-\infty, +\infty)$, where (Ω, p) is a Riemann domain over a locally convex space E, is plurisubharmonic if it is upper semi-continuous and for all $x \in \Omega$, $a \in E$ and $r > 0$ such that $\overline{D_\Omega(x, a, r)} \subset \Omega$ we have

$$u(x) \leq \frac{1}{2\pi} \int_0^{2\pi} u(x + re^{i\theta} a)\, d\theta. \tag{5.18}$$

We let $PSH(\Omega)$ denote the set of all plurisubharmonic functions on Ω.

Condition (5.18) is the classical inequality used to define subharmonic functions on open subsets of $\mathbb{C}$ and plurisubharmonic functions on open subsets of $\mathbb{C}^n$. To check that a given function is plurisubharmonic one must verify upper semi-continuity and condition (5.18) which is determined on the finite dimensional sections of the domain. At times this makes it easy to lift or extend results to infinite dimensional spaces. For example, if $f \in H(\Omega)$ then $|f| \in PSH(\Omega)$ and if $u_\alpha \in PSH(\Omega)$ for all α then $u := \sup_\alpha u_\alpha \in PSH(\Omega)$ if and only if $u(x) < \infty$ for every x in Ω and u is upper semi-continuous. It is also easily verified for Riemann domains (Ω_1, p_1) and (Ω_2, p_2) that $u \in PSH(\Omega_2)$ and $f \in H(\Omega_1, \Omega_2)$ imply $u \circ f \in PSH(\Omega_1)$.

Definition 5.25 A Riemann domain (Ω, p) spread over a locally convex space E is pseudo-convex if $-\log d_\Omega$ is a plurisubharmonic function on the Riemann domain $(\Omega \times E, p \times Id_E)$.

If (Ω, p) is spread over E and F is a subspace of E let

$$\Omega_F = p^{-1}\left(p(\Omega) \cap F\right) = \{x \in \Omega : p(x) \in F\}.$$

It is clear that $(\Omega_F, p|_{\Omega_F})$ is a Riemann domain over F and we have the commutative diagram of continuous mappings

$$
\begin{array}{ccc}
\Omega_F & \xrightarrow{\; i_{\Omega_F}(x)=x \;} & \Omega \\
{\scriptstyle p|_{\Omega_F}} \downarrow & & \downarrow {\scriptstyle p} \\
F & \xrightarrow{\; i_F(x)=x \;} & E
\end{array}
$$

Hence i_{Ω_F} is an i_F-morphism of Riemann domains. We call Ω_F the F-section of Ω. If $\dim(F) < \infty$ we call Ω_F a finite dimensional section of Ω.

We have already noted that d_Ω is lower semi-continuous on $\Omega \times E$ and hence $-\log d_\Omega$ is upper semi-continuous. This means that Ω is pseudo-convex if and only if the restriction of $-\log d_\Omega$ to the finite dimensional sections of Ω satisfy (5.18). This observation proves the following result.

Proposition 5.26 *A Riemann domain is pseudo-convex if and only if all of its finite dimensional sections are pseudo-convex.*

Next we define the *plurisubharmonic hull* of a set. If A is a subset of a Riemann domain (Ω, p) let

$$\widehat{A}_{PSH(\Omega)} = \{x \in \Omega \; : \; u(x) \leq \sup_{y \in A} u(y), \text{ for all } u \in PSH(\Omega)\}$$

and recall (Chapter 3) that

$$\widehat{A}_{H(\Omega)} = \{x \in \Omega : |f(x)| \leq \|f\|_A \text{ for all } f \in H(\Omega)\}.$$

We call $\widehat{A}_{PSH(\Omega)}$ and $\widehat{A}_{H(\Omega)}$ the *plurisubharmonic* and *holomorphic hull* of A respectively. Since $|f| \in PSH(\Omega)$ if $f \in H(\Omega)$ we have $\widehat{A}_{PSH(\Omega)} \subset \widehat{A}_{H(\Omega)}$. The holomorphic hull of any set is closed and if Ω is a domain in E the Hahn–Banach Theorem implies that $\widehat{A}_{H(\Omega)}$ is contained in the closed convex hull of A. We confine the following definition to quasi-complete spaces to avoid any confusion that might arise between the notions of precompact and relatively compact sets.

A Riemann domain (Ω, p) spread over a quasi-complete locally convex space is PSH-convex (respectively holomorphically convex) if the plurisubharmonic hull (respectively holomorphic hull) of each compact subset of Ω is also a compact subset of Ω.

In the following proposition we collect a number of important finite dimensional results. This proposition shows that *all* the concepts we have introduced coincide *for finite dimensional Riemann domains*. The equivalence of (iv) and (vi) is known as the *Cartan–Thullen Theorem* and the problem of showing that (i) and (vi) are equivalent is known as the *Levi problem*.

Proposition 5.27 *Let* (Ω, p) *denote a Riemann domain over a finite dimensional space* E *and let* α *denote a norm on* E. *The following are equivalent:*

(i) Ω *is pseudo-convex,*

(ii) Ω *is PSH-convex,*

(iii) $d_{\Omega}^{\alpha}(\widehat{K}_{H(\Omega)}) > 0$ *(or* $d_{\Omega}^{\alpha}(\widehat{K}_{PSH(\Omega)}) > 0$) *for each compact subset* K *of* Ω,

(iv) Ω *is holomorphically convex,*

(v) Ω *is a domain of holomorphy,*

(vi) Ω *is a domain of existence.*

Moreover, if any of these conditions are satisfied, then

(a) $\widehat{K}_{PSH(\Omega)} = \widehat{K}_{H(\Omega)}$ *for each compact subset* K *in* Ω,

(b) *if* K *is a compact subset of* Ω *then each* $f \in H(\widehat{K}_{H(\Omega)})$ *can be uniformly approximated on* K *by functions from* $H(\Omega)$,

(c) *if* U *is an open subset of* Ω *then* $\widehat{K}_{H(\Omega)} \subset U$ *for each compact subset* K *in* U *if and only if* U *is pseudo-convex and* $H(\Omega)$ *is dense in* $(H(U), \tau_0)$,

(d) $\mathcal{S}_c(\Omega) = \Omega$,

(e) Ω *is holomorphically separated.*

Investigations aimed at generalizing the results of Proposition 5.27 to infinite dimensional spaces have played an important role in the development of infinite dimensional holomorphy over the last thirty years. In particular, we recall our introduction of the τ_ω topology in Section 3.2 and the motivation provided by the implication *(iv)* $\Longrightarrow$ *(v)*. It is not difficult to show that a proper open subset U of a locally convex space is holomorphically convex if and only if for each sequence $(x_n)_n$ in U which converges to a boundary point of U there exists $f \in H(U)$ such that $\sup_n |f(x_n)| = \infty$. Hence, if U is a holomorphically convex open subset of a locally convex space E then U is a domain of holomorphy if E is a $\mathcal{DFM}$ space, since $(H(U), \tau_0)$ is barrelled (Example 3.20), or if U is a balanced subset of a Banach space with Schauder basis since $(H(U), \tau_\omega)$ is barrelled (Corollary 4.18). A considerable part of the remainder of this chapter deals with extending Proposition 5.27 to separable Fréchet spaces with the bounded approximation property and to $\mathcal{DFM}$ spaces with the approximation property (we require the $\mathcal{DFM}$ result in Chapter 6).

Proposition 5.26 and the equivalence of (i) and (ii) in Proposition 5.27 show that PSH-convex Riemann domains over locally convex spaces are pseudo-convex, while the equivalence of (i) and (iv) (for finite dimensional spaces) and the inclusion $\widehat{K}_{PSH(\Omega)} \subset \widehat{K}_{H(\Omega)}$ show that holomorphically convex Riemann domains are pseudo-convex. We have already noted that domains of existence are domains of holomorphy and so $(vi) \Longrightarrow (v)$.

The implications in the previous paragraph are those that we have proved and will use. They do not include some other general results that we do not use but now state for the sake of completeness (we will, however, prove some of these for the special spaces mentioned above). For a Riemann domain (Ω, p) spread over a locally convex space we have $(i) \iff (ii)$ and $(v) \Longrightarrow (iv)$, i.e.

Ω is pseudo-convex $\iff$ Ω is PSH-convex,
if Ω is a domain of holomorphy then Ω is holomorphically convex.

Motivated by (c) of Proposition 5.27 we introduce the following useful definition. The reader should take care in distinguishing between the concepts of $PSH(\Omega)$-convex and PSH-convex domains.

Definition 5.28 If (Ω, p) is a Riemann domain over a locally convex space E and U is an open subset of Ω then U is $PSH(\Omega)$-convex if $\widehat{K}_{PSH(\Omega)} \subset U$ whenever K is compact in U.

The following lemma is easily proved.

Lemma 5.29 *Let (Ω, p) denote a Riemann domain over a locally convex space E. Then*

(a) *Ω is $PSH(\Omega)$-convex,*
(b) *the finite intersection of $PSH(\Omega)$-convex sets is $PSH(\Omega)$-convex,*
(c) *if $f \in H(\Omega_1; \Omega_2)$ and $U \subset \Omega_2$ is $PSH(\Omega_2)$-convex then $f^{-1}(U)$ is $PSH(\Omega_1)$-convex,*
(d) *a convex open subset of a locally convex space E is $PSH(E)$-convex,*
(e) *if V is a convex open subset of E then $p^{-1}(V)$ is a $PSH(\Omega)$-convex subset of Ω,*
(f) *If (Ω, p) is pseudo-convex and $U \subset \Omega$ is $PSH(\Omega)$-convex then $(U, p|_U)$ is a pseudo-convex Riemann domain.*

Proposition 5.30 *If (Ω, p) is a pseudo-convex Riemann domain spread over a locally convex space and $u \in PSH(\Omega)$ then for any $c \in \mathbb{R}$, $V_c := \{x \in \Omega : u(x) < c\}$ is pseudo-convex and $PSH(\Omega)$-convex.*

Proof. If u is a constant function or V_c is empty then the result is trivial and so we assume that neither of these conditions apply. If K is a compact subset of V_c then, since non-constant plurisubharmonic functions attain their maximum on compact sets and do not attain a maximum on open sets, there exists $c' < c$ such that $u(x) \leq c'$ for all $x \in K$. Hence

$$\sup_{x \in \widehat{K}_{PSH(\Omega)}} u(x) = \sup_{x \in K} u(x) \le c' < c$$

and $\widehat{K}_{PSH(\Omega)} \subset V_c$. This shows that V_c is $PSH(\Omega)$-convex.

If F is a finite dimensional subspace of E then Ω_F is pseudo-convex and $u|_{\Omega_F} \in PSH(\Omega_F)$. If K is a compact subset of $V_F := V_c \cap \Omega_F$ then

$$\widehat{K}_{PSH(V_F)} \subset \widehat{K}_{PSH(\Omega_F)} \cap \{x \in \Omega : u(x) \le c''\}$$

for some $c'' < c$. Since $(\Omega_F, p|_F)$ is a pseudo-convex domain over a finite dimensional space, $\widehat{K}_{PSH(\Omega_F)}$ is a relatively compact – in fact compact – subset of Ω_F. Hence $\widehat{K}_{PSH(V_F)}$ is a compact subset of $V_F \cap \{x \in \Omega : u(x) \le c''\} \subset V_F$. This shows that V_F is pseudo-convex for each finite dimensional subspace F of E and, by Propositions 5.26 and 5.27, V_c is pseudo-convex. This completes the proof.

From Proposition 5.27 we derive the following important property of the spectrum.

Proposition 5.31 *If (Ω, p) is a Riemann domain over a quasi-complete locally convex space E then $(S_c(\Omega), q)$ and $(\widehat{\Omega}_c, q)$ are pseudo-convex Riemann domains over E.*

Proof. Since a Riemann domain is pseudo-convex if and only if each of its connected components is pseudo-convex it suffices to prove the result for $S_c(\Omega)$. Let F denote a finite dimensional subspace of E and let K_F denote a compact subset of $S_c(\Omega)_F$. By Proposition 5.21(d) there exists K compact in Ω such that

$$\|\tilde{f}\|_{K_F} \le \|f\|_K$$

for all $f \in H(\Omega)$. If $\phi \in (\widehat{K_F})_{H(S_c(\Omega))}$ then

$$|\tilde{f}(\phi)| \le \|\tilde{f}\|_{K_F} \le \|f\|_K.$$

Hence $\phi \prec K$ and

$$\|\tilde{f}\|_{\widehat{(K_F)}_{H(S_c(\Omega))}} \le \|f\|_K$$

for all $f \in H(\Omega)$. Let α denote a continuous semi-norm on E such that $d_\Omega^\alpha(K) > 0$ and $\alpha|_F$ is a norm. By (5.16)

$$d_{S_c(\Omega)}^\alpha \left((\widehat{K_F})_{H(S_c(\Omega))} \right) \ge d_\Omega^\alpha(K) > 0.$$

Since

$$(\widehat{K_F})_{PSH(S_c(\Omega)_F)} \subset (\widehat{K_F})_{PSH(S_c(\Omega))} \subset (\widehat{K_F})_{H(S_c(\Omega))}$$

and α is a norm on F we have shown

$$d^{\alpha|F}_{\mathcal{S}_c(\Omega)_F} \left(\widehat{(\widehat{K_F})}_{PSH(\mathcal{S}_c(\Omega)_F)} \right) > 0.$$

By Proposition 5.27, $\mathcal{S}_c(\Omega)_F$ is pseudo-convex and, by Proposition 5.26, $\mathcal{S}_c(\Omega)$ is pseudo-convex. This completes the proof.

This completes the second stage of our investigation. The final, more technically demanding, stage consists of examining the structure of holomorphically convex and pseudo-convex domains over certain locally convex spaces. We do not aim at a comprehensive treatment, since excellent partial accounts already exist in book form (see Section 5.4), but present fully some non-trivial results. As a general guide to combinations that have yielded positive results we refer to Table 5.1.

We could discuss the problems defined by any choice of one entry from each row using columns (a) or (b). The entries in column (b) lead to problems that are more general or difficult than the entries on the same row in column (a). We discuss the simplest and most difficult options suggested by the above scheme. These are the Cartan–Thullen Theorem for domains in $\mathcal{DFM}$ spaces (all arising from column (a)) and the Levi problem for Riemann domains over Fréchet spaces with the bounded approximation property (all arising in column (b)). We also require, for Example 6.11, a solution to the Levi problem for $\mathcal{DFN}$ spaces and we prove this later (Corollary 5.45).

Table 5.1

	(a)	(b)
Riemann domain (Ω, p) over E	one-sheeted domain (or open subset of E)	arbitrary Riemann domains
locally convex space E	some class of DF spaces	some class of Fréchet spaces
problem	extend Cartan–Thullen Theorem	solve the Levi problem

Let E denote a $\mathcal{DFM}$ space. Then E'_β is Fréchet–Montel and contains a dense sequence $(\phi_j)_j$. For x, $y \in E$ let

$$\rho(x, y) = \sum_{j=1}^{\infty} \frac{1}{2^j} \frac{|\phi_j(x - y)|}{1 + |\phi_j(x - y)|}.$$

Then ρ defines a continuous metric on E which induces the original topology on each compact subset of E. Moreover, ρ is translation invariant, i.e. $\rho(x, y) = \rho(x - y, 0)$ for $x, y \in E$ and, using the strict convexity of $f(s) := s/(1 + s)$, we see that $\rho(\alpha x, 0)$ is a strictly increasing function of $|\alpha|$ for all $x \in E$. Let Ω denote a proper open subset of E and let $(K_m)_{m=1}^\infty$ denote a fundamental sequence of compact convex subsets of E such that $K_m \cap \Omega$ and $K_m \setminus \Omega$ are both non-empty for all m. For each $m \in \mathbb{N}$ and $\xi \in K_m \cap \Omega$ let

$$\rho_m(\xi) := \rho(\xi, K_m \setminus \Omega) := \inf\{\rho(\xi, x) : x \in K_m \setminus \Omega\}$$

and let

$$B_m(\xi) = \{x \in K_m : \rho(\xi, x) < \rho_m(\xi)\}.$$

Since $\xi \in K_m \cap \Omega$ it is clear that $B_m(\xi) \subset K_m \cap \Omega$. Choose $\eta \in K_m \setminus \Omega$ so that $\rho_m(\xi) = \rho(\xi, \eta)$ – this is possible since ρ is continuous and $K_m \setminus \Omega$ is compact. By the convexity of K_m, $x_t := \xi + t(\eta - \xi) \in K_m$ for all t, $0 < t < 1$, and

$$\rho(\xi, x_t) = \rho(\xi - x_t, 0) = \rho(t(\xi - \eta), 0) < \rho(\xi - \eta, 0) = \rho(\xi, \eta) = \rho_m(\xi).$$

Hence $x_t \in B_m(\xi)$ for $0 < t < 1$. Since $\lim_{t \to 1^-} x_t = \eta \notin \Omega$ we have proved $\overline{B_m(\xi)} \not\subset \Omega$ and $B_m(\xi) \not\subset L$ for any compact subset L of Ω.

Proposition 5.32 (Cartan–Thullen Theorem) *A holomorphically convex open subset Ω of a $\mathcal{DFM}$ space E is a domain of existence.*

Proof. We use the above notation and assume, without loss of generality, that Ω is connected. Thus $(K_m)_{m=1}^\infty$ denotes a fundamental sequence of convex compact subsets of E such that $K_m \cap \Omega$ and $K_m \setminus \Omega$ are non-empty for all m. Let $(L_m)_{m=1}^\infty$ denote an increasing fundamental sequence of compact subsets of Ω satisfying $(\widehat{L_m})_{H(\Omega)} = L_m$ for all m. For each m let D_m denote a countable dense subset of $K_m \cap \Omega$ and let $D = \{(m, \xi) : m \in \mathbb{N}, \xi \in D_m\}$. Now choose a sequence $(m_j, \xi_j)_j$ using elements of D as entries such that each point in D appears infinitely often in the sequence. Since $B_{m_j}(\xi_j) \not\subset L_k$ for all j and k we can, if necessary on replacing $(L_m)_m$ by a subsequence, choose a sequence $(\eta_j)_j \subset \Omega$ such that $\eta_j \in B_{m_j}(\xi_j) \cap (L_{j+1} \setminus L_j)$ for all j. Since $L_j = (\widehat{L_j})_{H(\Omega)}$ we can find for each j, $g_j \in H(\Omega)$, such that $|g_j(\eta_j)| > \|g_j\|_{L_j}$. If $\|g_j\|_{L_j} < \varepsilon_j < |g_j(\eta_j)|$ then

$$\lim_{s \to \infty} \left\| \left(\frac{g_j}{\varepsilon_j} \right) \right\|_{L_j}^s = \lim_{s \to \infty} \left(\frac{\|g_j\|_{L_j}}{\varepsilon_j} \right)^s = 0$$

and

$$\lim_{s \to \infty} \left| \left(\frac{g_j}{\varepsilon_j} \right)^s (\eta_j) \right| = \lim_{s \to \infty} \left(\frac{|g_j(\eta_j)|}{\varepsilon_j} \right)^s = \infty.$$

Choose s_j sufficiently large so that, for $f_j := g_j^{s_j}$, $\|f_j\|_{L_j} \leq 1/2^j$ and

$$|f_j(\eta_j)| \geq j + 1 + \left| \sum_{k<j} f_k(\eta_j) \right|$$

for all j. The series $\sum_j f_j$ converges uniformly on the compact subsets of Ω and hence, by Example 3.8(e), defines a holomorphic function f on Ω. It is also clear that $|f(\eta_j)| > j$ for all j.

We claim that Ω is the domain of existence for f. Let $p(x) = x$ for all $x \in \Omega$ and let $\tau: (\Omega, p) \to (\widetilde{\Omega}, \widetilde{p})$ denote a holomorphic extension of f, i.e. τ is a morphism and there exists a unique $\widetilde{f} \in H(\widetilde{\Omega}, \widetilde{p})$ such that $\widetilde{f} \circ \tau = f$. Since $\widetilde{p}(\tau(x)) = x$ for all $x \in \Omega$, τ is injective. If τ is not surjective there exists $a \in \partial(\tau(\Omega))$ in $\widetilde{\Omega}$. Let W denote a chart about a in $(\widetilde{\Omega}, \widetilde{p})$ such that $\widetilde{p}(W)$ is a convex subset of E and $\|\widetilde{f}\|_W < \infty$. Let $b \in \tau(\Omega) \cap W$. Since $\widetilde{p}(W)$ is convex

$$\widetilde{p}_W^{-1}\left(\widetilde{p}(b) + t(\widetilde{p}(a) - \widetilde{p}(b))\right) \in W$$

for $0 \leq t \leq 1$. Hence there exists ε, $0 < \varepsilon \leq 1$, such that

$$\widetilde{p}_W^{-1}\left(\widetilde{p}(b) + t(\widetilde{p}(a) - \widetilde{p}(b))\right) \in \tau(\Omega) \cap W$$

for $0 \leq t < \varepsilon$ and

$$c := \widetilde{p}_W^{-1}\left(\widetilde{p}(b) + \varepsilon(\widetilde{p}(a) - \widetilde{p}(b))\right) \notin \tau(\Omega).$$

This implies $c \in \partial(\tau(\Omega))$ in $\widetilde{\Omega}$, $\widetilde{p}(c) \in \partial\Omega$ in E and, by uniqueness of analytic continuation, f is bounded on the connected component U of $\Omega \cap \widetilde{p}(W)$ which contains $\widetilde{p}(b)$. We complete the proof by showing that this leads to a contradiction.

Choose m sufficiently large so that $\widetilde{p}(b)$ and $\widetilde{p}(a)$ belong to K_m. Since K_m is convex $\widetilde{p}(b) + t(\widetilde{p}(a) - \widetilde{p}(b)) \in \Omega \cap K_m \cap \widetilde{p}(W)$ for $0 \leq t < \varepsilon$. Let U_m denote the connected component of $\Omega \cap K_m \cap \widetilde{p}(W)$ containing $\widetilde{p}(b)$. Clearly $U_m \subset U$ and $\widetilde{p}(c) \in \partial U_m \cap \partial\Omega \cap K_m$. Since $\widetilde{p}(W) \cap K_m$ is a neighbourhood of $\widetilde{p}(c)$ in K_m and ρ generates the topology on each K_m we can choose $r > 0$ such that

$$B_m\left(\widetilde{p}(c), 2r\right) := \{x \in K_m : \rho(\widetilde{p}(c), x) < 2r\} \subset K_m \cap \widetilde{p}(W).$$

Now choose $\xi \in D_m \cap U_m \cap B_m\left(\widetilde{p}(c), r\right)$. Since $\widetilde{p}(c) \in K_m \setminus \Omega$, $\rho_m(\xi) < r$ and

$$B_m(\xi) \subset B_m(\xi, r) \subset B_m\left(\widetilde{p}(c), 2r\right) \subset K_m \cap \widetilde{p}(W).$$

Since $B_m(\xi) \subset \Omega$ and $B_m(\xi)$ is connected we have $B_m(\xi) \subset U_m$. By the definition of the sequence $(m_j, \xi_j)_j$ there exists an infinite sequence $(j_k)_k$ such that $\eta_{j_k} \in B_m(\xi)$ for all k. Since $f(\eta_{j_k}) \to \infty$ as $k \to \infty$ we conclude that f is unbounded on $B_m(\xi)$ and hence on U_m and U. This completes the proof.

We have arrived at a delicate point in our investigation as the remaining material we present is rather technical. The results we consider are known to hold for separable Fréchet spaces with the bounded approximation property. We have noted in the remarks after Definition 2.7 that such spaces are complemented subspaces of Fréchet spaces with a Schauder basis and, in the literature, many results are first proved using a Schauder basis and afterwards extended by complementation (see for instance Corollary 4.18). Since these results are also known to be false for arbitrary Banach spaces (Example 5.55) some hypotheses are necessary. A second technical obstacle arises if the Fréchet space does not admit a continuous norm. This difficulty can frequently be overcome, although not always in a routine fashion, with the aid of the following lemma.

Lemma 5.33 *(a) Let (Ω, p) denote a pseudo-convex Riemann domain over a locally convex space E and let $\alpha \in \mathrm{cs}(E)$. If there exists $x \in \Omega$ such that $d_\Omega^\alpha(x) > 0$ then for all $x \in \Omega$ and $y \in E$, $\alpha(y) = 0$,*

$$D_\Omega(x, y, \infty) \subset \Omega.$$

(b) Let E denote a Fréchet space with Schauder basis $(e_n)_n$ and let $\alpha \in \mathrm{cs}(E)$ satisfy

$$\alpha\Big(\sum_{n=1}^{\infty} x_n e_n\Big) = \sup_m \alpha\Big(\sum_{n=1}^{m} x_n e_n\Big)$$

for all $\sum_1^\infty x_n e_n \in E$. If $Z^\alpha = \{n \in \mathbb{N} : \alpha(e_n) = 0\}$ then $E^\alpha := \{\sum_1^\infty x_n e_n \in E : x_n = 0 \text{ for all } n \in Z^\alpha\}$ has a Schauder basis and E is a topological direct sum of E^α and $\alpha^{-1}(0)$.

We have chosen to state our main results for separable Fréchet spaces with the bounded approximation property while restricting the proofs to the case where we have a Schauder basis and continuous norm. This reduces somewhat the technical details but they are still considerable.

Let E denote a Fréchet space with Schauder basis $(e_n)_n$ which admits a continuous norm. If (Ω, p) is a Riemann domain over E then our general approach is to project onto the finite dimensional sections generated by the Schauder basis and to use the finite dimensional results presented in Proposition 5.27. We construct in $H(\Omega, \Omega)$ a sequence of mappings $(\tau_n)_{n=1}^\infty$ which behave like projections. However, the lack of a global linear structure on Ω means that we must first use p to go from Ω to E, then use projections on E_n and finally transfer back to Ω using local inverses of p. In Chapter 4 we chose a generating family of semi-norms so that the canonical projections onto the span of the basis mapped the unit ball of each semi-norm into itself. We require a similar property here both in order to obtain suitable estimates and to ensure that the compositions of certain mappings are well defined. The mappings δ_U^n, defined shortly, help us to manage locally the sequence $(\tau_n)_n$ while the functions d_Ω^α allow us to maintain a safe distance from the boundary and to approach it at a controlled rate. We require nine technical lemmata and maintain the same notation, that we now start introducing, in all of them.

Let π_n denote the canonical projection of E onto the span of the first n coordinates E_n of E and let $\pi^n = 1_E - \pi_n$. We fix once and for all an increasing fundamental sequence of continuous *norms* on E, $(\alpha_n)_{n=1}^{\infty}$, such that $\alpha_{n+1} \geq 2\alpha_n$ for all n and $\alpha_n = \sup_k \alpha_n \circ \pi_k$ and let $V_n = \{x \in E : \alpha_n(x) < 1\}$. For each n let $\Omega_n := \Omega_{E_n} = p^{-1}(p(\Omega) \cap E_n)$. If (Ω, p) is pseudo-convex then $(\Omega_n, p\big|_{\Omega_n})$ is a pseudo-convex domain over the finite dimensional space E_n. If U is an open subset of Ω and n is a positive integer let

$$\delta_U^n(x) = \inf_{k \geq n} d_U\left(x, \pi^k(p(x))\right)$$

for all $x \in U$.

Lemma 5.34 *If U is an open subset of (Ω, p) then $-\log \delta_U^n$ is upper semi-continuous and, if $(U, p\big|_U)$ is pseudo-convex, $-\log \delta_U^n$ is plurisubharmonic on U.*

Proof. If $x \in \Omega$, $\pi^k(p(x)) \to 0$ as $k \to \infty$. Hence $\liminf_{k \to \infty} d_U\left(x, \pi^k(p(x))\right) > 0$ and $\delta_U^n(x) > 0$ for all $x \in U$. To show that δ_U^n is lower semi-continuous, and hence that $-\log \delta_U^n$ is upper semi-continuous, we must show that $\{x \in U : \delta_U^n(x) > c\}$ is open for all $c \in \mathbb{R}$. If $\delta_U^n(x) > c$ and $c < r < \delta_U^n(x)$, Lemma 5.12 implies

$$L := \overline{\bigcup_{k \geq n} D\left(x, \pi^k(p(x)), r\right)}$$

is a compact subset of U. Hence we can choose $\alpha \in \mathrm{cs}(E)$ such that $d_U^\alpha(L) > 0$. Choose $s > 0$ such that $(1 + 2r)s < d_U^\alpha(L)$. If $y \in B_U^\alpha(x, s)$, $|\lambda| \leq r$ and $k \geq n$ then

$$\alpha\left(p(y) + \lambda\pi^k(p(y)) - p(x) - \lambda\pi^k(p(x))\right)$$
$$< (1 + |\lambda|)\,\alpha\left(p(y) - p(x)\right) + |\lambda|\alpha\left(\pi_k\left(p(y) - p(x)\right)\right)$$
$$\leq (1 + r)s + rs < d_U^\alpha(L).$$

Since $x + \lambda\pi^k(p(x)) \in L$ for $|\lambda| \leq r$ this implies $d_U\left(y, \pi^k(p(y))\right) \geq r$ for all $k \geq n$ and $\delta_U^n(y) \geq r > c$ for all $y \in B_U^\alpha(x, s)$. We have proved that $-\log \delta_U^n$ is upper semi-continuous.

If U is pseudo-convex then $-\log d_U$ is plurisubharmonic. Since the composition of a holomorphic function and a plurisubharmonic function is plurisubharmonic the function

$$x \in U \longmapsto -\log d_U\left(x, \pi^k(p(x))\right)$$

is plurisubharmonic for each k. Finally, since $-\log \delta_U^n(x) < +\infty$ for all x and the supremum of plurisubharmonic functions is plurisubharmonic if and only if it is upper semi-continuous and does not take the value $+\infty$, $-\log \delta_U^n$ is plurisubharmonic and the proof is complete.

For each positive integer n let

$$A_n = \{x \in \Omega \ : \ \delta_\Omega^n(x) > 1\} = \{x \in \Omega \ : \ -\log \delta_\Omega^n < 0\}$$
$$D_x = D_\Omega\big(x, \pi^n(p(x)), \delta_\Omega^n(x)\big) \quad \text{for } x \in A_n$$

and

$$\tau_n(x) = p_{D_x}^{-1} \circ \pi_n \circ p(x) \quad \text{for } x \in A_n.$$

For simplicity of notation we use D_x in place of the more appropriate $D_{x,n}$ – the context specifies the integer n. Since δ_Ω^n is increasing with n, A_n is increasing and $\bigcup_n A_n = \Omega$ and, as $\delta_\Omega^n(x) > 1$ for all $x \in A_n$, the mapping τ_n is well defined.

Lemma 5.35 *If (Ω, p) is a Riemann domain over a Fréchet space with Schauder basis and continuous norm then*

(a) $\tau_n \in H(A_n; \Omega_n)$ for all n,
(b) $\tau_n\big|_{\Omega_n} = $ identity,
(c) $\tau_n \circ \tau_{n+1} = \tau_{n+1} \circ \tau_n = \tau_n$ on A_n for all n,
(d) each τ_n is a π_n-morphism, i.e.

$$
\begin{array}{ccc}
A_n & \xrightarrow{\ \tau_n\ } & \Omega_n \\
{\scriptstyle p|_{A_n}}\downarrow & & \downarrow{\scriptstyle p|_{\Omega_n}} \\
E & \xrightarrow{\ \pi_n\ } & E_n
\end{array}
$$

is a commutative diagram,
(e) if K is a compact subset of Ω, V is a convex balanced neighbourhood of zero in E and $K + V \subset \Omega$ then there exists a positive integer n such that $K \subset A_n$ and $\tau_k(x) \in x + V$ for all $x \in K$ and all $k \geq n$,
(f) if (Ω, p) is pseudo-convex then each A_n is pseudo-convex.

Proof. (a) is obvious. If $x \in \Omega_n$ then $\pi_k(p(x)) = p(x)$ for all $k \geq n$. Hence $\delta_\Omega^n(x) = \infty$ and $\tau_n(x) = p_{D_x}^{-1} \circ p(x) = x$ for all $x \in \Omega_n$. This proves (b) and (c) follows from the relationships $\pi_n \circ \pi_{n+1} = \pi_{n+1} \circ \pi_n = \pi_n$ on E. Each τ_n is a π_n-morphism since $p \circ p_{D_x}^{-1} \circ \pi_n \circ p = p \circ \tau_n = \pi_n \circ p$ on A_n and this proves (d). Since $(A_n)_n$ is an increasing open cover of Ω, $K \subset A_n$ for all n sufficiently large. Since $\pi^k \to 0$ uniformly on compact subsets of E as $k \to \infty$ there exists a positive integer k_0 such that $\tau_k(x) \in x + V$ for all $x \in K$ and all $k \geq k_0$. This proves (e), and (f) follows from the plurisubharmonicity of $-\log \delta_\Omega^n$ and Proposition 5.30. This completes the proof.

Lemma 5.35 tells us that the sequence $(\tau_k)_k$ behaves on Ω in many ways like the sequence $(\pi_k)_k$ on E. In particular, we note that (e) implies $\tau_k(x) \to$

x as $k \to \infty$ uniformly on compact subsets of Ω. In the following lemma we construct, using a diagonal process, fundamental sequences of relatively compact sets in each Ω_k and an increasing countable open cover of Ω, each with special convexity properties. We keep the same notation.

Lemma 5.36 *Let (Ω, p) denote a connected Riemann domain over a Fréchet space with basis and continuous norm. There exist two countable increasing open covers of Ω, $(B_n)_n$ and $(C_n)_n$ and a neighbourhood basis at the origin $(W_n)_n$ such that*

(a) $C_n \subset B_n \subset A_n$ and $C_n + W_n \subset C_{n+1}$ for all n,
(b) $\overline{B_n \cap \Omega_k}$ is a compact subset of $A_n \cap \Omega_k$ for all k and n,
(c) $\tau_k(C_n) \subset B_n \cap \Omega_k$ for all $k \geq n$,
(d) if (Ω, p) is pseudo-convex then $(\widehat{B_n \cap \Omega_k})_{H(\Omega_k)}$ is a compact subset of $A_n \cap \Omega_k$ for all k and n.

Proof. Let $X_n = \{x \in A_n : d_{A_n}^{\alpha_n}(x) > 1\}$ and $Y_n = \{x \in X_n : \delta_{X_n}^n(x) > 1\}$. Clearly $(X_n)_n$ and $(Y_n)_n$ are increasing open covers of Ω. If $x \in Y_n$ and $k \geq n$, then $x + \lambda \pi^k(p(x)) \in X_n$ for $|\lambda| \leq 1$. Hence, letting $\lambda = -1$, we have

$$\tau_k(x) = p_{D_x}^{-1} \circ \pi_k \circ p(x)$$
$$= p_{D_x}^{-1}\big((1_E - \pi^k)(p(x))\big)$$
$$= p_{D_x}^{-1}\big(p(x) + \lambda \pi^k(p(x))\big)$$
$$= x + \lambda \pi^k(p(x))$$

for $k \geq n$ and $\tau_k(Y_n) \subset X_n$ for all $k \geq n$. Since $Y_n \subset Y_k \subset A_k$ and $\tau_k(A_k) \subset \Omega_k$, by Lemma 5.35(a), this implies $\tau_k(Y_n) \subset X_n \cap \Omega_k$ for all $k \geq n$. Hence $\tau_n(Y_n) \subset X_n \cap \Omega_n = Y_n \cap \Omega_n$ for every n.

We construct a sequence of continuous functions $\gamma_n : Y_n \to [0, +\infty]$ such that $\{x \in Y_n \cap \Omega_n : \gamma_n(x) < m\}$ is relatively compact in Ω_n for all n and m. Suppose, without loss of generality, that $Y_1 \cap \Omega_1 \neq \emptyset$. Fix $x_1 \in Y_1 \cap \Omega_1$. Let $\Gamma_n(x)$ denote the set of all finite sequences $(x_1, \ldots, x_s, x_{s+1})$ in Y_n with $x_{s+1} = x$ such that $D_\Omega(x_r, p(x_{r+1}) - p(x_r), 1 + q_r) \subset Y_n$ for some $q_r > 0$. Let

$$\gamma_n(x) = \inf\Big\{\sum_{r=1}^s \alpha_1\big(p(x_{r+1}) - p(x_r)\big) : (x_1, \ldots, x_{s+1}) \in \Gamma_n(s)\Big\}$$

if $\Gamma_n(x) \neq \emptyset$ and let $\gamma_n(x) = \infty$, otherwise. Clearly $\Gamma_n(x) \neq \emptyset$ if and only if x_1 and x belong to the same connected component of Y_n. In fact, $\Gamma_n(x)$ consists of the centres of a chain of discs in Y_n joining x_1 and x and γ_n may be considered a *length function*. If $x \in Y_n$, $\gamma_n(x) < \infty$ and $d_{Y_n}^\alpha(x) > 1$ for some continuous norm $\alpha \geq \alpha_1$ on E then there exists $\varepsilon > 0$ such that $x + w \in Y_n$ for $\alpha(w) \leq 1 + \varepsilon$. If $y \in B_{Y_n}^\alpha(x, 1/2)$ and $z \in B_{Y_n}^\alpha(y, 1/2)$ then $z = x + z_1 + z_2$

where z_1, $z_2 \in E$ and $\alpha(z_i) < 1/2$ for $i = 1,\ 2$. Moreover, $p(y) = p(x) + z_1$ and $p(z) = p(x) + z_1 + z_2$. Hence

$$x + \lambda\big(p(y) - p(x)\big) = x + \lambda z_1 \in Y_n$$

for $|\lambda| \leq 2(1 + \varepsilon)$ and

$$y + \lambda\big(p(z) - p(y)\big) = y + \lambda z_2 \in Y_n$$

for $|\lambda| \leq 1 + \varepsilon$. If $(x_1, \ldots, x_s, x) \in \Gamma_n(x)$ then $(x_1, \ldots, x_s, x, y) \in \Gamma_n(y)$ and $\gamma_n(x) < \infty$ implies $\gamma_n(y) < \infty$. If $(y_1, \ldots, y_r, y) \in \Gamma_n(y)$ then $(y_1, \ldots, y_r, y, z) \in \Gamma_n(z)$ and

$$\gamma_n(z) \leq \gamma_n(y) + \alpha_1\big(p(z) - p(y)\big). \tag{$*$}$$

Since $x \in B_{Y_n}^{\alpha}(y, 1/2)$ and $\alpha_1 \leq \alpha$ we may let $z = x$ in $(*)$ to obtain

$$\gamma_n(x) \leq \gamma_n(y) + \alpha\big(p(x) - p(y)\big)$$

and, on letting $y = x$ and $z = y$ in $(*)$, we get

$$\gamma_n(y) \leq \gamma_n(x) + \alpha\big(p(x) - p(y)\big).$$

Hence

$$|\gamma_n(y) - \gamma_n(x)| \leq \alpha_1\big(p(x) - p(y)\big)$$

for all $y \in B_{Y_n}^{\alpha}(x, 1/2)$ and γ_n is continuous for all n.

Now fix n and define inductively a sequence of sets K_m as follows:

$$K_1 := Y_n \cap B_{\Omega_n}^{\alpha_n}(x_1, 1/2),$$

$$K_{m+1} := Y_n \cap \bigcup\{B_{\Omega_n}^{\alpha_n}(x, 1/2) \ : \ x \in K_m\}.$$

Since each α_n is a norm K_1 is a relatively compact subset of Ω_n. If K_m is relatively compact in Ω_n then there exist points $y_1, \ldots, y_s \in K_m$ such that

$$K_m \subset \bigcup_{r=1}^{s} B_{\Omega_n}^{\alpha_n}(y_r, 1/2)$$

and

$$K_{m+1} \subset \bigcup_{r=1}^{s} B_{\Omega_n}^{\alpha_n}(y_r, 1).$$

By induction K_m is a relatively compact subset of Ω_n for all m.

Since each α_n is a norm and $E_n = \pi_n(E)$ is finite dimensional there exists $c > 1$ such that $\alpha_n \leq c\alpha_1$ on E_n. For each positive integer m let

$$L_m = \Big\{x \in Y_n \cap \Omega_n \ : \ \gamma_n(x) < \frac{m}{2c}\Big\}.$$

Our next step is to show, by induction, that $L_m \subset K_m$ for all m. If $\gamma_n(x) < 1/2c$ then there exists $\{x_1, \ldots, x_s, x_{s+1} = x\} \subset \Gamma_n(x)$ such that

$$\alpha_n\big(p(x_1) - p(x)\big) \leq \sum_{r=1}^{s} c\,\alpha_1\big(p(x_{r+1}) - p(x_r)\big) < \frac{c}{2c}.$$

Hence $x \in K_1$ and $L_1 \subset K_1$. Suppose $L_m \subset K_m$. Let $x \in L_{m+1}$. If $\gamma_n(x) < m/2c$ then $x \in L_m \subset K_m \subset K_{m+1}$. If $m/2c \leq \gamma_n(x) < (m+1)/2c$ choose $\{x_1, \ldots, x_{s+1}\} \subset \Gamma_n(x)$ such that

$$\sum_{r=1}^{s} \alpha_1\big(p(x_{r+1}) - p(x_r)\big) < \frac{m+1}{2c}. \qquad (**)$$

Since $\tau_n(Y_n) \subset Y_n \cap \Omega_n$, $D_\Omega(x, a, r) \subset Y_n$ implies

$$D_\Omega\big(\tau_n(x), \pi_n(a), r\big) \subset Y_n \cap \Omega_n.$$

Hence, on replacing x_r by $\tau_n(x_r)$ in $(**)$ if necessary, we may suppose

$$D_\Omega\big(x_r, p(x_{r+1}) - p(x_r), 1 + q_r\big) \subset Y_n \cap \Omega_n.$$

Let $t \leq s$ denote the first integer such that

$$\sum_{r=1}^{t} \alpha_1\big(p(x_{r+1}) - p(x_r)\big) \geq \frac{m}{2c}.$$

By the Intermediate Value Theorem we can find y in the line segment $[x_t, x_{t+1}]$ such that

$$\sum_{r<t} \alpha_1\big(p(x_{r+1}) - p(x_r)\big) + \alpha_1\big(p(y) - p(x_t)\big) < \frac{m}{2c}$$

and

$$\alpha_1\big(p(x_{t+1}) - p(y)\big) + \sum_{r>t} \alpha_1\big(p(x_{r+1}) - p(x_r)\big) < \frac{1}{2c}.$$

Then $\gamma_n(y) < m/2c$ and $y \in Y_n \cap \Omega_n$. By our induction hypothesis $y \in K_m$. Since $x_{s+1} = x$ and

$$\alpha_n\big(p(y) - p(x)\big) \leq c\alpha_1\big(p(y) - p(x)\big)$$

$$\leq c\alpha_1\big(p(y) - p(x_{t+1})\big) + c\sum_{r>t} \alpha_1\big(p(x_r) - p(x_{r+1})\big) < \frac{1}{2}$$

$x \in K_{m+1}$ and, by induction, $L_m \subset K_m$ for all m. The relative compactness of K_m in Ω_n implies that $\{x \in Y_n \cap \Omega_n : \gamma_n(x) < m\}$ is a relatively compact subset of Ω_n for all m and n.

Let $B_n = \{x \in Y_n : \gamma_n(x) < n\}$ for all n. Then $B_n \cap \Omega_k$ is relatively compact in $A_n \cap \Omega_k$ for all n and k and (b) holds. Since Ω is connected, $\Omega = \bigcup_{n=1}^{\infty} B_n$ and $\gamma_{n+1} \leq \gamma_n$ on Y_n implies that $(B_n)_n$ is an increasing sequence. If (Ω, p) is pseudo-convex then (d) follows from (b), Proposition 5.27 ((i) $\Longleftrightarrow$ (iv)) and Lemma 5.35(f).

Let $Z_n = \{x \in B_n : \delta^n_{B_n}(x) > 1\}$ and $C_n = \{z \in Z_n : d^{\alpha_n}_{Z_n}(x) > 1\}$. We have

$$C_n \subset Z_n \subset B_n \subset Y_n \subset X_n \subset A_n$$

which implies $\tau_k(C_n) \subset \tau_k(Z_n) \cap \Omega_k$ for $k \geq n$. If $z \in Z_n$ then

$$D_\Omega\big(z, \pi^k(p(z)), 1 + \varepsilon_k\big) \subset B_n$$

for all $k \geq n$ and some positive sequence $(\varepsilon_k)_{k \geq n}$. Hence $z - \pi^k(p(z)) = \tau_k(z) \in B_n$ and $\tau_k(Z_n) \subset B_n \cap \Omega_k$ for $k \geq n$ which proves (c). A routine calculation, using $2\alpha_n \leq \alpha_{n+1}$, shows that $C_n + W_n \subset C_{n+1}$ where $W_n := \{x \in E : \alpha_{n+1}(y) < 1\}$. This completes the proof of (a) and of the lemma.

To proceed we require the use of the finite dimensional $\overline{\partial}$ operator. We do not give a formal definition but instead list the properties that we need. We let (Ω, p) denote a Riemann domain over a finite dimensional space. The $\overline{\partial}$ operator is defined on

$$\bigcup_{p \geq 0, \, q \geq 0} \mathcal{C}^\infty_{(p,q)}(\Omega)$$

where $\mathcal{C}^\infty_{(p,q)}(\Omega)$ denotes the space of smooth (p, q)-forms on Ω. The space $\mathcal{C}^\infty_{(0,0)}(\Omega)$ is identified with $\mathcal{C}^\infty(\Omega)$ – the space of smooth functions on Ω. The $\overline{\partial}$ operator is a first order differential operator and obeys the product rule for differentiation.

Lemma 5.37 *Let (Ω, p) denote a Riemann domain over a finite dimensional space. Then*

(a) if $g \in \mathcal{C}^\infty(\Omega)$ then $\overline{\partial} g = 0 \iff g \in H(\Omega)$,
(b) if $g \in \mathcal{C}^\infty(\Omega)$ then $\overline{\partial}(\overline{\partial} g) = 0$,
(c) if U is an open subset of (Ω, p), $g \in \mathcal{C}^\infty(\Omega)$ and $g\big|_U \in H(U)$ then
$$\overline{\partial} g\big|_U = 0,$$
(d) if (Ω, p) is pseudo-convex, $v \in \mathcal{C}^\infty_{(0,1)}(\Omega)$ and $\overline{\partial} v = 0$ then there exists $u \in \mathcal{C}^\infty(\Omega)$ such that $\overline{\partial} u = v$.

An important and partially solved problem in infinite dimensional holomorphy is the so called "$\overline{\partial}$ problem" (see Section 4.6). Note that (d) above says that the finite dimensional problem has a solution on pseudo-convex domains.

Our next result shows that we can extend holomorphic functions from finite dimensional sections of a pseudo-convex Riemann domain to the whole space. Alternatively, it shows that the restriction mappings to the finite dimensional sections are surjective. Both points of view are useful.

Lemma 5.38 *Let (Ω, p) denote a connected pseudo-convex Riemann domain spread over a Fréchet space E with Schauder basis and continuous norm.*

Then, with our previous notation, for each $f_n \in H(\Omega_n)$ and each $\varepsilon > 0$ there exists $f \in H(\Omega)$ such that

(a) $f|_{\Omega_n} = f_n$,

(b) $\|f - f_n \circ \tau_n\|_{C_n} \leq \varepsilon$,

(c) $\|f\|_{C_j} < \infty$ *for all j.*

Proof. We first extend f_n to Ω_{n+1} and then proceed inductively. Let $X = \Omega_n \cup (B_n \cap \widehat{\Omega_{n+1}})_{H(\Omega_{n+1})}$ and $Y = \Omega_{n+1} \setminus A_n$. By Lemma 5.36(d), X and Y are disjoint closed subsets of Ω_{n+1}. Since connected Riemann domains over finite dimensional spaces are second countable manifolds (see Lemma 5.41) there exists $\psi \in C^\infty(\Omega_{n+1})$ such that $\psi \equiv 1$ on an open neighbourhood W_1 of X and $\psi \equiv 0$ on an open neighbourhood W_2 of Y. Let $\xi_{n+1} = \pi_{n+1} - \pi_n$ on E, i.e. ξ_{n+1} is evaluated at the $(n+1)^{\text{th}}$ coordinate of the basis and let

$$v(x) = \begin{cases} 0 & \text{if } x \in X \cup Y, \\ \dfrac{f_n(\tau_n(x))}{\xi_{n+1}(p(x))} \, \bar{\partial}\psi(x) & \text{if } x \in \Omega_{n+1} \setminus (X \cup Y). \end{cases}$$

The $(0,1)$-form v is well defined since $\bar{\partial}\psi \equiv 0$ on $W_1 \cup W_2$ and $\xi_{n+1}(p(x)) \neq 0$ when $x \in \Omega_{n+1} \setminus \Omega_n$.

By Lemma 5.37(b) and (c), $\bar{\partial}v = 0$ and, by Lemma 5.37(d), there exists $u \in C^\infty(\Omega_{n+1})$ such that $\bar{\partial}u = v$. Since τ_n is defined on A_n and $\psi = 0$ on $\Omega_{n+1} \setminus A_n$ the function

$$g := \psi \cdot f_n \circ \tau_n - (\xi_{n+1} \circ p) \cdot u$$

is well defined on Ω_{n+1}. As $f_n \circ \tau_n$ and $\xi_{n+1} \circ p$ are holomorphic, Lemma 5.37(a) implies

$$\bar{\partial}g = (f_n \circ \tau_n) \cdot \bar{\partial}\psi - (\xi_{n+1} \circ p) \cdot \bar{\partial}u = (f_n \circ \tau_n) \cdot \bar{\partial}\psi - (\xi_{n+1} \circ p) \cdot v = 0$$

on $\Omega_{n+1} \setminus (X \cup Y)$ while on $X \cup Y$,

$$\bar{\partial}g = -(\xi_{n+1} \circ p) \cdot \bar{\partial}u = -(\xi_{n+1} \circ p) \cdot v = 0.$$

By Lemma 5.37(a), $g \in H(\Omega_{n+1})$. Since $\bar{\partial}u - v$, $\bar{\partial}u = 0$ on $W_1 \cup W_2$ and u is holomorphic on a neighbourhood of $(B_n \cap \widehat{\Omega_{n+1}})_{H(\Omega_{n+1})}$. By Proposition 5.27(b) there exists $g_n \in H(\Omega_{n+1})$ such that $\|g_n - u\|_{B_n \cap \Omega_{n+1}} \leq \varepsilon/2c$ where $c = \|\xi_{n+1} \circ p\|_{B_n \cap \Omega_{n+1}}$. Let

$$h_{n+1} = g + (\xi_{n+1} \circ p) \cdot g_n = \psi \cdot (f_n \circ \tau_n) + \xi_{n+1} \circ p \cdot (g_n - u).$$

On Ω_n, $\psi \equiv 1$, $\tau_n(x) = x$ (Lemma 5.35(b)) and $\xi_{n+1}(p(x)) = 0$. This implies $h_{n+1}|_{\Omega_n} = f_n$. If $x \in B_n \cap \Omega_{n+1}$ then $\psi(x) = 1$. For all $x \in B_n \cap \Omega_{n+1}$

$$|h_{n+1}(x) - f_n(\tau_n(x))| = |\xi_{n+1}(p(x)) \cdot (g_n(x) - u(x))|$$
$$\leq \frac{|\xi_{n+1}(p(x))|\,\varepsilon}{2\|\xi_{n+1} \circ p\|_{B_n \cap \Omega_{n+1}}} \leq \frac{\varepsilon}{2}$$

and

$$\|h_{n+1} - f_n \circ \tau_n\|_{B_n \cap \Omega_{n+1}} \leq \frac{\varepsilon}{2}. \tag{5.19}$$

If $x \in C_n$ then $\tau_{n+1}(x) \in B_n \cap \Omega_{n+1}$, by Lemma 5.36(c), and $\overline{B_n \cap \Omega_{n+1}}$ is a compact subset of $A_n \cap \Omega_{n+1}$ by Lemma 5.36(b). By (5.19) and Lemma 5.35(c), and since $\tau_{n+1} \in H(A_{n+1}; \Omega_{n+1})$, we have

$$\left| h_{n+1}(\tau_{n+1}(x)) - f_n(\tau_n(x)) \right| \leq \frac{\varepsilon}{2} \tag{5.20}$$

for all $x \in C_n$. Proceeding inductively we can define $(h_j)_{j=n+1}^{\infty}$ such that $h_j \in H(\Omega_j)$ for all j, $h_{j+1}\big|_{\Omega_j} = h_j$ and

$$\|h_{j+1} \circ \tau_{j+1} - h_j \circ \tau_j\|_{C_j} \leq \frac{\varepsilon}{2^{j-n+1}}$$

for all $j \geq n$. Hence

$$\|h_k \circ \tau_k - h_j \circ \tau_j\|_{C_j} \leq \frac{\varepsilon}{2^{j-n}}$$

whenever $k \geq j \geq n$ and the sequence $(h_j \circ \tau_j)$ converges uniformly on each C_j to a function $f \in H(\Omega)$. Clearly $f = h_j$ on Ω_j for all j and, in particular, $f\big|_{\Omega_n} = f_n$. Moreover,

$$\|f - h_j \circ \tau_j\|_{C_j} \leq \frac{\varepsilon}{2^{j-n}} \tag{5.21}$$

for $j \geq n$. Taking $j = n$ in (5.21), we obtain

$$\|f - f_n \circ \tau_n\|_{C_n} \leq \varepsilon.$$

By Lemma 5.36(c), $\tau_j(C_j) \subset B_j \cap \Omega_j$ and, since $\overline{B_j \cap \Omega_j}$ is compact in Ω_j for all j (Lemma 5.36(b)), it follows from (5.21) that $\|f\|_{C_j} < \infty$ for each j. This completes the proof.

Lemma 5.39 *Let (Ω, p) denote a connected pseudo-convex Riemann domain over a Fréchet space with Schauder basis and continuous norm and let U denote a $PSH(\Omega)$-convex open subset of Ω. Then each $g \in H(U)$ is the limit in $(H(U), \tau_0)$ of a sequence $(f_n)_n \subset H(\Omega)$. Furthermore, there is a countable open cover of U, $(C_k(U))_k$, such that, if $g \in H^\infty(U)$, then the sequence $(f_n)_n$ can be chosen so that for each k, $\sup_n \|f_n\|_{C_k(U)} < \infty$.*

Proof. Since U is $PSH(\Omega)$-convex and Ω is pseudo-convex U and $U \cap \Omega_n$, $n = 1, 2 \ldots$ are pseudo-convex for all n. Let $g \in H(U)$ be given. Let

$$A_n(U) = \{x \in U \; : \; \delta_U^n(x) > 1\},$$
$$B_n(U) = \{x \in B_n \cap A_n(U) \; : \; d_{A_n(U)}^{\alpha_n}(x) > 1\},$$
$$C_n(U) = \{x \in C_n \cap B_n(U) \; : \; \delta_{B_n(U)}^n(x) > 1\}$$

where A_n, B_n and C_n are defined in Lemma 5.36. Since $B_n(U) \cap \Omega_n \subset B_n \cap \Omega_n$, Lemma 5.36(d) implies that $\big(B_n\widehat{(U)} \cap \Omega_n\big)_{H(\Omega_n)}$ is a compact subset of Ω_n and, as Ω_n is a finite dimensional pseudo-convex domain, Proposition 5.27(a) shows that

$$\big(B_n\widehat{(U)} \cap \Omega_n\big)_{H(\Omega_n)} = \big(B_n\widehat{(U)} \cap \Omega_n\big)_{PSH(\Omega)}.$$

Moreover, $B_n(U) \cap \Omega_n$ is contained in $U \cap \Omega_n$ and bounded away from the boundary of $U \cap \Omega_n$. Hence $B_n(U) \cap \Omega_n$ is a relatively compact subset of $U \cap \Omega_n$. Since U is $PSH(\Omega)$-convex, $\big(B_n\widehat{(U)} \cap \Omega_n\big)_{H(\Omega_n)}$ is a compact subset of $U \cap \Omega_n$. By Proposition 5.27(c), $H(\Omega_n)$ is τ_0-dense in $H(U \cap \Omega_n)$. Let $(h_n)_n \subset H(\Omega_n)$ be chosen so that

$$\|h_n - g\|_{B_n(U) \cap \Omega_n} \leq \frac{1}{n} \tag{5.22}$$

for all n. If $x \in C_n(U)$ then $D_\Omega\big(x, \pi^n(p(x)), 1 + \varepsilon\big) \subset B_n(U)$ for some $\varepsilon > 0$ and, in particular, $\tau_n(x) \in B_n(U)$. Since $\tau_n(C_n) \subset \Omega_n$ for all n we have $\tau_n(C_n(U)) \subset B_n(U) \cap \Omega_n$ and, by (5.22),

$$\|h_n \circ \tau_n - g \circ \tau_n\|_{C_n(U)} \leq \frac{1}{n} \tag{5.23}$$

for all n. For each n there exists, by Lemma 5.38, $f_n \in H(\Omega)$ such that

$$\|f_n - h_n \circ \tau_n\|_{C_n} \leq \frac{1}{n}. \tag{5.24}$$

Since $C_n(U) \subset C_n$, (5.23) and (5.24) imply

$$\|f_n - g \circ \tau_n\|_{C_n(U)} \leq \frac{2}{n} \tag{5.25}$$

for all n. If K is a compact subset of U and $\varepsilon > 0$ is arbitrary then, since g is continuous, we can find a convex balanced neighbourhood of 0 in E, V, such that $K + V \subset U$ and

$$|g(y) - g(x)| \leq \varepsilon$$

for $x \in K$ and $y \in x + V$. By Lemma 5.35(e) we can find $n_0 \geq 1/\varepsilon$ such that $K \subset C_{n_0}(U)$ and

$$\|g \circ \tau_n - g\|_K \leq \varepsilon$$

for all $n \geq n_0$. By (5.25),

$$\|f_n - g\|_K \leq 3\varepsilon$$

for all $n \geq n_0$. If $\|g\|_U < \infty$ then (5.25) implies that the sequence $(f_n)_n$ is uniformly bounded on each $C_k(U)$. This completes the proof.

The previous two results approximate holomorphic functions by functions which are uniformly bounded on an increasing countable open cover of Ω. We can, see for instance the proof of Proposition 5.17, endow this collection with the structure of a locally m-convex Fréchet algebra and apply the approach, suggested in our introduction of the τ_ω topology in Section 3.2, to study holomorphic convexity on (Ω, p). We need to show that the pseudo-convex Riemann domain (Ω, p) admits *sufficiently many suitable* $PSH(\Omega)$-convex subsets.

We continue with our previous notation and as usual (Ω, p) will denote a connected pseudo-convex Riemann domain over E where E is a Fréchet space with continuous norm and Schauder basis $(e_n)_n$. We recall that $(\alpha_n)_n$ is a fundamental system of monotone norms for E chosen so that $2\alpha_n \leq \alpha_{n+1}$ and $V_n = \{x \in E : \alpha_n(x) < 1\}$ for all n. Consider the locally m-convex Fréchet algebra (see Proposition 5.17)

$$H_C(\Omega) := \{f \in H(\Omega) : \|f\|_{C_n} < \infty \text{ for all } n\},$$

where $C := (C_n)_n$ denotes the increasing countable open cover of Ω defined in Lemma 5.36. If $A \subset \Omega$ let

$$\widehat{A}_C = \{x \in \Omega : |f(x)| \leq \|f\|_A \text{ for all } f \in H_C(\Omega)\}$$

and call $\widehat{A}_C$ the $H_C(\Omega)$-hull of A. By Lemma 5.36(a), $C_n + W_n \subset C_{n+1}$ for all n where $W_n := \{x \in E : \alpha_{n+1}(x) < 1\} = V_{n+1}$. Using this inclusion and the Cauchy inequalities we see that $\dfrac{\widehat{d^n f_a}}{n!} \in H_C(\Omega)$ when $f \in H_C(\Omega)$, $n \in \mathbb{N}$ and $a \in E$. If $x \in \widehat{(C_n)}_C$, $y \in W_n$ and $f \in H_C(\Omega)$, then

$$|f(x+y)| = \left| \sum_{m=0}^{\infty} \frac{\widehat{d^m f_y}}{m!}(x) \right| \leq \sum_{m=0}^{\infty} \left\| \frac{\widehat{d^m f_y}}{m!} \right\|_{C_n}$$

$$\leq \sum_{m=0}^{\infty} \frac{1}{\rho^m} \|f\|_{C_n + W_n} \leq \frac{\rho}{\rho - 1} \|f\|_{C_{n+1}},$$

where $\rho > 1$ is chosen so that $\rho y \in W_n$. Since $H_C(\Omega)$ is an algebra we can use the sequence $(f^j)_j$ in $H_C(\Omega)$ to deduce that $|f(x+y)| \leq \|f\|_{C_{n+1}}$ for all $f \in H_C(\Omega)$ and $y \in W_n$. Hence

$$\widehat{(C_n)}_C + W_n \subset \widehat{(C_{n+1})}_C \tag{5.26}$$

for all n.

Lemma 5.40 *Let (Ω, p) denote a connected pseudo-convex Riemann domain over a Fréchet space with Schauder basis and continuous norm. With the above notation*

(a) $\widehat{(C_n)}_C \cap \Omega_k \subset (\widehat{B_n \cap \Omega_k})_{H(\Omega_k)}$ for $k \geq n$,

(b) $\widehat{(C_n)}_C \cap \Omega_k$ is relatively compact in Ω_k for all k and n,

(c) $\widehat{(C_n)}_C + V_n \subset \Omega$ for every n.

Proof. Let $x \in \widehat{(C_n)}_C \cap \Omega_k$, $\varepsilon > 0$, $k \geq n$ and $f_k \in H(\Omega_k)$. By Lemma 5.38 there exists $f \in H_C(\Omega)$ such that $f\big|_{\Omega_k} = f_k$ and $\|f - f_k \circ \tau_k\|_{C_k} < \varepsilon$. By Lemma 5.36(c), $\tau_k(C_n) \subset B_n \cap \Omega_k$ for $k \geq n$. Hence

$$|f_k(x)| = |f(x)| \leq \|f\|_{C_n} \leq \|f_k \circ \tau_k\|_{C_n} + \varepsilon \leq \|f_k\|_{B_n \cap \Omega_k} + \varepsilon \tag{5.27}$$

and as ε and f_k were arbitrary this implies $x \in (\widehat{B_n \cap \Omega_k})_{H(\Omega_k)}$ when $k \geq n$. We have proved (a). By (a) and Lemma 5.36(d) we obtain (b) for $k \geq n$ and this implies (b) for all k and n.

We now prove (c). We keep the above notation, except we now suppose that $x \in \widehat{(C_n)}_C \cap C_k$. For $k \geq n$,

$$|f_k(\tau_k(x))| \leq |f(x)| + \varepsilon, \quad \text{since } x \in C_k,$$
$$\leq \|f\|_{C_n} + \varepsilon, \quad \text{since } x \in \widehat{(C_n)}_{H_C(\Omega)},$$
$$\leq \|f_k \circ \tau_k\|_{C_n} + 2\varepsilon, \quad \text{since } k \geq n \text{ implies } C_n \subset C_k,$$
$$\leq \|f_k\|_{B_n \cap \Omega_k} + 2\varepsilon, \quad \text{by Lemma 5.36(c).}$$

Since $\varepsilon > 0$ was arbitrary this implies $\tau_k(x) \in (\widehat{B_n \cap \Omega_k})_{H(\Omega_k)}$ for all $k \geq n$. By construction $B_n \subset A_n \subset \{x \in \Omega : d_\Omega^{\alpha_n}(x) > 1\}$. Hence

$$B_n \cap \Omega_k \subset \{x \in \Omega_k : d_{\Omega_k}^{\alpha_n}(x) > 1\}.$$

By Proposition 5.27(a) and the pseudo-convexity of Ω_k we obtain

$$\tau_k(x) \in (\widehat{B_n \cap \Omega_k})_{H(\Omega_k)} = (\widehat{B_n \cap \Omega_k})_{PSH(\Omega)}$$
$$\subset \{x \in \Omega_k : d_{\Omega_k}^{\alpha_n}(x) \geq 1\} \subset \{x \in \Omega : d_\Omega^{\alpha_n}(x) \geq 1\}.$$

Since $\lim_{k \to \infty} \tau_k(x) = x$ this implies

$$\widehat{(C_n)}_C \cap C_k \subset \{x \in \Omega : d_\Omega^{\alpha_n}(x) \geq 1\}$$

and

$$\widehat{(C_n)}_C = \bigcup_k \widehat{(C_n)}_C \cap C_k \subset \{x \in \Omega : d_\Omega^{\alpha_n}(x) \geq 1\}.$$

Hence $\widehat{(C_n)}_C + V_n \subset \Omega$. This proves (c) and completes the proof.

In order to adopt a sequential approach to compact sets in (Ω, p) we show, in our next lemma, that connected Riemann domains over separable Fréchet spaces are not too large. A topological space which admits a countable basis of open sets is said to be second countable.

Lemma 5.41 *A connected Riemann domain* (Ω, p) *over a separable Fréchet space is second countable, Lindelöf and separable.*

Proof. To show that (Ω, p) is second countable it suffices to find a countable covering of Ω consisting of second countable open sets. Fix x_1 in Ω and suppose, without loss of generality, that $d_\Omega^{\alpha_1}(x_1) > 1$. For each n let ω_n denote the connected component of $\{x \in \Omega : d_\Omega^{\alpha_n}(x) > 1/n\}$ which contains x_1. Since Ω is connected it is easily seen that $\Omega = \bigcup_{n=1}^\infty \omega_n$ and hence it suffices to prove that each ω_n is second countable. Let $\omega_{n,2} = B_\Omega^{\alpha_n}(x_1, \frac{1}{2n})$ and for each integer $m \geq 2$ let

$$\omega_{n,m} = \left\{ x_m \in \Omega_n : \text{ there exists } (x_2, \dots, x_{m-1}) \in \omega_n \text{ such that}\right.$$
$$\left. x_i \in B_\Omega^{\alpha_n}\left(x_{i-1}, \frac{1}{2n}\right) \text{ for } i = 2, \dots, m \right\}.$$

Again it is easily seen that $\omega_n = \bigcup_{m=2}^\infty \omega_{n,m}$ and it suffices to show that each $\omega_{n,m}$ is second countable. Since $\omega_{n,2} = B_\Omega^{\alpha_n}(x_1, \frac{1}{2n}) \cong B_E^{\alpha_n}(0, \frac{1}{2n})$, $\omega_{n,2}$ is second countable. Now suppose $\omega_{n,m}$ is second countable. Hence, $\omega_{n,m}$ is Lindelöf and there exists $(y_j)_{j=1}^\infty \subset \omega_{n,m}$ such that

$$\omega_{n,m} \subset \bigcup_{j=1}^\infty B_\Omega^{\alpha_n}\left(y_j, \frac{1}{2n}\right).$$

If $x \in \omega_{n,m+1}$ then $x \in B_\Omega^{\alpha_n}(y, \frac{1}{2n})$ for some $y \in \omega_{n,m}$ and hence

$$\omega_{n,m+1} \subset \bigcup_{j=1}^\infty B_\Omega^{\alpha_n}\left(y_j, \frac{1}{n}\right).$$

Since each $B_\Omega^{\alpha_n}(y_j, \frac{1}{n})$ is second countable we have proved that (Ω, p) is second countable. The Lindelöf property and separability follow from second countability and this completes the proof.

As a consequence of Lemma 5.41 we see that $p^{-1}(x)$ is finite or countable for each x in a connected Riemann domain (Ω, p) over a separable Fréchet space. Moreover, as a topological space, (Ω, p) is metrizable.

Lemma 5.42 *Let* (Ω, p) *denote a connected pseudo-convex Riemann domain over a Fréchet space E with Schauder basis and continuous norm. If K is a compact subset of E then*

$$L := p^{-1}(K) \cap \widehat{(C_n)}_c$$

is a compact subset of Ω for all n.

Proof. By Lemma 5.41 it suffices to show that each sequence in L has a convergent subsequence. Let $(x_j)_{j=1}^{\infty} \subset L$. By taking, if necessary, a subsequence we may suppose $p(x_j) \to a \in K$ as $j \to \infty$. Choose j_0 and $b \in \bigcup_{j=1}^{\infty} E_j = \bigcup_{j=1}^{\infty} \pi_j(E)$ such that $b - p(x_j) \subset V_{n+4}$ for all $j \geq j_0$. Suppose $b \in E_k$. By Lemma 5.40(c), $x_j + V_{n+4} \subset \Omega$ for all j. If $j \geq j_0$, then $y_j := x_j + (b - p(x_j)) \in \Omega$. Since $p(y_j) = b \in E_k$, $y_j \in \Omega_k$. Hence, by (5.26),

$$y_j \in ((\widehat{C_n})_C + V_{n+4}) \cap \Omega_k \subset (\widehat{C_{n+1}})_C \cap \Omega_k$$

for all $j \geq j_0$. By Lemma 5.40(b), $(y_j)_j$ contains a convergent subsequence, $(y_{n_j})_j$ with $y_{n_j} \to y \in (\widehat{C_{n+1}})_C \cap \Omega_k$ as $j \to \infty$. Hence $(y_{n_j})_{j=1}^{\infty} \cup \{y\}$ is a compact subset of Ω and Lemma 5.40(c) implies

$$\{(y_{n_j})_{j=1}^{\infty} \cup \{y\}\} + V_{n+3} \subset (\widehat{C_{n+1}})_C + V_{n+3} \subset \Omega.$$

Since $(p(x_{n_j}) - b)_{j=1}^{\infty}$ is a convergent sequence in V_{n+3}, Lemma 5.12 implies $x_{n_j} = y_{n_j} + p(x_{n_j}) - b \to w \in \Omega$ as $j \to \infty$. Since $p(w) = a \in K$ and $(\widehat{C_n})_C$ is a closed subset of Ω this implies $w \in L$ and completes the proof.

Proposition 5.43 *Let (Ω, p) be a pseudo-convex Riemann domain over a Fréchet space E with Schauder basis. Then every bounding subset of Ω is relatively compact.*

Proof. (When E has a continuous norm.) We may assume, without loss of generality, that Ω is connected. Let B denote a bounding subset of Ω. Then $p(B)$ is a bounding subset of E and, by Example 3.20(c), $\overline{p(B)}$ is a compact subset of E. The set $\{f \in H_C(\Omega) : \|f\|_B \leq 1\}$ is a closed convex balanced absorbing subset of $H_C(\Omega)$ and, as $H_C(\Omega)$ is Fréchet and hence barrelled, there exists a positive integer n_0 and $c > 0$ such that

$$\|f\|_B \leq c\|f\|_{C_{n_0}}$$

for all $f \in H_C(\Omega)$. Since $H_C(\Omega)$ is an algebra we can use powers of f to show

$$\|f\|_B \leq \|f\|_{C_{n_0}}$$

for all $f \in H_C(\Omega)$. Hence $B \subset (\widehat{C_{n_0}}) \cap p^{-1}(\overline{p(B)})$. By Lemma 5.41, B is a relatively compact subset of Ω. This completes the proof.

Proposition 5.44 (Levi Problem) *If (Ω, p) is a pseudo-convex Riemann domain over a separable Fréchet space with the bounded approximation property then Ω is holomorphically separated, holomorphically convex and a domain of existence.*

Proof. (For Fréchet spaces with Schauder basis and continuous norm.) We suppose, without loss of generality, that Ω is connected. If K is a compact subset of (Ω, p) then

$$\|f\|_{\widehat{K}_{H(\Omega)}} \leq \|f\|_K < \infty$$

for all $f \in H(\Omega)$. Hence $\widehat{K}_{H(\Omega)}$ is a closed bounding subset of Ω. By Proposition 5.43, $\widehat{K}_{H(\Omega)}$ is a compact subset of Ω and (Ω, p) is holomorphically convex.

Let $(\tau_k)_k$ and $\mathcal{C} = (C_n)_n$ denote the sequences constructed in Lemmata 5.35 and 5.36 respectively. Since $(C_n)_n$ is an increasing open cover of Ω and $\tau_k(x) \to x$ for all $x \in \Omega$ we can choose for a given x, $y \in \Omega$, $x \neq y$, a positive integer n such that $\tau_n(x) \in C_n$, $\tau_n(y) \in C_n$ and $\tau_n(x) \neq \tau_n(y)$. By Proposition 5.27 pseudo-convex Riemann domains over finite dimensional spaces are holomorphically separated. Hence there exists $f \in H(\Omega_n)$ such that $|f(\tau_n(x)) - f(\tau_n(y))| = \varepsilon > 0$. By Lemma 5.38 there exists $g \in H_{\mathcal{C}}(\Omega)$ such that $g\big|_{\Omega_n} = f$ and $\|g - f \circ \tau_n\|_{C_n} < \varepsilon/3$. Hence

$$|g(x) - g(y)| \geq \big|f(\tau_n(x)) - f(\tau_n(y))\big| - \big|g(x) - f(\tau_n(x))\big|$$
$$- \big|f(\tau_n(y)) - g(y)\big| > \frac{\varepsilon}{3}$$

and $H_{\mathcal{C}}(\Omega)$ separates the points of Ω. This shows that Ω is holomorphically separated.

Let D denote a countable dense subset of E, let $R = p^{-1}(D)$ and

$$S = \{(x, y) \in R \times R : x \neq y, \, p(x) = p(y)\}.$$

By Lemma 5.41, R, and hence S, is countable. Let $R =: (x_l)_{l=1}^{\infty}$. Since $H_{\mathcal{C}}(\Omega)$ separates the points of Ω, $S_{x,y} := \{g \in H_{\mathcal{C}}(\Omega) : g(x) \neq g(y)\}$ is non-empty for $(x, y) \in S$. Clearly $S_{x,y}$ is open. If $g \in S_{x,y}$, $f \in H_{\mathcal{C}}(\Omega)$ and $f \notin S_{x,y}$ then $f + (1/n)g \in S_{x,y}$ for $n \neq 0$ and $S_{x,y}$ is a dense open subset of the Fréchet space $H_{\mathcal{C}}(\Omega)$. Since S is countable the Baire Category Theorem implies

$$H_S(\Omega) := \bigcap_{(x,y) \in S} S_{x,y} \neq \emptyset.$$

For each triple of positive integers l, k and m choose $\xi_{l,k,m} \in E$ such that $\alpha_k(\xi_{l,k,m}) < d_{\Omega}^{\alpha_k}(x_l) + 1/m$ and $d_{\Omega}(x_l, \xi_{l,k,m}) \leq 1$ (see Exercise 5.78). Let $T = (\xi_{l,k,m})_{l,k,m=1}^{\infty}$ and let $(z_j)_j$ denote a sequence, with entries from T, such that each term appears infinitely often. By (5.26) and Lemma 5.40(c) we can, if necessary replacing $(C_j)_j$ with a subsequence, find $y_j \in C_j \setminus C_{j-1}$ such that, if $z_j = \xi_{l,k,m}$, then $y_j \in D_{\Omega}(x_l, \xi_{l,k,m}, d_{\Omega}(x_l, \xi_{l,k,m}))$. Let $h \in H_S(\Omega)$. Now choose inductively $(f_j)_j \in H_{\mathcal{C}}(\Omega)$ such that $\|f_j\|_{C_j} \leq 1/2^j$ and

$$|f_j(y_j)| > |h(y_j)| + j + 1 + \left|\sum_{i=1}^{j-1} f_i(y_j)\right|$$

for all j. The series $\sum_{j=1}^{\infty} f_j$ converges uniformly on each C_j and defines an element $f \in H_{\mathcal{C}}(\Omega)$. Since the quotients $(f(x) - f(y))/(h(x) - h(y))$, $(x, y) \in S$, are countable we can find $\theta \in (0, 1) \subset \mathbb{R}$ such that

$$f(x) - f(y) \neq \theta\big(h(x) - h(y)\big)$$

for all $(x, y) \in S$. Let $w = f - \theta h$. Then $w \in H_C(\Omega)$. We have $w(x) \neq w(y)$ for all x, $y \in S$ and $|w(y_j)| > j$ for all j.

To complete the proof we show that (Ω, p) is a domain of existence for w. Let $j: (\Omega, p) \to (\widetilde{\Omega}, \widetilde{p})$ denote a morphism for which there exists a unique $\widetilde{w} \in H(\widetilde{\Omega})$ satisfying $\widetilde{w} \circ j = w$. We first show that j is injective. If x, $y \in \Omega$ and $p(x) \neq p(y)$ then $\widetilde{p}(j(x)) = p(x) \neq p(y) = \widetilde{p}(j(y))$ and $j(x) \neq j(y)$. If $x \neq y$, $p(x) = p(y)$ and $j(x) = j(y)$ then, for all $n \in \mathbb{N}$ and all $z \in E$, we have

$$\frac{\widehat{d^n \widetilde{w}_z}}{n!}\big(j(x)\big) = \frac{\widehat{d^n \widetilde{w}_z}}{n!}\big(j(y)\big)$$

and hence

$$\frac{\widehat{d^n w_z}}{n!}(x) = \frac{\widehat{d^n w_z}}{n!}(y).$$

Since $x \neq y$ we can find $\alpha \in \mathrm{cs}(E)$ and $r > 0$ such that $d_\Omega^\alpha(x) > r$, $d_\Omega^\alpha(y) > r$ and $B_\Omega^\alpha(x, r) \cap B_\Omega^\alpha(y, r) = \emptyset$. Let $p_1 = p\big|_{B_\Omega^\alpha(x,r)}$ and $p_2 = p\big|_{B_\Omega^\alpha(y,r)}$. If $z \in E$ and $\alpha(z) < r$ then

$$w \circ p_1^{-1}\big(p(x) + z\big) = \sum_{n=0}^{\infty} \frac{\widehat{d^n w_z}}{n!}(x)$$

$$= \sum_{n=0}^{\infty} \frac{\widehat{d^n w_z}}{n!}(y)$$

$$= w \circ p_2^{-1}\big(p(x) + z\big).$$

Since D is dense in E we can choose z_0, $\alpha(z_0) < r$, such that $p(x) + z_0 = p(y) + z_0 \in D$. If $x_0 = p_1^{-1}\big(p(x) + z_0\big)$ and $y_0 = p_2^{-1}\big(p(y) + z_0\big)$ then $(x_0, y_0) \in S$. Since $w(x_0) = w(y_0)$ this is impossible and hence j is injective.

If $j(\Omega) \neq \widetilde{\Omega}$ then there exists a point a which lies in the boundary of $j(\Omega)$ in $\widetilde{\Omega}$. Suppose $B_{\widetilde{\Omega}}^{\alpha_k}(a, 2r) \subset \widetilde{\Omega}$ and $\|\widetilde{w}\|_{B_{\widetilde{\Omega}}^{\alpha_k}(a, 2r)} < \infty$. We can find $x_s \in R$ such that $j(x_s) \in B_{\widetilde{\Omega}}^{\alpha_k}(a, r)$. Hence $d_{\widetilde{\Omega}}^{\alpha_k}(x_s) < r$ and there exists m_0 such that $d_\Omega(x_s, \xi_{s,k,m}) < r$ for all $m \geq m_0$. In particular,

$$j\big(D_\Omega(x_s, \xi_{s,k,m_0}, d_\Omega(x_s, \xi_{s,k,m_0}))\big) = D_{\widetilde{\Omega}}\big(j(x_s), \xi_{s,k,m_0}, d_\Omega(x_s, \xi_{s,k,m_0})\big)$$
$$\subset B_{\widetilde{\Omega}}^{\alpha_k}(a, 2r).$$

By construction there exists an infinite sequence $(j_k)_k$ such that $z_{j_k} = \xi_{s,k,m_0}$ for all j_k. Hence $y_{j_k} \in D_\Omega(x_s, \xi_{s,k,m_0}, d_\Omega(x_s, \xi_{s,k,m_0}))$ and $j(y_{j_k}) \subset B_{\widetilde{\Omega}}^{\alpha_k}(a, 2r)$. Since $\left|\widetilde{w}(j(y_{j_k}))\right| = |w(y_{j_k})| > j$ this implies

$$\|\widetilde{w}\|_{B_{\widetilde{\Omega}}^{\alpha_k}(a, 2r)} = \infty.$$

This contradicts our choice of a and r. Hence $j(\Omega) = \widetilde{\Omega}$ and (Ω, p) is a domain of existence for w. This completes the proof.

Corollary 5.45 *A pseudo-convex open subset U of a $\mathcal{DFM}$ space E with the approximation property is a domain of existence.*

Proof. (For $\mathcal{DFM}$ spaces with Schauder basis.) The case $U = E$ is easily handled and we suppose $U \neq E$. By Proposition 5.32, it suffices to show that U is holomorphically convex. Since every $\mathcal{DFM}$ space is barrelled and admits a continuous norm the topology on E is generated by a family of norms $\mathcal{A}$ such that each $E_\alpha := (E, \alpha)$ has a Schauder basis. By Example 3.11, we can choose $\alpha \in \mathcal{A}$ such that U is an α-open subset of E and, by Proposition 5.26, $U_\alpha := (U, \alpha)$ is a pseudo-convex open subset of E_α. The mapping $x \in U_\alpha \mapsto d_U^\alpha(x)$ is continuous and, since $d_U^\alpha(x) = \inf\limits_{\substack{a \in E \\ \alpha(a) \leq 1}} d_U(x, a)$ and $-\log d_U(\cdot, \cdot)$ is plurisubharmonic, $\theta := -\log d_U^\alpha \in PSH(U_\alpha)$.

For each $x \in U_\alpha$ the *radius of boundedness* of θ at x, $r_{\alpha, \theta}(x)$, is defined by letting

$$r_{\alpha, \theta}(x) = \{\sup r > 0 \; : \; B_E^\alpha(x, r) \subset U \text{ and } \sup_{y \in B_E^\alpha(x, r)} \theta(y) < \infty\}.$$

Since $U \neq E$ it is easily seen that

$$|r_{\alpha, \theta}(x) - r_{\alpha, \theta}(y)| \leq \alpha(x - y) \tag{$*$}$$

for all $x, y \in U$ and $\omega := -\log r_{\alpha, \theta} \in PSH(U_\alpha)$.

We define $\widehat{U} = \{x' \in \widehat{E}_\alpha \; : \; \alpha(x - x') < d_U^\alpha(x) \text{ for some } x \in U\}$. Clearly $\widehat{U}$ is an open subset of $\widehat{E}_\alpha$ and $\widehat{U} \cap E = U$. Define $\widetilde{r}_{\alpha, \theta} : \widehat{U} \longrightarrow [-\infty, +\infty)$ by letting

$$\widetilde{r}_{\alpha, \theta}(x') = \limsup_{x \in U \to x'} r_{\alpha, \theta}(x).$$

Since $|\widetilde{r}_{\alpha, \theta}(x') - \widetilde{r}_{\alpha, \theta}(y')| \leq \alpha(x' - y')$ for all $x', y' \in \widehat{U}, \widetilde{r}_{\alpha, \theta}$ and $\widetilde{\omega} := -\log \widetilde{r}_{\alpha, \theta}$ are continuous functions on $\widehat{U}$. If $a \in \widehat{U}$ and $\varepsilon > 0$ are arbitrary we can choose by continuity, a neighbourhood V of 0 in $\widehat{E}_\alpha$, such that

$$\omega(a' + b' e^{i\theta}) \leq \widetilde{\omega}(a + b e^{i\theta}) + \varepsilon$$

for all $\theta \in [0, 2\pi]$, $a' \in (a + V) \cap E$, $b \in V$ and $b' \in V \cap E$. Hence

$$\widetilde{\omega}(a) = \limsup_{a' \in E \to a} \omega(a')$$

$$\leq \limsup_{a' \in E \to a} \frac{1}{2\pi} \int_0^{2\pi} \omega(a' + b' e^{i\theta}) \, d\theta$$

$$\leq \frac{1}{2\pi} \int_0^{2\pi} \widetilde{\omega}(a + b e^{i\theta}) \, d\theta + \varepsilon.$$

Since $\varepsilon > 0$ was arbitrary, $\tilde{\omega} \in PSH(\widehat{U})$. Moreover, $\tilde{\omega}(x) \to \infty$ as $x \in \widehat{U} \to \partial\widehat{U}$. This implies that $\widehat{U}$ is PSH-convex and hence a pseudo-convex open subset of $\widehat{E}_\alpha$. Since $\widehat{E}_\alpha$ is a Banach space with Schauder basis, Proposition 5.43 implies that $\widehat{U}$ is holomorphically convex. If K is a compact subset of U then K is also a compact subset of $\widehat{U}$ and, since $\widehat{K}_{H(U)} \subset \widehat{K}_{H(\widehat{U})}$, we have $d_U^\alpha(\widehat{K}_{H(U)}) > 0$. Hence $\widehat{K}_{H(U)}$ is bounded away from the boundary of U. Since $\widehat{K}_{H(U)}$ is contained in the convex hull of K this implies that $\widehat{K}_{H(U)}$ is a compact subset of U and completes the proof.

Proposition 5.46 *Let (Ω, p) be a pseudo-convex Riemann domain spread over a Fréchet space with Schauder basis and continuous norm. If K is compact in Ω and $K = \widehat{K}_{H(\Omega)}$ then K contains a neighbourhood basis consisting of $PSH(\Omega)$-convex open sets.*

Proof. We suppose, without loss of generality, that Ω is connected. Let $\mathcal{C} = (C_n)_n$ denote the covering of Ω constructed in Lemma 5.36 and let U denote an open subset of Ω which contains K. If $\overline{\Gamma}(p(K))$ is the closed convex hull of $p(K)$, Lemma 5.42 implies $L := p^{-1}\big(\overline{\Gamma}(p(K))\big) \cap \widehat{(C_n)}_{\mathcal{C}}$ is a compact subset of Ω for all n. Fix n so that $K \subset C_n$. It follows that $K \subset L$. Since $\widehat{K}_{H(\Omega)} = K$ there exists for each $x \in L\backslash U$, $f_x \in H(\Omega)$, such that $|f_x(x)| > 1 > \|f_x\|_K$ and, as $L\backslash U$ is compact, we can find $f_1, \ldots, f_m \in H(\Omega)$ such that $\sup_j \|f_j\|_K < 1$ and $L\backslash U \subset \bigcup_{j=1}^m \{x \,:\, |f_i(x)| > 1\}$. Hence $L \cap \{x \in \Omega \,:\, |f_j(x)| \le 1, j = 1, \ldots m\} \subset U$.

Let $u = \sup(|f_1|, \ldots, |f_m|)$. Then $u \in PSH(\Omega)$ and

$$p^{-1}\big(\overline{\Gamma}(p(K))\big) \cap \widehat{(C_n)}_{\mathcal{C}} \cap \{x \in \Omega : u(x) \le 1\} \subset U. \qquad (5.28)$$

By Lemma 5.36, $C_n + W_n \subset C_{n+1}$ for all n. We claim there exists a positive integer k such that

$$p^{-1}\left(\overline{\Gamma}(p(K)) + W_k\right) \cap \widehat{(C_n)}_{\mathcal{C}} \cap \{x \in \Omega : u(x) \le 1\} \subset U.$$

Otherwise, we can find sequences $(x_k)_{k>n}$ and $(b_k)_{k>n}$ such that

$$x_k \in p^{-1}\left(\overline{\Gamma}(p(K)) + W_k\right) \cap \widehat{(C_n)}_{\mathcal{C}} \cap \{x \in \Omega : u(x) \le 1\} \backslash U,$$

$b_k \in \overline{\Gamma}(p(K))$ and $p(x_k) - b_k \in W_k$ for all k. By (5.26), $y_k := x_k + b_k - p(x_k) \in \widehat{(C_n)}_{\mathcal{C}} + W_k \subset \widehat{(C_{n+1})}_{\mathcal{C}}$ for all k. Since $p(y_k) = b_k \in \overline{\Gamma}(p(K))$ we have

$$y_k \in p^{-1}\left(\overline{\Gamma}(p(K))\right) \cap \widehat{(C_{n+1})}_{\mathcal{C}}$$

for all k. By Lemma 5.42 this set is compact and, taking a subsequence if necessary, we may suppose $y_k \to y$ as $k \to \infty$. Since $p(x_k) - b_k \to 0$ as $k \to \infty$ and $x_k = y_k + p(x_k) - b_k$ it follows that $\lim_k x_k = y$. Hence $y \in$

$p^{-1}\left(\overline{\Gamma}(p(K))\right) \cap \widehat{(C_n)}_C \cap \{x \in \Omega : u(x) \le 1\} \setminus U$. This contradicts (5.28) and establishes our claim. Hence there exists a convex balanced neighbourhood of 0, W, such that

$$V := p^{-1}\left(\overline{\Gamma}(p(K)) + W\right) \cap \text{Int } \left(\widehat{(C_n)}_C\right) \cap \{x \in \Omega : u(x) < 1\} \subset U.$$

Since $u \in PSH(\Omega)$ the set $\{x \in \Omega : u(x) < 1\}$ is $PSH(\Omega)$-convex by Proposition 5.30. The set $p^{-1}\left(\overline{\Gamma}(p(K)) + W\right)$ is $PSH(\Omega)$-convex by the Hahn–Banach Theorem since p is holomorphic and $\overline{\Gamma}(p(K)) + W$ is convex. If J is a compact set contained in the interior of $\widehat{(C_n)}_C$ then there exists a convex balanced neighbourhood of 0, W, such that $J + W \subset \widehat{(C_n)}_C$. By using Cauchy estimates, as we did to establish (5.26), we obtain $\widehat{J}_C + W \subset \widehat{(C_n)}_C$. Hence

$$\widehat{J}_{PSH(\Omega)} \subset \widehat{J}_C \subset \text{Interior } \left(\widehat{(C_n)}_C\right)$$

and the interior of $\widehat{(C_n)}_C$ is $PSH(\Omega)$-convex. Since the finite intersection of $PSH(\Omega)$-convex sets is $PSH(\Omega)$-convex, V is $PSH(\Omega)$-convex. This completes the proof.

Proposition 5.47 (Oka–Weil Theorem) *Let (Ω, p) denote a pseudoconvex Riemann domain over a Fréchet space with Schauder basis and continuous norm. Let K be a compact subset of Ω such that $\widehat{K}_{H(\Omega)} = K$. For each open set U, $K \subset U \subset \Omega$, there is an open set V, $K \subset V \subset U$, such that each $g \in H(U)$ is the limit in $\left(H(V), \tau_0\right)$ of a sequence $(f_n)_n \subset H(\Omega)$. Furthermore, whenever $g \in H^\infty(U)$ we can choose $(f_n)_n$ such that $\sup_n \|f_n\|_V < \infty$.*

Proof. By Proposition 5.46, there exists a $PSH(\Omega)$-convex open subset W of Ω such that $K \subset W \subset U$. If $\mathcal{C}_W := \left(C_k(W)\right)_{k=1}^\infty$ is the increasing open cover of W constructed in Lemma 5.39 and $K \subset C_{k_0}(W)$ then the set $V := C_{k_0}(W)$ has the required properties.

Our aim now is to show that all the natural maximal extensions agree for domains spread over certain Fréchet spaces. We require the following lemma which bears comparison with Lemma 3.25.

Lemma 5.48 *If (Ω, p) is a connected Riemann domain over a separable Fréchet space, $\mathcal{F}$ is an equicontinuous or locally bounded subset of $H(\Omega)$ and $\phi \in S_b(\Omega)$ then $\phi|_{\mathcal{F}}$ is τ_p-continuous.*

Proof. Since $\tau_\delta \ge \tau_p$ it suffices to show the following: if $\varepsilon > 0$, $(f_\alpha)_\alpha \subset \mathcal{F}$ and $f \in \mathcal{F}$, $f_\alpha(x) \to f(x)$ for all $x \in \Omega$ as $\alpha \to \infty$, then there exists α_0 such that $|\phi(f_\alpha) - \phi(f)| \le \varepsilon$ for all $\alpha \ge \alpha_0$.

For each $x \in \Omega$ we can find a neighbourhood of 0 in E, V_x, such that $x + V_x \subset \Omega$ and $|g(x + y) - g(x)| < \varepsilon/3$ for all $y \in V_x$ and all $g \in \mathcal{F}$. By

Lemma 5.41, (Ω, p) is Lindelöf. Hence there exists a countable set $(x_i)_i$ in Ω such that $\Omega \subset \bigcup_{i=1}^{\infty} x_i + V_{x_i}$. Since ϕ is τ_δ-continuous there exists $c > 0$ and m a positive integer such that

$$|\phi(h)| \leq c\|h\|_{\bigcup_{i=1}^{m} x_i + V_{x_i}}$$

for all $h \in H(\Omega)$. For all n and all h we have

$$|\phi(h)| = |\phi(h^n)|^{1/n} \leq c^{1/n}\|h\|_{\bigcup_{i=1}^{m} x_i + V_{x_i}}$$

and letting $n \to \infty$ we see that we can take $c = 1$. Now choose α_0 such that

$$|f_\alpha(x_i) - f(x_i)| \leq \varepsilon/3$$

for all $\alpha \geq \alpha_0$ and $i = 1, \ldots, m$. For $\alpha \geq \alpha_0$,

$$|\phi(f_\alpha) - \phi(f)| \leq \sup_{i \leq m} \|f_\alpha - f\|_{x_i + V_i}$$

$$\leq \sup_{i \leq m} |f_\alpha(x_i) - f(x_i)| + \frac{2\varepsilon}{3}$$

$$\leq \varepsilon.$$

This completes the proof.

Proposition 5.49 *If (Ω, p) is a connected pseudo-convex Riemann domain over a separable Fréchet space E with the bounded approximation property then each non-zero τ_δ-continuous multiplicative functional on $H(\Omega)$ is a point evaluation at some point of Ω.*

Proof. (When E has Schauder basis and continuous norm.) Let $h \in \mathcal{S}_b(\Omega)$. If $\mathcal{C}$ denotes the open covering constructed in Lemma 5.36 then there exists a positive integer n_0 and $c > 0$ such that

$$|h(f)| \leq c\|f\|_{C_{n_0}}$$

for all $f \in H(\Omega)$. Since h is multiplicative we may suppose $c = 1$. Fix $n \geq n_0$ and $g \in H(\Omega_n)$. By Lemma 5.38 there exists a sequence $(f_k)_k \subset H_{\mathcal{C}}(\Omega)$ such that

$$\lim_{k \to \infty} \|f_k - g \circ \tau_n\|_{C_n} = 0 \tag{5.29}$$

where $(\tau_k)_k$ are the mappings defined in Lemma 5.35. Since $n \geq n_0$, (5.29) implies

$$\lim_{k, j \to \infty} \|f_k - f_j\|_{C_{n_0}} = 0$$

and $(h(f_k))_k$ is a Cauchy sequence in $\mathbb{C}$. Let

$$h_n(g) = \lim_{k \to \infty} h(f_k)$$

for all $g \in H(\Omega_n)$ and all $n \geq n_0$. It is clear that h_n does not depend on the approximation used and defines a non-zero multiplicative linear functional on $H(\Omega_n)$. By Lemma 5.36(c),

$$|h_n(g)| \leq \limsup_{k \to \infty} |h(f_k)| \leq \limsup_{k \to \infty} \|f_k\|_{C_{n_0}}$$

$$\leq \|g \circ \tau_n\|_{C_{n_0}} = \|g\|_{\tau_n(C_{n_0})} \leq \|g\|_{B_{n_0} \cap \Omega_n}$$

for all $g \in H(\Omega_n)$. By Lemma 5.36(d), $B_{n_0} \cap \Omega_n$ is a relatively compact subset of the finite dimensional pseudo-convex domain Ω_n and there exists, by Proposition 5.27(d), $a_n \in \Omega_n$ such that $h_n(g) = g(a_n)$ for all $g \in H(\Omega_n)$.

Now fix $f \in H(\Omega)$. By the above argument there exists, for $n \geq n_0$, $a_n \in \Omega_n$ and a sequence $(f_{n,k})_k$ in $H(\Omega)$ such that $\lim_{k \to \infty} \|f_{n,k} - f\big|_{\Omega_n} \circ \tau_n\|_{C_n} = 0$ and $\lim_{k \to \infty} h(f_{n,k}) = h_n(f\big|_{\Omega_n}) = f\big|_{\Omega_n}(a_n) = f(a_n)$. For each $n \geq n_0$ choose k_n such that

$$\|f_{n,k} - f\big|_{\Omega_n} \circ \tau_n\|_{C_n} \leq \frac{1}{n}$$

and

$$|h(f_{n,k}) - f(a_n)| \leq \frac{1}{n} \tag{$*$}$$

for all $k \geq k_n$. Since $\tau_n(C_n) \subset \Omega_n$, $f\big|_{\Omega_n} \circ \tau_n(x) = f \circ \tau_n(x)$ for all $x \in C_n$ and

$$\|f_{n,k} - f\big|_{\Omega_n} \circ \tau_n\|_{C_n} = \|f_{n,k} - f \circ \tau_n\|_{C_n}$$

for all n and k. We claim that $\lim_{n \to \infty} f_{n,k_n} = f$ in $(H(\Omega), \tau_0)$. Indeed, if K is compact in Ω and $\varepsilon > 0$ are arbitrary then we can find a neighbourhood of 0 in E, W, such that $K + W \subset \Omega$ and $|f(y) - f(x)| \leq \varepsilon$ whenever $x \in K$ and $y \in x + W$. By Lemma 5.35(e) we can find $n_1 \geq n_0$, $n_1 \geq 1/\varepsilon$ such that $K \subset C_n$, and $\tau_n(x) \in x + W$ whenever $x \in K$ and $n \geq n_1$. For $n \geq n_1$

$$\|f_{n,k_n} - f\|_K \leq \|f_{n,k_n} - f \circ \tau_n\|_{C_n} + \|f \circ \tau_n - f\|_K \leq 2\varepsilon$$

and we have established our claim. We have also shown that the sequence $(f_{n,k_n})_n$ is uniformly bounded on compact subsets of Ω. Hence $(f_{n,k_n})_n$ is an equicontinuous or locally bounded subset of $H(\Omega)$. By Lemma 5.48 and $(*)$

$$h(f) = \lim_{n \to \infty} h(f_{n,k_n}) = \lim_{n \to \infty} f(a_n)$$

for every $f \in H(\Omega)$. Hence the sequence $(a_n)_n$ is bounding and, by Proposition 5.43, contains a convergent subsequence $(a_{n_j})_j$. If $a_{n_j} \to a \in \Omega$ as $j \to \infty$ then $h(f) = f(a)$ for all $f \in H(\Omega)$ and this completes the proof.

Corollary 5.50 *If (Ω, p) is a connected holomorphically separated Riemann domain spread over a separable Fréchet space E with the bounded approximation property then*

$$\Sigma(\Omega) = \mathcal{S}_b(\Omega) = \mathcal{S}_c(\Omega) = \widehat{\Omega}_c$$

and $\widehat{\Omega}_c$ is a pseudo-convex, holomorphically separated domain of existence spread over E.

Proof. (When E has a Schauder basis and continuous norm.) If $h \in \mathcal{S}_b(\Omega)$ we define $\widetilde{h}$ on $H(\widehat{\Omega}_c)$ by letting $\widetilde{h}(f) := h\big(\delta^*(f)\big)$. By Proposition 5.22, $\widetilde{h}$ defines a non-zero multiplicative linear functional on $H(\widehat{\Omega}_c)$. If $(W_n)_{n=1}^\infty$ is an increasing countable open cover of $\widehat{\Omega}_c$ then $\Big(\delta^{-1}\big(W_n \cap \delta(\Omega)\big)\Big)_{n=1}^\infty$ is an increasing countable open cover of Ω. Hence there exists $c > 0$ and n_0 such that

$$|h(f)| \le c\|f\|_{\delta^{-1}\left(W_{n_0} \cap \delta(\Omega)\right)}$$

for all $f \in H(\Omega)$. This implies

$$|\widetilde{h}(f)| = \big|h\big(\delta^*(f)\big)\big| \le c\|\delta^*(f)\|_{\delta^{-1}\left(W_{n_0} \cap \delta(\Omega)\right)}$$
$$= c\|f\|_{W_{n_0} \cap \delta(\Omega)} = c\|f\|_{W_{n_0}}$$

for all $f \in H(\widehat{\Omega}_c)$, and hence $\widetilde{h} \in \mathcal{S}_b(\widehat{\Omega}_c)$. By Proposition 5.31, $\widehat{\Omega}_c$ is a connected pseudo-convex Riemann domain spread over E and, by Proposition 5.49, there exists x_h in $\widehat{\Omega}_c$ such that $\widetilde{h}(f) = f(x_h)$ for all $f \in H(\widehat{\Omega}_c)$. Since $\delta^*(\widetilde{f}|_{\widehat{\Omega}_c}) = f$ for all $f \in H(\Omega)$, where $\widetilde{f}(\phi) = \phi(f)$ for all $\phi \in \mathcal{S}_c(\Omega)$ (see Proposition 5.22), we have

$$h(f) = \widetilde{h}\big((\delta^*)^{-1}(f)\big) = \widetilde{h}(\widetilde{f}|_{\widehat{\Omega}_c}) = \widetilde{f}|_{\widehat{\Omega}_c}(x_h) = \widetilde{f}(x_h)$$

and $\mathcal{S}_b(\Omega) \subset \widehat{\Omega}_c$. Since $\widehat{\Omega}_c \subset \mathcal{S}_c(\Omega) \subset \mathcal{S}_b(\Omega)$ this implies $\widehat{\Omega}_c = \mathcal{S}_c(\Omega) = \mathcal{S}_b(\Omega)$. By Proposition 5.17, $\Sigma(\Omega) \subset \mathcal{S}_b(\Omega)$ and, by Proposition 5.22, $\widehat{\Omega}_c \subset \Sigma(\Omega)$. This completes the proof.

The following proposition completes our infinite dimensional extensions of Proposition 5.27.

Proposition 5.51 *If (Ω, p) is a pseudo-convex Riemann domain spread over a Fréchet space with Schauder basis and continuous norm and K is a compact subset of Ω then*

$$\widehat{K}_{PSH(\Omega)} = \widehat{K}_{H(\Omega)}.$$

Proof. We always have $\widehat{K}_{PSH(\Omega)} \subset \widehat{K}_{H(\Omega)}$ and, by Proposition 5.44, $\widehat{K}_{H(\Omega)}$ is a compact subset of Ω. If $a \in \widehat{K}_{H(\Omega)} \setminus \widehat{K}_{PSH(\Omega)}$ then there exists $u \in PSH(\Omega)$ such that $\sup_{x \in K} u(x) < 0 < u(a)$.

Our aim is to use the corresponding finite dimensional result (Proposition 5.27(a)). In order to do so, we must "holomorphically embed" $\widehat{K}_{H(\Omega)}$

in a finite dimensional section so that the restriction of u still separates K and a. We only require the approximation property to achieve this end but require Proposition 5.47 to complete the proof. In certain cases, e.g. $\mathcal{DFC}$ spaces, the approximation property alone suffices.

Since u is upper semi-continuous and $\widehat{K}_{H(\Omega)}$ is compact we can find $\alpha \in$ cs(E) and $\delta > 0$ such that $d_{\Omega}^{\alpha}(\widehat{K}_{H(\Omega)}) > 3\delta$ and $u < 0$ on $B_{\Omega}^{\alpha}(K, 3\delta)$. Let $T: E \rightarrow E$ denote a finite rank continuous linear operator such that $\alpha(T(p(x)) - p(x)) < \delta$ for all $x \in \widehat{K}_{H(\Omega)}$ and let S denote the continuous affine operator defined by $S(y) = T(y) + p(a) - T(p(a))$ for all $y \in E$. If $x \in \widehat{K}_{H(\Omega)}$ then

$$S(p(x)) = p(x) + T(p(x)) - p(x) + p(a) - T(p(a)). \tag{5.30}$$

Hence $S(p(a)) = p(a)$, $S(p(\widehat{K}_{H(\Omega)})) \subset B_{E}^{\alpha}(p(\widehat{K}_{H(\Omega)}), 2\delta)$ and $S(p(K)) \subset B_{E}^{\alpha}(p(K), 2\delta)$. Let

$$U = \{x \in \Omega \; : \; \alpha(S(p(x)) - p(x)) < d_{\Omega}^{\alpha}(x)\}.$$

If $x \in \widehat{K}_{H(\Omega)}$ then $d_{\Omega}^{\alpha}(x) > 3\delta$ and $\alpha(S(p(x)) - p(x)) < \delta$. Hence $\widehat{K}_{H(\Omega)} \subset U$. Let M denote the finite dimensional subspace of E generated by $S(E)$ and let

$$r(x) = \left(p|_{B_{\Omega}^{\alpha}(x, d_{\Omega}^{\alpha}(x))}\right)^{-1} \circ S \circ p(x)$$

for all $x \in U$. Since $p|_{\Omega_M} \circ r = S \circ p$, $r \in H(U, \Omega_M)$. If $x \in K$ then (5.30) implies $r(x) \in B_{\Omega}^{\alpha}(x, 2\delta)$ and hence $u(r(x)) < 0$. Since $u(r(a)) = u(a) > 0$ it follows that $r(a) \notin \widehat{r(K)}_{PSH(\Omega_M)}$ and as Ω_M is a pseudo-convex domain over a finite dimensional space, Proposition 5.27(a) implies, $r(a) \notin \widehat{r(K)}_{H(\Omega_M)}$. Hence there exists $g \in H(\Omega_M)$ such that

$$\|g\|_{r(K)} < |g(r(a))|.$$

If $h = g \circ r$ then $h \in H(U)$ and $\|h\|_K < |h(a)|$. Since $\widehat{K}_{H(\Omega)} \subset U$ the Oka–Weil Theorem (Proposition 5.47) implies there exists $f \in H(\Omega)$ such that

$$\|f\|_K < |f(a)|.$$

This contradicts the fact that $a \in \widehat{K}_{H(\Omega)}$ and completes the proof.

Finally we return to the problem of equality between the τ_0 and τ_ω topologies. Proposition 5.9 extends to Riemann domains and we obtain

$$(H(\Omega), \tau_0) = \varprojlim_{K \subset \Omega} (H(K), \tau_0) = \varprojlim_{K \subset \Omega} (H(K), \tau_\omega)$$

for any domain (Ω, p) spread over a Fréchet–Schwartz space. To obtain $\tau_0 = \tau_\omega$ on $H(\Omega)$ we must, by Corollary 5.23 and the above, show that

$$\big(H(\Omega), \tau_\omega\big) = \varprojlim_K \big(H(K), \tau_\omega\big)$$

when (Ω, p) is a connected pseudo-convex Riemann domain. This is the content of our next proposition.

Proposition 5.52 *If (Ω, p) is a connected pseudo-convex Riemann domain spread over a Fréchet–Schwartz space E with the bounded approximation property then*

$$\big(H(\Omega), \tau_\omega\big) = \varprojlim_K \big(H(K), \tau_\omega\big)$$

Proof. (For Fréchet spaces with Schauder basis and continuous norm.) Clearly the canonical identity mapping from $\big(H(\Omega), \tau_\omega\big)$ into $\varprojlim_K \big(H(K), \tau_\omega\big)$ is a continuous bijection. Let $(f_\alpha)_{\alpha \in \Gamma}$ denote a net in $H(\Omega)$ and suppose $[f_\alpha]_K \to 0$ in $\big(H(K), \tau_\omega\big)$ for all K compact in Ω. It suffices to show $q(f_\alpha) \to 0$ for each τ_ω-continuous semi-norm q on $H(\Omega)$. We may suppose, by Proposition 5.44, that q is ported by a compact subset K of Ω for which $K = \widehat{K}_{H(\Omega)}$. Let U denote an open subset of Ω with $K \subset U \subset \Omega$. By Proposition 5.46 there exists a $PSH(\Omega)$-convex open set V such that $K \subset V \subset U$.

Let $f \in H^\infty(U)$. The Oka–Weil Theorem implies there exists a sequence $(f_n)_n$ in $H(\Omega)$, uniformly bounded on V, such that $f_n\big|_V \to f\big|_V$ as $n \to \infty$ in $\big(H(V), \tau_0\big)$. By Corollary 5.10, extended to Riemann domains, there exists a neighbourhood W of K such that $\|f_n - f\|_W \to 0$. Hence $\|f_n - f_m\|_W \to 0$ as $n, m \to \infty$. If $c(W) > 0$ satisfies $q(f) \le c(W)\|f\|_W$ for all $f \in H(\Omega)$ then

$$|q(f_n) - q(f_m)| \le q(f_n - f_m) \le c(W)\|f_n - f_m\|_W \to 0$$

as $n, m \to \infty$. Clearly if $(g_n)_n \in H(\Omega)$ and $\|g_n - f\|_{W'} \to 0$ for some neighbourhood W' of K then $\lim_{n\to\infty} q(f_n) = \lim_{n\to\infty} q(g_n)$. Hence the function $\widetilde{q}(f) := \lim_{n\to\infty} q(f_n)$ is well defined for $f \in H^\infty(U)$. By letting U range over all neighbourhoods of K we obtain a semi-norm $\widetilde{q}$ on $H(K)$ such that $\widetilde{q}(f) = q(f)$ for all $f \in H(\Omega)$. We claim that $\widetilde{q}$ is τ_ω-continuous on $H(K)$. If not, there exists a neighbourhood W of K and $(f_n)_n \subset H^\infty(W)$, $\|f_n\|_W = 1$ and $\widetilde{q}(f_n) > n$ for all n. By our construction above there exists a neighbourhood $\widetilde{W}$ of K, $K \subset \widetilde{W} \subset W$, and $g_n \in H(\Omega) \cap H^\infty(\widetilde{W})$ such that $\|f_n - g_n\|_{\widetilde{W}} \le 1/n$ and $|\widetilde{q}(f_n) - q(g_n)| \le 1/n$ for all n. This implies $\lim_{n\to\infty} q(g_n) = \infty$ and $\|g_n\|_{\widetilde{W}} \le 2$ – a contradiction. Hence, if $(f_\alpha)_{\alpha \in \Gamma}$ is a net in $H(\Omega)$ and $[f_\alpha]_K \to 0$ in $\big(H(K), \tau_\omega\big)$ for all K compact in Ω then $\widetilde{q}([f_\alpha]_K) = q(f_\alpha) \to 0$ as $\alpha \to \infty$. This completes the proof.

Corollary 5.53 *If (Ω, p) is a holomorphically separating Riemann domain over a Fréchet–Schwartz space E with the bounded approximation property then $\tau_0 = \tau_\omega$ on $H(\Omega)$.*

Corollary 5.54 *If U is an open subset of a Fréchet–Schwartz space with the bounded approximation property then $\tau_0 = \tau_\omega$ on $H(U)$.*

In solving the Levi Problem we used, and were heavily dependent on, approximations of one kind or another and this required restrictions on the underlying space. Such restrictions cannot be removed without some sort of replacement. If E is a locally convex space we let $\mathcal{P}(E)$ denote the set of all continuous polynomials on E. If Ω is an open subset of E and $K \subset \Omega$ let

$$\widehat{K}_{\mathcal{P}(E)} = \{z \in \Omega \ : \ |P(z)| \leq \|P\|_K \text{ for all } P \in \mathcal{P}(E)\}.$$

An open set U is *polynomially convex* if the polynomial hull, $\widehat{K}_{\mathcal{P}(E)}$, of K is compact for each K compact in Ω. Since $\mathcal{P}(E) \subset H(E)$ polynomially convex sets are holomorphically convex. There exists an example of a connected polynomially convex open subset Ω of $c_0(\Gamma)$, Γ uncountable, which is not a domain of holomorphy. In particular, Ω is pseudo-convex and holomorphically convex and the *Levi problem does not always have a positive solution* over infinite dimensional Banach spaces. From this we obtain a further negative answer to the topologies problem (see Corollary 4.52).

Example 5.55 There exists an open subset Ω of $c_0(\Gamma)$, Γ uncountable, such that $\tau_\omega \neq \tau_\delta$ on $H(\Omega)$. In fact, if Ω denotes the counterexample mentioned above and $\Sigma(\Omega)$ is its envelope of holomorphy then Proposition 5.17 implies $\Sigma(\Omega) \subset S_b(\Omega)$. The set Ω can be identified with a proper subset of a connected component of $\Sigma(\Omega)$. Hence there exist $(x_j)_j \in \Omega$, $x_j \to x \in \partial\Omega$ as $j \to \infty$ and $p(f) := \sup_j |f(x_j)| < \infty$ for all $f \in H(\Omega)$. If p was τ_ω-continuous and ported by the compact subset K of Ω then, using the algebra structure of $H(\Omega)$, we would have

$$|f(x_j)| \leq \|f\|_K$$

for all j and $(x_j)_j \subset \widehat{K}_{H(\Omega)}$. This is impossible since Ω is holomorphically convex and shows that p is not τ_ω-continuous. Since τ_δ is a barrelled topology and each $\delta_{x_j} \in S_b(\Omega)$, p is τ_δ-continuous and we conclude that $\tau_\omega \neq \tau_\delta$ on $H(\Omega)$.

5.3 Exercises

Exercise 5.56* Let (Ω, p) denote a Riemann domain spread over a locally convex space E. If $f \in H(\Omega)$ show that $\rho_f^\alpha(x) := \sup\Big\{r > 0 :$

$$\sum_{n=0}^{\infty} \left\|\frac{\widehat{d^n} f(x)}{n!}\right\|_{B_E^\alpha(0,1)} r^n < \infty\Big\} = \left[\limsup_{n\to\infty}\left(\left\|\frac{\widehat{d^n} f(x)}{n!}\right\|_{B_E^\alpha(0,1)}^{1/n}\right)\right]^{-1}.$$ Show that

$x \to \rho_f^\alpha(x)$ is continuous for all $\alpha \in cs(E)$ and that $x \to -\log\rho_f^\alpha(x)$ is plurisubharmonic.

Exercise 5.57 If (Ω, p) is a connected Riemann domain over a Banach space $(E, \| \cdot \|)$ and there exists $a \in E$ such that $d_\Omega^{\| \cdot \|}(a) = \infty$ show that $d_\Omega^{\| \cdot \|}(x) = \infty$ for all $x \in E$ and that $p : \Omega \to E$ is a surjective homomorphism.

Show that a locally convex space is a domain of holomorphy and a domain of existence.

Exercise 5.58 If (Ω, p) is a Riemann domain over a locally convex space E and (Ω_1, p_1) and (Ω_2, p_2) are two envelopes of holomorphy of (Ω, p) show that $(\Omega_1, p_1) \simeq (\Omega_2, p_2)$. Show that $\sum(\Omega) = \bigcup_{i \in A} \sum(\Omega_i)$ where $\Omega_i, i \in A$, denote the connected components of Ω. Show that $\mathcal{S}_c(\Omega) = \bigcup_i \mathcal{S}_c(\Omega_i)$.

Exercise 5.59* Let (Ω, p) denote a Riemann domain over a locally convex space E and suppose Ω is not holomorphically separated. Let $x \sim y, x, y \in \Omega$, if $f(x) = f(y)$ for all $f \in H(\Omega)$. Let $\widetilde{\Omega} = \Omega_{/\sim}$. Show that p induces a mapping $\widetilde{p} : \widetilde{\Omega} \to E$ such that $(\widetilde{\Omega}, \widetilde{p})$ is a Riemann domain over E. Show that $\pi : (\Omega, p) \to (\widetilde{\Omega}, \widetilde{p})$, where π is the quotient mapping, is a non-injective holomorphic extension of (Ω, p).

Exercise 5.60 Show that every holomorphic extension of a holomorphically separated Riemann domain is holomorphically separated.

Exercise 5.61* Let $j : (\Omega, p) \to (\widetilde{\Omega}, \widetilde{p})$ denote a morphism of Riemann domains over a locally convex space E. Define $j^* : H(\widetilde{\Omega}) \to H(\Omega)$ by $j^*(f) = f \circ j$ all $f \in H(\widetilde{\Omega})$. Show that

$$j^* \left(\frac{\widehat{d^n} f_a}{n!} \right) = \frac{\widehat{d^n}(j^*(f))_a}{n!}$$

for all $f \in H(\widetilde{\Omega})$, all $a \in E$ and all $n \in \mathbb{N}$.

Exercise 5.62 If (Ω, p) is a Riemann domain over a locally convex space E and $f \in H(\Omega)$ show that $|f| \in PSH(\Omega)$. If $(u_\alpha)_\alpha \subset PSH(\Omega)$, $u = \sup_\alpha u_\alpha$ and $u(x) < \infty$ for all $x \in \Omega$, show that $u \in PSH(\Omega)$ if and only if u is upper semi-continuous. If $(\widetilde{\Omega}, \widetilde{p})$ is a Riemann domain over a locally convex space F, $g \in H(\Omega, \widetilde{\Omega})$ and $u \in P(\widetilde{\Omega})$ show that $g \circ u \in PSH(\Omega)$.

Exercise 5.63 A function $u : X \to [-\infty, \infty)$ where X is a topological space is said to be upper semi-continuous if $\{x \in X : u(x) < c\}$ is open for all $c \in \mathbb{R}$. We say that $v : X \to (-\infty, \infty]$ is lower semi-continuous if $-v$ is upper semi-continuous. Show that an upper semi-continuous function achieves its maximum on each compact subset of its domain. If X is metrizable show that each upper semi-continuous function is the pointwise limit of a decreasing sequence of continuous functions.

Exercise 5.64* Find an example of a proper open subset Ω of $\mathbb{C}$, with boundary containing a dense sequence $(x_n)_n$, and $f \in H(\Omega)$ such that for each n there exists a sequence in Ω, which converges to x_n, on which f is unbounded but such that Ω is not the domain of existence of f. Show that Ω is a domain of holomorphy.

Exercise 5.65 Let $f(z) = \sqrt{z}$ on $\mathbb{C} \setminus \{x \in \mathbb{R} : x \leq 0\}$. Construct the envelope of holomorphy of f, (Ω, p), and show that $\left|p^{-1}(x)\right| = 2$ for all $x \in \mathbb{C} \setminus \{0\}$ and $\left|p^{-1}(0)\right| = 0$.

Exercise 5.66* Let U be a balanced open subset of a quasi-complete locally convex space E. For each compact subset K of E let $\widehat{K}_{\mathcal{P}(E)} = \{x \in E : |P(x)| \leq \|P\|_K$ for all P in $\mathcal{P}(E)\}$. Let

$$\widehat{U} = \bigcup_{\substack{K \subset U \\ K \text{ compact}}} \widehat{K}.$$

Show that $\widehat{U}$ is a pseudo-convex polynomially convex balanced open subset of E. If $f = \sum_{n=0}^{\infty} \frac{\widehat{d}^n f(0)}{n!} \in H(U)$ show that $\sum_{n=0}^{\infty} \frac{\widehat{d}^n f(0)}{n!}$ converges and defines a holomorphic function $\widetilde{f}$ on $\widehat{U}$. Show that the mapping $f \in (H(U), \tau) \longrightarrow \widetilde{f} \in (H(\widehat{U}), \tau)$ is a linear topological isomorphism for $\tau = \tau_0$, τ_ω.

Exercise 5.67* If $\mathcal{F}$ is a collection of $\mathbb{C}$-valued holomorphic functions on the open subset U of the Banach space show that the mapping

$$F_{\mathcal{F}} : x \in U \to \left(f(x)\right)_{f \in \mathcal{F}} \subset \mathbb{C}^{\mathcal{F}}$$

is holomorphic when $\mathbb{C}^{\mathcal{F}}$ is endowed with the product topology. If U is a domain of holomorphy show that U is a domain of existence for $F_{H(U)}$.

Exercise 5.68* Let (Ω, p) denote a Riemann domain over a Fréchet space and let K denote a *locally connected* compact subset of Ω. Show that the topology of $(H(K), \tau_0)$ is generated by all semi-norms which satisfy (5.1).

Exercise 5.69* If (Ω, p) is a Riemann domain over a Fréchet–Schwartz space show that $(H(K), \tau_\omega)$ is a $\mathcal{DFS}$ space for each compact subset K of Ω.

Exercise 5.70* If (Ω, p) is a Riemann domain over a Fréchet space show that $(H(\Omega), \tau_0)$ and $(H(\Omega), \tau_\omega)$ are locally m-convex algebras.

Exercise 5.71 If U is an open polydisc in a fully nuclear space with basis show that $\mathcal{S}_b(U) \cong U$.

Exercise 5.72* If $\pi: E \to F$ is a continuous linear surjection, U is a pseudo-convex open subset of E and V is an open subset of F such that $\pi^{-1}(V) \subset U$ show that $\pi(U)$ is a pseudo-convex open subset of F and $U = \pi^{-1}(\pi(U))$.

Exercise 5.73* Let U denote a proper open subset of the Banach space E and let $H_b(U) = \{f \in H(U) : \|f\|_V < \infty$ for each U-bounded set $V\}$. Show that $\widehat{V}_{H_b(U)}$ is a U-bounded subset of U for each U-bounded set $V \subset U$ if and only if there does not exist V_1, V_2 open and connected in E, $\phi \neq V_2 \subset V_1 \cap U$, $V_1 \not\subset U$ such that for each $f \in H_b(U)$ there exists $f_1 \in H_b(V_1)$ satisfying $f|_{V_2} = f_1|_{V_2}$.

Exercise 5.74* If U and U' are two pseudo-convex open subsets of a Banach space with the bounded approximation property, $U \subset U'$, show that the following are equivalent:

(a) $H(U')$ is dense in $(H(U), \tau_0)$,

(b) $\widehat{K}_{H(U)} = \widehat{K}_{H(U')}$ for every compact subset K of U,

(c) $\widehat{K}_{PSH_c(U')}$ is a compact subset of U for every K compact in U.

 (PSH_c denotes continuous plurisubharmonic functions.)

Exercise 5.75* A subset U of a Banach space with unconditional basis $(e_j)_j$ is Reinhardt if $\sum_{j=1}^{\infty} x_j \lambda_j e_j \in U$ whenever $|\lambda_j| = 1$ for all j and $\sum_{j=1}^{\infty} x_j e_j \in U$. If U is an open Reinhardt domain containing the origin show that $(H(U), \tau_\pi) = (H(U), \tau_\omega)$.

Exercise 5.76* Consider the following conditions on a locally convex space E.

(A) E admits a fundamental directed set of semi-norms $\mathcal{A}$ and for each $\alpha \in \mathcal{A}$ there exists a bounded subset B_α of E such that $\pi_\alpha(B_\alpha)$ is a neighbour-hood of zero in E_α.

(B) E has a fundamental directed set of semi-norms Γ such that $(\mathcal{P}(^m E), \tau_\omega)$ induces the norm (i.e. $\|\cdot\|_\beta$) topology on $\mathcal{P}(^m E_\beta)$ for all $\beta \in \Gamma$.

(C) E is an open surjective limit of normed linear spaces (in particular if E is Fréchet then E is a quojection).

(N) E has a fundamental directed set of semi-norms $\mathcal{N}$ such that E_α is a Banach space for each $\alpha \in \mathcal{N}$.

 Show that (A) $\Longrightarrow$ (B), (A) $\Longrightarrow$ (C), (N) $\Longrightarrow$ (C). If E is Fréchet show that (A) $\Longleftrightarrow$ (C) $\Longleftrightarrow$ (N) $\Longrightarrow$ (B). If E is a distinguished Fréchet space show that all these conditions are equivalent.

Exercise 5.77* Let K denote a compact metric subset of a locally convex space E and suppose

(a) $(H(0_E), \tau_\omega)$ is regular,

and either

(b) if V is a convex balanced open subset of E and $(f_n)_n \subset H(V)$, $f_n \neq 0$ for each n, then there exists a bounded sequence in V, $(x_n)_{n=1}^{\infty}$, such that $f_n(x_n) \neq 0$ for all n

or

(c) $H(0_E)$ does not contain a non-trivial very strongly convergent sequence.

Show that $(H(K), \tau_\omega) = \varinjlim_U (H^\infty(U), \|\cdot\|)$ is a regular inductive limit.

Exercise 5.78* If (Ω, p) is a Riemann domain over a locally convex space E and $\alpha \in cs(E)$ show that d_Ω^α is continuous and d_Ω is lower semi-continuous. Show that $d_\Omega^\alpha(x) = \sup\{r > 0 \ : \ \alpha(a) < r \Rightarrow d_\Omega(x, a) > 1\}$ and if $d_\Omega^\alpha(x) > 0$, show that $d_\Omega^\alpha(x) = \inf\{d_\Omega(x, a) \ : \ \alpha(a) \leq 1\}$.

Exercise 5.79 If $\tau \colon (X, p) \to (\widetilde{X}, \widetilde{p})$ is a morphism of Riemann domains and $\tau^*(g) = g \circ \tau$ for all $g \in H(\widetilde{X})$ show that τ is a holomorphic extension if and only if $\tau^* \colon H(\widetilde{X}) \to H(X)$ is an algebra homomorphism.

Exercise 5.80* For (Ω, p), a fixed Riemann domain over a locally convex space E, let $\Omega/D(E)$ denote the category whose objects are the morphisms $j \colon (\Omega, p) \to (\Sigma, q)$ where (Σ, q) is a Riemann domain over E. A $\Omega/D(E)$-morphism between $j \colon (\Omega, p) \to (\Sigma, q)$ and $j' \colon (\Omega, p) \to (\Sigma', q')$ is a morphism $\phi \colon (\Sigma, q) \to (\Sigma', q')$ of Riemann domains such that $j' = \phi \circ j$. Prove that the category $\Omega/D(E)$ admits arbitrary products and projective limits. Let $\mathcal{A} \subset H(\Omega)$. For each $f \in \mathcal{A}$ let (Σ_f, p_f) denote the envelope of holomorphy of f and let $j_f \colon (\Omega, p) \to (\Sigma_f, p_f)$ denote the f-extension of (Ω, p). Show that the $\mathcal{A}$-envelope of holomorphy of (Ω, p) is the projective limit of (Σ_f, p_f), $f \in \mathcal{A}$, in $\Omega/D(E)$.

Exercise 5.81* Let $j \colon (\Omega, p) \to (\widetilde{\Omega}, \widetilde{p})$ denote a holomorphic extension where (Ω, p) and $(\widetilde{\Omega}, \widetilde{p})$ are Riemann domains spread over a Fréchet space and let F denote a sequentially complete locally convex space. Show that

$$j_F^* \colon H(\Omega, F) \to H(\widetilde{\Omega}, F), \quad j_F^*(f) = f \circ j$$

is surjective.

Exercise 5.82* Let U denote an open subset of a locally convex space with the approximation property. Show that U is polynomially convex if and only if all finite dimensional sections of U are polynomially convex.

Exercise 5.83 Show that the extension $j \colon (\Omega, p) \to (\widetilde{\Omega}, \widetilde{p})$ is an $\mathcal{A}$-envelope of holomorphy if and only if $(\widetilde{\Omega}, \widetilde{p})$ is an $\mathcal{A}$-domain of holomorphy.

Exercise 5.84 If U is a proper open subset of a locally convex space show that U is a domain of holomorphy if and only if for each sequence $(x_n)_n \subset U$ which converges to a point $x_0 \in \partial U$ there exists $f \in H(U)$ such that $\sup_n |f(x_n)| = \infty$.

Exercise 5.85* Let (Ω, p) denote a finitely sheeted Riemann domain spread over a locally convex space E, i.e. $\sup_{x \in E} p^{-1}(x) < \infty$. Show that Ω is holomorphically convex if and only if $d_\Omega^\alpha(\widehat{K}_{H(\Omega)}) = d_\Omega^\alpha(K)$ for all K compact in Ω and all $\alpha \in \mathrm{cs}(E)$.

Exercise 5.86* Let E be a separable Banach space with open unit ball B and let D denote the open unit disc in $\mathbb{C}$. Show that there exists $f \in H(D; B)$ such that $B \subset \overline{f(D)}$.

Exercise 5.87* Let F denote a reflexive Fréchet space and let

$$\mathcal{K} = \{K \ : \ K \subset E \text{ (Fréchet nuclear) and } H(K)'_\beta \text{ has property } (\widetilde{\Omega})\}.$$

Show that F has (DN) if and only if all F-valued weakly holomorphic germs on K, $K \in \mathcal{K}$, are holomorphic.

5.4. Notes

The two principal topics discussed in this chapter are holomorphic germs on compact subsets of Fréchet spaces and the Levi problem (first posed for finite dimensional spaces in 1911 by E.E. Levi [549]) for Riemann domains spread over Fréchet spaces. These were eventually combined to prove $\tau_0 = \tau_\omega$ for the space of holomorphic functions on an *arbitrary* open subset of a Fréchet–Schwartz space with the bounded approximation property (Proposition 5.52). The ideas and methods in the two topics are quite different and reflect our slow but persistent drift to non-linear aspects of infinite dimensional holomorphy. This change of emphasis, which continues in the next chapter, led to the two practically self-contained sections of text presented here.

Two of the important ideas associated with holomorphic germs emerged during 1970. A. Hirschowitz [458], while investigating problems of analytic continuation, holomorphic completions, etc. (see Section 6.1) needed to show that the inductive limit $\varinjlim_{K \subset U} (H^\infty(U), \| \cdot \|)$ was regular for K a compact subset of a Banach space. In the process of correcting his original proof Hirschowitz introduced the semi-norms (5.1) and (5.2). At the same time C.B. Chae [213, 214] was considering the following problem: if U is an open subset of a Banach space is $(H(U), \tau_\omega)$ complete? He approached this problem by localizing it to compact sets and introduced, as did A. Hirschowitz [458] independently, the τ_π topology on $H(U)$ (see Lemma 4.28) by letting

$$\left(H(U), \tau_\pi\right) = \varprojlim_{K \subset U} \left(H(K), \tau_\omega\right).$$

The general idea was to show $\tau_\pi = \tau_\omega$ and to prove that $\left(H(K), \tau_\omega\right)$ was complete for all K compact in U. Chae proceeded by defining a compact set K to be U-Runge if $H(U)$ is sequentially dense in $\left(H(K), \tau_\omega\right)$ and said that $H(U)$ had the *Runge property* if it contained a fundamental system of compact sets consisting of U-Runge sets. He showed $\tau_\pi = \tau_\omega$ on $H(U)$ and that $(H(U), \tau_\omega)$ was complete whenever U has the Runge property and noted that this was the case for balanced domains. J. Mujica [639, 640, 644] afterwards proved that $(H(U), \tau_\omega)$ was complete for any open subset of a Banach space and, as previously noted, this result has now been extended to arbitrary open subsets of Fréchet spaces.

On replacing τ_π by τ_0 and Banach spaces by Fréchet–Schwartz spaces one sees (e.g. consider Proposition 5.52) that this approach contains the essential ingredients of the overall strategy adopted in this chapter. To bridge the gap between the results of Chae and Hirschowitz required a deeper analysis, due mainly to J. Mujica and M. Schottenloher.

In his 1974 thesis [644] J. Mujica studied holomorphic germs on compact subsets of a Fréchet space and showed

$$\left(H(U), \tau_\omega\right) = \varprojlim_{K \subset U} \left(H(K) \cap (H(U), \tau_\omega)\right)$$

for arbitrary open sets and, in Fréchet spaces satisfying (B) of Exercise 5.76,

$$\left(H(U), \tau_\omega\right) = \varprojlim_{K \subset U} \left(H(K), \tau_\omega\right)$$

when $H(U)$ has the Runge property (these results are extended to Banach-valued holomorphic germs in [220]). A distinguished Fréchet space has condition (B) if and only if it is a quojection (see Exercise 5.76). He established the completeness of $\left(H(U), \tau_\omega\right)$ in certain cases and motivated further investigation of the Runge property, the τ_π topology and completeness of $\left(H(K), \tau_\omega\right)$. Further progress was made in 1976 by K.-D. Bierstedt–R. Meise [120, 121] who investigated holomorphic germs on compact subsets of Fréchet–Schwartz and Fréchet nuclear spaces. They showed that $\left(H(K), \tau_\omega\right)$ was $\mathcal{DFS}$ (respectively $\mathcal{DFN}$) for any K compact in the Fréchet space E if and only if E'_β is $\mathcal{DFS}$ (respectively $\mathcal{DFN}$). This was carried a stage further in 1976–77 by P. Avilés–J. Mujica [80] who obtained analogous results for quasi-normable Fréchet spaces. A comprehensive and unified survey on the state of development achieved by 1977–78 is given in K.-D. Bierstedt–R. Meise [122]. In this article the local or sheaf-like nature of the τ_π topology is highlighted and the authors show that τ_π is the (unique) sheaf topology associated with τ_ω. This was followed by S. Dineen [300] who proved, using the topology generated by the semi-norms (5.1) and (5.2), that $\big(H(K), \tau_\omega\big)$ is complete for any compact subset K of a Fréchet space (we discussed abstract linear developments associated with this result in Section 3.6).

In her thesis, published in 1981, O. Nicodemi [688] considered algebra homomorphisms between spaces of holomorphic germs and introduced the τ_0 topology on $H(K)$. The final step in obtaining the main results in Section 5.1 was taken by J. Mujica [646] who obtained Lemmata 5.1 and 5.2, Propositions 5.3, 5.4 and 5.9 and Corollaries 5.5 and 5.6 for univalent domains in 1981 and for Riemann domains the following year. Using Corollary 5.5, J.M. Ansemil–S. Ponte [28] obtained Proposition 5.7 which led to the first example of a Fréchet–Montel space E, which was not Fréchet–Schwartz, such that $\tau_0 = \tau_\omega$ on $H(E)$ (see Section 4.6). Corollary 5.8 is a special case of results in S. Dineen–M. Lindström [317] (see also F. Blasco [131, 132]). Corollary 5.10 is due to K.-D. Bierstedt–R. Meise [121, Theorem 7]. R.L. Soraggi [807, 809, 811] made extensive use of the semi-norms (5.1) and (5.2) in his examination of the regularity of $H(K)$, K compact in a non-metrizable locally convex space (see the remarks on Exercise 5.77). More recently, K.-D. Bierstedt–J. Bonet–A. Peris [118], J. Bonet–P. Domański–J. Mujica [160] and C. Boyd [183] investigated the regularity and completeness of spaces of holomorphic vector-valued germs on compact subsets of certain Fréchet spaces. The methods in [160] include linearization techniques (see Section 3.2).

With hindsight, and keeping in mind the developments that led to the results in Section 5.1 and other similar examples, it could be argued that holomorphic problems on Fréchet spaces go through five phases before reaching a satisfactory resolution (we hesitate to use the word solution since there are usually many unresolved problems at the end of the fifth phase). These phases are characterized by different collections of Fréchet spaces (Table 5.2), each more inclusive than the previous phase.

Table 5.2

Phase 1	Phase 2	Phase 3	Phase 4	Phase 5
$\mathbb{C}^{\mathbb{N}}$	Banach spaces Fréchet nuclear spaces	quojections Fréchet–Schwartz spaces	quasi-normable spaces Fréchet–Montel spaces	Fréchet spaces with splitting conditions

For example:
- $\tau_b = \tau_\omega$ on $H(U)$, U balanced, has gone through all five phases (Section 4.3),

-   $\tau_0 = \tau_\omega$ on $H(U)$, U arbitrary, phase 2 is complete (Section 5.2, if E is a quojection then $\tau_0 = \tau_\omega$ on $H(E)$ $\iff$ $E = \mathbb{C}^{\mathbb{N}}$),
-   $\tau_\omega = \tau_\delta$ on $H(E)$, phase 1 is complete, partial results in phase 2 are given in Section 4.2.

The pairs of spaces in phases 2, 3 and 4 have little in common ($\mathbb{C}^{\mathbb{N}}$ in the case of phases 3 and 4) and come from opposite ends of the spectrum. The approach at each phase is often quite different for each of the two spaces but the combined solutions often point the way forward. At the end of phase 4 the required splitting condition may have revealed itself (see the introduction to Chapter 4).

To the reader unfamiliar with the material in Section 5.2, we strongly recommend the books [235, 446, 651, 696] all of which were written from an infinite dimensional point of view. The books of Ph. Noverraz [696] and G. Coeuré [235] were written in 1972 and 1973, respectively, when the basic concepts they discuss, plurisubharmonic functions and pseudo-convex domains in the case of [696], Riemann domains and analytic continuation in [235], were just being developed for infinite dimensional spaces. Time has shown that both cover fundamental ideas and techniques in an efficient and clear fashion. The book of M. Hervé [446] appeared in 1989 and contains a comprehensive treatment of plurisubharmonic functions and pseudo-convex sets over open subsets of infinite dimensional spaces. J. Mujica's 1984 book [651], deals with all aspects of the Levi problem for Riemann domains over finite and infinite dimensional Banach spaces. The basic ideas are presented so clearly and carefully in [651] that it is eminently suitable as a text for an introductory graduate level course in several complex variables. Our goal is different, but not too different, to those in [651] and [446], and as the approach we follow is a development and refinement of both, the reader will find useful additional background information by consulting them.

During 1968–1988 a number of authors including V. Aurich, W. Bogdanowicz, G. Coeuré, J.F. Colombeau, S. Dineen, L. Gruman, Y. Hervier, A. Hirschowitz, J.M. Isidro, G. Katz, C.O. Kiselman, E. Ligocka, M.L. Lourenço, M.C. Matos, L.A. Moraes, J. Mujica, Ph. Noverraz, R. Pomes, N. Popa, C.E. Rickart, M. Schottenloher and L. Waelbroeck wrote on different aspects of the Levi problem, the Cartan–Thullen Theorem and the Oka–Weil Theorem over infinite dimensional spaces. The literature includes the following: [74, 75, 76, 77, 141, 234, 235, 236, 248, 277, 281, 283, 287, 288, 292, 293, 294, 316, 325, 410, 411, 412, 447, 453, 454, 455, 456, 457, 458, 460, 473, 487, 503, 504, 552, 553, 564, 565, 574, 575, 576, 578, 624, 625, 641, 642, 643, 645, 648, 649, 650, 652, 693, 694, 695, 696, 697, 698, 699, 700, 701, 702, 703, 708, 735, 736, 738, 744, 745, 777, 778, 779, 780, 781, 783, 784, 785, 786, 789, 793, 794, 850]. Techniques and ideas from these papers motivated, influenced and contributed to the final results presented here. We *confine* our remarks, for the most part, to articles of *immediate relevance* to the results presented here. Our overall strategy and technical approach in Section 5.2, including Lemmata 5.35, 5.36, 5.38, 5.39, 5.40, 5.42 and Proposition 5.43 is modelled on J. Mujica [649]. To achieve his objectives in [649] Mujica was obliged to modify and sharpen a number of technical results in [785] and [412], e.g. Lemma 5.35 is due to M. Schottenloher [785], Lemmata 5.36 and 5.38 are due to J. Mujica but inspired by results in [785] and [412] respectively. Lemma 5.41 is taken from [651, p.334] (see also P. Berner [108]). To avoid excessive technicalities we have confined our proofs (the technical lemmata) to Fréchet spaces with basis and continuous norms and indicated the known results in the statement of the propositions.

We now discuss how the results in Section 5.2 evolved. H.J. Bremermann [197] initiated the study of pseudo-convex domains, domains of holomorphy and plurisubharmonic functions over infinite dimensional spaces in 1956. He defined a domain U, in a Banach space, to be pseudo-convex if $-\log d_U$ is plurisubharmonic (d_U is the distance to the boundary) and showed (Proposition 5.26) that this was equivalent to U having finite dimensional pseudo-convex sections ([198, 199, 200]). The next step was taken in 1968 by H. Alexander, a student of H.J. Bremermann, whose interest lay in the existence and construction of envelopes of holomorphy for domains spread over Banach spaces. In his thesis [13], Alexander endowed the compact-open spectrum $\mathcal{S}_c(\Omega)$, Ω a Riemann domain over a Banach space, with a complex structure so that it became an analytic extension of Ω and proved Propositions 5.19, 5.20, 5.21, 5.22 and Corollary 5.23 (the finite dimensional case was discussed a few years previously by H. Rossi [750]). He was unable to show that $\mathcal{S}_c(\Omega)$ was the envelope of holomorphy of Ω and, indeed, a 1973 counterexample of B. Josefson [487] showed that this was *not* always the case for Banach spaces (Example 5.55). Alexander posed the problem of finding a natural topology τ_Ω on $H(\Omega)$ such that *any* holomorphic extension $j \colon (\Omega, p) \to (\widetilde{\Omega}, \widetilde{p})$ induces a linear topological isomorphism $j^* \colon \big(H(\widetilde{\Omega}), \tau_{\widetilde{\Omega}}\big) \to \big(H(\Omega), \tau_\Omega\big)$, $j^*(f) = f \circ j$ (see also [783, Problem 5]). J.M. Exbrayat [344] has written a brief survey of some of the results in [13].

The next five years were ones of rapid growth and development. To overcome the shortcomings of the compact open topology described in [13], G. Coeuré [232] introduced the τ_δ topology on $H(\Omega)$ for a domain Ω spread over a *separable* Banach space and showed that analytic extensions led to τ_δ isomorphisms (Proposition 5.17). This result was extended to arbitrary Banach spaces by A. Hirschowitz [460, 455], to Riemann domains spread over metrizable locally convex spaces by M. Schottenloher [779] (see also V. Aurich [74] and P. Berner [108]) and to domains spread over Baire locally convex spaces by K. Rusek–J. Siciak [756]. Using germs of holomorphic functions, A. Hirschowitz [460, 455], proved the existence of envelopes of holomorphy for Riemann domains spread over Banach spaces. M. Schottenloher [777, 781, 783] gave alternative proofs, in the more general setting of domains spread over quasi-complete locally convex spaces, by extending the classical method of intersections (Exercise 5.80) and by considering a suitable subset of the set of *all* multiplicative linear functionals on $H(\Omega)$. K. Rusek–J. Siciak [756] obtained the maximal analytic extension as the domain of existence of $\delta : \Omega \to (H(\Omega), \tau)'$ where τ is an "admissable" topology. During this period variations of the above were considered, e.g. vector-valued and weak holomorphic extensions were studied in W. Bogdanowicz [141], G. Coeuré [232], A. Hirschowitz [455, 456, 460], E. Ligocka–J. Siciak [553], M. Schottenloher [777, 779, 784, 789, 791] and L. Waelbroeck [850], analytic continuation of $\mathcal{G}$-holomorphic, LF-holomorphic and $\mathcal{A}$-holomorphic functions were examined in C.E. Rickart [744, 745] and M. Schottenloher [779, 784], simultaneous analytic extensions of subspaces and subalgebras (of the space of all holomorphic functions) were investigated in G. Coeuré [235], M. Schottenloher [779, 789], K. Rusek–J. Siciak [756] and M.C. Matos [575, 576, 578], and the envelope of holomorphy for domains over $\mathbb{C}^A$ was characterized by V. Aurich [74].

A. Hirschowitz [460, Théoreme 1.6] showed that domains of holomorphy in Banach spaces are holomorphically convex and this result was extended to locally convex spaces by Ph. Noverraz [696, Théoreme 3.5] (see also J. Mujica [645, Lemmata 2.2 and 2.3]). On the negative side A. Hirschowitz [453] (see [456] for details) showed that the open unit ball of $\mathcal{C}([0, \Omega])$, Ω the first uncountable ordinal, is not a domain of existence (by the Hahn–Banach Theorem any convex open set is a domain of holomorphy) and the counterexample of B. Josefson [487], mentioned previously, shows that there are holomorphically convex (even polynomially convex) domains which are not domains of holomorphy. Both these counterexamples involve non-separable Banach spaces. Since holomorphic convexity involves compact sets which, in a Banach space lie in a separable subspace, and as we are considering $\mathbb{C}$-valued holomorphic functions, it is not surprising that countability considerations enter the picture. The situation changes when we move to vector-valued holomorphic functions (see Exercise 5.67). In the DF direction Ph. Noverraz [695, 697, 702] proved that holomorphically convex open subsets of $\mathcal{DFS}$ spaces are domains of holomorphy and this result was extended to $\mathcal{DFM}$ spaces by S. Dineen [292, 295]. The Cartan–Thullen Theorem for E'_c, E a Fréchet space, is due independently to J. Mujica [645] and M. Schottenloher [780, 793]. This result contains Proposition 5.32 as a special case and we have followed the proof in [645].

Parallel to these developments on analytic continuation and holomorphic convexity, the theory of plurisubharmonic functions over infinite dimensional spaces was being constructed. We refer to [446] for a comprehensive treatment of plurisubharmonic functions on locally convex spaces and note that important contributions are due to P. Lelong [535, 536, 537, 538, 539], who pioneered the study of these functions over both finite and infinite dimensional spaces, G. Coeuré [230, 231, 232], M. Hervé [445], C.O. Kiselman [514] and Ph. Noverraz [690, 691, 692]. It is convenient at this point to mention some of the particular results from this area used in Section 5.2. G. Coeuré [232] and Ph. Noverraz [695; 696, Corollaire 2.37] proved that pseudo-convex $\Longleftrightarrow$

PSH-convex, and this includes Proposition 5.26. In his fundamental contribution to the subject, M. Schottenloher [785] (see also [649, Lemma 2.5]) introduced the δ_U^n functions and proved Lemma 5.34. Lemma 5.33 is due to S. Dineen [292]. J. Mujica [649] introduced "$PSH(\Omega)$-convex sets" (Definition 5.28) under the name "sets with Property (P)" and proved Lemma 5.29 and Proposition 5.30. The proof of Corollary 5.45 contains the following result of Ph. Noverraz [700]:

> if U is a pseudo-convex open subset of a locally convex space E with completion $\widehat{E}$, then there exists a pseudo-convex open subset $\widehat{U}$ of $\widehat{E}$ such that $\widehat{U} \cap E = U$.

We now consider the Levi problem and refer to L. Hormander [466] and J. Mujica [649, 651] for the finite dimensional results used, Proposition 5.27 and Lemma 5.37 (the first part of Proposition 5.27 is a solution to the Levi problem for finite dimensional spaces). In 1969, A. Hirschowitz [454] gave a positive solution to the Levi problem for pseudo-convex open subsets of $\mathbb{C}^{\mathbb{N}}$, a result afterwards extended to Riemann domains over $\mathbb{C}^{\mathbb{N}}$ by M.C. Matos [574] and to Riemann domains over $\mathbb{C}^A$, A uncountable, by V. Aurich [74, 75]. L. Gruman [410, 411] gave the first complete solution to the Levi problem for infinite dimensional Banach spaces. He used the finite dimensional $\bar{\partial}$ operator and an inductive process to show that pseudo-convex domains in separable Hilbert spaces are domains of existence. The methods used in [410] influenced most later researchers on the Levi problem. L. Gruman–C.O. Kiselman [412] then solved the Levi problem for open subsets of Banach spaces with a Schauder basis, Y. Hervier [447, 448] extended this result to Riemann domains and Ph. Noverraz [695] replaced the Schauder basis hypothesis with the bounded approximation property. M. Schottenloher [780, 785] defined admissible coverings and regular classes, modified a key result of [412] to prove his Main Lemma [785, p.223] and solved the Levi problem for separable Fréchet spaces with the bounded approximation property (Proposition 5.44 and the preliminary result Proposition 5.43). In fact, using the Lindelöf property and uniformly open sets M. Schottenloher [785] obtained a unified solution for $\mathcal{DFS}$ and Fréchet spaces with the bounded approximation property (see also R. Pomes [735, 736]) and J.F. Colombeau–J. Mujica [248] obtained a solution for arbitrary $\mathcal{DFN}$ spaces. The Levi problem is solved for F'_c, F a separable Fréchet space with the approximation property, in [645, Theorem 11.1] and [793]. This includes Corollary 5.45 as a special case (see also [690]). The first part of [645] surveys the basic properties of domains of holomorphy in locally convex spaces. M.L. Lourenço [565] extended a number of these results to Riemann domains over $\mathcal{DFC}$ spaces and, in particular, extended Proposition 5.51 to pseudo-convex Riemann domains spread over $\mathcal{DFC}$ spaces with continuous norm and the approximation property.

In [794] M. Schottenloher proved that $\mathcal{S}_c(\Omega) = \widehat{\Omega}_c = \Sigma(\Omega) = \widehat{\Omega}_b$ for a Riemann domain Ω spread over a separable Fréchet space with the bounded approximation property ($\widehat{\Omega}_b$ is the connected component of $\mathcal{S}_b(\Omega)$) and J. Mujica [652] completed this line of research by proving Lemma 5.48, Proposition 5.49 and Corollary 5.50. This corollary was the final result of a short sequence of earlier results on the same topic, e.g. in [642] (respectively [643]) J. Mujica proved that $S_c(U) = U$ (respectively $S_b(U) = U$) if U is a polynomially convex domain in a Fréchet space with the approximation property (respectively with the bounded approximation property). The final unknown general implication is whether or not pseudo-convex Riemann domains are holomorphically convex. We conjecture that this is the case when the Riemann domain is spread over a locally convex space with the approximation property. Our observations above and Proposition 5.26 suggests that non-separability will not play a role in this problem.

The Oka–Weil Approximation Theorems (Propositions 5.46 and 5.47) are due to J. Mujica [649] and are the culmination of similar results due to C. Matyszczyk [593, 594], J. Mujica [648, 650], Ph. Noverraz [696, 701] and M. Schottenloher [783, 785, 786, 793]. In contrast to the majority of results in Section 5.2 the continuous norm hypothesis cannot be removed without some replacement in Proposition 5.47 since C. Matyszczyk [594] has given an example which shows that the result is not true for certain domains in $\mathbb{C}^{\mathbb{N}}$. We do not know if the same remark applies to Propositions 5.46 and 5.51 (see the final remarks in [565]). Proposition 5.52 and Corollaries 5.23, 5.53 and 5.54 are due to J. Mujica [649]. In an earlier paper [648], Mujica obtained Corollary 5.54 for pseudo-convex open subsets of a Fréchet space with the bounded approximation property and indicated there that the problem for arbitrary open sets could be obtained by going to envelopes of holomorphy (an approach followed, as we have noted previously, by M. Schottenloher [787] for domains over $\mathbb{C}^{\mathbb{N}}$). Approximation theorems for Riemann domains over certain $\mathcal{DFC}$ spaces have been obtained by M.L. Lourenço [565].

In 1952, E.A. Michael [618] asked if every multiplicative linear functional on a locally m-convex Fréchet algebra was continuous (or equivalently if every multiplicative linear functional on a locally m-convex topological algebra was bounded). Surprisingly the problem for real algebras proved to be much less difficult then the complex problem which was only recently solved positively by B. Stensones [814] using several complex variables. A number of researchers in infinite dimensional holomorphy became interested in this result and reduced the general problem to various spaces of holomorphic functions over infinite dimensional spaces. For further details we refer the reader to D. Clayton [229], J. Mujica [651, §33; 652, 656] and M. Schottenloher [792].

The results in this book are *independent* of this problem and its solution. We could have used this result to shorten the proof of Proposition 5.17 but felt that the proof given here is instructive in view of later developments and we also wished to give a self-contained treatment of infinite dimensional holomorphy. In Chapter 6 the solution to the Michael Problem shows that all multiplicative linear functionals on $H_b(E)$ are continuous but we make no use of this fact.

Chapter 6. Extensions

Extensions of polynomials and holomorphic mappings arose in most of the previous chapters of this book, e.g. in estimating the polarization constant of the bidual (Corollary 1.52), in the definition of Q-reflexive Banach spaces (Definition 2.44), in studying the strong localization property for holomorphic germs (Proposition 4.37), in passing the relationship $\tau_0 = \tau_\omega$ to closed subspaces of fully nuclear spaces (Proposition 4.54) and, in Chapter 5, while proving the equality $\tau_0 = \tau_\omega$ on arbitrary open subsets of a Fréchet–Schwartz space with the bounded approximation property (Corollary 5.53) we used the holomorphic extension obtained in Lemma 5.38. In all these cases extensions were a means to an end. In this chapter we make them our primary concern. We have extended holomorphic functions from *open* sets to the envelope of holomorphy in Chapter 5. In the first two sections of this chapter we investigate:

(a) *extending holomorphic functions from E to F where E is a dense subspace of F,*

(b) *extending holomorphic functions from E to F where E is a closed subspace of F.*

Topic (a) does not arise in several complex variables theory since no finite dimensional space contains a proper dense subspace and does not arise in linear functional analysis since all continuous linear functions can easily be extended to the completion (provided the range is complete). Moreover, since continuous polynomials can also be readily extended in this setting we are dealing with a problem that is exclusive to infinite dimensional holomorphy.

Topic (b) may be regarded as a holomorphic approach to the Hahn–Banach Theorem. Motivated by results on analytic continuation in Chapter 5 and guided by developments in Section 6.2 we discuss multiplicative linear functionals on spaces of holomorphic functions of bounded type in Section 6.3 and discover close connections between this topic and the theory of holomorphic extensions discussed in the previous chapter.

6.1 Holomorphic Extensions from Dense Subspaces

Let E denote a locally convex space with completion $\widehat{E}$. We consider the problem of extending holomorphic functions on E holomorphically to $\widehat{E}$. This is not always possible but we shall see that each locally convex space has a holomorphic completion $E_{\mathcal{O}}$, i.e. a maximal *subspace* of $\widehat{E}$ to which all holomorphic functions on E admit a (necessarily unique) holomorphic extension.

If $f \in H(E)$, let A_f denote the set of all pairs (x, V), where $x \in E$ and V is a convex balanced open subset of $\widehat{E}$ such that

$$\sum_{n=0}^{\infty} \left\| \frac{\widehat{d^n} f(x)}{n!} \right\|_V < \infty.$$

In defining A_f we used the fact that continuous n-homogeneous polynomials have unique extensions as n-homogeneous polynomials to $\widehat{E}$. If $(x, V) \in A_f$ then

$$f_x(y) := \sum_{n=0}^{\infty} \frac{\widehat{d^n} f(x)}{n!} (y - x)$$

defines a holomorphic function on $x + V$. If (x, V) and (y, W) belong to A_f and $(x + V) \cap (y + W)$ is non-empty, then $E \cap (x + V) \cap (y + W)$ is dense in $(x + V) \cap (y + W)$ and hence, by continuity, f_x and f_y agree on the overlap. By analytic continuation it follows that the function

$$g := \begin{cases} f_x & \text{on } x + V \\ f_y & \text{on } y + W \end{cases}$$

is holomorphic on $(x + V) \cup (y + W)$.

Let $\Omega_f = \bigcup \{x + V : (x, V) \in A_f\}$. Clearly Ω_f is an open subset of $\widehat{E}$ containing E and since E is a connected dense subset of $\widehat{E}$, Ω_f is connected. From these observations we deduce that there exists a (unique) $\widetilde{f} \in H(\Omega_f)$ which coincides with f on E. We now show that the envelope of holomorphy of $\widetilde{f}$ is univalent, i.e. it lies in $\widehat{E}$. Let (ω, p) denote a connected Riemann domain over $\widehat{E}$ and let $j : (\Omega_f, i_{\Omega_f}) \longrightarrow (\omega, p)$ denote a morphism of Riemann domains. We suppose that there exists a unique $g \in H(\omega)$ which extends $\widetilde{f}$, i.e. we have the following commutative diagram where i_{Ω_f} is the inclusion mapping from Ω_f into $\widehat{E}$:

$$(6.1)$$

If $j(\Omega_f) \neq \omega$ then there exists $\xi \in \partial(j(\Omega_f))$, $\alpha \in \mathrm{cs}(\widehat{E}) \cong \mathrm{cs}(E)$, and $\rho > 0$ such that $B_\omega^\alpha(\xi, 2\rho) \subset \omega$ and $\|g\|_{B_\omega^\alpha(\xi, 2\rho)} < \infty$. Let $q = \left(p|_{B_\omega^\alpha(\xi, 2\rho)}\right)^{-1}$. We have $g \circ q \in H^\infty(B_{\widehat{E}}^\alpha(p(\xi), 2\rho))$. Since $\xi \in \partial(j(\Omega_f))$ we can find $\eta \in E$ such that $j(\eta) \in B_\omega^\alpha(\xi, \rho)$ and, using the Cauchy inequalities, we see that

$$\sum_{n=0}^\infty \left\| \frac{\widehat{d}^n(g \circ q)}{n!}(p(j(\eta))) \right\|_{B_{\widehat{E}}^\alpha(0, \rho)} < \infty.$$

By (6.1), $p(j(\eta)) = \eta$ and

$$\frac{\widehat{d}^n(g \circ q)}{n!}(p(j(\eta))) = \frac{\widehat{d}^n \widetilde{f}(\eta)}{n!} = \frac{\widehat{d}^n f(\eta)}{n!}$$

for all n. Hence

$$\sum_{n=0}^\infty \left\| \frac{\widehat{d}^n f(\eta)}{n!} \right\|_{B_{\widehat{E}}^\alpha(0, \rho)} < \infty$$

and $(\eta, B_{\widehat{E}}^\alpha(0, \rho)) \in A_f$. This implies that $\xi \in j(\Omega_f)$ and j is an isomorphism. Hence Ω_f is the domain of existence of $\widetilde{f}$. We have proved the following proposition.

Proposition 6.1 *Let E denote a locally convex space with completion $\widehat{E}$ and let $f \in H(E)$. Then f can be extended as a holomorphic function $\widetilde{f}$ to an open subset of $\widehat{E}$ and the domain of existence of $\widetilde{f}$ can be realized as an open subset of $\widehat{E}$.*

Definition 6.2 Let E denote a locally convex space with completion $\widehat{E}$. We define $E_{\mathcal{O}}$, the *holomorphic completion* of E, to be the intersection of all open domains of existence in $\widehat{E}$ which contain E.

Proposition 6.3 *The holomorphic completion of a locally convex space E is a subspace of $\widehat{E}$.*

Proof. Let $\mathcal{E}$ denote the set of all open subsets of $\widehat{E}$ which are domains of existence and contain E. If Ω is an open subset of $\widehat{E}$ and $x \in \widehat{E}$ then, using the function $f_x(y) := f(y - x)$ for $y \in x + \Omega$, we see that Ω is a domain of existence if and only if $x + \Omega$ is also a domain of existence for all $x \in \widehat{E}$.

If $x \in E$ then $E = x + E$ and, if the domain of existence Ω contains E, then $x + \Omega$ is also a domain of existence containing E. Hence

$$E_{\mathcal{O}} = \bigcap \{\Omega \; : \; \Omega \in \mathcal{E}\} = \bigcap \{x + \Omega \; : \; \Omega \in \mathcal{E}\} = x + E_{\mathcal{O}}$$

and $E + E_{\mathcal{O}} = E_{\mathcal{O}}$.

If $\lambda \neq 0$, $\lambda \in \mathbb{C}$, then $\Omega \in \mathcal{E}$ if and only if $\lambda\Omega \in \mathcal{E}$. Hence $\lambda E_{\mathcal{O}} = E_{\mathcal{O}}$ for all $\lambda \neq 0$.

Let $y \in E_{\mathcal{O}}$. If $\Omega \in \mathcal{E}$ then $y + \Omega \supset y + E + E_{\mathcal{O}} \supset E$, since $-y \in E_{\mathcal{O}}$. Hence $y + \Omega \in \mathcal{E}$. Conversely, if $y + \Omega \in \mathcal{E}$ then $-y + (y + \Omega) = \Omega \in \mathcal{E}$ and we obtain

$$E_{\mathcal{O}} = \bigcap \{\Omega \ : \ \Omega \in \mathcal{E}\} = \bigcap \{y + \Omega \ : \ \Omega \in \mathcal{E}\} = y + E_{\mathcal{O}}$$

Hence $E_{\mathcal{O}} = E_{\mathcal{O}} + E_{\mathcal{O}}$ and $E_{\mathcal{O}}$ is a subspace of $\widehat{E}$. This completes the proof.

Our next proposition justifies the terminology 'holomorphic completion'.

Proposition 6.4 *If E is a locally convex space with holomorphic completion $E_{\mathcal{O}}$ then each $f \in H(E)$ has a unique holomorphic extension to $E_{\mathcal{O}}$. Moreover if F is any subspace of $\widehat{E}$ containing E and each holomorphic function on E has a holomorphic extension to F then $F \subset E_{\mathcal{O}}$.*

Proof. By our construction each $f \in H(E)$ can be extended uniquely to a holomorphic function on a domain of existence contained in $\widehat{E}$ and as $E_{\mathcal{O}}$ is the intersection of all such domains of existence, it follows that f can be holomorphically continued to $E_{\mathcal{O}}$. If $f \in H(E)$, f_F is its holomorphic extension to F and $x \in F$, then there exists a convex balanced neighbourhood of 0 in $\widehat{E}$, V, and $g \in H^{\infty}(x + V)$ such that $g|_{(x+V) \cap F} = f_F|_{(x+V) \cap F}$ (it suffices to apply the construction, used in Proposition 6.1 for $f \in H(E)$, to f_F). If $y \in (x + \frac{1}{4}V) \cap E$ then $g|_{y+\frac{3}{4}V} \in H^{\infty}(y + \frac{3}{4}V)$ and $(y, \frac{1}{2}V) \in A_f$. Hence $x \in \Omega_f$. Since $E_{\mathcal{O}} = \bigcap \{\Omega_f \ : \ f \in H(E)\}$ it follows that $F \subset E_{\mathcal{O}}$. This completes the proof.

In certain circumstances we may use *bounding sets* to describe the holomorphic completion. If E is a locally convex space with completion $\widehat{E}$ we let

$$E_B = \bigcup \{\overline{A} \ : \ A \subset E \text{ bounding, closure taken in } \widehat{E}\}.$$

We refer to Example 3.11 and the proof of Proposition 5.17 for techniques similar to those employed in part (c) of the following proposition.

Proposition 6.5 *Let E denote a locally convex space with completion $\widehat{E}$ and holomorphic completion $E_{\mathcal{O}}$.*

(a) *If τ_0 and τ_δ define the same bounded subsets of $H(E)$ then E_B is a vector subspace of $\widehat{E}$.*

(b) *If E is sequentially dense in $\widehat{E}$ then $E_{\mathcal{O}} \subset E_B$.*

(c) *If for each $f \in H(E)$ there exists a sequence of continuous semi-norms $(p_n)_n$ on E such that f is continuous with respect to the locally convex topology generated by $(p_n)_n$ then $E_B \subset E_{\mathcal{O}}$.*

Proof. (a) If A is a bounding subset of $H(E)$ and $\lambda \in \mathbb{C}$ then λA is bounding and hence $\lambda E_{\mathcal{B}} \subset E_{\mathcal{B}}$ for all $\lambda \in \mathbb{C}$. Hence, in order to show that $E_{\mathcal{B}}$ is a vector space, it suffices to prove that the vector sum of the bounding subsets A_1 and A_2 of E is also bounding. By Example 3.20(b), the semi-norms $p_{A_i}(f) := \|f\|_{A_i}$, $i = 1, 2$, are τ_δ continuous on $H(E)$. If $f \in H(E)$ and $y \in E$ let $f_y \in H(E)$ denote the function defined by $f_y(x) = f(x + y)$ for all $x \in E$. Since the vector sum of compact sets is again compact, $(f_y)_{y \in K}$ is a τ_0, and hence τ_δ, bounded subset of $H(E)$ for any fixed compact subset K of E. Hence

$$\sup_{x \in A_1} \|f_x\|_K = \sup_{y \in K} \|f_y\|_{A_1} = \sup_{y \in K, \, x \in A_1} |f(x + y)| < \infty$$

and $(f_x)_{x \in A_1}$ is τ_0 and, by hypothesis, τ_δ bounded. This implies

$$\sup_{x \in A_1, \, y \in A_2} |f(x + y)| = \sup_{x \in A_1} p_{A_2}(f_x) < \infty.$$

Hence $A_1 + A_2$ is a bounding subset of $H(E)$ and we have proved (a).

  (b) We prove the following (possibly) slightly stronger result: if $(x_n)_n$ is a sequence in E which converges to $x \in E_{\mathcal{O}}$ then $(x_n)_n$ is a bounding subset of E. Let $f \in H(E)$. Then f admits a holomorphic extension to $E_{\mathcal{O}}$ and hence $(f(x_n))_n$ is a convergent sequence. In particular $\sup_n |f(x_n)| < \infty$ for all $f \in H(E)$ and $\{x_n\}_{n=1}^\infty$ is a bounding subset of E. Since $x \in \{x_n\}_n$ we have shown $x \in E_{\mathcal{B}}$ and hence $E_{\mathcal{O}} \subset E_{\mathcal{B}}$.

  (c) Let A denote a bounding subset of E. We consider a fixed f in $H(E)$. By our hypothesis there exists a strictly increasing sequence of continuous semi-norms $(p_n)_n$ such that f is continuous with respect to the locally convex topology on E generated by $(p_n)_n$. For each positive integer n let

$$U_n = \{x \in E : \|f\|_{x + 2\{y \, : \, p_n(y) \le 2/n\}} \le n\}$$

and let

$$\tilde{U}_n = U_n + \{x \in E : p_n(x) < 1/n\}.$$

If $V_n = \bigcup_{i \le n} \tilde{U}_i$ and $W_n = \bigcap_{i \le n} \{x \in E : p_i(x) < 1/i\}$ then $\|f\|_{V_n + W_n} =: M_n < \infty$ and $(V_n)_n$ is an increasing countable open cover of E. Since $\|\cdot\|_A$ defines a τ_δ continuous semi-norm on $H(E)$ there exists a positive integer n_0 and $c > 0$ such that

$$\|g\|_A \le c\|g\|_{V_{n_0}} \tag{6.2}$$

for all $g \in H(E)$. Let $x_0 \in \overline{A}$. Now choose y_0 in A such that $x_0 \in y_0 + \frac{1}{6}\overline{W}_{n_0}$ where the closure is taken in $\hat{E}$. If $x \in E$ and n is a positive integer then the mapping

$$y \in E \longrightarrow \frac{\hat{d}^n f(y)}{n!}(x)$$

is holomorphic and, by the Cauchy inequalities (3.12),

$$\sup_{y \in V_{n_0}} \left\| \frac{\widehat{d^n} f(y)}{n!} \right\|_{\frac{1}{3}\overline{W}_{n_0}} \leq \frac{1}{2^n} \|f\|_{V_{n_0}} + (2/3)W_{n_0} \leq \frac{M_{n_0}}{2^n}$$

for $n \geq 0$. Since $y_0 \in A$, (6.2) implies

$$\sum_{n=0}^{\infty} \left\| \frac{\widehat{d^n} f(y_0)}{n!} \right\|_{\frac{1}{3}\overline{W}_{n_0}} \leq c \sum_{n=0}^{\infty} \frac{M_{n_0}}{2^n} < \infty.$$

If W is the interior of $\frac{1}{3}\overline{W}_{n_0}$ (in $\widehat{E}$) then we have shown, in the notation of Proposition 6.1, that $(y_0, W) \in A_f$ and hence $x_0 \in \Omega_f$. Since f was arbitrary this implies

$$x_0 \in \bigcap \{\Omega_f : f \in H(E)\} = E_{\mathcal{O}}.$$

Hence $\overline{A} \subset E_{\mathcal{O}}$ and $E_{\mathcal{B}} \subset E_{\mathcal{O}}$. This completes the proof.

The first three parts of the following example follow immediately from Proposition 6.5. In the remainder of this section we use positive solutions to the Levi problem (Chapter 5), plurisubharmonic functions and polar sets to examine the holomorphic completion of various locally convex spaces. Part (d) illustrates a number of the techniques that we employ in later examples.

Example 6.6 (a) If E is a metrizable locally convex space then $E_{\mathcal{O}} = E_{\mathcal{B}}$.

(b) If $f \in H(E)$ where E is a dense subspace of a $\mathcal{DFM}$ space then, as we have previously noted, f can be extended to an open subset Ω of $\widehat{E}$. Example 3.11 and Proposition 6.5(c) imply that $E_{\mathcal{B}} \subset E_{\mathcal{O}}$. If, in addition, E is sequentially dense in E – in particular if $\widehat{E}$ has a Schauder basis and E contains the vector space spanned by the basis – then Proposition 6.5(b) implies $E_{\mathcal{B}} = E_{\mathcal{O}}$.

(c) If E is Lindelöf – in particular if E has a countable algebraic basis – then $E_{\mathcal{B}} \subset E_{\mathcal{O}}$.

(d) Consider a metrizable locally convex space of countable dimension E, with completion $\widehat{E}$, which admits a continuous norm $\|\cdot\|$. Let $(x_n)_{n=1}^{\infty}$ denote an algebraic basis for E. If $(e_n)_{n=1}^{\infty}$ denotes the standard unit vector basis for l_1, let l_1^0 denote the vector subspace of l_1 spanned by $(e_n)_{n=1}^{\infty}$. Now fix $x_0 \in \widehat{E} \setminus E$, $\|x_0\| = 1$. For each positive integer n define a linear mapping

$$T_n : E_n := \text{Span}\{x_0, \ldots, x_n\} \longrightarrow \mathbb{C}$$

by letting $T_n(\sum_{i=0}^{n} \lambda_i x_i) = \lambda_0 + \lambda_n$. Since T_n is linear and E_n is finite dimensional, T_n is continuous. By the Hahn–Banach Theorem, there exists $\widetilde{T}_n \in (\widehat{E}, \|\cdot\|)'$ such that $\widetilde{T}_n|_E = T_n$.

Let $T : \widehat{E} \longrightarrow l_1$ be defined by $T(x) := \left(\frac{\widetilde{T}_n(x)}{3^n \|\widetilde{T}_n\|} \right)_{n=1}^{\infty}$. Since

$$\sum_{n=1}^{\infty} \sup_{\|x\| \le 1} \left| \frac{\tilde{T}_n(x)}{3^n \|\tilde{T}_n\|} \right| \le \sum_{n=1}^{\infty} \frac{1}{3^n},$$

T is a continuous linear mapping. If $x = \sum_{i=1}^{n} \lambda_i x_i$ then $\tilde{T}_k(x) = 0$ for $k > n$ and hence $T(E) \subset l_1^0$. Since $T(x_m) = 3^{-m} \|\tilde{T}_m\|^{-1} e_m$, we have $T(E) = l_1^0$ and, as $T_n(x_0) = 1$ for all n, $T(x_0)$ does not belong to l_1^0.

Let $T(x_0) = (\beta_n)_{n=1}^{\infty}$ and let $\alpha_n = -(n^2 \log \beta_n)^{-1}$. Clearly $\alpha_n > 0$ for all n. Since the mapping

$$\sum_{n=1}^{\infty} z_n e_n \in l_1 \longrightarrow \sum_{n=1}^{m} \alpha_n \log |z_n|$$

is plurisubharmonic for all m, the function $\theta : l_1 \longrightarrow [-\infty, +\infty)$ defined by

$$\theta \left(\sum_{n=1}^{\infty} z_n e_n \right) := \sum_{n=1}^{\infty} \alpha_n \log |z_n|$$

is locally the decreasing limit of a sequence of plurisubharmonic functions and hence is plurisubharmonic on l_1. We have

$$\theta \left(\sum_{n=1}^{\infty} \beta_n e_n \right) = \sum_{n=1}^{\infty} \frac{-\log \beta_n}{n^2 \log \beta_n} = -\sum_{n=1}^{\infty} \frac{1}{n^2} =: c \in \mathbb{R}.$$

The set $U := \{z \in l_1 : \theta(z) < c\}$ is a pseudo-convex open subset of l_1 and, since $\theta \left(\sum_{n=1}^{m} z_n e_n \right) = -\infty$ for all m, $l_1^0 \subset U$. Since E is a dense subspace of the metrizable space $\hat{E}$ we can find a sequence $(\xi_n)_n$ in E such that $\xi_n \to x_0$ as $n \to \infty$. We have $T(\xi_n) \in l_1^0 \subset U$ and $T(\xi_n) \to T(x_0) = \sum_{m=1}^{\infty} \beta_m e_m$ as $n \to \infty$. By Proposition 5.44, U is holomorphically convex and thus there exists $f \in H(U)$ such that $\sup_n |f(T(\xi_n))| = \infty$. Since $f \circ T \in H(T^{-1}(U))$ and the sequence $(\xi_n)_n$ is not a bounding subset of E, the proof of Proposition 6.5(b) implies that $x_0 \notin E_{\mathcal{O}}$. Hence $E_{\mathcal{O}} \subset E$ and, as we always have $E \subset E_{\mathcal{O}}$, we have shown $E = E_{\mathcal{O}}$. The hypothesis of a continuous norm is necessary (see Exercise 6.61).

We now turn to the relationship between $E_{\mathcal{O}}$ and $\hat{E}$. If $E_{\mathcal{O}} \ne \hat{E}$ then E may be considered a "small" (although dense) subspace of $\hat{E}$. We introduce *polar sets*, which reflect both the complex and linear structures of E (and $\hat{E}$), and use them to obtain a further measure of the size of E in $\hat{E}$.

Definition 6.7 If U is an open subset of a locally convex space E then $A \subset U$ is a polar (respectively strictly polar) subset of U if there exists a plurisubharmonic (respectively negative plurisubharmonic) function ν on U, $\nu \not\equiv -\infty$, such that $A \subset \{x \in U : \nu(x) = -\infty\}$.

A key feature of Example 6.6(d) may be rephrased as follows; *a countable dimensional subspace of a Fréchet space E with continuous norm is a polar subset of $\widehat{E}$.*

This theme is developed in the following proposition.

Proposition 6.8 *Let E denote a locally convex space with completion $\widehat{E}$. If each pseudo-convex open subset of $\widehat{E}$ is a domain of existence of a holomorphic function, then the following are equivalent:*

(a) E is a polar subset of $\widehat{E}$,

(b) $E_\mathcal{O} \neq \widehat{E}$,

(c) $E_\mathcal{O}$ is a polar subset of $\widehat{E}$.

Proof. If $\nu \in \mathrm{PSH}(\widehat{E})$, $E \subset \{x : \nu(x) = -\infty\}$ and $\nu(a) \neq -\infty$ for some a in $\widehat{E}$ then $U := \{x \in \widehat{E} : \nu(x) < \nu(a)\}$ is a proper open subset of $\widehat{E}$ containing E. Since U is pseudo-convex, it is a domain of existence. Hence $E_\mathcal{O} \subset U \subsetneq \widehat{E}$ and (a) $\Longrightarrow$ (b).

If $E_\mathcal{O} \neq \widehat{E}$ then there exists a domain of existence U in $\widehat{E}$ such that $E_\mathcal{O} \subset U \subsetneq \widehat{E}$. Let $a \in \widehat{E} \setminus U$. Since domains of existence are pseudo-convex, the function $-\log d_U$ is plurisubharmonic on $U \times \widehat{E}$. Hence the mapping

$$\nu : x \in \widehat{E} \longrightarrow -\log d_U(0,x) = \inf\{-\log r : \lambda x \in U, |\lambda| \leq r\}$$

is plurisubharmonic on $\widehat{E}$. As $E_\mathcal{O} \subset U$ it follows that $\nu(x) = -\infty$ for all $x \in E_\mathcal{O}$ and since $\nu(a) > -\infty$ we have $\nu \not\equiv -\infty$. Hence $E_\mathcal{O}$ is a polar subset of $\widehat{E}$ and (b) $\Longrightarrow$ (c).

The implication (c) $\Longrightarrow$ (a) follows immediately from the fact that subsets of polar sets are themselves polar. This completes the proof.

To motivate our next set of examples we introduce the concept of *control set* for plurisubharmonic functions.

A subset A of an open subset U in a locally convex space E, is called a control set for plurisubharmonic functions on U if there exists a function $\gamma : U \longrightarrow (0,1]$ such that

$$\nu(x) \leq \gamma(x) \sup_{y \in A} \nu(y) + (1 - \gamma(x)) \sup_{y \in U} \nu(y) \tag{6.3}$$

for all $x \in U$ and all $\nu \in \mathrm{PSH}(U)$. The function γ is called a control function.

By (6.3) control sets and functions estimate the growth of plurisubharmonic functions – a classical example is the Hadamard Three Circles Theorem. A set A will not be a control set if it is *"too small"* and will yield only trivial estimates if it is *"too large"*, e.g. if A is a dense open subset of U then $\sup_{y \in A} \nu(y) = \sup_{y \in U} \nu(y)$ and (6.3) provides no new information. How small is "too small" and how large is "too large"? A set A is a control set if and only if it is not *strictly polar*. At the other extreme a more subjective choice is necessary and it is reasonable to say, especially in the case of infinite dimensional Banach and Fréchet spaces that compact sets are not very large. We are thus led to the following problem:

to classify the locally convex spaces which admit compact non-polar subsets.

A complete locally convex space E will contain a compact non-polar set if and only if it contains a convex balanced compact non-polar set. Since polar subsets of $\mathbb{C}$ have empty interiors it is easily seen that K, compact convex balanced, is non-polar in E if and only if $E_K := \bigcup_{n>0} nK$ is non-polar. By Proposition 6.8, this will occur if and only if $(E_K)_{\mathcal{O}} = E$. In many cases, e.g. if E is Fréchet or $\mathcal{DFS}$, the open mapping theorem can be used to show that $E_K \neq E$ and if E_K is not dense in E then the Hahn–Banach Theorem and functions of the form $\log |\varphi|$, $\varphi \in E'$, can be used to show that $(E_K)_{\mathcal{O}} \neq \widehat{E}$. If K is polar in E then K will admit a neighbourhood basis of open sets $(V_\alpha)_\alpha$ with K strictly polar in each V_α. Thus we see that the existence of control sets and the problem of whether or not holomorphic completions coincide with linear completions are closely related.

We confine ourselves to certain fully nuclear spaces with basis. The basis hypothesis is not necessary (see Section 6.5) but its inclusion leads to more transparent proofs. Nuclearity, although not necessary, cannot be omitted without some replacement (see Exercise 6.36). If $E = \Lambda(P)$ is a fully nuclear space with basis and $(\alpha_n)_n \in E$ we let

$$[(\alpha_n)_n] = \{(z_n)_n \in \Lambda(P) \ : \ |z_n| \leq |\alpha_n|, \ \forall n\}.$$

It is clear that the polydiscs $[(\alpha_n)_n]$ form a fundamental system of compact subsets of E as $(\alpha_n)_n$ ranges over $\Lambda(P)$.

Proposition 6.9 *Let $E = \Lambda(P)$ denote a fully nuclear space with basis and suppose*

(a) pseudo-convex open subsets of E are domains of existence for holomorphic functions,

(b) if $f \in H_{HY}(E)$ then the set of points of continuity of f is an open and closed subset of E.

If $(\alpha_n)_n \in E$, $\alpha_n \neq 0$ all n, then $[(\alpha_n)_n]$ is a compact non-polar subset of E if and only if for each $(\beta_n)_n \in E$ and each $(p_n)_n \in P$ there exists $c > 0$ and $d > 0$ such that

$$|\beta_n|^{1+d} p_n^d \leq c|\alpha_n| \tag{6.4}$$

for all n.

Proof. We first suppose that (6.4) holds. By Proposition 6.8 it suffices to show that each $f \in H(U)$, U open in E containing $E_{[(\alpha_n)_n]}$, can be extended holomorphically to E. If $f \in H(U)$ then f has a monomial expansion which converges absolutely on some neighbourhood of the origin and on a neighbourhood of $k[(\alpha_n)_n]$ for each positive real number k. Let

$$f(z) = \sum_{m \in \mathbb{N}^{(\mathbb{N})}} a_m z^m.$$

Choose $p = (p_n)_n \in P$ so that

$$\sup_{m \in \mathbb{N}^{(\mathbb{N})}} \|a_m z^m\|\{(z_n)_n \ : \ |z_n|p_n \le 1\} = M < \infty,$$

i.e.

$$|a_m| \le Mp^m \tag{6.5}$$

for all $m \in \mathbb{N}^{(\mathbb{N})}$. For each positive number k we have

$$\sup_{m \in \mathbb{N}^{(\mathbb{N})}} \|a_m z^m\|\{(z_n)_n \ : \ |z_n| \le k\alpha_n\} =: a(k) < \infty,$$

i.e.

$$|a_m|k^{|m|}|\alpha^m| \le a(k) \tag{6.6}$$

holds for all $m \in \mathbb{N}^{(\mathbb{N})}$. If $(\beta_n)_n \in E$ then, by full nuclearity, we can choose $(\gamma_n)_n \in E$, $|\gamma_n| > |\beta_n|$ all n, such that $\sum_{n=1}^{\infty} \left|\dfrac{\beta_n}{\gamma_n}\right| < \infty$. By our hypothesis there exists $c > 0$ and $d > 0$ such that

$$|\gamma_n|^{1+d} p_n^d \le c|\alpha_n| \tag{6.7}$$

for all n. If $m \in \mathbb{N}^{(\mathbb{N})}$ then

$$(|a_m||\gamma^m|)^{1+d} \le |a_m||\gamma^m|^{1+d} M^d (p^m)^d, \text{ by (6.5)},$$
$$\le M^d c^{|m|} |a_m||\alpha^m|, \text{ by (6.7)},$$
$$\le M^d a(c), \text{ by (6.6)}.$$

Hence

$$\sum_{m \in \mathbb{N}^{(\mathbb{N})}} \|a_m z^m\|_{[(\beta_n)_n]} \le \sum_{m \in \mathbb{N}^{(\mathbb{N})}} \left|\left(\frac{\beta}{\gamma}\right)^m\right| \|a_m z^m\|_{[(\gamma_n)_n]}$$

$$\le M^{\frac{d}{1+d}} a(c)^{\frac{1}{1+d}} \sum_{m \in \mathbb{N}^{(\mathbb{N})}} \left|\left(\frac{\beta}{\gamma}\right)^m\right|$$

$$= \frac{M^{\frac{d}{1+d}} a(c)^{\frac{1}{1+d}}}{\prod_{n=1}^{\infty} \left(1 - \left|\frac{\beta_n}{\gamma_n}\right|\right)}$$

$$< \infty.$$

Since f is continuous at the origin and $f \in H_{HY}(E)$, (b) implies that $f \in H(E)$. We have shown that (6.4) is sufficient for non-polarity.

Conversely, suppose $[(\alpha_n)_n]$ is a non-polar subset of E. If (6.4) is not satisfied then we can find $\beta = (\beta_n)_n \in E$ and $p = (p_n)_{n=1}^{\infty} \in P$ such that for all $d > 0$

$$A(d) := \{n \, : \, |\beta_n|^{1+d} p_n^d \geq |\alpha_n|\}$$

is an infinite set. Since $\sum_{n=1}^{\infty} |\beta_n p_n| < \infty$ we may suppose, without loss of generality, that $|\beta_n p_n| < 1$ for all n. Since $|\beta_n|^{1+d} p_n^d = |\beta_n p_n|^d p_n$ we see that $A(d') \subset A(d)$ if $d' > d$. Using a diagonal process we can choose a strictly increasing sequence of positive integers $(n_j)_j$ such that

$$|\beta_{n_j}|^{1+j^2} p_{n_j}^{j^2} \geq |\alpha_{n_j}| \tag{6.8}$$

for all j. Let

$$\nu((z_n)_n) = \sum_{j > j_0} \frac{\log(|z_{n_j}| p_{n_j})}{-\log(|\alpha_{n_j}| p_{n_j})}$$

where j_0 is chosen so that $|\alpha_{n_j}| p_{n_j} < 1$ for all $j > j_0$. Since ν is locally the limit of a decreasing sequence of plurisubharmonic functions on E it is plurisubharmonic. If $(z_n)_n \in [(\alpha_n)_n]$ then, $|z_n| \leq |\alpha_n|$ all n,

$$\nu((z_n)_n) \leq \sum_{j > j_0} \frac{\log(|\alpha_{n_j}| p_{n_j})}{-\log(|\alpha_{n_j}| p_{n_j})} \leq \sum_{j > j_0} (-1) = -\infty$$

and $\nu\big|_{[(\alpha_n)_n]} = -\infty$. By (6.8),

$$|\beta_{n_j}|^{1+j^2} p_{n_j}^{1+j^2} \geq |\alpha_{n_j}| p_{n_j}$$

for all j. Hence

$$(1 + j^2) \log(|\beta_{n_j}| p_{n_j}) \geq \log(|\alpha_{n_j}| p_{n_j})$$

and

$$\frac{\log(|\beta_{n_j}| p_{n_j})}{-\log(|\alpha_{n_j}| p_{n_j})} \geq \frac{-1}{1 + j^2}$$

for all j. This implies

$$\nu((\beta_n)_n) = \sum_{j > j_0} \frac{\log(|\beta_{n_j}| p_{n_j})}{-\log(|\alpha_{n_j}| p_{n_j})} \geq -\sum_{j > j_0} \frac{-1}{1 + j^2} > -\infty.$$

Hence $[(\alpha_n)_n]$ is a polar subset of E. This contradicts our hypothesis, shows that (6.4) is necessary and completes the proof.

Condition (a) of Proposition 6.9 is satisfied by Fréchet nuclear spaces with the bounded approximation property (Proposition 5.44) and by $\mathcal{DFN}$ spaces

(Corollary 5.45). Condition (b) is satisfied by Fréchet spaces and $\mathcal{DFS}$ spaces (Example 3.8(d)).

Example 6.10 Let $E = \lambda(A)$ denote a Fréchet nuclear space with basis. We suppose $0 \le a_{j,k} \le a_{j,k+1}$ $\forall j, k$ and $\sum_j \frac{a_{j,k}}{a_{j,k+1}} < \infty$ for all k where $\frac{a}{0} := \infty$ if $a > 0$ and $\frac{0}{0} := 0$. A fundamental system of compact subsets of E is given by $c\left[\left(\frac{1}{a_{j,n_j}}\right)_j\right]$ where $c > 0$, $(n_j)_j$ is increasing to infinity and $a_{j,n_j} > 0$ $\forall j$. By Proposition 6.9, $\left[\left(\frac{1}{a_{j,n_j}}\right)_j\right]$ is a compact non-polar subset of E if and only if for every increasing to infinity sequence of positive integers $(l_j)_j$ and for every positive k there exists $c > 0$ and $d > 0$ such that the following equivalent conditions are satisfied:

$$a_{j,l_j}^{-1-d} a_{j,k}^d \le \frac{c}{a_{j,n_j}} \tag{a}$$

$$a_{j,n_j}^{\frac{1}{1+d}} a_{j,k}^{\frac{d}{1+d}} \le c^{\frac{1}{1+d}} a_{j,l_j} \tag{b}$$

$$\sup_j \frac{a_{j,n_j}^{\frac{1}{1+d}} a_{j,k}^{\frac{d}{1+d}}}{a_{j,l_j}} < \infty. \tag{c}$$

From the above we see that E contains a compact non-polar set

$\Longleftrightarrow \exists (n_j)_j, \; \forall (l_j)_j, \; \forall k \; \exists d$ such that

$$\sup_j \frac{a_{j,n_j}^{\frac{1}{1+d}} a_{j,k}^{\frac{d}{1+d}}}{a_{j,l_j}} < \infty \tag{*}$$

$\Longleftrightarrow \exists (n_j)_j, \; \forall k \; \exists l \in \mathbb{N}, \; d > 0, c > 0$ such that

$$a_{j,n_j}^{\frac{1}{1+d}} a_{j,k}^{\frac{d}{1+d}} \le c a_{j,l} \text{ for all } j \tag{**}$$

$\Longleftrightarrow \exists (n_j)_j, \; \forall k \; \exists l \in \mathbb{N}, \; d > 0$, such that

$$\sup_j \frac{a_{j,n_j}}{a_{j,l}} \left(\frac{a_{j,k}}{a_{j,l}}\right)^d < \infty$$

$\Longleftrightarrow \forall k \; \exists l \in \mathbb{N}, \; d > 0$, such that $\forall m$

$$\sup_j \frac{a_{j,m}}{a_{j,l}} \left(\frac{a_{j,k}}{a_{j,l}}\right)^d < \infty \tag{6.9}$$

$\Longleftrightarrow \forall k \; \exists l \in \mathbb{N}, \; d > 0$, such that $\forall m \; \exists c > 0$ satisfying

$$a_{j,m} a_{j,k}^d \le c a_{j,l}^{1+d} \text{ for all } j. \tag{6.10}$$

Indeed, $(*)$ is condition (c) and $(**)$ is easily seen to imply (b). Suppose $(**)$ does not hold for the positive integer k. For all $l > k$ and $d > 0$ let

$$N(l, d) = \{j \; : \; n_j \ge k, \; a_{j,n_j}^{\frac{1}{1+d}} a_{j,k}^{\frac{d}{1+d}} \ge a_{j,l}\}.$$

Since $a_{j,l'} \geq a_{j,l}$ we have $N(l,d) \supset N(l',d)$ for $l' \geq l$. Moreover,

$$a_{j,n_j}^{\frac{1}{1+d}} a_{j,k}^{\frac{d}{1+d}} = a_{j,n_j} \left(\frac{a_{j,k}}{a_{j,n_j}} \right)^{\frac{d}{1+d}} \geq a_{j,n_j} \left(\frac{a_{j,k}}{a_{j,n_j}} \right)^{\frac{d'}{1+d'}}$$

if $d < d'$ since $k < n_j$ and $\phi(x) := \frac{x}{1+x}$ is an increasing function of $x > 0$.
This implies $N(l',d) \supset N(l',d')$ if $d < d'$. Since $(**)$ does not hold we have
$|N(l,d)| = \infty$ for all l and d. By using a diagonal process on the sequence
of sets $(N(j,j))_j$ we can choose an increasing to infinity sequence of positive
integers $(s_j)_j$ such that

$$a_{s_j,n_{s_j}}^{\frac{j}{1+j}} a_{s_j,k}^{\frac{1}{1+j}} \geq a_{s_j,j}$$

for all j. Hence

$$\frac{a_{s_j,n_{s_j}}^{\frac{1}{1+d}} a_{s_j,k}^{\frac{d}{1+d}}}{a_{s_j,j-1}} \geq \frac{a_{s_j,n_{s_j}}^{\frac{1}{1+j}} a_{s_j,k}^{\frac{j}{1+j}}}{a_{s_j,j-1}} \geq \frac{a_{s_j,j}}{a_{s_j,j-1}} \to \infty$$

as $j \to \infty$ and, by interpolation, we obtain a sequence which does not satisfy
$(*)$. Hence $(*) \iff (**)$. The other equivalences are obtained in a similar
fashion.

*A Fréchet space E is said to have property $(\widetilde{\Omega})$ if there exists a compact
subset K of E such that for every zero neighbourhood U there exists a
zero neighbourhood V, $c > 0$ and $d > 0$ such that*

$$V \subset \frac{c}{r} U + r^d K \tag{6.11}$$

for all $r > 0$.

Condition (6.11) is a splitting condition similar to those discussed in the
introduction to Chapter 4 and a Fréchet nuclear space with basis $\lambda(A)$ has
property $(\widetilde{\Omega})$ if and only if it satisfies (6.10). By Proposition 6.9 a Fréchet
nuclear space with basis contains a compact non-polar set if and only if it
has property $(\widetilde{\Omega})$. The basis hypothesis in Proposition 6.9 may be replaced
by the bounded approximation property (see Section 6.5).

We now use Proposition 6.9 to identify compact non-polar subsets of
power series spaces. We use the notation of Example 4.13. First suppose
$E = \Lambda_\infty(\alpha)$ where $\alpha = (\alpha_j)_j$. The compact polydisc $\left[(q^{\alpha_j n_j})_j \right]$, $n_j \to \infty$, is
non-polar if and only if for all $l_j \to \infty$ and all k there exists $d > 0$ such that

$$\sup_j \frac{(q^{\alpha_j l_j})^{1+d} (q^{-\alpha_j k})^d}{q^{\alpha_j n_j}} = \sup_j \left(q^{l_j(1+d)-kd-n_j} \right)^{\alpha_j} < \infty.$$

Since $\alpha_j \to +\infty$ and $0 < q < 1$, we see that this is never possible, e.g. take l_j to be the integer part of $n_j^{1/2}$ for all j. Hence all compact subsets of a power series space of infinite type are polar.

Let $E = \Lambda_1(\alpha)$ and let $(a_j)_j \in \Lambda_1(\alpha)$, $a_j > 0$, all j. Then $[(a_j)_j]$ is compact non-polar if and only if for all $l_j \to \infty$ and all k there exists $d > 0$ such that

$$\sup_j \left(\frac{\left(p_{l_j}^{-\alpha_j}\right)^{1+d} \left(p_k^{\alpha_j}\right)^d}{|a_j|} \right) = \sup_j \left(\frac{1}{|a_j|^{1/\alpha_j} p_{l_j} \left(p_{l_j} p_k^{-1}\right)^d} \right)^{\alpha_j} < \infty$$

where $0 < p_k < 1$, $p_{l_j} \to 1^-$ and $\alpha_j \to \infty$ as $j \to \infty$. Clearly this will occur if and only if $\liminf_{j \to \infty} |a_j|^{1/\alpha_j} > 0$.

If $E = H(D)$, $D = \{z \in \mathbb{C} : |z| < 1\}$ and $f = \sum_{n=0}^{\infty} a_n z^n \in H(D)$ then $[f]$ is compact non-polar if and only if $\liminf_{n \to \infty} |a_n|^{1/n} > 0$. In particular, if $f \in H(\mathbb{C})$ then $[f]$ is polar in $H(D)$.

Example 6.11 We consider the $\mathcal{DFN}$ space $E = \lambda(A)'_\beta$ where $\lambda(A)$ is a Fréchet nuclear space with basis. We use the notation of the previous example. In this case $(w_j)_j \in E$ if and only if there exists $c > 0$ and $k \in \mathbb{N}$ such that $|w_j| \le c a_{j,k}$ for all j and, moreover, we have $E \cong \Lambda(P)$ where $P = \left(\frac{1}{a_{j,n_j}}\right)_j$ and $(n_j)_j$ ranges over all strictly increasing sequences with $a_{j,n_j} > 0$. By Proposition 6.9, E contains a compact non-polar subset

$$\Longleftrightarrow \exists k, \ \forall l, \ \forall (n_j)_j, \ n_j \to \infty, \ \exists d > 0 \text{ such that}$$
$$\sup_j \frac{a_{j,l}^{1+d}}{a_{j,k} a_{j,n_j}^d} < \infty$$
$$\Longleftrightarrow \exists k, \ \forall l, \ \exists m \in \mathbb{N}, \ d > 0 \text{ such that}$$
$$\sup_j \frac{a_{j,l}^{1+d}}{a_{j,k} a_{j,m}^d} < \infty$$
$$\Longleftrightarrow \exists k, \ \forall l, \ \exists m \in \mathbb{N}, \ d > 0, \ c > 0 \text{ such that}$$
$$a_{j,l}^{1+d} \le c a_{j,k} a_{j,m}^d. \tag{6.12}$$

This leads us to the following definition.

A Fréchet space with fundamental system of semi-norms $(\|\cdot\|_n)_{n=1}^{\infty}$ *has property* (DN) *if it admits a continuous norm* $\|\cdot\|$ *such that for all l there exists $m \in \mathbb{N}$, $c > 0$ and $d > 0$ such that*

$$\|\cdot\|_l^{1+d} \le c \|\cdot\| \|\cdot\|_m^d.$$

If $E = \lambda(A)$ is a Fréchet nuclear space with basis then E has (<u>DN</u>) if and only if A satisfies (6.12). This can be expressed as a splitting condition since a Fréchet nuclear space has (<u>DN</u>) if and only if for every fundamental sequence of convex balanced compact subsets of E'_β, $(B_n)_n$, we can choose k_0 such that for every $k \in \mathbb{N}$ there exists $n \in \mathbb{N}$, $c > 0$ and $d > 0$ such that

$$B_k \subseteq \frac{c}{r} B_n + r^d B_{k_0} \tag{6.13}$$

for all $r > 0$.

Condition (6.13) may be compared with (6.11) and also with condition (DN) of Chapter 4. Nuclear power series spaces of finite type (Example 4.13(b)) have (<u>DN</u>) while the space $\lambda(A)$ where $a_{j,k} = \exp\left(\sum_{j=1}^{k} \exp(j^{1/n})\right)$ for all j and k does not have (<u>DN</u>). Since the basis hypothesis can be removed in this example (see Section 6.5) the following holds: if E is a Fréchet nuclear space then E'_β contains a compact non-polar subset if and only if E has (<u>DN</u>).

6.2 Holomorphic Extensions from Closed Subspaces

In this section we consider the problem of extending holomorphic functions from a closed subspace of a locally convex space to the whole space as a holomorphic function. We begin with some simple observations which illustrate both the difficulties that must be overcome and ideas which prove useful.

If E is a closed subspace of F, $P \in \mathcal{P}(^nE)$, $f \in H(F)$ and $f\big|_E = P$ then $Q := \dfrac{\hat{d}^n f(0)}{n!} \in \mathcal{P}(^nF)$ and $Q\big|_E = P$. Hence if each holomorphic function on E can be extended as a holomorphic function to F then each continuous n-homogeneous polynomial on E can be extended as a continuous n-homogeneous polynomial on F. Thus, in contrast to the situation encountered in the previous section, we see from our results in Chapter 1, that the extension problem we are considering is non trivial at the homogeneous polynomial level. At the same time this suggests that, if we extend each homogeneous polynomial, we may be able to combine the extensions to form the Taylor series expansion of a function which gives the required holomorphic extension. This turns out to be the case but we require uniform bounds on the norms of the polynomial extensions. We next present an example which illustrates a non-polynomial obstruction to the construction of extensions and afterwards give a fresh look at how we extend polynomials.

Example 6.12 Propositions 1.51 and 1.53 both show that continuous n-homogeneous polynomials on c_0 admit norm preserving extensions to l_∞ as n-homogeneous polynomials. If $A := (e_n)_n$ is the standard unit vector basis for c_0 then A is a closed non-compact, and by Example 3.20(c), non-bounding subset of c_0. Hence there exists $f \in H(c_0)$ such that $\|f\|_A = \infty$. By Proposition 4.51, A is a bounding subset of l_∞ and thus any holomorphic extension $\tilde{f}$ of f to l_∞ must satisfy $\|f\|_A = \|\tilde{f}\|_A < \infty$. This is clearly impossible and shows that holomorphic functions on a Banach space do not necessarily extend holomorphically to the bidual. However, if $f \in H_b(c_0)$ (see Section 1.3) then, by the Cauchy inequalities, $\lim_{n\to\infty} \left\| \dfrac{\widehat{d^n f(0)}}{n!} \right\|^{1/n} = 0$. The polynomial extension property implies that for each n there exists $P_n \in \mathcal{P}(^n l_\infty)$ such that $\|P_n\| = \left\| \dfrac{\widehat{d^n f(0)}}{n!} \right\|$ and $P_n|_{c_0} = \dfrac{\widehat{d^n f(0)}}{n!}$. Hence $\tilde{f} := \sum_{n=0}^\infty P_n \in H_b(l_\infty)$ and $\tilde{f}\Big|_{c_0} = f$. We have thus shown that each $f \in H_b(c_0)$ has a holomorphic extension to the bidual as a holomorphic function of bounded type and that there exist non-extendible $f \in H(c_0) \setminus H_b(c_0)$. Later we develop this example into a general result. At this time we are obliged, in the Banach space setting, to restrict ourselves to holomorphic functions of bounded type since we do not know of any pair of Banach spaces $\{E, F\}$, where E is a closed non-complemented subspace of F, such that each holomorphic function on E can be extended to a holomorphic function on F (see Example 6.33(b)).

We return now to a more detailed examination of polynomial extensions. Various approaches to extensions have been proposed and developed in recent years and, whether explicit or implicit, each successful effort employs one or both of the canonical mappings $J_E : E \longrightarrow E''$, $J_{E'} : E' \longrightarrow E'''$. In particular, we see that the Aron–Berner extension, defined in Section 1.3, may be considered an extension of $J_{E'}$, i.e. $(AB)_1 = J_{E'}$, and, moreover, $[AB_1(\varphi)](J_E x) = \varphi(x)$ for $x \in E$ and $\varphi \in E'$. We concentrate here on the Aron–Berner extension operator as it leads to the results we require although we could also have used the principle of local reflexivity and an ultraproduct extension in some cases. The Aron–Berner extension is more general than it initially appears and, although defined in topological terms, is essentially an *algebraic* extension, which, however, respects and preserves *topological* environments. This can be put more clearly in focus by examining alternative methods of introducing it. In particular, (1.58) in Section 1.5 shows that the topological content of this extension is derived from the mapping $J_{E'}$.

We recall the definition of AB_n and extend it to the vector-valued setting (we retain the notation AB_n in this case). In doing so we will also extend elements of $\mathcal{L}(^n E; F)$ to define elements of $\mathcal{L}(^n E''; F'')$ and use $\mathcal{E}_n$ to denote the operator which realizes this multilinear extension. Thus if E is a Banach space and $L \in \mathcal{L}(^n E)$, then $\mathcal{E}_n(L)$ or $\tilde{L}$ denotes the n-linear mapping on $(E'')^n$ defined by

$$\tilde{L}(x_1'', \ldots, x_n'') = \lim_{\alpha_1} \lim_{\alpha_2} \cdots \lim_{\alpha_n} L(x_{\alpha_1}, \ldots, x_{\alpha_n}) \tag{6.14}$$

where (x_{α_j}) is a net in E such that $J_E x_{\alpha_j} \to x_j''$ in the $\sigma(E'', E')$ topology as $\alpha_j \to \infty$ for all j. It is important to note the order, right to left, in which the limits are taken in (6.15). A different ordering may give rise to a different extension.

For each j, $1 \leq j \leq n$, $(x_i)_{i=1}^{j} \in E^j$ and $(x_i'')_{i=j+2}^{n} \in (E'')^{n-j-1}$ the mapping

$$x'' \in E'' \longrightarrow \widetilde{L}(J_E x_1, \ldots, J_E x_j, x'', x_{j+2}'', \ldots, x_n'')$$

is $\sigma(E'', E')$ continuous. From the proof of Proposition 1.53 we have $\widetilde{L} \in \mathcal{L}(^n E'')$ and $\|L\| = \|\widetilde{L}\|$. If $\widehat{L} = P \in \mathcal{P}(^n E)$ then $[AB_n(P)](z) = \mathcal{E}_n(L)(z^n) = \widetilde{L}(z^n)$ for all $z \in E''$. For convenience of notation we sometimes write $\overset{\circ}{P}$ in place of $AB_n(P)$.

We now consider the vector-valued situation. If E and F are Banach spaces and $L \in \mathcal{L}(^n E; F)$, let $(\widetilde{L}(x_1'', \ldots, x_n''))(\varphi) = \varphi \circ L(x_1'', \ldots, x_n'')$ for $x_i'' \in E''$, $1 \leq i \leq n$, and $\varphi \in F'$. Since $\|\widetilde{\varphi \circ L}\| = \|\varphi \circ L\| \leq \|\varphi\| \, \|L\|$, the mapping $\varphi \in F' \longrightarrow (\widetilde{L}(x_1'', \ldots, x_n''))(\varphi)$ is a continuous linear mapping for each fixed $(x_1'', \ldots, x_n'') \in (E'')^n$ and hence $(x_1'', \ldots, x_n'') \longrightarrow \widetilde{L}(x_1'', \ldots, x_n'') \in F''$ defines a continuous n-linear F''-valued form on E''. Moreover, $\|L\| = \|\widetilde{L}\|$ and the mapping

$$\mathcal{E}_n : L \in \mathcal{L}(^n E; F) \longrightarrow \mathcal{E}_n(L) =: \widetilde{L} \in \mathcal{L}(^n E''; F'')$$

is a linear isometric extension operator.

If $P \in \mathcal{P}(^n E; F)$ we define $\overset{\circ}{P}$ or $AB_n(P)$ by

$$\overset{\circ}{P}(z) := [AB_n(P)](z) := [\mathcal{E}_n(\overset{\vee}{P})](z^n)$$

for all $z \in E''$. From the $\mathbb{C}$-valued case (Proposition 1.53) and the Hahn–Banach Theorem we see that

$$AB_n : \mathcal{P}(^n E; F) \longrightarrow \mathcal{P}(^n E''; F'')$$

is an isometric linear extension operator.

First we investigate algebraic properties of this extension. If $P \in \mathcal{P}(^n E; F)$, $x \in E$, $y'' \in E''$, $0 \leq k \leq n$ and $(y_\alpha)_\alpha$ is a net in E such that $(J_E y_\alpha)_\alpha$ converges $\sigma(E'', E')$ to y'', then, if $(y_{\alpha_j})_{\alpha_j} = (y_\alpha)_\alpha$ for all j,

$$\left[\left(\frac{\widehat{d^k} AB_n(P)}{k!} \right) (J_E x) \right] (y'') = \binom{n}{k} \overset{\vee}{AB}_n(P)((J_E x)^{n-k}, (y'')^k)$$

$$= \binom{n}{k} \lim_{\alpha_1} \lim_{\alpha_2} \cdots \lim_{\alpha_k} \overset{\vee}{P}(x^{n-k}, y_{\alpha_1}, \ldots, y_{\alpha_k})$$

$$= \lim_{\alpha_1} \lim_{\alpha_2} \cdots \lim_{\alpha_k} \left(\frac{\widehat{d^k} P(x)}{k!} \right) (y_{\alpha_1}, \ldots, y_{\alpha_k})$$

$$= \left[AB_k \left(\frac{\widehat{d^k} P(x)}{k!} \right) \right] (y'').$$

Hence the Aron–Berner extension commutes with differentiation and we may write symbolically $\dfrac{\widehat{d^k}}{k!} \circ AB_n = AB_k \circ \dfrac{d^k}{k!}$.

Next we consider multiplicative properties of $(AB)_n$ when $F = \mathbb{C}$. If L_1 is k-linear, L_2 is $(n-k)$-linear and σ is a permutation of $\{1,\dots,n\}$ let

$$(L_1 \cdot L_2)_\sigma(x_1,\dots,x_n) = L_1(x_{\sigma(1)},\dots,x_{\sigma(k)})L_2(x_{\sigma(k+1)},\dots,x_{\sigma(n)}).$$

It is easily seen that

$$s(L_1 \cdot L_2) = \frac{1}{n!} \sum_{\sigma \in S_n} (L_1 \cdot L_2)_\sigma$$

where s denotes the symmetrization operator on n-linear forms. Using the notation of (6.14) we see that

$$
\begin{aligned}
&\mathcal{E}_n((L_1 \cdot L_2)_\sigma)(x_1'',\dots,x_n'') \\
&= \lim_{\alpha_1}\lim_{\alpha_2}\cdots\lim_{\alpha_n} L_1(x_{\alpha_{\sigma(1)}},\dots,x_{\alpha_{\sigma(k)}})L_2(x_{\alpha_{\sigma(k+1)}},\dots,x_{\alpha_{\sigma(n)}}) \\
&= \mathcal{E}_k(L_1)(x_{\sigma(1)}'',\dots,x_{\sigma(k)}'')\mathcal{E}_{n-k}(L_2)(x_{\sigma(k+1)}'',\dots,x_{\sigma(n)}'') \\
&= (\mathcal{E}_k(L_1)\cdot\mathcal{E}_{n-k}(L_2))_\sigma(x_1'',\dots,x_n'')
\end{aligned}
$$

and hence

$$
\begin{aligned}
\mathcal{E}_n(s(L_1 \cdot L_2)) &= \mathcal{E}_n\left(\frac{1}{n!}\sum_{\sigma \in S_n}(L_1\cdot L_2)_\sigma\right) \\
&= \frac{1}{n!}\sum_{\sigma \in S_n}\mathcal{E}_n((L_1\cdot L_2)_\sigma) \\
&= \frac{1}{n!}\sum_{\sigma \in S_n}(\mathcal{E}_k(L_1)\cdot\mathcal{E}_{n-k}(L_2))_\sigma \\
&= s(\mathcal{E}_k(L_1)\cdot\mathcal{E}_{n-k}(L_2)).
\end{aligned}
$$

If $P \in \mathcal{P}(^k E)$, $Q \in \mathcal{P}(^{n-k}E)$ and $x'' \in E''$ then

$$
\begin{aligned}
\overset{\circ}{\overline{PQ}}(x'') &= \mathcal{E}_n(\widehat{\overset{\vee}{PQ}})(x'') \\
&= \mathcal{E}_n(\widehat{s(\overset{\vee}{P}\overset{\vee}{Q})})(x'') \\
&= s(\mathcal{E}_k(\overset{\vee}{P})\mathcal{E}_{n-k}(\overset{\vee}{Q}))(x'') \\
&= \mathcal{E}_k(\overset{\vee}{P})(x'')^k\mathcal{E}_{n-k}(\overset{\vee}{Q})(x'')^{n-k} \\
&= \widehat{\mathcal{E}_k(\overset{\vee}{P})}(x'')\widehat{\mathcal{E}_{n-k}(\overset{\vee}{Q})}(x'') \\
&= \overset{\circ}{\overline{P}}(x'')\overset{\circ}{\overline{Q}}(x'').
\end{aligned}
$$

If we define $AB : \mathcal{P}(E) \longrightarrow \mathcal{P}(E'')$ by $AB\left(\sum_{j=0}^{n} P_j\right) = \sum_{j=0}^{n} AB_j(P_j)$ where $P_j \in \mathcal{P}(^j E)$ for $j > 0$ and $AB_0 = $ identity on $\mathcal{P}(^0 E) := \mathbb{C}$, then the above shows that $\overset{\circ}{PQ} = \overset{\circ}{P}\,\overset{\circ}{Q}$ for $P, Q \in \mathcal{P}(E)$ and AB is a multiplicative linear function from $\mathcal{P}(E)$ into $\mathcal{P}(E'')$.

Next, we consider the question of permuting the order of taking limits in (6.14). By the Hahn–Banach Theorem it suffices to consider the case $F = \mathbb{C}$ and by permuting limits, two at a time, we see that it suffices to look at the $n = 2$ case. If $L \in \mathcal{L}(^2 E)$ and the order of taking limits in (6.14) can be interchanged then we say that L is *Arens regular* and if every $L \in \mathcal{L}(^2 E)$ is Arens regular we say that E is *regular*. Clearly, if E is regular then $\mathcal{E}_n(\mathcal{L}^s(^n E, F)) \subset \mathcal{L}^s(^n E'', F'')$ for every Banach space F. To characterize regular Banach spaces it is convenient to consider the following alternative construction of the Aron–Berner extension. We only require the $n = 2$ case but the procedure can be carried out for all n. If $A \in \mathcal{L}(^2 E)$ we define $T \in \mathcal{L}(E; E')$ (see Proposition 1.1) by the formula $[T(x)](y) = A(x, y)$ for all $(x, y) \in E^2$. The double transpose of T, T^{**}, maps E'' into E''' and $\mathcal{E}_2(A)(x'', y'') = \tilde{A}(x'', y'') = [T^{**}(x'')](y'')$.

The following characterization leads to examples which show that A symmetric does not necessarily imply $\mathcal{E}_2(A)$ is symmetric (see Exercise 6.50).

Proposition 6.13 *If E is a Banach space and $A \in \mathcal{L}^s(^2 E)$ then the following are equivalent:*

(a) $\tilde{A} = \mathcal{E}_2(A) \in \mathcal{L}^s(^2 E'')$,

(b) $\mathcal{E}_2(A)$ *is separately weak* continuous in both variables*,

(c) $T \in \mathcal{L}(E; E')$ *is weakly compact (i.e. T maps the unit ball of E into a weakly relatively compact subset of E').*

Proof. (a) $\Longrightarrow$ (b). Since $\tilde{A}$ is always weak* continuous in the first variable, symmetry implies weak* continuity in the second variable.

(b) $\Longrightarrow$ (c). Fix $x'' \in E''$. Since

$$y'' \in E'' \longrightarrow \tilde{A}(x'', y'') = [T^{**}(x'')](y'')$$

is weak* continuous we have $T^{**}(E'') \subset J_{E'}(E')$ and hence T is weakly compact (see Exercise 1.82).

(c) $\Longrightarrow$ (a). If $x, y \in E$ then

$$[(T^* \circ J_E)(x)](y) = [T^*(J_E(x))](y)$$
$$= (J_E(x))(T(y))$$
$$= (T(y))(x)$$
$$= A(y, x).$$

Since A is symmetric, this implies

$$[(T^* \circ J_E)(x)](y) = A(y, x) = (T(y))(x) = (T(x))(y)$$

and $T^* \circ J_E = T$. Hence, if $x \in E$ and $x'' \in E''$, then

$$
\begin{aligned}
[T^*(x'')](x) &= x''(Tx) \\
&= x''(T^* \circ J_E(x)) \\
&= (x'' \circ T^*)(J_E(x)) \\
&= (T^{**}(x'') \circ J_E)(x)
\end{aligned}
$$

and

$$T^*(x'') = T^{**}(x'') \circ J_E \qquad\qquad (*)$$

for all $x'' \in E''$. If (c) holds then $T^{**}(E'') \subset J_{E'}(E')$. To prove $T^{**} = J_{E'} \circ T^*$ it suffices, since E separates the points of E', to show $(T^{**}(x''))(J_E(x)) = (T^*(x''))(x)$ for $x'' \in E''$ and $x \in E$. This, however, is just $(*)$. Hence, if x'' and y'' belong to E'' then

$$
\begin{aligned}
\widetilde{A}(x'', y'') &= [T^{**}(x'')](y'') \\
&= [(J_{E'} \circ T^*)(x'')](y'') \\
&= [J_{E'}(T^*(x''))](y'') \\
&= y''(T^*(x'')) \\
&= (y'' \circ T^*)(x'') \\
&= [T^{**}(y'')](x'') \\
&= \widetilde{A}(y'', x'').
\end{aligned}
$$

Hence, $\widetilde{A}$ is symmetric, (c) $\Longrightarrow$ (a) and the proof is complete.

Condition (c) in Proposition 6.13 is quite useful in finding both positive and negative examples of symmetric bilinear forms with symmetric extensions to the bidual. For instance, it is well known that any continuous linear mapping $T : C(K) \longrightarrow C(K)'$, K a compact Hausdorff space, factors through a Hilbert space. Since Hilbert spaces are reflexive and have weakly compact unit balls, T is weakly compact and the associated bilinear form has a symmetric extension to $C(K)''$. The Banach space l_1 is a negative example (see Exercise 6.50 for details).

The problem of extending bilinear forms to the bidual has been considered in the literature from another important perspective that we now briefly describe. The set $\mathcal{L}(^2E; E)$ may be regarded as the collection of all *normed algebra structures* on E. Thus each product $*$ on E may be identified with the bilinear form L_* on $E \times E$ by letting $L_*(x, y) = x * y$ and conversely each $L \in \mathcal{L}(^2E; E)$ defines a product $*_L$ where $x *_L y = L(x, y)$. In this identification the symmetric bilinear forms correspond to the commutative algebra structures and continuity of L corresponds to joint continuity of the product. Extensions to the bidual as bilinear forms correspond to extending the product from E to E''. The extension operator $\mathcal{E}_2$ achieves this purpose but iteration of the limits in (6.14) may give rise to two different algebra structures on E''. With our previous terminology we say that the original product is Arens regular if both extensions agree. Since any product is the sum of a commutative and a non-commutative product, the proof of Proposition 6.13 can be modified to prove the following result.

Proposition 6.14 *If E and G are Banach spaces then the following are equivalent:*

(a) each $L \in \mathcal{L}(^nE; G)$ is Arens regular,

(b) if $L \in \mathcal{L}(^nE; G)$ then $\mathcal{E}_n(L)$ is separately weak* continuous in each variable,

(c) each continuous linear mapping from E to E' is weakly compact.

Every C^* algebra is regular and, in particular, the usual product on $\mathcal{C}(K)$ extends to endow $\mathcal{C}(K)''$ with the structure of a commutative C^* algebra.

Extensions from E to E'' of scalar-valued mappings were based on the duality $(J_Ex)(\varphi) = \varphi(x)$. By generalizing this duality to G-valued mappings we obtain a natural domain for G-valued extensions. Given Banach spaces E and G let

$$J(E; G) := \mathcal{L}(\mathcal{L}(E; G); G).$$

We embed E in $J(E; G)$ by letting $(j_{E,G}(x))(\varphi) = \varphi(x)$ for $x \in E$ and $\varphi \in \mathcal{L}(E; G)$. If $G = \mathbb{C}$ then $\mathcal{L}(\mathcal{L}(E; \mathbb{C}); \mathbb{C}) = \mathcal{L}(E'; \mathbb{C}) = E''$ and $j_{E,G} = J_E$. The space $J(E; G)$ is a Banach space when we endow $\mathcal{L}(E; G)$ and $\mathcal{L}(\mathcal{L}(E; G); G)$ with the topologies of uniform convergence on bounded sets. We define a G-valued extension in two steps. If $\phi \in \mathcal{L}^s(^nE, G)$ and $T \in J(E; G)$ let $T^\Diamond(\phi) \in \mathcal{L}^s(^{n-1}E, G)$ be defined by

$$[T^\Diamond(\phi)](x_1, \ldots, x_{n-1}) := T[x_n \longrightarrow \phi(x_1, \ldots, x_n)]$$

for $x_i \in E$, $1 \leq i \leq n$. Clearly the mapping

$$T \in J(E; G) \longrightarrow T^\Diamond \in \mathcal{L}(\mathcal{L}^s(^nE, G); \mathcal{L}^s(^{n-1}E, G))$$

is linear and $\|T^\Diamond\| \leq \|T\|$. Moreover, if $x \in E$, $\phi \in \mathcal{L}^s(^nE, G)$ and $x_i \in E$ for $1 \leq i < n$ then

$$[(j_{E,G}(x))^\Diamond(\phi)](x_1 \ldots, x_{n-1}) = j_{E,G}(x)[x_n \longrightarrow \phi(x_1, \ldots, x_n)]$$
$$= \phi(x_1, \ldots, x_{n-1}, x).$$

The second step consists in repeating the first step n times; i.e. if $\phi \in \mathcal{L}^s(^nE, G)$ we define $\phi^\Diamond \in \mathcal{L}^s(^nJ(E; G), G)$ by

$$\phi^\circ(T_1, \ldots, T_n) = (T_1^\circ \circ T_2^\circ \circ \cdots \circ T_n^\circ)(\phi)$$

for $T_i \in J(E; G)$, $1 \le i \le n$. It is easily seen that $\|\phi^\circ\| \le \|\phi\|$ for all $\phi \in \mathcal{L}^s(^n E, G)$ but, in general, ϕ° may not be symmetric. A linear extension operator Z_n on $\mathcal{P}(^n E; G)$ is obtained by letting

$$[Z_n(P)](T) = \left(\overset{\vee}{P}\right)^\circ (T, \ldots, T)$$

for $T \in J(E; G)$. We have $\|Z_n(P)\| \le \|P\|$ for all $P \in \mathcal{P}(^n E; G)$ and all n and $[Z_n(P)](j_{E,G}(x)) = P(x)$ for all $x \in E$.

Example 6.15 (a) If $G = \mathbb{C}$ then $J(E; \mathbb{C}) = E''$ and $Z_n = AB_n$ for all n.

(b) If $E = \mathbb{C}$ then $J(\mathbb{C}; G) = \mathcal{L}(\mathcal{L}(\mathbb{C}; G); G) = \mathcal{L}(G; G)$ is the space of bounded linear operators from G into G. Each $P \in \mathcal{P}(^n \mathbb{C}; G)$ has the form $P(z) = az^n$ for some $a \in G$ and $[Z_n(P)](T) = T \circ T \circ \cdots \circ T(a) = T^n(a)$ for all $T \in \mathcal{L}(G; G)$ and all n.

We now return to holomorphic functions of bounded type. Any of AB_n, Z_n or the extension in Proposition 1.51 could be used to prove the following proposition.

Proposition 6.16 *If E and G are Banach spaces then there exists a continuous linear mapping*

$$T : H_b(E; G) \longrightarrow H_b(E''; G'')$$

such that $T f\big|_E = f$. If $F = \mathbb{C}$ then T is multiplicative.

Proof. If $f \in H(E; G)$ then, by the Cauchy inequalities, $f \in H_b(E; G)$ if and only if $\lim_{n \to \infty} \left\| \dfrac{\widehat{d^n} f(0)}{n!} \right\|_B^{1/n} = 0$ where B denotes the unit ball of E. Moreover, the Fréchet topology on $H_b(E; G)$ of uniform convergence on bounded subsets of E is generated by

$$p_r \left(\sum_{n=0}^{\infty} \frac{\widehat{d^n} f(0)}{n!} \right) := \sum_{n=0}^{\infty} r^n \left\| \frac{\widehat{d^n} f(0)}{n!} \right\|_B,$$

$r = 1, 2, \ldots$. If $f \in H_b(E; G)$ let $T(f) := \sum_{n=0}^{\infty} AB_n \left(\dfrac{\widehat{d^n} f(0)}{n!} \right)$. Since $AB_n : \mathcal{P}(^n E; G) \longrightarrow \mathcal{P}(^n E''; G'')$ and $\|AB_n\| \le 1$ it follows that T is a continuous linear operator and clearly $T(f)\big|_E = f$. We have already noted that AB_n is multiplicative on $\mathcal{P}(E)$ and, by continuity, T is multiplicative on $H_b(E)$. This completes the proof.

Since the bounded subsets of c_0 are bounding subsets of l_∞ (see Section 4.6) Example 6.12 and Proposition 6.16 imply the following corollary.

Corollary 6.17 *A holomorphic function f on c_0 has a holomorphic extension to l_∞ if and only if $f \in H_b(c_0)$.*

Our next result extends Proposition 6.16 into a complete characterization. To stay with the same range space it is natural, in view of our extensions to F'', to consider Banach spaces which are complemented in their bidual, for example dual Banach spaces.

Proposition 6.18 *If E is a subspace of a Banach space F then the following are equivalent.*

(1) *If the Banach space G is complemented in G'' then there exists a sequence of continuous linear operators $(T_n)_n$,*

$$T_n : \mathcal{P}(^nE; G) \longrightarrow \mathcal{P}(^nF; G)$$

such that $T_n(P_n)\big|_E = P_n$ for all $P_n \in \mathcal{P}(^nE; G)$ and all n and

$$\sup_n \|T_n\|^{1/n} < \infty.$$

(2) *If the Banach space G is complemented in G'' then there exists a continuous linear operator $T : H_b(E; G) \longrightarrow H_b(F; G)$ such that $T(f)\big|_E = f$ for all $f \in H_b(E; G)$.*

(3) *There exists a continuous linear operator $T : H_b(E) \longrightarrow H_b(F)$ such that $T(f)\big|_E = f$ for all $f \in H_b(E)$.*

(4) *There exists a continuous linear operator $T_1 : E' \longrightarrow F'$ such that $T_1(\phi)\big|_E = \phi$ for all $\phi \in E'$.*

(5) *There exists a continuous linear operator $S : F \longrightarrow E''$ which extends J_E.*

Proof. The equivalence of (1) and (2) is easily established using the methods employed in the proof of Proposition 6.16. In particular, if (1) holds let $T(f) = \sum_{n=0}^{\infty} T_n\left(\dfrac{\widehat{d^n} f(0)}{n!}\right)$ for all $f \in H_b(E; G)$. This gives the required linear extension of holomorphic functions. If (2) holds then the mappings $T_n(P_n) := \dfrac{\widehat{d^n}(T(P_n))}{n!}(0)$ for all n and all $P \in \mathcal{P}(^nE; G)$ are easily shown to imply (1).

Clearly (2) implies (3). If (3) holds then the mapping $T_1 : E' \longrightarrow F'$ defined by $T_1(\phi) = d(T(\phi))(0)$ is a continuous linear operator satisfying (4).

Hence (3) $\implies$ (4). If (4) holds then T_1^*, the transpose of T_1, maps F'' into E''. Let $S = T_1^*\big|_F$. If $x \in E$ and $\phi \in E'$ then

$$[S(x)](\phi) = [T_1^*\big|_F(x)](\phi) = [T_1^*(x)](\phi)$$
$$= [T_1(\phi)](x)$$
$$= [d(T(\phi))(0)](x)$$
$$= \phi(x)$$

and S has the required properties. Hence (4) $\implies$ (5).

Now suppose (5) holds. Let G denote a Banach space which is complemented in its bidual and let π denote a continuous projection from G'' onto G. For each n and $P_n \in \mathcal{P}(^nE; G)$ let $T_n(P_n) = \pi \circ AB_n(P_n) \circ S$. Then T_n is linear, $T_n(P_n)\big|_E = P_n$ and

$$\|T_n\| = \sup_{\|P_n\| \leq 1} \|\pi \circ AB_n(P_n) \circ S\|$$
$$\leq \|\pi\| \sup_{\|P_n\| \leq 1,\ x \in F,\ \|x\| \leq 1} \|AB_n(P_n)(Sx)\|$$
$$\leq \|\pi\| \|S\|^n \sup_{\|P_n\| \leq 1} \|AB_n(P_n)\|$$
$$\leq \|\pi\| \|S\|^n$$

for all n. Hence (5) $\implies$ (1) and this completes the proof.

Example 6.19 If E is complemented in its bidual and is a subspace of the Banach space F, and there exists a continuous linear operator

$$T : H_b(E) \longrightarrow H_b(F)$$

such that $T(f)\big|_E = f$ then Proposition 6.18 implies that $\mathrm{Id}_E \in H_b(E; E)$ extends to $\widetilde{\mathrm{Id}}_E \in H_b(F; E)$. Hence $d(\widetilde{\mathrm{Id}}_E)(0)$ defines a continuous projection from F onto E and E is a complemented subspace of F. If F is reflexive and there exists for each closed subspace E of F a linear extension operator from $H_b(E)$ into $H_b(F)$ then every closed subspace of F is complemented and F is isomorphic to a Hilbert space.

Example 6.20 The proof of Proposition 6.16 can be applied to show that

$$Z\Big(\sum_{n=0}^{\infty} \frac{\widehat{d^n} f(0)}{n!}\Big) := \sum_{n=0}^{\infty} Z_n\Big(\frac{\widehat{d^n} f(0)}{n!}\Big)$$

defines a linear extension operator from $H_b(E; G)$ into $H_b(J(E; G); G)$.

Let $E = \mathbb{C}$. Consider $f \in H(\mathbb{C}; G)$ which has the form $f = g \cdot a$ where $g \in H(\mathbb{C})$ and $a \in G$. If $g(z) = \sum_{n=0}^{\infty} \alpha_n z^n$ then, by Example 6.15,

$$[Z(g \cdot a)](T) = \sum_{n=0}^{\infty} \alpha_n T^n(a) = g(T)(a)$$

for all $T \in \mathcal{L}(G; G)$ and Z implements the *operator functional calculus* for holomorphic function of bounded type.

Since $H_b(E) \neq H(E)$ for any infinite dimensional Banach space E (Example 3.8(f)), the results we have obtained do not apply to $H(E)$. However, we can apply our approach to the space of all holomorphic functions on certain non-Banach locally convex spaces.

Proposition 6.21 *The following are equivalent conditions on a nuclear locally convex space E;*
(a) $H(E) = H_{ub}(E)$,

(b) if E is a subspace of a locally convex space F and F has a fundamental system of semi-norms consisting of inner products then each $f \in H(E)$ can be extended holomorphically to F.

Proof. (a) $\Longrightarrow$ (b). Let $\sum_{n=0}^{\infty} P_n$ denote the Taylor series expansion at the origin of $f \in H(E)$. Since $H(E) = H_{ub}(E)$ there exists a continuous semi-inner product α on F such that $\lim_{n \to \infty} \|P_n\|_{B_\alpha(1) \cap E}^{1/n} = 0$. By Lemma 3.9 there exists $Q_n \in \mathcal{P}(^n \widehat{E}_\alpha)$ such that $P_n = Q_n \circ (\pi_\alpha|_E)$ where $\widehat{E}_\alpha$ is the completion of E_α and π_α denotes the canonical quotient mapping from F into $\widehat{F}_\alpha$. Since $\widehat{E}_\alpha$ is a closed complemented subspace of $\widehat{F}_\alpha$ there exists a continuous projection π from $\widehat{F}_\alpha$ onto $\widehat{E}_\alpha$. Hence $Q_n \circ \pi \in \mathcal{P}(^n \widehat{F}_\alpha)$ for all n and, moreover,

$$\|Q_n \circ \pi\|_{\pi_\alpha(B_\alpha(1))} = \|P_n\|_{B_\alpha(1) \cap E}.$$

We have $Q_n \circ \pi \circ \pi_\alpha \in \mathcal{P}(^n F)$, $\|Q_n \circ \pi \circ \pi_\alpha\|_{B_\alpha(1)} = \|P_n\|_{B_\alpha(1) \cap E}$ and $Q_n \circ \pi \circ \pi_\alpha|_E = Q_n \circ (\pi_\alpha|_E) = P_n$. Hence $\lim_{n \to \infty} \|Q_n \circ \pi \circ \pi_\alpha\|_{B_\alpha(1)}^{1/n} = \lim_{n \to \infty} \|P_n\|_{B_\alpha(1) \cap E}^{1/n} = 0$, $g := \sum_{n=0}^{\infty} Q_n \circ \pi \circ \pi_\alpha \in H(F)$ and $g|_E = f$. We have shown that (a) $\Longrightarrow$ (b).

(b) $\Longrightarrow$ (a). The nuclear space E can be realized as a subspace of F^I where I is some indexing set and F is a $\mathcal{DFS}$ space with topology generated by a family of semi-inner products. If (b) holds and $f \in H(E)$ then there exists $g \in H(F^I)$ such that $g|_E = f$. By Example 3.10(b), $g = h \circ \pi_J$ for some finite subset J of I and $h \in H(F^J)$ where π_J is the canonical projection from F^I onto F^J. Since F^J is a $\mathcal{DFS}$ space, Example 3.11 shows that $h \in H_{ub}(F^J)$ and thus there exists a neighbourhood V of 0 in F^J such that $\|h\|_{cV} < \infty$ for all $c \in \mathbb{R}$. Hence $\|g\|_{c\pi_J^{-1}(V)} < \infty$ and $\|f\|_{c\pi_J^{-1}(V) \cap E} < \infty$ for all $c \in \mathbb{R}$. This proves $f \in H_{ub}(E)$ and completes the proof.

We did not use the full set of hypothesis in either direction in the previous proposition and hence using the same methods we obtain part (a) of the following example.

Example 6.22 (a) If E is a $\mathcal{DFM}$ subspace of a locally convex space F and the topology on F is generated by semi-inner products then each holomorphic function on E admits a holomorphic extension to F (see also Exercises 4.76 and 4.77).

(b) In Example 3.12 we proved $H(H(\mathbb{C})) \neq H_{ub}(H(\mathbb{C}))$. By Proposition 6.21 this implies that holomorphic functions on $H(\mathbb{C})$ cannot always be extended to superspaces of $H(\mathbb{C})$ (see Exercise 3.120).

Our aim now is to show that $H(E) = H_{ub}(E)$ where E is a Fréchet nuclear space with basis and property $(\widetilde{\Omega})$ (Example 6.10). By Proposition 6.21 this can be viewed as an extension result for holomorphic functions. Our method of proof is similar to the proof of Proposition 6.9.

Proposition 6.23 *If $\lambda(A)$ is a Fréchet nuclear space with property $(\widetilde{\Omega})$ then $H(\lambda(A)) = H_{ub}(\lambda(A))$.*

Proof. Let $f \in H(\lambda(A))$ have monomial expansion $\sum_{m\in\mathbb{N}^{(\mathbb{N})}} a_m z^m$. Since f is bounded on a neighbourhood of the origin there exists a positive integer k such that

$$\|f\|_{\{(z_j)_j \ : \ |z_j|\alpha_{j,k} \le 1\}} =: M < \infty. \qquad (*)$$

If $N_1 = \{j \ : \ \alpha_{j,k} > 0\}$ and $N_2 = N \setminus N_1$ then it is easily seen that $\lambda(A) = \lambda(A_1) \times \lambda(A_2)$ where $\lambda(A_i) = \{(z_j)_j \ : \ z_j = 0 \text{ all } j \in N_i\}$ and $f(x + y) = f(x)$ for all $x \in \lambda(A_1)$, $y \in \lambda(A_2)$. Clearly it suffices to show $f|_{\lambda(A_1)} \in H_{ub}(\lambda(A_1))$ and thus we may suppose, from now on, that $\alpha_{j,k} \neq 0$ for all j. Consequently we may suppose that $\alpha_{j,k} < \alpha_{j,k+1}$ for all j and k and, by the Grothendieck–Pietsch criterion, $\sum_j \frac{\alpha_{j,k}}{\alpha_{j,k+1}} < \infty$. Using property $(\widetilde{\Omega})$ and (6.9) and taking, if necessary, a subsequence of the rows of A we may suppose there exists $d > 0$ such that for all m there exists $c(m) > 0$ such that

$$\left(\frac{\alpha_{j,k}}{\alpha_{j,k+1}}\right)^d \le c(m)\frac{\alpha_{j,k+1}}{\alpha_{j,k+m}}$$

for all j. Since $\sum_j \frac{\alpha_{j,k+m}}{\alpha_{j,k+m+1}} < \infty$ we can choose for each m, j_m, such that

$c(m + 1)\frac{\alpha_{j,k+m}}{\alpha_{j,k+m+1}} \le 1$ all $j \ge j_m$. Hence

$$\left(\frac{\alpha_{j,k}}{\alpha_{j,k+1}}\right)^d \le c(m + 1)\frac{\alpha_{j,k+1}}{\alpha_{j,k+m+1}} \le c(m + 1)\frac{\alpha_{j,k+m}}{\alpha_{j,k+m+1}}\frac{\alpha_{j,k+1}}{\alpha_{j,k+m}} \le \frac{\alpha_{j,k+1}}{\alpha_{j,k+m}}$$

for all $j \geq j_m$. We suppose, without loss of generality, that $(j_m)_m$ is strictly increasing. Let

$$n_j = \begin{cases} k+2 & \text{for } j < j_2, \\ k+m & \text{for } j_m \leq j < j_{m+1}, m \geq 2. \end{cases}$$

By construction there exists $\gamma > 0$ such that

$$\left(\frac{\alpha_{j,k}}{\alpha_{j,k+1}}\right)^d \leq \gamma \frac{\alpha_{j,k+1}}{\alpha_{j,n_j}} \tag{$**$}$$

for all j. For any $\beta > 0$ the set $\{(z_j)_j \; : \; |z_j| \leq \frac{\beta}{\alpha_{j,n_j}} \text{ all } j\}$ is a compact subset of E. Hence, by the Cauchy inequalities, there exists for each $\delta > 0$, $c(\delta) > 0$, such that

$$|a_m|\delta^{|m|} \leq c(\delta)\alpha_{\bullet,n_\bullet}^{m_\bullet}. \tag{6.15}$$

for all $m \in \mathbb{N}^{(\mathbb{N})}$. In (6.15) we have used $\alpha_{\bullet,n_\bullet}^{m_\bullet}$ to denote $\prod_{j=1}^{\infty} \alpha_{j,n_j}^{m_j}$, i.e. a black dot replaces the variable j and we take the product over all j. We adopt a similar convention in the following calculations. By $(*)$, and the Cauchy inequalities,

$$|a_m| \leq M\alpha_{\bullet,k}^{m_\bullet} \tag{6.16}$$

for all $m \in \mathbb{N}^{(\mathbb{N})}$. If $\beta \in \mathbb{R}^+$ and $m \in \mathbb{N}^{(\mathbb{N})}$ then

$$\left(\frac{|a_m|\beta^{|m|}}{\alpha_{\bullet,k+1}^{m_\bullet}}\right)^{1+d} \leq \frac{|a_m|}{(\alpha_{\bullet,k+1}^{m_\bullet})^{1+d}}(\beta^{1+d})^{|m|}M^d(\alpha_{\bullet,k}^{m_\bullet})^d, \text{ by (6.16)}$$

$$= \frac{M^d(\beta^{1+d})^{|m|}|a_m|}{\alpha_{\bullet,k+1}^{m_\bullet}}\left(\frac{\alpha_{\bullet,k}^{m_\bullet}}{\alpha_{\bullet,k+1}^{m_\bullet}}\right)^d$$

$$\leq \frac{M^d(\beta^{1+d})^{|m|}}{\alpha_{\bullet,k+1}^{m_\bullet}}|a_m|\gamma^{|m|}\left(\frac{\alpha_{\bullet,k+1}}{\alpha_{\bullet,n_\bullet}}\right)^{m_\bullet}, \text{ by } (**)$$

$$= M^d(\beta^{1+d}\gamma)^{|m|}\frac{|a_m|}{\alpha_{\bullet,n_\bullet}^{m_\bullet}}$$

$$\leq M^d c(\gamma\beta^{1+d}), \text{ by (6.15)}.$$

Hence

$$\sum_{m \in \mathbb{N}^{(\mathbb{N})}} \|a_m z^m\|_{\{(z_j)_j \; : \; |z_j\alpha_{j,k+2}| \leq \beta\}} = \sum_{m \in \mathbb{N}^{(\mathbb{N})}} \frac{|a_m|\beta^{|m|}}{\alpha_{\bullet,k+1}^{m_\bullet}}\left(\frac{\alpha_{\bullet,k+1}}{\alpha_{\bullet,k+2}}\right)^{m_\bullet}$$

$$\leq M^{\frac{d}{1+d}}c(\gamma\beta^{1+d})^{\frac{1}{1+d}}\prod_{j=1}^{\infty}\left(1 - \frac{\alpha_{j,k+1}}{\alpha_{j,k+2}}\right)^{-1}$$

$$< \infty$$

and f is bounded on every multiple of the zero neighbourhood

$$\{(z_j)_j \; : \; |z_j|\alpha_{j,k+2} \le 1\}.$$

This completes the proof.

6.3 Holomorphic Functions of Bounded Type

In this section we examine the *spectrum* of $H_b(E)$ where E is a Banach space and $H_b(E)$ is given the topology of uniform convergence on the bounded subsets of E. Endowed with this topology $H_b(E)$ is a locally m-convex Fréchet algebra. We denote by $\mathcal{S}(H_b(E))$ the set of all continuous *non-zero* multiplicative linear functionals on $H_b(E)$ and call $\mathcal{S}(H_b(E))$ the spectrum of $H_b(E)$. In Chapter 5 we examined the spectrum of $(H(\Omega), \tau_0)$, $\mathcal{S}_c(\Omega)$, for Ω a Riemann domain spread over a Fréchet space. Our analysis here follows a somewhat similar pattern but there are subtle differences since we are using a stronger topology which frequently generates new elements.

In the previous chapter the formula $\delta_x(f) = f(x)$ allowed us to identify, using point evaluations, the set Ω with a subset of $\mathcal{S}_c(\Omega)$ and, for Fréchet spaces with the bounded approximation property, we constructed the envelope of holomorphy $\Sigma(\Omega)$ so that *all elements* of $\mathcal{S}_c(\Omega)$ became point evaluations (Corollary 5.50). We encountered a similar situation in the previous section when we considered the duality $J_E x(\phi) = \phi(x)$ for $x \in E$ and $\phi \in E'$. These mappings are examples of the same basic *"adjoint"* or *"transpose"* construction that occurs in many parts of mathematics and are combined here in our examination of $\mathcal{S}(H_b(E))$. Our initial motivation is to describe $\mathcal{S}(H_b(E))$ as a collection of point evaluations and to examine its structure from this perspective. We concentrate on positive results and, in almost all cases, require the Banach space E to be *symmetrically regular*. This is a slightly weaker version of the concept of regularity discussed in Section 6.2 (see Exercises 6.48 and 6.49). It can be shown (see Section 6.5) that a symmetrically regular hypothesis is both necessary and sufficient for many of our results.

Definition 6.24 A Banach space E is symmetrically regular if each continuous symmetric bilinear form on E is Arens regular.

Proposition 6.13 can be read as a characterization of symmetrically regular Banach spaces. Since $[dP(x)](y) = 2\check{P}(x, y)$ for $P \in \mathcal{P}(^2E)$, condition (v) of Proposition 2.6 implies (c) of Proposition 6.13 and, combining these two results, we obtain the following result:

if E is a Banach space and $\mathcal{P}(^2E) = \mathcal{P}_w(^2E)$ then E is symmetrically regular.

If E is a Banach space then clearly we have $E \subset \mathcal{S}(H_b(E))$. By Proposition 6.16 each $f \in H_b(E)$ can be extended to $\overset{\circ}{f} \in H_b(E'')$. Moreover the mapping $f \in H_b(E) \longrightarrow \overset{\circ}{f} \in H_b(E'')$ is continuous, linear and multiplicative. Hence the mapping $f \to \overset{\circ}{f}(x'')$, $x'' \in E''$, allows us to identify elements of E'' with points in $\mathcal{S}(H_b(E))$. We may continue the process and extend $\overset{\circ}{f} \in H_b(E'')$ to $\overset{\circ\circ}{f} := \overset{\overset{\circ}{\circ}}{f} \in H_b(E^{(iv)})$ where $E^{(iv)} = (E'')''$. This also gives rise to elements of $\mathcal{S}(H_b(E))$ and we now show that this second extension does not lead to any *new* elements of $\mathcal{S}(H_b(E))$ when E is symmetrically regular. We will use the following well-known decomposition of $E^{(iv)}$,

$$E^{(iv)} = J_{E''}(E'') \oplus \{J_{E'}(E')\}^{\perp}$$

and let ρ denote the canonical projection from $E^{(iv)}$ onto $J_{E''}(E'')$.

Proposition 6.25 *If E is a symmetrically regular Banach space, $y \in E''$, $z \in E^{(iv)}$ and $J_{E''}(y) = \rho(z)$ then*

$$\overset{\circ\circ}{f}(z) = \overset{\circ\circ}{f}(\rho(z)) = \overset{\circ}{f}(y) \tag{6.17}$$

for all $f \in H_b(E)$.

Proof. Since the extension in Proposition 6.16 is linear and continuous it suffices to show, using Taylor series expansions, that (6.17) holds for continuous n-homogeneous polynomials.

If $P \in \mathcal{P}(^nE)$ and $A := \overset{\vee}{P}$ then $\overset{\circ\circ}{P}(z) = \overset{\approx}{A}(z^n)$ for all $z \in E^{(iv)}$. In general, the bidual of a symmetrically regular Banach space will not be symmetrically regular (see Exercise 6.55) but we do know, by extending Proposition 6.13 to n linear forms, the following about $\widetilde{A} = \mathcal{E}_n(A) \in \mathcal{L}(^nE'')$ and $\overset{\approx}{A} = \mathcal{E}_n(\mathcal{E}_n(A)) \in \mathcal{L}(^nE^{(iv)})$,

$\widetilde{A} \in \mathcal{L}(^nE'')$ and $\widetilde{A}$ is separately $\sigma(E'', E')$ continuous
in each variable. $\tag{6.18}$

If $y_1, \ldots, y_j \in E''$ and $z_{j+2}, \ldots, z_n \in E^{(iv)}$, where $1 \leq j < n$,
then $z \in E^{(iv)} \longrightarrow \overset{\approx}{A}(J_{E''}y_1, \ldots, J_{E''}y_j, z, z_{j+2}, \ldots, z_n)$ is
$\sigma(E^{(iv)}, E''')$ continuous. $\tag{6.19}$

Now let $z_1, \ldots, z_n \in E^{(iv)}$, $y_1, \ldots, y_n \in E''$ and suppose $\rho(z_j) = J_{E''}y_j$ for all j. By the definition of $\widetilde{A}$

$$\overset{\approx}{A}(J_{E''}y_1, \ldots, J_{E''}y_{n-1}, z_n) = \lim_{\alpha} \widetilde{A}(y_1, \ldots, y_{n-1}, z_{\alpha})$$

where $(z_{\alpha})_{\alpha}$ is any net in E'' such that $(J_{E''}z_{\alpha})_{\alpha}$ converges $\sigma(E^{(iv)}, E''')$ to z_n as $\alpha \to \infty$. Then $(z_{\alpha})_{\alpha}$ converges $\sigma(E'', E')$ to y_n and, by (6.18),

$$\widetilde{\widetilde{A}}(J_{E''}y_1, \ldots, J_{E''}y_{n-1}, z_n) = \widetilde{A}(y_1, \ldots, y_{n-1}, y_n)$$
$$= \widetilde{\widetilde{A}}(J_{E''}y_1, \ldots, J_{E''}y_{n-1}, \rho(z_n)).$$

By (6.19)

$$\widetilde{\widetilde{A}}(J_{E''}y_1, \ldots, J_{E''}y_{n-2}, z_{n-1}, z_n) = \lim_\beta \widetilde{\widetilde{A}}(J_{E''}y_1, \ldots, J_{E''}y_{n-2}, J_{E''}z_\beta, z_n)$$

where $(z_\beta)_\beta$ is any net in E'' such that $(J_{E''}z_\beta)_\beta$ converges $\sigma(E^{(iv)}, E''')$ to z_{n-1} as $\beta \to \infty$. Since

$$\widetilde{\widetilde{A}}(J_{E''}y_1, \ldots, J_{E''}y_{n-2}, J_{E''}z_\beta, z_n) = \widetilde{A}(y_1, \ldots, y_{n-2}, z_\beta, y_n),$$

(6.18) implies

$$\widetilde{\widetilde{A}}(J_{E''}y_1, \ldots, J_{E''}y_{n-2}, z_{n-1}, z_n) = \widetilde{A}(y_1, \ldots, y_{n-2}, y_{n-1}, y_n)$$
$$= \widetilde{\widetilde{A}}(y_1, \ldots, y_{n-2}, \rho(z_{n-1}), \rho(z_n)).$$

Continuing in this way we obtain

$$\widetilde{\widetilde{A}}(z_1, \ldots, z_n) = \widetilde{A}(y_1, \ldots, y_n) = \widetilde{\widetilde{A}}(\rho(z_1), \ldots, \rho(z_n)).$$

Letting $z_i = z$ for all i we obtain

$$\overset{\infty}{P}(z) = \widetilde{\widetilde{A}}(z^n) = \widetilde{A}(y^n) = \overset{\circ}{P}(y) = \widetilde{\widetilde{A}}(\rho(z)^n) = \overset{\infty}{P}(\rho(z))$$

for all $z \in E^{(iv)}$. This completes the proof.

Proposition 6.25 shows that point evaluations in the fourth dual do not define new points in $\mathcal{S}(H_b(E))$ when E is symmetrically regular and a similar argument shows that the higher order even duals also do not define *new* multiplicative linear functionals on $H_b(E)$ when E is symmetrically regular. This is not a contradiction since the Aron–Berner extension is generally not surjective. If E is not symmetrically regular it can be shown that certain points in $E^{(iv)}$ give rise to *new* elements in the spectrum.

Our aim is to show, when E is symmetrically regular, that $\mathcal{S}(H_b(E))$ can be endowed with the structure of a Riemann domain over E'' and to extend each $f \in H_b(E)$ to a holomorphic function on $\mathcal{S}(H_b(E))$. This is clearly analogous to the programme we carried out in the previous chapter for $\mathcal{S}_c(U) = \mathcal{S}((H(U), \tau_0))$. We use similar methods here. It is interesting to compare the two cases at the linear level by noting that $(E', \tau_0)' = E$ and $(E', \|\cdot\|)' = E''$. If $m \in \mathcal{S}(H_b(E))$ let $\pi(m) = m\big|_{E'} \in (E')' = E''$. This defines a surjective mapping $\pi : \mathcal{S}(H_b(E)) \longrightarrow E''$. Since E'' may be identified with a subset of $\mathcal{S}(H_b(E))$, by letting $x''(f) = \overset{\circ}{f}(x'')$ for all $x'' \in E''$

and $f \in H_b(E)$, we can also consider π as a mapping from $S(H_b(E))$ into itself and with this interpretation π is a projection, i.e. $\pi^2 = \pi$.

Moreover, as

$$m(\phi^n) = (m(\phi))^n = (\pi(m)(\phi))^n = (\overset{\circ}{\phi}(\pi(m)))^n = (\overset{\circ}{\phi})^n(\pi(m))$$

for all $\phi \in E'$, we have $m(P) = \overset{\circ}{P}(\pi(m))$ for any approximable polynomial P. If the approximable polynomials are dense in $H_b(E)$ then π is bijective. Combining these observations with Proposition 2.8 we obtain the following result.

Proposition 6.26 *If E is a Banach space, E' has the approximation property and all continuous homogeneous polynomials on E are weakly continuous on bounded sets then $\pi : S(H_b(E)) \longrightarrow E''$ is bijective and every multiplicative linear functional m on $H_b(E)$ has the form $m(f) = \overset{\circ}{f}(\pi(m))$.*

Proposition 6.26 applies, in particular, to c_0, T^* and T_J^*. We thus have the following *set theoretic* identifications,

$$S(H_b(c_0)) \cong l_\infty, \quad S(H_b(T^*)) \cong T^*, \quad S(H_b(T_J^*)) \cong (T_J^*)''.$$

Our next result shows that weak continuity on bounded sets is also necessary in Proposition 6.26.

Proposition 6.27 *If there exists a continuous polynomial P on the Banach space E which is not weakly continuous on bounded sets then there exists $\theta \in S(H_b(E))$ such that $\theta \neq \pi(\theta)$.*

Proof. We may suppose without loss of generality that P is n-homogeneous. By hypothesis there exists a bounded net $(x_\alpha)_{\alpha \in \Gamma}$ which is weakly convergent to the point $x_0 \in E$ such that

$$\lim_\alpha P(x_\alpha) \neq P(x_0).$$

By taking a subnet and multiplying by a scalar, if necessary, we can suppose

$$|P(x_\alpha) - P(x_0)| \geq 1$$

for all $\alpha \in \Gamma$. On Γ take the filter base $\mathcal{B} := \{\alpha \in \Gamma : \alpha \geq \alpha_0\}_{\alpha_0 \in \Gamma}$ and let $\mathcal{U}$ be an ultrafilter such that $\mathcal{U} \supset \mathcal{B}$. Let

$$\theta(f) = \lim_\mathcal{U} f(x_\alpha)$$

for all $f \in H_b(E)$. Since $(x_\alpha)_{\alpha \in \Gamma}$ is bounded $\lim_\mathcal{U} f(x_\alpha)$ converges for all $f \in H_b(E)$ and clearly defines an element of $S(H_b(E))$. We claim that $\theta \neq \pi(\theta)$. If $x' \in E'$ then $x'(x_\alpha) \to x'(x_0)$ as $\alpha \to \infty$ and hence $\pi(\theta)(x') = \theta(x') = \lim_\mathcal{U} x'(x_\alpha) = x'(x_0)$. This shows that $(\pi(\theta))(Q) = \overset{\circ}{Q}(Jx_0) = Q(x_0)$ since

$x_0 \in E$. On the other hand $\theta(P) = \lim_{\mathcal{U}} P(x_\alpha) \neq P(x_0)$ since $|P(x_\alpha) - P(x_0)| \geq 1$ for all α. Hence $\pi(\theta) \neq \theta$ and this completes the proof.

We endow $\mathcal{S}(H_b(E))$ with the structure of a complex manifold so that it may be regarded as the envelope of "bounded" holomorphy of E. We keep the notation $\overset{\circ}{f}$ to denote the Aron–Berner extension of $f \in H_b(E)$ to E'' and identify, as above, E'' with a subset of $\mathcal{S}(H_b(E))$.

If $z \in E''$ let $\tau_z(x) = J_E x + z$ for all $x \in E$. This induces a mapping $\tau_z^* : H_b(E) \longrightarrow H_b(E)$ where

$$(\tau_z^* f)(x) = \overset{\circ}{f}(J_E x + z) = \overset{\circ}{f} \circ \tau_z(x).$$

Clearly

$$\tau_z^*(f) = \sum_{n=0}^{\infty} \tau_z^* \left(\frac{\widehat{d}^n f(0)}{n!} \right) \tag{6.20}$$

for all $f \in H_b(E)$.

Lemma 6.28 *If E is a symmetrically regular Banach space then*

$$\tau_{z+w}^* = \tau_z^* \circ \tau_w^*$$

for all $z, w \in E''$.

Proof. It suffices, by (6.20), to show $\tau_{z+w}^*(P) = \tau_z^* \circ \tau_w^*(P)$ for all $P \in \mathcal{P}(^n E)$, n arbitrary. Let $A = \overset{\vee}{P}$. We have

$$\overset{\circ}{P}(J_E x + w) = \tilde{A}((J_E x + w)^n) = \sum_{k=0}^{n} \binom{n}{k} \tilde{A}(J_E x^k, w^{n-k})$$

for all $x \in E$ and $w \in E''$. Let P_k, $1 \leq k \leq n$, be defined by

$$P_k(x) := \tilde{A}((J_E x)^k, w^{n-k})$$

for all $x \in E$. Then $P_k \in \mathcal{P}(^k E)$ and if $A_k = \overset{\vee}{P_k}$ then $A_k(x_1, \ldots, x_k) = \tilde{A}(J_E x_1, \ldots, J_E x_k, w^{n-k})$ for $x_1, \ldots, x_k \in E$. Since E is symmetrically regular $\tilde{A}$ is separately $\sigma(E'', E')$ continuous in each variable. This implies

$$\overset{\circ}{P_k}(x'') = \tilde{A}_k((x'')^k) = \tilde{A}((x'')^k, w^{n-k})$$

for all $x'' \in E''$. Hence

$$\overline{\tau_w^*(P)}(x'') = \sum_{k=0}^{n} \binom{n}{k} \tilde{A}((x'')^k, w^{n-k}) = \overset{\circ}{P}(x'' + w)$$

for all $x'' \in E''$. If $x \in E$ then

$$(\tau_z^*(\tau_w^*(P)))(x) = \overline{\tau_w^*(P)}^{\circ}(J_E x + z)$$
$$= \overset{\circ}{P}(J_E x + z + w)$$
$$= (\tau_{z+w}^* P)(x).$$

Hence $\tau_z^* \circ \tau_w^* = \tau_{z+w}^*$ and this completes the proof.

We suppose, from now on, that E is symmetrically regular. If $z \in E''$ and $\phi \in E'$ then

$$[\tau_z^*(\phi)](x) = \overset{\circ}{\phi}(J_E x + z) = \phi(x) + z(\phi)$$

and $\tau_z^*(\phi) = z(\phi) + \phi$ where $z(\phi)$ is the constant mapping on E. Note that $\phi \in E'$ acts, by definition, on E and by the classical duality, $J_E x(\phi) = \phi(x)$, also on E'' – this, in fact, is the definition of J_E. In functional analysis we often do not distinguish between these two roles of ϕ, i.e. whether it acts on E or E''. Here, for the sake of applying a consistent notation to $E' = \mathcal{P}(^1 E)$ and $\mathcal{P}(^n E)$, we do and let $\overset{\circ}{\phi}$ denote the mapping ϕ acting on E''.

Since $\mathcal{S}(H_b(E))$ consists of non-zero multiplicative forms, $m(1) = 1$ and $m(z(\phi)) = z(\phi)$ for all $m \in \mathcal{S}(H_b(E))$. Hence, if $m \in \mathcal{S}(H_b(E))$ and $z \in E''$ then clearly $m \circ \tau_z^* \in \mathcal{S}(H_b(E))$ and, for all $\phi \in E'$, we have

$$m \circ \tau_z^*(\phi) = m(z(\phi) + \phi) = z(\phi) + m(\phi).$$

This shows that $\pi(m \circ \tau_z^*) = \pi(m) + z$ for all $m \in \mathcal{S}(H_b(E))$ and all $z \in E''$.

If $m \in \mathcal{S}(H_b(E))$ and $\varepsilon > 0$ let

$$V_{m,\varepsilon} = \{m \circ \tau_z^* : z \in E'', \|z\| < \varepsilon\}.$$

We claim that $\mathcal{V}_m := (V_{m,\varepsilon})_{\varepsilon>0}$ forms a neighbourhood basis at m for a Hausdorff topology on $\mathcal{S}(H_b(E))$.

If $m_1 \in V_{m,\varepsilon}$ then $m_1 = m \circ \tau_z^*$ for some $z \in E''$, $\|z\| < \varepsilon$. Let $\delta = \varepsilon - \|z\|$. If $w \in E''$, $\|w\| < \delta$, Lemma 6.28 implies, since associativity is easily checked, that

$$m_1 \circ \tau_w^* = (m \circ \tau_z^*) \circ \tau_w^* = m \circ (\tau_z^* \circ \tau_w^*) = m \circ \tau_{z+w}^* \in V_{m,\varepsilon}$$

and hence $V_{m_1,\delta} \subset V_{m,\varepsilon}$. This shows that we have defined a topology on $\mathcal{S}(H_b(E))$. We show that this topology is Hausdorff. Consider $m, m_1 \in \mathcal{S}(H_b(E))$. Let $2\varepsilon = \|\pi(m) - \pi(m_1)\|$ if $\pi(m) \neq \pi(m_1)$ and suppose $w \in V_{m,\varepsilon} \cap V_{m_1,\varepsilon}$. We have $w = m \circ \tau_z^* = m_1 \circ \tau_{z_1}^*$ for some $z, z_1 \in E''$, $\|z\| < \varepsilon$ and $\|z_1\| < \varepsilon$. Hence

$$\pi(m) + z = \pi(m \circ \tau_z^*) = \pi(m_1 \circ \tau_{z_1}^*) = \pi(m_1) + z_1$$

and

$$2\varepsilon = \|\pi(m) - \pi(m_1)\| = \|z - z_1\| < 2\varepsilon.$$

This contradiction shows that $V_{m,\varepsilon} \cap V_{m_1,\varepsilon} = \emptyset$.

Now suppose $\pi(m) = \pi(m_1)$. If $m_2 \in V_{m,r} \cap V_{m_1,s}$ for r, s positive then $m_2 = m \circ \tau_z^* = m_1 \circ \tau_{z_1}^*$ and

$$\pi(m) + z = \pi(m \circ \tau_z^*) = \pi(m_1 \circ \tau_{z_1}^*) = \pi(m_1) + z_1.$$

Since $\pi(m) = \pi(m_1)$ this implies $z = z_1$ and

$$m = m \circ \tau_z^* \circ \tau_{-z}^* = m_1 \circ \tau_{z_1}^* \circ \tau_{-z}^* = m_1 \circ \tau_{z_1-z}^* = m_1.$$

This proves that the topology on $\mathcal{S}(H_b(E))$ is Hausdorff. Moreover, the above also shows, if $\pi(m) = \pi(m_1)$, that either $V_{m,r} \cap V_{m_1,s} = \emptyset$ or $m = m_1$, in which case, $V_{m,r} \cap V_{m_1,s} = V_{m,\min(r,s)}$.

The mapping π restricted to $V_{m,\varepsilon}$ has the form $\pi(m \circ \tau_z^*) = \pi(m) + z$ and maps $V_{m,\varepsilon}$ homeomorphically onto $B_{E''}(\pi(m),\varepsilon)$. Hence π is a local homeomorphism and $(\mathcal{S}(H_b(E)), \pi)$ is a Riemann domain spread over E''.

Clearly, $V_{m,\infty} := \bigcup_{\varepsilon>0} V_{m,\varepsilon}$ is mapped homeomorphically onto E'' and hence is an open subset of $\mathcal{S}(H_b(E))$. It is also closed since, if $m \circ \tau_{z_n} \to m_1 \in \mathcal{S}(H_b(E))$ as $n \to \infty$, then $\pi(m) + z_n \to \pi(m_1)$ as $n \to \infty$ and $(z_n)_n$ converges to some point z in E''. Hence $m \circ \tau_z = \lim_n m \circ \tau_{z_n} = m_1$ and $m_1 \in V_{m,\infty}$.

We have located the connected components of $\mathcal{S}(H_b(E))$ and may visualize $\mathcal{S}(H_b(E))$ as a set of disjoint copies of E'' lying above E''. Combining a number of these results with earlier results we obtain a proof of the first part of the following proposition.

Proposition 6.29 *If E is a symmetrically regular Banach space then $(\mathcal{S}(H_b(E)), \pi)$ is a Riemann domain spread over E''.*

If E is a Banach space and E' has the approximation property then the following are equivalent:

(1) E is symmetricaly regular and $\mathcal{S}(H_b(E))$ is connected,

(2) E is symmetricaly regular and $\pi : \mathcal{S}(H_b(E)) \longrightarrow E''$ is a bijective homeomorphism,

(3) each continuous n-homogeneous polynomial on E is weakly continuous on bounded sets.

Proof. It suffices to show that the three conditions are equivalent. Let E denote a symmetrically regular Banach space. Since $\mathcal{S}(H_b(E))$ is connected if and only if it contains a single connected component and as π maps each connected component of $\mathcal{S}(H_b(E))$ homeomorphically onto E'', (1) and (2) are equivalent. By Proposition 6.27, $\sim$(3) $\Longrightarrow$ $\sim$(2) and hence (2) $\Longrightarrow$ (3). For these implications we did not need the approximation property. If E' has the approximation property then Proposition 6.26 shows that (3) $\Longrightarrow$ (2). This completes the proof.

We next consider the problem of extending holomorphic functions of bounded type on E to holomorphic functions of "bounded" type on $S(H_b(E))$. As in Chapter 5 we identify E with a subset of $S(H_b(E))$ by using the Dirac delta function, i.e. $\delta : E \longrightarrow S(H_b(E))$ is defined by letting $\delta(x)(f) = \delta_x(f) = f(x)$ for all $x \in E$ and all $f \in H_b(E)$. We extend f to $S(H_b(E))$ by letting $f''(m) = m(f)$ for all $m \in S(H_b(E))$. We now have two extensions of $f \in H_b(E)$ – the Aron–Berner extension $\overset{\circ}{f}$ to E'' and the extension f'' to $S(H_b(E))$. Let μ denote the mapping which identifies E'' with a subset of $S(H_b(E))$. If $f \in H_b(E)$ and $x'' \in E''$ then

$$f''(\mu(x'')) = \mu(x'')(f) = \overset{\circ}{f}(x'')$$

and f'' extends $\overset{\circ}{f}$. This is further evidence of the canonical behaviour of the Aron–Berner extension. Our next proposition shows that f'' is also holomorphic.

Proposition 6.30 *If E is a symmetrically regular Banach space and $f \in H_b(E)$ then $f'' \in H(S(H_b(E)))$ and we have the following commutative diagram:*

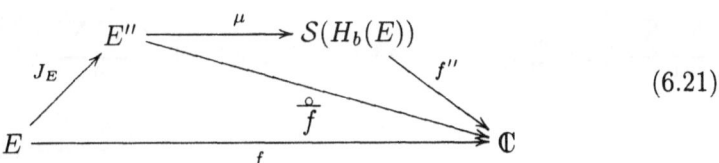

$$(6.21)$$

Proof. By our remark above the diagram is commutative. We show that f'' is holomorphic. Since $\pi\big|_{V_{m,\infty}}$ is bijective it suffices to show that $f'' \circ (\pi\big|_{V_{m,\infty}})^{-1} \in H(E'')$ for each $m \in S(H_b(E))$. Our previous calculations show that

$$f'' \circ (\pi\big|_{V_{m,\infty}})^{-1}(\pi(m) + z) = f''(m \circ \tau_z^*) = m \circ \tau_z^*(f) = m(\tau_z^*(f))$$

for all $z \in E''$ and we are required to prove that the mapping

$$z \in E'' \longrightarrow m(\tau_z^*(f))$$

is holomorphic for all $m \in S(H_b(E))$. Fix m and suppose $|m(f)| \le \|f\|_{rB}$ for all $f \in H_b(E)$ where B is the unit ball of E.

Let $\sum_{n=0}^{\infty} P_n$, $P_n \in \mathcal{P}(^nE)$, denote the Taylor series expansion of f at the origin. For each n let $A_n = \overset{\vee}{P_n}$ and for $z \in E''$ and $0 \le k \le n$ let $P_{n,k,z}$ denote the mapping

$$x \in E \longrightarrow \tilde{A}_n(J_E x^{n-k}, z^k).$$

Clearly $P_{n,k,z} \in \mathcal{P}_a(^{n-k}E)$ and

$$\|P_{n,k,z}\| \le \|z\|^k \sup_{z_i \in E'', \|z_i\| \le 1} |\tilde{A}_n(z_1, \ldots, z_n)|$$

$$= \|z\|^k \|A_n\|. \tag{6.22}$$

Now

$$(\tau_z^* f)(x) = \overset{\circ}{f}(J_E x + z) = \sum_{n=0}^{\infty} \overset{\circ}{P}_n(J_E x + z)$$

$$= \sum_{n=0}^{\infty} \left(\sum_{k=0}^{n} \binom{n}{k} \tilde{A}_n(J_E x^{n-k}, z^k) \right)$$

$$= \sum_{n=0}^{\infty} \left(\sum_{k=0}^{n} \binom{n}{k} P_{n,k,z} \right)(x).$$

Since $f \in H_b(E)$, $\lim_{n \to \infty} \|P_n\|^{1/n} = 0$. By the Polarization Inequality (Proposition 1.8) and Stirling's formula (4.37) this implies $\lim_{n \to \infty} \|A_n\|^{1/n} = 0$.

Suppose $\|x\| \le s$ and $\|z\| \le t$. Choose $\varepsilon > 0$ such that $\varepsilon(s+t) < 1$ and choose $c > 0$ such that $\|A_n\| \le c\varepsilon^n$ for all n. By (6.22)

$$\sum_{n=0}^{\infty} \sum_{k=0}^{n} \binom{n}{k} \sup_{\|z\| \le t} \|P_{n,k,z}\| \{x \in E : \|x\| \le s\} \le \sum_{n=0}^{\infty} \sum_{k=0}^{n} \binom{n}{k} t^k s^{n-k} \|A_n\|$$

$$\le c \sum_{n=0}^{\infty} \sum_{k=0}^{n} \binom{n}{k} t^k s^{n-k} \varepsilon^n$$

$$= c \sum_{n=0}^{\infty} \varepsilon^n (t+s)^n$$

$$< \infty$$

and for each z, $\sum_{n=0}^{\infty} \sum_{k=0}^{n} \binom{n}{k} P_{n,k,z}$ converges absolutely to $\tau_z^* f$ on the bounded subsets of E and, moreover, the series is absolutely uniformly convergent as z ranges over any fixed bounded subset of E''. Hence

$$m(\tau_z^* f) = \sum_{n=0}^{\infty} \sum_{k=0}^{n} \binom{n}{k} m(P_{n,k,z})$$

$$= \sum_{k=0}^{\infty} \sum_{n \ge k} \binom{n}{k} m(P_{n,k,z}). \tag{6.23}$$

The mapping

$$P_{n,k} : z \in E'' \longrightarrow m(P_{n,k,z})$$

is a k-homogeneous polynomial for all n and (6.22) implies

$$|m(P_{n,k,z})| \le \sup_{\|x\| \le r} \|P_{n,k,z}(x)\| \le \|z\|^k r^{n-k} \|A_n\|$$

for all n, all $k \leq n$ and all $z \in E''$. Let $\alpha > 0$ be arbitrary. Choose $\delta > 0$ such that $\delta r < 1$ and $\frac{\delta \alpha}{1 - r\delta} < 1$ and then choose $c' > 0$ such that $\|A_n\| \leq c' \delta^n$ for all n. If $g(z) = \sum_{n=0}^{\infty} z^n$, $|z| < 1$, then $g \in H(D)$ and

$$\limsup_{k \to \infty} \left(\frac{g^{(k)}(r\delta)}{k!} \right)^{1/k} \leq \frac{1}{1 - r\delta}.$$

With the above choice of δ we have

$$\sum_{n \geq k} \binom{n}{k} \sup_{\|z\| \leq \alpha, z \in E''} |P_{n,k}(z)| = \sum_{n \geq k} \binom{n}{k} \sup_{\|z\| \leq \alpha, z \in E''} |m(P_{n,k,z})|$$

$$\leq \sum_{n \geq k} \binom{n}{k} \alpha^k r^{n-k} \|A_n\|$$

$$\leq c' \sum_{n \geq k} \binom{n}{k} \alpha^k r^{n-k} \delta^n$$

$$= \frac{c'(\delta \alpha)^k}{k!} \sum_{n \geq k} n(n-1) \cdots (n-k+1)(r\delta)^{n-k}$$

$$= \frac{c'(\delta \alpha)^k}{k!} g^{(k)}(r\delta).$$

Hence the series $\displaystyle\sum_{n \geq k} \binom{n}{k} \sup_{\|z\| \leq \alpha, z \in E''} |P_{n,k}(z)|$ is convergent and

$$Q_k := \sum_{n \geq k} \binom{n}{k} P_{n,k} \in \mathcal{P}(^k E'').$$

Moreover

$$\sum_{k=0}^{\infty} \sup_{\|z\| \leq \alpha, z \in E''} \|Q_k(z)\| \leq \sum_{k=0}^{\infty} \sum_{n \geq k} \binom{n}{k} \sup_{\|z\| \leq \alpha, z \in E''} |P_{n,k}(z)|$$

$$\leq c' \sum_{k=0}^{\infty} (\delta \alpha)^k \frac{g^{(k)}(r\delta)}{k!}$$

$$< \infty$$

since

$$\limsup_{k \to \infty} \left(c'(\delta \alpha)^k \frac{g^{(k)}(r\delta)}{k!} \right)^{1/k} \leq \frac{\delta \alpha}{1 - r\delta} < 1.$$

Hence $\sum_{k=0}^{\infty} Q_k \in H_b(E'')$ and, by (6.23), $m(\tau_z^* f) = \sum_{k=0}^{\infty} Q_k(z)$ for all $z \in E''$. We have shown that f'' is a holomorphic function on $\mathcal{S}(H_b(E))$ and this completes the proof.

We also proved in Proposition 6.30 that the extension f'' of f is a holomorphic function of "bounded" type on each sheet of the Riemann domain $\mathcal{S}(H_b(E))$. The connected component of $\mathcal{S}(H_b(E))$, which is isomorphic to E'', may thus be regarded, in the symmetrically regular case, as the envelope of holomorphy of E (to make this rigorous a modified definition of envelope of holomorphy is required).

The above results show a remarkable synthesis of ideas and techniques from linear functional analysis and several complex variable theory and gives insight into both areas. For instance, we may regard the bidual of a Banach space as the *"envelope of linearity"* while the space $H_b(\mathcal{S}(H_b(E)))$ behaves in many ways like a *"holomorphic bidual"* of E'. Furthermore, we have seen that the extension f'' gives on each *sheet* of $\mathcal{S}(H_b(E))$ a holomorphic function which may, via π, be regarded as a holomorphic extension of f to E''. This partially explains the appearance of different extensions to the bidual and why, for instance, different ultrafilters may lead to different extensions. The result in Proposition 6.30 indicates the special position occupied by the Aron–Berner extension. Further interesting results, e.g. the study of $\mathcal{S}(H^\infty(U))$ and $\mathcal{S}(H_b(U))$, have appeared in the literature (see Section 6.5) and it is not surprising that a number of the results in this section can be obtained by alternative routes. For the reader's convenience and also for the sake of coherence we have followed the approach used for domains of holomorphy in Chapter 5 and sketch now briefly an elegant alternative approach to some of the above results.

If $f \in H_b(E)$ let $f_x(y) = f(x + y)$ for all $x, y \in E$. If $\phi, \theta \in \mathcal{S}(H_b(E))$ then the mapping

$$x \in E \longrightarrow \theta(f_x)$$

defines a holomorphic function of bounded type on E. Repeating this process we see that the mapping

$$f \in H_b(E) \longrightarrow \phi(x \longrightarrow \theta(f_x))$$

defines an element of $\mathcal{S}(H_b(E))$ that we denote by $\theta * \phi$ and call the *convolution* of θ and ϕ. The convolution product is easily seen to be associative and endows $\mathcal{S}(H_b(E))$ with the structure of a *semi-group* (which will not in general be a group). The identity of this semi-group is point evaluation at the origin and E, identified with a subset of $\mathcal{S}(H_b(E))$, is a subgroup lying in the centre of $\mathcal{S}(H_b(E))$. Since $\pi(\phi * \theta) = \pi(\phi) + \pi(\theta)$ the convolution $*$ may, intuitively, be regarded as an "extension" to the Riemann domain $\mathcal{S}(H_b(E))$ of the operation of addition on E''. Furthermore, if $z_1, \ldots, z_n \in E''$, $P \in \mathcal{P}(^nE)$, $A = \overset{\vee}{P}$, then

$$(\delta_{z_1} * \delta_{z_2} * \cdots * \delta_{z_n})(P) = n!\widetilde{A}(z_1, \ldots, z_n).$$

We let $z^{*n} := \delta_z * \delta_z * \cdots * \delta_z$ and thus we have, symbolically,

$$\delta_z = \sum_{n=0}^{\infty} \frac{1}{n!} z^{*n} := \exp(z^*)$$

for all $z \in E''$. The correspondence $z \longrightarrow \delta_z * \theta$ is the inverse of π restricted to $V_{\theta,\infty}$ for all $\theta \in S(H_b(E))$ and, if $f \in H_b(E)$, then

$$(\delta_z * \theta)(f) = f'' \circ \left(\pi\big|_{V_{\theta,\infty}}\right)^{-1} \tag{6.24}$$

is the restriction of f to the θth sheet of the Riemann domain $S(H_b(E))$.

We complete this chapter by obtaining function space representations of the biduals of spaces of holomorphic functions. We draw on results from Chapters 2, 3, 4 and 5 and these representations, obtained using the linear theory, complement and illuminate the results derived above from the multiplicative structure.

We consider holomorphic functions on rB, $0 < r \le \infty$, where B is the open unit ball of the Banach space E (note that $rB = E$ when $r = \infty$). From Chapter 2 we recall that $\mathcal{P}_w(^nE)$ and $\mathcal{P}_{w*}(^nE'')$ denote the subspaces of $\mathcal{P}(^nE)$ and $\mathcal{P}(^nE'')$ which are weakly uniformly continuous and weak* continuous, respectively, on bounded sets. Let

$$H_{wu}(rB) := \left\{ f \in H_b(rB) \ : \ \frac{\widehat{d^n} f(0)}{n!} \in \mathcal{P}_w(^nE) \text{ all } n \right\}$$

and

$$H_{w*u}(rB'') := \left\{ f \in H_b(rB'') \ : \ \frac{\widehat{d^n} f(0)}{n!} \in \mathcal{P}_{w*}(^nE'') \text{ all } n \right\}.$$

These spaces are endowed with the induced topologies from $H_b(rB)$ and $H_b(rB'')$ respectively. If $f \in H(rB)$ then

$$f \in H_{wb}(rB) \iff f\big|_{\rho B} \text{ is weakly uniformly continuous on } \rho B \text{ for all } \rho < r$$

and, if $f \in H(rB'')$, then

$$f \in H_{w*b}(rB'') \iff f\big|_{\rho B''} \text{ is weak* continuous on } \rho B'' \text{ for all } \rho < r.$$

The sequences $\{\mathcal{P}_w(^nE)\}_{n=0}^\infty$ and $\{\mathcal{P}_{w*}(^nE'')\}_{n=0}^\infty$ are S-absolute decompositions (Definition 3.32) for $H_{wu}(rB)$ and $H_{w*u}(rB'')$ respectively.

If E is a locally convex space we write the canonical mapping into the bidual as follows:

$$J_E : E \longrightarrow J_E(E) \subset E''$$

i.e. the space in the middle identifies the image of the canonical mapping while the space on the far right is the bidual of E. A locally convex space E is infrabarrelled if and only if J_E is an isomorphism onto its range.

Proposition 6.31 *If E is a Q-reflexive Banach space and E'' has the approximation property and the Radon–Nikodým Property then*

$$J_{H_b(rB)} : H_b(rB) \longrightarrow H_{\omega^* u}(rB'') \subset H_b(rB'')$$

and $J_{H_b(rB)} = \overset{a}{f}$.

Proof. Applying Proposition 3.35 twice we see that $\{\mathcal{P}(^n E)''\}_{n=0}^{\infty}$ is an $\mathcal{S}$-absolute decomposition for $H_b(rB)'' := ((H_b(rB)'_\beta)'_\beta$. Since $H_b(rB)$ is infrabarrelled the strongly bounded and equicontinuous subsets of $H_b(rB)'_\beta$ coincide. Hence, Lemma 3.33 implies that the sets

$$\left\{ \sum_{n=0}^{\infty} T'_n \; : \; T'_n \in \mathcal{P}(^n E)', \|T'_n\| \le c \cdot C^n \text{ all } n \right\}$$

$c = 1, 2, \ldots, 0 < C < r$, form a fundamental system of bounded sets in $H_b(rB)'_\beta$ and thus

$$H_b(rB)'' = \left\{ \sum_{n=0}^{\infty} T''_n \; : \; T''_n \in \mathcal{P}(^n E)'', \sum_{n=0}^{\infty} \rho^n \|T''_n\| < \infty \text{ for } 0 < \rho < r \right\} \quad (*)$$

with topology generated by

$$p_r \left(\sum_{n=0}^{\infty} T''_n \right) := \sum_{n=0}^{\infty} \rho^n \|T''_n\| \qquad (**)$$

where $0 < \rho < r$. Moreover, $J_{H_b(rB)} = \sum_{n=0}^{\infty} J_{\mathcal{P}(^n E)}$. Since E is Q-reflexive and E'' has both the approximation property and the Radon–Nikodým Property, Proposition 2.49 implies that for any compact balanced convex subset K of E,

$$J_{\mathcal{P}(^n E)} : (\mathcal{P}(^n E), \| \cdot \|_{K+B}) \to (\mathcal{P}_{\omega^*}(^n E''), \| \cdot \|_{K+B''}) \subset (\mathcal{P}(^n E), \| \cdot \|_{K+B''}).$$

$$(6.25).$$

Moreover, $J_{\mathcal{P}(^n E)}(P)$ is the extension of P to E'' obtained by using either the Aron–Berner extension or weak uniform continuity. Combining $(*)$, $(**)$ and (6.25) with $K = \{0\}$, we obtain the required representation. This completes the proof.

Now suppose $f \in H(rB)$ where $rB \subset E$, E is Q-reflexive and E'' has the approximation property. If $f = \sum_{n=0}^{\infty} \dfrac{\hat{d}^n f(0)}{n!}$ and K is a compact convex balanced subset of E then here exists $\varepsilon := \varepsilon(K, f) > 0$ such that

$$\sum_{n=0}^{\infty} \left\| \frac{\hat{d}^n f(0)}{n!} \right\|_{K+\varepsilon B} < \infty.$$

This estimate implies that each $f \in H(rB)$ has a holomorphic local weak* continuous extension $\overset{\circ}{f}$ to an open subset of E'' containing $J_E(rB)$ and leads us to holomorphic germs on *closed* sets (see Section 4.3). If A is a closed subset of an open subset U of a locally convex space E, let

$$H_U(A) = \bigcup_{V \text{open}, \, A \subset V \subset U} H(V)/\sim$$

where $f \sim g$ if f and g coincide on a neighbourhood of A in U. For $0 < r \le \infty$ let

$$H_{rB'',\omega^*}(J_E(rB)) = \left\{ f \in H_{rB''}(J_E(rB)) \; : \; \frac{\widehat{d^n f}(0)}{n!} \in \mathcal{P}_{\omega^*}(^nE'') \text{ all } n \right\}.$$

We endow both $H_{rB''}(J_E(rB))$ and $H_{rB'',\omega^*}(J_E(rB))$ with the topology τ_ω generated by the semi-norms

$$p_{K,(\beta_n)_n}\left(\sum_{n=0}^{\infty} \frac{\widehat{d^n f}(0)}{n!} \right) := \sum_{n=0}^{\infty} \left\| \frac{\widehat{d^n f}(0)}{n!} \right\|_{J_E(K)+\beta_n B''} \qquad (6.26)$$

where K ranges over the convex balanced compact subsets of rB and $(\beta_n)_{n=0}^{\infty}$ over c_0 (see Proposition 3.47).

Proposition 6.32 *If E is a separable Q-reflexive Banach space with the bounded approximation property and E'' has both the approximation property and the Radon–Nikodým Property then, for $0 < r \le \infty$,*

$$J_{H(rB)} : (H(rB), \tau_\omega) \longrightarrow H_{rB'',\omega^*}(J_E(rB)) \subset (H_{rB''}(J_E(rB)), \tau_\omega)$$

and $J_{H(rB)}(f) = \overset{\circ}{f}$ is obtained by using either the Aron–Berner extension or weak continuity.

Proof. By Proposition 3.47 the τ_ω topology on $H(rB)$ is generated by the semi-norms $p_{K,(\beta_n)_n}$ where K ranges over the compact convex balanced subsets of rB and $(\beta_n)_n$ over c_0 and, by Corollary 4.18, $(H(rB), \tau_\omega)$ is an infrabarrelled locally convex space. To complete the proof it suffices to use the method of the previous proposition and to apply the isometric formula in (6.25) to the spaces $(\mathcal{P}(^nE), \| \cdot \|_{K+\beta_n B})_{n=0}^{\infty}$.

Example 6.33 (a) The Tsirelson–James space T_j^* is an exampe of a non-reflexive space which satisfies all the hypotheses of Proposition 6.32.

(b) If $x \in E''$, E a separable Banach space with the bounded approximation property, and

$$\sum_{n=0}^{\infty} \left[AB_n\left(\frac{\widehat{d^n f}(0)}{n!} \right) \right](x) < \infty$$

for all $f \in H(E)$ then the mapping

$$\phi : f \longrightarrow \sum_{n=0}^{\infty} \left[AB_n \left(\frac{\widehat{d^n} f(0)}{n!} \right) \right](x)$$

defines a multiplicative linear functional on $H(E)$, which is, as the pointwise limit of a sequence of τ_δ continuous linear functionals, τ_δ continuous. By Corollary 4.18, ϕ is τ_ω continuous and, using the sequence $(f^n)_n$ we see that it is τ_0 continuous. The Hahn–Banach Theorem now implies that $x \in E$. This shows that $(H(rB), \tau_\omega)'' \cong H(rB'')$ (as sets) under the hypotheses of Proposition 6.32 if and only if E is reflexive. In this case $(H(rB), \tau_\omega)$ is itself reflexive. An example of an infinite dimensional Banach space of this kind is the original Tsirelson space T^* (see Example 4.21).

6.4 Exercises

Exercise 6.34* For a given fixed positive ρ and $n \in \mathbb{N}$, let

$$E(\rho) = \{f \in H(\mathbb{C}^n) \ : \ \exists c > 0 \text{ such that } |f(z)| \le c e^{\|z\|^\rho} \text{ for all } z \in \mathbb{C}^n\}.$$

Show that $E(\rho)$ endowed with the norm

$$\|f\|_\rho = \inf\{c > 0 \ : \ |f(z)| \le c e^{\|z\|^\rho} \text{ for all } z \in \mathbb{C}^n\}$$

is a Banach space and that $\bigcup_{\alpha < \rho} E(\alpha)$ is a polar subset of $E(\rho)$.

Exercise 6.35 Show that a countable union of polar subsets in a Fréchet space has non-empty interior. Show that this result is not true for $\mathbb{C}^{(\mathbb{N})}$.

Exercise 6.36* Show that a Fréchet space which contains a compact non-polar subset is Schwartz.

Exercise 6.37* Let B denote a balanced convex bounded subset of a Banach space E and let $(n_j)_j$ denote a strictly increasing sequence of positive integers. If $P_j \in \mathcal{P}(^{n_j}E)$ for all j and

(a) $\sup_j \sup_{\|x\| \le 1} |P_j(x)| \le 2^{-n_j}$,

(b) there exists $z \in E$, $\|z\| < 1$, such that $\inf_{j \in \mathbb{N}} |P_j(z)|^{1/n_j} > 0$,

(c) $\sup_{x \in B} |P_j(x)|^{1/n_j} \le e^{-j^2}$ for every $j \in \mathbb{N}$,

show that $\theta(x) := \sum_{j=1}^{\infty} \frac{1}{j^2 n_j} \log |P_j(x)|$ defines a plurisubharmonic function on E satisfying $\theta\big|_B \equiv -\infty$ and $\theta(z) > -\infty$.

Exercise 6.38* A subset B of a locally convex space is *uniformly polar* if there exists a convex balanced neighbourhood of zero U in E and a plurisubharmonic function θ on $\widehat{E}_U$ such that $\theta \circ \pi_U$ is not identically $-\infty$ and

$\theta(\pi_U(B)) = -\infty$ (π_U is the canonical quotient mapping from E onto E_U). Using the plurisubharmonic function from the previous exercise, show that convex balanced polar subsets of a locally convex space are uniformly polar if all pseudo-convex open subsets of E are domains of existence and the set of points of continuity of $\mathcal{G}$-holomorphic functions is open and closed.

Exercise 6.39* Let $E = \lambda(A)'_\beta$ denote a $\mathcal{DFS}$ space with absolute basis. Show that E contains a compact non-polar set if and only if E satisfies (6.9), i.e. if and only if E'_β has property $(\widetilde{\Omega})$.

Exercise 6.40* Let E denote a locally convex space with completion $\widehat{E}$. Show that the following are equivalent: (a) E is a polar subset of $\widehat{E}$, (b) there exists U pseudo-convex open in $\widehat{E}$, $E \subset U \subsetneq \widehat{E}$, (c) E is not determining for plurisubharmonic functions on $\widehat{E}$.

Exercise 6.41* If for all n, E_n is a proper closed subspace of a Banach space E with the bounded approximation property show that $(\bigcup_n E_n)_O \neq E$.

Exercise 6.42* If E and F are metrizable locally convex spaces, show that $(E \times F)_O = E_O \times F_O$.

Exercise 6.43* Let E denote a Fréchet space with strictly increasing fundamental sequence of semi-norms $(p_n)_{n=1}^\infty$. If there exists a function μ on $E \times \mathbb{C}$ with $(x, z) \longmapsto \mu(x, |z|)$ plurisubharmonic and an unbounded strictly increasing sequence of positive real numbers $(t_n)_n$ such that $\log p_n(x) = \mu(x, t_n)$ for all $x \in E$ and all n, show that E is either a Banach space or every bounded subset B of E is strictly polar in some open set U, $B \subset U$.

Exercise 6.44* If E is a locally convex space and $H_{ub}(E) = H(E)$, show that $H_{ub}(F) = H(F)$ for any quotient F of E. If, in addition, E is stable show that $H_{ub}(E^{\mathbb{N}}) = H(E^{\mathbb{N}})$.

Exercise 6.45* If E is a Fréchet space show that $H_b(E) = H_{ub}(E)$ if and only if $(H_{ub}(E), \tau_b^{bor}) = \varinjlim_{U \in \mathcal{U}} (H_b(E_U), \tau_b)$ and the inductive limit is regular ($\mathcal{U}$ is a fundamental system of convex balanced zero neighbourhoods).

Exercise 6.46* If P is an n-homogeneous integral polynomial on the Banach space E show that $AB_n(P)$ is an integral polynomial on E''.

Exercise 6.47* A polynomial $P \in \mathcal{P}(^nE)$, E a Banach space, is called *extendible* if for every Banach space F containing E as a subspace there exists $P_F \in \mathcal{P}(^nF)$ such that $P_F|_E = P$. Show that every extendible polynomial is weakly sequentially continuous and that every integral polynomial is extendible. If $P \in \mathcal{P}(^nE)$ is extendible let

$$\|P\|_e = \inf\{c > 0 : \quad \text{for all } F, \ E \hookrightarrow F \text{ there exists } P_F \in \mathcal{P}(^n F)$$
$$\text{which extends } P \text{ and such that } \|P_F\| \le c\|P\|\}.$$

Show that the space of all extendible n-homogeneous polynomials on E, $\mathcal{P}_e(^n E)$, endowed with $\| \cdot \|_e$ is a dual Banach space.

Exercise 6.48* If E_1 and E_2 are Banach spaces show that $E_1 \times E_2$ is regular if and only if all continuous linear mappings from E_i into E'_j, $i, j = 1$ or 2, are weakly compact.

Exercise 6.49* Show that a stable Banach space is regular if and only if it is symmetrically regular.

Exercise 6.50* Let $T : l_1 \longrightarrow l_\infty$ be given by $T((x_i)_i) = (y_i)_i$ where $y_i = (-1)^{i+1} \sum_{j \le i} x_j + \sum_{j > i} (-1)^{j+1} x_j$. Show that $T \in \mathcal{L}(l_1; l_\infty)$ but that $\overline{T(B_{l_1})}$ is not a weakly compact subset of l_∞. Show that the bilinear form associated with T is symmetric.

Exercise 6.51* If E is a Banach space, $Q \in \mathcal{P}(^n E'')$ and $P \in \mathcal{P}(^n E)$ show that $Q = (AB)_n(P)$ if and only if $Q\big|_E = P$, $dQ(x) \in E'$ is weak* continuous for each $x \in E$ and $dQ(x'')(x_\alpha) \to dQ(x'')(x'')$ for all $x'' \in E''$ and all $(x_\alpha)_\alpha \subset E$ which converges weak* to x''.

Exercise 6.52* If E is a Banach space and $f \in H_b(E)$ show that the following are equivalent:

(a) $df : E \longrightarrow E'$ is weakly compact,

(b) there exists a reflexive Banach space Y, $T \in \mathcal{L}(E; Y)$ and $\tilde{f} \in H_b(Y)$ such that $f = \tilde{f} \circ T$,

(c) if $g := AB(f)$ is the Aron–Berner extension of f to $H_b(E'')$ then $dg : E'' \longrightarrow E'''$ is weakly compact.

Exercise 6.53* Show that T_J^* (Example 2.43) is a regular Banach space.

Exercise 6.54* Let A denote a continuous symmetric bilinear form on the Banach space E and let T denote the associated linear mapping from E to E'. If $x'', y'' \in E''$ and $(y_\beta)_\beta$ is a net in E such that $J_E y_\beta \to y''$, in $(E'', \sigma(E'', E'))$, as $\beta \to \infty$ show that $\tilde{A}(x'', y'') = T^{**}x''(y'')$ and $\tilde{A}(y'', x'') = \lim_\beta T^{**}x''(J_E y_\beta)$. Hence deduce, for fixed x'' in E'', that $T^{**}(x'') \in J_{E'}(E')$ if and only if $A(x'', y'') = A(y'', x'')$ for all $y'' \in E''$. Using this result show that point evaluations from $E^{(iv)}$ gives homomorphisms of $H_b(E)$ which do not arise from point evaluations in E'' if E is not symmetrically regular.

Exercise 6.55* Show that $E := c_0(\{l_1^n\}_n) := \{(x_n)_n \ : \ x_n \in l_1^n$ and $\|x_n\| \to$ 0 as $n \to \infty\}$ endowed with the norm $\|(x_n)_n\| = \sup_n \|x_n\|$ is regular but that $E'' = l_\infty(\{l_1^n\}_n)$ is not symmetrically regular.

Exercise 6.56* If E is a Banach space, $f \in H_b(E)$ and $x \in E$ let $f_x(y) = f(x + y)$. Show that $f_x \in H_b(E)$. If $\phi \in H_b(E)'$, show that the mapping

$$x \in E \longrightarrow \phi(f_x)$$

defines a holomorphic function of bounded type on E.

Exercise 6.57* If E is a Banach space and $f \in H(E)$, show that there exists an open subset Ω of E'', $E \subset \Omega$, and $\tilde{f} \in H(\Omega)$ such that $\tilde{f}\big|_E = f$.

Exercise 6.58* Let E denote a Banach space and let

$$\theta : H_{\omega u}(E) \longrightarrow H_{\omega u}(E)$$

denote a continuous multiplicative linear function. If $\theta \circ \tau_z^* = \tau_z^* \circ \theta$ for all $z \in E$, that is, θ commutes with all translation operators, we call θ a convolution operator. Show that τ_z^* is a convolution operator on $H_{\omega u}(E)$ for all $z \in E''$ and that the converse is true (i.e. all convolution operators have this form) when E has the approximation property.

Exercise 6.59* If E and F are Banach spaces, $E' \cong F'$ and E is regular, show that F is regular and $(H_b(E), \tau_b) \cong (H_b(F), \tau_b)$.

Exercise 6.60* If E is a Banach space, F is a uniform algebra (i.e. a closed subalgebra of $\mathcal{C}(K)$, K compact Hausdorff) with identity and E' has the approximation property, show that $\mathcal{S}(H_b(E; F)) \cong E'' \times \mathcal{S}(F)$ if and only if $\mathcal{P}(^n E) = \mathcal{P}_\omega(^n E)$ for all n.

Exercise 6.61* If E is a locally convex space, with completion $\widehat{E}$, let

$$E_* = \{x \in \widehat{E} \ : \ \exists (x_\alpha)_\alpha \subset E, \text{ such that } \lambda_\alpha(x_\alpha - x) \to 0 \text{ as } \alpha \to \infty$$
$$\text{for } \textit{any} \text{ net } (\lambda_\alpha)_\alpha \text{ of scalars}\}.$$

If E has countable algebraic dimension, show that $E_* = E_{\mathcal{O}}$.

Exercise 6.62* Let $F : H(\mathbb{C}^n) \longrightarrow H(\mathbb{C}^n)$ be given by $F(f) = e^f$. Show that $F(H(\mathbb{C}^n))$ is a polar but not an analytic subset of $H(\mathbb{C}^n)$.

Exercise 6.63* If E is a Fréchet space show that each $f \in H_b(E)$ has an extension to a holomorphic function $\overset{\circ}{\tilde{f}} \in H_b(E'')$.

Exercise 6.64* If E is a distinguished closed subspace of a Fréchet space F show that the following are equivalent:

(a) there exists a continuous linear mapping $\pi : F \to E''$ such that $\pi\big|_E$ is the identity,

(b) there exists a continuous linear extension mapping $\rho : E' \to F'$,

(c) there exists a continuous linear extension mapping $T : H_b(E) \to H_b(F)$,

(d) there exists a continuous linear extension mapping $T_G : H_b(E; G'_\beta) \to H_b(F; G'_\beta)$ for every locally convex space G such that G'_β is quasi-complete.

Exercise 6.65* If E is a Banach space show that $(H_b(E), \tau_b)$ is reflexive if and only if $(\mathcal{P}(^nE), \|\cdot\|)$ is reflexive for all n.

Exercise 6.66* If U is a convex balanced open subset of T^*, the original Tsirelson space, and $f_1, \ldots, f_n \in H(U)$ do not have a common zero show that there exist $g_1, \ldots g_n \in H(U)$ such that $\sum_{i=1}^n f_i g_i = 1$.

Exercise 6.67* If E is a locally convex space, $\nu \in PSH(E)$, $A := \{x \in E : \nu(x) = -\infty\}$ and μ is a Gaussian measure on E, show that $\mu(A) = 0$ or 1. Show that both possibilities may arise.

6.5 Notes

Holomorphic completions (Definition 6.2) of normed linear spaces were introduced by A. Hirschowitz [458, 460] who obtained Propositions 6.3 and 6.4 and Example 6.6(a). These results were extended to certain locally convex spaces (including metrizable spaces) in [292, §6] and [778]. Proposition 6.5 can be found in [285] (spaces which do not satisfy the hypothesis of Proposition 6.5(c) are discussed in [418, 419, 790]). A. Hirschowitz [460] gave the first examples of normed linear spaces which are not holomorphically complete by showing that

$$(l_p, \|\cdot\|_q) \subsetneq \bigcap_{r>p} l_r =: l_{p+} \subseteq (l_p, \|\cdot\|_q)_{\mathcal{O}}$$

where $1 \le p < q < \infty$ (a structural investigation of l_{p+} is undertaken by G. Metafune–V.B. Moscatelli in [617] and the same space appears in an example, due to J.M. Ansemil–F. Blasco–S. Ponte [22], showing that $(BB)_2 \not\Rightarrow (BB)_3$). Ph. Noverraz [696] showed that the subspace spanned by the elements of a Schauder basis in a Banach space is holomorphically complete and noted [696, p.172] a connection between polar subspaces and holomorphic completions. Example 6.6 (see also Exercise 6.32) is due to S. Dineen [285]. Proposition 6.8 is due to G. Coeuré [237] and Ph. Noverraz [699, Théorème 5; 697]. Using Gaussian measures S. Dineen–Ph. Noverraz [323, 324] proved that every infinite dimensional Fréchet space E contained a *proper dense hyperplane* H such that $H_{\mathcal{O}} = E$. In [778], M. Schottenloher showed that each $f_i \in H(E; F)$, E and F metrizable locally convex spaces, or ω-spaces (see Exercise 6.42), admits a holomorphic extension $f_O \in H(E_O; F_O)$.

Control sets for plurisubharmonic functions on infinite dimensional spaces were investigated by P. Lelong [541, 542] and C.O. Kiselman [518]. In [542] P. Lelong showed that all compact subsets of $H(\mathbb{C}^n)$ were polar (Example 6.10 and Exercise 6.43). The remaining results in Section 6.1, i.e. Proposition 6.9 and Examples 6.10 and 6.11 are due to S. Dineen–R. Meise–D. Vogt [319, 320, 321]. For the sake of a shorter and more accessible exposition we assumed the existence of a basis, although this hypothesis is not necessary (see [321]), and this obliged us to construct new or modified proofs. The invariants $(\underline{DN})$ and $(\widetilde{\Omega})$ were introduced and linearly characterized, as in Examples 6.10 and 6.11, by D. Vogt in [838] and [839], respectively, and $(\widetilde{\Omega})$ appears in the holomorphic classification of infinite dimensional polydiscs [606, 774]. In the last ten years, these invariants and others introduced by D. Vogt have played a key role in the examination of uniform holomorphicity, weak holomorphic extensions, etc. on Fréchet nuclear, Fréchet–Schwartz, Fréchet–Montel and Fréchet spaces by L.M. Hai, N.V. Khue and their co-workers. We refer to [427, 428, 429, 433, 434, 435, 506, 507, 508, 509, 510, 740] for details. The final space mentioned in Example 6.11 first appeared in [610] (see also [321]). Pseudo-convex completions are investigated by Ph. Noverraz in [698, 700].

R. Arens [31, 32] initiated the study of extending bilinear mappings and products to the bidual using the transpose operator and obtained essentially Propositions 6.13 and 6.14 and showed that l_1 is not regular (see [53, 741, 747, 860]). R.M. Aron–P. Galindo [57] defined m-regular Banach spaces for multilinear mappings and characterized polynomials with weakly compact first derivative (see Exercise 6.52). Extensions of holomorphic functions from closed subspaces and the connection between bounding sets and extensions (Example 6.12) are outlined in [285]. A detailed study of extensions of polynomials and holomorphic functions to the bidual, which included Propositions 6.16 and 6.18, Corollary 6.17 and Example 6.19, was undertaken by R.M. Aron–P. Berner in [47]. The space $J(E; F)$ and the extension operator Z_n were introduced and investigated by I. Zalduendo [856]. The paper [856] contains a proof of Proposition 6.16, Example 6.20 and the characterization of the Aron–Berner extension given in Exercise 6.51. A detailed discussion of the Aron–Berner extension for polynomials is given in the remarks on Chapter 1. Extensions of holomorphic functions to the bidual of Fréchet and other locally convex spaces are due to N.V. Khue [506, 507], L.A. Moraes [628, 629, 632], L.A. Moraes–P.A. Burlandy [634] and P. Galindo–D. García–M. Maestre [371] (we refer to the remarks on Exercises 6.63 and 6.64). The dual lifting problem is considered in [66] (see the remarks on Exercise 4.67). Recent results on polynomial extensions may be found in [48, 202, 205, 372, 373, 512]. Proposition 6.21 and Example 6.22(a) are due to R. Meise–D. Vogt [608] and the embedding theorem used in Proposition 6.21 is also proved in [608]. An earlier version of Example 6.22(a) (for closed subspaces of $\mathcal{DFN}$ spaces) is

due to P. Boland [149] (see Exercise 4.77) and this result has been extended to
$\mathcal{DFC}$ spaces in [366] (see also [369, Proposition 8; 252, 432]). Example 6.22(b)
is the now classical example of L. Nachbin [670, 679] that we encountered previously in Example 3.12 and Exercise 3.120. In [607], R. Meise–D. Vogt show
that every infinite dimensional Fréchet nuclear space, with the exception of
$\mathbb{C}^{\mathbb{N}}$, contains a closed subspace F such that $H_{ub}(F) \neq H(F)$ (see also [371,
610] and Exercises 3.69, 3.78 and 3.80). Proposition 6.23, without the basis
hypothesis, is due to R. Meise–D. Vogt [610]. L.M. Hai–T.T. Quang [435]
consider the vector-valued case and show that $H(E;F) = H_u(E;F)$ if E is a
Fréchet nuclear space with $(\widetilde{\Omega})$ and F is a Fréchet space with (DN).

The dominant influence and motivation for the results given in Section
6.3 is the paper [51] by R.M. Aron–B.J. Cole–T.W. Gamelin. The focus of
attention in that article was the spectrum of $H^{\infty}(B)$, where B is the open
unit ball of a Banach space. The mathematical literature (see for instance
[376] and [382]) contains abundant evidence of the rich interplay between
complex function theory on the one-dimensional disc and Banach algebra
theory. Thus it is no surprise that in infinite dimensions the topic is rich in
interesting ideas and results. To analyse the structure of $H^{\infty}(B)$ the authors
of [51] found it convenient to begin by examining $\mathcal{S}(H_b(E))$. The convolution
$*$ is defined and its basic properties developed in [51] and the important
role of symmetrically regular Banach spaces uncovered, for example it is
shown that $\delta_z * \delta_w = \delta_w * \delta_z$ if and only if E is symmetrically regular. The
extension result (6.24), Proposition 6.26 and Lemma 6.28 are taken from [51].
We did not find it convenient to include an examination of $H^{\infty}(B)$. Three
important papers devoted to suggestions and problems arising in [51] are [53,
58 and 347]. Extensions to the bidual of holomorphic functions which are
weakly continuous on bounded sets are discussed in [53] and the converse of
Proposition 6.26, which combines with results in [51] to complete the proof of
Proposition 6.29, is proved in [53] (a vector-valued version of Proposition 6.26
is given in L.A. Moraes–P. Burlandy [634]). A brief partial survey of the
results in [51, 53] is given in [52] and a more extensive introduction may be
found in the lecture notes of T.W. Gamelin [377]. Propositions 6.25, 6.27
and 6.29 are due to R.M. Aron–P. Galindo–D. García–M. Maestre [58] (see
also [532, Remark 3] for a further proof of Proposition 6.25). More generally,
the authors in [58] have shown that $\mathcal{S}(H_b(U))$ is a Riemann domain over E''
when E is a symmetrically regular Banach space and U is an arbitrary open
subset of E. The proof when $U \neq E$ is much more technical and the authors
point out in [58] the simplifications available when $U = E$. We have followed
their suggestions and presented the simple case. The fact that $\mathcal{S}(H_b(E))$
is an analytic extension of "bounded type" of E (Proposition 6.30) in the
sense of Chapter 5 has not, apparently, previously appeared in the literature.
Homomorphisms and convolution operators on spaces of weak* continuous
holomorphic functions on a dual Banach space are discussed by R.M. Aron–

P. Rueda [69] and the set $S(H_b(U, F))$, where E is a Banach space and F is a Banach algebra, is studied in [378] and [634].

C.P. Gupta [421, 422, 423] and L. Nachbin–C.P. Gupta [684] (see Section 4.3) studied convolution operators on Banach spaces (i.e. linear operators which commute with the translation operators) in the late 1960s and in doing so were led to develop a *preduality* theory for $H_b(E)$ and $H(E)$ (see Sections 2.6 and 4.6). Topological *duality* of $H_b(E; F)$, E and F Banach, was investigated by J.M. Isidro [472] and J.M. Isidro–J.P. Méndez [476] and their sequence space representation of $H_b(E)'_\beta$ was extended by J.M. Ansemil–S. Ponte [25] to a similar result for the bidual. The Fréchet space $H_{\omega u}(E)$ was introduced by R.M. Aron [43]. J.M. Ansemil–S. Ponte [25, Corollary 3] show that $H_b(E)$ is reflexive if and only if $\mathcal{P}(^nE)$ is reflexive for all n and further equivalent conditions are given in M. González–J. Gutiérrez [395, Theorem 4]. In [739], A. Prieto showed that $(H_b(T^*), \tau_b)$ is reflexive and used S-absolute decompositions to show $(H_{\omega u}(E), \tau_b)'' \cong (H_b(E), \tau_b) \iff \mathcal{P}(^nE) = \mathcal{P}_{wu}(^nE)''$ all n (this result is correct but the statement and proof are somewhat ambiguous since the equality between the spaces of homogeneous polynomials must be implemented in a canonical way, say by I_n, and one obtains an equivalence if and only if $\sup_n[\|I_n\| \cdot \|I_n^{-1}\|]^{1/n} < \infty$). This was clarified in the thesis of P. Rueda [753] and in P. Galindo–M. Maestre–P. Rueda [374] and the results of Prieto are generalized in [831, 832]. Proposition 6.32 is obtained in [753, Teorema 3.22] and [374, Corollary 16] for Banach spaces such that $l_1 \not\hookrightarrow \mathcal{P}_{w^*}(^nX')$ all n. Proposition 6.32 is due to S. Dineen [313]. The semi-norms (6.36) are easily seen to generate a complete locally convex space structure on both spaces (see R.M. Aron [36, Theorem 2] for a similar result on $H(K)$, K compact balanced in a Banach space). Example 6.33(a) is stated in [55] and rescued from ambiguity by results in [374]. Holomorphic functions of bounded type on a Banach space are discussed and vector-valued extensions of Proposition 6.31 obtained in [477] by J.A. Jaramillo–L.A. Moraes. In [379] D. García–M. Maestre–P. Rueda study Schauder decompositions, duality, preduality and reflexivity of weighted spaces of holomorphic functions on balanced domains in Banach spaces.

Appendix. Remarks on Selected Exercises

Chapter 1

1.61 Polynomials on finite dimensional spaces are easily seen to be continuous. This fact together with the method used in Example 1.25 can be combined to show that $\mathcal{P}_a(\mathbb{C}^{(\mathbb{N})}) = \mathcal{P}(\mathbb{C}^{(\mathbb{N})})$. The converse was announced by L. Nachbin [677] and two proofs, each assuming the continuum hypothesis, are given in [92] by J.A. Barroso–M.C. Matos–L. Nachbin. To prove the converse, it suffices to show, if E is a vector space which contains an algebraic or Hamel basis of uncountable cardinality I, that there exists a 2-homogeneous polynomial on E which is not continuous when E is endowed with its finest locally convex topology. If $(e_i)_{i \in I}$ is an algebraic basis for E then the polynomial

$$P\left(\sum_{i \in I} \alpha_i e_i\right) := \sum_{i,j \in I} \alpha_i \alpha_j r(b_i, bj)$$

has the required properties when r is the function defined in Example 2.19. The continuum hypothesis was removed in [482] and [682]. See also [446, p. 37] and [162, Example 3.7].

This exercise is related to the result of S. Kakutani–V. Klee [498] (see also [446, Proposition 2.3.2]) which says that the finite open topology (Section 3.1) on a vector space is locally convex if and only if E has a countable algebraic basis. Various topologies, which lie between the finite open topology and locally convex topologies that arise in the study of plurisubharmonic functions on infinite dimensional spaces, are considered in P. Lelong [537, 538, 539]. See also the remarks on Exercise 3.55.

1.62 Let $(e_\alpha)_{\alpha \in \Gamma}$ denote an algebraic basis for E. For each $\alpha \in \Gamma$ let $e_\alpha^* \in E^*$ denote evaluation at the αth coordinate. Let $\Gamma^{(N)}$ denote all sets of non-negative integers indexed by Γ of which at most a finite number are non-zero. If $m = (m_i)_{i \in \Gamma} \in \Gamma^{(N)}$ let $|m| = \sum_i |m_i|$ and let z^m denote the monomial $\prod_{i \in \Gamma}(e_i^*)^{m_i}$. It is clear, and well known, that z^m is a $|m|$-homogeneous polynomial and if $|\Gamma| < \infty$ then $(z^m)_{m \in \Gamma^{(N)}, |m|=n}$ is a basis for $\mathcal{P}_a(^n E)$. If $P : E \to F$ has the property that $P|_G \in \mathcal{P}(^n G; F)$ for each finite dimensional subspace G of E then it is easily seen that

$$P\big((z_i)_{i\in\Gamma}\big) = \sum_{\substack{m\in\Gamma^{(N)}\\|m|=n}} b_m z^m$$

where $b_m \in F$ for all m. It is now easy to construct an n-linear mapping L from E into F such that $\widehat{L} = P$ and hence $P \in \mathcal{P}_a(^nE; F)$. See Section 3.1 and Exercise 3.109.

1.63 See Example 3.8(g).

1.64 This method of differences was used by M. Fréchet [357] (see also [820]) to define polynomials on an abstract space. For locally convex spaces over $\mathbb{R}$ this is equivalent to our definition but this is not the case for complex spaces, e.g. let $P(z) = Re(z)$ for $z \in \mathbb{C}$. If, in addition, one supposes that the mapping is everywhere complex Gâteaux differentiable then we get the usual definition (this is due to I.E. Highberg [449] for Banach spaces and to D.H. Hyers [471] for locally convex spaces).

1.65 This result is due to S. Mazur–W. Orlicz [599]. A function which is continuous when restricted to the complement of a set of first category is called a B-continuous function. These functions arise in measure theory and are useful since the pointwise limit of a sequence of B-continuous functions on a Baire space is B-continuous. For general results on B-continuous functions on a Baire space we refer to H. Hahn [431] and J. C. Oxtoby [710]. Generalizations and applications are given in S. Mazur–W. Orlicz [599, 600], M.A. Zorn [862] and A. Alexiewicz–W. Orlicz [14]. Closely related but less general are the functions of Baire first class. These functions are the pointwise limit of a *sequence* of continuous functions and have played an important role in the recent history of Banach space theory. We use properties of Baire first class functions in Example 3.8(a).

1.67 This is a polynomial version of the Banach–Dieudonné Theorem due to J. Mujica [643, Theorem 2.1]. His proof is modelled on the classical proof given in J. Horváth [467, p. 245]. Alternatively, one may apply, after some identification of topologies, the linear Banach–Dieudonné theorem. First, as we have already noted, if E is Fréchet then τ_0 on $\mathcal{P}(^nE)$ coincides with the topology of uniform convergence on the compact subsets of $\widehat{\bigotimes}_{n,s,\pi} E$. Next, mappings of the form $P \to \sum_{j=1}^{\infty} \lambda_j P(x_j)$, $P \in \mathcal{P}(^nE)$, where $(\lambda_j)_j \in l_1$ and $(x_j)_j$ is a null sequence in E generate the $\sigma\big(\mathcal{P}(^nE), \widehat{\bigotimes}_{n,s,\pi} E\big)$ topology on $\mathcal{P}(^nE)$. Hence the topology of pointwise convergence coincides with the weak* topology on equicontinuous or, equivalently, locally bounded subsets of $\mathcal{P}(^nE)$. This gives the proof when E is Fréchet and can be adjusted for the metrizable case. P. Rueda considers extensions of the Banach–Dieudonné

Theorem to $H_b(E)$, E Banach, and to $H(E)$, E Fréchet, in [752, 753]. See the remarks in Exercise 3.108.

1.68 The result for arbitrary E is due to L.A. Harris [438] – note that $(*)$ is satisfied, if $p = 1$, by any set of unit vectors and on taking $m = k$ and $n_1 = n_2 = \cdots = n_k = 1$ the estimate reduces to the classical polarization inequality with bound $m^m/m!$. An Example in [767], due to A. Tonge, shows that the same inequality does not hold for real Banach spaces. The result for $L^p(\mu)$ is due to Y. Sarantopoulos [765] who also shows that the result is best possible. Alternative proofs are given in [64, Corollary 8] and in [441] using generalized Rademacher functions.

1.69 Use the Baire Category Theorem and Proposition 1.11. Alternatively, the second method in Exercise 1.67 shows that $\mathcal{F} \subset \mathcal{P}(^nE)$ is pointwise bounded if and only if it is $\sigma\big(\mathcal{P}(^nE), \widehat{\bigotimes}_{n,s,\pi} E\big)$ bounded and one can now apply the uniform boundedness principle to show that $\mathcal{F}$ is locally bounded. Refined versions of this Banach–Steinhaus Theorem for polynomials are given in P. Lelong [537] and [446, Theorem 3.18]. S. Mazur–W. Orlicz [600, p.182] and A.E. Taylor [820, p.311] prove that the pointwise limit of a sequence of continuous n-homogeneous polynomials on a Banach space is itself a continuous n-homogeneous polynomial. See [92, Proposition 40] and [534; §IV, Proposition 1-2].

1.70 See [295]. The holomorphic analogue for $\mathcal{DFS}$-spaces is given in Example 3.8(d) but we do not know if this extends to arbitrary $\mathcal{DFM}$-spaces [318].

1.71 Consider the mapping

$$(\lambda_m)_{m=1}^\infty \longrightarrow \Phi\big((\lambda_m)_{m=1}^\infty\big)\Big[P \longrightarrow \sum_{m=1}^\infty \lambda_m P(e_m) f_m\Big].$$

1.72 The space $c_0(\Gamma)$, Γ uncountable, is a useful counterexample space in infinite dimensional holomorphy (see for instance C. Boyd–R.A. Ryan [191] B. Josefson [487, 489], Ph. Noverraz [706], J. Globevnik [389], S. Dineen [292] and Example 5.55). The theory of surjective limits (discussed in Section 3.6) partially explains the behaviour of $c_0(\Gamma)$ while the geometry of the unit ball (essentially a product of discs) is also important. The first part of this exercise is easy. The second part is due to R.M. Aron [42]. See also A. Pełczyński–Z. Semadini [718] and Exercise 3.74.

1.73 See S. Banach [85] and W. Bogdanowicz [140].

1.74 See Y. Sarantopoulos [765].

1.75 This result is due to S. Banach [85], and generalizes to symmetric n-linear forms the well known linear result that a self-adjoint compact operator from a Hilbert space into itself has an eigenvalue (characteristic value) whose absolute value is equal to the norm of the operator. A characterization of polynomials in $L_\mu^2(M)$, M a locally compact space, which can be represented by means of L^2 kernels is given in T.A.W. Dwyer [338].

1.77 The result is due to F. Blasco [129, 130].

1.78 The Banach space case is due to R.M. Aron–M. Schottenloher [70, 71] while the result for arbitrary E and the vector-valued analogue are given in F. Blasco [129, 130]. Typical applications occur in [860, §2] and [661]. See also Proposition 3.22.

1.79 Show that the sequence $(Q_m := \sum_{n=1}^m P_n)_{m=1}^\infty$, where $(P_n)_n$ is the sequence defined in Example 1.27, is τ_0 Cauchy but does not converge. This result is related to results in Proposition 2.16 and Examples 2.19 and 3.50(d). See also [153].

1.80 This is just a rephrasing of (1.55).

1.81 We are concerned in this book with polynomials over complex spaces and have neglected for the most part any reference to real polynomials. This theory contains interesting (and surprising) intrinsic results and also material which extends our knowledge of complex polynomials. This exercise deals with one of the most basic questions – the extension of polynomials from a real Banach space to its complexification. As the exercise shows this involves constructing a *complex norm* (i.e. one which satisfies property (a)) on the complex space. Property (b) says that the norm extends the norm of the underlying Banach space while (c) is "desirable" ([511]) and (d) "reasonable" ([662]). Norms on $E_{\mathbb{C}}$ satisfying all four properties can be constructed in many different ways (e.g. using averaging techniques, tensor products, etc.) and a number of obvious candidates for complex norms, e.g. $\|x + iy\| = (\|x\|^2 + \|y\|^2)^{1/2}$ for $x, y \in E$, fail to satisfy all four. This question was considered in the period 1938–1943 by A.E. Taylor [820, 821] and A.D. Michal–M. Wyman [622]. In the 1950s A. Alexiewicz–W. Orlicz [14] developed a theory of *real analytic functions* on Banach spaces by extending mappings, including polynomials, from E to $E_{\mathbb{C}}$.

The results in this exercise were obtained independently and at the same time by P. Kirwan [511] and G. Muñoz–Y. Sarantopoulos–A. Tonge [662]. These papers contain interesting new results and examples and give a comprehensive account of the interplay between real and complex Banach spaces and the geometry of Banach spaces (see also J. Bochnak [135] and J. Bochnak–J. Siciak [136, 137, 138]). Extremal polynomials on real Banach spaces are discussed in [441, 513] and the relationship between the real and complex norms of a polynomial are investigated in [46], [63] and [820, p. 314]. See also Exercise 2.80.

1.82 The adjoint of P, $P^* \in \mathcal{L}(F'; \mathcal{P}(^n E))$, is given by $P^*(\psi) = \psi \circ P$ where $\psi \in F'$ (see [71]). The results in this exercise are due to R.A. Ryan [758] (the reflexivity result is discussed in Section 2.4). The special case $E = \mathcal{C}(K)$ is considered in [715]. Related results on the polynomial Dunford–Pettis property can be found in [757]. See also J. Mujica [653, Proposition 3.4]. The equivalence of (i), (ii) and (iii) for linear mappings is Gantmacher's Theorem [270, p.21], a holomorphic analogue is given in [401].

1.83 In the real case, extreme points have the form $\pm x \otimes \cdots \otimes x$. This is due to R.A. Ryan–B. Turett [761] when $\dim(E) < \infty$ and to C. Boyd–R.A. Ryan [191] for reflexive Banach spaces. Since $\underset{1,s,\pi}{\bigotimes} E = E$ the result cannot hold for all E. Extreme points for spaces of polynomials are also discussed in [223, 226, 227].

1.86 Let $P(y + y') = y'(y)$. Extend P to $P_l \in \mathcal{P}(^2(X \times Y'))$ using the Polynomial Extension Property. Extend $\overset{\vee}{P_l}$ to a bilinear form Q on $X'' \times Y'''$ using weak* continuity as in the proof of Proposition 1.53. The mapping π defined by $[\pi(x^{**})](y') = 2Q(x^{**}, y')$ is a linear projection from X'' onto Y''. This implication is taken from [307, Proposition 4.18]. A number of other equivalent conditions can be found in [860, Theorem 2.8].

If X is a Banach space then the 2-homogeneous polynomial $P(y, y') = y'(y)$ on $X \times X'$ is weakly sequentially continuous if and only if X has the Dunford–Pettis Property (Definition 2.33). The Aron–Berner extension of P to $X'' \times X'''$ has the form

$$AB_2(P)(x'', x''') = \frac{1}{2}\left(x'''(x'') + x''(\rho(x'''))\right)$$

where $\rho: X''' \to X'$ is the restriction (i.e. the transpose of the natural inclusion $J_X: X \to X''$). This example is used by S. Lassalle–I. Zalduendo [532] to show that $l_1 \hookrightarrow X \times X'$ if X is an infinite dimensional Banach space with the Dunford–Pettis Property and by P. Galindo in [365] to construct a counterexample.

1.87 This exercise shows that $\mathbb{C}^I$, I uncountable, does not satisfy (1.39). By [90] the space $\mathcal{P}(^n \mathbb{C}^I)$ has a τ_ω-neighbourhood basis at the origin consisting of τ_0-closed sets but τ_0 and τ_ω are not compatible on $\mathcal{P}(^n \mathbb{C}^I)$, $n > 1$ (see Example 2.19). This result shows that (1.39) cannot be removed, without some replacement, as a hypothesis in Proposition 1.35. The result in this exercise is due to C. Boyd [184] where further pairs of spaces, e.g. $((\mathbb{C}^{(\mathbb{N})})^{\mathbb{N}}, (\mathbb{C}^{(\mathbb{N})})^{\mathbb{N}})$ and $(\mathcal{D}', \mathcal{D}')$, that fail (1.39), can be found. See the remarks on Exercises 1.94 and 2.69, and [187].

1.89 A projective limit $E = \varprojlim_{n} E_n$ is *reduced* if the canonical mappings
from E into E_n have dense ranges. This exercise is a characterization of
distinguished Fréchet spaces. We refer to J. Bonet–S. Dierolf [159] for a survey
on distinguished Fréchet spaces (see also [21] and Exercise 6.64).

1.90 This result is not difficult ([370]). An unpublished proof using tensor
products is due to R.A. Ryan.

1.91 A sequence $(x_j)_j \subset E$ such that $P(x_j) \to 0$ as $j \to \infty$ for all $P \in$
$\mathcal{P}(^n E)$ all $n \in \mathbb{N}$ is called a *weak polynomially null sequence*. T.K. Carne–
B.J. Cole–T.W. Gamelin [207] define Λ-spaces as those Banach spaces in
which weak polynomially null and norm null sequences coincide. The result
of this exercise, proved in [207] for $p \geq 2$ and in [478] for $1 < p < 2$, shows
that l_p is a Λ-space. See also [210], [400] and Section 2.6.

1.92 This result, is due to L. Harris [438, remark after Corollary 5] when
$\dim(H) = 1$ and to Y. Sarantopoulos [767, Proposition 3], for arbitrary H.
It shows that polarization constants cannot be used to characterize complex
Hilbert spaces isometrically, in contrast to the case for real Banach spaces
[98, Proposition 2.8].

1.93 Use Proposition 1.51. An application is given in Example 4.14 of [307].

1.94 By Exercise 1.61 there exists $P \in \mathcal{P}_a(^n \mathbb{C}^{(I)}) \backslash \mathcal{P}(^2 \mathbb{C}^{(I)})$ for I uncountable.
The set $(P \circ \pi_J)_{J \subset I \text{ (finite)}}$, where π_J is the canonical projection onto $\mathbb{C}^{(J)}$,
is Cauchy but does not converge [92, Example 7]. Analogous questions for
uncountable products have positive answers since the τ_0-bounded subsets of
$(H(\mathbb{C}^I), \tau_0)$ are locally bounded and $(H(\mathbb{C}^I), \tau)$ is semi-Montel when $\tau = \tau_0$
([93]) and $\tau = \tau_\omega$ or τ_δ ([184]). The τ_δ topology on $H(\mathbb{C}^I)$ (see Chapter 3)
is the bornological topology associated with both τ_0 and τ_ω ([93]) and the
space $(H(\mathbb{C}^I), \tau_0)$ is complete [90]. Note that semi-Montel spaces are quasi-
complete but not necessarily complete. The holomorphic results imply, by
Proposition 3.22, the polynomial results. The space $\mathbb{C}^I$, I uncountable, is the
classical example of a complete non-metrizable Baire space and this property
may be used to prove some of the above, see for instance [92, Proposition 37],
and to extend some of the results in Example 3.8.

1.95 This example is due to M. Schottenloher and appears in [47] (the first
example of the failure of the Hahn–Banach Theorem for polynomials is due
to B. Grünbaum [417]). These show that (1.52) does not follow from (1.51),
as is the case for linear mappings by the Hahn–Banach Theorem, and leads to
the following problem considered by C. Benítez–M.C. Otero: *characterize the
Banach spaces such that each 2-homogeneous polynomial on each subspace
has a norm preserving extension to the whole space.* The authors in [97] show

that the problem for complex Banach spaces can be reduced to the real case and give an example of a three-dimensional space (this is extended to finite dimensional spaces in [554]), whose unit ball is the intersection of two ellipsoids but which is not an inner product space, with this property. An interesting geometrical interpretation of this result is given in [97]. The same problem for n-homogeneous polynomials may prove interesting.

In [598], P. Mazet shows that each $P \in \mathcal{P}(^2H)$ has an extension $\tilde{P} \in \mathcal{P}(^2E)$, H a hyperplane in a Banach space E, satisfying $\|\tilde{P}\| \leq c\|P\|$ where $c = 2$ is best possible for real Banach spaces and where the best possible constant for complex Banach spaces lies between $7/3$ and $2\sqrt{2}$.

1.96 These results are due to J. Mujica [645]. It is not known if the separability hypothesis is necessary. Earlier S. Dineen [295] had obtained similar results for $\mathcal{DFM}$ spaces.

1.97 This is Markov's Inequality for real Banach spaces and is due to Y. Sarantopoulos [765] – see also O.D. Kellogg [505], A. Tonge [829] and L.A. Harris [439, 440, 441]. The result follows for Hilbert spaces from Exercise 1.106.

1.98 This result, due to C. Boyd [184, Corollary 3.6], shows that the following result does not extend to $\mathcal{DFN}$ spaces: if $\{E_n\}_n$ and $\{F_m\}_m$ are two sequences of Fréchet spaces and each pair (E_n, F_m) has the (BB)-property then $(\prod_{n=1}^{\infty} E_n, \prod_{m=1}^{\infty} F_m)$ has the (BB)-property (see [523, p.194] and [816]).

1.99 This result is given in C. Boyd [184]. The proof is somewhat indirect and uses the results in Proposition 2.16 and Exercise 2.57.

1.100 This is a polynomial version of Propositions 3.26 and 3.27 and these are due to M. Schottenloher [784], P. Mazet [597], R.A. Ryan [758] and J. Mujica–L. Nachbin [660].

1.101 This was obtained by J.H. Grace and can be deduced from the following result in [407]: if A and B are zeros of the polynomial of one complex variable and degree n, P, and C is the mid-point of the line segment joining A and B then P' has a zero in the circle with centre C and radius $\frac{1}{2}|AB| \cot(\pi/n)$. See Exercise 1.107.

1.102 The topic in this exercise – reverse polynomial inequalities – was first considered in an infinite dimensional setting by R.A. Ryan–B. Turrett [761]. The result for arbitrary Banach spaces is due to C. Benítez–Y. Sarantopoulos–A. Tonge [99]. To see that this result is generally best possible consider $E = l_1$, $P(\sum_{i=1}^{\infty} x_i e_i) = x_1 \cdots x_k$ and $Q(\sum_{i=1}^{\infty} x_i e_i) = x_{k+1} \cdots x_{k+l}$. The result for Hilbert spaces is due to C. Boyd–R.A. Ryan [192]. Extremal values for the constants are used in [99] to isometrically characterize

l_1^n and real Hilbert spaces. There is extensive literature on the corresponding finite dimensional problem (see [46, 99]).

1.104 Proposition 1.58 and (1.56) consider summability properties along the diagonal for homogeneous polynomials while this exercise deals with summability over all terms for n-linear forms (and hence also for homogeneous polynomials). The case $n = 2$ is due to J.E. Littlewood [562, Theorem 1(i)] and the fact that the exponent is best possible in this case is due to O. Toeplitz [826] (see the additional note in [437]). The result for arbitrary n is due to H. F. Bohnenblust–E. Hille [142] (see also A.M. Davie [255], Y.S.Choi et al. [228] and G.H. Hardy–J.E. Littlewood [437]). Since these results are best possible it follows that the monomials *do not* form an absolute basis for $(\mathcal{P}(^n c_0), \|\cdot\|)$ for $n > 1$ – see Proposition 4.4 and Sections 4.4 and 4.6). If one considers the monomial expansion of a holomorphic function on c_0 which is bounded on the unit ball, or indeed an arbitrary polynomial, then the series of coefficients will lie in l_2 (and this is best possible). This result is due to H. Bohr [143, p.462] and was afterwards rediscovered by B. Josefson [487]. Related results for sets which lie between the diagonal and the full matrix are considered by I. Zalduendo [859].

1.105 This result arose in the proof of Proposition 1.34 [266]. See also [24] and [34].

1.106 This result is due to L.A. Harris [438, Theorem 5]. It extends and refines inequalities of J.G. van der Corput–G. Schaake [254, Sätz 6] and S. Bernstein [110, (63)]. For information on derivatives see Section 3.1 and, in particular, Example 3.4.

1.107 This result is due to L. Hörmander [465] who was motivated by the result of J.H. Grace [407] quoted in the remarks on Exercise 1.101. In a recent article L.A. Harris [440] provides a simple proof, based on the fundamental theorem of algebra, and uses it to derive and extend several classical inequalities.

1.108 The results in this exercise are due to M. González–J. Gutiérrez [395]. Note that the first part follows directly from (1.55).

1.109 We are, of course, assuming, as we do throughout this book, that E is Hausdorff. Any $\theta \in \bigotimes_{n,s} E$ has a representation of the form $\theta = \sum_i^k x_i \otimes \cdots \otimes x_i$ where the finite set $(x_i)_{i=1}^k$ consists of linearly independent vectors. By the Hahn–Banach Theorem we can find $x' \in E'$ such that $x'(x_1) \neq 0$ and $x'(x_i) = 0$ for $i > 1$. If $P = (x')^n$ then $\langle P, \theta \rangle = x'(x_1)^n \neq 0$.

Chapter 2

2.50 This example is due to R. Alencar [5].

2.51 See F. Bombal [156].

2.52 See [279, Lemma 13].

2.53 See [275] and [405, Corollary 2.9].

2.54 This result is due to R. Gonzalo–J.A. Jaramillo [405].

2.55 This result is due to R. Alencar [7]. The proof uses the Bartle integral and the Singer representation of $\mathcal{C}(K, E)'$ to identify the dual of $\mathcal{P}_A({}^nE', F)$. The polynomial $P((x_i)_i) = (x_i^n)_i$, $(x_i)_i \in l_p$, $nq = p$, is used to show $\mathcal{P}_A({}^nl_p, l_q) \neq \mathcal{P}({}^nl_p, l_q)$, and Exercise 1.77 can be applied if $nq > p$. This result on l_p can also be deduced from Exercise 2.65.

A number of authors have investigated vector-valued versions of the results presented in Chapter 2. In [339], T.A.W. Dwyer proves $\mathcal{P}_N({}^nE; F)' = \mathcal{P}({}^nE'; F')$ (isometrically) when E' and F' have the approximation property (see Proposition 2.10), R. Alencar [7] shows that $\mathcal{P}_A({}^nE; F)' = \mathcal{P}_I({}^nE'; F')$ (isometrically) when E and F are reflexive Banach spaces (see Proposition 2.24), $\mathcal{P}_N({}^nE; F) = \mathcal{P}_I({}^nE; F)$ (isomorphically) when E' has the Radon–Nikodým Property (see Propositions 2.27 and 2.47), and $\mathcal{P}({}^nE; F)$ is reflexive, E and F reflexive with the approximation property, if and only if $\mathcal{P}_A({}^nE; F) = \mathcal{P}({}^nE; F)$ (see Proposition 2.31). These results have been generalized in D. Carando–V. Dimant [203] and J.A. Jaramillo–L.A. Moraes [477]. The authors of [477] also consider vector-valued polynomial and holomorphic Q-reflexivity (see Sections 2.4 and 6.3).

2.57 This result is due to L.A. Moraes [626]. The proof is technical and involves ideas similar to Exercise 2.66. By [269] the equivalent conditions in the exercise do not imply that E admits a continuous norm. See the remarks on Exercises 3.87 and 4.74 and [111], [627] and [638].

2.58 These results may be obtained using surjective limits (see Section 3.6) and appeared in [290, 292].

2.59 See [55, Theorem 5(a)]. The lower q estimate and the approximation property imply that n-homogeneous polynomials of finite type are dense in $\mathcal{P}({}^nE)$. It suffices to show that any sequence $\{\phi_j^n\}$, $\phi_j \in E'$ for all j, generates a Banach space isomorphic to a subspace of a separable dual space.

2.60 Clearly $l_1 \hookrightarrow E$ implies $(l_1)_{\mathbb{R}} \hookrightarrow E_{\mathbb{R}}$. The converse is non-trivial. Suppose $l_1 \not\hookrightarrow E$. Let $(x_n)_n$ denote a bounded sequence in $E_{\mathbb{R}}$. Then $(x_n)_n$ is also a bounded sequence in E and, by a result of L.E. Dor [331], it contains a weak Cauchy sequence $(x_{n_j})_j$. If $\phi \in E'_{\mathbb{R}}$ then $\tilde{\phi}(x) := \phi(x) - i(\phi(ix))$ defines an element in E' and hence $(\phi(x_{n_j}))_j = (\mathrm{Re}\ \tilde{\phi}(x_{n_j}))_j$ is a Cauchy sequence of real numbers. Hence $(x_{n_j})_j$ is a weak Cauchy sequence in $E_{\mathbb{R}}$ and a result of H.P. Rosenthal [749, Theorem 1] implies $(l_1)_{\mathbb{R}} \not\hookrightarrow E_{\mathbb{R}}$. This result allows one to extend (2.26), a result required in the proof of Proposition 2.47, to complex Banach spaces.

2.61 This result is due to M. González–J. Gutiérrez [397, Theorem 14 and Corollary 15]. The extension in (c) is obtained by using the Aron–Berner extension on each coordinate. See Exercise 2.71 and [365].

2.62 The polynomial results are due to S. Dineen–M. Lindström [317]. The result that quotient mappings from quasi-normable spaces onto l_p spaces do not lift bounded sets is due to M. Miñarro [623]. Lifting also arises in the study of holomorphic functions on quotient spaces (see Corollary 1.56, Exercises 4.67 and 4.88, [131, 132, 262, 392]).

2.63 C.P. Gupta [421, Lemma 4] proved that

$$\|P\|_N \le \frac{n^n}{n!} \|\overset{\vee}{P}\|_N$$

for $P \in \mathcal{P}_N(^nE)$ and using this T. Abuabara [1, Corollary 2] deduced that

$$\|PQ\|_N \le \frac{(m+n)^{m+n}}{(m+n)!} \|P\|_N \|Q\|_N$$

for $P \in \mathcal{P}_N(^nE)$ and $Q \in \mathcal{P}_N(^mE)$. To obtain the improved estimate in this exercise proceed as follows:

(a) First show that it suffices to obtain the estimate for $P = \phi^n$ and $Q = \psi^m$ where $\phi, \psi \in E'$.

(b) Show, using the binomial theorem, that

$$\binom{n+m}{n} \phi^n \psi^m = \frac{1}{2\pi} \int_0^{2\pi} e^{-im\theta} (\phi + e^{i\theta}\psi)^{n+m}\, d\theta$$

and hence deduce that

$$\|\phi^n \psi^m\|_N \le \frac{n!m!}{(n+m)!} \sup_{\theta \in R} \|\phi + e^{i\theta}\psi\|^{n+m}.$$

(c) By first considering the case $a^n b^m = 1$, $a, b > 0$ show, using Lagrange multipliers, that $(a+b)^{n+m} \le a^n b^m (n+m)^{n+m}/m^m n^n$.

(d) By (b) and (c) deduce

$$\|\phi^n \psi^m\|_N \le \frac{n!m!}{(n+m)!} (\|\phi\| + \|\psi\|)^{n+m}$$

$$\le \frac{n!m!}{(n+m)!} \frac{(n+m)^{n+m}}{m^m n^n} \|\phi\|^n \|\psi\|^m.$$

A weaker estimate for nuclear polynomials on a Hilbert space is given in [279, Lemma 15] (see also [688]). The inequality $\|t_n \otimes t_m\| \leq \|t_n\| \cdot \|t_m\|$ for $t_i \in \widehat{\bigotimes}_{n,\pi} E$, $i = n, m$, where E is a locally convex space, is used to endow the space $l_1\left(\{\widehat{\bigotimes}_{n,\pi} E\}\right)$ with the structure of a (tensor) Banach algebra in A. Colojoară [240]. This space has the following universal property:

If A is a Banach algebra with identity and $L: E \to A$ is a linear contraction satisfying $L(x)L(y) = L(y)L(x)$ then there exists a unique Banach algebra homomorphism $\widetilde{L}$ such that the following diagram commutes:

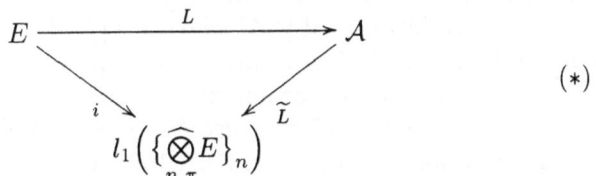

$$(*)$$

where i is the natural identification of E with the diagonal in $\widehat{\bigotimes}_{i,\pi} E$

and, moreover, $\widetilde{L}$ is also a contraction.

The *symmetric* algebra of a locally convex space is a complete *commutative* locally m-convex algebra $S(E)$ with unit together with a continuous injection $i: E \to S(E)$ with the following universal property:

for any continuous linear mapping ϕ of E into a complete locally m-convex algebra A there exists a unique continuous algebra homomorphism $\Phi: S(E) \to A$ with $\phi = \Phi \circ i$.

The symmetric algebra exists and is unique within the category of locally m-convex algebras. It is natural to look for a concrete realization of $S(E)$ by replacing arbitrary tensors in $(*)$ by symmetric tensors. The author in [240] follows this approach but incorrectly assumes $\|PQ\|_N \leq \|P\|_N \|Q\|_N$ for nuclear polynomials. A successful study of this topic is carried out in [241] for Fréchet nuclear spaces. The topic is further developed by M. Börgens–R. Meise–D. Vogt [170, Theorem 2.7] (see also [299, Theorem 6.55]) who prove the following result (see Sections 3.3 and 4.3 for notation):

if $\Lambda(P)$ is a reflexive nuclear space with absolute basis then

$$S(\Lambda(P)) \cong \left(H_{HY}(\Lambda(P)'_\beta), \tau_0\right)$$

algebraically and topologically (this implies in particular that $S(s) \cong s$).

2.64 This exercise deals with homogeneous polynomials which preserve or improve certain kinds of summability. Useful general results of this kind are

(1.18) and (1.55). The result of this exercise, which appears in [589, Proposition 3.1], is an integral domination theorem, of the Grothendieck–Pietsch type, for n-homogeneous polynomials which map weakly r-summable series onto (r/n)-summable series. Polynomials which map weakly r-summing series onto s absolutely summing series are called (r,s)-summing polynomials. In [172], G. Botelho proves that all $\mathbb{C}$-valued 2-homogeneous polynomials on $\mathcal{C}(K), K$ a compact Hausdorff space, are $(1,2)$ summing. Further results of this kind, and in some cases holomorphic analogues, may be found in [12], [173], [174], [228], [354], [386], [496], [587], [588], [590], [612].

2.65 The equivalence of (a) and (b) for $n = 2$ and scalar range is due to H.R. Pitt [732]. The condition $\sum_{i=1}^{m} p_i^{-1} < 1$ occurs in the work of G.H. Hardy–J.E. Littlewood [437] who call $\prod_{i=1}^{m} l_{p_i}$ a space "of type β" and also in [228] and [859]. The result for $p_i = p$ all i is due to A. Pełczyński [713] while the multilinear version given here is due to R. Alencar–K. Floret [10] and T.W. Gamelin [377] (see also A. Arias–J. D. Farmer [34], R.M. Aron et al. [64], R. Bonic–J. Frampton [167, 168], V. Dimant–I. Zalduendo [275], J.R. Holub [462] and N.J. Kalton [502]).

2.66 This result is due to P.J. Boland–S. Dineen [155] and can be used to prove Exercise 2.72. This general method of constructing polynomials and holomorphic functions on direct sums and countable inductive limits has been used in other situations (see [299, Exercises 3.85 and 3.86], [303], [307], [322]).

2.67 This result is due to E. Toma [827]. Further proofs are given in [57], [65], [203], [204] and [633]. Polynomials satisfying the condition of the exercise are called K-bounded in [204]. The approximation of K-bounded polynomials by polynomials of finite type and extensions to the bidual are discussed in [204] and [633].

2.68 This result is due to M. Schottenloher [784, Example 1.5]. The author also shows in [784] that E'_σ is not a k-space if $E'_c \neq E'_\sigma$.

2.69 For $\mathcal{DFM}$ spaces this appears in [295] (and for $\mathcal{DFS}$ spaces in [697]). These are special cases of the result in the previous exercise and also of the result that open and closed subsets of $\mathcal{DFC}$ spaces are k-spaces [645]. See [90] for results on $\mathbb{C}^I$.

2.71 See [309, Lemma 5]. A related result is given in [397, Theorem 14] and a further proof appears in [191].

2.72 The result is due to P.J. Boland–S. Dineen [154]. See also Example 3.24(b) and the remarks on Exercises 1.66, 2.66, 3.123 and 4.75. The topologies problem for $H(\mathcal{D})$ and $H(\mathcal{D}')$ is fully solved in [154] and [155] and we have $\tau_0 = \tau_\omega = \tau_\delta$ on $H(\mathcal{D})$ while $\tau_0 \neq \tau_\omega \neq \tau_\delta = \tau_0^{bor} = \tau_\omega^{bor}$ on $H(\mathcal{D}')$. P. Berner [107] shows that $\big(H(\mathcal{D}'),\tau_\delta\big)$ is a strict LF-Montel space and in [169] and [170, Corollary 6.4(a)] M. Börgens–R. Meise–D. Vogt show that $\big(H(\mathcal{D}'),\tau_\delta\big) \simeq \mathcal{D}$.

2.73 This result is due to J.M.F. Castillo [211].

2.74 For $n = 1$ this is due to R.S. Phillips [723] and for arbitrary n to R.M. Aron [42]. The proof uses induction and a variant of Proposition 1.1. A vector-valued version is given in [633, Proposition 8].

2.75 For reflexive spaces this result is due to R.A. Ryan [758, Proposition 5.3]. The approximation property was inadvertently omitted as a hypothesis in [758]. The result is also given in M. González–J. Gutiérrez [401, Lemma 1] and used to show the following: if all weakly null sequences in E' are norm null then all weakly null sequences in $\mathcal{P}(^nE)$ are norm null. (i.e. E Schur implies $\widehat{\bigotimes}_{n,s,\pi} E$ Schur). P. Galindo [365, Corollary 2(ii)] shows that the approximation property is not required if the sequence $(x_j)_j$ is limited (see Section 3.6). In [401, Lemma 1], M. González–J.M. Gutiérrez prove that $\{x\otimes\cdots\otimes x : \|x\| \leq 1\}$ is norm closed in $\widehat{\bigotimes}_{n,s,\pi} E$ when E is a Banach space with the approximation property, while in [761, Proposition 11], R.A. Ryan–B. Turett show that the approximation property is not necessary when $n = 2$. See also Exercise 2.51.

2.76 This result is due to J.A. Jaramillo–A. Prieto–I. Zalduendo [479].

2.77 $\mathcal{C}\big(\beta(\mathbb{N}^n)\big) \cong \mathcal{C}_b(\mathbb{N}^n)$. If $f \in \mathcal{C}_b(\mathbb{N}^n)$ and $\sigma \in P_n$ let $f_\sigma(x_1,\ldots,x_n) = f(x_{\sigma(1)},\ldots,x_{\sigma(n)})$. A measure ν on $\beta(\mathbb{N}^n)$ is symmetric if $\nu(f) = \nu(f_\sigma)$ for all $f \in \mathcal{C}_b(\mathbb{N}^n)$ and all $\sigma \in P_n$. This result is due to R.M. Aron–B.J. Cole–T.W. Gamelin [51, Lemma 7.1].

2.78 This result is due to T.K. Carne–B.J. Cole–T.W. Gamelin [207]. See also [479, Corollary 2].

2.79 It is not clear what role the approximation property will eventually play in infinite dimensional holomorphy. In pursuit of positive results we have found it convenient, whenever necessary, to assume such a hypothesis (and the same remarks apply to the existence of a Schauder basis and finite dimensional decomposition). Clarification will appear someday. It is not known (see [661]) if the approximation property is necessary for either of the results in this exercise and also for Proposition 2.31. It is known that the compact approximation property is strictly stronger than the approximation property for reflexive Banach spaces ([853]). The approximation property when it appears naturally is much more likely – see Proposition 2.8 – to be necessary and sufficient in vector-valued situations while the effort required to remove it in scalar-valued cases may be highly non-trivial. We refer to J. Mujica [651, Theorem 28.1] and J. Mujica–M. Valdivia [661, Proposition 2.2] for the results of this exercise. See also Exercise 3.124.

2.80 F. Blasco develops the theory of polynomials over X-Köthe spaces in [131, 132]. In the process, he proved the result in this exercise. If $A = (a_m)_m = ((a_{m,n})_n)_m$ is a Köthe matrix and X is a Banach space with unconditional basis $(e_n)_n$ then

$$\lambda_X(A) := \left\{ (x_n)_n : \sum_n a_{m,n} x_n e_n \in X \text{ all } m \right\}$$

is endowed with a Fréchet space structure by the sequence of semi-norms $\|(x_n)_n\|_m := \|\sum_n a_{m,n} x_n e_n\|$, $m = 1, 2, \ldots$. Reflexive and Montel $\lambda_X(A)$ spaces are discussed by Blasco who shows that for certain A the space $\lambda_{T^*}(A)$ does not satisfy the density condition. Holomorphic functions on $\lambda^p(A)$ appear in Corollary 5.8.

Chapter 3

3.55 In one direction use the fact that $\mathbb{C}^{(\mathbb{N})} = \varinjlim_n \mathbb{C}^n$ in the category of topological spaces and continuous mappings. For the converse use Exercise 1.61 or the following ([287, Proposition 2.3]):

if $(e_\alpha)_\alpha \in \Gamma$ is an algebraic basis for E then

$$f\left(\sum_\alpha z_\alpha e_\alpha \right) := \sum_{n=1}^\infty n^{2n} \left\{ \sum_{\substack{l_1,\ldots,l_n \in \Gamma \\ l_j \neq l_k \text{ for } j \neq k}} z_{l_1} \cdots z_{l_n} \right\}$$

is a G-holomorphic function which is locally bounded if and only if Γ is countable.

3.56 The composition of Gâteaux holomorphic mappings is not always Gâteaux holomorphic, see for instance M. Hervé [446, Proposition 2.3.10].

3.57 See J.M. Isidro–J.P. Méndez [476].

3.58 "Generalized" in the sense that the "basis" may have arbitrary cardinality.

3.59 This result is due to M. Estéves–C. Hervés [342, 343]. They show more generally that one only need assume that f is universally measurable. See also Ph. Noverraz [708].

3.60 This result is due to L. Nachbin [673]. It shows that conditions on the range of a $\mathcal{G}$-holomorphic function can provide information about its continuity. A similar type of weak implies strong holomorphic result is due to N. Dunford [336, p.354] who only requires weak holomorphy with respect to a determining manifold. A.E. Taylor, [819] and [821], proves a weaker result, along the same lines, on the analytic dependence of an operator-valued function on a parameter (background information is discussed in [824, p.334]). See also Example 3.8(g) and [141, 434, 435, 510, 553, 675, 676, 729, 850]. A different type of examination of the range (how to densely approximate a given set by the range of a holomorphic function) is discussed in the remarks on Exercise 5.86.

3.62 This result is due to E. Bishop [123]. The proof uses a monomial expansion and an Auerbach basis and the result is applied to obtain vector-valued versions of Cartan's theorems A and B.

3.64 A.E. Taylor [818, pp. 474–475] was the first to note that entire functions on certain Banach spaces may have finite radii of uniform convergence. Subsequently, L. Nachbin [666] gave the definition of radius of boundedness and proved the result in this exercise. The fact (Example 3.8(f)) that $H_b(E) \neq H(E)$ or that bounding sets are nowhere dense in E for any infinite dimensional Banach space E are both equivalent to the result that every infinite dimensional Banach space admits a holomorphic function f such that $r_f \not\equiv +\infty$ (see Exercise 5.56).

3.65 This result is due to R.M. Aron–J. Cima [50]. See A.E. Taylor [821, Theorem 3], [818], [584] and [676, Proposition 16] for related results.

3.66 This is the holomorphic analogue of Exercise 2.66 and is given in [154]. The construction bears comparison with that given in Example 3.8(f). See also Exercise 3.86 and [307, §4].

3.68 See [162, Theorem 1.2]. Further infinite dimensional generalizations of Hartogs' Theorem can be found in L.M. Graves [408, p.648 and p.653], A.E. Taylor [818], M.A. Zorn [863], J. Sebastiaõ e Silva [796, 799], H. Alexander [13], E. Ligocka [551], J. Bochnak–J. Siciak [138], J.F. Colombeau [242], S. Dineen [290, 291, 292, 299], D. Lazet [534], M.C. Matos [573, 580, 581], Ph. Noverraz [690], S. Dineen–M. Lourenço [318] and D. Pisanelli [726, 727, 728, 729]. A. Hirschowitz [459] and M.C. Matos [581] give examples which show that we do not have a Hartogs' theorem for an arbitrary pair of locally convex spaces. See also Example 3.8(c) and Exercise 3.84.

3.69 (a) See M. Bianchini [112], (c) M. Lindström [557], the remarks on Exercise 3.78 and [163].

3.71 This exercise deals with local uniform convergence of the Taylor series expansion of *vector*-valued holomorphic functions. We refer to L. Nachbin–J.A. Barroso [683] for details.

3.73 The space $\mathbb{C}^{\mathbb{N}}$ has many different structures and consequently the function space $H(\mathbb{C}^{\mathbb{N}})$ may be studied from many different angles. One approach is to note (Example 3.10(b)) that each holomorphic function on $\mathbb{C}^{\mathbb{N}}$ factors through a finite dimensional space. This has been used by J.M. Ansemil [15], V. Aurich [74, 75], C.E. Rickart [743], J.A. Barroso [88], P. Berner [106, 107] and H. Hirschowitz [454]. An alternative approach, as we have seen in Section 3.3, is to view $\mathbb{C}^{\mathbb{N}}$ as a Fréchet nuclear space with absolute basis. Either approach can be used to obtain a solution to this exercise. A full discussion of this exercise, including a representation of analytic functionals on $\mathbb{C}^{\mathbb{N}}$ by functions of exponential type (see Proposition 4.49), is given in J.M. Ansemil [15]. P. Berner [107] also gives a general result which includes a solution to this exercise.

3.74 Use Exercise 1.72. If we let

$$\left(c_0(\Gamma), \tau\right) = \varprojlim_{\Gamma_1 \subset \Gamma \text{ countable}} \left(c_0(\Gamma_1), \|\cdot\|\right)$$

then this shows that $H\left(c_0(\Gamma), \|\cdot\|\right) = H\left(c_0(\Gamma), \tau\right)$ – a result of B. Josefson [487] (see also [92, 221, 292, 706]).

3.76 If X is completely regular Hausdorff space then

$$\left(\mathcal{C}(X), \tau_0\right) = \varprojlim_{K \subset X, \text{ compact}} \left(\mathcal{C}(K), \|\cdot\|_K\right)$$

is an open and compact surjective limit (see Section 3.6) of Banach spaces by the open mapping theorem and the Tietze Extension Theorem. The result of this exercise may be found in [290, 292]. Holomorphic germs on compact subsets of $\mathcal{C}(X)$ are discussed by R.L. Soraggi [811] (see also [299, Proposition 6.29]).

3.77 The equivalence of (a) and (b) is the well known Nachbin [663]–Shirota [800] Theorem. Condition (b) involves a concept of bounding set for continuous functions. Condition (c) can be rephrased, using the terminology of [91, 92], as "$\mathcal{C}(X)$ is holomorphically barrelled" – see Section 3.6 and Example 3.28(a). The equivalence of (a) or (b) with (c) is due to J. Bonet–P. Galindo–D. García–M. Maestre [162, Corollary 2.2]. A similar result for $\left(\mathcal{C}(X), \tau_0\right)$ holomorphically infrabarrelled is given in S. Dineen [292].

3.78 A fairly comprehensive account of the topological theory of holomorphic functions on DF and gDF spaces may be found in [369] (see also [26, 27,

371, 472, 474, 557 and 758]). The equality $H(E) = H_{ub}(E)$ has been studied
for $\mathcal{DFM}$ spaces by J.F. Colombeau–J. Mujica [249], for $\mathcal{DFC}$ spaces by
P. Galindo–D. García–M. Maestre [370] and L.M. Hai–N.V. Khue [433], for
Fréchet nuclear spaces by R. Meise–D. Vogt [608], N.M. Ha–N.V. Khue [428],
L.M. Hai [432] and L.M. Hai–T.T. Quang [435] and for quasi-normable spaces
by M. Lindström [557]. We discuss the Fréchet nuclear space case in Section
6.2. Non-nuclear results include: [557], if E is a quasi-normable Fréchet space
then $H(E) = H_{ub}(E) \Rightarrow E$ is Schwartz and, [369], if E is a complete gDF-
space then $H(E) = H_b(E) \Longleftrightarrow E$ is a $\mathcal{DFC}$ space (and in this case $H(E) =$
$H_{ub}(E)$). See Exercises 3.69, 3.80 and 6.45.

3.79 The method used for n-homogeneous polynomials on bornological DF
spaces (Example 1.25) or for holomorphic functions on $\mathcal{DFM}$ spaces (Example 3.8 (e)) can be extended to obtain the first part which is due to P. Galindo–
D. García–M. Maestre [369, Proposition 15] – see also [26, Proposition 3; 107;
128; 307, Example 4.10(c); 475; 758].

3.80 The first part is due to R. Meise–D. Vogt [610] and the second to
P. Galindo–D. García–M. Maestre [371, Theorem 2.2]. The first part and the
fact that every quojection is a quotient of a product of Banach spaces ([165])
reduces the final part to the case of a product of Banach spaces. See also
L.M. Hai [432] and L.M. Hai–T.T. Quang [435].

3.81 This is a result of L. Nachbin [679].

3.83 This result is due to R. Pomes [736]. See also [286, footnote page 42]
and [707, Proposition 7].

 The topology τ_e generated by the inductive limit of *all open covers* is
discussed by S. Bjon–M. Lindström [127]. They show that $\tau_\omega \leq \tau_e \leq \tau_\delta$ and
$\tau_e = \tau_\delta$ if U is Lindelöf.

3.84 See S. Dineen [286] and the remarks on Exercise 3.68.

3.85 H. Alexander [13] introduced these topologies while investigating the
envelope of holomorphy of a domain spread over a Banach space (see Sections
5.2 and 5.4) and proved the result given here for τ_n, $n = 0, 1, \ldots, \infty$. Alexan-
der also showed $\mathcal{S}\big(H(U), \tau_0\big) = \mathcal{S}\big(H(U), \tau_n\big) = \mathcal{S}\big(H(U), \tau_\infty\big)$ for any positive
integer n. L. Nachbin [665, 666] obtained the same result, independently, and
at this time defined the topology τ on $H(U)$ generated by the semi-norms
ported by the compact sets. Nachbin showed, for infinite dimensional Banach
spaces, that $\tau_0 \underset{\neq}{<} \ldots \underset{\neq}{<} \tau_n \underset{\neq}{<} \ldots \underset{\neq}{<} \tau_\infty \underset{\neq}{<} \tau$ and, in view of the well ordering of
this sequence of topologies and the traditional use of ω to denote infinite
ordinal numbers, introduced the notation τ_ω for the *ported* or Nachbin topology. The word *"ported"* is a corruption of supported (see Section 3.2, [669,
footnote (1)] and [671, p.20]).

3.86 This result is due to S. Dineen [307, Proposition 4.8].

3.87 The result is due to M.A. Simões [803] and generalizes [111, Lemma 2 and Theorem 2], [501, Proposition 2.3] as well as part of [292, Proposition 7.5(a)]. M.A. Simões also shows in [803] that E'_β contains $\mathbb{C}^{(\mathbb{N})}$ as a complemented subspace if E (not necessarily sequentially complete) contains a non-trivial very strongly convergent sequence – this extends the remainder of Proposition 7.5(a) and also Proposition 7.4(c) in [292]. In [158], J. Bonet shows that if E is a non-normable Fréchet space, F is a locally complete space and $L(E; F) = LB(E; F)$ then F does not admit non-trivial very strongly convergent sequences. Very strongly convergent sequences also appear in R.L. Soraggi's investigation [807, 809, 811] of regularity and completeness for spaces of holomorphic germs (see Exercise 5.77) and in L.A. Moraes' investigation of holomorphic functions on strict inductive limits of reflexive Fréchet spaces [626].

3.88 See [803] and Exercise 6.61.

3.89 This result is due to A. Hirschowitz [457, Lemma 4].

3.91 See [162, Proposition 4.5].

3.92 See J. Mujica [647, 656], J. Mujica–L. Nachbin [660], K.-D. Bierstedt–J. Bonet [117], [299, p. 417] and Section 3.6.

3.93 See [299, Proposition 3.10].

3.94 This result is due to C. Boyd [179]. Using this characterization and a result of P. Avilés–J. Mujica [80], Boyd also showed, for balanced domains in Fréchet spaces, that the space $\big(H(U), \tau_\delta\big)$ satisfies the strict Mackey convergence criterion if and only if E is quasi-normable. The balanced hypothesis was removed in D. García–J. Mujica [380] when E is separable. Related results concerning the predual $G(U)$ are given in the above references and [177, 178, 372, 373, 597, 653, 654, 655, 660, 758].

3.95 Apply Corollary 3.37(a) for a direct proof. This result is proved in Ph. Noverraz [705, 707] (countably barrelled spaces are called d-barrelled spaces in [705] and $\aleph_0$-barrelled spaces in [720]). See P. Berner [105] for related results.

3.98 The equivalence of the first three conditions is due to Ph. Noverraz [707, Proposition 6] while the remaining equivalences are given in [286] and [299, Proposition 3.30].

3.99 See Ph. Noverraz [707, 705].

3.100 Determining sets are used in Example 3.24(b). See J. Ansemil–S. Dineen [23], P. Boland–S. Dineen [154], J. Chmielowski [218], J. Chmielowski–G. Łubczonok [219], S. Dineen [300], P. Galindo–D. García–M. Maestre [367], L.A. Moraes [626, 627], L. Waelbroeck [850] and Exercise 4.81.

3.101 Use transfinite induction as in the proof of Proposition 3.34(b). See [299, Proposition 3.5].

3.102 See [602].

3.103 See [307, Theorem 1.8].

3.105 See the final paragraph in Section 3.6 and [307].

3.106 See B.L. Sanders [764] and T.A. Cook [253]. A similar result holds for boundedly complete, shrinking Schauder decompositions (N.J. Kalton [500, Theorem 3.2]). We use this exercise in Corollary 4.19. Note that S-absolute decompositions are both shrinking and boundedly complete.

3.107 See [314, 315, 660]. Example 3.28(a) states that a locally convex space E is holomorphically barrelled if and only if $G_0(U)$ is barrelled. An analogous result is proved for holomorphically Mackey spaces in [660] but, the results of this exercise applied to $\mathbb{C}^{\mathbb{N}} \times \mathbb{C}^{(\mathbb{N})}$ produce examples showing that similar results do not hold for holomorphically bornological and infrabarrelled spaces ([314, 315]). The role of locally bounded sets of holomorphic mappings in infinite dimensional holomorphy is discussed in J.A. Barroso–M.C. Matos–L. Nachbin [92], J. Bonet–P. Galindo–D. García–M. Maestre [162] and C. Boyd–S. Dineen [187].

3.108 This result is due to P. Rueda [752, 753] and shows that we do not have a Banach–Dieudonné Theorem for holomorphic functions of bounded type on a Banach space. A positive result for polynomials is given in J. Mujica [643] (see Exercise 1.67).

3.109 Even though $\mathbb{C}^I$ does not have a Schauder basis the monomials do give a topological decomposition of $\mathcal{P}(^n\mathbb{C}^I)$ and the methods used for Fréchet nuclear spaces with basis can be applied with obvious modifications. See [90] and Exercises 1.62 and 3.58.

3.110 By Exercise 1.96, F is separable and $H(U)$ is a Fréchet space. If $\mathcal{F} \subset H(U)$ is τ_0-bounded and $(f_n)_n \subset \mathcal{F}$ then for each n and each x in F there exists a neighbourhood of 0, V_n, such that $\|f_n - f_n(x)\|_{x+V_n} \le 1$. By [645, Corollary 7.9], there exists $\lambda_n > 0$ such that $\cap \lambda_n V_n$ is a neighbourhood of zero. Hence $(f_n)_n$ is locally bounded and, by Exercise 3.91, $\mathcal{F}$ is locally bounded.

3.112 This example is due to J.M. Ansemil–S. Ponte [26, Proposition 5]. In fact τ_0 is an ultrabornological topology and $\tau_0 \underset{\neq}{\leq} \tau_\omega$.

3.114 See [303], [162, Example 3.7] and Example 3.8(g). This example shows that $\mathbb{C}^{(A)}$, A uncountable, is not holomorphically Mackey. In [94], J.A. Barroso–L. Nachbin prove that $\mathbb{C}^{(A)}$, with the topology induced by $\mathbb{C}^A$, is holomorphically bornological.

3.115 If $(e_n)_n$ is an absolute basis for E and $(\delta_n)_n$ is the sequence defining A-nuclearity show that

$$\left\{ \sum_{n=1}^{\infty} \delta_n a_n e_n : \text{ there exists } \sum_{n=1}^{\infty} b_n e_n \in B \text{ such that } |a_n| \leq |b_n| \text{ all } n \right\}$$

is bounded in the completion of E whenever B is bounded in E. This can be used to show that (E, τ^{bor}) is an A-nuclear space and that $(e_n)_n$ is again an absolute basis for (E, τ^{bor}). The same ideas work for $(E, \tau)'_\beta$ ([301]). Use the same sequence $(\delta_n)_n$ to show A-nuclearity.

A locally convex space (E, τ) is a Mackey space $\iff \tau_1 \geq \tau$, τ_1 locally convex, and $(E, \tau)' = (E, \tau_1)'$ imply $\tau = \tau_1 \iff \tau$ coincides with the topology τ_m of uniform convergence on the weakly compact convex subsets of $(E, \tau)'$. For arbitrary τ, $\tau \leq \tau_m \leq \tau^{bor}$. From the first part $(e_n)_n$ is an absolute basis for (E, τ_m). It suffices to show that the modular hull of a weakly compact convex subset of $(E, \tau)'$ is weakly compact and convex.

The above result can be extended to $(H(U), \tau_0)$ and $(H(U), \tau_\omega)$, U an open polydisc in an A-nuclear space, using germs of holomorphic functions (note that we do not know if these spaces are always A-nuclear). We refer to [154] for details.

3.116 First show that for any sequence $(u^m)_{m=1}^\infty$, $u^m := (u_n^m)_{n=1}^\infty$, in l_1^+ there exists a sequence $(u_n)_{n=1}^\infty$ in l_1^+ such that $(u_n^m / u_n)_{n=1}^\infty \in c_0$ for all m. For further details, see P.J. Boland–S. Dineen [154].

3.117 This result is due to J. Mujica–M. Valdivia [661]. The proof proceeds by first showing that the inductive limit $H(K) = \varinjlim_n H^\infty(U_n)$ is weakly compact and hence totally reflexive for K compact in E. The hypotheses are satisfied by Tsirelson's space, T^*, considered in Section 2.4. Previously C. Boyd [176, 177] had shown that $H(K)$ was reflexive. The article [661] also discusses the current role of the approximation property and the compact approximation property as hypotheses in various results on polynomials which are weakly continuous on bounded sets.

3.118 See S. Dineen [301] and [299, Corollary 5.26]. If p is a τ_δ-continuous semi-norm on $H(U)$ then

$$p'\Big(\sum_{m\in\mathbb{N}^{(\mathbb{N})}} a_m z^m\Big) := \sup_{\substack{J\subset\mathbb{N}^{(\mathbb{N})}\\ J\text{ finite}}} p\Big(\sum_{m\in J} a_m z^m\Big)$$

is also τ_δ-continuous.

3.119 See S. Dineen [301].

3.120 This example is due to R. Meise–D. Vogt [610]. The proof is an extension of the method outlined in Example 3.12. See Section 6.2.

3.122 This result is due to M. Börgens–R. Meise–D. Vogt [170] (see also [169] and [299, §6.4]). In [170] the authors show that $H\big(\Lambda_\infty(\alpha)'_\beta\big) \cong \Lambda_\infty(\delta)$ where $\delta = (\alpha|m))_{m\in\mathbb{N}^{(\mathbb{N})}}$ and using the functions

$$N(t) := \big|\{m \in \mathbb{N}^{(\mathbb{N})} : (\alpha|m) \le t\}\big|$$

and

$$F(t) = \sup_n \frac{t^n}{\alpha_1 \cdots \alpha_n n!}$$

prove that the sequence δ is equivalent to $\big(F^{-1}(n)\big)_{n=1}^\infty$ (in that both define the same power series space of infinite type) if either (a) $(\alpha_n n^{-p})_n$ is increasing for some $p > 0$ or (b) $1 < \inf_n(\alpha_{kn}/\alpha_n) \le \sup_n(\alpha_{kn}/\alpha_n) < \infty$ for some positive integer k. If (b) is satisfied then $\big(H(\Lambda_\infty(\alpha)'_\beta), \tau_0\big) \cong \Lambda_\infty(\bar\delta)$ where $\bar\delta_n = [\log(n+1)]\alpha_{[\log(n+1)]}$ and $[\cdot]$ denotes integer part. As a particular example they show

$$\Big(H\big(H(\mathbb{C}^m)'_\beta\big), \tau_0\Big) \cong \Lambda_\infty\Big(\log(n+1)^{1+\frac{1}{m}}\Big).$$

3.123 This result is due to M. Börgens–R. Meise–D. Vogt [170] and can be used to show that $\big(H(\mathcal{D}'), \tau_\delta\big) \cong \mathcal{D}$. See the remarks on Exercise 2.72.

3.124 A comprehensive study – which includes the results of this exercise – of the approximation property for spaces of holomorphic functions on Banach spaces is undertaken in R.M. Aron–M. Schottenloher [70, 71]. The result in (c) follows from (a) and (b) and from the fact that there exists a Banach space E with the approximation property such that E'_β fails to have the approximation property. An example is given in [71] of a reflexive Banach space with basis E such that $\mathcal{P}(^2E)$ does not have the approximation property – if H is a separable Hilbert space, then $\mathcal{B}(H)$, the bounded linear operators from H into itself, does not have the approximation property. Since H is stable, it follows from Proposition 1.34 that $\mathcal{P}(^2H)$ does not have the approximation property. See J. Mujica [653] for similar results on the metric approximation property, K.-D. Bierstedt–R. Meise [121] for the Fréchet–Schwartz space case and O.T.W. Paques [711] for results on Silva holomorphic functions.

3.125 By Example 3.8(f), $(\phi_n)_n$ is a weak* null sequence in E'. The results in this exercise are due to C.O. Kiselman [515].

3.126 P. Lelong [535] proved that r_f is a Lip 1 function and that $-\log r_f$ is plurisubharmonic (Definition 5.14) and these results motivated C.O. Kiselman ([515, 516, 517]) to undertake a detailed investigation of the radius of uniform convergence. The remaining results in this exercise may be found in [517]. C.O. Kiselman posed the problem of prescribing the radius of uniform convergence and the following results have been obtained:

> If E is an infinite dimensional separable Banach space and $\phi : E \to R^+$ satisfies
> (i) $|\phi(x) - \phi(y)| \le \|x - y\|$ for all $x, y \in E$,
> (ii) $\log \phi$ is plurisubharmonic
>   then
>
>   (a) if $E = l_1$ there exist $f \in H(E)$ such that $r_f = \phi$,
>   (b) if E has a Schauder basis then there exists $f \in H(E)$ such that $\phi/3 \le r_f \le \phi$.

Result (a), when ϕ only depends on a finite number of coordinates, is due to C.O. Kiselman and the general case is due to G. Coeuré [238]. This result does not extend to arbitrary l_p spaces. Result (b) for l_p and c_0, when ϕ depends on a finite number of variables, is also due to C.O. Kiselman and the general result is due again to G. Coeuré [238]. It is unknown if (b) holds for arbitrary separable Banach spaces. More accessible proofs of these results are given in M. Schottenloher [788]. N. Cherfaoui [217] studies the radius of boundedness of plurisubharmonic functions on Orlicz spaces and M. Schottenloher [789] has written a very readable survey article on bounding sets and the radius of boundedness.

Chapter 4

4.60 Use the Grothendieck–Pietsch criterion for nuclearity and Proposition 3.47 (see [273]).

4.61 This example is given in V. Dimant–S. Dineen [273]. The proof uses Proposition 3.37(c). Both Exercises 4.60 and 4.61 suggest that Fréchet subspaces of $(H(E), \tau_\omega)$, E a Banach space, are sums of Banach spaces and Fréchet nuclear spaces. It might be interesting to follow this line of investigation. See also the notes on the final part of Section 4.2.

4.62 See the remarks in Section 3.6 on bounding sets and M. Schottenloher [782, 789].

4.63 This may be regarded as a τ_ω version of Corollary 4.43. For details see [153] and [299, Proposition 5.37].

4.64 See V. Dimant–I. Zalduendo [275] for this and related results.

4.65 See A. Hirschowitz [458].

4.66 This example is taken from [299, Example 3.47] and is related to an example of R.M. Aron given in R.L. Soraggi [804]. The technique of finding n such that $f\big|_{\mathbb{C}^n} \equiv 0 \Longrightarrow p(f) = 0$ has been developed and found quite useful in dealing with holomorphic functions on strict inductive limits – see for instance [154], [322] and [307, Proposition 4.11]. The fact that the inductive limit is not regular also follows from Corollary 4.44 and is equivalent to the result $\tau_0 \neq \tau_\delta$ on $H(\mathbb{C}^{\mathbb{N}})$ (Example 3.24(a)).

One can use this result and $\mathcal{S}$–absolute decompositions to show that the τ_ω topology on $H(0_{\mathbb{C}^{(\mathbb{N})}})$ is generated by the semi-norms

$$p_{K,(\alpha_n)_n}\Big(\sum_{n=0}^{\infty} \frac{\hat{d}^n f(0)}{n!}\Big) := \sum_{n=0}^{\infty} |\alpha_n| \Big\| \frac{\hat{d}^n f(0)}{n!} \Big\|_K$$

where K ranges over the compact subsets of $\mathbb{C}^{(\mathbb{N})}$ and $(\alpha_n)_n$ over all sequences of scalars such that $|\alpha_n|^{1/n} \to 0$. See [299, Example 3.47].

4.67 The result for Banach spaces is due to R.M. Aron–L. Moraes–R.A. Ryan [67] and for Fréchet–Schwartz spaces to S. Ponte [737]. The τ_ω result cannot be extended to Fréchet–Montel spaces which are not Fréchet–Schwartz [737, Theorem 3] and [19, Theorem 2.6]. Analogous problems are discussed for holomorphic germs on compact sets in [19, 29, 67, 603], for holomorphic mappings of bounded type in [18] and [262], and for bounded holomorphic functions in [655]. For arbitrary (i.e. non-balanced) open sets the problem becomes more interesting and leads to the construction of several complicated examples and a number of open problems [18]. The associated linear problem, which deals with the classical isomorphism $(E/F)' \simeq F^\perp$, appears in Lemma 4.53. Liftings may be considered dual to extensions and R.M. Aron–L.A. Moraes–O.W. Paques [66] have constructed a universal lifting space analogous to that constructed by I. Zalduendo [856] for extensions (see Example 6.20).

As noted in the final remarks of [759] the result shows that here exists for any Banach space E an indexing set I such that $(H(E), \tau_\omega)$ is isomorphic to a closed subspace of $(H(l_1(I)), \tau_\omega)$. Hence $(H(l_1(I)), \tau_\omega)$ may be regarded as a universal space for the τ_ω topology on Banach spaces (see also the final remarks of [67]). Since the identification carries through at the homogeneous polynomial level we have a simple proof of Corollary 1.56: if l_p is a quotient of E then $\mathcal{P}(^n E)$ is a subspace of $\mathcal{P}(^n l_p)$ and l_∞ is contained in $\mathcal{P}(^n l_p)$ whenever $n \geq p$.

4.68 The results in this exercise are due to A. Benndorf [100, 101, 102]. If E is a Fréchet spaces with the bounded approximation property then, by [102], E is isomorphic to a complemented subspace of F where F has a finite dimensional decomposition and, if E is Schwartz, F is isomorphic to a subspace of $E \widehat{\otimes}_\pi s$. We have the following composition of mappings:

$$E \xrightarrow{\;\;i\;\;} F \xrightarrow{\;\;j\;\;} E \widehat{\otimes}_\pi s \qquad (*)$$

where i and j are isomorphisms onto their ranges and $i(E)$ is complemented in F. The space s may be replaced by $\lambda^1(A)$ where A is a Köthe matrix such that $a_{j,1} \geq 1$ for all j (in particular $\lambda^1(A)$ admits a continuous norm) and if E admits a continuous norm F can be chosen so that $i(E) = F$. This result refines the following proposition of A. Pełczyński [716] and A. Pełczyński–P. Wojtaszczyk [719]: *every separable Fréchet space with the bounded approximation property (and continuous norm) is isomorphic to a complemented subspace of a Fréchet space with a finite dimensional decomposition (and continuous norm)*. In particular, using $(*)$, we see that properties, preserved by subspaces and stable under $\widehat{\otimes}_\pi s$, such as nuclearity, the (DN)-property (use Theorem 2.5 of [842]) and the existence of a continuous norm transfer from E to F. One needs to be careful with these type of results since every Fréchet nuclear space with the bounded approximation property is isomorphic *both* to a complemented subspace of a Fréchet space with basis *and* to a complemented subspace of a Fréchet nuclear space with finite dimensional decomposition but may not, simultaneously, be isomorphic to a complemented subspace of a Fréchet nuclear space with basis (see [334]). The characterization given in the exercise is used by J.C. Díaz [264] to study non-primary Fréchet–Schwartz spaces. An example of a Banach space with the bounded approximation property and without a basis is given in S.J. Szarek [815].

4.69 We already noted, when introducing the ε-tensor product (Section 2.1), that $E \otimes_\varepsilon F = E \otimes_\pi F$ if E is nuclear and F is any locally convex space (see Section 4.6 and [261, 265, 326, 329, 484, 486]). The estimate (4.49) is a special case of the following result due to A. Pełczyński–C. Schütt [717];

if

$$b(p, q) = \begin{cases} \frac{1}{p} + \frac{1}{q} & \text{for } 1 \leq p, q \leq 2 \\ (\min(p, q))^{-1} & \text{for } \max(p, q) > 2 \end{cases}$$

then there exist positive constants c_1 and c_2 such that

$$c_1 n^{b(p,q)} \leq \left\| \sum_{k,l=1}^{n} e^{-2\pi i k l / n} e_k \otimes e_l \right\|_{l_p^n \otimes_\varepsilon l_q^n} \leq c_2 n^{b(p,q)}.$$

4.70 The concept of weak holomorphic convergence was introduced by Y. Petunin–V. Savkin [722] who proved that weak holomorphic and norm sequential convergence coincide in weakly compactly generated Banach spaces. In [49, Example 1.7] the authors show that this result does not extend to l^∞.

Comparison of these results with results (positive and negative) on bounding sets (see the notes on Chapter 3 and Exercise 4.62) led P. Galindo–L.A. Moraes–J. Mujica [375] to the result of this exercise which generalizes the results quoted from [722] and [49]. See also [365].

4.71 The results of this exercise are due to R.A. Ryan [758, Proposition 5.2]. The relationship between this result and Proposition 4.4 is similar to the relationship observed between Definition 2.7 and (2.4). See also Proposition 4.56.

4.72 See the remarks on Exercise 3.78 and [369].

4.73 If $\mathcal{P}_\omega({}^nE) \neq \mathcal{P}({}^nE)$ show that there exists $P \in \mathcal{P}({}^nE)$ and a disjointly supported sequence of vectors $(w_j)_j$ such that $|P(w_j)| \geq \varepsilon$ for all j and apply Proposition 2.41. See [273].

4.74 This result is due to S. Dineen [298; 299, Lemma 5.43 and Exercise 5.59] and has been extended in stages by K. Floret [353] and K. Floret–B.V. Moscatelli [355, 356] (see E. Dubinsky [335, p.4]) who eventually arrived at the following result:

if E is a Fréchet space with an unconditional basis then either E admits a continuous norm or E is isomorphic to a product of a sequence of non-zero Fréchet spaces each of which has a continuous norm and an unconditional basis.

This implies in particular that a twisted quojection (Example 3.10(b)) cannot have an unconditional basis (although by [614, 615] it may have a Schauder basis and even an unconditional finite dimensional decomposition consisting of two-dimensional spaces). Combined with the existence of twisted quojections this leads to examples of strict inductive limits of Fréchet nuclear spaces, $E = \varinjlim_{n} E_n$, where each E_n has a continuous norm but E does not. See Exercise 2.57 and [269]. If E is a quojection then $\widehat{\bigotimes}_{n,s,\pi} E$ is a prequojection and $(\mathcal{P}({}^nE), \tau_\omega)'_\beta$ is a quojection, see H. Hüser [470, Korollar 3.2.7]. Background information on quojections may be found in [616] and [638]. Holomorphic functions on direct sums and strict inductive limits are discussed in [94, 180, 187, 286, 298, 303, 322, 626, 627].

4.75 This result is due to S. Dineen [296, 298] while the special case $E \simeq s^{(\mathbb{N})} \simeq \mathcal{D}(\Omega)$, Ω open in $\mathbb{R}^n$, is due to P.J. Boland–S. Dineen [154, 155]. See also the remarks on Exercise 2.72. One may use a similar approach and Corollary 4.16 to prove the following which extends a result in [283]; if E is a countable direct sum of separable Banach spaces, each of which has the bounded approximation property, then $\tau_\omega = \tau_\delta$ on $H(U)$, U balanced in E.

4.76 See S. Dineen [298] and [299, Corollary 5.49].

4.77 This, the first infinite dimensional holomorphic Hahn–Banach theorem, is due to P.J. Boland [149] and is a special case of Example 6.22(a). Proofs based on Example 3.11 are due to J.F. Colombeau–J. Mujica [249]. Exercise 4.76 may also be used to prove this result ([298, Corollary 5.50]). The result is extended to $\mathcal{DFS}$ spaces and Fréchet-valued holomorphic functions in J.F. Colombeau–B. Perrot [252] and in R. Meise–D.Vogt [608], to DF-valued mappings in N.V. Khue [506, 507] (see also [252, 366, 369]) and to holomorphic functions of bounded nuclear type in [47]. Holomorphic extensions from closed subspaces are investigated in Sections 6.2 and 6.5.

4.78 Use $\widehat{\bigotimes\limits_{n,s,\pi}} s \cong s$ and the fact that E has (DN) if and only if E is isomorphic to a subspace of s. This result is due to E. Mangino [567, Lemma 3.29] and also follows from the following result of R. Meise–D. Vogt [606, Lemma 4.2(a)]; a Fréchet nuclear space E has (DN) if and only if $(H(E'_\beta), \tau_0)$ has (DN).

The results in [606] have been extended by N.M. Ha–L.M. Hai [427] who obtained the following results for a Fréchet space E:

if E has (DN) then $H_b(E')$ has (DN),

if E' has an absolute basis and E has (Ω) then $H_b(E')$ has (Ω),

if E is Montel with topology generated by semi-inner-products and E has (Ω) then $(H(E'), \tau_0)$ has (Ω).

Property $(\widetilde{\Omega})$ is defined in Example 6.10 and also appears in Proposition 6.23. Property (Ω), which characterizes within the collection of Fréchet nuclear spaces the quotients of s, is related to $(\widetilde{\Omega})$ and is an important Fréchet space invariant.

In a recent paper, [508], N.V. Khue–P.T. Danh extend these results to the space of holomorphic germs on compact subsets of certain Fréchet spaces.

4.79 See Example 6.12.

4.80 See Section 6.2 and [47].

4.81 This result is due to L. Waelbroeck [850] and involves an application of the closed graph theorem (see also W. Bogdanowicz [141], L.M. Hai–N.V. Khue–N.T. Nga [434], N.V. Khue–B.D. Tac [510], and E. Ligocka–J. Siciak [553]).

4.82 Use the result that $\mathcal{C}([0,1])$ has the polynomial Dunford–Pettis property ([757]) and the fact that the standard unit vector basis for l_2 is weakly null. See R.M. Aron [45] for further details.

4.83 Any continuous linear mapping from c_0 to l_1 is compact. See R.M. Aron [45].

4.84 These results are due to C. Boyd [177]. Tsirelson's space, T^*, considered in Sections 2.4 and 4.1, is an example of the required kind for the final part of the exercise since $\tau_0 \neq \tau_\omega$ for any infinite dimensional Banach space and $(P(^nT^*), \|\cdot\|)$ is reflexive for all n. It is rather surprising that τ_0 and τ_ω are compatible but do not coincide. A further interesting feature of this result is that we derive results about the τ_0 and τ_ω topologies on $H(U)$ and yet a crucial feature in the proof is the predual of $(H(U), \tau_\delta)$.

In [182], C. Boyd considered the vector-valued case and showed that τ_0 and τ_ω are compatible on $H(E; F)$ when E is Fréchet–Schwartz and F is a reflexive Banach space.

4.85 This result is due to R.A. Ryan [758]. See Proposition 2.20 for the corresponding polynomial result.

4.87 Apply Proposition 4.22 and the fact that $(\mathcal{P}(^nE), \tau_\omega)$ is a dual Banach space when E is a Banach space. See [273].

4.88 This result, the analogue of Proposition 4.56 for holomorphic functions of bounded type, is due to R.A. Ryan [759, Theorem 3.3] and is used, together with the vector-valued version of Proposition 4.56 for $r = \infty$ (proved also in [759]), to lift F-valued holomorphic mappings of bounded type on l_1 to G-valued holomorphic mappings when there exists a bounded linear surjection from G onto F. See the remarks on Exercises 2.62 and 4.67. A number of other finite dimensional coefficient characterizations of this type have appeared in the literature, for example, from results in C.G. Astashina [73] one can deduce that $\sum_{m,n=0}^{\infty} a_{n,m} z^n w^m \in H(B_{l_1^2})$ if and only if for all λ, $0 \leq \lambda \leq \infty$,

$$\limsup_{n+m \to \infty, \frac{n}{m} \to \lambda} \frac{|a_{n,m}|^{\frac{1}{n+m}} n^{\frac{n}{n+m}} m^{\frac{m}{m+n}}}{n+m} \leq \frac{\lambda^{\lambda(1+\lambda)}}{1+\lambda}$$

(this easily follows from Corollary 4.57). See also O.D. Kellogg [505, §3].

4.89 The monomials do *not* form a basis for $(H(l_1), \tau_\omega)$ since $(l_1', \|\cdot\|) = (l_\infty, \|\cdot\|)$ is a closed complemented subspace of $(H(l_1), \tau_\omega)$. Nevertheless the monomials, which do form a basis for the weaker τ_0 topology (Proposition 4.56), can be used, as this example shows, to generate a fundamental system of semi-norms for the τ_ω topology. The result in this exercise is due to R.A. Ryan [759, Theorem 4.12].

4.90 Use Exercise 2.63.

4.91 These are special cases of results due to M. Schottenloher [784], R.M. Aron [38], and R.M. Aron–M. Schottenloher [70, 71]. An identification, $H(U \times V) = H(U) \widehat{\otimes}_\varepsilon H(V)$ is called a *product formula* while $H(U \times V) = H(U; H(V))$ is called an *exponential law*. Exponential laws for holomorphic functions are also given in S. Bjon [125, 126], S. Bjon–M. Lindström [128], C. Boyd [181], M.C. Matos [581] and J. Mujica [655]. In [181], the author obtains an exponential law for the τ_ω topology when E and F are Fréchet spaces with Qno (quasi-normable by operators) and U and V are balanced. Boyd shows that Qno is necessary. The condition Qno is related to the *strong localization property* (Definition 4.36) and is defined as follows; E has Qno if for every neighbourhood V of 0, there exists a 0-neighbourhood W such that for all $\varepsilon > 0$, there is a $P \in L(E; E)$ with $P(W)$ bounded and $(I - P)(W) \subset \varepsilon V$. Fréchet–Schwartz spaces with the bounded approximation property and Banach spaces have Qno. The formula $H(U; F) = H(U) \widehat{\otimes}_\varepsilon F$ (see [784]) is often used to extend the validity of scalar-valued results to the vector-valued case. We have essentially used these formulae in Example 3.12 and Proposition 4.55.

4.92 This exercise is a basis free version of Proposition 4.55. By the previous exercise, $H(D \times E), \tau_0) = (H(D), \tau_0) \widehat{\otimes}_\pi (H(E), \tau_0)$. Since E'_c is complemented in $(H(E), \tau_0)$, $H(D) \widehat{\otimes}_\pi E'_c$ is complemented in $(H(D \times E), \tau_0)$. Hence, if $\tau_0 = \tau_\delta$ on $H(D \times E)$ then $H(D) \widehat{\otimes}_\pi E'_c$ is bornological and, by D. Vogt [839; 841, Theorem 4.9], E has (DN) (note that $H(D) \widehat{\otimes}_\pi E'_c = L_b(E; H(D))$ for any Fréchet nuclear space E). Conversely, if E has (DN) then E can be identified with a subspace of s and $D \times E$ with a subset of the polydisc $D \times s$ in $\mathbb{C} \times s \cong s$. By Example 4.40(e), $\tau_0 = \tau_\delta$ on $H(D \times s)$. If p is a τ_δ continuous semi-norm on $H(D \times E)$ then $\tilde{p}(f) := p(f|_{D \times E})$ is τ_δ and hence τ_0 continuous on $H(D \times s)$ (see the proof of Proposition 4.54). Hence there exists r, $0 < r < 1$, K compact in s and $c > 0$ such that $\tilde{p}(f) \le c\|f\|_{\{z \in \mathbb{C}: |z| \le r\} \times K}$. By the remarks prior to Lemma 4.53, each $P \in \mathcal{P}(^n\mathbb{C} \times E)$ can be extended to $\tilde{P} \in \mathcal{P}(^n\mathbb{C} \times s)$. Hence

$$p(P_n) \le c \inf\{\|Q_n\|_{\{z \in \mathbb{C}: |z| \le r\} \times K} : Q_n \in \mathcal{P}(^n\mathbb{C} \times s), Q_n|_{\mathbb{C} \times E} = P_n\}$$

for all $P_n \in \mathcal{P}(^n\mathbb{C} \times E)$ and all n. By Lemma 4.53, there exists a compact convex balanced subset L of E such that for all n and all $P \in \mathcal{P}(^nE)$

$$\inf\{\|Q\|_K : Q \in \mathcal{P}(^ns), Q|_E = P\} \le \|P\|_L.$$

If $P_n \in \mathcal{P}(^n\mathbb{C} \times E)$ then $P_n(\lambda, y) = \sum_{j=0}^n \lambda^j P_{n,j}(y)$, for $\lambda \in \mathbb{C}$ and $y \in E$, where $P_{n,j} \in \mathcal{P}(^{n-j}E), 0 \le j \le n$. Hence

$$p(P_n) \le c \inf\left\{\sum_{j=o}^n r^j \|Q_{n,j}\|_K : Q_{n,j} \in \mathcal{P}(^{n-j}s), Q_{n,j}|_E = P_{n,j}\right\}$$

$$= c \left\{\sum_{j=o}^n r^j \inf \|Q_{n,j}\|_K : Q_{n,j} \in \mathcal{P}(^{n-j}s), Q_{n,j}|_E = P_{n,j}\right\}$$

$$\le c \left\{\sum_{j=o}^n r^j \|P_{n,j}\|_L : P_n(\lambda, y) = \sum_{j=0}^n \lambda^j P_{n,j}(y), \lambda \in \mathbb{C}, y \in E\right\}$$

since the $Q_{n,j}$ can be chosen independently of one another. By (3.5) or Proposition 3.1, $\sum_{j=0}^{n} r^j \|P_{n,j}\|_L \le n \|P_n\|_{\{z:|z|<r\}\times L}$ and

$$p(f) \le \sum_{n=0}^{\infty} p\left(\frac{\hat{d}^n f(0)}{n!}\right) \le c \sum_{n=0}^{\infty} n \left\|\frac{\hat{d}^n f(0)}{n!}\right\|_{\{z:|z|<r\}\times L}$$

for all $f \in H(D \times E)$. By Proposition 3.36, p is τ_0 continuous. Hence $\tau_0 = \tau_\delta$ on $H(D \times E)$.

We have provided in this exercise a proof of a result announced in [299, p.396].

Chapter 5

5.56 See M. Schottenloher [779, Lemma 1.6], M. Hervé [446, p.144] and the remarks on Exercise 3.126.

5.59 See [651, Lemma 52.3].

5.61 We refer to [651, Exercise 47 I].

5.64 At first it is surprising that such examples exist and indeed several writers, including the author, have made the erroneous assumption that this was not possible. For further details we refer to [645, p.522] and Exercise 5.84. Note that every open subset of $\mathbb{C}$ is a domain of existence.

5.66 See [278] and [696, §4.4].

5.67 This result is due to M. Schottenloher [779; p.225, 3°] and shows that a $\mathbb{C}$-valued domain of holomorphy is a vector-valued domain of existence. A. Hirschowitz [453, 456, 457] gives examples of domains of holomorphy and domains of existence of Banach-valued holomorphic functions which are not domains of existence of $\mathbb{C}$-valued holomorphic functions (see Example 5.55).

5.68 For finite dimensional spaces this led to some interesting investigations by A. Baernstein [81], J.R. Rogers, Jr W.R. Zame [748], K. Rusek [755] and W.R. Zame [861]. The result in this exercise is due to J. Mujica [646]. One may also consider the question of when the τ_ω topology on $H(K)$ is generated by

$$\sum_{n=0}^{\infty} \sup_{x \in K} p\left(\frac{\hat{d}^n f(x)}{n!}\right)$$

where p ranges over the continuous semi-norms on $H(0)$.

5.69 This result is due to K.-D. Bierstedt–R. Meise [121] for domains in E (see also [80]) and to J. Mujica [646] when Ω is spread over E.

5.70 This result is due to J. Mujica [639, 640, 644]. See also J.M. Isidro [473].

5.72 This result is useful in removing the continuous norm assumption in various situations and is due to A. Hirschowitz [454] when $E = \mathbb{C}^{\mathbb{N}}$ and, in general, independently, to S. Dineen [282; 287, Lemma 1.1] and Ph. Noverraz [693; 695; 696, Theoréme 2.1.7].

5.73 This is a Cartan–Thullen Theorem for holomorphic functions of bounded type (see [277]) and the finite dimensional proof (see [466] and Section 3.2) works since $H_b(U)$ is a locally m-convex Fréchet algebra. For finite dimensional spaces, the definition given, involving V_1 and V_2, is the classical definition that the univalent domain U is a domain of holomorphy. Similar results for other Fréchet subalgebras of $H(U)$ are given in M.C. Matos [575, 576, 578] and M. Schottenloher [777, 779].

5.74 See Ph. Noverraz [694, 696, 698, 701].

5.75 Reinhardt domains in Banach spaces with an unconditional basis are investigated in G. Katz [503] and M.C. Matos–L. Nachbin [592]. See also M.C. Matos [586].

5.76 Condition (A) is due to R.M. Aron (unpublished) – see [644, p.17]. Condition (B) was introduced by J.A. Barroso [87, Theorem 3.1] while investigating topologies on spaces of holomorphic mappings and implies that the inductive limit $\varinjlim_{n} \mathcal{P}(^m E_{\beta_n})$ is strict. A related condition, which is satisfied by all Fréchet–Schwartz spaces, is considered in [121, Proposition 4]. Condition (N) was defined by L. Nachbin [670] to obtain factorization theorems for holomorphic mappings. The different implications in this exercise are due to J. Mujica [644, Theorem 4.1] (see also [80, 122, 220, 221, 679, 680, 681]).

5.77 The main interest here is when E is non-metrizable and, consequently, the inductive limit uncountable. Such limits are, in general, very badly behaved and regularity and quasi-completeness may be difficult to verify but, by Corollary 4.43, we have positive results. Regularity at the origin (condition (a)) implies Cauchy estimates on the compact set and so the main difficulty is to construct sufficiently many semi-norms, having the form (5.2), to obtain coherence of the Taylor series expansions. The results of this exercise are due to R.L. Soraggi [811] (see also [300, 807, 808]). R.L. Soraggi also shows in [811] that condition (b) is satisfied if and only if $\big(H(E), \tau_0\big)$ does not contain a non-trivial very strongly convergent sequence. Further results on the same topic can be found in R.L. Soraggi [804, 805, 806] and J. Mujica [646]. Positive results are obtained by assuming that K is locally connected.

It is not known if condition (a) alone suffices for regularity. Vector-valued holomorphic germs are studied in [118, 160, 182, 183, 189].

5.78 Let $s = \sup\{r \geq 0 : \alpha(a) \leq r \Rightarrow d_\Omega(x,a) > 1\}$. By definition $s \geq 0$, hence $d_\Omega^\alpha(x) \leq s$ if $d_\Omega^\alpha(x) = 0$. If $d_\Omega^\alpha(x) > 0$ then clearly $d_\Omega(x,a) > 1$ for all $a \in E$, $\alpha(a) < d_\Omega^\alpha(x)$, and $d_\Omega^\alpha(x) \leq s$. To prove $d_\Omega^\alpha(x) \geq s$, we can suppose $s > 0$. For $0 < r < s$, let $Z = \bigcup\{D_\Omega(x,a,1) : a \in E, \alpha(a) < r\}$. By the definition of Z, $p\big|_Z : Z \to B_E^\alpha(p(x),r)$ is bijective. To show that $p\big|_Z$ is open, which implies $d_\Omega^\alpha(x) \geq r$, it suffices to show that Z is open. If $\alpha(a) < r$ then $D_\Omega(x,a,1)$ is relatively compact in Ω. Hence there exists a neighbourhood U of $D_\Omega(x,a,1)$ in Ω such that U is mapped homeomorphically onto an absolutely convex neighbourhood $p(x)+V$ of $D_E(x,a,1)$ which lies in $B_E^\alpha(p(x),r)$. If $b \in V$ then $p\big|_{D(x,b,1)}$ and $p\big|_U$ are homeomorphisms and $x \in D_\Omega(x,b,1) \cap U$. Hence $D_\Omega(x,b,1) \subset U$ and $\bigcup\{D_\Omega(x,b,1) : b \in B\} \subset Z$. This shows that Z is open and $s = d_\Omega^\alpha(x)$.

This result, for normed linear spaces, is due to G. Coeuré [232, Proposition 9.2] and, for locally convex spaces, to M. Schottenloher [779, Lemma 1.3]. According to [783], the infimum result is not generally true when $d_\Omega^\alpha(x) = 0$.

5.80. This exercise can be used to construct the envelope of holomorphy and domains of existence (see Definition 5.14) and is taken from the excellent survey article of M. Schottenloher [783] (see also G. Coeuré [235]). Briefly one proceeds as follows: given $\{j_k : (\Omega, p) \to (\Omega_k, p_k)\}_k \subset \Omega/D(E)$ let

$$X = \{(x_k)_k : p_k(x_k) = p_l(x_l) \text{ and } \exists \text{ charts } (W_k, p_k)$$
$$\text{such that } p_k(W_k) = p_l(W_l)\} \subset \prod_k \Omega_k\}$$

and take sets of the form $\prod_k W_k$ as neighbourhood systems to define a topology on X. Let $\widetilde{p} : X \to E$, $\widetilde{p}((x_k)_k) = p_k(x_k)$. The set of connected components of X which intersect $j(\Omega)$, where $j : \Omega \to X$ is given by $j(x) = (j_k(x))_k \subset X$, is denoted by $\widehat{\Omega}$, and $(\widehat{\Omega}, \widehat{p} := \widetilde{p}\big|_{\widehat{\Omega}}) \in \Omega/D(E)$ is the maximal extension of all j_k (in a sense defined analogously to Definition 5.14). The Riemann domain $(\widehat{\Omega}, \widehat{p})$ is called the intersection of $(\Omega_k, p_k)_k$ (in the category – $\Omega/D(E)$ – of domains below Ω and above E).

5.81 This result is due to A. Hirschowitz [455, 460] for domains spread over Banach spaces (the result for $j : (\Omega, p) \to (S_c(\Omega), q)$ is given in [13] and other Banach space results are given in [232] and [777]) and the general result is due to M. Schottenloher [784, Theorem 4.3]. A key result in vector-valued holomorphic extensions is due to W. Bogdanowicz [141] who proved that Gâteaux holomorphic extension into sequentially complete spaces always exist. This result was combined with Example 3.8(b) and (d), by M. Schottenloher [779, 791], to obtain Fréchet holomorphic extensions. The completeness requirement on the range is weakened in [292] and [779] using holomorphic completions (Definition 6.2). Further results on this topic may be found in [13, 232, 292, 551, 553, 675, 696, 779, 783].

5.82 This result is due to M. Schottenloher [786]. Earlier the same result has been obtained for Banach spaces with Schauder basis in [278], for Banach spaces with the bounded approximation property in [693, 695], and for various other spaces, including nuclear spaces, by using surjective limits in [282, 287, 693, 695, 696, 702]. This exercise leads to Oka–Weil and Runge approximation theorems.

5.85 A Riemann domain Ω over a locally convex space E such that $d_\Omega^\alpha(\widehat{K}_{H(\Omega)})$ $= d_\Omega^\alpha(K)$ for all K compact in Ω and all $\alpha \in cs(E)$ is called metrically holomorphically convex (see [783]). Domains of holomorphy and holomorphically convex domains are metrically holomorphically convex [696, p.62] while metrically holomorphically convex domains are pseudo-convex.

5.86 This is known as Patil's Problem. It was first posed by D. Patil at the Kentucky conference on infinite dimensional holomorphy during June of 1973. Extensive work has been undertaken on this problem by a number of authors, notably J. Globevnik. With this problem as motivation R.M. Aron–J. Globevnik–M. Schottenloher [60] studied interpolation sequences and found new proofs of a number of classical theorems. The problem for separable ranges was solved independently by R.M. Aron [41] (who first reduced the problem to the c_0 case and then used cluster sets and Blaschke products), by J. Globevnik [388] (whose approach involved a generalization of the Rudin–Carleson interpolation theorem to vector-valued functions) and by W. Rudin [751]. A counterexample which shows that the result does not extend to non-separable Banach spaces is given in B. Josefson [489]. Josefson proved the following result:

> Let B_0 be the open unit ball of $c_0(\Gamma)$, Γ uncountable, and suppose $f \in H\big(B_0; c_0(\Gamma)\big)$, then there exists an open connected bounded subset U of $c_0(\Gamma)$ such that if $U \subset \overline{f(B_0)}$ then $f(B_0) \not\subset U + \frac{1}{10}B_0$.

Further extensions of this counterexample are due to J. Globevnik [389, 390] who showed that if a Banach space contains a non-separable analytic image of the unit ball of $c_0(\Gamma)$, Γ uncountable, then it contains an isomorphic copy of $c_0(\Gamma')$ where Γ' is uncountable. The range of polynomial and holomorphic mappings on certain subsets of c_0 are discussed in [59].

5.87 This result is due to L.M. Hai–T.T. Quang [435] and concerns a topic previously discussed by N.V. Khue–B.D. Tac [510], E. Ligocka–J. Siciak [553] and L. Waelbroeck [850].

Chapter 6

6.34 See C.O. Kiselman [518].

6.36 See [321, Corollary 8(b)]. Alternatively one can proceed directly as follows. Suppose K is compact non-polar in E Fréchet. Then K is contained in the closed convex hull of a null sequence and thus in a separable subspace F of E. If $\phi \not\equiv 0$ and $\phi\big|_F = 0$ then $\log|\phi|(F) = -\infty$ and $\log|\phi| \not\equiv -\infty$. Hence E must be separable. If E is not Schwartz then E contains a non-Montel quotient $\pi : E \longrightarrow E_0$ and $\pi(K)$ is non-polar in E_0. Since E_0 is separable its closed *limited* subsets are compact and one can choose a bounded sequence B in E_0 and a weak* null sequence $(\phi_n)_n \in E_0'$ such that $\|\phi_n\|_B = 1$ for all n. If $x_n \in B$ and $|\phi_n(x_n)| > 1/2$ all n, there exists a sequence of scalars $(\alpha_n)_n$ such that $x_0 := \sum_{n=1}^{\infty} \alpha_n x_n \in E_0$ and a sequence of scalars $(\beta_n)_n$ such that

$$\nu := \sum_{n=1}^{\infty} \beta_n \log|\phi_n| \text{ is plurisubharmonic, } \nu(\pi(K)) = -\infty \text{ and } \nu(x_0) > -\infty.$$

6.37 See [321, Lemma 2.1]. Using this result one can show easily that bounding subsets of a Banach space are polar.

6.38 See [321, Theorem 2.2].

6.39 In [320, Corollary 14] the authors show; E has property $(\widetilde{\Omega}) \iff E$ contains a compact non-uniformly polar subset. In [321] these conditions are shown to be equivalent to the condition that E contains a compact non-polar subset.

6.40 The pseudo-convex completion of a locally convex space is investigated by Ph. Noverraz in [696]. His results include this exercise and also the following: plurisubharmonic functions on E extend to open subsets of the completion $\widehat{E}$ – this result is contained within the proof of Corollary 5.45.

6.41 This result is due to Ph. Noverraz [697].

6.42 See Ph. Noverraz [698, 700] and M. Schottenloher [778]. This result holds for open surjective limits of metrizable spaces and ω-spaces [790] and is conjectured to hold for all locally convex spaces. A locally convex space E is an ω-space if each $f \in H(E)$ depends only on a countable set of continuous semi-norms on E ([285]). Lindelöf spaces are ω-spaces and examples of spaces which are not ω-spaces are given in [418, 419, 790].

6.43 This result is due to P. Lelong [541, 542] and was applied in [542] to show that all compact subsets of $H(\mathbb{C})$ are polar (Example 6.10). See also M. Hervé [446, Corollary 4.5.14 and Example 4.5.14].

6.44 This result is due to R. Meise–D. Vogt [608].

6.45 See [610, Proposition 4.1; 371, Proposition 2.5] and the remarks on Exercises 3.69, 3.78 and 3.80.

6.46 See [205].

6.47 Extendible polynomials were introduced by P. Kirwan–R.A. Ryan [512]. Use the embedding $E \longrightarrow \mathcal{C}(B_{E'}, \sigma(E', E))$, the Dunford–Pettis Property and Proposition 2.3.4 to show extendible polynomials are weakly sequentially continuous. D. Carando–I. Zalduendo [205] proved that integral polynomials are extendible. The norm $\| \cdot \|_e$ is defined and the Banach space structure of $\mathcal{P}_e(^nE)$ investigated in [512]. Preduals of $(\mathcal{P}_e(^nE), \| \cdot \|_e)$ are constructed in [202] and [512].

6.48 This result is due to R.M. Aron–P. Galindo–D. García–M. Maestre [58]. As corollaries one obtains (a) E regular $\Longleftrightarrow$ $E \times E$ symmetrically regular, (b) $E \times E'$ is not regular if E is non-reflexive.

6.49 Use the previous exercise. D. Leung [548] has shown that J' is symmetrically regular but not regular (J is the classical James space – see [555, Example 1.d.2]).

6.50 This example is due to J. Rennison [741] (see also [53] and [860]).

6.51 This result is due to I. Zalduendo [856] and characterizes homogeneous polynomials on the bidual which are Aron–Berner extensions of homogeneous polynomials on E. The result in [856] is given for holomorphic functions and is applied to show that the Aron–Berner extension is an algebra morphism from $H_b(E)$ into $H_b(E'')$.

R.M. Aron–C. Boyd–Y.S. Choi [48] obtain a different type of characterization for c_0 and show that $P \in \mathcal{P}(^n l_\infty)$ is an Aron–Berner extension of an element in $\mathcal{P}(^n c_0)$ if and only if

$$\left\| \frac{\widehat{d^j} P(x)}{j!} \right\|_{l_\infty} = \left\| \left\| \frac{\widehat{d^j} P(x)}{j!} \right\|_{c_0} \right\|_{c_0}$$

for all $x \in c_0$ and all j, $1 \leq j \leq n$. They also show that the Aron-Berner extension is the unique norm preserving extension for nuclear polynomials when E is an M-ideal in its bidual.

6.52 This result, due to R.M. Aron–P. Galindo [57], extends to holomorphic mappings of bounded type the following well known linear result; weakly compact linear mappings between Banach spaces factor through reflexive Banach spaces. Factorization results for multilinear mappings, polynomials and holomorphic mappings may be found in [65, 386, 399, 402, 493, 495, 758, 760]. The factorizations in [386], [402] and [495] use operator ideals.

Compact, weakly compact, bounding and limited holomorphic mappings are considered in the above references and also in [38, 70, 71, 558, 560]. M. Lindström–R.A. Ryan show in [560] that the composition of three (respectively two) limited holomorphic mappings is compact (respectively weakly compact). Asplund holomorphic mappings are studied in [746] by N. Robertson.

6.53 See R.M. Aron–P. Galindo [57].

6.54 This exercise shows that (6.17) holds for all $z \in E^{(iv)}$ if and only if E is symmetrically regular and is due to R.M. Aron–P. Galindo–D. García–M. Maestre [58].

6.55 This example, due to P. Harmand, is given in [58, Remark 1.4(d)]. See [548] for other examples.

6.56 This result is due to R.M. Aron–B.J. Cole–T.W. Gamelin [51, Theorem 6.1]. The result for $\phi \in \mathcal{S}(H_b(E))$ is given in Proposition 6.30.

6.57 This result is due to R.M. Aron–P. Berner [47]. Further proofs can be found in [560] and [856]. In [256] the authors prove that each $f \in H^\infty(B)$ admits a holomorphic norm preserving extension to $B^{\circ\circ}$ where B is the open unit ball of a Banach space and $B^{\circ\circ}$ is the open unit ball of its bidual (other proofs are given in [328] and [371]). See the remarks on Exercises 6.63 and 6.64.

6.58 See R.M. Aron–P. Rueda [69] and P. Rueda [753].

6.59 The results in this exercise are due independently to F. Cabello Sánchez–J.M.F. Castillo–R. García [201] and S. Lassalle–I. Zalduendo [532] and are a response to the following question posed in [266]: if E and F are Banach spaces and $E' \simeq F'$ does this imply $(\mathcal{P}(^n E), \|\cdot\|) \cong (\mathcal{P}(^n F), \|\cdot\|)$? The authors in [201] provide a positive solution to this question, using Nicodemi extension operators (see [372, 373, 688] and Section 1.5), when E and F are stable Banach spaces and give a number of interesting and unexpected examples, e.g. there exist a separable Banach space E and a non-separable Banach space F such that $(\mathcal{P}(^n E), \|\cdot\|) \cong (\mathcal{P}(^n F), \|\cdot\|)$ for all n. In [532] the authors show that any continuous linear operator $s : E' \to F'$ leads to an extension operator which can be expressed using s' and the Aron–Berner extension. Using this they obtain analogous results for integral polynomials and for polynomials which are weakly continuous on bounded sets.

The following simple proof is due to B. Grecu. If E is symmetrically regular and $\phi : E' \to F'$ is a linear isomorphism with transpose ϕ^* then ϕ^* is easily seen to induce an isomorphism ϕ_n^* between $\mathcal{L}^s(^nE'')$ and $\mathcal{L}^s(^nF'')$. If $\tilde{L}$ denotes the extension of $L \in \mathcal{L}^s(^nE)$ to the bidual, obtained using the Aron–Berner extension, then $L \in \mathcal{L}^s(^nE) \to \phi_n^*(\tilde{L})\big|_F \in \mathcal{L}^s(^nF)$ is the required isomorphism.

6.60 This result when $F = \mathbb{C}$ is due to R.M. Aron–P. Galindo–D. García–M. Maestre [58] and for F a uniform algebra to D. García–M.L. Lourenço–L. Moraes–O.W. Paques [378]. A generalization is given in [634].

6.61 See [285, 292] and Exercises 3.87, 3.88 and 5.77. By using this result and by considering the subspace of $\mathbb{C}^{\mathbb{N}}$ spanned by the standard basis one sees that the continuous norm hypothesis in Example 6.6(d) is necessary.

6.62 This result is due to P. Lelong [540]. A subset A of an open subset Ω of a locally convex space is analytic if for each $\xi \in A$ there exists an open set ω_ξ, $\xi \in \omega_\xi \subset \Omega$ and $f \in H(\omega_\xi)$ such that $A \cap \omega_\xi \subseteq \{z \in \omega_\xi : f(z) = 0\}$. If $f \in H(\Omega)$ then $\log|f| \in \mathrm{PSH}(\Omega)$ and hence analytic sets are polar. A deep analysis of analytic sets in locally convex spaces is given in [597] by P. Mazet.

6.63 The Aron–Berner extension (see Section 1.3) can be obtained either directly or by using the Factorization Lemma for polynomials (Lemma 1.13) to extend each continuous homogeneous polynomial to the bidual. By Proposition 1.53

$$\|AB(P)\|_{B^{\circ\circ}} = \|P\|, \quad P \in \mathcal{P}(^nE)$$

for B convex balanced. By Goldstine's Theorem, applied to E_B, B is $\sigma(E'', E')$ dense in $B^{\circ\circ}$. If $f = \sum_{n=0}^{\infty} \dfrac{\widehat{d}^n f(0)}{n!} \in H_b(E)$ then

$$\sum_{n=0}^{\infty} \left\| AB_n \left(\frac{\widehat{d}^n f(0)}{n!} \right) \right\|_{B^{\circ\circ}} = \sum_{n=0}^{\infty} \left\| \frac{\widehat{d}^n f(0)}{n!} \right\|_B < \infty$$

and, since $E'' = \bigcup \{B^{\circ\circ} : B \text{ bounded in } E\}$, the results in Example 3.8 (a) and (d) show that $AB(f) := \sum_{n=0}^{\infty} AB_n \left(\dfrac{\widehat{d}^n f(0)}{n!} \right) \in H(E'')$. However, if E is *not* distinguished the collection $\{B^{\circ\circ} : B \text{ bounded}\}$ may not form a fundamental system of bounded subsets of E and further analysis is required. It suffices to show that $AB(f)$ is bounded on each *countable* bounded subset of E'' and such sets are equicontinuous in E'' [467, p. 293] and hence contained in sets of the form $B^{\circ\circ}$. Hence $AB(f) \in H_b(E'')$. The result is due to N.V. Khue [507] who, however, only used the estimate

$$\left\| AB_n \left(\frac{\widehat{d}^n f(0)}{n!} \right) \right\| \leq \frac{n^n}{n!} \left\| \frac{\widehat{d}^n f(0)}{n!} \right\|$$

which suffices for entire functions (due to R.M. Aron–P. Berner [47] for Banach spaces). The improved estimate, Proposition 1.53, used above, due to A.M. Davie–T.W. Gamelin [256] for Banach spaces and extended to locally convex spaces by P. Galindo–D. García–M. Maestre [371] can be used to obtain a similar result for $H_b(U)$, U a convex balanced open subset of E. Analogous results for holomorphic functions which are weakly uniformly continuous on bounded sets are given in L.A. Moraes [628, 629, 632].

6.64 This result is due to N.V. Khue [506, 507] and extends Proposition 6.18. See also the remarks on the previous exercise, Proposition 3.48, Sections 1.5, 6.5 and Exercises 1.88 and 1.89.

The result of this exercise can be extended to a convex balanced open subset U of a distinguished Fréchet space E. For this one requires

$$\text{Int}(U^{\circ\circ}) = \bigcup_{B \subset U,\ B \text{ bounded}} B^{\circ\circ}$$

(a result of P. Galindo–D. García–M. Maestre [371] – see also [308, 632, 633]).

6.65 This result is due to J.M. Ansemil–S. Ponte [25] with proof relying on a characterization of $H_b(E)'_\beta$ due to J.M. Isidro [472] (see also [739] and Section 6.5). A proof is also possible using S-absolute decompositions (see Proposition 3.35). A vector-valued version of this result is given in [401].

6.66 This result, which shows that finitely generated ideals in $H(U)$ are either contained in a closed maximal ideal or generate the whole space (and thus $H(U)$ contains no proper finitely generated dense ideals), is due to J. Mujica [659] and is currently the only known infinite dimensional Fréchet space example of this type (see also [642]). M. Schottenloher [793] obtains the same result for domains of holomorphy in $\mathcal{DFC}$ spaces and a similar result for $H_b(U)$ is given in [659].

6.67 This zero-one law for *complete* polar subsets is due to S. Dineen [297]. Polar subsets of finite-dimensional spaces are of Lebesgue measure zero and in [324] the authors obtain the following infinite-dimensional translation-invariant substitute: if A is a circled (i.e. $e^{i\theta} A \subset A$ all $\theta \in \mathbb{R}$) complete polar subset of the locally convex space E then there exists a Gaussian measure μ on E such that $\mu(x + A) = 0$ for all $x \in E$.

References

[1] *T. Abuabara*, A version of the Paley-Weiner-Schwartz theorem in infinite dimensions, Advances in Holomorphy, Ed. J.A. Barroso, North-Holland Math. Stud., **34**, 1979, 1–29.

[2] *M.D. Acosta, F.J. Aguirre, R. Payá*, There is no bilinear Bishops-Phelps theorem, Israel J. Math., **93**, 1996, 221–227.

[3] *J. Alaminos, Y.S. Choi, S.G. Kim, R. Payá*, Norm attaining bilinear forms on spaces of continuous functions, Glasgow Math. J., **40**, 1998, 359–365.

[4] *R. Alencar*, Aplicações nucleares e integrais e a propriedade de Radon-Nikodým, Thesis, Universidade de São Paulo, 1982.

[5] *R. Alencar*, Multilinear mappings of nuclear and integral type, Proc. Amer. Math. Soc., **94**, 1, 1985, 33–38.

[6] *R. Alencar*, On reflexivity and basis for $P(^mE)$, Proc. Royal Irish Acad. Sect. A, **85**, 2, 1985, 131–138.

[7] *R. Alencar*, An application of Singer's theorem to homogenous polynomials, Contemp. Math., **144**, 1993, 1–8.

[8] *R. Alencar, R.M. Aron, S. Dineen*, A reflexive space of holomorphic functions in infinitely many variables, Proc. Amer. Math. Soc., **90**, 1984, 407–411.

[9] *R. Alencar, R.M. Aron, G. Fricke*, Tensor products of Tsirelson's space, Illinois J. Math., **31**, 1, 1987, 17–23.

[10] *R. Alencar, K. Floret*, Weak-strong continuity of multilinear mappings and the Pełczyński-Pitt Theorem, J. Math. Anal. Appl., **206**, 1997, 532–546.

[11] *R. Alencar, K. Floret*, Weak continuity of multilinear mappings on Tsirelson's space, Quaestiones Math., **21**, 1998, 177–186.

[12] *R. Alencar, M.C. Matos*, Some classes of multilinear mappings between Banach spaces, Publicaciones del Departamento de Análisis Matemático, Universidad Complutense de Madrid, **12**, 1989.

[13] *H. Alexander*, Analytic functions on Banach spaces, Thesis, University of California, Berkeley, 1968.

[14] *A. Alexiewicz, W. Orlicz*, Analytic operations in real Banach spaces, Studia Math., **14**, 1953, 57–81.

[15] *J.M. Ansemil,* Topological duality on the function space $H(\mathbb{C}^{\mathbb{N}})$, J. Math. Anal. Appl., **14**, 1979, 188–197.

[16] *J.M. Ansemil,* Relations between τ_0 and τ_ω on spaces of holomorphic functions, Advances in the Theory of Fréchet Spaces, Ed. T. Terzioğlu, Kluwer Academic Publishers, Series C, **287**, 1989, 173–180.

[17] *J.M. Ansemil,* On the quasi-normability of $H_b(U)$, Extracta Math., **9**, 1, 1994, 71–74.

[18] *J.M. Ansemil, R.M. Aron, S. Ponte,* Embeddings of spaces of holomorphic functions of bounded type, J. London Math. Soc., (2), **46**, 1992, 482–490.

[19] *J.M. Ansemil, R.M. Aron, S. Ponte,* Spaces of holomorphic functions and germs on quotients, Progress in Functional Analysis, Ed. K.-D. Bierstedt et al., North-Holland Math. Stud., **170**, 1992, 163–177.

[20] *J.M. Ansemil, F. Blasco, S. Ponte,* Quasi-normability and topologies on spaces of polynomials, J. Math. Anal. Appl., **213**, 1997, 534–539.

[21] *J.M. Ansemil, F. Blasco, S. Ponte,* On the "Three-Space Problem" for spaces of polynomials, Proceedings of the Second International Workshop on Functional Analysis at Trier, 1997, Note Mat. (to appear).

[22] *J.M. Ansemil, F. Blasco, S. Ponte,* (BB) properties on Fréchet spaces, Ann. Acad. Sc. Fenn. Math. (to appear).

[23] *J.M. Ansemil, S. Dineen,* Locally determining sequences in infinite dimensional spaces, Note Mat., **7**, 1987, 41–45.

[24] *J.M. Ansemil, K. Floret,* The symmetric tensor product of a direct sum of locally convex spaces, Studia Math., **129**, 3, 1998, 285–295.

[25] *J.M. Ansemil, S. Ponte,* An example of a quasi-normable Fréchet function space which is not a Schwartz space, Functional Analysis, Holomorphy and Approximation Theory. Rio de Janeiro, 1978. Ed. S. Machado, Lecture Notes in Math. **843**, 1981, 1–8.

[26] *J.M. Ansemil, S. Ponte,* Topologies associated with the compact open topology on $H(U)$, Proc. Royal Irish Acad. Sect. A, **82**, 1982, 121–128.

[27] *J.M. Ansemil, S. Ponte,* The barrelled topology associated with the compact-open topology on $H(U)$ and $H(K)$, Portugal. Math., **43**, 4, 1985, 429–438.

[28] *J.M. Ansemil, S. Ponte,* The compact open topology and the Nachbin ported topology on spaces of holomorphic functions, Arch. Math. (Basel), **51**, 1988, 65–70.

[29] *J.M. Ansemil, S. Ponte,* Spaces of holomorphic germs on quotients, J. Math. Anal. Appl., **172**, 1, 1993, 33–38.

[30] *J.M. Ansemil, J. Taskinen,* On a problem of topologies in infinite dimensional holomorphy. Arch. Math. (Basel), **54**, 1990, 61–64.

[31] *R. Arens,* Operations induced in function classes, Monatsh. Math., **55**, 1951, 1–19.

[32] *R. Arens,* The adjoints of a bilinear operation, Proc. Amer. Math. Soc., **2**, 1951, 839–848.

[33] *R. Arens, J.L. Kelley,* Characterizations of the space of continuous functions over a compact Hausdorff space, Trans. Amer. Math. Soc., **62**, 1947, 499–508.

[34] *A. Arias, J.D. Farmer,* On the structure of tensor products of l_p spaces, Pacific J. Math., **175**, 1996, 13–37.

[35] *R.M. Aron,* Sur la topologie bornologique pour l'espace d'applications holomorphes, C. R. Acad. Sci. Paris, Sér. A, **272**, 1971, 872–873.

[36] *R.M. Aron,* Holomorphic functions on balanced subsets of a Banach space, Bull. Amer. Math. Soc., **78**, 1972, 624–627.

[37] *R.M. Aron,* Holomorphy types for open subsets of Banach spaces, Studia Math., **45**, 1973, 273–289.

[38] *R.M. Aron,* Tensor products of holomorphic functions, Indag. Math., **35**, 3, 1973, 192–202.

[39] *R.M. Aron,* The bornological topology on the space of holomorphic mappings on a Banach space, Math. Ann., **202**, 1973, 256–272.

[40] *R.M. Aron,* Entire functions of unbounded type on a Banach space, Boll. Un. Mat. Ital., (4), **9**, 1974, 28–31.

[41] *R.M. Aron,* The range of vector-valued holomorphic mappings, Ann. Polon. Math., Conference on Analytic Functions, **33**, 1976, 17–20.

[42] *R.M. Aron,* Compact polynomials and compact differentiable mappings between Banach Spaces, Séminaire P. Lelong, 1974–75, Lecture Notes in Math., **524**, 1976, 213–222.

[43] *R.M. Aron,* Weakly uniformly continuous and weakly sequentially continuous entire functions, Advances in Holomorphy, Ed. J.A. Barroso, North-Holland Math. Stud., **34**, 1979, 47–66.

[44] *R.M. Aron,* Polynomial approximation and a question of G.E. Shilov, Approximation Theory and Functional Analysis, Ed. J.B. Prolla, North-Holland Math. Stud., **35**, 1979, 1–12.

[45] *R.M. Aron,* Extension and lifting theorems for analytic mappings, Functional Analysis: Surveys and Recent Results II, Ed. K.-D. Bierstedt, B. Fuchssteiner, North-Holland Math. Stud., **38**, 1980, 257–267.

[46] *R.M. Aron, B. Beauzamy, P. Enflo,* Polynomials in many variables, real vs. complex norms, J. Approx. Theory, **74**, 2, 1993, 181–198.

[47] *R.M. Aron, P.D. Berner,* A Hahn-Banach extension theorem for analytic mappings, Bull. Soc. Math. France, **106**, 1978, 3–24.

[48] *R.M. Aron, C. Boyd, Y.S. Choi,* Unique Hahn-Banach extensions of spaces of homogeneous polynomials (preprint).

[49] *R.M. Aron, Y.S. Choi, J. Llavona,* Estimates by polynomials, Bull. Austral. Math. Soc., **52**, 3, 1995, 475–486.

[50] *R.M. Aron, J. Cima,* A theorem on holomorphic mappings into Banach spaces, Proc. Amer. Math. Soc., **36**, 1, 1972, 289–292.

[51] *R.M. Aron, B.J. Cole, T.W. Gamelin,* Spectra of algebras of analytic functions on a Banach space, J. Reine Angew. Math., **415**, 1991, 51–93.

[52] *R.M. Aron, B.J. Cole, T.W. Gamelin*, Spectra of analytic functions in infinite dimensions, Rev. Un. Mat. Argentina, **37**, 1991, 5–9.

[53] *R.M. Aron, B.J. Cole, T.W. Gamelin*, Weak-star continuous analytic functions, Canad. J. Math., **47**, 4, 1995, 673–683.

[54] *R.M. Aron, J. Diestel, A.K. Rajappa*, Weakly continuous functions on Banach spaces containing l_1, Banach Spaces, Ed. N.J. Kalton and E. Saab, Lecture Notes in Math., **1166**, 1985, 1–3.

[55] *R.M. Aron, S. Dineen*, Q-reflexive Banach spaces, Rocky Mountain J. Math., **27**, 4, 1997, 1009–1025.

[56] *R.M. Aron, C. Finet, E. Werner*, Some remarks on norm attaining n-linear forms, Functions Spaces, Proceedings Edwardsville, 1994, Lecture Notes in Pure and Appl. Math., **172**, 1995, 19–28.

[57] *R.M. Aron, P. Galindo*, Weakly compact multilinear mappings, Proc. Edinburgh Math. Soc., **40**, 1997, 181–192.

[58] *R.M. Aron, P. Galindo, D. García, M. Maestre*, Regularity and algebras of analytic functions in infinite dimensions, Trans. Amer. Math. Soc., **348**, 2, 1996, 543–559.

[59] *R.M. Aron, J. Globevnik*, Analytic functions on c_0, Rev. Mat. Univ. Complut. Madrid, **2**, 1989, 27–33.

[60] *R.M. Aron, J. Globevnik, M. Schottenloher*, Interpolation by vector-valued analytic functions, Rend. Mat. Appl., (6) **9**, 2, 1976, 347–364.

[61] *R.M. Aron, C. Hervés*, Weakly sequentially continuous analytic functions on a Banach space, Functional Analysis, Holomorphy and Approximation Theory II, Ed. G.I. Zapata, North-Holland Math. Stud., **86**, 1984, 23–38.

[62] *R.M. Aron, C. Hervés, M. Valdivia*, Weakly continuous mappings on Banach spaces, J. Funct. Anal., **52**, 1983, 189–204.

[63] *R.M. Aron, M. Klimek*, Supremum norms for quadratic polynomials (preprint).

[64] *R.M. Aron, M. Lacruz, R.A. Ryan, A.M. Tonge,* The generalized Rademacher functions, Note Mat., **12**, 1992, 15–25.

[65] *R.M. Aron, M. Lindström, W. Ruess, R.A. Ryan*, Uniform factorization for compact sets of operators, Proc. Amer. Math. Soc., **127**, 4, 1999, 1119–1125.

[66] *R.M. Aron, L.A. Moraes, O.W. Paques*, Lifting of holomorphic mappings, Proc. Royal Irish Acad. Sect. A, **94**, 1, 1994, 119–126.

[67] *R.M. Aron, L.A. Moraes, R.A. Ryan*, Factorization of holomorphic mappings in infinite dimensions, Math. Ann., **277**, 1987, 617–628.

[68] *R.M. Aron, J.B. Prolla*, Polynomial approximation of differentiable functions on Banach spaces, J. Reine Angew. Math., **313**, 1980, 195–216.

[69] *R.M. Aron, P. Rueda*, Homomorphisms on spaces of weakly continuous holomorphic functions, Arch. Math. (Basel) (to appear).

[70] *R.M. Aron, M. Schottenloher,* Compact holomorphic mappings on Banach spaces and the approximation property, Bull. Amer. Math. Soc., **80**, 6, 1974, 1245–1249.

[71] *R.M. Aron, M. Schottenloher,* Compact holomorphic mappings on Banach spaces and the approximation property, J. Funct. Anal., **21**, 1976, 7–30.

[72] *R.M. Aron, I. Zalduendo,* Polynomial norms and coefficients (preprint).

[73] *O.G. Astashina,* The completeness of a multi-dimensional system in the space of functions which are analytic in an unbounded multicircular domain, Izv. Vyssh. Uchrbn. Zaved. Mat., **30**, 2, 1986, 50–52.

[74] *V. Aurich,* The spectrum as envelope of holomorphy of a domain over an arbitrary product of lines, Proc. on Infinite Dimensional Holomorphy, Ed. T.L. Hayden, T.J. Suffridge, Lecture Notes in Math., **364**, 1974, 109–122.

[75] *V. Aurich,* Fonctions méromorphes sur $\mathbb{C}^{\wedge}$, Infinite Dimensional Holomorphy and Applications, Ed. M.C. Matos, North-Holland Math. Stud., **12**, 1977, 19–30.

[76] *V. Aurich,* Das meromorphe Levi Problem in unendlichdimensionalen Banachräumen, Bayer Akad. Wiss. Math. Natur. Kl. Sitzungsber, **5**, 1979, 35–42.

[77] *V. Aurich,* Der invariante Kontinuitätssatz für meromorphe Funktionen, Manuscripta Math., **31**, 1980, 149–166.

[78] *V.I. Averbukh, O.G. Smolyanov,* The theory of differentiation in linear topological spaces, Russian Math. Surveys, **22**, 4, 1967, 201–258.

[79] *V.I. Averbukh, O.G. Smolyanov,* The various definitions of the derivative in linear topological spaces, Russian Math. Surveys, **23**, 4, 1968, 67–113.

[80] *P. Avilés, J. Mujica,* Holomorphic germs and homogeneous polynomials on quasi-normable metrizable spaces, Rend. Math. (6), **10**, 1, 1977, 117–127.

[81] *A. Baernstein,* Representation of holomorphic functions by boundary integrals, Trans. Amer. Math. Soc., **160**, 1971, 27–37.

[82] *R.R. Baldino,* Aplicações holomorfas em produtos cartesianos, Thesis, I.M.P.A., Rio de Janeiro, 1972.

[83] *S. Banach,* Théorie des Opérations Linéaires, Warsaw, 1932. (Republished by Chelsea Publishing Company, New York, 1978.)

[84] *S. Banach,* Über n-lineare symmetrische Formen, Ann. Polon. Math., **12**, 1933, 116–117.

[85] *S. Banach,* Über Homogene Polynome in (L^2), Studia Math., **7**, 1938, 36–44.

[86] *J.A. Barroso,* Topologies sur les espaces d'applications holomorphes entre des espaces localement convexes, C. R. Acad. Sci. Paris, Sér. A, **271**, 1970, 264–265.

[87] *J.A. Barroso,* Topologias nos espaços de aplicações holomorfas entre espaços localmente convexos, An. Acad. Brasil. de Ciênc., **43**, 1971, 527–546.

[88] *J.A. Barroso,* Comparaison de topologies sur les espaces d'applications holomorphes, Séminaire P. Lelong/H. Skoda, 1978/79, Lecture Notes in Math., **822**, 1980, 18–32.

[89] *J.A. Barroso,* Introduction to Holomorphy, North-Holland Math. Stud., **106**, 1985.

[90] *J.A. Barroso, S. Dineen,* Holomorphic Functions on $\mathbb{C}^I$, I uncountable, Note Mat., **X**, suppl. no. 1, 1990, 65–71.

[91] *J.A. Barroso, M.C. Matos, L. Nachbin,* On bounded sets of holomorphic mappings, Proc. on Infinite Dimensional Holomorphy, Ed. T.L. Hayden, T.J. Suffridge, Lecture Notes in Math., **364**, 1974, 123–134.

[92] *J.A. Barroso, M.C. Matos, L. Nachbin,* On holomorphy versus linearity in classifying locally convex spaces, Infinite Dimensional Holomorphy and Applications, Ed. M.C. Matos, North-Holland Math. Stud., **12**, 1977, 31–74.

[93] *J.A. Barroso, L. Nachbin,* Some topological properties of holomorphic mappings in infinitely many variables, Advances in Holomorphy, Ed. J.A. Barroso, North-Holland Math. Stud., **34**, 1979, 67–91.

[94] *J.A. Barroso, L. Nachbin,* A direct sum is holomorphically bornological with the topology induced by the Cartesian product, Portugal. Math., **40**, 2, 1981, 252–256.

[95] *A. Bayoumi,* Bounding subsets of some metric vector spaces, Ark. Mat., **18**, 1, 1980, 13–17.

[96] *S. Bellenot, E. Dubinsky,* Fréchet spaces with nuclear Köthe quotients, Trans. Amer. Math. Soc., **273**, 1982, 579–594.

[97] *C. Benítez, M.C. Otero,* On the extension of continuous 2-polynomials in normed linear spaces, Geometric Aspects of Banach Spaces, London Math. Soc. Lecture Notes Ser., **140**, 1989, 125–132.

[98] *C. Benítez, Y. Sarantopoulos,* Characterization of real inner product spaces by means of symmetric bilinear forms, J. Math. Anal. Appl., **180**, 1993, 207–220.

[99] *C. Benítez, Y. Sarantopoulos, A. Tonge,* Lower bounds for norms of products of polynomials, Math. Proc. Cambridge Philos. Soc., **124**, 1998, 395–408.

[100] *A. Benndorf,* Über die Beziehungen zwischen einigen beschränkten Approximationseigenschaften in nuklearen Fréchet Räumen, Dissertation, Darmstadt, 1980.

[101] *A. Benndorf,* On bounded approximation properties of spaces of holomorphic functions on certain open subsets of strong duals of nuclear Fréchet spaces, Arch. Math. (Basel), **38**, 3, 1982, 248–257.

[102] *A. Benndorf*, On the relation of the bounded approximation property and a finite dimensional decomposition in nuclear Fréchet spaces, Studia Math., **75**, 1983, 103–119.

[103] *G. Bennett, V. Goodman, C. Newman*, Norms of random matrices, Pacific J. Math., **59**, 1975, 359–365.

[104] *J.A. Berezanskiĭ*, Inductively reflexive locally convex spaces, Soviet Math. Dokl., **9**, 1968, 1080–1082.

[105] *P. Berner*, Sur la topologie de Nachbin de certains espaces de fonctions holomorphes, C. R. Acad. Sci. Paris, Sér. A, **280**, 1975, 431–433.

[106] *P. Berner*, A global factorization property for holomorphic functions spread over a surjective limit, Séminaire P. Lelong, 1974–1975, Lecture Notes in Math., **524**, 1976, 130–155.

[107] *P. Berner*, Holomorphy on surjective limits of locally convex spaces, Atti. Accad. Naz. Lincei (8), LX, 6, 1976, 760–762.

[108] *P. Berner*, Topologies on spaces of holomorphic functions on certain surjective limits, Infinite Dimensional Holomorphy and Applications, Ed. M.C. Matos, North-Holland Math. Stud., **12**, 1977, 75–92.

[109] *P. Berner*, Convolution operators and surjective limits, Advances in Holomorphy, Ed. J.A. Barroso, North-Holland Math. Stud., **34**, 1979, 93–102.

[110] *S.N. Bernstein*, Leçons sur les Propriétes Extremales et la Meilleure Approximation des Fonctions Analytiques d'une Variable Réelle, Gauthier-Villars, Paris, 1926.

[111] *C. Bessaga, A. Pełczyński*, On a class of B_0 spaces, Bull. Polish Acad. Sci. Math., **5**, 4, 1957, 375–377.

[112] *M. Bianchini*, $H(E)$-bounded subsets of a locally convex space, Advances in Holomorphy, Ed. J.A. Barroso, North-Holland Math. Stud., **34**, 1979, 103–110.

[113] *M. Bianchini*, Silva-holomorphy types, Borel transforms and partial differential operators, Functional Analysis, Holomorphy and Approximation Theory, Ed. S. Machado, Lecture Notes in Math., **843**, 1980, 55–92.

[114] *M. Bianchini, O.W. Paques, M.C. Zaine*, On the strong compactported topology for spaces of holomorphic mappings, Pacific J. Math., **77**, 1, 1978, 33–49.

[115] *K.-D. Bierstedt*, An introduction to locally convex inductive limits, Functional Analysis and its Applications, World Sci. Publ., Singapore, 1988, 35–133.

[116] *K.-D. Bierstedt, J. Bonet*, Stefan Heinrich's density condition for Fréchet spaces and the characterisation of the distinguished Köthe spaces, Math. Nachr., **135**, 1988, 149–180.

[117] *K.-D. Bierstedt, J. Bonet*, Biduality in Fréchet and (LB)-spaces, Progress in Functional Analysis, Ed K.-D. Bierstedt et al., North-Holland Math. Stud., **170**, 1992, 113–133.

[118] *K.-D. Bierstedt, J. Bonet, A. Peris,* Vector-valued holomorphic germs on Fréchet-Schwartz spaces, Proc. Royal Irish Acad. Sect. A, **94**, 1, 1994, 31–46.

[119] *K.-D. Bierstedt, B. Gramsch, R. Meise,* Approximationseigenschaft, Lifting und Kohomologie bei localkonvexen Produkt-garben, Manuscripta Math., **19**, 1976, 319–364.

[120] *K.-D. Bierstedt, R. Meise,* $H(K)$ et $\bigl(H(U), \tau_\omega\bigr)$ sur les espaces métrisables nucléaires ou de Schwartz, C. R. Acad. Sci. Paris, Sér. A, **283**, 1976, 325–327.

[121] *K.-D. Bierstedt, R. Meise,* Nuclearity and the Schwartz property in the theory of holomorphic functions on metrizable locally convex spaces, Infinite Dimensional Holomorphy and Applications, Ed. M.C. Matos, North-Holland Math. Stud., **12**, 1977, 93–129.

[122] *K.-D. Bierstedt, R. Meise,* Aspects of inductive limits in spaces of germs of holomorphic functions on locally convex spaces and applications to the study of $\bigl(H(U), \tau_\omega\bigr)$, Advances in Holomorphy, Ed. J.A. Barroso, North-Holland Math. Stud., **34**, 1979, 111–178.

[123] *E. Bishop,* Analytic functions with values in a Fréchet space, Pacific J. Math., **12**, 1962, 1177–1192. (Reprinted in E. Bishop, Selected Papers, World Scientific, 1986.)

[124] *P. Biström, J.A. Jaramillo, M. Lindström,* Polynomial compactness in Banach spaces, Rocky Mountain J. Math., **28**, 4, 1998, 1203–1226.

[125] *S. Bjon,* Differentiation under the integral sign and holomorphy, Math. Scand., **60**, 1987, 77–95.

[126] *S. Bjon,* On the exponential law for spaces of holomorphic mappings, Math. Nachr., **131**, 1987, 201–204.

[127] *S. Bjon, M. Lindström,* A general approach to infinite-dimensional holomorphy, Monatsh. Math., **101**, 1986, 11–26.

[128] *S. Bjon, M. Lindström,* Algebras of holomorphic functions, J. Math. Anal. Appl., **128**, 1, 1987, 207–213.

[129] *F. Blasco,* Complementación, casinormabilidad y tonelación en espacios de polinomios, Thesis, Universidad Complutense de Madrid, 1996.

[130] *F. Blasco,* Complementation in spaces of symmetric tensor products and polynomials, Studia Math., **123**, 2, 1997, 165–173.

[131] *F. Blasco,* Polynomials on Köthe echelon spaces, Arch. Math. (Basel), **70**, 1998, 147–152.

[132] *F. Blasco,* On X-Köthe echelon spaces and applications, Math. Proc. Royal Irish Acad., **98**A, 2, 187–199.

[133] *H.P. Boas,* The football player and the infinite series, Notices Amer. Math. Soc., **44**, 11, 1997, 1430–1435.

[134] *H.P. Boas, D. Khavison,* Bohr's power series theorem in several variables, Proc. Amer. Math. Soc., **125**, 10, 1997, 2975–2979.

[135] *J. Bochnak,* Analytic functions in Banach spaces, Studia Math., **35**, 1970, 273–292.

[136] *J. Bochnak, J. Siciak,* Fonctions analytiques dans les espaces vectoriels topologiques réel ou complexes, C. R. Acad. Sci. Paris, Sér. A, **270**, 1970, 643–646.

[137] *J. Bochnak, J. Siciak,* Polynomials and multilinear mappings in topological vector spaces, Studia Math., **39**, 1971, 59–76.

[138] *J. Bochnak, J. Siciak,* Analytic functions in topological vector spaces, Studia Math., **39**, 1971, 77–112.

[139] *W. Bogdanowicz,* On the weak continuity of the polynomials functions defined on the space c_0, Bull. Polish Acad. Sci. Math., **5**, 1957, 243–246.

[140] *W. Bogdanowicz,* Integral representation of multilinear continuous operators from the space of Lebesgue–Bochner summable functions into any Banach space, Bull. Amer. Math. Soc., **72**, 1966, 317-321.

[141] *W. Bogdanowicz,* Analytic continuation of holomorphic functions with values in a locally convex space, Proc. Amer. Math. Soc., **22**, 1969, 660–666.

[142] *H.F. Bohnenblust, E. Hille,* On the absolute convergence of Dirichlet series, Ann. of Math. (2), **32**, 1931, 600–622.

[143] *H. Bohr,* Über die Bedeutung der Potenzreihen unendlich vieler Variablen in der Theorie der Dirichletschen Reihen $\sum \frac{a_n}{n^s}$, Nachrichten von der Königlichen Gesellschaft der Wissenschaften zu Göttingen, 1913, 441–488.

[144] *H. Bohr,* A theorem concerning power series, Proc. London Math. Soc. (2), **13**, 1914, 1–5.

[145] *P.J. Boland,* Espaces pondérés de fonctions entières et de fonctions entières nucleaires sur des espaces de Banach, C. R. Acad. Sci. Paris, Sér. A, **275**, 1972, 587–590.

[146] *P.J. Boland,* Malgrange theorem for entire functions on nuclear spaces, Proc. on Infinite Dimensional Holomorphy, Ed. T.L. Hayden, T.J. Suffridge, Lecture Notes in Math., **364**, 1974, 38–60.

[147] *P.J. Boland,* Some spaces of entire and nuclearly entire functions on a Banach space I, J. Reine Angew. Math., **270**, 1974, 38–60.

[148] *P.J. Boland,* Some spaces of entire and nuclearly entire functions on a Banach space II, J. Reine Angew. Math., **271**, 1974, 8–27.

[149] *P.J. Boland,* Holomorphic functions on nuclear spaces. Trans. Amer. Math. Soc., **209**, 1975, 275–281.

[150] *P.J. Boland,* An example of a nuclear space in infinite dimensional holomorphy, Ark. Mat., **15**, 1977, 87–91.

[151] *P.J. Boland,* Duality and spaces of holomorphic functions, Infinite Dimensional Holomorphy and Applications, Ed. M.C. Matos, North-Holland Math. Stud., **12**, 1977, 131–138.

[152] *P.J. Boland, S. Dineen,* Fonctions holomorphes sur des espaces pleinement nucléaires, C. R. Acad. Sci. Paris, Sér. A, **286**, 1978, 1235–1237.

[153] *P.J. Boland, S. Dineen,* Holomorphic functions on fully nuclear spaces, Bull. Soc. Math. France, **106**, 1978, 311–336.

[154] *P.J. Boland, S. Dineen,* Duality theory for spaces of germs and holomorphic functions on nuclear spaces, Ed. J.A. Barroso, North-Holland Math. Stud., **34**, 1979, 179–207.

[155] *P.J. Boland, S. Dineen,* Holomorphy on spaces of distributions, Pacific J. Math., **92**, 1, 1981, 27–34.

[156] *F. Bombal,* On polynomial properties in Banach spaces, Atti. Sem. Mat. Fis. Univ. Modena, **44**, 1, 1996, 135–146.

[157] *J. Bonet,* The countable neighbourhood property and tensor products, Proc. Edinburgh Math. Soc., **28**, 1985, 207–215.

[158] *J. Bonet,* On the identity $L(E; F) = LB(E; F)$ for pairs of locally convex spaces E and F, Proc. Amer. Math. Soc., **69**, 2, 1987, 249–255.

[159] *J. Bonet, S. Dierolf,* On distinguished Fréchet spaces, Progress in Functional Analysis, Ed. K.-D. Bierstedt et al., North–Holland Math. Stud., **170**, 1992, 201–214.

[160] *J. Bonet, P. Dománski, J. Mujica,* Complete spaces of vector-valued holomorphic germs, Math. Scand., **75**, 1995, 250–260.

[161] *J. Bonet, A. Galbis,* The identity $L(E; F) = LB(E; F)$, tensor products and inductive limits, Note Mat., **9**, 2, 1989, 195–216.

[162] *J. Bonet, P. Galindo, D. García, M. Maestre,* Locally bounded sets of holomorphic mappings, Trans. Amer. Math. Soc., **309**, 2, 1988, 609–620.

[163] *J. Bonet, M. Lindström,* Convergent sequences in duals of Fréchet spaces, Functional Analysis, Essen 1991, Lecture Notes in Pure and Appl. Math., **150**, 1994, 391–404.

[164] *J. Bonet, M. Lindstöm, M. Valdivia,* Two theorems of Josefson-Nissenzweig type for Fréchet spaces, Proc. Amer. Math. Soc., **117**, 1993, 363–364.

[165] *J. Bonet, M. Maestre, G. Metafune, V.B. Moscatelli, D. Vogt,* Every quojection is the quotient of a countable product of Banach spaces, Advances in the Theory of Fréchet Spaces, Ed. T. Terzioğlu, Kluwer Academic Publishers, Ser. C, 1989, 355–356.

[166] *J. Bonet, A. Peris,* On the injective tensor product of quasi-normable spaces, Results Math., **20**, 1991, 431–443.

[167] *R. Bonic, J. Frampton,* Differentiable functions on certain Banach spaces, Bull. Amer. Math. Soc., **71**, 1965, 393–395.

[168] *R. Bonic, J. Frampton,* Smooth functions on Banach manifolds, J. Math. Mech., **15**, 1966, 877–898.

[169] *M. Börgens, R. Meise, D. Vogt,* Functions holomorphes sur certaines espaces échelonnés et λ-nucléarité, C. R. Acad. Sci. Paris, Sér. A, **290**, 1980, 229–232.

[170] *M. Börgens, R. Meise, D. Vogt,* Entire functions on nuclear sequence spaces, J. Reine Angew. Math., **322**, 1981, 196–220.

[171] *M. Börgens, R. Meise, D. Vogt,* $\Lambda(\alpha)$-nuclearity in infinite dimensional holomorphy, Math. Nachr., **106**, 1982, 129–146.

[172] *G. Botelho,* Cotype and absolute summing multilinear mappings and homogeneous polynomials, Proc. Royal Irish Acad. Sect. A, **97**, 2, 1997, 145–153.

[173] *G. Botelho,* Type, cotype and the generalised Rademacher functions, Rocky Mountain J. Math. (to appear).

[174] *G. Botelho,* Almost summing polynomials (preprint).

[175] *J. Bourgain, J. Diestel,* Limited operators and strict cosingularity, Math. Nachr., **11**, 1984, 55–58.

[176] *C. Boyd,* Preduals of the space of holomorphic functions on a Fréchet space, Thesis, University College Dublin, National University of Ireland, 1992.

[177] *C. Boyd,* Montel and reflexive preduals of spaces of holomorphic functions on Fréchet spaces, Studia Math., **107**, 3, 1993, 305–315.

[178] *C. Boyd,* Distinguished preduals of spaces of holomorphic functions, Rev. Mat. Univ. Complut. Madrid, **6**, 2, 1993, 221–231.

[179] *C. Boyd,* Some topological properties of preduals of spaces of holomorphic functions, Proc. Royal Irish Acad. Sect. A, **94**, 2, 1994, 167–178.

[180] *C. Boyd,* Holomorphic functions on $\mathbb{C}^{(I)}$, I uncountable, Results Math. (to appear).

[181] *C. Boyd,* Exponential laws for the Nachbin ported topology, Canadian Math. Bull. (to appear).

[182] *C. Boyd,* Preduals of spaces of vector-valued holomorphic functions (preprint).

[183] *C. Boyd,* Holomorphic germs on Schwartz spaces (preprint).

[184] *C. Boyd,* Holomorphic functions and the BB-property on product spaces (preprint).

[185] *C. Boyd,* Duality and reflexivity of spaces of approximable polynomials on locally convex spaces (preprint).

[186] *C. Boyd,* Polynomially significant properties and the bounded approximation property (preprint).

[187] *C. Boyd, S. Dineen,* Locally bounded subsets of holomorphic functions, Comp. Appl. Math., **13**, 3, 1994, 189–194.

[188] *C. Boyd, S. Dineen,* Compact sets of holomorphic functions, Math. Nachr., **193**, 1998, 27–36.

[189] *C. Boyd, A. Peris,* A projective description of the Nachbin-ported topology, J. Math. Anal. Appl., **197**, 3, 1996, 635–657.

[190] *C. Boyd, R.A. Ryan,* Bounded weak continuity of homogeneous polynomials at the origin, Arch. Math. (Basel), **71**, 3, 1998, 211–218.

[191] *C. Boyd, R.A. Ryan,* Geometric theory of spaces of integral polynomials and symmetric tensor products (preprint).

[192] *C. Boyd, R.A. Ryan,* Inequalities for the factor of a polynomial in infinite dimensions (preprint).

[193] *R.W. Braun,* Linear topological structure of closed ideals in certain F-algebras, Proc. Royal Irish. Acad. Sect. A, **8**, 1, 1987, 35–44.

[194] *H.A. Braunss,* On holomorphic mappings of Schatten class type, Arch. Math. (Basel), **59**, 1992, 450–456.

[195] *H.A. Braunss, H. Junek,* Bilinear mappings and operator ideals, Proc. of the 13th Winter School on Abstract Analysis (Srni, 1985), Rend. Circ. Mat. Palermo (2), Suppl. (1985), **10**, 1986, 25–35.

[196] *H.A. Braunss, H. Junek,* On types of polynomials and holomorphic functions on Banach spaces, Note Mat., **X**, 1, 1990, 47–58.

[197] *H.J. Bremermann,* Complex convexity, Trans. Amer. Math. Soc., **82**, 1956, 17–51.

[198] *H.J. Bremermann,* Holomorphic functionals and complex convexity in Banach spaces, Pacific J. Math., **7**, 1957, 811–831.

[199] *H.J. Bremermann,* The envelope of holomorphy of tube domains in Banach spaces, Pacific J. Math., **10**, 1960, 1149–1153.

[200] *H.J. Bremermann,* Pseudo-convex domains in linear topological spaces, Proc. Conf. Complex Analysis, Minneapolis 1964, Springer, Berlin, 1965, 182–186.

[201] *F. Cabello Sánchez, J.M.F. Castillo, R. García,* Polynomials on dual isomorphic spaces, Arch. Math (Basel) (to appear).

[202] *D. Carando,* Extendible polynomials on Banach spaces (preprint).

[203] *D. Carando, V. Dimant,* Duality in spaces of nuclear and integral polynomials, J. Math. Ann. Appl. (to appear).

[204] *D. Carando, V. Dimant, B. Duarte, S. Lassalle,* K-bounded polynomials, Math. Proc. Royal Irish. Acad., **98**A, 2, 1998, 159–171.

[205] *D. Carando, I. Zalduendo,* A Hahn-Banach theorem for integrable polynomials, Proc. Amer. Math. Soc., **127**, 1999, 241–250.

[206] *B. Carl, A. Defant,* Asymptotic estimates for approximation quantities of tensor product identities, J. Approx. Theory, **88**, 2, 1997, 228–256.

[207] *T.K. Carne, B.J. Cole, T.W. Gamelin,* A uniform algebra of analytic functions on a Banach space, Proc. Amer. Math. Soc., **314**, 1989, 639–659.

[208] *H. Cartan,* Some applications of the new theory of Banach analytic spaces, J. Lond. Math. Soc., **41**, 1, 1966, 70–78.

[209] *P. Casazza,* Review of "Classical sequences in Banach spaces" by S. Guerre-Delabrière, Bull. Amer. Math. Soc., **30**, 1, 1994, 117–124.

[210] *J.M.F. Castillo,* Extraction of subsequences in Banach spaces, Extracta Math., **7**, 2, 1992, 77–88.

[211] *J.M.F. Castillo,* private communication, 1995.

[212] *J.M.F. Castillo, R. García, R. Gonzalo,* Banach spaces in which all multilinear forms are weakly sequentially continuous, Studia Math. (to appear).

[213] *S.B. Chae,* Sur les espaces localement convexes de germs holomorphes, C. R. Acad. Sci. Paris, Sér. A, **271**, 1970, 990–991.

[214] *S.B. Chae,* Holomorphic germs on Banach spaces, Ann. Inst. Fourier (Grenoble) **21**, 1971, 107–141.

[215] *S.B. Chae,* Holomorphy and calculus in normed spaces, Marcel Dekker, 1985.

[216] *S.D. Chatterji,* Continuous functions representable as sums of independent random variables, Z. Wahrsch., **13**, 1969, 338–341.

[217] *N. Cherfaoui,* Rayon de bornologie des fonctions plurisousharmoniques sur les espaces de Banach, Thése de 3eme cycle, Lille, 1980.

[218] *J. Chmielowski,* Constructions des ensembles déterminants pour les fonctions analytiques, Studia Math., **54**, 1976, 141–146.

[219] *J. Chmielowski, G. Lubczonok,* A property of determining sets for analytic functions, Studia Math., **40**, 1977, 285–288.

[220] *Y.S. Choi,* Spaces of holomorphic mappings on locally convex spaces, Portugal. Math., **45**, 3, 1988, 273–293.

[221] *Y.S. Choi,* Concerning conditions for uniform factorization and uniform holomorphy, J. Math. Anal. Appl., **135**, 2, 1988, 611–614.

[222] *Y.S. Choi,* Norm attaining bilinear forms on $L^1[0,1]$, J. Math. Anal. Appl., **211**, 2, 1997, 295–300.

[223] *Y.S. Choi, H. Ki, S.G. Kim,* Extreme polynomials and multilinear forms on l_1, J. Math. Ann. Appl., **228**, 1998, 467–484.

[224] *Y.S. Choi, S.G. Kim,* Polynomial properties of Banach spaces, J. Math. Anal. Appl., **190**, 1995, 203–210.

[225] *Y.S. Choi, S.G. Kim,* Norm or numerical radius attaining multilinear mappings and polynomials, J. London Math. Soc., (2), **54**, 1996, 135–147.

[226] *Y.S. Choi, S.G. Kim,* Smooth points of the unit ball of the space $\mathcal{P}(^2l_1)$, Results Math. (to appear).

[227] *Y.S. Choi, S.G. Kim,* The unit ball of $\mathcal{P}(^2l_2^2)$, Arch. Math. (Basel), **71**, 1998, 472–480.

[228] *Y.S. Choi, S.G. Kim, Y. Meléndez, A. Tonge,* Estimates for absolutely summing norms of polynomials and multilinear mappings (preprint).

[229] *D. Clayton,* A reduction of the continuous homomorphism problem for F-algebras, Rocky Mountain J. Math., **5**, 1975, 337–344.

[230] *G. Coeuré,* Le théorème de convergence sur les espaces localement convexes complexes, C. R. Acad. Sci. Paris, Sér. A, **264**, 1967, 287–288.

[231] *G. Coeuré,* Fonctions plurisousharmoniques et fonctions ℂ-analytiques a une infinite de variables, C. R. Acad. Sci. Paris, Sér. A, **267**, 1968, 440–442, 473–476, erratum, 816.

[232] *G. Coeuré*, Fonctions plurisousharmoniques sur les espaces vectoriels topologiques et applications a l'etude des fonctions analytiques, Ann. Inst. Fourier (Grenoble), **20**, 1, 1970, 361–432.

[233] *G. Coeuré*, Fonctionnelles analytiques sur certains espaces de Banach, Ann. Inst. Fourier (Grenoble), **21**, 2, 1971, 15–21.

[234] *G. Coeuré*, Propriétè de Runge et enveloppe d'holomorphie de certaines variétés analytiques de dimensions infinies, Bull. Soc. Math. France, **102**, 1974, 281–288.

[235] *G. Coeuré*, Analytic functions and manifolds in infinite dimensional spaces, North-Holland Math. Stud., **11**, 1974.

[236] *G. Coeuré*, Prolongement analytique en dimension infinie, Functional Analysis and Applications, Lecture Notes in Math., **384**, 1974, 1–19.

[237] *G. Coeuré*, O-completion of normed spaces, Analyse fonctionnelle et applications, Ed. L. Nachbin, Hermann, 1975, 91–93.

[238] *G. Coeuré*, Sur le rayon de bornologie des fonctions holomorphes, Journées de Fonctions Analytiques, Toulouse 1976, Lecture Notes in Math., **578**, 1977, 183–194.

[239] *G. Coeuré*, L'equation $(\bar{\partial}u = F)^*$ en dimension infinie, Journées Bruxelles-Lille-Mons d'Analyse Fonctionnelle et Équations aux dérivées partielles, Université de Lille Publications Internes, **131**, 1978, 6–9.

[240] *A. Colojoară*, Sur l'algébra symétrique de certains espaces de Banach, Rev. Roumaine Math. Pures Appl., **18**, 9, 1973, 1345–1369.

[241] *A. Colojoară*, Algébra symétrique du dual d'un espace nucléaire DF, Rev. Roumaine Math. Pures Appl., **23**, 1978, 1317–1339.

[242] *J.F. Colombeau*, Sur les applications G-analytiques et analytiques en dimension infinie, Seminaire P. Lelong, 1971/72, Lecture Notes in Math., **332**, 1973, 48–58.

[243] *J.F. Colombeau*, Holomorphy in locally convex spaces and operators on the Fock spaces, Séminaire P. Lelong/H. Skoda, 1978/79, Lecture Notes in Math., **822**, 1980, 46–60.

[244] *J.F. Colombeau, M.C. Matos*, Convolution equations in spaces of infinite dimensional entire functions, Indag. Math., **42**, 1980, 375–389.

[245] *J.F. Colombeau, M.C. Matos*, Convolution equations in infinite dimensions: brief survey, new results and proofs, Functional Analysis, Holomorphy and Approximation Theory, Ed. J.A. Barroso, North-Holland Math. Stud., **71**, 1982, 131–178.

[246] *J.F. Colombeau, R. Meise*, Strong nuclearity in spaces of holomorphic mappings, Advances in Holomorphy, Ed. J.A. Barroso, North-Holland Math. Stud., **34**, 1979, 233–248.

[247] *J.F. Colombeau, R. Meise, B. Perrot*, A density result in spaces of Silva holomorphic mappings, Pacific J. Math., **84**, 1, 1979, 35–42.

[248] *J.F. Colombeau, J. Mujica*, The Levi problem in nuclear Silva spaces, Ark. Mat., **18**, 1, 1980, 117–123.

[249] *J.F. Colombeau, J. Mujica,* Holomorphic and differentiable mappings of uniform bounded type, Functional Analysis, Holomorphy and Approximation Theory, Ed. J.A. Barroso, North-Holland Math. Stud., **71**, 1982, 179–200.

[250] *J.F. Colombeau, B. Perrot,* Reflexivity and kernels in infinite dimensional holomorphy, Portugal. Math., **36**, 3–4, 1977, 291–300.

[251] *J.F. Colombeau, B. Perrot,* Une caractérisation de la nucléarité des espaces de fonctions holomorphes en dimension infinie, C. R. Acad. Sci. Paris, Sér. A, **284**, 1977, 1275–1278.

[252] *J.F. Colombeau, B. Perrot,* Transformation de Fourier-Borel et réflexivité dans les espaces d'applications Silva analytiques à valeurs vectorielles; applications, C. R. Acad. Sci. Paris, Sér. A, **285**, 1977, 19–21.

[253] *T.A. Cook,* Schauder decompositions and semi-reflexivity, Math. Ann., **182**, 1969, 232–235.

[254] *J.G. van der Corput, G. Schaake,* Ungleichungen für Polynome und Trigonometrische Polynome, Compositio Math., **2**, 1935, 321–361. Corrections, **3**, 1936, 128.

[255] *A.M. Davie,* Quotient algebras of uniform algebras, J. London Math. Soc., **7**, 1973, 31–40.

[256] *A.M. Davie, T.W. Gamelin,* A theorem on polynomial-star approximation, Proc. Amer. Math. Soc., **106**, 1989, 351–356.

[257] *A. Defant,* The local Radon-Nikodým property for duals of locally convex spaces, Bull. Soc. Royal Sci. Liège, **53**, 5, 1984, 233–246.

[258] *A. Defant, K. Floret,* Tensor norms and operator ideals, North-Holland Math. Stud., **176**, 1993.

[259] *A. Defant, W. Govaerts,* Tensor products and spaces of vector-valued continuous functions, Manuscripta Math., **55**, 1986, 433–449.

[260] *A. Defant, M. Maestre,* Property (*BB*) and holomorphic functions on Fréchet-Montel spaces, Math. Proc. Cambridge Philos. Soc., **115**, 2, 1994, 305–313.

[261] *A. Defant, L.A. Moraes,* On unconditional bases in spaces of holomorphic functions in infinite dimensions, Arch. Math. (Basel), **56**, 1991, 163–173.

[262] *J.C. Díaz,* A note on holomorphic functions of bounded type in Fréchet spaces, J. Math. Anal. Appl., **177**, 1, 1993, 308–313.

[263] *J.C. Díaz,* On 2-homogeneous polynomials on some non-stable Banach and Fréchet spaces, J. Math. Anal. Appl., **20**, 6, 1997, 322–331.

[264] *J.C. Díaz,* On non-primary Fréchet-Schwartz spaces, Studia Math., **126**, 3, 1997, 291–307.

[265] *J.C. Díaz,* On unconditional bases in tensor products of Köthe echelon spaces, Michigan Math. J., **44**, 1997, 409–415.

[266] *J.C. Díaz, S. Dineen,* Polynomials on stable Banach spaces, Ark. Mat., **36**, 1, 1998, 87–96.

[267] *J.C. Díaz, M.A. Miñarro,* On Fréchet Montel spaces and their projective tensor product, Math. Proc. Cambridge Philos. Soc., **11**, 1993, 335–341.

[268] *P. Dienes,* The Taylor Series, Oxford, 1931, Reprint Dover, New York, 1957.

[269] *S. Dierolf, K. Floret,* Über die Fortsetzberkeit stetiger Normen, Ark. Mat., **35**, 1980, 149–154.

[270] *J. Diestel,* Sequences and series in Banach spaces, Springer-Verlag, Graduate Texts in Math., **92**, 1984.

[271] *J. Diestel, H. Jarchow, A. Tonge,* Absolutely summing operators, Cambridge University Press, **43**, 1995.

[272] *J. Diestel, J.J. Uhl,* Vector measures, Amer. Math. Soc. Mathematical Surveys, **15**, 1979.

[273] *V. Dimant, S. Dineen,* Banach subspaces of spaces of holomorphic mappings and related topics, Math. Scand., **83**, 1998, 142–160.

[274] *V. Dimant, R. Gonzalo,* Block diagonal polynomials (preprint).

[275] *V. Dimant, I. Zalduendo,* Bases in spaces of multilinear forms over Banach spaces, J. Math. Anal. Appl., **200**, 1996, 548–566.

[276] *S. Dineen,* Holomorphic functions on a Banach space, Bull. Amer. Math. Soc., **76**, 4, 1970, 883–886.

[277] *S. Dineen,* The Cartan-Thullen theorem for Banach Spaces, Ann. Sci. École Norm. Sup. Pisa, **24**, 1970, 883–886.

[278] *S. Dineen,* Runge domains in Banach spaces, Proc. Royal Irish Acad. Sect. A, **71**, 1971, 85–89.

[279] *S. Dineen,* Holomorphy types on a Banach space, Studia Math., **39**, 1971, 241–288.

[280] *S. Dineen,* Bounding subsets of a Banach space, Math. Ann., **192**, 1971, 61–70.

[281] *S. Dineen,* Convexité holomorphe en dimension infinie, Séminaire P. Lelong 1970/71, Lecture Notes in Math., **275**, 1972, 177–181.

[282] *S. Dineen,* Fonctions analytiques dans les espaces vectoriels topologiques localement convexes, C. R. Acad. Sci. Paris, Sér. A, **274**, 1972, 544–546.

[283] *S. Dineen,* Holomorphic functions on (c_0, X_b)-modules, Math. Ann., **196**, 1972, 106–116.

[284] *S. Dineen,* Unbounded holomorphic functions on a Banach space, J. London Math. Soc. (2), **4**, 1971/1972, 461–465.

[285] *S. Dineen,* Holomorphically complete locally convex topological vector spaces, Séminaire P. Lelong (Analyse), 1971–1972, Lecture Notes in Math., **332**, 1973, 77–111.

[286] *S. Dineen,* Holomorphic functions on locally convex topological vector spaces I, Locally convex topologies on $H(U)$, Ann. Inst. Fourier (Grenoble), **23**, 1, 1973, 19–54.

[287] *S. Dineen*, Holomorphic functions on locally convex topological vector spaces II, Pseudo-convex domains, Ann. Inst. Fourier (Grenoble), **23**, 3, 1973, 155–185.

[288] *S. Dineen*, Sheaves of holomorphic functions on infinite dimensional spaces, Math. Ann., **202**, 1973, 337–345.

[289] *S. Dineen*, Holomorphically significant properties of topological vector spaces, Coll. Int. du CNRS, 1972, Fonctions Analytiques de Plusieurs Variables et Analyse Complexe. Agora Math., Gauthier-Villars, **1**, 1974, 25–34.

[290] *S. Dineen*, Holomorphic functions and surjective limits, Proc. on Infinite Dimensional Holomorphy, Ed. T.L. Hayden, T.J. Suffridge, Lecture Notes in Math., **364**, 1974, 1–12.

[291] *S. Dineen*, Equivalent definitions of holomorphy, Séminaire P. Lelong, 1973/74, Lecture Notes in Math., **474**, 1975, 114–122.

[292] *S. Dineen*, Surjective limits of locally convex spaces and their application to infinite dimensional holomorphy. Bull. Soc. Math. France, **103**, 1975, 441–509.

[293] *S. Dineen*, Cousin's first problem on certain locally convex topological vector spaces, An. Acad. Brasil. Ciênc., **48**, 1, 1976, 11–12.

[294] *S. Dineen*, Growth properties of pseudo-convex domains and domains of holomorphy in locally convex spaces, Math. Ann., **226**, 1977, 229–236.

[295] *S. Dineen*, Holomorphic functions on strong duals of Fréchet-Montel spaces, Infinite Dimensional Holomorphy and Applications, Ed. M.C. Matos, North-Holland Math. Stud., **12**, 1977, 147–166.

[296] *S. Dineen*, Holomorphic functions on nuclear sequence spaces, Functional Analysis: Surveys and Recent Results II, Ed . K.-D. Bierstedt, B. Fuchssteiner, North-Holland Math. Stud., **38**, 1979, 239–256.

[297] *S. Dineen*, Zero-one laws for probability measures on locally convex spaces, Ark. Mat., **17**, 1979, 217–233.

[298] *S. Dineen*, Topological properties inherited by certain subsets of holomorphic functions, Mathematical Analysis and Applications, Part A, Adv. in Math. Supp. Stud., **7A**, 1981, 317–326.

[299] *S. Dineen*, Complex Analysis on Locally Convex Spaces, North-Holland Math. Stud., **57**, 1981.

[300] *S. Dineen*, Holomorphic germs on compact subsets of locally convex spaces, Functional Analysis, Holomorphy and Approximation Theory, Ed. S. Machado, Lecture Notes in Math., **843**, 1981, 247–263.

[301] *S. Dineen*, Analytic functionals on fully nuclear spaces, Studia Math., **73**, 1982, 11–32.

[302] *S. Dineen*, Entire functions on c_0, J. Funct. Anal., **52**, 1983, 205–218.

[303] *S. Dineen*, Holomorphic functions on inductive limits of $\mathbb{C}^{\mathbb{N}}$, Proc. Royal Irish Acad. Sect. A, **86**, 2, 1986, 143–146.

502   References

[304] *S. Dineen*, Polar subsets of infinite dimensional spaces – small sets in large spaces, Contemp. Math., **54**, 1986, 9–16.

[305] *S. Dineen*, Monomial expansions in infinite dimensional holomorphy. Advances in the Theory of Fréchet spaces, Ed. T. Terzioğlu, Kluwer Academic Publishers, Series C, **287**, 1989, 155–171.

[306] *S. Dineen*, Holomorphic functions on Fréchet Montel spaces, J. Math. Anal. Appl., **163**, 1992, 581–587.

[307] *S. Dineen*, Quasi-normable spaces of holomorphic functions, Note Mat., **12**, 1, 1993, 155–195.

[308] *S. Dineen*, Holomorphic functions and the (*BB*)-property, Math. Scand., **74**, 1994, 215–236.

[309] *S. Dineen*, A Dvoretzky theorem for polynomials, Proc. Amer. Math. Soc., **123**, 9, 1995, 2817–2821.

[310] *S. Dineen*, Holomorphic functions and Banach-nuclear decompositions of Fréchet spaces, Studia Math., **113**, 1995, 43–54.

[311] *S. Dineen*, Holomorphic functions on Fréchet spaces with Schauder basis, Extracta Math., **10**, 2, 1995, 164–167.

[312] *S. Dineen*, Complex analysis in an infinite dimensional setting, Geometric Complex Analysis, Ed. J. Noguchi, World Scientific Publishing Company, 1996, 183–194.

[313] *S. Dineen*, Canonical mappings for polynomials and holomorphic mappings on Banach spaces (preprint).

[314] *S. Dineen, P. Galindo, D. García, M. Maestre*, Linearity and holomorphy on fully nuclear spaces with a basis, C. R. Acad. Sci. Paris, Sér. A, **314**, 1992, 715–718.

[315] *S. Dineen, P. Galindo, D. García, M. Maestre*, Linearization of holomorphic mappings on fully nuclear spaces with a basis, Glasgow Math. J., **36**, 1994, 201–208.

[316] *S. Dineen, A. Hirschowitz*, Sur le théorème de Levi banachique, C. R. Acad. Sci. Paris, Sér. A, **272**, 1971, 1245–1247.

[317] *S. Dineen, M. Lindström*, Spaces of homogeneous polynomials containing c_0 or l^∞, Functional Analysis, Ed. S. Dierolf, S. Dineen, P. Domanski, Walter de Gruyter & Co, Berlin, 1996, 183–194.

[318] *S. Dineen, M.L. Lourenço*, Holomorphic functions on strong duals of $\mathcal{DFM}$ spaces II, Arch. Math. (Basel), **53**, 1989, 590–598.

[319] *S. Dineen, R. Meise, D. Vogt*, Caractérisation des espaces de Fréchet nucléaires dans lesquels tous les bornés sont pluripolaires, C. R. Acad. Sci. Paris, Sér. I Math., **295**, 1982, 385–388.

[320] *S. Dineen, R. Meise, D. Vogt*, Characterisation of nuclear Fréchet spaces in which every bounded set is polar, Bull. Soc. Math. France, **112**, 1984, 41–68.

[321] *S. Dineen, R. Meise, D. Vogt*, Polar subsets of locally convex spaces, Aspects of Mathematics and its Applications, Ed. J.A. Barroso, North-Holland Math. Library, **34**, 1986, 295–319.

[322] *S. Dineen, L.A. Moraes,* Holomorphic functions on strict inductive limits of Banach spaces, Rev. Mat. Univ. Complut. Madrid, **5**, 2, 1992, 177–183.

[323] *S. Dineen, Ph. Noverraz,* Mesure gaussienne des ensembles polaires en dimension infinie, C. R. Acad. Sci. Paris, Sér. A, **287**, 1978, 225–228.

[324] *S. Dineen, Ph. Noverraz,* Gaussian measures and polar sets in locally convex spaces, Ark. Mat., **17**, 1979, 217–223.

[325] *S. Dineen, Ph. Noverraz, M. Schottenloher,* Le problème de Levi dans certains espaces vectoriels topologiques localement convexes, Bull. Soc. Math. France, **104**, 1976, 87–97.

[326] *S. Dineen, R.M. Timoney,* Absolute bases, tensor products and a theorem of Bohr, Studia Math., **94**, 1989, 227–234.

[327] *S. Dineen, R.M. Timoney,* On a problem of H. Bohr, Bull. Soc. Roy. Sci. Liège, **60**, 6, 1991, 401–404.

[328] *S. Dineen, R.M. Timoney,* Complex geodesics on convex domains, Progress in Functional Analysis, Ed. K.-D. Bierstedt et al., North-Holland Math. Stud., **170**, 1992, 333–365.

[329] *S. Dineen, M. Yoshida,* Tensor products and projections, Bull. Irish Math. Soc., **35**, 1995, 4–11.

[330] *J. Dixmier,* Sur un théorème de Banach, Duke Math. J., **151**, 1948, 1057–1071.

[331] *L.E. Dor,* On sequences spanning a complex l_1 space, Proc. Amer. Math. Soc., **47**, 2, 1975, 515–516.

[332] *A. Douady,* Le problème des modules pour les sous-espaces analytiques compacts d'un espace analytique donné, Ann. Inst. Fourier (Grenoble), **16**, 1, 1966, 1–95.

[333] *E. Dubinsky,* Nuclear Fréchet spaces without the bounded approximation property, Studia Math., **71**, 1981, 85–105.

[334] *E. Dubinsky,* Approximation properties in nuclear Fréchet spaces, Functional analysis, Holomorphy and Approximation Theory, Ed. J.A. Barroso, North-Holland Math. Stud., **71**, 1982, 215–234.

[335] *E. Dubinsky,* Approximation properties of nuclear Fréchet spaces, Advances in the Theory of Fréchet spaces, Ed. T. Terzioğlu, Kluwer Academic Publishers, Series C, **287**, 1989, 1–10.

[336] *N. Dunford,* Uniformity in linear spaces, Trans. Amer. Math. Soc., **44**, 1938, 305–356.

[337] *N. Dunford, J. Schwartz,* Linear Operators, Part I, Interscience, New York, 1976.

[338] *T.A.W. Dwyer III,* Partial differential equations in Fisher-Fock spaces for the Hilbert-Schmidt holomorphy type, Bull. Amer. Math. Soc., **77**, 1971, 725–730.

[339] *T.A.W. Dwyer III,* Convolution equations for vector-valued entire functions of nuclear bounded type, Trans. Amer. Math. Soc., **217**, 1976, 105–119.

[340] *T.A.W. Dwyer III*, Vector-valued convolution equations for the nuclear holomorphy type, Proc. Royal Irish Acad. Sect. A, **76**, 1976,
101–110.

[341] *T.A.W. Dwyer III*, Dualité des espaces de fonctions entiéres en dimension infinie, Ann. Inst. Fourier, **26**, 4, 1976, 151–195.

[342] *M. Estévez, C. Hervés*, Mesures Gaussiennes complexes et fonctions
plurisousharmoniques en dimension infinie, C. R. Acad. Sci. Paris, Sér.
A, **287**, 1978, 417–420.

[343] *M. Estévez, C. Hervés*, Application de mesures Gaussiennes complexes
aux fonctions plurisousharmoniques et analytiques, Bull. Sci. Math.,
105, 1981, 73–83.

[344] *J.M. Exbrayat*, Fonctions analytiques dans un espace de Banach
(d'après Alexander), Seminaire P. Lelong, 1969, Lecture Notes in
Math., **116**, 1970, 30–38.

[345] *L. Fantappié*, I funzionali analitici, Memorie della Academia Nazionale
dei Lincei, Ser. VI, **3**, 11, 1930, 453–683.

[346] *J.D. Farmer*, Polynomial reflexivity in Banach spaces, Israel J. Math.,
87, 1994, 257–273.

[347] *J.D. Farmer*, Fibres over the sphere of a uniformly convex Banach
space, Michigan Math. Jour., **45**, 1998, 211–226.

[348] *J.D. Farmer, W.B. Johnson*, Polynomial Schur and polynomial
Dunford-Pettis properties, Contemp. Math., **144**, 1993, 95–105.

[349] *J. Ferrera*, Mosco convergence of sequences of homogeneous polynomials, Rev. Mat. Univ. Complut. Madrid, **11**, 1998, 31–41.

[350] *J. Ferrera*, Norm-attaining polynomials and differentiability
(preprint).

[351] *J. Ferrera, J. Gómez Gil, J.G. Llavona*, On completion of spaces of
weakly continuous functions, Bull. Lond. Math. Soc., **15**, 1983, 260–
264.

[352] *K. Floret*, Über den Dualraüm eines lokalkonvexen Unterräumes,
Arch. Math. (Basel), **25**, 6, 1974, 646–548.

[353] *K. Floret*, Continuous norms on locally convex inductive limits, Math.
Z., **188**, 1984, 75–88.

[354] *K. Floret, M.C. Matos*, Application of a Khintchine inequality to holomorphic mappings, Math. Nachr., **176**, 1995, 65–72.

[355] *K. Floret, V.B. Moscatelli*, On bases in strict inductive and projective
limits of locally convex spaces, Pacific J. Math., **119**, 1, 1985, 103–113.

[356] *K. Floret, V.B. Moscatelli*, Unconditional bases in Fréchet-spaces,
Arch. Math. (Basel), **47**, 1986, 129–130.

[357] *M. Fréchet*, Une définition fontionnelle des polynômes, Nouv. Ann.
Math., **9**, 1909, 145–162.

[358] *M. Fréchet*, Sur les fonctionnelles continues, Ann. Sci. École Norm.
Sup., (3), **27**, 1910, 193–216.

[359] *M. Fréchet*, Sur les fonctionnelles bilinéaires, Trans. Amer. Math. Soc., **16**, 1915, 215–234.

[360] *M. Fréchet*, Les transformations ponctuelles abstraites, C. R. Acad. Sci. Paris, Sér. A, **180**, 1925, 1816–1817.

[361] *M. Fréchet*, La notion de différentielle dans l'analyse génerale, Ann. Sci. École Norm. Sup., (3), **42**, 1925, 293–323.

[362] *M. Fréchet*, Les polynômes abstraits, J. Math. Pures Appl., (9), **8**, 1929, 71–92.

[363] *A. Frölicher, W. Bucher*, Calculus in vector spaces without norm, Lecture Notes in Math., **30**, 1966.

[364] *Y. Fujimoto*, Analytic functional on a countably infinite dimensional topological vector space, Tokyo J. Math., **3**, 2, 1980, 271–289.

[365] *P. Galindo*, Polynomials and limited sets, Proc. Amer. Mat. Soc., **124**, 1996, 1481–1488.

[366] *P. Galindo, D. García, M. Maestre*, Holomorphically ultrabornological spaces and holomorphic inductive limits, J. Math. Anal. Appl., **124**, 1, 1987, 15–26.

[367] *P. Galindo, D. García, M. Maestre*, A remark on locally determining sequences in infinite dimensional spaces, Note Mat., **X**, 2, 1990, 267–272.

[368] *P. Galindo, D. García, M. Maestre*, The coincidence of τ_0 and τ_ω for spaces of holomorphic functions on some Fréchet-Montel spaces, Proc. Royal Irish Acad. Sect. A, **91**, 2, 1991, 137–143.

[369] *P. Galindo, D. García, M. Maestre*, Holomorphic mappings of bounded type on (DF)-spaces, Progress in Functional Analysis, Ed. K.-D. Bierstedt et al., North-Holland Math. Stud., **170**, 1992, 135–148.

[370] *P. Galindo, D. García, M. Maestre*, Holomorphic mappings of bounded type, J. Math. Anal. Appl., **166**, 1, 1992, 236–246.

[371] *P. Galindo, D. García, M. Maestre*, Entire functions of bounded type on Fréchet spaces, Math. Nachr., **161**, 1993, 185–198.

[372] *P. Galindo, D. García, M. Maestre, J. Mujica*, Extension of multilinear mappings to the bidual of a Banach space, Sem. Brasileiro de Análise, 1993, 311–322.

[373] *P. Galindo, D. García, M. Maestre, J. Mujica*, Extension of multilinear mappings on Banach spaces, Studia Math., **108**, 1, 1994, 55–76.

[374] *P. Galindo, M. Maestre, P. Rueda*, Biduality in spaces of holomorphic functions, Math. Scand. (to appear).

[375] *P. Galindo, L.A. Moraes, J. Mujica*, Weak holomorphic convergence and bounding sets in Banach spaces, Math. Proc. Royal Irish Acad., **98**A, 2, 153–157.

[376] *T.W. Gamelin*, Uniform Algebras, second edition, Chelsea Pub. Co., New York, 1984.

506   References

[377] *T. W. Gamelin*, Analytic functions in Banach spaces, Complex Function Theory, Ed. Gauthier and Sabidussi, Kluwer Academic Publishers, Amsterdam, 1994, 187–223.

[378] *D. García, M.L. Lourenço, L. Moraes, O.W. Paques*, The spectra of
some algebras of analytic mappings, Indag. Math. (to appear).

[379] *D. García, M. Maestre, P. Rueda*, Weighted spaces of holomorphic
functions on Banach spaces, Studia Math. (to appear).

[380] *D. García, J. Mujica*, Quasi-normable preduals of spaces of holomorphic functions, J. Math. Anal. Appl., **268**, 1997, 171–180.

[381] *L. Gårding*, An inequality for hyperbolic polynomials, J. Math. Mech.,
8, 1959, 957–965.

[382] *J. Garnett*, Bounded analytic functions, Academic Press, London-New
York, 1981.

[383] *R. Gâteaux*, Sur les fonctionnelles continues et les fonctionnelles analytiques, C. R. Acad. Sci. Paris, Sér. A, **157**, 1913, 325–327.

[384] *R. Gâteaux*, Fonctions d'une infinité des variables indépendantes, Bull.
Soc. Math. France, **47**, 1919, 70–96.

[385] *R. Gâteaux*, Sur diverses questions de calcul functionnel, Bull. Soc.
Math. France, **50**, 1922, 1–37.

[386] *S. Geiss*, Ein Faktorisierungssatz für multilineare Funktionale, Math.
Nachr., **134**, 1987, 149–159.

[387] *B.R. Gelbaum, J.G. de Lamadrid*, Bases of tensor products of Banach
spaces, Pacific J. Math., **11**, 1961, 1281–1286.

[388] *J. Globevnik*, Analytic functions whose range is dense in a ball, J.
Funct. Anal., **22**, 1976, 32–28.

[389] *J. Globevnik*, On the range of analytic maps on $c_0(\Gamma)$, Boll. Un. Mat.
Ital. A (5), **17**, 1980, 149–155.

[390] *J. Globevnik*, Separability of analytic images of some Banach spaces,
Boll. Un. Mat. Ital. A., (5), **17**, 1980, 149–155.

[391] *H. Goldman*, Uniform Fréchet Algebras, North-Holland Math. Stud.,
162, 1990.

[392] *J. Gómez, J.A. Jaramillo*, Interpolation by weakly differentiable functions on Banach spaces, J. Math. Anal. Appl., **182**, 2, 1994, 501–515.

[393] *M. González*, Remarks on Q-reflexive Banach spaces, Proc. Royal Irish
Acad. Sect. A, **96**, 2, 1996, 195–201.

[394] *M. González, J.M. Gutiérrez*, Weakly continuous mappings on Banach
spaces with the Dunford-Pettis property, J. Math. Anal. Appl., **173**,
2, 1993, 460–482.

[395] *M. González, J.M. Gutiérrez*, Unconditionally converging holomorphic
mappings between Banach spaces, Acta Univ. Carolin. Math. Phys.,
35, 2, 1994, 13–22.

[396] *M. González, J.M. Gutiérrez*, When every polynomial is unconditionally converging, Arch. Math. (Basel), **63**, 1994, 145–151.

[397] *M. González, J.M. Gutiérrez*, Polynomial Grothendieck properties, Glasgow Math. J., **37**, 1995, 211–219.

[398] *M. González, J.M. Gutiérrez*, Unconditionally converging polynomials on Banach spaces, Math. Proc. Cambridge Philos. Soc., **117**, 1995, 321–331.

[399] *M. González, J.M. Gutiérrez*, Factorization of weakly continuous holomorphic mappings, Studia Math., **118**, 2, 1996, 117–133.

[400] *M. González, J.M. Gutiérrez*, Schauder type theorems for differentiable and holomorphic mappings, Monatsh. Math., **122**, 1996, 325–343.

[401] *M. González, J.M. Gutiérrez*, Gantmacher type theorems for holomorphic mappings, Math. Nachr., **186**, 1997, 131–145.

[402] *M. González, J.M. Gutiérrez*, Injective factorization of holomorphic mappings, Proc. Amer. Math. Soc. (to appear).

[403] *M. González, J.M. Gutiérrez, J.G. Llavona*, Polynomial continuity on l_1, Proc. Amer. Math. Soc., **125**, 5, 1997, 1349–1353.

[404] *R. Gonzalo*, Multilinear forms, subsymmetric polynomials and spreading models on Banach spaces, J. Math. Anal. Appl., **202**, 1996, 379–397.

[405] *R. Gonzalo, J. Jaramillo*, Compact polynomials between Banach spaces, Proc. Royal Irish Acad. Sect. A, **95**, 1995, 213–226.

[406] *R. Gonzalo, J. Jaramillo*, Smoothness and estimates of sequences in Banach spaces, Israel J. Math., **89**, 1995, 321–341.

[407] *J.H. Grace*, On the zeros of a polynomial, Math. Proc. Cambridge Philos. Soc., **11**, 1900–1902, 352–357.

[408] *L.M. Graves*, Topics in the functional calculus, Part I, The theory of functionals, Bull. Amer. Math. Soc., **41**, 1935, 641–662.

[409] *M.M. Grinblyum*, On the representation of a space of type B in the form of a direct sum of subspaces, Doklady Akad. Nauk., SSRS (N.S.), **70**, 1950, 749–752.

[410] *L. Gruman*, The Levi problem in certain infinite dimensional vector spaces, Illinois J. Math., **18**, 1974, 20–26.

[411] *L. Gruman*, Le problème de Levi en dimension infinie, Coll. Int. du CNRS, 1972, Fonctions Analytiques de Plusieurs Variables et Analyse Complexe, Agora Math., Gauthier-Villars, **1**, 1974, 65–69.

[412] *L. Gruman, C.O. Kiselman*, Le problème de Levi dans les espaces de Banach à base, C. R. Acad. Sci. Paris, Sér. A, **274**, 1972, 821–824.

[413] *A. Grothendieck*, Sur certains espaces de fonctions holomorphes, J. Reine Angew. Math., **192**, 1953, Part I, 35–44, Part II, 77–95.

[414] *A. Grothendieck*, Produits tensoriels topologiques et espaces nucléaires, Mem. Amer. Math. Soc., **16**, 1955.

[415] *A. Grothendieck*, Résumé de la théorie métrique des produits tensoriels topologiques, Bol. Soc. Mat. São Paulo, **8**, 1953, 1–79.

508   References

[416] *A. Grothendieck*, Topological Vector Spaces, Gordon and Breach, New York, 1973.

[417] *B. Grünbaum*, Two examples in the theory of polynomial functionals, Reveon Lematematika, **11**, 1957, 56–60.

[418] *E. Grusell*, An example of a locally convex space which is not an ω-space, Ark. Mat., **12**, 1974, 213–216.

[419] *E. Grusell*, ω-spaces and σ-convex spaces, Infinite Dimensional Holomorphy and Applications, Ed. M.C. Matos, North-Holland Math. Stud., **12**, 1977, 211–216.

[420] *S. Guerre-Delabrière*, Classical sequences in Banach spaces, Pure and Applied Mathematics, **166**, Marcel-Dekker Inc., 1992.

[421] *C.P. Gupta*, On the Malgrange theorem for nuclearly entire functions of bounded type on a Banach space, Notas de Matemática, **37**, IMPA, Rio de Janeiro, 1968.

[422] *C.P. Gupta*, Convolution operators and holomorphic mappings of bounded type on a Banach space, Séminaire d'Analyse Moderne, No. 2, Dept. Math. Université Sherbrooke, Quebec, 1969.

[423] *C.P. Gupta*, On the Malgrange theorem for nuclearly entire functions of bounded type on a Banach space, Indag. Math., **32**, 1970, 356–358.

[424] *J.M. Gutiérrez*, Weakly continuous functions on Banach spaces not containing l_1, Proc. Amer. Math. Soc., **119**, 1993, 147–152.

[425] *J.M. Gutiérrez, J.A. Jaramillo, J.G. Llavona*, Polynomials and Geometry of Banach spaces, Extracta Math., **10**, 2, 1995, 79–114.

[426] *J.M. Gutiérrez, J.G. Llavona*, Polynomially continuous operators, Israel J. Math., **102**, 1997, 179–187.

[427] *N.M. Ha, L.M. Hai*, Linear topological invariants of spaces of holomorphic functions in infinite dimensions, Publ. Mat., **39**, 1, 1995, 71–88.

[428] *N.M. Ha, N.V. Khue*, The property (H_u) and $(\tilde{\Omega})$ with the exponential representation of holomorphic functions, Publ. Mat., **38**, 1, 1994, 37–49.

[429] *N.M. Ha, N.V. Khue*, The property (DN) and the exponential representation of holomorphic functions, Monatsh. Math., **120**, 3-4, 1995, 281–288.

[430] *J. Hadamard*, Le développement et le rôle scientifique du calcul fonctionnel. Atti de Congresso Internazionale dei Matematici, Bologna, (VI), Tomo I, 1928, 143–161.

[431] *H. Hahn*, Reelle Funktionen, Erster Teil, Leipzig, 1934.

[432] *L.M. Hai*, Holomorphic maps of uniform type, Colloq. Math., **69**, 1, 1995, 81–86.

[433] *L.M. Hai, N.V. Khue*, Holomorphic and meromorphic maps on (DFC)-spaces, Southeast Asian Bull. Math., **20**, 2, 1996, 89–94.

[434] *L.M. Hai, N.V. Khue, N.T. Nga*, Weak meromorphic extensions, Colloq. Math., **64**, 1, 1993, 65–70.

[435] *L.M. Hai, T.T. Quang*, Weak extension of Fréchet-valued holomorphic functions on compact sets and linear topological invariants, Acta Math. Vietnam, **21**, 2, 1996, 183–199.

[436] *P. Hájek, J.G. Llavona*, *P*-continuity on classical Banach spaces, Proc. Amer. Math. Soc. (to appear).

[437] *G.H. Hardy, J. E. Littlewood*, Bilinear forms bounded in space [*p, q*], Quart. J. Math., Oxford, **5**, 1934, 241–254.

[438] *L.A. Harris*, Bounds on the derivatives of holomorphic functions of vectors, Analyse fonctionnelle et applications, Ed. L. Nachbin, Hermann, 1975, 145–163.

[439] *L.A. Harris*, Problems 73 and 74, The Scottish Book, Ed. R.D. Mauldin, Birkhäuser, Basel, 1981, 143–150.

[440] *L.A. Harris*, Bernstein's polynomial inequalities and functional analysis, Bull. Irish Math. Soc., **36**, 1996, 19–33.

[441] *L.A. Harris*, A Bernstein-Markov theorem for normed spaces, J. Math. Anal. Appl., **208**, 1997, 476–480.

[442] *R. Haydon*, Some more characterizations of Banach spaces containing l_1, Math. Proc. Cambridge Philos. Soc., **80**, 1976, 269–276.

[443] *R. Haydon*, Non-separable Banach spaces, Functional Analysis, Surveys and Recent results, Ed. K.-D. Bierstedt, B. Fuchssteiner, North-Holland Math. Stud., **38**, 1979, 19–30.

[444] *R. Haydon*, A non-reflexive Grothendieck space that does not contain l_∞, Israel J. Math., **40**, 1981, 65–73.

[445] *M. Hervé*, Analytic and plurisubharmonic functions in finite and infinite dimensional spaces, Lecture Notes in Math., **198**, 1971.

[446] *M. Hervé*, Analyticity in Infinite Dimensional Spaces, de Gruyter Studies in Mathematics, **10**, 1989.

[447] *Y. Hervier*, Sur le problème de Levi pour les espaces étales banachiques, C. R. Acad. Sci. Paris, Sér. A, **275**, 1972, 821–824.

[448] *Y. Hervier*, On the Weierstrass problem in Banach spaces, Proc. on Infinite Dimensional Holomorphy, Ed. T. L. Hayden, T.J. Suffridge, Lecture Notes in Math., **364**, 1974, 157–167.

[449] *I.E. Highberg*, A note on abstract polynomials in complex spaces, J. Math. Pures Appl. (9), **16**, 1937, 307–314.

[450] *D. Hilbert*, Wesen und Zieleiner Analysis der unendlich vielen unabhängigen Variabeln, Rend. Circ. Mat. Palermo, **27**, 1909, 59–74.

[451] *E. Hille*, Functional analysis and semigroups, Amer. Math. Soc. Colloq. Publ., **31**, 1948.

[452] *E. Hille, R.S. Phillips*, Functional analysis and semi-groups, Amer. Math. Soc. Colloq. Pub., **31**, 1957. (This is a revised version of [451].)

[453] *A. Hirschowitz*, Sur le non-prolongement des variétés analytiques banachiques réelles, C. R. Acad. Sci. Paris, Sér. A, **269**, 1969, 844–846.

[454] *A. Hirschowitz*, Remarks sur les ouverts d'holomorphie d'un produit dénombrable de droites, Ann. Inst. Fourier (Grenoble), **19**, 1, 1969, 219–229.

[455] *A. Hirschowitz*, Prolongement analytiques en dimension infinie, C. R. Acad. Sci. Paris, Sér. A, **270**, 1970, 1736–1737.

[456] *A. Hirschowitz*, Sur les suites de fonctions analytiques, Ann. Inst. Fourier (Grenoble), **20**, 2, 1970, 403–413.

[457] *A. Hirschowitz*, Diverses notions d'ouverts d'analyticité en dimensions infinie, Séminaire Pierre Lelong 1969–70. Lecture Notes in Math., **205**, 1971, 11–20 (and typed correction).

[458] *A. Hirschowitz*, Bornologie des espaces de fonctions analytiques en dimension infinie, Séminaire P. Lelong (Analyse) 1969–1970, Lecture Notes in Math., **205**, 1971, 21–33.

[459] *A. Hirschowitz*, Sur un theoreme de M.A. Zorn, Arch. Math. (Basel), **23**, 1972, 1–13.

[460] *A. Hirschowitz*, Prolongement des fonctions analytiques, Ann. Inst. Fourier (Greboble), **22**, 1972, 255–292.

[461] *R.B. Holmes*, Geometric Functional Analysis and its Applications, Springer-Verlag Graduate Texts in Mathematics, **24**, 1975.

[462] *J.R. Holub*, Tensor product mappings, Math. Ann., **188**, 1970, 1–12.

[463] *T. Honda, M. Miyagi, M. Nishihara, M. Yosida*, On Polynomials of weak type on locally convex spaces, Proceedings 4th International conference on Finite and Infinite Dimensional Holomorphy, Fukuoka Univ. Sci. Rep., **27**, 1997, 9–16.

[464] *T. Honda, M. Miyagi, M. Nishihara, M. Yosida*, On the extension of holomorphic mappings of weak type, Fukuoka Univ. Sci. Rep., **27**, 2, 1997, 55–63.

[465] *L. Hörmander*, On a theorem of Grace, Math. Scand., **2**, 1954, 55–64.

[466] *L. Hörmander*, An Introduction to Complex Analysis in Several Variables, Van Nostrand, 1966.

[467] *J. Horváth*, Topological Vector Spaces and Distributions, Vol. I, Addison-Wesley, Reading Massachusetts, 1966.

[468] *J. Horváth*, The life and works of Leopoldo Nachbin, Aspects of Math. and its Applications, Ed. J.A. Barroso, North-Holland Math. Library, **34**, 1986, 1–75.

[469] *J. Horváth*, The late works of Leopoldo Nachbin, Comp. Appl. Math., **13**, 3, 1994, 175–188.

[470] *H. Hüser*, Lokalkonvexe Topologien auf Räumen n-linearer Abbildungen und n-homogener Polynome, Ph. D. Dissertation, Universität Trier, 1994.

[471] *D.H. Hyers*, A note on Fréchet's definition of "Polynômes abstrait", Houston J. Math., **4**, 3, 1978, 359–362.

[472] *J.M. Isidro*, Topological duality on the function space $\left(H_b(U;F), \tau_b\right)$, Proc. Royal Irish Acad. Sect. A, **79**, 1979, 115–130.

[473] *J.M. Isidro,* Characterisation of the spectrum of some topological algebras of holomorphic functions, Advances in Holomorphy, Ed. J.A. Barroso, North-Holland Math. Stud., **34**, 1979, 407–416.

[474] *J.M. Isidro,* On the distinguished character of the function spaces of holomorphic functions of bounded type, J. Funct. Anal., **38**, 2, 1980, 139–145.

[475] *J.M. Isidro,* Quasi-normability of some spaces of holomorphic mappings, Rev. Mat. Univ. Complut. Madrid, **3**, 1, 1990, 13–17.

[476] *J.M. Isidro, J.P. Méndez,* Topological duality on the function space $(H(U;F),\tau_\delta)$, J. Math. Anal. Appl., **67**, 1, 1979, 239–248.

[477] *J.A. Jaramillo, L.A. Moraes,* Duality and reflexivity in spaces of polynomials, Arch. Math. (Basel) (to appear).

[478] *J.A. Jaramillo, A. Prieto,* Weak polynomial convergence on a Banach space, Proc. Amer. Math. Soc., **118**, 2, 1993, 463–468.

[479] *J.A. Jaramillo, A. Prieto, I. Zalduendo,* The bidual of a space of polynomials on a Banach space, Math. Proc. Cambridge. Philos. Soc., **122**, 1997, 457–471.

[480] *H. Jarchow,* Locally Convex Spaces, B.G. Teubner, Stuttgart, 1981.

[481] *H. Jarchow, K. John,* Bilinear forms and nuclearity, Czechoslovak Math. J., **44**, 119, 1994, 367–373.

[482] *T.J. Jech,* On a problem of L. Nachbin, Proc. Amer. Math. Soc., **79**, 2, 1980, 341–342.

[483] *M. Jiménez-Sevilla, R. Payá,* Norm attaining multilinear mappings and preduals of Lorentz sequence spaces, Studia Math., **127**, 2, 1998, 99–112.

[484] *K. John,* On tensor product characterizations of nuclear spaces, Math. Ann., **257**, 1981, 341–353.

[485] *K. John,* Counterexample to a conjecture of Grothendieck, Math. Ann., **265**, 1983, 169–179.

[486] *K. John,* Tensor products and nuclearity in Banach space theory and its applications, Lecture Notes in Math., **991**, 1993, 124–129.

[487] *B. Josefson,* A counterexample in the Levi problem, Proc. Infinite Dimensional Holomorphy, Ed. T.L. Hayden, T. J. Suffridge, Lecture Notes in Math., **364**, 1974, 169–177.

[488] *B. Josefson,* Weak sequential convergence in the dual of a Banach space does not imply norm convergence, Ark. Mat., **13**, 1975, 79–89.

[489] *B. Josefson,* Some remarks on Banach valued polynomials on $c_0(A)$, Infinite Dimensional Holomorphy and Applications, Ed. M.C. Matos, North-Holland Math. Stud., **12**, 1977, 231–238.

[490] *B. Josefson,* Bounding subsets of $l^\infty(A)$, J. Math. Pures e Appl., **57**, 1978, 397–421.

[491] *B. Josefson,* A Banach space containing non-trivial limited sets but no non-trivial bounding sets, Israel J. Math., **71**, 1990, 321–327.

[492] *B. Josefson,* Uniform bounds for limited sets and applications to bounding sets (preprint).

[493] *H. Junek,* Holomorphic functions on intermediate spaces, Wiss. Z. d. Päd., Hochsch. Potsdam, **31**, 1, 1987, 131–137.

[494] *H. Junek,* Polynomials and holomorphic functions on interpolation spaces, Acta Univ. Carolin. Math. Phys., **30**, 2, 1989, 69–75.

[495] *H. Junek,* Factorization of operator ideals and the *BB*-property, Funtional Analysis, Essen 1991, Lecture Notes in Pure and Appl. Math., **150**, 1994, 203–216.

[496] *H. Junek, M.C. Matos,* Unconditionally *p*-summing polynomials, Arch. Math. (Basel), **70**, 1, 1998, 41–51.

[497] *W. Kaballo,* Lifting-Sätze für Vektorfunktionen und das ϵ-Tensor produkt, Habilitationsschrift, Kaiserslauten, 1976.

[498] *S. Kakutani, V. Klee,* The finite topology of a linear space, Arch. Math. (Basel), **14**, 1963, 55–58.

[499] *N.J. Kalton,* Schauder decompositions in locally convex spaces, Math. Proc. Cambridge Philos. Soc., **68**, 1970, 377–392.

[500] *N.J. Kalton,* Schauder decompositions and completeness, Bull. London Math. Soc., **2**, 1970, 34–36.

[501] *N.J. Kalton,* Normalization properties of Schauder bases, Proc. London Math. Soc., (3) **22**, 1971, 91–105.

[502] *N.J. Kalton,* Spaces of compact operators, Math. Ann., **208**, 1974, 267–278.

[503] *G. Katz,* Analytic continuation in infinite dimensions and Reinhardt sets, Thesis, University of Rochester, 1974.

[504] *G. Katz,* Domains of existence in infinite dimensions, Infinite Dimensional Holomorphy and Applications, Ed. M. C. Matos, North-Holland Math. Stud., **12**, 1977, 239–247.

[505] *O.D. Kellogg,* On bounded polynomials in several variables, Math. Z., **27**, 1928, 55–64.

[506] *N.V. Khue,* On the extension of holomorphic maps on a locally convex space, Ann. Polon. Math., **43**, 3, 1983, 267–282.

[507] *N.V. Khue,* On the extension of holomorphic functions on locally convex spaces with values in Fréchet spaces, Ann. Polon. Math., **44**, 2, 1984, 163–175.

[508] *N.V. Khue, P.T. Danh,* Structure of spaces of germs of holomorphic functions, Pub. Mat., **41**, 2, 1997, 467–480.

[509] *N.V. Khue, N.T. Nga,* Lifting vector-valued meromorphic functions in infinite dimensions, Proc. Amer. Math. Soc., **122**, 1, 1994, 91–96.

[510] *N.V. Khue, B.D. Tac,* Extending holomorphic maps from compact sets in infinite dimensions, Studia Math., **95**, 3, 1990, 263–272.

[511] *P. Kirwan,* Complexification of multilinear and polynomial mappings on normed spaces, Thesis, University College Galway, 1997.

[512] *P. Kirwan, R.A. Ryan,* Extendibility of homogenous polynomials on Banach spaces, Proc. Amer. Math. Soc., **126**, 4, 1998, 1023–1029.

[513] *P. Kirwan, Y. Sarantopoulos, A. Tonge,* Extremal homogeneous polynomials on real Banach spaces, J. Approx. Theory (to appear).

[514] *C.O. Kiselman,* On entire functions of exponential type and indicators of analytic functionals, Acta Math., **117**, 1967, 1–35.

[515] *C.O. Kiselman,* On the radius of convergence of an entire function in a normed space, Ann. Polon. Math., **33**, 1976, 39–55.

[516] *C.O. Kiselman,* Construction de fonctions entières à rayon de convergence donné, Journèes sur les Fonctions Analytiques, Toulouse 1976, Ed. P. Lelong, Lecture Notes in Math., **578**, 1977, 246–253.

[517] *C.O. Kiselman,* Geometric aspects of the theory of bounds for entire functions in normed spaces, Infinite Dimensional Holomorphy and Applications, Ed. M.C. Matos, North-Holland Math. Stud., **12**, 1977, 249–275.

[518] *C.O. Kiselman,* Croissance des fonctions plurisousharmoniques en dimension infinie, Ann. Inst. Fourier (Grenoble), **34**, 1, 1984, 155–183.

[519] *J. Kopeć, J. Musielak,* On the estimation of the norm of the *n*-linear symmetric operation, Studia Math., **15**, 1955, 29–30.

[520] *H. von Koch,* Sur les système d'ordre infinie d'equations différentielles, Öfversigt af Kongl. Svenska Vetenskaps-Akademiens Forhandlinger, **61**, 1899, 395–411.

[521] *G. Köthe,* Dualität in der Funktionentheorie, J. Reine Angew. Math., **191**, 1953, 30–49.

[522] *G. Köthe,* Topological Vector Spaces I, Springer-Verlag, 1969.

[523] *G. Köthe,* Topological Vector Spaces II, Springer-Verlag, 1979.

[524] *Y. Kōmura,* On linear topological spaces, Kumamoto J. Sci. Math., **5A**, 1962, 148–157.

[525] *B. Kramm, DFN-*analytic spaces, Stein algebras and a "universal" holomorphic functional calculus, Séminaire P. Lelong/H. Skoda, 1978/79, Lecture Notes in Math., **822**, 1980, 109–128.

[526] *B. Kramm,* Analytische Struktur in Spektren ein Zugang über die ∞—dimensionale Holomorphie, J. Funct. Anal., **37**, 3, 1980, 249–270.

[527] *B. Kramm,* Nuclearity and function algebras – a survey, Functional Analysis: Surveys and Recent Results III, Ed. K.-D. Bierstedt and B. Fuchssteiner, North-Holland Math. Stud., **90**, 1984, 233–252.

[528] *P. Kreé,* Produits tensoriels topologiques multiples, Seminaire P. Kreé, Equations aux dérivées partielles en dimension infinie, 1975/1976.

[529] *P. Kreé,* Méthodes holomorphes et méthodes nucléaires en analyse de dimension infinie et en théorie quantiques des champs, Vector Space Measures and Applications I, Ed. R.M. Aron, S. Dineen, Lecture Notes in Math., **644**, 1978, 212–254.

[530] *A. Kriegl, P.W. Michor,* The convenient setting for global analysis, American Math. Soc., Math. Surveys Monogr., **53**, 1997.

[531] *A. Kriegl, L.D. Nel,* A convenient setting for holomorphy, Cahiers de Topologie et Geometrie Différentielle Catégoriques, **26**, 3, 1985, 273–309.

[532] *S. Lassalle, I. Zalduendo,* To what extent does the dual Banach space E' determine the polynomials over E? (preprint).

[533] *D. Lazet,* Sur la différentiabilité des applications analytiques en dimension infinite, C. R. Acad. Sci. Paris, Sér. A, **273**, 1971, 155–157.

[534] *D. Lazet,* Applications analytiques dans les espaces bornologiques, Seminaire P. Lelong, 1971–1972, Lecture Notes in Math., **332**, 1973, 1–47.

[535] *P. Lelong,* Fonctions plurisousharmoniques dans les espaces vectoriels topologiques, Séminaire P. Lelong, 1967/68, Lecture Notes in Math., **71**, 1968, 167–189.

[536] *P. Lelong,* Fonctions plurisousharmoniques et ensembles polaires dans les espaces vectoriels topologiques, C. R. Acad. Sci. Paris, Sér. A, **267**, 1968, 916–918.

[537] *P. Lelong,* Théorème de Banach-Steinhaus pour les polynômes; applications entières d'espaces vectoriels complexes, Séminaire P. Lelong 1970, Lecture Notes in Math., **205**, 1971, 87–112.

[538] *P. Lelong,* Topologies semi-vectorielles et topologies pseudo-convex sur un espace vectoriel complexe, Séminaire P. Lelong 1973/1974, Lecture Notes in Math., **474**, 1975, 1–15.

[539] *P. Lelong,* Topologies semi-vectorielles, Applications a l'analyse complexe, Ann. Inst. Fourier (Grenoble), **25**, 1975, 381–407.

[540] *P. Lelong,* Sur l'application exponentielle dans l'espaces des fonctions entières, Infinite Dimensional Holomorphy and Applications, Ed. M.C. Matos, North-Holland Math. Stud., **12**, 1977, 297–311.

[541] *P. Lelong,* Ensembles de contrôle de croissance pour l'analyse complexe dans les espaces de Fréchet, C. R. Acad. Sci. Paris, Sér. A, **287**, 1978, 1097–1100.

[542] *P. Lelong,* A class of Fréchet spaces in which the bounded sets are C-polar, Functional Analysis, Holomorphy and Approximation Theory, Ed. J.A. Barroso, North-Holland Math. Stud., **71**, 1982, 255–272.

[543] *L. Lempert,* The Dolbeault complex in infinite dimensions I, J. Amer. Math. Soc., **11**, 1998, 485–520.

[544] *L. Lempert,* The Cauchy Riemann equation in infinite dimensions, Journées équations aux dérivées partielles, CNRS, Nantes, 1998.

[545] *L. Lempert,* The Dolbeault complex in infinite dimensions II, J. Amer. Math. Soc. (to appear).

[546] *L. Lempert,* Approximation de fonctions holomorphes d'un nombre infinie de variables, Ann. Inst. Fourier (Grenoble) (to appear).

[547] *L. Lempert,* The Dolbeault complex in infinite dimensions III (preprint).

[548] *D.H. Leung,* Some remarks on regular Banach spaces, Glasgow Math. J., **38**, 1996, 243-248.

[549] *E.E. Levi,* Sulle ipersuperfici dello spazio a 4 dimensioni che passono essere frontiera del campo di esistanza di una funzioni analitica di due variabili complesse, Ann. Math. Pura Appl., (3), **18**, 1911, 69-79.

[550] *P. Lévy,* Leçons d'analyse fonctionnelle, Gauthier-Villars, Paris, 1922 (reprinted 1950).

[551] *E. Ligocka,* A local factorization of analytic functions and its applications, Studia Mat., **47**, 1973, 239-252.

[552] *E. Ligocka,* Levi forms, differential forms of type (0,1) and pseudoconvexity in Banach spaces, Conference on Analytic Functions, Ann. Polon. Math., **33**, 1976, 63-69.

[553] *E. Ligocka, J. Siciak,* Weak analytic continuation, Bull. Acad. Pol. Sci. Math., **20**, 6, 1972, 461-466.

[554] *P.K. Lin,* On the extension of 2-polynomials, J. Math. Anal. Appl., **193**, 3, 1995, 795-799.

[555] *J. Lindenstrauss, L. Tzafriri,* Classical Banach Spaces I, Sequence Spaces, Ergeb. Math. Grenzgeb., Springer-Verlag, **92**, 1977.

[556] *J. Lindenstrauss, T. Tzafriri,* Classical Banach Spaces II, Function Spaces, Ergeb. Math. Grenzgeb., Springer-Verlag, **97**, 1979.

[557] *M. Lindström,* A characterization of Schwartz spaces, Math. Z., **198**, 1988, 423-430.

[558] *M. Lindström,* On compact and bounding holomorphic mappings, Proc. Amer. Math. Soc., **105**, 1989, 356-361.

[559] *M. Lindström, R.A. Ryan,* Limited holomorphic mappings on Banach spaces, Proc. Royal Irish. Acad. Sect. A, **91**, 1, 1991, 57-62.

[560] *M. Lindström, R.A. Ryan,* Applications of ultraproducts to infinite dimensional holomorphy, Math. Scand., **71**, 1992, 229-242.

[561] *M. Lindström, T. Schlumprecht,* A Josefson-Nissenzweig theorem for Fréchet spaces, Bull. London Math. Soc., **25**, 1, 1993, 55-58.

[562] *J.E. Littlewood,* On bounded bilinear forms in an infinite number of variables, Quart. J. Math. Oxford Ser., **1**, 1930, 164-174.

[563] *J.G. Llavona,* Approximation of continuously differentiable functions, North-Holland Math. Stud., **130**, 1986.

[564] *M.L. Lourenço,* A projective limit representation theory of DFC spaces with the approximation property, J. Math. Anal. Appl., **115**, 2, 1986, 422-433.

[565] *M.L. Lourenço,* Riemann domains over (DFC)-spaces, Complex Variable Theory Appl., **10**, 1, 1988, 67-82.

[566] *B. Malgrange,* Existence et approximation des solutions des équations aux dérivées partielles et des équations de convolution. Ann. Inst. Fourier (Grenoble), **6**, 1955-56, 271-355.

[567] *E. Mangino,* Productos tensoriales de espacios (LF), (DF) y de Fréchet, Thesis, Universitat de València, 1997.

[568] *A.M. Mantero, A. Tonge*, The Schur multiplication on tensor algebras, Studia Math., **68**, 1980, 1–24.

[569] *R.S. Martin*, Contributions to the theory of functionals, Thesis, University of California, 1932 (unpublished).

[570] *A. Martineau*, Sur les fonctionnelles analytiques et la transformation de Fourier-Borel, J. Anal. Math., **11**, 1963, 1–164.

[571] *A. Martineau*, Sur la topologie des espaces de fonctions holomorphes, Math. Ann., **163**, 1966, 62–88.

[572] *A. Martineau*, Les supports des fonctionnelles analytiques, Séminaire P. Lelong 1969, Lecture Notes in Math., **116**, 1970, 175–195.

[573] *M.C. Matos*, Sur les applications holomorphes définies dans des espaces vectoriels topologiques de Baire, C. R. Acad. Sci. Paris, Sér. A, **271**, 1970, 599.

[574] *M.C. Matos*, Sur l'enveloppe d'holomorphie des domains de Riemann sur un produit dénombrable de droites, C. R. Acad. Sci. Paris, Sér. A, **271**, 1970, 727–728.

[575] *M.C. Matos*, Sur les ouvertes de τ holomorphy dans les espaces de Banach separables, C. R. Acad. Sci. Paris, Sér. A, **271**, 1970, 1165–1166.

[576] *M.C. Matos*, Domains of τ-holomorphy in a separable Banach space, Math. Ann., **195**, 1972, 273–278.

[577] *M.C. Matos*, Sur le théorème d'approximation et d'existence de Malgrange-Gupta, C. R. Acad. Sci. Paris, Sér. A, **195**, 1972, 273–278.

[578] *M.C. Matos*, On the Cartan-Thullen theorem for some subalgebras of holomorphic functions in a locally convex space, J. Reine Angew. Math., **270**, 1974, 7–14.

[579] *M.C. Matos*, On holomorphy and compactness in Banach spaces, Ark. Mat., **12**, 4, 1974, 235–238.

[580] *M.C. Matos*, Holomorphically bornological spaces and infinite dimensional versions of Hartogs' theorem, J. London Math. Soc., (2), **17**, 1978, 363–368.

[581] *M.C. Matos*, On separately holomorphic and Silva holomorphic mappings, Advances in Holomorphy, Ed. J.A. Barroso, North-Holland Math. Stud., **34**, 1979, 509–520.

[582] *M.C. Matos*, Convolution operators in spaces of uniform nuclear entire functions, Functional Analysis, Holomorphy and Approximation Theory, Ed. G.I. Zapata, Lecture Notes in Pure and Appl. Math., **83**, 1979, 207–231.

[583] *M.C. Matos*, On the Fourier–Borel transformation on spaces of entire functions in a normed space, Functional Analysis, Holomorphy and Approximation Theory II, Ed. G. I. Zapata, North-Holland Math. Stud., **86**, 1984, 139–169.

[584] *M.C. Matos*, A characterization of holomorphic mappings in Banach spaces with a Schauder basis, Atas do 19° Seminário Brasileiro de Análise, Soc. Brasil. de Math., São Paulo, 1984.

[585] *M.C. Matos*, On convolution operators in spaces of entire functions of a given type and order, Complex Analysis, Functional Analysis and Approximation Theory, Ed. J. Mujica, North-Holland Math. Stud., **125**, 1986, 129–171.

[586] *M.C. Matos*, On pseudo-convex polycircular domains in Banach spaces, Proc. Royal Irish Acad. Sect. A, **90**, 2, 1990, 235–240.

[587] *M.C. Matos*, On multilinear mappings of nuclear type, Rev. Mat. Univ. Complut. Madrid, **6**, 1993, 61–81.

[588] *M.C. Matos*, On the question of the coincidence between spaces of multilinear absolutely summing mappings and Hilbert-Schmidt mappings, 41° Seminário Brasileiro de Análise, 1995.

[589] *M.C. Matos*, Absolutely summing holomorphic mappings, An. Acad. Brasil. Ciênc., **68**, 1, 1996, 1–13.

[590] *M.C. Matos*, Strictly absolutely summing multilinear mappings (preprint).

[591] *M.C. Matos, L. Nachbin*, Silva holomorphy types, Functional Analysis, Holomorphy and Approximation Theory, Ed. S. Machado, Lecture Notes in Math., **843**, 1981, 437–487.

[592] *M.C. Matos, L. Nachbin*, Reinhardt domains of holomorphy in Banach spaces, Adv. Math., **92**, 1992, 266–278.

[593] *C. Matyszczyk*, Approximation of analytic operators by polynomials in complex B_0 spaces with bounded approximation property, Bull. Polish Acad. Sci. Math., **20**, 10, 1972, 833–836.

[594] *C. Matyszczyk*, Approximation of analytic and continuous mappings by polynomials in Fréchet spaces, Studia Math., **60**, 3, 1977, 223–238.

[595] *R.D. Maudlin*, The Scottish Book, Mathematics from the Scottish Café, Birkhauser, 1981.

[596] *P. Mazet*, Définition d'une application universelle sur un espace analytique de dimension finie, Seminaire P. Lelong, 1974/75, Lecture Notes in Math., **524**, 1976, 67–78.

[597] *P. Mazet*, Analytic Sets in Locally Convex Spaces, North-Holland Math. Stud., **89**, 1984.

[598] *P. Mazet*, A Hahn-Banach theorem for quadratic forms, Math. Proc. Royal Irish Acad. (to appear).

[599] *S. Mazur, W. Orlicz*, Grundlegende Eigenschaften der polynomischen Operationen I, Studia Math., **5**, 1934, 50–68.

[600] *S. Mazur, W. Orlicz*, Grundlegende Eigenschaften der polynomischen Operationen II, Studia Math., **5**, 1934, 179–189.

[601] *C.W. McArthur, J.R. Rutherford*, Uniform and equicontinuous Schauder bases of subspaces, Canad. J. Math., **17**, 1965, 207–212.

[602] *L.R. Medeiros*, Absolute basis in the space of holomorphic mappings, Proc. Royal Irish Acad. Sect. A, **90**, 2, 1990, 247–249.

[603] *R. Meise*, A remark on the ported and the compact-open topology for spaces of holomorphic functions on nuclear Fréchet spaces, Proc. Royal Irish Acad. Sect. A, **81**, 1981, 217–223.

[604] *R. Meise, D. Vogt*, An interpretation of τ_w and τ_0 as normal topologies of sequence spaces, Functional Analysis, Holomorphy and Approximation Theory, Ed. J.A. Barroso, North-Holland Math. Stud., **71**, 1982, 273–285.

[605] *R. Meise, D. Vogt*, Analytic isomorphisms of infinite dimensional polydiscs and an application, Bull. Soc. Math. France, **111**, 1983, 3–20.

[606] *R. Meise, D. Vogt*, Structure of spaces of holomorphic functions on infinite-dimensional polydiscs, Studia Math., **75**, 1983, 235–252.

[607] *R. Meise, D. Vogt*, Counterexamples in holomorphic functions on nuclear Fréchet Spaces, Math. Z., **182**, 1983, 167–177.

[608] *R. Meise, D. Vogt*, Extension of entire functions on nuclear locally convex spaces. Proc. Amer. Math. Soc., **92**, 4, 1984, 495–500.

[609] *R. Meise, D. Vogt*, A characterization of quasi-normable Fréchet spaces, Math. Nachr., **122**, 1985, 141–150.

[610] *R. Meise, D. Vogt*, Holomorphic functions of uniformly bounded type on nuclear Fréchet spaces, Studia Math., **83**, 1986, 147–166.

[611] *R. Meise, D. Vogt*, Holomorphic functions on nuclear sequence spaces, Lecture Notes, Dept. de Teoría de Funciones, Universidad Complutense de Madrid, 1986.

[612] *Y. Meléndez, A. Tonge*, Polynomials and the Pietsch Domination Theorem, Math. Proc. Royal Irish Acad. (to appear).

[613] *P. Mellon*, The polarization constant for JB^* triples. Extracta Math., **9**, 3, 1994, 160–163.

[614] *G. Metafune*, Quojections and finite-dimensional decompositions, Arch. Math. (Basel), **57**, 6, 1991, 609–616.

[615] *G. Metafune, V.B. Moscatelli*, A twisted Fréchet space with basis, Monatsh. Math., **105**, 1988, 127–129.

[616] *G. Metafune, V.B. Moscatelli*, Quojections and prequojections, Advances in the theory of Fréchet spaces, Ed. T. Terzioğlu, Kluwer Academic Publishers, Series C, **287**, 1989, 235–254.

[617] *G. Metafune, V.B. Moscatelli*, On the space $l^{p^+} = \bigcap_{q>p} l^q$, Math. Nachr., **147**, 1990, 7–12.

[618] *E.A. Michael*, Locally multiplicatively-convex topological algebras, Mem. Amer. Math. Soc., **11**, 1952.

[619] *A.D. Michal*, Le calcul différential dans les espaces de Banach, Gauthier-Villars, Paris, 1958.

[620] *A.D. Michal, A.H. Clifford*, Fonctions analytiques implicites dans des espaces vectoriels abstraits, C. R. Acad. Sci. Paris, Sér. A, **197**, 1933, 735–737.

[621] *A.D. Michal, R.S. Martin,* Some expansions in vector spaces, J. Math. Pures Appl. (9), **13**, 1934, 69–91.

[622] *A.D. Michal, M. Wyman,* Characterization of complex couple spaces, Ann. of Math., **42**, 1941, 247–250.

[623] *M. Miñarro,* A characterization of quasi-normable Köthe sequence spaces, Proc. Amer. Math. Soc., **123**, 4, 1995, 1207–1212.

[624] *L.A. Moraes,* Theorems of the Cartan-Thullen type and Runge domains, Advances in Holomorphy, Ed. J.A. Barroso, North-Holland Math. Stud., **34**, 1979, 521–561.

[625] *L.A. Moraes,* Envelopes for types of holomorphy, Functional Analysis, Holomorphy and Approximation Theory, Ed. S. Machado, Lecture Notes in Math., **843**, 1981, 488–499.

[626] *L.A. Moraes,* Holomorphic functions on strict inductive limits, Resultate Math., **4**, 1981, 201–212.

[627] *L.A. Moraes,* Holomorphic functions on holomorphic inductive limits and on the strong duals of strict inductive limits, Functional Analysis, Holomorphy and Approximation Theory II, Ed. G.I. Zapata, North-Holland Math. Stud., **86**, 1984, 297–310.

[628] *L.A. Moraes,* The Hahn-Banach extension theorem for some spaces of n-homogeneous polynomials, Functional Analysis: Surveys and Recent Results III, Ed. K.-D. Bierstedt, B. Fuchssteiner, North-Holland Math. Stud., **90**, 1984, 265–274.

[629] *L.A. Moraes,* A Hahn-Banach extension theorem for some holomorphic functions, Complex Analysis, Functional Analysis and Approximation Theory, Ed. J. Mujica, North-Holland Math. Stud., **125**, 1986, 205–220.

[630] *L.A. Moraes,* Quotients of spaces of holomorphic functions on Banach spaces, Proc. Royal Irish Acad. Sect. A, **87**, 2, 1987, 181–186.

[631] *L.A. Moraes,* On some relations between linear and topological mappings on Banach spaces, Note Mat., **9**, 2, 1989, 153–163.

[632] *L.A. Moraes,* Extension of holomorphic mappings from E to E'', Proc. Amer. Math. Soc., **118**, 2, 1993, 455–461.

[633] *L.A. Moraes,* Weakly continuous holomorphic mappings, Proc. Royal Irish Acad. Sect. A, **97**, 2, 1997, 139–144.

[634] *L.A. Moraes, P.A. Burlandy,* The spectrum of an algebra of weakly continuous holomorphic mappings (preprint).

[635] *L.A. Moraes, O.W. Paques, M.C.F. Zaine,* F-Quotients and envelope of F-holomorphy, J. Math. Anal. Appl., **163**, 2, 1992, 393–405.

[636] *L.A. Moraes, O.W. Paques, M.C.F. Zaine,* Factorization of uniformly holomorphic functions, Ann. Polonici Math., **61**, 1, 1995, 1–11.

[637] *V.B. Moscatelli,* Fréchet spaces without continuous norms and without basis, Bull. London Math. Soc., **12**, 1980, 63–66.

[638] *V.B. Moscatelli,* Strict inductive and projective limits, twisted spaces and quojections, Rend. Circ. Mat. Palermo, (2), **10**, 1985, 119–131.

[639] J. Mujica, On the Nachbin topology in spaces of holomorphic germs, Bull. Amer. Math. Soc., **81**, 5, 1975, 904–906.

[640] J. Mujica, Holomorphic germs on infinite dimensional spaces, Infinite Dimensional Holomorphy and Applications, Ed. M. C. Matos, North-Holland Math. Stud., **12**, 1977, 313–321.

[641] J. Mujica, The Oka-Weil theorem in locally convex spaces with the approximation property, Séminaire P. Krée 1977/1978, Paris, Exp. 3., 1979.

[642] J. Mujica, Ideals of holomorphic functions on Fréchet spaces, Advances in Holomorphy, Ed. J.A. Barroso, North-Holland Math. Stud., **34**, 1979, 563–567.

[643] J. Mujica, Complex homomorphisms of the algebras of holomorphic functions on Fréchet spaces, Math. Ann., **241**, 1979, 73–82.

[644] J. Mujica, Spaces of germs of holomorphic functions, Adv. Math. Suppl. Stud., **4**, Academic Press, 1979, 1–41.

[645] J. Mujica, Domains of holomorphy in (DFC)-spaces, Functional Analysis, Holomorphy and Approximation Theory, Ed. S. Machado, Lecture Notes in Math., **843**, 1981, 500–533.

[646] J. Mujica, A Banach-Dieudonné theorem for germs of holomorphic functions, J. Funct. Anal., **57**, 1, 1984, 31–48.

[647] J. Mujica, A completeness criterion for inductive limits of Banach spaces, Functional Analysis, Holomorphy and Approximation Theory II, Ed. G.I. Zapata, North-Holland Math. Stud., **86**, 1984, 319–329.

[648] J. Mujica, Holomorphic approximation in Fréchet spaces with basis, J. London Math. Soc., (2), **29**, 1984, 113–126.

[649] J. Mujica, Holomorphic approximation in infinite-dimensional Riemann domains, Studia Math., **82**, 1985, 107–134.

[650] J. Mujica, Polynomial approximation in nuclear Fréchet spaces, Aspects of Math. and its Applications, Ed. J.A. Barroso, Elsevier Sc. Publishers, 1986, 601–606.

[651] J. Mujica, Complex Analysis in Banach Spaces, North-Holland Math. Stud., **120**, 1986.

[652] J. Mujica, Spectra of algebras of holomorphic functions on infinite dimensional Riemann domains, Math. Ann., **276**, 1987, 317–322.

[653] J. Mujica, Linearization of bounded holomorphic mappings on Banach spaces, Trans. Amer. Math. Soc., **324**, 2, 1991, 867–887.

[654] J. Mujica, Linearization of holomorphic mappings on infinite dimensional spaces, Rev. Un. Mat. Argentina, **57**, 1991, 127–134.

[655] J. Mujica, Linearization of holomorphic mappings of bounded type, Progress in Functional Analysis, Ed. K.-D. Bierstedt et al., North-Holland Math. Stud., **170**, 1992, 149–162.

[656] J. Mujica, Mittag–Leffler Methods in Analysis, Rev. Mat. Univ. Complut. Madrid, **8**, 2, 1995, 309–325.

[657] J. Mujica, Spaces of holomorphic mappings on Banach spaces with a Schauder basis, Studia Math., **122**, 2, 1997, 139–151.

[658] J. Mujica, Separable quotients of Banach spaces, Rev. Mat. Univ. Complut. Madrid, **10**, 2, 1997, 299–330.

[659] J. Mujica, Ideals of holomorphic functions on Tsirelson's space, (preprint).

[660] J. Mujica, L. Nachbin, Linearization of holomorphic mappings on locally convex spaces, J. Math. Pures Appl., **71**, 1992, 543–560.

[661] J. Mujica, M. Valdivia, Holomorphic germs on Tsirelson's space, Proc. Amer. Math. Soc., **123**, 5, 1995, 1379–1386.

[662] G. Muñoz, Y. Sarantopoulos, A. Tonge, Complexification of real Banach spaces, polynomials and multilinear mappings, Studia Math., **134**, 1999, 1–33.

[663] L. Nachbin, Topological vector spaces of continuous functions, Proc. Nat. Acad. Sci. U.S.A., **40**, 1954, 471–474.

[664] L. Nachbin, On spaces of holomorphic functions of a given type, Functional Analysis, Irvine, 1966, Ed. B. R. Gelbaum, Academic Press, 1967, 50–70.

[665] L. Nachbin, On the topology of the space of all holmorphic functions on a given open set, Indag. Math., **29**, 1967, 366–368.

[666] L. Nachbin, Topology on spaces of holomorphic mappings, Ergeb. Math. Grenzgeb., **47**, Springer–Verlag, Berlin, 1969.

[667] L. Nachbin, Convolution operators in spaces of nuclearly entire functions on a Banach space, Proc. Conf. on Functional Analysis and Related Fields, University of Chicago, 1968, Ed. F.E. Browder, Springer–Verlag, Berlin, 1970, 167–171.

[668] L. Nachbin, Concerning holomorphy types for Banach spaces, International Colloquium on Nuclear Spaces and Ideals in Operator Algebras, Studia Math., **38**, 1970, 407–412.

[669] L. Nachbin, Sur les espaces vectoriels topologiques d'applications continues, C. R. Acad. Sci. Paris, Sér. A, **271**, 1970, 596–598.

[670] L. Nachbin, Uniformité d'holomorphie et type exponential, Séminaire P. Lelong, 1970, Lecture Notes in Math., **205**, 1971, 216–224.

[671] L. Nachbin, Concernente a espaçõs de aplicações holomorfas, Atas da 3ª Quinzena de Análise Functional e Equações Differenciais Parciais, São José dos Campos, 1970, Sociedade Brasileira de Matemática, **2**, 1971, 1–47.

[672] L. Nachbin, Convoluções en funções inteiras nucleares, Atas da 1ª e 2ª Quinzenas de Análise Functional e Equações Diferenciais Parciais, São Jose dos Campos, 1967 and 1969, Vol. 3, Rio de Janeiro, 1972.

[673] L. Nachbin, Weak holomorphy, Parts I and II, unpublished manuscript, 1972.

[674] L. Nachbin, Recent developments in infinite dimensional holomorphy, Bull. Amer. Math. Soc., **79**, 1973, 625–640.

[675] *L. Nachbin*, On vector-valued versus scalar-valued analytic continuation, Indag. Math., **35**, 4, 1973, 352–354.

[676] *L. Nachbin*, A glimpse at infinite dimensional holomorphy, Proc. on Infinite Dimensional Holomorphy, Ed. T. L. Hayden, T. J. Suffridge, Lecture Notes in Math., **364**, 1974, 69–79.

[677] *L. Nachbin,* Some holomorphically significant properties of locally convex spaces, Functional Analysis, Ed. J. Figuereido, Marcel Dekker, Lecture Notes in Pure Appl. Math., **18**, 1976, 251–277.

[678] *L. Nachbin,* Some problems in the application of functional analysis to holomorphy, Advances in Holomorphy, Ed. J.A. Barroso, North-Holland Math. Stud., **34**, 1979, 577–583.

[679] *L. Nachbin,* On pure uniform holomorphy in spaces of holomorphic germs, Results Math., **8**, 1985, 117–122.

[680] *L. Nachbin,* Some aspects and problems in holomorphy, Extracta Mat., **1**, 2, 1986, 57–86.

[681] *L. Nachbin,* A glance at holomorphic factorization and uniform holomorphy, Complex Analysis, Functional Analysis and Approximation Theory, Ed. J. Mujica, North-Holland Math. Stud., **125**, 1986, 221–245.

[682] *L. Nachbin,* When does finite holomorphy imply holomorphy?, Portugal. Math., **51**, 4, 1994, 525–528.

[683] *L. Nachbin, J. A. Barroso,* Sur certaines propriétés bornologiques des espaces d'applications holomorphes, Troisième Colloque sur l'Analyse Fonctionnelle, Liège, 1970, Centre Belge de Rech. Math. (Vander, Louvain), 1971, 47–55.

[684] *L. Nachbin, C.P. Gupta,* Malgrange theorem for nuclearly entire functions on a Banach space, Unpublished Manuscript, IMPA, 1966, Abstract 642-98, Notices Amer. Math. Soc., 1967.

[685] *E. Nelimarkka,* On spaces of holomorphic functions on locally convex spaces defined by an operator ideal. Notes on Funct. Anal. II, Ed. L. Holmström, Univ. of Helsinki, 1980, 25–35.

[686] *E. Nelimarkka,* Spaces of holomorphic germs on locally convex spaces defined by an operator ideal, Report HTKK–MAT–A, **175**, Helsinki University of Technology, 1980.

[687] *K.F. Ng,* On a theorem of Dixmier, Math. Scand., **29**, 1971, 297–280.

[688] *O. Nicodemi,* Homomorphisms of algebras of germs of holomorphic functions, Functional Analysis, Holomorphy and Approximation Theory, Ed. S. Machado, Lecture Notes in Math., **843**, 1981, 534–546.

[689] *A. Nissenzweig, w** sequential convergence, Israel J. Math., **22**, 1975, 266–272.

[690] *Ph. Noverraz,* Fonctions plurisousharmoniques et analytiques dans les espaces vectoriels topologiques, Ann. Inst. Fourier (Grenoble), **19**, 2, 1969, 419–493.

[691] *Ph. Noverraz*, Fonctions analytiques et théorème de prolongement dans les espaces vectoriels complexes, Séminaire P, Lelong 1969/70, Lecture Notes in Math., **116**, 1970, 21–29.

[692] *Ph. Noverraz*, Un théorème de Hartogs et théorèmes de prolongement dans les espaces vectoriels topologiques complexes, C. R. Acad. Sci. Paris, Sér. A, **266**, 1970, 806–808.

[693] *Ph. Noverraz*, Sur la convexité fonctionelle en dimension infinie, C. R. Acad. Sci. Paris, Sér. A, **274**, 1972, 313–316.

[694] *Ph. Noverraz*, Convexité fonctionelle dans les espaces de Banach a base, Séminaire P. Lelong, 1970–1971, Lecture Notes in Math., **275**, 1972, 166–176.

[695] *Ph. Noverraz*, Sur le pseudo-convexité et la convexité polynomiale en dimension infinie, Ann. Inst. Fourier (Grenoble), **23**, 1, 1973, 113–134.

[696] *Ph. Noverraz*, Pseudo-convexite, convexité polynomiale et domaines d'holomorphie, North-Holland Math. Stud., **3**, 1973.

[697] *Ph. Noverraz*, Sur le theoreme de Cartan-Thullen-Oka en dimension infinie, Séminaire P. Lelong, 1971/1972, Lecture Notes in Math., **332**, 1973, 59–68.

[698] *Ph. Noverraz*, Prolongement, complétion pseudo-convexe et approximation en dimension infinie, C. R. Acad. Sci. Paris, Sér. A, **276**, 1973, 1601–1604.

[699] *Ph. Noverraz*, Pseudo-Convexité et complétion holomorphe en dimension infinie, Coll. Int. du CNRS, 1972, Fonctions Analytiques de Plusieurs Variables et Analyse Complexe, Agora Math., Gauthier-Villars, **1**, 1974, 172–179.

[700] *Ph. Noverraz*, Pseudo-convex completion of locally convex topological vector spaces, Math. Ann., **208**, 1974, 59–69.

[701] *Ph. Noverraz*, Approximation of holomorphic or plurisubharmonic functions in certain Banach spaces, Proc. on Infinite Dimensional Holomorphy, Ed. T.L. Hayden, T.J. Suffridge, Lecture Notes in Math., **364**, 1974, 178–185.

[702] *Ph. Noverraz*, Pseudo-convexité et base de Schauder dans les espaces localement convexes, Séminaire P. Lelong, 1973/1974. Lecture Notes in Math., **474**, 1975, 63–82.

[703] *Ph. Noverraz*, Le problème de Levi en dimension infinie, Journés sur la Géometrie de Dimension Infinie, Bull. Soc. Math. France, **46**, 1976, 73–82.

[704] *Ph. Noverraz*, Sur la topologie tonnelée bornologique associée à la topologie de Nachbin, C. R. Acad. Sci. Paris, Sér. A, **279**, 1976, 459–462.

[705] *Ph. Noverraz*, On topologies associated with Nachbin topology, Proc. Royal Irish Acad. Sect. A, **77**, 1977, 85–95.

[706] *Ph. Noverraz*, On a particular case of surjective limit, Infinite Dimensional Holomorphy and Applications, Ed. M.C. Matos, North-Holland Math. Stud., **12**, 1977, 323–331.

[707] *Ph. Noverraz*, Topologies associated with Nachbin topology, Advances in Holomorphy, Ed. J.A. Barroso, North-Holland Math. Stud., **34**, 1979, 609–628.

[708] *Ph. Noverraz*, Approximation of plurisubharmonic functions, Approximation Theory and Functional Analysis, Ed. J.B. Prolla, North-Holland Math. Stud., **35**, 1979, 343–349.

[709] *M.L. Ostrovskii*, On complemented subspaces of sums and products of Banach spaces, Proc. Amer. Math. Soc., **129**, 7, 1996, 2005–2012.

[710] *J.C. Oxtoby*, Measure and Category, Springer-Verlag Graduate Texts in Math., **2**, 1971.

[711] *O.T.W. Paques*, Tensor products of Silva holomorphic functions, Advances in Holomorphy, Ed. J. A. Barroso, North-Holland Math. Stud., **34**, 1979, 629–700.

[712] *R. Payá*, Norm attaining operators versus bilinear forms, Extracta Math., **12**, 2, 1997, 179–183.

[713] *A. Pełczyński*, A property of multilinear operations, Studia Math., **16**, 1957, 173–182.

[714] *A. Pełczyński*, On weakly compact polynomial operators on B-spaces with Dunford-Pettis property, Bull. Polish Acad. Sci. Math., **11**, 6, 1963, 371–378.

[715] *A. Pełczyński*, A theorem of Dunford–Pettis type for polynomial operators, Bull. Polish Acad. Sci. Math., **11**, 6, 1963, 379–386.

[716] *A. Pełczyński*, Any separable Banach space with the bounded approximation property is a complemented subspace of a Banach space with a basis, Studia Math., **40**, 1971, 239–243.

[717] *A. Pełczyński, C. Schütt*, Factoring the natural injection $i^{(n)} : L_n^\infty \to L_n^1$ through finite dimensional Banach spaces and geometry of finite dimensional unitary ideals, Mathematical Analysis and Applications, Part B, Adv. in Math. Suppl. Stud., **7b**, 1981, 653–683.

[718] *A. Pełczyński, Z. Semandini*, Spaces of continuous functions III, Spaces $C(\Omega)$ for Ω without perfect subsets, Studia Math., **18**, 1959, 211–222.

[719] *A. Pełczyński, P. Wojtaszczyk*, Banach spaces with finite dimensional expansions of identity and universal bases of finite dimensional subspaces, Studia Math., **40**, 1971, 91–108.

[720] *P. Pérez Carreras, J. Bonet*, Barrelled Locally Convex Spaces, North-Holland Math. Stud., **131**, 1987.

[721] *A. Peris*, Productos tensoriales de espacios localmente convexos y otras clases realacionadas, Thesis, Universitat de València, 1992.

[722] Y.P. Petunin, V. Savkin, Convergence generated by analytic functionals and isomorphisms of algebras of analytic functions, Ukranian Math., J., **40**, 1988, 676–679.

[723] R.S. Phillips, On linear transformations, Trans. Amer. Math. Soc., **48**, 1940, 516–541.

[724] A. Pietsch, Operator Ideals, North-Holland Math. Library, **20**, 1980.

[725] A. Pietsch, Ideals of multilinear functionals, Forschungsergebnisse, Friedrich Schiffer Universität, Jena, 1983.

[726] D. Pisanelli, Sui funzionali transcendenti interi dello spazio L. N., Boll. Un. Mat. Ital., 8, **19**, 1964, 106–109.

[727] D. Pisanelli, Sur les applications analytiques en dimension infinie, C. R. Acad. Sci. Paris, Sér. A, **274**, 1972, 760–762.

[728] D. Pisanelli, Applications analytiques en dimension infinie, Bull. Sci. Math. (2), **96**, 1972, 181–191.

[729] D. Pisanelli, Sur la LF-analiticité, Analyse fonctionnelle et Applications, Ed. L. Nachbin, Hermann, 1975, 215–224.

[730] G. Pisier, Counterexamples to a conjecture of Grothendieck, Acta Math., **151**, 1983, 181–218.

[731] G. Pisier, Factorization of linear operators and geometry of Banach spaces, Amer. Math. Soc. Reg. Conf. Ser. Math., **60**, 1986.

[732] H.R. Pitt, A note on bilinear forms, J. London Math. Soc., **11**, 1936, 174–180.

[733] E. Plewnia, Homogeneous polynomials on Banach spaces, University of Potsdam, Institut für Mathematik, 1995.

[734] E. Plewnia, Approximationstheorie homogener Polynome auf Banachräumen, Dissertation, Universität Potsdam, 1997.

[735] R. Pomes, Solution du problème de Levi dans les espaces de Silva à base, C. R. Acad. Sci. Paris, Sér. A, **278**, 1974, 707–710.

[736] R. Pomes, Ouverts pseudo-convexes et domaines d'holomorphie en dimension infinie, Thése de $3^{\text{ième}}$ cycle, Université de Bordeaux I, 1977.

[737] S. Ponte, A remark about the embedding $\big(H(E/F), \tau\big) \to \big(H(E), \tau\big)$ with $\tau = \tau_0, \tau_w$ in Fréchet spaces, Note Mat., **IX**, 2, 1989, 217–220.

[738] N. Popa, Sur le problème de Levi dans les espaces de Silva à base, C. R. Acad. Sci. Paris, Sér. A, **277**, 1973, 211–214.

[739] A. Prieto, The bidual of spaces of holomorphic functions in infinitely many variables, Proc. Royal Irish Acad. Sect. A, **92**, 1, 1992, 1–8.

[740] T.T. Quang, Holomorphic functions of uniform type with values in Riemann domains (preprint).

[741] J. Rennison, A note on the extension of associative products, Proc. Amer. Math. Soc., **17**, 6, 1966, 1375–1377.

[742] G. Restrepo, An infinite dimensional version of a theorem of Bernstein, Proc. Amer. Math. Soc., **23**, 1969, 193–198.

[743] C.E. Rickart, Analytic functions of an infinite number of complex variables, Duke Math. J., **39**, 1969, 581–597.

[744] *C.E. Rickart*, Plurisubharmonic functions and convexity properties for general function algebras, Trans. Amer. Math. Soc., **169**, 1972, 1–24.

[745] *C.E. Rickart*, Holomorphic extensions and domains of holomorphy for general function algebras, Proc. on Infinite Dimensional Holomorphy, Ed. T.L. Hayden, T.J. Suffridge, Lecture Notes in Math., **364**, 1974, 80–91.

[746] *N. Robertson*, Asplund operators and holomorphic mappings, Manuscripta Math., **75**, 1992, 25–34.

[747] *A. Rodríguez Palacios*, A note on Arens' regularity, Quart. J. Math. Oxford Ser. (2), **38**, 1987, 91–93.

[748] *J.T. Rogers Jr., W.R. Zame*, Extensions of analytic functions and the topology in spaces of analytic functions, Indiana Univ. Math. J., **31**, 6, 1982, 809–818.

[749] *H.P. Rosenthal*, Some recent discoveries in the isomorphic theory of Banach spaces, Bull. Amer. Math. Soc., **84**, 5, 1978, 803–831.

[750] *H. Rossi*, On envelopes of holomorphy, Comm. Pure Appl. Math., **16**, 1963, 9–17.

[751] *W. Rudin*, Holomorphic maps of discs into F-spaces, Complex Analysis, Kentucky, 1976, Lecture Notes in Math., **599**, 1977, 104–108.

[752] *P. Rueda*, On the Banach-Dieudonné theorem for spaces of holomorphic functions, Quaestiones Math., **19**, 1996, 341–352.

[753] *P. Rueda*, Algunos problemas sobre holomorfía en dimension infinita, Thesis, Universitat de València, 1997.

[754] *K. Rusek*, Remarks on H-bounded subsets of Banach spaces, Zeszyty Nauk. Univ. Jagiellonski, Prace Mat., **356**, 16, 1974, 55–59.

[755] *K. Rusek*, A new topology in the space of germs of holomorphic functions on a compact set in $\mathbb{C}^n$, Bull. Polish Acad. Sci. Math. **25**, 1977, 1227–1232.

[756] *K. Rusek, J. Siciak*, Maximal analytic extensions of Riemann domains over topological vector spaces, Infinite Dimensional Holomorphy and Applications, Ed. M.C. Matos, North-Holland Math. Stud., **12**, 1977, 347–377.

[757] *R.A. Ryan*, Dunford-Pettis properties, Bull. Acad. Polon. Sci. Ser. Sci. Math., **27**, 5, 1979, 373–379.

[758] *R.A. Ryan*, Applications of topological tensor products to infinite dimensional holomorphy, Thesis, Trinity College Dublin, 1980.

[759] *R.A. Ryan*, Holomorphic mappings on l_1, Trans. Amer. Math. Soc., **302**, 7, 1987, 797–811.

[760] *R.A. Ryan*, Weakly compact holomorphic mappings on Banach spaces, Pacific J. Math., **131**, 1, 1988, 179–190.

[761] *R.A. Ryan, B. Turett*, Geometry of spaces of polynomials, J. Math. Anal. Appl., **221**, 1998, 698–711.

[762] E. Saab, Une characterization des convexes $\sigma(E', E)$-compactes possedent le propriete de Radon-Nikodým, C. R. Acad. Sci. Paris, Sér. A, **286**, 1978, 45–48.

[763] E. Saab, A characterization of w^*-compact convex sets having the Radon–Nikodým property, Bull. Sci. Math., (2), **104**, 1980, 79–88.

[764] B.L. Sanders, Decompositions and reflexivity in Banach spaces, Proc. Amer. Math. Soc., **16**, 1965, 204–208.

[765] Y. Sarantopoulos, Estimates for polynomial norms on $L^p(\mu)$-spaces, Math. Proc. Cambridge Philos. Soc., **99**, 1986, 263–271.

[766] Y. Sarantopoulos, Extremal multilinear forms on Banach spaces, Proc. Amer. Math. Soc., **99**, 2, 1987, 340–346.

[767] Y. Sarantopoulos, Polynomials on certain Banach spaces, Bull. Greek Math. Soc., **28**, 1987, 89–102.

[768] Y. Sarantopoulos, Bounds on the derivatives of polynomials on Banach spaces, Math. Proc. Cambridge Philos. Soc., **110**, 1991, 307–312.

[769] Y. Sarantopoulos, A. Tonge, Norm attaining polynomials (preprint).

[770] H.H. Schaefer, Topological Vector Spaces, McMillan Co. New York, 1966.

[771] R. Schatten, A theory of cross products, Ann. of Math. Stud., **26**, Princeton University Press, 1950.

[772] M. Scheve, Räume holomorpher Funktionen auf unendlich-dimensionalen Polyzylindern, Dissertation, Düsseldorf, 1988.

[773] M. Scheve, Isomorphic spaces of entire functions on non-isomorphic (DFN)-spaces, Proc. Royal Irish Acad. Sect. A, **90**, 1990, 241–244.

[774] M. Scheve, Isomorphy classes of spaces of holomorphic functions on open polydiscs in dual nuclear power series spaces, Studia Math., **101**, 1, 1991, 83–104.

[775] T. Schlumprecht, A limited set which is not bounding, Proc. Royal Irish Acad. Sect. A, **90**, 1990, 125–129.

[776] B. Schneider, On absolutely p-summing and related multilinear mappings, Wissenschaftliche Brandenburg Zeitschrift, Brandenburg Lands., **35**, 1991, 105–117.

[777] M. Schottenloher, Über analytische Forsetzung in Banachraümen, Math. Ann., **199**, 1972, 313–336.

[778] M. Schottenloher, Holomorphe Vervollständigung metrisiebarer lokalkonvexer Räume, Bayer Akad. d. Wiss., Math. Nat. Sitz.-bar, 1973, 57–66.

[779] M. Schottenloher, Analytic continuation and regular classes in locally convex Hausdorff spaces, Portugal. Math., **33**, 1974, 219–250.

[780] M. Schottenloher, Das Leviproblem in unendlichdimensionalen Räumen mit Schauderzerlegung, Habilitationsschrift, Universität Munchen, 1974.

[781] *M. Schottenloher*, The envelope of holomorphy as a functor, Coll. Int. du CNRS, 1972, Fonctions Analytiques de Plusieurs Variables et Analyse Complexe. Agora Math., Gauthier-Villars, **1**, 1974, 221–230.

[782] *M. Schottenloher*, Bounding sets in Banach spaces and regular classes of analytic functions, Functional Analysis and Applications, Recife 1972, Lecture Notes in Math., **383**, 1974, 109–122.

[783] *M. Schottenloher*, Riemann domains: Basic results and open problems, Proc. on Infinite Dimensional Holomorphy, Ed. T.L. Hayden, T.J. Suffridge, Lecture Notes in Math., **364**, 1974, 196–212.

[784] *M. Schottenloher*, ϵ-product and continuation of analytic mappings, Analyse Fonctionnelle et Applications, Ed. L. Nachbin, Hermann, Paris, 1975, 261–270.

[785] *M. Schottenloher*, The Levi problem for domains spread over locally convex spaces with a finite dimensional decomposition, Ann. Inst. Fourier (Grenoble), **26**, 4, 1976, 207–237.

[786] *M. Schottenloher*, Polynomial approximation on compact sets, Infinite Dimensional Holomorphy and Applications, Ed. M.C. Matos, North-Holland Math. Stud., **12**, 1977, 379–391.

[787] *M. Schottenloher*, $\tau_\omega = \tau_0$ for domains in $\mathbb{C}^{\mathbb{N}}$, Infinite Dimensional Holomorphy and Applications, Ed. M.C. Matos, North-Holland Math. Stud., **12**, 1977, 393–395.

[788] *M. Schottenloher*, Holomorphe Funktionen auf Gebieten über Banach-Räumen zu Vorgegebenen Konvergenzradien, Manuscripta Math., **21**, 4, 1977, 315–327.

[789] *M. Schottenloher*, Richness of the class of holomorphic functions on an infinite dimensional space. Functional Analysis: Surveys and Recent Results, Ed. K.-D. Bierstedt, B. Fuchsteiner, North-Holland Math. Stud., **27**, 1977, 209–225.

[790] *M. Schottenloher*, An example of a locally convex space which is not an ω-space, Advances in Holomorphy, Ed. J.A. Barroso, North-Holland Math. Stud., **34**, 1979, 735–744.

[791] *M. Schottenloher*, Théorème de Zorn et prolongement analytique en dimension infinie, unpublished manuscript, 1976.

[792] *M. Schottenloher*, Michael problem and algebras of holomorphic functions, Arch. Math. (Basel), **37**, 1981, 241–247.

[793] *M. Schottenloher*, A Cartan-Thullen theorem for domains spread over $\mathcal{DFM}$-spaces, J. Reine Angew. Math., **345**, 1983, 201–220.

[794] *M. Schottenloher*, Spectrum and envelope of holomorphy for infinite dimensional Riemann domains, Math., Ann., **263**, 1983, 213–219.

[795] *L. Schwartz,* Proprieté de Radon-Nikodým, Séminare Maurey-Schwartz, 1974–1975, Exp. V-VI.

[796] *J. Sebastião e Silva,* As funções analíticas e a análise functional, Portugal. Math., **9**, 1950, 1–130.

[797] *J. Sebastião e Silva,* Sui fondamenti della teoria dei funzionali analitici, Portugal. Math., **12**, 1953, 1–46.

[798] *J. Sebastião e Silva,* Le calcul différentielle et intégral dans les espaces localement convexes, réels ou complexes, Atti. Acad. Naz. Lincei, I, **20**, 1956, 743–750 and II, **21**, 1956, 40–46.

[799] *J. Sebastião e Silva,* Conceitos de função diferenciável em espaços localmente convexos, Publicações do Centro de Estudos Matemáticos de Lisboa, 1957.

[800] *T. Shirota,* On locally convex vector spaces of continuous functions, Proc. Japan Acad. Ser. A, Math. Sci., **30**, 1954, 294–298.

[801] *C.L. da Silva Diàs,* Sobre o conceito de funcional analítico, An. Acad. Brasil. Ciênc., 1943, 1–9.

[802] *C.L. da Silva Diàs,* Espaços vectoriais topologicos e suas applicações nos espaços functionais analíticos, Bol. Soc. Mat. São Paulo, **5**, 1952, 1–58.

[803] *M.A. Simões,* Very strongly and very weakly convergent sequences in locally convex spaces, Proc. Royal Irish Acad. Sect. A, **84**, 2, 1984, 125–132.

[804] *R.L. Soraggi,* Partes limitadas nos espaços de germes de aplicações holomorfas, An. Acad. Brasil. Ciênc., **49**, 1, 1977, 21–46.

[805] *R.L. Soraggi,* On bounded sets of holomorphic germs, Proc. Japan Acad. Ser. A Math. Sci., **53**, 1977, 198–201.

[806] *R.L. Soraggi,* Bounded sets in spaces of holomorphic germs, Advances in Holomorphy, Ed. J.A. Barroso, North-Holland Math. Stud., **34**, 1979, 745–766.

[807] *R.L. Soraggi,* Quasi-completeness on the spaces of holomorphic germs, Atti. Acad. Naz. Lincei, (8), **70**, 1981, 243–345.

[808] *R.L. Soraggi,* On entire functions of nuclear type, Proc. Royal Irish Acad. Sect. A, **83**, 2, 1983, 171–177.

[809] *R.L. Soraggi,* A remark on the regularity of spaces of germs, Houston J. Math., **10**, 3, 1984, 445–452.

[810] *R.L. Soraggi,* Analytic Functionals, Note Mat., **VI**, 1986, 1–34.

[811] *R.L. Soraggi,* Holomorphic germs on certain locally convex spaces, Ann. Math. Pura Appl., (4), **144**, 1986, 1–22.

[812] *R.L. Soraggi,* The Cauchy–Riemann operator in infinite dimensional spaces, Rev. Mat. Univ. Complut. Madrid, **3**, 2, 1990, 143–150.

[813] *R.L. Soraggi,* The $\overline{\partial}$-problem for a $(0, 2)$-form in a D.F.N. space, J. Funct. Anal., **98**, 2, 1991, 380–403.

[814] *B. Stensones,* A proof of the Michael conjecture (preprint).

[815] *S.J. Szarek,* A Banach space without a basis which has the bounded approximation property, Acta. Math., **159**, 1987, 81–98.

[816] *J. Taskinen,* Counterexamples to "Problème des topologies" of Grothendieck, Ann. Acad. Sci. Fenn. Math. Diss., **63**, 1986.

530   References

[817] A.E. Taylor, Analytic functions in general analysis, Ann. Scuola
      Norm. Sup. Pisa Cl. Sci., (2), **6**, 1937, 277–292.
[818] A.E. Taylor, On the properties of analytic functions in abstract spaces,
      Math. Ann., **115**, 1938, 466–484.
[819] A.E. Taylor, Linear operators which depend analytically on a param-
      eter, Ann. of Math., **39**, 1938, 574–593.
[820] A.E. Taylor, Additions to the theory of polynomials in normed linear
      spaces, Tohoku Math. J., **44**, 1938, 302–318.
[821] A.E. Taylor, Analysis in complex Banach spaces, Bull. Amer. Math.
      Soc., **49**, 1943, 652–669.
[822] A.E. Taylor, Review of a paper of Kopeć and Musielak, Math. Re-
      views, **17**, 1956, 512.
[823] A.E. Taylor, Historical notes on analyticity as a concept in functional
      analysis, Problems in Analysis, Ed. R.C. Gunning, Princeton Math.
      Ser., **31**, 1970, 325–343.
[824] A.E. Taylor, Notes on the history of the uses of analyticity in operator
      theory, Amer. Math. Monthly, **78**, 1971, 331–342.
[825] A.E. Taylor, The differential: nineteenth and twentieth century devel-
      opments, Arch. Hist. Exact Sci., **12**, 4, 1974, 355–383.
[826] O. Toeplitz, Über eine bei den Dirichletschen Reihen auftretende
      Aufgabe aus der Theorie der Potenzreihen von unendlich vielen
      Veränderlichen, Nachrichten von der Königlichen Gesellchaft der Wis-
      senschaften zu Göttingen, 1913, 417–432.
[827] E. Toma, Aplicações holomorfas e polinómios τ-contínuous, Thesis,
      Univ. Federal do Rio de Janeiro, 1993.
[828] A.M. Tonge, Polarization and the two dimensional Grothendieck in-
      equality, Math. Proc. Cambridge Philos. Soc., **95**, 1984, 313–318.
[829] A.M. Tonge, The failure of Bernstein's Theorem for polynomials on
      $\mathcal{C}(K)$ spaces, J. Approx. Theory, **51**, 2, 1987, 160–162.
[830] B.S. Tsirelson, Not every Banach space contains an imbedding of l_p
      or c_0, Funct. Anal. Appl. **8**, 1974, 138–144.
[831] M. Valdivia, Banach spaces of polynomials without copies of l_1, Proc.
      Amer. Math. Soc. **123**, 10, 1995, 3143–3150.
[832] M. Valdivia, Fréchet spaces of holomorphic functions without copies
      of l_1, Math. Nachr., **181**, 1996, 277–287.
[833] M. Valdivia, Some properties of spaces of multilinear functions and
      spaces of polynomials, Math. Proc. Royal Irish Acad., **98**A, 1, 1998,
      87–106.
[834] N. Th. Varopoulos, Some remarks on Q-algebras, Ann. Inst. Fourier
      (Grenoble), **22**, 1972, 1–11.
[835] D. Vogt, Vektorwertige Distributionen als Randverteilunger holomor-
      pher Functionen, Manuscripta Math., **17**, 1975, 267–299.
[836] D. Vogt, Charakterisierung der Unterräume von s, Math. Zeit., **155**,
      1977, 109–117.

[837] *D. Vogt*, Subspaces and quotients of (s), Functional Analysis: Surveys and Recent Results, Ed. K.-D. Bierstedt, B. Fuchssteiner, North-Holland Math Stud., **27**, 1977, 167–187.

[838] *D. Vogt*, Charakterisierung der Unterräume eines nuklearen stabilen Potenzreihenraumes von endlichen Typ, Studia Math., **71**, 1982, 251–270.

[839] *D. Vogt*, Frécheträume, zwischen denen jede stetige lineare Abbildung beschränkt ist, J. Reine Angew. Math., **345**, 1983, 182–200.

[840] *D. Vogt*, An example of a nuclear Fréchet space without the bounded approximation property, Math. Z., **182**, 1983, 265–267.

[841] *D. Vogt*, Some results on continuous linear mappings between Fréchet spaces, Functional Analysis, Surveys and Recent Results III, Ed. K.-D. Bierstedt, B. Fuchssteiner, North-Holland Math. Stud., **90**, 1984, 349–381.

[842] *D. Vogt*, On two classes of (F)-spaces, Arch. Math. (Basel), **45**, 3, 1985, 255–266.

[843] *V. Volterra*, Sopra le funzioni che dependone da altre funzioni, Nota I, Rend. Acad. Lincei, (4), **3**, 1887, 97–105.

[844] *V. Volterra*, Sopra le funzioni che dependone da altre funzioni, Nota II, Rend. Acad. Lincei, (4), **3**, 1887, 141–146.

[845] *V. Volterra*, Sopra le funzioni che dependone da altre funzioni, Nota III, Rend. Acad. Lincei, (4), **3**, 1887, 153–158.

[846] *V. Volterra,* Sopra le funzioni dependenti da linee, Nota I, Rend. Acad. Lincei, (4), **3**, 1887, 225–230.

[847] *V. Volterra*, Sopra le funzioni dependenti da linee, Nota II, Rend. Acad. Lincei, (4), **3**, 1887, 274–281.

[848] *V. Volterra*, Leçons sur les fonctions de lignes, Collection de Monographies sur le théorie des Fonctions, Gauthier-Villars, Paris, 1913.

[849] *L. Waelbroeck*, Duality and the injective tensor product, Math. Ann., **163**, 1966, 122–126.

[850] *L. Waelbroeck*, Weak analytic functions and the closed graph theorem, Proc. on Infinite Dimensional Holomorphy, Ed. T. L. Hayden, T.J. Suffridge, Lecture Notes in Math., **364**, 1974, 97–100.

[851] *L. Waelbroeck*, The nuclearity of $\mathcal{O}(U)$, Infinite Dimensional Holomorphy and Applications, Ed. M.C. Matos, North-Holland Math. Stud., **12**, 1977, 425–436.

[852] *N. Weiner*, Note on a paper of M. Banach, Fund. Math., **4**, 1923, 136–143.

[853] *G. Willis*, The compact approximation property does not imply the approximation property, Studia Math., **103**, 1992, 99–108.

[854] *A.J.M. Wanderley*, Germes de aplicações holomorfas em espaços localmente convexos, Thesis, Universidade Federal do Rio de Janeiro, 1974.

[855] *S. Yamamuro,* Differential calculus in topological linear spaces, Lecture Notes in Math., **374**, 1974.

[856] *I. Zalduendo,* A canonical extension for analytic functions on Banach spaces, Trans. Amer. Math. Soc., **320**, 1990, 747–763.

[857] *I. Zalduendo,* Extensiones de formas multilineales y funciones analíticas sobre espacios de Banach, Publ. Centro Latinoamericano de Matemáticas e Informática, **17**, 1992, 79–85.

[858] *I. Zalduendo,* Dualidad y extensiones en holomorfía infinita. Rev. Colombiana Mat., **27**, 1993, 131–135.

[859] *I. Zalduendo,* An estimate for multilinear mappings on l^p spaces, Proc. Royal Irish Acad. Sect. A, **93**, 1, 1993, 137–142.

[860] *I. Zalduendo,* Extending polynomials – a survey, Publ. del Departamento de Análisis Matemático de la Univ. Complut. de Madrid, **41**, 1998.

[861] *W.R. Zame,* Extendibility, boundedness and sequential convergence in spaces of holomorphic functions, Pacific J. Math., **57**, 2, 1975, 619–625.

[862] *M.A. Zorn,* Gâteaux differentiability and essential boundedness, Duke Math. J., **12**, 1945, 579–583.

[863] *M.A. Zorn,* Characterization of analytic functions in Banach spaces, Ann. of Math., **46**, 1945, 585–593.

[864] *M.A. Zorn,* Derivatives and Fréchet differentials, Bull. Amer. Math. Soc., **52**, 1946, 133–137.

Index